Singularities, Algebraic Geometry, Commutative Algebra, and Related Topics

Gert-Martin Greuel • Luis Narváez Macarro •
Sebastià Xambó-Descamps
Editors

Singularities, Algebraic Geometry, Commutative Algebra, and Related Topics

Festschrift for Antonio Campillo on the Occasion of his 65th Birthday

Springer

Editors
Gert-Martin Greuel
Fachbereich Mathematik
Universität Kaiserslautern
Kaiserslautern, Germany

Luis Narváez Macarro
Departamento de Álgebra
Instituto de Matemáticas (IMUS)
Universidad de Sevilla
Sevilla, Spain

Sebastià Xambó-Descamps
Departament de Matemàtiques
Universitat Politècnica de Catalunya
Barcelona, Spain

ISBN 978-3-030-07258-2 ISBN 978-3-319-96827-8 (eBook)
https://doi.org/10.1007/978-3-319-96827-8

Mathematics Subject Classification (2010): 14-XX, 32SXX, 11AXX, 13PXX, 19DXX, 37FXX, 51EXX, 94BXX

This Springer imprint is published by the registered company Springer Nature Switzerland AG
The registered company address is: Gewerbestrasse 11, 6330 Cham, Switzerland

Preface

The decision to prepare a volume of contributions in honor of Antonio Campillo was taken at the University of Valladolid (Spain) during 19–23 June 2017 on the occasion of the 4th Joint Congress of the Spanish and Mexican Mathematical Societies.

The Congress paid homage to Campillo by including two plenary lectures to present his contributions to mathematics and his service and leadership in a variety of institutions, two special sessions on Singularities and Algebraic Geometry, and a special scientific day to celebrate more closely his mathematical work.

The announcement of this book project was sent to the list of all participants at the Congress enlarged with the names of other colleagues with whom Campillo has had significant interactions in his professional career.

This volume collects the refereed papers that were submitted following that announcement. After the opening article, which is meant to be a first approach to Campillo's biography, we have grouped the papers in five categories: Singularities (ten papers), Algebraic Geometry (five), Commutative Algebra (five), Algebraic Codes (three), and Other Topics (three). The variety of topics correlates well with the very broad research interests of Antonio Campillo.

The grouping is of course imperfect, as some papers would easily fit in more than one category, but we hope that the classification respects nevertheless the main area in which each paper is ascribed. Within each category, papers that explicitly connect with published work of Campillo are placed first, and these, as well as the others, are ordered alphabetically by the family name of the first author.

We are most grateful to all the authors (two per paper on average) for their positive response, their patience with the long refereeing process, and above all, for their scientific contribution to the volume. Thanks also to the generosity and keen judgments of the many anonymous referees.

Going back to the beginning, we are thankful to the Mexican and Spanish mathematical societies for having embraced the idea of paying tribute to Antonio Campillo on the occasion of their 4th joint meeting, and to the University of Valladolid, in particular to its Mathematical Institute (IMUVA), for providing excellent organization for all activities in that week.

Antonio Campillo has been our colleague and friend for many years, in a variety of circumstances, and it is a pleasant duty to express our acknowledgment to him for his productive career, his dedication to causes that promote the welfare of institutions and professional communities, his vision regarding academic progress and research organization, his engagement in increasing public awareness of mathematics, and of course for all his achievements in mathematics.

We wish you, Antonio, a most happy 65th birthday and that your career may continue to be, for many years to come, as productive and inspirational as it has been hitherto.

Kaiserslautern, Germany — Gert-Martin Greuel
Sevilla, Spain — Luis Narváez Macarro
Barcelona, Spain — Sebastià Xambó-Descamps

Contents

Contributors

Argimiro Arratia Department of Computer Science and Barcelona Graduate School of Mathematics, Universitat Politècnica de Catalunya, Barcelona, Spain

Enrique Artal Bartolo Departamento de Matemáticas, IUMA, Universidad de Zaragoza, Zaragoza, Spain

Rafael G. Campos División de Ciencias e Ingeniería, Departamento de Ciencias, Universidad de Quintana Roo, Chetumal, México

Pierrette Cassou-Noguès Institut de Mathématiques de Bordeaux, (UMR 5251), Université de Bordeaux, Talence Cedex, France

José Ignacio Cogolludo-Agustín Departamento de Matemáticas, IUMA, Universidad de Zaragoza, Zaragoza, Spain

G. Cortiñas Department of Matemática-Inst. Santaló, FCEyN, Universidad de Buenos Aires, Buenos Aires, Argentina

Laura Costa Departament de Matemàtiques i Informàtica, Facultat de Matemàtiques i Informàtica, Barcelona, Spain

Steven Dale Cutkosky Department of Mathematics, University of Missouri, Columbia, MO, USA

Daniel Duarte CONACYT, Universidad Autónoma de Zacatecas, Zacatecas, Mexico

Gioia Failla Università Mediterranea di Reggio Calabria, DIIES, Reggio Calabria, Italy

J. I. Farrán Universidad de Valladolid, Valladolid, Spain

Juan Bosco Frías-Medina Instituto de Matemáticas, Universidad Nacional Autónoma de México. Área de la Investigación Científica. Circuito Exterior, Ciudad Universitaria, Coyoacán, Mexico

Unidad Académica de Matemáticas, Universidad Autónoma de Zacatecas, Zacatecas, México

Carlos Galindo Departamento de Matemáticas, Instituto Universitario de Matemáticas y Aplicaciones de Castellón, Universitat Jaume I, Castelló, Spain

Alejandra García Facultad de Ciencias UNAM, Ciudad Universitaria México, México City, México

Evelia R. García Barroso Departamento de Matemáticas, Estadística e I.O. Sección de Matemáticas, Universidad de La Laguna, La Laguna, Tenerife, España

Philippe Gimenez Facultad de Ciencias, Instituto de Investigación en Matemáticas de la Universidad de Valladolid (IMUVA), Valladolid, Spain

M. R. Gonzalez-Dorrego Departamento de Matemáticas, Universidad Autónoma de Madrid, Madrid, Spain

Pedro D. González Pérez Instituto de Ciencias Matemáticas (CSIC-UAM-UC3M-UCM), Madrid, Spain

Departamento de Álgebra, Geometría y Topología, Facultad de Ciencias Matemáticas, Universidad Complutense de Madrid, Madrid, Spain

Gerardo Gonzalez Sprinberg Facultad de Ciencias, Centro de Matemática, UdelaR, Montevideo, Uruguay

Université Grenoble Alpes, Grenoble, France

Manuel González Villa Centro de Investigación en Matemáticas, A.C., Guanajuato, México

Departamento de Álgebra, Geometría y Topología, Facultad de Ciencias Matemáticas, Universidad Complutense de Madrid, Madrid, España

Gert-Martin Greuel Fachbereich Mathematik, Universität Kaiserslautern, Kaiserslautern, Germany

C. Haesemeyer School of Mathematics and Statistics, University of Melbourne, Melbourne, VIC, Australia

Fernando Hernando Departamento de Matemáticas, Instituto Universitario de Matemáticas y Aplicaciones de Castellón, Universitat Jaume I, Castelló, Spain

Gary Kennedy Ohio State University at Mansfield, Mansfield, OH, USA

Mustapha Lahyane Instituto de Física y Matemáticas (IFM), Universidad Michoacana de San Nicolás de Hidalgo (UMSNH), Morelia, México

Santiago López de Medrano Instituto de Matemáticas, Universidad Nacional Autónoma de México, México City, México

José Martínez-Bernal Departamento de Matemáticas, Centro de Investigación y de Estudios Avanzados del IPN, México City, México

Jorge Martín-Morales Centro Universitario de la Defensa-IUMA, Academia General Militar, Zaragoza, Spain

Laura Felicia Matusevich Mathematics Department, Texas A&M University, College Station, TX, USA

Lee J. McEwan Ohio State University at Mansfield, Mansfield, OH, USA

Rosa Maria Miró-Roig Departament de Matemàtiques i Informàtica, Facultat de Matemàtiques i Informàtica, Barcelona, Spain

Francisco Monserrat Department of Applied Mathematics, Instituto Universitario de Matemática Pura y Aplicada (IUMPA), Universidad Politécnica de Valencia, Valencia, Spain

Luis Narváez Macarro Departamento de Álgebra & Instituto de Matemáticas (IMUS), Facultad de Matemáticas, Universidad de Sevilla, Sevilla, Spain

Ignacio Ojeda Departamento de Matemáticas, Universidad de Extremadura, Badajoz, Spain

Jorge Olivares Centro de Investigación en Matemáticas, A.C., Guanajuato, México

Lourdes Palacios Universidad Autónoma Metropolitana Iztapalapa, México City, México

Ruud Pellikaan Discrete Mathematics, Technische Universiteit Eindhoven, Eindhoven, MB, The Netherlands

Arkadiusz Płoski Department of Mathematics and Physics, Kielce University of Technology, Kielce, Poland

Patrick Popescu-Pampu Département de Mathématiques, Université de Lille, Villeneuve d'Ascq Cedex, France

Michel Raibaut Université Grenoble Alpes, Grenoble, France

CNRS, LAMA, Université Savoie Mont Blanc, Cedex, France

Brenda Leticia De La Rosa-Navarro Facultad de Ciencias, Universidad Autónoma de Baja California, Ensenada, México

Diego Ruano IMUVA (Mathematics Research Institute), University of Valladolid, Valladolid, Spain

Department of Mathematical Sciences, Aalborg University, Aalborg, Denmark

Carlos Signoret Universidad Autónoma Metropolitana Iztapalapa, México City, México

Aron Simis Departamento de Matemática, Universidade Federal de Pernambuco, Recife, Brazil

Diego Sulca CIEM-FAMAF, Universidad Nacional de Córdoba, Córdoba, Argentina

Bernard Teissier Institut Mathématique de Jussieu-Paris Rive Gauche, Paris, France

Daniel Green Tripp Universidad Nacional Autónoma de México, México City, México

Rosanna Utano Dipartimento di Scienze Matematiche e Informatiche. Scienze Fisiche e Scienze della Terra, Università di Messina, Messina, Italy

Orlando Villamayor ICMAT-Universidad Autónoma de Madrid, Madrid, Spain

Rafael H. Villarreal Departamento de Matemáticas, Centro de Investigación y de Estudios Avanzados del IPN, México City, México

Carlos E. Vivares Departamento de Matemáticas, Centro de Investigación y de Estudios Avanzados del IPN, México City, México

M. E. Walker Department of Mathematics, University of Nebraska – Lincoln, Lincoln, NE, USA

C. A. Weibel Department of Mathematics, Rutgers University, New Brunswick, NJ, USA

Sebastià Xambó-Descamps Mathematics Department, Universitat Politècnica de Catalunya, Barcelona, Spain

Acronyms

AMS	American Mathematical Society
ANECA	Agencia Nacional de Evaluación de la Calidad y Acreditación
ANEP	Agencia Nacional de Evaluación y Prospective
BBVA	Banco de Bilbao Vizcaya Argentaria
BMS	Belgium Mathematical Society, SMB
CDM	Conferencia de Decanos de Matemáticas
CEMAT	Comité Español de Matemáticas
CIMPA	Centre International de Mathématiques Pures et Appliquées
EMS	Europcan Mathcmatical Society
FESPM	Federación Española de Sociedades de Profesores de Matemáticas
FPI	Formación de Personal Investigador
HNE	Hamburger-Noether expansion
ICIAM	International Congress on Industrial and Applied Mathematics
IMU	International Mathematical Union
MCyT	Ministerio de Ciencia y Tecnología
MO	Mathematical Olympiad
MPE	Mathematics of Planet Earth
MR	Mathematical Reviews
MSP	Minero Siderúrgica de Ponferrada
RSEF	Real Sociedad Española de Física
RSEQ	Real Sociedad Española de Química
RSME	Real Sociedad Matemática Española
SCM	Societat Catalana de Matemàtiques
SEIO	Sociedad de Estadística e Investigación Operativa
SEMA	Sociedad Española de Matemática Aplicada
SEN	Sociedad Española de Neurociencias
SIMAI	Società Italiana di Matematica Applicata e Industriale
SMB	Société Mathématique de Belgique
SML	Société Mathématique du Luxembourg
SMM	Sociedad Matemática Mexicana
UCM	Universidad Complutense de Madrid

UMI	Unione Matematica Italiana
US	Universidad de Sevilla
UVA	Universidad de Valladolid

Antonio Campillo

A Portrayal of His Life and Work

Francisco Monserrat and Sebastià Xambó-Descamps

Abstract The main purpose is to present a biographical portrayal of Antonio Campillo López. In addition to the most relevant aspects of his life and academic career, we offer a panoramic view of his scientific work in fields such as singularity theory, commutative algebra, algebraic geometry, or coding theory. We also stress that his endeavors have been closely linked to an important work of research training, from which a thriving school has emerged. Finally, his vision and influence will be considered through his numerous institutional and scientific policy responsibilities at all levels.

1 Prelude

The University of Valladolid (UVA) hosted the 4th Joint Meeting of the Real Sociedad Matemática Española (RSME) and the Sociedad Matemática Mexicana (SMM) from June 19 to 22, 2017. The meeting, and a special scientific day on June 23, had the character of a tribute to Antonio Campillo. This tribute was acknowledged from the table of the opening ceremony[1] and further stressed by the fact that the opening plenary lecture had been planned to offer an outline of his

[1]Constituted by Daniel Miguel San José, rector of the UVA; Pablo Raphael de la Madrid, Director of the Cultural Institute of the Embassy of Mexico in Spain; Gelasio Salazar Anaya, President of the SMM; Francisco Marcellán Español, President of the RSME; and by Edgar Martínez Moro, President of the Organizing Committee of the meeting.

F. Monserrat (✉)
Department of Applied Mathematics, Universidad Politécnica de Valencia, Valencia, Spain
e-mail: framonde@mat.upv.es

S. Xambó-Descamps
Mathematics Department, Universitat Politècnica de Catalunya, Barcelona, Spain
e-mail: sebastia.xambo@upc.edu

G.-M. Greuel et al. (eds.), *Singularities, Algebraic Geometry, Commutative Algebra, and Related Topics*, https://doi.org/10.1007/978-3-319-96827-8_1

life and achievements.[2] To a good extend, this paper is an elaboration of that lecture aimed at providing a wider and deeper coverage of several aspects, most particularly of his mathematical research.

The tribute befitted the occasion on many counts, as we will point out in due time, but especially because Antonio Campillo was the promoter, in the two terms as President of the RSME (2009–2012 and 2012–2016), of the first four joint meetings of the two societies: Oaxaca (2009), Torremolinos (2012), Zacatecas (2014), and Valladolid (2017).

Let us also mention, for the effect it has had on the the gestation of this work, and also on the scientific relations between the two mathematical communities, that one of the most outstanding projects Antonio Campillo set out in his first mandate was the celebration of the (first) centenary of the RSME (2011). It entailed a vast mobilization in which many people were involved in a variety of initiatives, and one of them was the launching of the *RSME-Universia Gallery of Mathematics, Science and Technology* (colloquially ArbolMat).[3] The mission of this initiative is "to publish scientific profiles of personalities from Latin America, including Spain and Portugal, distinguished by their eminent research in Mathematics or in the use of Mathematics (in science, technology, economics, ...), for their high generativity and influence, and for their ability to inspire younger generations." Fifteen profiles were published in November 2011, 24 in November 2014, and 29 in January 2017. Among the latter there was Antonio Campillo's profile,[4] whose inclusion was decided after finishing in November 2015 his second term as President of the RSME. From the Mexican side, the profiles that have been produced so far are those of José Antonio de la Peña, Xavier Gómez Mont, Santiago López de Medrano, and Raimundo Bautista Ramos, and at the time of this writing those of Francisco González Acuña and Alberto Verjovsky are in the pipeline.

The picture for the ArbolMat portal (Fig. 1) was taken on the occasion of the joint meeting (another really outstanding organizational feat) of the Unione Matematica Italinana (UMI), RSME, Catalan Mathematical Society (SCM), Sociedad Española de Matemática Aplicada (SEMA), and Società Italiana per la Matematica Applicata e Industriale (SIMAI), which was held at the Universidad del País Vasco/Euskal Herriko Unibertsitatea (Bilbao) from 30 June to 4 July 2014. We believe that it captures and transmits, at least to those that know him, characteristic features of his scientific, academic, institutional and organizational personality.

Antonio Campillo was born on 26 November, 1953, in the industrious Ponferrada, the capital of El Bierzo, a region in the North-Western of the León province, which today belongs to the Autonomous Community of Castilla-León, Spain. Founded in the eleventh century, today Ponferrada is the largest city crossed by the

[2] "Antonio Campillo, bosquejos sobre su vida y obra", by S. Xambó (https://mat-web.upc.edu/people/sebastia.xambo/Bios/AC/2017-Xambo--s-AC.pdf)

[3] http://www.arbolmat.com/. Since February 2018, however, ArbolMat belongs solely to RSME and its name is *RSME Gallery of Mathematics, Science and Technology*.

[4] http://www.arbolmat.com/antonio-campillo/

Fig. 1 A. Campillo's picture at the ArbolMat portal. With permission from the author, S. Xambó

rapid Sil River, and the last major town on the Camino de Santiago before reaching Santiago de Compostela. To facilitate the crossing of the river to Santiago pilgrims, the initial wooden bridge was commissioned to be fortified with iron by Osmundo, Bishop of Astorga (1082–1098), and hence the name *Pons Ferrata*, or *Iron Bridge*. León itself is the old *Legio VI* of the Romans, for whom the whole area turned out to be the largest open mining site across the Empire, particularly the gold mines of Las Médulas (now a Unesco World Heritage Site), and an excellent producer of many other goods, notably food and wine.

The modern Ponferrada also owes much to mining, but this time of high quality coal (anthracite) and iron mines that began to be exploited after World War I by the mining and steel company MSP (Minero Siderùrgica de Ponferrada). The MSP built a railway that provided services beyond its use for ore transportation, and a thermal power plant (Central Térmica) that was powered by coal extracted from the mines and popularly known as the "Light Factory" (La Fábrica de Luz) because it supplied electrical power to the city. At its height in over half a century ago, the MSP provided employments and services, including housing for workers, and a school, but two decades later it underwent a deep crisis that led to it being absorbed by the Asturian steel industry. Early in this century, La Fábrica de Luz was rehabilitated, preserving the old structure, into the Ponferrada Energy Museum (Fig. 2).

These are only a few glimpses of a rich cultural heritage that is much cherished by the inhabitants of those lands and cities and a real joy for the outsiders that visit them. Antonio Campillo grew up in this resourceful environment, where he spend his childhood and youth. He attended the MSP school in the period 1960–1963. Being the brightest student, the school endowed him with a grant to attend the public high school Gil y Carrasco, in the City Hall square, where he completed the whole secondary education cycle in the period 1964–1971. At age of 17 he entered the University of Oviedo, where he spent the term 1971–1972 as a student of the first course of the science faculties and engineering schools, which at that time was mandatory for all science and engineering careers. His early decision to pursue Mathematics lead him to the University of Valladolid, which is about 200 km South-East of Ponferrada, to pursue the remaining four terms to get the "Licenciado

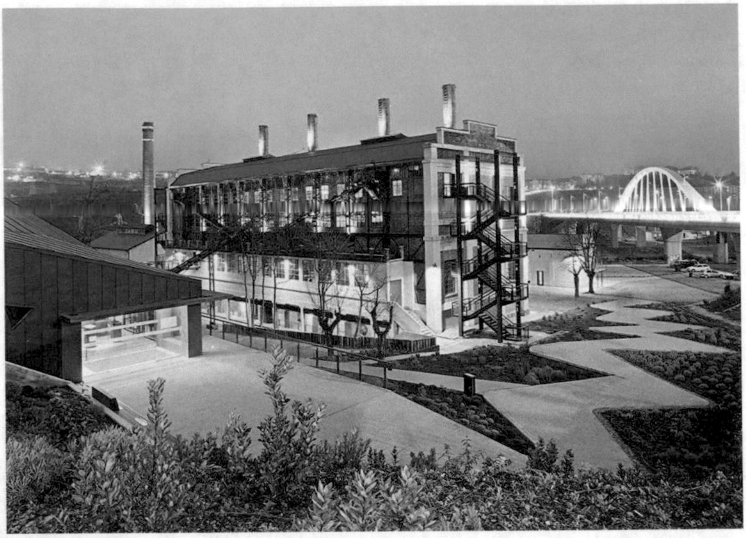

Fig. 2 Ponferrada's "La Fábrica de Luz. Energy Museum", an award-winning rehabilitation of the one-century old thermal power plant. Published with the kind permission of the author, José Hevia, through the good offices of Míriam Fernández Cuevas

en Matemáticas" degree, which in todays terms would be equivalent to a Master's degree in Mathematics.

Let us end this prelude with a short description of the sections in which we have divided what lies ahead:

- Sect. 2: A timeline of Antonio Campillo's main academic landmarks.
- Sect. 3: A relation of his PhD students with a brief summary of their theses.
- Sect. 4: A short report on his main collaborators.
- Sect. 5: A selection of his most representative works, with comments, often with quotations from other sources. When appropriate, we emphasize questions and observations that are likely seeds for future research opportunities. At the end we provide a summary of his main contributions in a headlines format.
- Sect. 6: An overview of his services in several fronts, and an assessment of his influence in important aspects of the mathematical and scientific communities.
- 7: References. The full list of Campillo's publications as of this writing (January 2018). References to works of others are given in footnotes inserted at the spot in the text where they are required.

2 Mathematical Education and Academic Positions

Antonio Campillo obtained a 5-year bachelor's degree in Mathematics[5] from the University of Valladolid (UVA) in 1976. In the last three terms of these studies, he won a salary-grant and he finished the degree as the first of his promotion and was the first nation-wide among the granted students in Mathematics. In the next two terms he was Assistant Professor in the same university and was awarded an FPI[6] grant that culminated with a PhD thesis that was defended in 1978.

The thesis advisor was José M. Aroca, a former student (1970) of Pedro Abellanas and Heisuke Hironaka, and the thesis title was *Singularidades de curvas algebroides planas en característica positiva*.[7] It was a remarkable memoir, for its extension and contents, and a no less remarkable feat that the author could complete it in a period that by all standards was very short.

It was published, with an important additional chapter (we will return to this later on), as a Springer Lecture Notes in Mathematics [1]. The main theme is the extension, and theoretical development, of the notion of equisingularity (introduced by Oscar Zariski in the 1960s) for plane curves over an algebraically closed field of any characteristic. Instead of the classical Puiseux expansions, the technical tools used are the parameterizations given by Hamburger-Noether expansions.

The PhD period coincided with those of Ignacio Luengo and Julio Castellanos. With both he has maintained a substantial collaboration along the years. In the period 1978–1982, he had a position of Associate Professor *ad interim* in the UVA. It is the time in which the Sevilla school of Algebraic Geometry and Singularities begins its own journey, inspired by José Luis Vicente Córdoba, another former student of Pedro Abellanas (1973), and that Antonio Campillo visits on several ocasions. In this way he formed a solid friendship with that school, particularly with Luis Narváez, Emilio Briales, Francisco Castro, Ramón Piedra, Javier Herrera, and of course Vicente Córdoba. It was also in this connection that he also met, as early as 1980, Lê Dũng Tráng, Andrei Todorov, Manfred Hermann, Gert-Martin Greuel, Bernard Teissier, and Jean Giraud.

The year 1981 of that period is especially intense, as he and Ignacio Luengo made postdoctoral stays at Columbia University and Harvard University. These allowed him to meet Oscar Zariski and several of his students, including Shreeran Abhyankar (PhD 1956), Heisuke Hironaka (1960), David Mumford (1961), and Joseph Lipman (1965). The exchanges with these researchers, and the motivation provided by their works, boosted his interest in the study of singularities in geometry and soon reached the maturity of being able to propose approaches and algebraic models for their study in any characteristic. Fundamentally, he focused on "classifying singularities from associated algebraic structures, mainly arcs and semigroups, and

[5]Licenciatura en Matemáticas. It was equivalent, roughly speaking, to a master's degree in Mathematics in todays terms.

[6]Formación de Personal Investigador, i.e. Training of Research Personel.

[7]*Singularities of algebroid plane curves in positive characteristic.*

on discovering how to read the geometry and topology directly in those structures" (AC). The thesis first, and the maturity reached in his postdoctoral stays afterwards, are clear signs of him having reached the proficiency and determination to carry on a creative and productive research career.

The academic consolidation came in the period 1982–1984, in which he had an appointment as Associate Professor of Geometry in the UVA (1982–1983), Full Professor of Geometry and Topology in the University of Sevilla (US) (1983–1984), and Full Professor of Algebra in the UVA since 1984. In the year 1983 he served as Director of the Algebra Department of the US, and after his settlement at the UVA, he was Director of the Department of Algebra, Geometry and Topology (1996–1999) and Dean of the Science Faculty (2000–2004). In 1991 he founded the Singacom research group and has been its coordinator ever since.

3 PhD Students

In a private communication about 2 weeks before the lecture referred to in the Foonote 2, Antonio Campillo described how he views himself in relation to scientific cooperation: "My idea has always been to feel inside a school of researchers, that is, to share with one another knowledge and ideas, including taking them to the classroom." This principle is clearly reflected in the ways he has carried out the orientation of doctoral students, to whom we dedicate this section, and in the collaboration with a rather long list of researchers, to whom we devote the next section.

- FÉLIX DELGADO DE LA MATA: *Invariantes numéricos de curvas algebroides con varias ramas. Una descripción explícita*. Universidad de Valladolid (UVA), 1986 (282 p).

Aimed at providing an explicit combinatorial description of several numerical invariants that characterize the equisingularity type of a plane reduced algebroid curve defined over an algebraically closed field, it essentially generalizes and extends the theory that was available in the irreducible case (Puiseux exponents, semigroup of values, indexes of maximal contact, dual graph, local topology, Whitney conditions, and so on) to the case of two or more branches, providing explicit descriptions of each kind of invariant in terms of the other kinds. In this case the equisingularity is defined in terms of the well-understood equisingularity of the branches and the intersection multiplicites of pairs of branches, and the value semigroup is a subsemigroup S of $\mathbb{N}^r$, where r is the number of branches. One of the many novel ideas introduced in this memoir is the concept of *gap* (laguna) relative to S, which is one of the keys for the main result of the thesis, namely the explicit description of the semigroup of values. The number of gaps is finite. This allows to compute the length of $\bar{R}/R$, where R denotes the local ring of the algebroid curve and $\bar{R}$ its integral closure in the total quotient ring of R, and so to characterize the Gorenstein condition of R in terms of the symmetries of the semigroup.

- CAROLINA ANA NÚÑEZ JIMÉNEZ: *Anillos saturados de dimensión uno: Clasificación, significado geométrico y aplicaciones.* UVA, 1986 (194 p).

In the Arcata Symposium "Singularities 1981", Antonio Campillo introduced a notion "absolute saturation" (he called it "presaturation"), see [2], and compared it with other analogous notions that had been introduced by Zariski, Teissier or Lipman. "This interesting paper illustrates the advantage of the use of Hamburger-Noether expansions instead of the Puiseux expansions in some problems over fields of arbitrary characteristic." (Doru Ştefănescu, MR713060). Ana Núñez' thesis is focused on the reduced and equicharacteristic one-dimensional complete local rings that are saturated according to Campillo's definition, providing a description of them, of their geometric interpretations, and their classification by means of a complete system of invariants. An outstanding contribution is the ring-theoretical formulation of Abhyankar's so-called inversion formulas for Puiseux exponents.

- JULEN SUSPERREGUI LESACA: *Invariantes determinantales de módulos sobre ciertos anillos no conmutativos.* UVA, 1987 (238 p).

Defines and studies the Fitting ideals of a finitely generated $\mathscr{D}$-module using the notion of determinant (introduced by Kossivi Adjamagbo) of a square matrix of elements of $\mathscr{D}$. For a summary, see the author's paper *Déterminant sur des anneaux filtrés* published by the Comptes Rendus de l'Académie de Sciences, Série I (Mathématiques), **293**/9 (1981), 447–449. A similar treatment was given for modules over skew-commutative graded algebras using a suitable definition of determinant for this case. In both cases he determined the geometrical invariants using the Fitting ideals.

- CARLOS MARIJUÁN LÓPEZ: *Una teoría birracional para grafos acíclicos.* UVA, 1988 (265 p).

A combinatorial study, in terms of oriented acyclic graphs, of processes, particularly the blowing-up along a subvariety, that concur in the algorithmic desingularization and birrational theory of algebraic curves. One of the main results is that each oriented acyclic graph G is birrationally equivalent to a unique oriented forest, and that this oriented forest classifies G up to birrational equivalence. He also obtained birrational contractions of an oriented forest with a minimal number of vertexes.

- JULIO GARCÍA DE LA FUENTE: *Geometría de los sistemas lineales de series de potencias en dos variables.* UVA, 1989 (118 p).

The topic of this thesis is the birrational reduction of the singularities of singular foliations on a smooth algebraic surface that have a meromorphic first integral, which means that the foliations are given as the tangents to the curves of a pencil $ag + bh = 0$, where $(a : b) \in \mathbb{P}^1$ and g and h are regular functions. The reduction turns out to be perfectly related to the elimination of the pencil base points, a result that has been used by Abhyankar, Artal and Luengo, among others, in their works on dicritical points.

- Carlos Galindo Pastor: *Desarrollos de Hamburger-Noether y equivalencia discreta de valoraciones*. UVA, 1991 (203 p).

The Hamburger-Noether expansion (HNE) of a plane curve germ is especially suitable for the study of its singularity in the case of positive characteristic (see [1]). In the thesis, the concept of HNE is extended to any plane valuation in any characteristic. This allows to refine Spivakovsky's classification of these valuations. In addition, it is proved that the HNE gives rise to parametric equations of the valuations with the consequent advantage from the computational point of view. The concept of the Poincaré series of a valuation is also introduced and an explicit calculation is given in the divisorial case. Finally, the concept of intersection number of germs of foliations and plane valuations is introduced, proving that the family of these values determines the birrational reduction of a foliation.

- Ana José Reguera López: *Proximidad, cúmulos e ideales completos sobre singularidades racionales de superficie*. UVA, 1993 (162 p). Codirigida con Monique Lejuene-Jalabert.

The geometric theory of complete ideals for rational singularities of an algebraic surface is established by means of a generalization, suitable for this context, of the Enriques-Zariski notion of proximity for infinitely near points associated to their resolution by blowing-up of points. A full dictionary is provided between the geometric theory and the combinatorial theory associated of its dual graph. This dictionary allows to understand the geometry of significant arcs traced on the rational surface singularities.

- José Ignacio Farrán Martín: *Construcción y decodificación de códigos álgebro-geométricos a partir de curvas planas: algoritmos y aplicaciones*. UVA, 1997 (142 p).

In the search of an efficient set-up for the algorithmic construction, coding and decoding of Goppa codes based on a smooth projective curve $\tilde{C}$ defined over a finite field, the main idea of this thesis is to cast all the relevant ingredients in terms of a *plane* model C by using a symbolic version of the Hamburger-Noether expansions to obtain $\tilde{C}$ (desingularization of C), the so-called embedded resolution, a basis for the space $L(G)$ used to construct a Goppa code, and the Weirstrass semigroup of a rational branch of C. The most satisfactory results are for the case $G = mP$ (point Goppa codes) and, more specifically, when P is the only branch at infinity of C.

- Eduardo Tostón Valdés: *Estudio y factorización de ideales completos en anillos locales*. UCM, 2002 (138 p).

The main contribution of this memoir is the discovery of a class of complete ideals in which the results of Zariski in dimension two can be generalized to arbitrary dimension. For this class of ideals, which is a wide subclass of toric complete ideals, it follows naturally that the semigroup they form is free, and that results of Zariski-Samuel (*Commutative Algebra*, volume II) and of Lipman (regular case) can be fully extended to this more general situation.

- FRANCISCO MONSERRAT DELPALILLO: *El cono de curvas asociado a una superficie racional. Poliedricidad.* Universidad Jaume I, 2003 (203 p). Codirigida con CARLOS GALINDO PASTOR.

To any projective surface X one can associate a series of convex cones (cone of curves, semi-ample cone, and characteristic cone) that provide information on the geometry of the surface. In this thesis the cone of curves associated to a rational and regular projective surface is studied. More specifically, conditions are established that involve the polyhedricity of that cone. These conditions are of two types: those that depend on the existence of certain effective divisors, and others that depend solely on obtaining the surface from a relatively minimal surface (which may be the projective plane or a Hirzebruch surface). The polyhedricity of the cone of curves has important geometric implications, such as the fact that the number of projective morphisms with connected fibers from X to another variety (contractions) is finite, and, consequently, that the number of (-1)-curves of X (that is, non-singular rational curves with self-intersection -1) is finite.

- DIEGO RUANO BENITO: *Estructura métrica de los códigos lineales. Códigos tóricos generalizados.* UVA, 2007 (100 p). Codirigida con JOSÉ IGNACIO FARRÁN MARTÍN.

A contribution to the theory of toric codes, that is, algebro-geometric codes on toric varieties obtained by evaluating rational functions at the torus points, and to some generalizations of toric codes. In particular he obtains, using intersection theory, lower bounds for the minimal distance. It also contributes to the study of linear codes in general by introducing structures of toric type that allow to understand and handle them in a way similar to the toric codes.

- ANN LEMAHIEU: *Poincaré series and zeta functions.* Katholieke Universiteit Leuven, 2007 (113 p). Codirigida con WILLEM VEYS.

One main result is to prove, in terms of the theory developed by Campillo, Gonzalez-Sprinberg and Lejeune-Jalabert [3], the monodromy conjecture for the Igusa function (Denef-Loeser theory) of surface singularities that are not degenerate with respect to a toric cluster in dimension 3, one of the few cases for which it has been possible to prove that conjecture. Another main result is to establish the theory, perform the corresponding calculations, and describe explicitly the geometrical meaning of the Poincaré series of toric varieties with respect to their natural filtractions.

- ROSA MARÍA DE FRUTOS MARÍN: *Perspectivas aritméticas para la Conjetura de Casas-Alvero.* UVA, 2013 (145 p).

Detailed study on the multiple facets and their interrelations of the Casas-Alvero conjecture (2001), according to which a complex univariate polynomial of degree n has only one root if it shares a root with each of its first $n-1$ derivatives. In particular it shows that the conjecture is valid if the polynomial has at most three monomials and, in general, it introduces a numerical discriminant, defined for each degree n, which is an integer, and whose non-vanishing is equivalent to the validity of the conjecture. On the other hand, since the validity of the conjecture follows from the validity of its reduction modulo some prime number p, it also considers six different reductions modulo p and proves that the validity of either of them for any given p is equivalent to the validity of the remaining five reductions for the same p. This result provides substantially simpler proofs of most known results.

- IRENE MÁRQUEZ CORBELLA: *Combinatorial Commutative Algebra Approach to Complete Decoding*. UVA, 2013 (222 p). Codirigida con EDGAR MARTÍNEZ-MORO.

It explores links between the algebraic structure of a linear code and the complete decoding process and develops an algebraic analysis of the decoding process by means of different mathematical structures associated with certain code families. Alternative algorithms and new techniques are also proposed that allow to relax the conditions of the decoding problem (in general NP-complete) and thereby reducing the resources of space and time necessary to handle it. The information concerning finite rings is represented in an original way by bit strings and their manipulation is based on toric geometry.

4 Collaborators

Antonio Campillo has created and led a dynamic research school composed not only by doctoral students and nearby colleagues, but also by pre and post doctoral students from numerous countries. All this activity has been carried out in step with his ability to attract, generate and manage appropriate resources. So it is not surprising that up till now he has had over 50 scientific collaborators.

Among the PhD students, we find the following list (the order of the names is by the year of the thesis defense, as in the previous section, and the order of the cited works is from oldest to newest):

- Félix DELGADO DE LA MATA: [4–31]
- Carlos MARIJUÁN LÓPEZ: [32–34]
- Julio GARCÍA DE LA FUENTE: [35, 36]
- Carlos GALINDO PASTOR: [35, 37–39]
- Ana José REGUERA LÓPEZ: [35, 40–42]
- José Ignacio FARRÁN MARTÍN: [43–47]
- Francisco MONSERRAT DELPALILLO: [48]
- Ann LEMAHIEU: [49]

Among the remaining collaborators, we consign a selection of those that seem to us to have more significance for the scientific endeavors of Campillo.

- François LOESER (PhD 1983, B. Teissier), Ignacio LUENGO (PhD 1979, J. M. Aroca), Claude SABBAH (PhD 1976, L. D. Tráng): [50]. See Fig. 3, left.

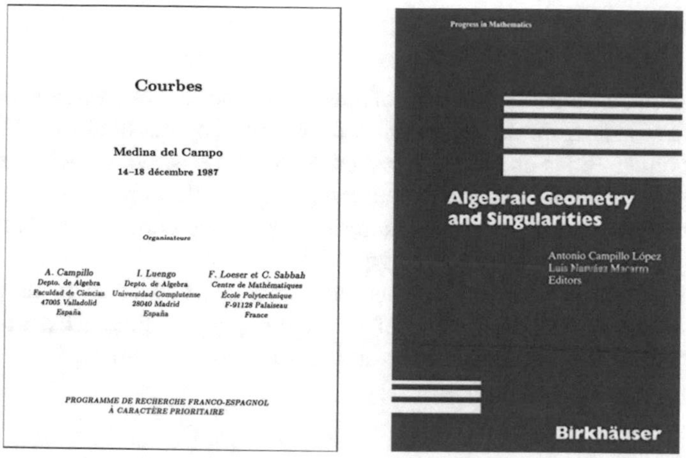

Fig. 3 Editorial work, I: *Courbes* (with permission of the organizers and editors), [50]; *Algebraic geometry and singularities*, [51]

- Monique LEJEUNE-JALABERT (PhD 1973, H. Hironaka): [3, 52–54]
- Pilar PISÓN (PhD 1991, J. L. Vicente): [33, 34, 55–60]
- Gerardo GONZALEZ-SPRINBERG (PhD 1982, J.-L. Verdier): [3, 48, 52–54, 61]
- Philippe THIERRY GIMENEZ (PhD 1993, M. Morales): [62, 63]
- Jorge Alberto GUCCIONE, Juan José GUCCIONE (PhD 1991, O. E. Villamayor), María Julia REDONDO (PhD 1991, O. E. Villamayor), Andrea SOLOTAR (PhD 1988, O. E. Villamayor), and Orlando E. VILLAMAYOR: [64]
- Luis NARVÁEZ (PhD 1984, Lê Dũng Tráng and José Luis Vicente Córdoba) [51]. See Fig. 3, right.
- Sabir GUSEIN-ZADE (PhD 1975, S. P. Novikov and V. I. Arnold): [5–31]
- Jorge OLIVARES (PhD 1994, X. Gómez-Mont): [65–70]
- Santiago ENCINAS (PhD 1996, O. E. Villamayor Uriburu): [71–73]
- Gert-Martin GREUEL (PhD 1973, E. V. Brieskorn), Christoph LOSSEN (PhD 1998, G.-M. Greuel): [74, 75]
- Patrick FITZPATRICK (PhD 1980, L. G. Kovács), Edgar MARTÍNEZ-MORO (PhD 2001, F. J. Galán Simón), Ruud PELLIKAAN (PhD 1985, D. Siersma): [76]
- Gabriel CARDONA (PhD 2002, J. Quer), Alejandro Melle-Hernández (PhD 1996, I. Luengo), Willem VEYS (PhD 1991, J. Denef), W. A. Zúñiga-Galindo (PhD 1996: K.-O. Stöhr) [77]
- Franz-Viktor KUHLMANN (PhD 1990, P. Roquette), Bernard TEISSIER (PhD 1973, H. Hironaka): [78]

5 Fruits and Seeds

The mathematical expertise domains of Antonio Campillo are *commutative algebra*, *singularities* and *algebraic geometry*, as represented by the books [1, 79] (Fig. 4). Among the main lines on which he has oriented his research, we will consider the following: equisingularity and deformations, particularly in positive characteristic; toric geometry; clusters, proximity, complete ideals and cones; algebraic geometry codes; algebraic foliations and polarity;

We also point out some questions and problems that seem to us promising research opportunities. At the end of the section, we include a list of headlines summarizing the major reseach achievements of Campillo so far.

5.1 Books

The memoir [1] essentially coincides with the thesis (1978) extended with a chapter on the singularities of space curves. "These notes intend to give a systematic development of the theory of equisingularity of irreducible algebroid curves over an algebraically closed field of arbitrary characteristic, using as main tool the Hamburger-Noether expansion instead of the Puiseux expansion which is usually employed in characteristic zero." (from the Introduction).

The review in the AMS Mathematical Reviews (MR) is signed by B. Teissier (1982), already a young giant in the theory of singularities by then, and it is worth quoting a few sentences from it: "This very clear book, defining all the concepts it uses, is an excellent introduction to the algebraic aspect of the classification of

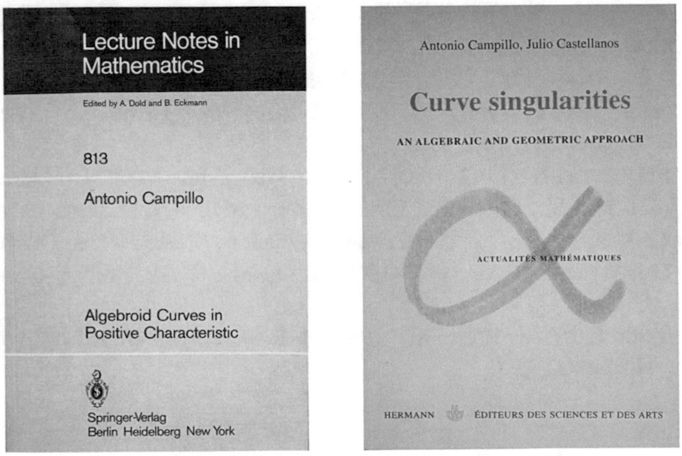

Fig. 4 Algebroid curves in positive characteristic, [1]; curve singularities, [79]

singularities of irreducible algebroid (or complex analytic) curves, in both positive and zero characteristics. [...] Here the central tool is the Hamburger-Noether expansion [...] In addition to the extension to positive characteristic, this book is useful because of the careful comparison that it develops among the different systems of invariants attached to a branch singularity: Hamburger-Noether, Puiseux, sequence of multiplicities of infinitely near points, semigroup of valuations, ... [...] There is also a very useful link, after Abhyankar's recent work, between the Hamburger-Noether expansion and the theory of the maximal contact of a branch with other singular branches [...] The last chapter studies three possible definitions of equisingularity of non-planar branch."[8]

It is worth mentioning that the Hamburger-Noether Expansions, which are the main tool used in this treatise, have been implemented by M. Lamm and C. Lossen as the package hnother.lib of the computer algebra system SINGULAR.[9]

The purpose of the book [79] is clearly described in its introduction: "One of the reasons for this book is that space curve singularities are still not classified by a good notion of equisingularity, so it becomes useful to clarify possible ways to focus on such a question. [...] Other reasons are that curve singularities also appear as a fundamental tool in recent developments concerning subjects such as arc spaces, resolution of singularities, complete ideals or valuations." In words of A. Dimca in the review MR1986115 (emphasis added), "The authors use the Arf closure relative to divisorial valuations *to define an extension of Puiseux exponents for general singularities*. Primary exponents come from valuation associated to arc space components. Secondary exponents are related to some Rees valuations." It may be worth remarking that part of this book amounts to a unified presentation of the papers [80–82] by the authors, the latter (*On Puiseux exponents for higher dimensional singularities*) included in the volumen in memory of Ruth Michler.

[8]Ce livre très clair, définissant tous les concepts qu'il utilise, constitue une excellente introduction à l'aspect algébrique de la classification des singularités des courbes algébroïdes (ou analytiques complexes) irréductibles, en charactéristique positive comme en caractéristique zéro. [...] Ici l'outil central est le développement de Hamburger-Noether [...] en plus de l'extension à la caractéristique positive, ce livre est utile par la comparison soigneuse qu'il contient entre les différents systèmes d'invariants attachés à une singularité de branche: Hamburger-Noether, Puiseux, suite de multiplicités de points infiniment voisins, semigroupe des valuations, [...] On trouve aussi le lien très utile après les travaux récents de Abhyankar entre le développement de Hamburger-Noether et la théorie du contact maximal d'une branche avec d'autres branches singulières [...] Le dernier chapitre étudie trois définitions possibles de l'équisingularité des branches non planes.

[9]W. Decker, G.-M. Greuel, G. Pfister, and H. Schönemann: SINGULAR 4-1-0 —*A computer algebra system for polynomial computations*, 2016. http://www.singular.uni-kl.de

5.2 Equisingularity and Deformations

Our general considerations above about the memoir [1] emphasized its pioneering character as a systematic study of equisingular families over fields of any characteristic and its status as a fundamental reference in the field. Here we will have a closer look on some of its more innovative features, and in particular on the three notions of equisingularity for space curves introduced in the last chapter.

The notion of equisingularity is characterized in terms of the completion of the local ring that is associated to the singularity (in particular it is invariant under change of coordinates and of parameterization), and numerically through several equivalent invariant systems: sequences of multiplicities of strict transforms of blowing-ups, characteristic exponents, maximal contact values, and semigroup of values.

In the case of twisted algebroid curves, three quite natural types of data are advanced as possible criteria for equisingularity:

1. The equisingularity class of a generic plane projection;
2. Sequences of multiplicities of strict transforms by blow-ups centered at closed points;
3. Semigroup of values.

More specifically, two irreducible algebroid curves are said to be, using the terminology of [1], *equisingular* E.s.1 if the equisingularity classes of their generic plane projections coincide [1, Definition 5.2.10]; E.s.2, if they have the same process of resolution (same multiplicity sequences) [1, Definition 5.3.1]; E.s.3, if their semigroups of values are equal [1, Definition 5.4.1]. For plane curves, these three definitions coincide, but they are different for space curves [1, Remark 5.4.4], and, as the author remarks, "none of them may be considered as a better definition than the other ones" [1, page 141]. On the other hand, the three notions of equisingularity extend naturally to curve singularities with several branches, but replacing the multiplicity sequence by the multiplicity tree.

Another important paper of A. Campillo concerning Hamburger-Noether expansions is [83]. In it, he defines a notion of Hamburger-Noether expansion over rings R (not only over fields), generalizing for them most of the results in [1] and obtaining some applications to equisingular deformation theory. In particular he proves that, for a fixed equisingularity type, there exists an equisingular family $X \to Y$ of $\mathbb{Z}$-schemes such that any equisingular family of algebroid curves (corresponding to that type) over any field K can be induced from $X \to Y$.

The notions of equisingularity E.s.1 and E.s.2 can be characterized by means of *saturation* and *Arf closure*, respectively. The concept of saturation was originally defined by Zariski in the 1960s using topological ideas (coverings); later, Pham and Teissier[10] gave an equivalent definition for the analytic case following the line of Grothendieck. In [2], Campillo introduced a third notion of saturation that is equivalent to the notions of Zariski and Pham-Teissier, but it is also valid in positive characteristic, fully extending and considerably simplifying the two previous theories. Both notions, saturation and Arf closure, have good algebraic characterizations in terms of the local algebra R of the singularity; more specifically, they can be translated in terms of the semigroup of values of R, leading up to two numerical equisingularity invariants of types E.s.1 and E.s.2, respectively: the semigroup of the saturation of R and the semigroup of the Arf closure of R. The paper [84] is devoted to develop arithmetical properties of the concept of saturation in arbitrary characteristic.

In [80], Arf closure and saturation are introduced with respect to a finite family of valuations. The paper [85] develops a geometrical theory of the Arf closure with respect to divisorial valuations for algebraic varieties of arbitrary dimension over arbitrary fields. Other papers related with Arf closure and Arf invariants associated with valuations are [81] and [82].

The aim of the paper [75] is to study equisingular deformations of plane curve singularities over an algebraically closed field of arbitrary characteristic. If this characteristic is zero it is known that the base space of an arbitrary deformation admits a unique maximal subspace over which the deformation is equisingular, that is, it admits a unique equisingularity stratum. However, if the field has positive characteristic, the situation may be different in special cases and they are described in this paper.

The paper [74] is computational and, in particular, an algorithm to compute the equisingularity stratum and its tangent space is given. About this paper, Arvid Siqveland says the following (MR2339830): "The way the deformation categories are described through the deformation functors is particularly interesting and illustrates a somewhat different technique. The proofs [...] illustrate beautiful and clever mathematics. Also, examples of occurring pathologies are given explicitly and the paper gives a really good introduction to the subject. The article should be read by anyone who is interested in a deep understanding of the geometry of deformation theory."

5.3 Toric Geometry

In *Syzygies of affine toric varieties* [63], a purely combinatorial method is given for computing the syzygies of the describing ring R of an affine toric variety. The working of the method is particularly illustrated in dimension one, and also on

[10] F. Pham, B. Teissier: *Fractions lipschitziennes d'un algèbre analytique complexe et saturation de Zariski.* Actes du Congrès International des Mathematiciens, Nice (1970), Tome 2, Gauthier-Villars, Paris (1971).

the problem of computing minimal generators of the defining ideal, and special attention is paid on the depth of R and Cohen-Macaulay semigroups. This paper was continued with [60], which presents a method for computing the syzigies.

In the paper *Toric mathematics from semigroup viewpoint* [59], toric geometry methods are developed which allow the study of non normal toric varieties by its close connection to semigroups and semigroup algebras. Here is a relevant quotation from the Introduction: "The purpose of this paper is to show how mathematics in toric geometry can be understood as the theory of appropriate classes of commutative semigroups with given generators. This viewpoint involves the description of various kinds of derived objects, as abelian groups and lattices, algebras and binomial ideals, cones and fans, affine and projective algebraic varieties, simplicial and cellular complexes, polytopes, and arithmetics. Our approach consists in showing the mathematical relations among the above objects and clarifying their possibilities for future developments in the area."

For other papers related to toric geometry, see [55, 56] (with P. Pisón); [33, 34] (with E. Briales, C. Marijuán and P. Pisón); [32] (with C. Marijuán); [86] (with M. A. Revilla); and [57, 58] (with E. Briales and P. Pisón). In the paper *Derivation algebras of toric varieties* [87] affine toric varieties are used for giving numerous examples of non normal varieties which are not determined by by their Lie algebras. Let us comment that affine normal varieties are determined by their Lie algebra, and those examples were the first that noticed that this does not hold in general.

5.4 Clusters, Proximity, Complete Ideals and Cones

Another relevant and fruitful topic appearing in the Campillo's work concerns clusters and proximity. These objects emerge with the study of singularities of plane curves, their desingularization by iterated point blowing-ups, and the problems of existence of curves having either assigned singularities or passing through a given set of points or infinitely near points with prescribed multiplicities. It goes back to Enriques and it is connected to the local study of (complete) linear systems of curves, carried out by Zariski, in the thirties, to define complete ideals on regular local rings and to investigate their structure and establish their theory in the smooth two dimensional case. Lipman continued the development of the theory of complete ideals for singular two dimensional cases and in higher dimension, establishing it for finitely supported complete ideals, i.e. for ideals supported at the closed point and such that there exist finite sequences of point blowing-ups (constellations or clusters) which make the ideals locally principal.

The paper [3] with G. González Sprinberg and M. Lejeune-Jalabert is devoted to establish a geometrical theory for finitely supported complete ideals in a regular local ring of arbitrary dimension, giving an explicit description of the monoid of monomial finitely supported complete ideals and starting the study of singularities naturally arising from these ideals.

In [40], with A. Reguera, the authors study morphisms given by composition of a sequence of point blow ups of smooth d-dimensional varieties in terms of combinatorial information coming from the d-ary intersection tensor on divisors with exceptional support. In particular, it is proved that this intersection tensor is an equivalent datum to the tree whose vertices are the points of the cluster that is associated to the sequence of blow ups, and whose edges join two points whenever one is proximate to the other.

The paper [61], with Gonzalez-Sprinberg, is devoted to the study of the relative cones of curves associated with morphisms as above. In particular, the authors consider clusters with enough few points such that the cone of curves (and its dual cone) is polyhedral (that is, finitely generated). The thesis of F. Monserrat (see the summary on page 9) was inspired by the results in this paper and identified a class of clusters for which the associated cones of curves are polyhedral.

In [41], with O. Piltant and A. Reguera, the surface obtained by blowing up the cluster of base (infinitely near) points of a pencil of curves of the projective plane with equation of the type $aF+bZ^r=0$ is considered, where F is an homogeneous polynomial of degree r defining a curve with only one place on the line at infinity ($Z=0$). It is proved that the cone of curves and the characteristic cone of that surface are polyhedral (in fact, regular) cones.

A more general case is studied in [42], again with O. Piltant and A. Reguera: they consider the surface obtained by blowing up the cluster of base (infinitely near) points of a pencil of curves of the projective plane with equation of type

$$aF+bZ^r=0$$

such that F is homogeneous of degree r and, moreover, it is a product of powers of polynomials defining curves with only one place on the line at infinity ($Z=0$). The results obtained are similar as those of [41], but in this case the cones are not necessarily regular.

The theory of finitely supported complete ideals, clusters and proximity has been a fruitful subject with many applications, some of them developed by A. Campillo and his collaborators and students. These include, among others: the study of singularities of polar curves by F. Delgado[11]; a geometric theory of complete ideals of rings associated with rational singularities (carried out by A. Reguera in her thesis; see page 8 for a summary); the Poincaré problem on the degree of projective integral curves of first order algebraic differential equations [35, 36, 88], and Galindo-Monserrat-2006[12]; the study of the cone of curves of rational surfaces

[11]A factorization theorem for the polar of a curve with two branches. *Composition Mathematica* **92**/3 (1994), 327–375.

[12]C. Galindo and F. Monserrat: Algebraic integrability of foliations of the plane, *Journal of Differential Equations* **231**/2 (2006), 611–632.

by Galindo-Monserrat-2005,[13] and Galindo-Monserrat-2004[14]; or the Harbourne–Hirschowitz conjecture for linear systems on rational surfaces Monserrat-2007.[15] The paper [48] is a survey on the subject where some of these applications can be found.

Nowadays, these techniques and results still constitute an open path for further research, specially considering, as starting point, the general framework established in [3], that allows to work in arbitrary dimension. Recall also what we said about Ann Lemahieu's thesis on page 9.

5.5 Algebraic Geometry Codes

Firstly we will look at three papers: [43] (with J. I. Farrán), [44] (with J. I. Farrán and C. Munuera) and [45] (with J. I. Farrán).

In [43] the authors give numerical expressions of the order (or Feng-Rao) bound on the minimum distance for sequences of duals of algebraic geometry codes defined over an algebraic curve C and using divisors mP, where P is a fixed rational point of C and m an element of the Weierstrass semigroup of C at P. It is worth mentioning that codes for suitable values of m in these sequences have been used by Tsfasman, Vlăduţ and Zink, and Garcia and Stichtenoth (among others) to show the existence of families of asymptotically good codes. The paper [44] is devoted to compute accurately the Feng-Rao bound for the codes of these sequences when the mentioned semigroup is Arf. The results developed for that purpose also provide the dimension of the improved geometric Goppa codes related to the considered algebraic geometry codes.

The paper [45] is an interesting work in computational algebra. On one hand, it is devoted to provide computational algorithms to determine, from the homogeneous equation of a projective plane curve defined over a perfect field, the singular points of the curve, parametrizations of the branches at these points (by means of Hamburger-Noether expansions) and equisingularity invariants of them. In particular, the desingularization of the curve (over the ground field) is determined. The algorithms are based on results in [1] and [83] and they have been implemented in the brnoeth library, which was produced by J. I. Farrán and Ch. Lossen, and distributed with versions 2.0 and higher of the SINGULAR computer algebra system. On the other hand, these algorithms are used to provide, when the ground field is finite, algorithms for coding and decoding of Algebraic Geometry codes defined

[13]The cone of curves associated to a plane configuration, *Commentarii Mathematici Helvectici* **80**/1 (2005), 75–93.

[14]On the cone of curves and of line bundles of a rational surface, *International Journal of Mathematics* **15**/4 (2004), 393–407.

[15]F. Monserrat: Curves having one place at infinity and linear systems on rational surfaces, *Journal of Pure and Applied Algebra* **211**/3 (2007), 685–701.

using general divisors over smooth curves for which a general plane model is known. The coding algorithm was essentially known,[16] but the decoding algorithm is new.

Here it is worth pointing out that no plane models are known for the curves of the explicit families providing asymptotically good codes. In fact, no decoding algorithms are known for them, but the procedures described in this paper will be ready to be used as soon as those models are obtained (AC).

Another interesting contribution to Coding Theory is given in [47], where the authors introduce *differential codes*. These are error-correcting codes defined by evaluating polynomials of degree less than m (for arbitrary m) at the singular points (in an affine chart) of an algebraic foliation of the projective plane (over a finite field) with reduced singularities. Applying results of [67], they obtain estimations of the parameters of these codes. This paper opens also interesting perspectives for future research: on the one hand, to extend the notion of differential codes using algebraic foliations defined in spaces of higher dimension and, on the other hand, to find efficient algorithms for decoding such codes.

It is worth adding here that the effort of A. Campillo to develop the theory of Algebraic Geometry codes has stimulated people to research on this subject, as, for example, his student Diego Ruano (see page 9 for a summary of his thesis), and also Fernando Hernando.

5.6 Algebraic Foliations and Polarity

In the papers [35, 88], and [36], the authors develop proximity and intersection inequalities for foliations (defined in surfaces in the case of the first two papers, and for higher dimensions in the last one). Moreover these inequalities are applied to obtain results concerning the so-called *Poincaré problem*, which was already considered by M. Carnicer for plane foliations having only non-dicritical singularities.[17]

Given a foliation of the plane having a rational (or meromorphic) first integral, the original Poincaré problem aims at bounding the degree of a general irreducible algebraic invariant curve in terms of the degree of the foliation. The interest in this problem stems from the following observation (Poincaré): if we have such a bound, then the problem of finding a rational first integral is reduced to purely algebraic computations. It turns out, however, that the problem has a negative answer in general, and as a consequence it has been rephrased as follows: to find a bound of the degree of the algebraic irreducible invariant curves of an arbitrary foliation of the plane *in terms of invariants of the foliation*. This problem, and the initial

[16] D. Le Brigand and J. J. Risler: Algorithme de Brill-Noether et codes de Goppa, *Bull. Soc. math. France*, **116**/2 (1988), 231–253.

[17] M. Carnicer: The Poincaré problem in the nondicritical case, *Annals of Mathematics* **140**/2 (1994), 289–294.

Poincaré's motivation (to find an algorithm for the computation of the rational first integral) are still open.

A 1-dimensional foliation $\mathscr{F}$ on the projective space $\mathbb{P}^k$ can be understood as a rational map $\mathbb{P}^k \cdots \to Gr(1,k)$ (where $Gr(1,k)$ denotes the Grassmannian of lines of $\mathbb{P}^k$) which satisfies the *polarity* condition, namely that image of every point P of $\mathbb{P}^k$ in the domain is a point of $Gr(1,k)$ that corresponds to a line going through P. The singular points of $\mathbb{F}$ are the points of indeterminacy of that rational map and the *singular subcheme* of $\mathscr{F}$, Sing($\mathscr{F}$), is the indeterminacy subscheme of the polarity map. In [66], A. Campillo and J. Olivares study this polarity map for plane foliations, establish a number of basic properties, and provide an explicit solution to the Poincaré problem for plane foliations and invariant curves having only simple singularities (corresponding to Dynkin diagrams A_n, D_n, E_6, E_7 and E_8).

In [67] it is proved that a plane foliation of degree greater than 1 is determined by its singular subscheme (roughly speaking, the singular subscheme is the set of singularities with multiplicities taken into account). Also, several conditions for subschemes of the projective plane to be the singular subscheme of a foliation are given. The proofs are based on properties of the polarity map and also on the Cayley-Bacharach Theorem.

In [68], the same authors (A. Campillo and J. Olivares) investigate the same problem studied in [67], but replacing $\mathbb{P}^2$ by an smooth projective variety of arbitrary dimension. In particular, they pinpoint a certain class of varieties (which includes projective spaces) for which foliations are determined by their singular subschemes. In this paper the results are not based on the Cayley-Bacharach Theorem and, therefore, they are not so explicit as those in [67]. An interesting challenge consists of finding more explicit proofs of these results using the Cayley-Bacharach theorem. In [69] such a proof is obtained in the case of $\mathbb{P}^3$, but the challenge is still open for the general case.

The paper [47] (considered before in the context of error-correcting codes) provides an interesting interplay between the theory of algebraic foliations and the theory of error-correcting codes. On one hand, foliations are used to define error-correcting codes (differential codes). On the other hand, the need to refine the estimation of the minimum distance of the given codes has given rise to an improvement of the conditions (given in [67]) that characterize the 0-dimensional schemes that are realized as singular subschemes of some plane foliation. This fact provides an unexpected application of error-correcting coding theory to the study of the geometry of the plane foliations. The following question, suggested by this paper, is still open: is it possible to construct differential codes using foliations in higher dimensional ambient spaces?

In the paper [70], A. Campillo and J. Olivares, applying results of [67], solve the problem of finding the subschemes (with the minimum possible degree) of the singular subscheme that determine a plane foliation. The framework provided by the results in this paper, and those in [47] and [67], suggest a genuine potential application to the design of secret sharing schemes in cryptography. Some current joint work of the same authors goes in this direction (AC).

5.7 *Poincaré Series and Zeta Functions*

A very remarkable part of A. Campillo's research consists of a series of papers on Poincaré series and zeta functions in several contexts that for the most part have been carried out in collaboration with F. Delgado and S. Gusein-Zade in the last 20 years. In this respect, the papers concerning the relationship between the Poincaré series and the topology of the singularities of plane curves have been seminal in the theory of Poincaré series. Nowadays, these series are key objects in a number of contexts such as the geometry and topology of singularities, Seiberg-Witten invariants, or the Floer homology, among others.

In [7] it is shown, by means of simple computations, that the Poincaré series of a complex plane curve singularity with only one branch coincides, up to a factor of the form $(1 - t)$, with the zeta function of the monodromy of the singularity. The surprising coincidence between combinatorial information coming from the resolution of singularities on one hand, and topological information on the other, is still a mystery because no topological or geometrical reason for this behaviour is known. In [8], an analogous result (which is proved in [13]) is announced for curve singularities with several branches. More specifically, in [13] a germ $C = \bigcup_{i=1}^{r} C_i \subset (\mathbb{C}^2, 0)$ of reduced plane curve (with an arbitrary number of branches) is considered. The multi-variable Alexander polynomial $\Delta^C(t_1, \ldots, t_r)$ is a complete topological invariant of the singularity $(C, 0)$, and it is shown that it coincides with the generalized Poincaré polynomial $P_C(t_1, \ldots, t_r)$ of the multi-index filtration induced by C on the ring $\mathcal{O}_{\mathbb{C}^2,0}$ of germs of functions in two variables. In the words of D. Matei in the review MR1962784, the proof is "a tour de force through the analytic, topological, and combinatorial characteristics of an embedded resolution of C". A different and simpler proof of this result, in the spirit of motivic integration, is given in [14]. Again, the intrinsic reason for this coincidence stands as a mystery many years after proving it (Fig. 5).

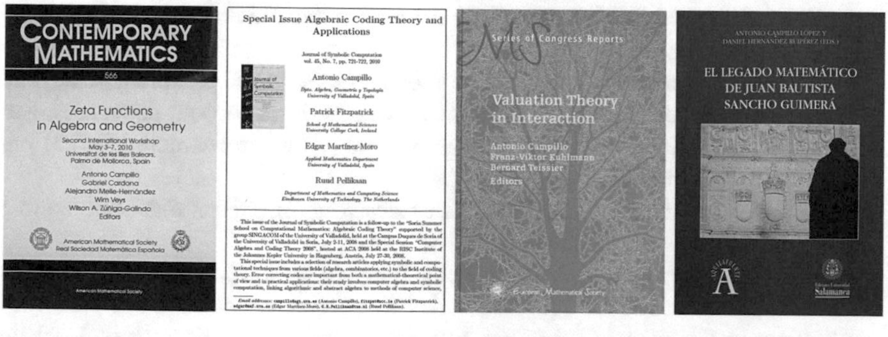

Fig. 5 Editorial work, II. From left to right: [77] *Zeta funtions in algebra and geometry*; [76] *algebraic coding theory and applications*; [78] *valuation theory and interactions*; and [89] *El legado matemático de Juan Bautista Sancho Guimerá*. (Images reproduced with the kind permissions by the American Mathematical Society, the Journal of Symbolic Computations, the EMS Publishing House, and Ediciones Universidad de Salamanca)

The result of [8] and [13] is considered paradigmatic in the theory of Poincaré series in several variables. The main reason is that one can use Poincaré series (a combinatorial object) in the comparison between the topology and the resolution of the singularity (via the A'Campo or Eisenbud-Newman formulas). As a consequence, these results gave rise to a wide program of research, still in full swing today, in which Poincaré series have replaced the topology. An evidence of the fruitfulness of this approach is given by a number of works of A. Campillo (most of them, as already said, in collaboration with F. Delgado and S. Gusein-Zade) concerning Poincaré series associated to different "singular" objects [6, 11, 15, 16, 18–31, 38, 49].

Among these works, [15] deserves a special comment. In it, the authors compute the Poincaré series of a natural multi-index filtration of the ring of germs of functions on a rational surface singularity and prove, among other results, that it is a rational function with a precisely given denominator.

The model of proof used to compute Poincaré series in the case of curves was conceptualized and formalized using *integration with respect to the Euler characteristic* (a notion that was introduced by O. Viro in the 1990s), providing a good framework for the computation of Poincaré series in more general cases [10, 12, 14].

5.8 Achivements in Headlines

- His main scientific interest has been the structure and classification of singularities in geometry and in any characteristic.
- His methodology is based on using associated algebraic structures, mainly arcs and semigroups, and on discovering how to read geometry and topology directly in those structures.
- He extended to a positive characteristic the theory of equisingularity for plane curves and showed that the moduli spaces are irreducible and independent of the base field.
- With G.-M. Greuel and C. Lossen, he proved that in positive characteristic, and contrary to expectations, all equisingularity strata have smooth properties, and that the singularities that present some pathology for the characteristic are not generic in their moduli spaces.
- In an arbitrary dimension, he searched extensively for objects that play the role of the semigroups of values in the case of curves.
- He introduced and developed, with Felix Delgado de la Mata and Sabir Gusein-Zade, a novel notion of Poincaré series in several variables associated with filtrations in the rings of functions. Numerous results were discovered, by them and other authors, in which relevant topological or geometric information of the singularity, such as the zeta function of the monodromy or the Seiberg-Witten invariants, can be read directly in a Poincaré series.

- A paradigmatic contribution has been to prove for germs of plane curves over the complex field that the Alexander polynomial of the knot entanglement that defines its topology is equal to the associated Poincaré series.
- He has applied his models to other problems of affine or projective geometry. For toric algebras, which are semigroup algebras, he described his minimal graded resolution in fully combinatorial terms. These works, with P. Pisón-Casares and P. T. Gimenez mainly, together with simultaneous advances by B. Sturmfels, are part of the development of the non-normal affine toric geometry motivated by its practical applications.
- He has created and implemented, with J. I. Farrán, computational algorithms for the coding and decoding of algebro-geometric codes constructed from plane projective curves over finite fields.
- He has contributed to developed, with G. González-Sprinberg and M. Lejeune-Jalabert, the theory of complete ideals (local linear systems) in cases of arbitrary dimension, as evisioned by Zariski.
- The same techniques have been useful to extend the theory of Mori to cases not covered by it and to bound the degree of invariant curves by foliations of dimension one in projective spaces.
- Has introduced, with J. Olivares, the polarity for such foliations, has proven that these are determined by their singular subscheme, thus extending the result of X. Gómez-Mont and G. Kempf in the generic case.
- He has developed applications for interpolation, coding and cryptography.

6 Services and Influence

Beyond his scholar achievements in research, research training and teaching, Antonio Campillo has always been willing to accept responsibilities in the service of community institutions of various sorts and levels, and quite often to undertake and lead projects aimed at their development. To see this side of his professional career at a glance, let us first compile a time-line for the most relevant of such endeavors.

6.1 University Administration

- 1983: Director of the Algebra Department of the University of Sevilla.
- 1996–1999: Director of the Department of Algebra, Geometry and Topology of the UVA.
- 2000–2004: Dean of the Faculty of Sciences of the UVA.
- 2003–2010: Member of the Advisory Board of the UVA.

6.2 Cooperation with Mathematical Societies

- 1997–2003: Spanish Corresponding Member of the French Mathematical Society.
- 2010–: Member of the International Center of Pure and Applied Mathematics (CIMPA-ICPAM).
- 2015–2016: President of Spain's Adhering Organization to the International Mathematical Union (IMU)

6.3 Scientific Policy

- 1999–2001: Member of the Physics and Mathematics Panel of the National Agency for Prospective Evaluation.[18]
- 2001–2002: Deputy to the ANEP Coordinating Committee for Mathematics.
- 2002–2003: Member of the Advisory Board for Physics and Mathematics (Area 1) of the Ministry of Education, Culture and Sports National Commission for the Evaluation of the Research Activity.
- 2003. Member of the Evaluation Commission of the National Plan for R+D+I of the Ministry of Science and Technology.[19]
- 2004: Member of the Committee for Physics and Mathematics of the Ministry of Science and Technology.[19]

6.4 Conference of Spanish Deans of Mathematics

- 2000–2001: Member of the Promoting Committe of the Conference of Spanish Deans of Mathematics.[20]
- 2001–2004: President of the CDM. In the period 2002–2004 was the Coordinator of Spain's National Quality Agency[21] White Book on University Mathematics Curriculae (see [90]).

[18] Agencia Nacional de Evaluación y Prospectiva, ANEP.

[19] Ministerio de Ciencia y Tecnología, MCyT. The Plan was coordinated by the Government Presidency.

[20] Conferencia de Decanos de Matemáticas, CDM

[21] Agencia Nacional de Evaluación de la Calidad y Acreditación, ANECA.

6.5 *Responsibilities in the RSME*

- 2003–2006: Member of the Governing Board.
- 2003–2007: President of the Scientific Commission.
- 2005: President of the Scientific Committee of the First Joint Meeing MAT.ES of the mathematical societies RSME, SCM, SEMA and SEIO.[22]
- 2009–2015: President of the RSME.
- 2009, 2012, 2014: Member of the Program Committee of the Joint Meetings of the SMM and the RSME.
- 2010: Promoter of the 2010 Joint Meeting of Mathematical Societies and CIMPA, sponsored by the Spanish Ministry of Science and Innovation.
- 2010–2016: President of CEMAT,[23] the Spanish Committee of Mathematics (its mission includes the coordination with IMU of Spanish activities in the international mathematical scene).
- 2010–: President of the RSME-Imaginary Committee.
- 2010: Head of Spain's Delegation at IMU's General Assembly in Hyderabad.
- 2010 Delegate at the EMS' Council held in Sophia.
- 2011: Promoter of the 4th Meeting of Presidents of National Member Societies of the EMS, held in Bilbao, and co-organizer of the same together with Marta Sanz-Solé, Stephen Huggett and Luis Vega.
- 2011: Coordinator of the Honorary Committee of the Centennial of the RSME.
- 2011–2015: President of the RSME-Universia Editorial Committee.[24]
- 2012: Member of the Scientific Committee of the First Imaginary Conference (Barcelona).
- 2012 Delegate at the EMS' Council held in Krakow.
- 2012, 2016: Member of the Program Committee of the Joint Conferences of the RSME, BMS/SMB and SML.[25]
- 2012–2013: Member of the Honorary Committee of the Turing Year in Spain.
- 2014: Member of Scientific and Organizing Committees of the Symposium in Memory of Juan Bautista Sancho Guimerá (cf. [89]).
- 2014: Head of Spain's Delegation at IMU's General Assembly in Seoul.
- 2014 Delegate at the EMS Council held in San Sebastián.
- 2014: Member of the Honorary Committee of 50 Years of the Mathematical Olympiad in Spain.
- 2016 Delegate at the EMS Council held in Berlin.
- 2018 Delegate at the EMS Council held in Prague.

[22]Societat Catalana de Matemàtiques, Sociedad Española de Matemática Aplicada, and Sociedad de Estadística e Investigación Operativa.

[23]Comité Español de Matemáticas, http://matematicas.uclm.es/cemat/

[24]http://www.arbolmat.com/ (cf. Footnote 3).

[25]Belgian Mathematical Society/Société Mathématique de Belgique, and Société Mathématique du Luxembourg.

There are several other relevant indicators of Campillo's influence and leadership that are worth recording here. For example, the following two tables summarize his supervision of predocs and postdocs:

Predocs		
Year	Name	Origin
1996	Patrick Popescu-Pampu	École Normale Supérieure, Paris, France
1999	Carles Bivia Ausina	Universidad Politécnica de Valencia, Spain
2002	Giovanni Molica	Università di Messina, Italy
2006	Ann Lemahieu	Katholieke Universiteit Leuven, Belgium
2007	M. Cruz Fernández Fernández	Universidad de Sevilla, Spain
2007	Gilherme Tizziotti	Universidade Estadual de Campinas, Brazil
2008	Javier Tordable	Google Seattle, USA
2008	Eugenia Ellis	Universidad de la República, Uruguay
2009	Irene Márquez Corbella	Universidad de La Laguna, Spain
2009	Ana Belén de Felipe	Universidad de La Laguna, Spain
2016	Wanderson Tenorio	Universidade Estadual de Campinas, Brazil

Postdocs		
Year	Name	Origin
1988	Zbigniew Hajto	Krakow University, Poland
1993	Philippe Gimenez	Institut Fourier de Grenoble, France
1996	Jorge Olivares Vázquez	CIMAT, Guanajuato, México
2001	Fernando Torres Orihuela	Universidade Estadual de Campinas, Brazil
2002	Mustapha Lahyane	ICTP, Trieste, Italy
2003	Thomas Brelivet	Université de Bordeaux, France
2004	Vincent Florens	Université de Strasbourg, France
2004	Adrian Varley	Warwick University, England
2004	Tristan Torrelli	Université de Nice, France

Another aspect showing the range of Campillo's influence is that he has been on the jury of PhD thesis defenses, often as chairman, in 108 occasions (at the time of this writing), 92 from Spanish universities (Valladolid, Madrid, Sevilla, Salamanca, Barcelona, La Laguna, Oviedo, Santander, Zaragoza, Granada, Málaga, Badajoz, Ciudad Real, Santiago de Compostela, and Valencia) and 16 in universities of other countries (Germany, France, Belgium, Finland, and Uruguay). Similarly, he has been on the jury for Full Professor (Associate Professor) openings[26] in 11 (20) occasions, again often as the chairman of the jury.

Some of the activities we have mentioned are related to Campillo's presence in committees of various sorts. If we now look only at the scientific committees in

[26]In recent years these include the habilitation and acreditation procedures.

which he has served, usually as chairman or main organizer, we find that he has been involved in about 30 conferences (including workshops, symposia, and networks' meetings), from the 1984 International Conference on Algebraic Geometry held at La Rábida (Huelva, Spain) to the 2018 VII Iberoamerican Congress on Geometry held in Valladolid[27]; in scientific days of the RSME with other scientific partners (one at a time: Bibao 2003 with SEMA; Barcelona 2003 with SCM; Elche 2004 with SEIO; Alicante 2005 with the Department of Mathematical Economy; Valladolid 2005 with RSEF; Sevilla 2006 with the SEN; and Sevilla 2011 with RSEQ); ten research schools, generally of a multidisciplinary character (from 2005 to 2010); and ten thematic seminars (from 2006 to 2010).

During the two terms as President of RSME (2009–2012 and 2012–2015), Antonio Campillo published a Letter of the President in LA GACETA of the RSME in 19 occasions.[28] Important as these documents are for a deeper appraisal of Campillo's steering thinking and achievements at the forefront of the RSME, here we can do not better than to offer a few general remarks about them aimed at highlighting points that have not been mentioned yet or that they provide relevant supplements to aspects already considered. As we will see, two letters (**14**/3 and **16**/2) are actually reproductions of speeches delivered on particularly significant occasions for the RSME mission, while the supposed space of the letter is ceded to other purposes for the issues **14**/1, **15**/4, and **16**/3.[29]

To a great extend, his first term was focused on the Centennial of the RSME. This is clearly underlined already in the first letter, **12**/4. Indeed, after thanking the ample support received, also the people that accepted going in his team (Henar Herrero, Joan Elías, Luis Vega and Santos González) and the three preceding presidents (Carlos Andradas, Antonio Martínez Naveira and Olga Gil), we find a broad picture of how the term ahead was envisioned:

> The Centenary of the RSME in the year 2011 is conceived as an exceptional occasion for the collective celebration of the advance of mathematics in all directions, as well as for the confluence of the various mathematical sectors and the reciprocal support between them. Research, university education, compulsory education and baccalaureate, publications and libraries, cooperation, applications and implications, or dissemination and culture, are, among others, priority directions for the RSME. I am pleased to invite this celebration to scientists, researchers, professionals, students, professors, authors, editors, artists, communicators, and those who feel interested in mathematics, to meet or associate with the Royal Spanish Mathematical Society to share the celebration.
>
> The celebration of the Centennial of the RSME will be in complete harmony with current trends in international mathematics, trying to promote the presence and external visibility of Spanish mathematicians and the RSME as an institution. In particular, the Centennial

[27] http://iberoamericangeometry2018.uva.es/

[28] 2009: **12**/4, 613–5. 2010: **13**/1, 5–6; **13**/2, 203–5; **13**/3, 405–7; **13**/4, 603–4. 2011: **14**/2, 209–10; **14**/3, 403–7; **14**/4, 611–2. 2012: **15**/1, 5–8; **15**/2, 233–4; **15**/3, 415–8. 2013: **16**/1, 5–8; **16**/2, 205–7; **16**/4, 605–9. 2014: **17**/1, 5–8; **17**/2, 203–6; **17**/3, 405–6; **17**/4, 611–3. 2015: **18**/, 7–10; **18**/2, 219–21.

[29] In this issue the piece is an interesting interview by Ann Lemahieu of María Pe Pereira, winner of the 2012 edition of the José Luis Rubio de Francia Prize.

Congress will be presented in scientific harmony with the two main international congresses between which it will be inserted: the 26th ICM of the IMU, from August 17 to 27, 2010, in Hyderabad, and the sixth EMS congress, from 2 to July 7, 2012, in Krakow.

The letter continues by announcing joint activities with half a dozen societies, the decisive support to important existing initiatives (like the Lluís Santaló and Miguel de Guzmán research schools, the Divulgamat portal, or the Women and Mathematics Committee) and the bolstering of others, particularly the publishing program and different ways of engaging the young. These promises were largely fulfilled, often beyond expectations, as attested by the facts recorded in this paper.

As hinted before, **14**/1 included, instead of the President's Letter, a long account (pages 5–24) of the rather solemn Centennial Open Ceremony. It features an introduction describing the unfolding of the ceremony, with a mention to the celebrated lecture *Olé las matemáticas*, by José Luis Fernández, and reproductions of the excellent speeches by the President of the Centennial Committee, María Jesús Carro (pages 10–14); by the Minister for Education, Ángel Gabilondo (pages 19–24); and by Antonio Campillo himself (pages 15–19), who in particular announced the publication of a History of the RSME.[30]

The letter in **14**/2 is an account of the good progress of the Centennial, while the issue **14**/3 reproduces two interventions of Campillo, one in the Opening of the Millenium Days (Barcelona, June 1st, 2011) and the other in the Opening of the Mathematical Days and the Life Sciences (Granada, June 6, 2011), both concerned with the ingredients needed for a flourishing research in the mathematical sciences. A reflection on the closing of the year, particularly focused on the Closing Ceremony in the Spanish Senate on November 29, 2011, is provided in the letter **14**/4.

Mathematics as an endurable value for education, research and culture will be the axis of the closing statement of the Centenary that, coordinated by Francisco Marcellán, will be presented at this ceremony. It will count, among institutional interventions, with a conference on "the eternal youth of Mathematics" by Sir Michael Atiyah.

In next letter, **15**/1, the first after the Centenary year, the RSME is reaffirmed "in its commitment to a growing dedication to education, research and mathematical culture in Spain, whose current challenges are formulated in the Centennial Closing Statement presented in the Senate on November 29."[31] In the same letter, we find the relation of the RSME honorary members at that moment: María Josefa Wonenburger, Antonio Martínez Naveira, Michael Atiyah, Haïm Brezis, Charles Fefferman, Gert-Martin Greuel, Heisuke Hironaka, Peter Lax, Yves Meyer, Jean Pierre Serre and Efim Zelmanov, noticing that all, except the first named, were nominated in the last months.

In the letter **15**/2, Divulgamat and two of the initiatives that were born in the Centennial Year, all alive and well at the time of this writing, are assessed in a broader context:

[30]Luis Español, *Historia de la RSME*. RSME, November 2011.

[31]http://www.rsme.es/org/DeclaracionRSME291111.pdf

> The experience acquired by the RSME, through Imaginary, Divulgamat or Arbolmat, among other initiatives, demonstrates how mathematical culture reinforces and stimulates both institutional relationships and communication. In particular, communication is one of the aspects that have stood out most during the celebration of the Centennial, and also continues to be valid both internally among the mathematical community, and externally in relation to the general public.

A report on the 2012 elections of the RSME is included in **15**/4 in place of the President's letter. Antonio Campillo was reelected as President, and the elected Governing Board was formed by two Vice-presidents (Francisco Marcellán and Adolfo Quirós), a Treasurer (Julio Bernués), a Secretary (Henar Herrero), a General Editor (Joan Elías) and ten additional members (vocales).

In the early months of Campillo's second term, a strategic plan 2013–2018 for the RSME was carefully devised. Its essence is described in the letter **16**/1:

> The need to channel the growing individual and collective contributions, the RSME has prepared its Strategic Plan 2013–2018, which is already in force. The objectives revolve around the axes of research, teaching, social responsibility, and professional image and its value. At the same time, the transversal axes of internationalization, interdisciplinarity, attraction and promotion of mathematical talent, and dissemination, will be constantly stimulated so that they permeate all the activity of the RSME.

In the next issue, **16**/2, the letter is replaced by the speech on the occasion of Prize Award Ceremony of the XLIX Spanish Mathematical Olimpiad (MO). It was given in Bilbao on April 6, 2013, and amounts to a detailed analysis and eloquent examples of the importance of the Mathematical Olympiad as an effective system "in the detection and training of young mathematical talents, very often those who in the future will be its most relevant scientists."

In the first months of the second term, scientific endeavors in Spain suffered an unprecedented setback. The letter **16/4** (the last of 2013) could express it louder, but not clearer:

> In difficult times for science, in which only one in ten vacant positions of professor and researcher can be replaced and the state fund for research has decreased by a third in two years of the legislature, the activity of scientific societies, and notably the RSME, becomes much more necessary. May these lines of complaint serve as a decry of the existing reality and as a rational allegation for its repair.

The letter continues by thanking many people who were helping, in institutions or individually, to maintain the RSME alive, and, quite fittingly, with a reproduction of Campillo's address to the ceremony organized to celebrate the 50th edition of the Mathematical Olympiad in Spain three months before it took place in Requena (Valencia, Spain). The letter also congratulates Ingrid Deaubechies and David Mumford for having been distinguished with the prestigious BBVA Foundations Frontiers of Knowledge Award[32] in the Basic Sciences. The event was an endearing landmark on several counts, including that it was the first time that Mathematics was the distinguished basic science.

---

[32] https://en.wikipedia.org/wiki/BBVA_Foundation_Frontiers_of_Knowledge_Award

The next six letters (4 in 2014 and 2 in 2015) reflect the decided cruise course of RSME, oriented by the strategic plan, in all fronts. Valuable as this may be, it is even more deserving to point out the achievements standing out of that course. Thus RSME became an associated member of the International Council on Industrial and Applied Mathematics (ICIAM), coinciding with the election of Maria Jesús Esteban to preside it for the period 2015–2019, and the decision to entrust SEMA of the organization of ICIAM 2019 in Valencia. "Our society promotes the role of mathematics as a science and supports the development of applied and industrial mathematics, as reflected in our participation in the international initiative Mathematics of Planet Earth (MPE), originated in 2013, and regularly in the schools, conferences and congresses that we organize" (**17**/1).

In **17**/2, the reported landmark is the launch of the "Vicent Caselles Research Awards" to annually distinguish, in remembrance of the scientific work of the teacher and colleague who left us in 2013 at a young age, six researchers in their first postdoctoral stage. The "RSME Medals" are also announced to distinguish annually, starting in 2014, three colleagues with outstanding career in the mathematical profession.

The 8th Miguel de Guzman School (9–11 July 2014) was organized in collaboration the Federation of Societies of Mathematics Teachers (FESPM) with the title "The mathematics we need: creativity and good practices in compulsory education". It was held at the UCM and "has had the support of the Ministry of Education, which, in addition to contributing to the activity through the National Institute for Educational Evaluation, has recognized the School for training purposes" (**17**/3). Moreover, "The active and numerous participation of teachers of all educational levels has represented an unprecedented success and has outlined a collective, generically young and with ample concurrence of women, capable of addressing the current challenges of teaching mathematics."

From **17**/4, let us single out two points. One is "The homage to exile celebrated in the Mexico's Spanish Athenaeum on September 5, by joint initiative with the SMM", one of the activities of the 3rd Joint Meeting RSME-SMM celebrated in Zacatecas (México), and which "allowed to testify and remember the work of Spanish mathematicians exiled 75 years ago, and had an important media impact". The other point concerns, besides the publication of own collections (authored books and facsimiles), the publishing agreements of RSME with several institutions (EMS-Publishing House, AMS, Springer-Verlag, SM and e-LectoLibris).

The two letters of 2015, particularly **18**/2, summarize the outcomes of past activities and express gratefulness to all people and institutions that have contributed to the achievement of this "unforgettable stage", ending with "But, above all, and of course in a very special way, Rosa, Elisa and Miguel" (see Fig. 6).

Fig. 6 Antonio Campillo's family: Rosa de Frutos (wife), Elisa (daughter, a journalist) and Miguel (son, junior student of Civil Engineering). (With the kind permission of Antonio Campillo)

6.6 Coda

Antonio Campillo was awarded the RSME-Medal in 2017 (see Fig. 7), a most deserved honor in our view, but we feel that we have to end this work by returning to research. Beyond the many mathematical contributions, we hope to have shown that we owe to Antonio Campillo a very enriched vision of the singularities and related areas that expands and considerably reinforces the crossroads position that these issues have in the whole of mathematics. Like the composer that starts fixing on the score the music he has been long listening in his mind, we take it as a snapshot of such a moment his March 3 talk on "Dual polyhedra and graphs" given in the Singacom Seminar with the following summary:

> Various alternative notions of dual polyhedra and dual graphs are presented for constellations of infinitely close points in arbitrary dimension, showing how all these notions have equivalent information, and that they extend to a higher dimension the notion of dual graph widely used in the description of the topology of the points of the surfaces. We analyze the role of the successions of blow ups and show how the difficulties brought about by the increase in dimension are offset by the richness of the geometry of the exceptional divisors.

Happy 65th birthday, Antonio, with the wish that you may continue to bring forth your inspirations for many years to come in the fields of your research an to the mathematical community in general!

Fig. 7 Francisco González (President of the Fundación BBVA), Carmen Vela (Secretary of State for Research), Antonio Campillo (awardee of the RSME Medal 2017), Francisco Marcellán (President of the RSME). (With the kind permission by the Fundación BBVA)

Acknowledgements Among the materials that we have had during the preparation of this paper, we are most grateful for a draft of notes by Antonio Campillo on his mathematical contributions at a much higher resolution than what we have been able to include in this work (the latest version of his notes has over 50 pages). It is like the transcription of music that a composer hears in his mind. We see it as a precise and precious mental map, with all the relevant concepts and connections in place, a generous tour-de-force on his part that will have positive and substantial repercussions in the investigation of the problems raised for many years. Our quotations from it are identified by the initials AC.

We also thank the students and collaborators of Antonio Campillo for providing us with all the answers to our questions and thus facilitating our task of improving and validating the informations in this work. With the understanding, of course, that only the authors are responsible for errors and misconceptions that may have remained.

References

1. Campillo, A.: Algebroid Curves in Positive Characteristic. Lecture Notes in Mathematics, 176pp. Springer, Heildelberg (1980)
2. Campillo, A.: On saturations of curve singularities (any characteristic). In: Orlik, P. (ed.) Singularities. Volume 40/1 of Proceedings of Symposia in Pure Mathematics, pp. 211–220. American Mathematical Society, Providence (1983). Proceedings of the Arcata Symposium "Singularities 1981"
3. Campillo, A., Gonzalez-Sprinberg, G., Lejeune-Jalabert, M.: Clusters of infinitely near points. Mathematische Annalen **306**(1), 169–194 (1996)
4. Campillo, A., Delgado, F., Kiyek, K.: Gorenstein property and symmetry for one-dimensional local Cohen-Macaulay rings. Manuscripta Mathematica **83**(3–4), 405–423 (1994)

5. Gusein-Zade, S., Delgado, F., Campillo, A.: The extended semigroup of a plane curve singularity. In: Proceedings of the Steklov Institute of Mathematics, vol. 221, pp. 139–156 (1998). From Trudy Matematicheskogo Instituta imeni V. A. Steklova, volume 221 (1998), 149–167. This volume was dedicated to V. I. Arnold on the occasion of his 60th birthday
6. Gusein-Zade, S., Delgado, F., Campillo, A.: On the monodromy at infinity of a plane curve and the Poincaré series of its ring of functions. In: Volume Dedicated to the 90th Birthday of Pontriagyn. Volume 68 of Contemporary Mathematics and Its Applications, pp. 49–54. Itogi Nauki i Tecniki, Moscow, 7
7. Gusein-Zade, S., Delgado, F., Campillo, A.: On the monodromy of a plane curve singularity and the Poincaré series of its ring of functions. Funct. Anal. Appl. **33**(1), 56–57 (1999)
8. Gusein-Zade, S., Delgado, F., Campillo, A.: The Alexander polynomial of a plane curve singularity and the ring of functions on the curve. Russ. Math. Surv. **54**(3), 634–635 (1999)
9. Campillo, A., Delgado, F., Gusein-Zade, S.M.: On generators of the semigroup of a plane curve singularity. J. LMS (2) **60**(2), 420–430 (1999)
10. Gusein-Zade, S., Delgado, F., Campillo, A.: Integration with respect to the Euler characteristic over a function space, and the Alexander polynomial of a plane curve singularity. Russ. Math. Surv. **55**(6), 1148–1149 (2000)
11. Gusein-Zade, S.M., Delgado, F., Campillo, A.: On the monodromy at infinity of a plane curve and the Poincaré series of its coordinate ring. J. Math. Sci. **105**(2), 1839–1842 (2001). English version of [6]
12. Gusein-Zade, S., Delgado, F., Campillo, A.: Integrals with respect to the euler characteristic over spaces of functions, and the alexander polynomial. Proc. Steklov Inst. Math. **238**(3), 134–147 (2002)
13. Campillo, A., Delgado, F., Gusein-Zade, S.: The Alexander polynomial of a plane curve singularity and the ring of functions on it. Duke Math. J. **117**(1), 125–156 (2003)
14. Campillo, A., Delgado, F., Gusein-Zade, S.: The Alexander polynomial of a plane curve singularity and integrals with respect to the Euler characteristic. Int. J. Math. **14**(1), 47–54 (2003)
15. Campillo, A., Delgado, F., Gusein-Zade, S.: On the zeta functions of a meromorphic germ in two variables. In: Buchstaber, V.M., Krichever, I.M. (eds.) Geometry, Topology, and Mathematical Physics (S. P. Novikov's Seminar 2002–2003). Volume 212 of Advances in Mathematical Sciences/Translations, pp. 67–74. American Mathematical Society, Providence (2004)
16. Campillo, A., Delgado, F., Gusein-Zade, S.: Poincaré series of a rational surface singularity. Inventiones Mathematicae **155**(1), 45–53 (2004)
17. Campillo, A., Delgado, F., Gusein-Zade, S.: On the monodromy of meromorphic germs of plane curves and the poincaré series of his ring of functions. In: Geometry, Topology, and Mathematical Physics (S. P. Novikov's Seminar 2002–2003). Volume 212 of Advances in Mathematical Sciences/Translations, pp. 234–243. American Mathematical Society, Providence (2004)
18. Campillo, A., Delgado, F., Gusein-Zade, S.: Poincaré series of curves on rational surface singularities. Commentarii Mathematici Helvetici **80**(1), 95–102 (2005)
19. Campillo, A., Delgado, F., Gusein-Zade, S.: Multi-index filtrations and generalized Poincaré series. Monatshefte fúr Mathematik **150**(3), 193–209 (2007)
20. Campillo, A., Delgado, F., Gusein-Zade, S.: On Poincaré series of filtrations on equivariant functions of two variables. Mosc. Math. J. **7**(2), 243–255 (2007)
21. Gusein-Zade, S., Delgado, F., Campillo, A.: Universal abelian covers of rational surface singularities, and multi-index filtrations. Funct. Anal. Its Appl. **42**(2), 83–88 (2008)
22. Campillo, A., Delgado, F., Gusein-Zade, S., Hernando, F.: Poincaré series of collections of plane valuations. Int. J. Math. **21**(11), 1461–1473 (2010)
23. Campillo, A., Delgado, F., Gusein-Zade, S.: The Poincaré series of divisorial valuations in the plane defines the topology of the set of divisors. Funct. Anal. Other Math. **3**(1), 39–46 (2010)
24. Gusein-Zade, S., Delgado, F., Campillo, A.: Alexander polynomials and Poincaré series of sets of ideals. Funct. Anal. Its Appl. **45**(4), 271–277 (2011)

25. Campillo, A., Delgado, F., Gusein-Zade, S.: Equivariant Poincaré series of filtrations. Revista Matemática Complutense **26**(1), 241–251 (2013)
26. Campillo, A., Delgado, F., Gusein-Zade, S.: Hilbert function, generalized Poincaré series and topology of plane valuations. Monatshefte für Mathematik **174**(3), 403–412 (2014)
27. Campillo, A., Delgado, F., Gusein-Zade, S.: Equivariant Poincaré series of filtrations and topology. Arkiv för Matematik **52**(1), 43–59 (2014)
28. Campillo, A., Delgado, F., Gusein-Zade, S.: On Poincaré series of filtrations. Azerbaijan J. Math. **5**(2), 125–139 (2015)
29. Campillo, A., Delgado, F., Gusein-Zade, S.: An equivariant Poincaré series of filtrations and monodromy zeta functions. Revista Matemática Complutense **28**(2), 449–467 (2015)
30. Campillo, A., Delgado, F., Gusein-Zade, S.M.: Equivariant Poincaré series and topology of valuations. Documenta Mathematica **21**, 271–286 (2016)
31. Campillo, A., Delgado, F., Gusein-Zade, S.: On the topological type of a set of plane valuations with symmetries. Mathematische Nachrichten **290**(13), 1925–1938 (2017)
32. Campillo, A., Marijuán, C.: Higher order relations for a numerical semigroup. Séminaire de Théorie des Nombres de Bordeaux **2**(3), 229–240 (1991)
33. Briales, E., Campillo, A., Marijuán, C., Pisón, P.: Minimal systems of generators of ideals of semigroups. J. Pure Appl. Algebra **124**(1), 7–30 (1998)
34. Briales, E., Campillo, A., Marijuán, C., Pisón, P.: Combinatorics and syzygies for semigroup algebras. Collectanea Mathematica **49**(2–3), 239–256 (1998). Special Volume in Memory of Ferran Serrano
35. Campillo, A., Galindo, C., García de la Fuente, J., Reguera, A.: Proximity and intersection inequalities for foliations on algebraic surfaces. In: Journés Singulières et Jacobiénes. Publications of the Institute Fourier de Grenoble, pp. 25–40. Institute Fourier (1994)
36. Campillo, A., Carnicer, M., García de la Fuente, J.: Curves invariant by vector fields on algebraic varieties. J. LMS (2) **62**(1), 56–70 (2000)
37. Campillo, A., Galindo, C.: On the graded ring relative to a valuation. Manuscripta Mathematica **92**, 173–189 (1997)
38. Campillo, A., Galindo, C.: The Poincaré series associated with finitely many monomial valuations. Math. Proc. Camb. Philos. Soc. **134**(3), 433–443 (2003)
39. Campillo, A., Galindo, C.: Toric structure of the graded algebra relative to a valuation. In: Christensen, C., Sathaye, A., Sundaram, G., Bajaj, C. (eds.) Algebra, Arithmetic and Geometry with Applications. Springer, Berlin/Heidelberg (2004)
40. Campillo, A., Reguera, A.: Combinatorial aspects of sequences of point blowing ups. Manuscripta Mathematica **84**(1), 29–46 (1994)
41. Campillo, A., Piltant, O., Reguera, A.: Cones of curves and of line bundles on surfaces associated with curves having one place at infinity. Proc. Lond. Math. Soc. **84**(3), 559–580 (2002)
42. Campillo, A., Piltant, O., Reguera, A.: Cones of curves and line bundles at infinity. J. Algebra **293**(2), 513–542 (2005)
43. Campillo, A., Farrán, J.I.: Computing Weierstrass semigroups and the Feng-Rao distance from singular plane models. Finite Fields Their Appl. **6**(1), 71–92 (2000)
44. Campillo, A., Farrán, J.I., Munuera, C.: On the parameters of codes coming from Arf semigroups. IEEE Trans. Inf. Theory **46**(7), 2634–2638 (2000)
45. Campillo, A., Farrán, J.I.: Construction of AG-codes from symbolic Hamburger-Noether expressions of plane curves. Math. Comput. **71**(240), 1759–1780 (2002)
46. Campillo, A., Farrán, J.I.: Adjoints and Codes. Rendiconti del Seminario Matematico Università e Politecnico de Torino **62**(2), 19–33 (2004)
47. Campillo, A., Farran, J.I., Pisabarro, M.J.: Evaluation codes at singular points of algebraic differential equations. Appl. Algebra Eng. Commun. Comput. **18**(1–2), 191–203 (2007)
48. Campillo, A., Gonzalez-Sprinberg, G., Monserrat, F.: Configurations of infinitely near points. Sao Paulo J. Math. **3**(1), 113–158 (2009)
49. Campillo, A., Lemahieu, A.: Poincaré series for filtrations defined by discrete valuations with arbitrary center. J. Algebra **377**, 66–75 (2013)

50. Campillo, A., Loeser, F., Luengo, I., Sabbah, C. (eds.): Courbes. Publications EPP. École Polytechnique Paris (1988)
51. Campillo, A., Narváez, L. (eds.): Algebraic Geometry and Singularities. Volume 134 of Progress in Mathematics. Birkhäuser, Boston/Basel/Berlin (1996)
52. Campillo, A., Gonzalez-Sprinberg, G., Lejeune-Jalabert, M.: Amas, ideaux complets et chaines toriques. Comptes Rendus de l'Académie de Sciences – Série I – Mathématiques **315**, 987–990 (1992)
53. Campillo, A., Gonzalez-Sprinberg, G., Lejeune-Jalabert, M.: Clusters, proximity inequalities and Zariski-Lipman complete ideal theory. Publicaciones Matemáticas Uruguay **6**, 7–46 (1995)
54. Campillo, A., Gonzalez-Sprinberg, G., Lejeune-Jalabert, M.: Enriques diagrams, resolutions and toric clusters. Comptes Rendus de l'Académie de Sciences – Série I – Mathématiques **320**, 329–334 (1995)
55. Campillo, A., Pisón, P.: Generators for a monomial curve and graphs for the associated semigroup. Bull. Belg. Math. Soc. Série A **45**(1–2), 45–58 (1993)
56. Campillo, A., Pisón, P.: L'ideal d'un semigroup de type fini. Comptes Rendus de l'Académie de Sciences – Série I – Mathématiques **316**, 1303–1306 (1993)
57. Briales, E., Campillo, A., Pisón, P.: On the equations defining toric projective varieties. In: Geometric and Combinatorial Aspects of Commutative Algebra (Messina, 1999). Volume 217 of Lecture Notes in Pure and Applied Mathematics, pp. 57–66. Dekker, New York (2001)
58. Briales, E., Campillo, A., Pisón, P., Vigneron, A.: Simplicial complexes and syzygies of lattice ideals. In: Symbolic Computation: Solving Equations in Algebra, Geometry, and Engineering (South Hadley, MA, 2000). Volume 286 of Contemporary Mathematics, pp. 169–183. American Mathematical Society, Providence (2001)
59. Campillo, A., Pisón, P.: Toric mathematics from semigroup viewpoint. In: Ring Theory and Algebraic Geometry (León, 1999). Volume 221 of Lecture Notes in Pure and Applied Mathematics, pp. 95–112. Dekker, New York (2001)
60. Briales, A., Campillo, A., Pisón, P., Vignerón, A.: Minimal resolution of lattices and integer linear programing. Revista Matemática Iberoamericana **19**(2), 287–306 (2003)
61. Campillo, A., Gonzalez-Sprinberg, G.: On characteristic cones, clusters and chains of infinitely near points. In: Brieskorn Anniversary Volume. Volume 162 of Progress in Mathematics, pp. 251–261. Birkhäuser, Basel (1996)
62. Campillo, A., Gimenez, P.: Graphes arithmetiques et syzygies. Comptes Rendus de l'Académie de Sciences – Série I – Mathématiques **324**, 313–316 (1997)
63. Campillo, A., Gimenez, P.: Syzygies of affine toric varieties. J. Algebra **225**(1), 142–161 (2000)
64. Campillo, A., et al.: A Hoschschild homology criterium for the smoothness of an algebra. Commentarii Mathematici Helvetici **69**, 163–168 (1994). The following people participated in the research of this paper under the acronym BACH (The Buenos Aires Cyclic Homology Group): Antonio Campillo, Jorge Alberto Guccione, Juan José Guccione, Maria Julia Redondo, Andrea Solotar and Orlando E. Villamayor
65. Campillo, A., Olivares, J.: Plane foliations of degree different from one are determined by their singular scheme. Comptes Rendus de l'Académie de Sciences – Série I – Mathématiques **328**, 877–882 (1999)
66. Campillo, A., Olivares, J.: Assigned conditions and geometry of foliations on the projective plane. In: Brasselet, J.-P., Suwa, T. (eds.) Singularities in Geometry and Topology 1998, Sapporo. Advanced Studies in Pure Mathematics, vol. 29, pp. 97–113. Mathematical Society of Japan, Amsterdam/Tokyo (2000)
67. Campillo, A., Olivares, J.: Polarity with respect to a foliation and Caylay-Bacharach theorems. Journal für die reine und angewandte Mathematik **534**, 95–118 (2001)
68. Campillo, A., Olivares, J.: On sections with isolated singualarities of twisted bundles and applications to foliations by curves. Math. Res. Lett. **10**(5–5), 651–658 (2003)
69. Campillo, A., Olivares, J.: On the polar linear system of a foliation by curves on the projective spaces. In: Muñoz-Porras, J.M., et al. (eds.) The Geometry of Riemann Surfaces and Abelian Varieties. Volume 397 of Contemporary Mathematics, pp. 7–14. American Mathematical

Society (2006). Proceedings of the "III Iberoamerican Congress on Geometry", in honor of Professor Sevín Recillas-Pishmish's 60th birthday, Salamanca, Spain, 8–12 June 2004

70. Campillo, A., Olivares, J.: Special subschemes of the scheme of singularities of a plane foliation. Comptes Rendus de l'Académie de Sciences – Série I – Mathématiques **344**(9), 581–585 (2007)
71. Campillo, A., Encinas, S.: Constructing nodes for two variable polynomial interpolation. ACM SIGSAM Bull. **35**(4), 17 (2001)
72. Campillo, A., Encinas, S.: Some applications of two dimensional complete ideals. In: Commutative Algebra (Grenoble/Lyon, 2001). Volume 331 of Contemporary Mathematics, pp. 55–74. American Mathematical Society (2003)
73. Campillo, A., Encinas, S.: Heisuke Hironaka. LA GACETA de la RSME **13**(1), 179–190 (2010)
74. Campillo, A., Greuel, G.M., Lossen, C.: Equisingular calculations of plane curve singularities. J. Symb. Comput. **42**(1–2), 89–114 (2007)
75. Campillo, A., Greuel, G.M., Lossen, C.: Equisingular deformations of plane curves in arbitrary characteristic. Compositio Mathematica **143**, 829–882 (2007)
76. Campillo, A., Fitzpatrick, P., Martínez-Moro, E., Pellikaan, R. (eds.): Special issue algebraic coding theory and applications. J. Symb. Comput. **45**(7), 721–824 (2010). From the editors Preface: This issue of the Journal of Symbolic Computation is a follow-up to the "Soria Summer School on Computational Mathematics: Algebraic Coding Theory" supported by the group SINGACOM of the University of Valladolid, held at the Campus Duques de Soria of the University of Valladolid in Soria, 2–11 July 2008 and the Special Session "Computer Algebra and Coding Theory 2008", hosted at ACA 2008 held at the RISC Institute of the Johannes Kepler University in Hagenberg, Austria, 27–30 July 2008
77. Campillo, A., Cardona, G., Melle-Hernández, A., Veys, W., Zúñiga-Galindo, W.A. (eds.): Zeta Functions in Algebra and Geometry. Contemporary Mathematics. American Mathematical Society (2010). The volume contains the proceedings of the "Second International Workshop on Zeta Functions in Algebra and Geometry" held 3–7 May 2010 at the Universitat de les Illes Balears, Palma de Mallorca, Spain
78. Campillo, A., Kulhmann, F.V., Teissier, B. (eds.): Valuation Theory in Interaction. EMS Series of Congress Reports. EMS Publishing House, Zürich (2014)
79. Campillo, A., Castellanos, J.: Curve Singularities. An Algebraic and Geometric Approach. Actualités Mathématiques, 126pp. Hermann, Paris (2005)
80. Campillo, A., Castellanos, J.: Arf closure relative to a divisorial valuation and transversal curves. Am. J. Math. **116**(2), 377–395 (1994)
81. Campillo, A., Castellanos, J.: Valuative Arf characteristic of singularities. Mich. Math. J. **49**(3), 435–450 (2001)
82. Campillo, A., Castellanos, J.: On Puiseux exponents for higher dimensional singularities. In: Topics in Algebraic and Noncommutative Geometry. Volume 324 of Contemporary Mathematics, pp. 91–102. American Mathematical Society, Providence (2003)
83. Campillo, A.: Hamburger-Noether expansions over rings. Trans. AMS **279**(1), 277–388 (1983)
84. Campillo, A.: Arithmetical aspects of saturation of singularities. Pol. Acad. Publ. **20**, 121–137 (1988)
85. Campillo, A.: Arf closure and saturation relative to divisorial valuations. Pubblicazioni Matematiche dell'Università di Pisa **742**, 11–30 (1993)
86. Campillo, A., Revilla, M.A.: Coin exchange algorithms and toric projective curves. Commun. Algebra **29**(7), 2985–2989 (2001)
87. Campillo, A., Grabowski, J., Müller, G.: Derivation algebras of toric varieties. Compositio Mathematica **116**(2), 119–132 (1999)
88. Campillo, A., Carnicer, M.: Proximity inequalities and bounds for the degree of curves invariant by foliations. Trans. AMS **349**(6), 2211–2228 (1997)
89. Campillo, A., Hernández-Ruipérez, D. (eds.): El legado matemático de Juan Bautista Sancho Guimerá. Ediciones Universidad de Salamanca (2015)
90. Campillo, A., et al.: Título de Grado en Matemáticas. Agencia Nacional de Evaluación y Acreditación (2004)

Singularities in Positive Characteristic: Equisingularity, Classification, Determinacy

Gert-Martin Greuel

Dedicated to Antonio Campillo on the occasion of his 65th birthday

Abstract In this survey paper we give an overview on some aspects of singularities of algebraic varieties over an algebraically closed field of arbitrary characteristic. We review in particular results on equisingularity of plane curve singularities, classification of hypersurface singularities and determinacy of arbitrary singularities. The section on equisingularity has its roots in two important early papers by Antonio Campillo. One emphasis is on the differences between positive and zero characteristic and on open problems.

2010 Mathematics Subject Classification 32S05, 32S15, 14B05, 14B07, 14H20

1 Historical Review

Singularity theory means in this paper the study of systems of polynomial or analytic or differentiable equations locally at points where the Jacobian matrix has not maximal rank. This means that the zero set of the equations at these points is not smooth. The points where this happens are called *singularities* of the variety defined by the equations. Singularities have been studied since the beginning of algebraic geometry, but the establishment of their own discipline arose about 50 years ago.

Singularity theory started with fundamental work of *Heisuke Hironaka* on the resolution of singularities (1964), *Oskar Zariski's* studies in equisingularity, (1965–1968), *Michael Artin's* paper on isolated rational singularities of surfaces (1966), and the work by *René Thom, Bernard Malgrange, John Mather,*... on singularities of differentiable mappings.

G.-M. Greuel (✉)
Fachbereich Mathematik, Universität Kaiserslautern, Kaiserslautern, Germany
e-mail: greuel@mathematik.uni-kl.de

G.-M. Greuel et al. (eds.), *Singularities, Algebraic Geometry, Commutative Algebra, and Related Topics*, https://doi.org/10.1007/978-3-319-96827-8_2

It culminated in the 1970s and 1980s with the work of *John Milnor*, who introduced what is now called the Milnor fibration and the Milnor number (1968), *Egbert Brieskorn's* discovery of exotic spheres as neighborhood boundaries of isolated hypersurface singularities (1966) and the connection to Lie groups (1971), *Vladimir Arnold's* classification of (simple) singularities (1973), and many others, e.g. *Andrei Gabrielov, Sabir Gusein-Zade, Ignacio Luengo, Antonio Campillo, C.T.C. Wall, Jonathan Wahl, Lê Dũng Tráng, Bernard Teissier, Dierk Siersma, Joseph Steenbrink*, ...

Besides the work of Artin, this was all in characteristic 0, mostly even for convergent power series over the complex or real numbers.

The first to study systematically "equisingular families" over a field of positive characteristic was *Antonio Campillo* in his thesis, published as Springer Lecture Notes in 1980.

2 Equisingularity

In the 1960s O. Zariski introduced the concept of equisingularity in order to study the resolution of hypersurface singularities by induction on the dimension. His idea can be roughly described as follows:

- To resolve the singularities of X consider a generic projection X to a smooth curve. We get a family of 1–codimensional subvarieties of X, which we know to resolve by induction on the dimension for each individual fibre.
- If the fibres are an "equisingular" family, then the resolution of a single fibre should resolve the nearby fibres simultaneously and then also the total space X.
- If the fibres are plane curves then "equisingular" means that the combinatorics of the resolution process of the fibre singularities is constant.

  Equivalently: the Puiseux pairs of each branch and the pairwise intersection numbers of the different branches are the same for each fibre, or the topological type of the fibre singularities is constant.
- Zariski's idea works if the fibres are plane curves, but not in general. Nevertheless, equisingularity has become an independent research subject since then.

2.1 Hamburger–Noether Expansions

Let me now describe Campillo's early contribution to equisingularity. There are two important papers by Antonio Campillo:

- *Algebroid Curves in Positive Characteristic* [4]
- *Hamburger–Noether expansion over rings* [5].

The first was Campillo's thesis and appeared as Springer Lecture Notes in 1980 and is now the standard reference in the field. The second paper is however less known but perhaps even more important.

For the rest of the paper let K **denote an algebraically closed field of characteristic** $p \geq 0$, unless otherwise stated.

We consider in this section reduced plane curve singularities defined over K. Over the complex numbers these are 1-dimensional complex germs $C \subset \mathbb{C}^2$ with isolated singularity at 0, given by a convergent power series $f \in \mathbb{C}\{x, y\}$ with C the germ of the set of zeros $V(f)$ of f. If K is arbitrary, a plane curve singularity is given by a formal power series $f \in K[[x, y]]$ with $C = V(f) = \operatorname{Spec} K[[x, y]]/\langle f\rangle$. C and f are also called **algebroid plane curves**.

A reduced and irreducible algebroid plane curve C can be given in two ways:

- by an *equation* $f = 0$, with f irreducible in the ring $K[[x, y]]$
- by a *parametrization* $x = x(t)$, $y = y(t)$ with $\langle f\rangle = \ker(K[[x, y]] \to K[[t]])$

Case $p = 0$

- In this case C has a special parametrization (the **Puiseux expansion**)

$$x = t^n \qquad : n = \operatorname{ord}(f) = \operatorname{mult}(C)$$
$$y = c_m t^m + c_{m+1} t^{m+1} + \cdots : m \geq n,\ c_i \in K$$

Here ord(f) is the **order** of f, also denoted the **multiplicity** of f, i.e. lowest degree of a non-vanishing term of the power series f.

If $f = f_1 \cdot \ldots \cdot f_r$ is reducible (but reduced) with f_i irreducible, we consider the parametrization of each **branch** f_i of f individually.

The Puiseux expansion determines the **characteristic exponents** of C (equivalently the **Puiseux pairs** of C).

- These data is called the **equisingularity type (es–type)** of the irreducible C.
- For a reducible curve C the **es–type** is defined by the **es–types of the branches** and the **pairwise intersection multiplicities** of different branches.
- Equivalently by the **system of multiplicities** of the reduced total transform in the resolution process of C by successive blowing up points.
- Two curves with the same es–type are called **equisingular**

For the case $K = \mathbb{C}$ and $f, g \in \mathbb{C}\{x, y\}$ we have the following nice topological interpretation of equisingularity:

- V(f) and V(g) are equisingular $\Leftrightarrow$ they are (embedded) **topologically equivalent**, i.e. there is a homeomorphism of triples $h : (B_\varepsilon, B \cap V(f), 0) \xrightarrow{\sim} (B_\varepsilon, B_\varepsilon \cap V(g), 0)$, with $B_\varepsilon \subset \mathbb{C}^2$ a small ball around 0 of radius ε (cf. Fig. 1).

Case $p > 0$

The resolution process of C by successive blowing up points exists as in the case $p = 0$. There exists also a parametrization of C, but a **Puiseux expansion does not exist** if $p|n$, n the multiplicity of C.

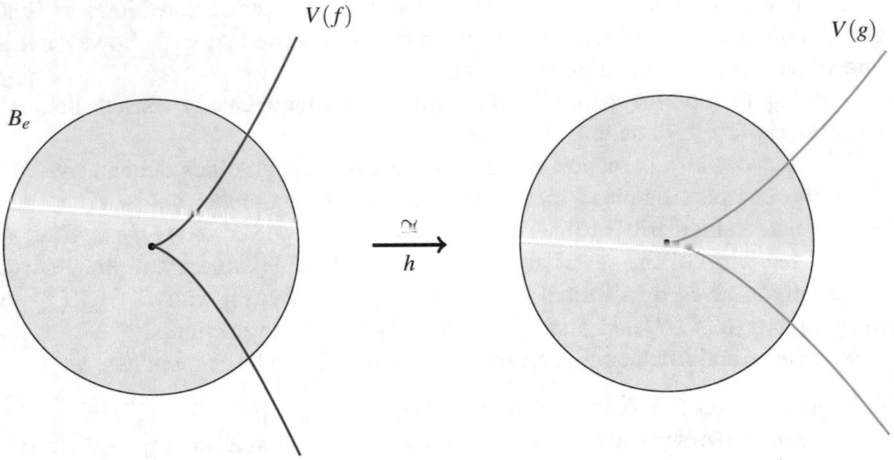

Fig. 1 Topological equivalence

- We define the **equisingularity type (es-type)** of C, by the **system of multiplicities** of the resolution as in characteristic 0.
- Instead of the Puiseux expansion another special parametrization exists and can be computed from any parametrization (or any equation) of C, the **Hamburger–Noether (HN) expansion**. It is determined by a chain of relations obtained from successive divisions by series of lower order (assume $ord(x) \leq ord(y)$) as follows:

$$
\begin{aligned}
y &= a_{01}x + \cdots + a_{0h}x^h + x^h z_1, \quad ord(z_1) < ord(x) \\
x &= a_{12}z_1^2 + \cdots + a_{1h_1}z_1h_1 + z_1^{h_1}z_2 \\
z_1 &= a_{22}z_2^2 + \cdots + a_{2h_2}z_2h_2 + z_2^{h_2}z_3 \qquad \textbf{(HN(C))} \\
&\vdots \\
z_{r-1} &= a_{r2}z_r^2 + \cdots \in K[[z_r]]
\end{aligned}
$$

We do not have Puiseux pairs, but we have characteristic exponents. Campillo defines the **characteristic exponents for** C from HN(C).

By substituting backwards, we get from HN(C) a parametrization of C:

$$x = x(z_r), \quad y = y(z_r) \in K[[z_r]]$$

Note that the uniformizing parameter z_r is a rational function of the coordinates x, y. It does not involve roots of unity as the uniformizing parameter for the Puiseux expansion. Moreover, computationally the Hamburger–Noether expansion is preferred to the Puiseux expansion as it needs the least number of field extensions if one wants to compute the es-type for an algebroid curve defined over a non algebraically closed field (such as $\mathbb{Q}$). This is implemented in SINGULAR [8].

For an arbitrary algebraically closed field K Campillo defines the complex model of C as follows:

- Let $F : K \to \mathbb{C}$ be any map with $F(a) = 0 \Leftrightarrow a = 0$ (e.g. $F(0) = 0$, $F(a) = 1$ for $a \neq 0$).

  A **complex model** $C_{\mathbb{C}}$ of the curve C is obtained from HN(C) by the HN-expansion

$$HN(C_{\mathbb{C}}) : \text{replace} a_{ij} \text{in HN}(C)\text{by} F(a_{ij}).$$

Theorem 2.1 (Campillo, [4, Chapter III])

(1) *The characteristic exponents of $C_{\mathbb{C}}$ do not depend on the complex model.*
(2) *They are a complete set of invariants of the es-type of C.*
(3) *They coincide with the characteristic exponents of $C_{\mathbb{C}}$ obtained from the Puiseux espansion.*

Note that the complex model $C_{\mathbb{C}}$ of C is defined over the integers if F has integer values (this is important for coding theory and cryptography).

We come now to the second paper of Campillo "Hamburger–Noether expansion over rings" mentioned above. For any ring A Campillo defines a

- **HN–expansion HN$_A$ over** A. HN$_A$ is similar to HN, but with $a_{ij} \in A$ and certain properties. It may be considered as a family over Spec(A) of parametrized curves with constant es-type. If A is the field K then HN$_K$ coincides with HN defined above.

If A is a local K–algebra with maximal ideal $\mathfrak{m}_A$ and $A/\mathfrak{m}_A = K$, then we may take residue classes of the HN-coefficients a_{ij} modulo $\mathfrak{m}_A$ and thus the HN-expansion over A induces a deformation

$$X = X(z_r), \quad Y = Y(z_r) \in A[[z_r]]$$

of the parametrization over Spec(A)

$$x = x(z_r), \quad y = y(z_r) \in K[[z_r]], \text{ with } x, y = X, Y \text{mod } \mathfrak{m}_A$$

of an irreducible plane algebroid curve C.

If A is irreducible, the es–types of the curve C parametrized by $x(z_r)$, $y(z_r)$ over K and of the curve defined by the parametrization $X(z_r)$, $Y(z_r)$ over the quotient field Quot(A) coincide.

We have the following important theorem, saying that for a fixed equisingularity type there exists some, in a sense "totally versal", equisingular family $X \to Y$ of $\mathbb{Z}$–schemes such that for any field K the following holds: any equisingular family of algebroid curves over K can be induced from $X \to Y$. More precisely, Campillo proves:

Theorem 2.2 (Campillo, [5, Teorem 5.1])

(1) *Let C be irreducible and E the es–type of C. Then there exists a morphism of $\mathbb{Z}$–schemes $\pi : X \to Y$ with section $\sigma : Y \to X$ s.t. for any algebraically closed field K the base change $\mathbb{Z} \to K$ induces a family $X^K \to Y^K$ with section σ^K such that the following holds:*

   (i) *Y^K is a* smooth, irreducible *affine algebraic variety over K.*

   (ii) *For any closed point $y \in Y^K$ the induced family*

$$Spec(\widehat{\mathcal{O}}_{X^K,\sigma^K(y)}) \to Spec(\widehat{\mathcal{O}}_{Y^K,y})$$

   *is a **total es–versal HN–expansion**, i.e.:*
   for any algebroid curve C' with $es(C') = E$ and any local K–algebra A s.t. HN_A induces $HN(C')$ mod $\mathfrak{m}_A$, there exists a morphism $\varphi : \widehat{\mathcal{O}}_{Y^K,y} \to A$ such that HN_A is induced from $HN_{\widehat{\mathcal{O}}_{Y^K,y}}$ via φ.

(2) *For a reducible curve C the construction is extended to finite sets of HN–expansions over A and the statement of* (1) *continues to hold.*

2.2 Equisingularity Strata

We note that Theorem 2.2 implies the existence of a semiuniversal equisingular deformation with smooth base space of the parametrization of a reduced curve C. For a different proof with an explicit formula see [6, Theorem 3.1] (where also a stronger versality property, the lifting of induced deformations over small extensions, is shown).

A Hamburger–Noether expansion over a ring A induces an **equisingular deformation of the parametrization** of an irreducible curve singularity C over Spec(A).

Such a deformation of the parametrization induces a deformation of the equation as follows:

- Let $X = X(z_r), Y = Y(z_r)$ be a HN–expansion over A. It is a deformation of the parametrization $x = x(z_r), y = y(z_r)$ of an algebroid curve $C = \text{Spec}(K[[x, y]]/\langle f\rangle)$.

  By elimination of z_r from $x - X(z_r)$ and $y - Y(z_r)$, we get a power series $F \in A[[x, y]]$ with $f = F \mod \mathfrak{m}_A$. F is a deformation of the curve $C = V(f)$ (in the usual sense) over Spec(A), also called a deformation of the equation. Since it is induced from an equisingular deformation of the parametrization, we call it an equisingular deformation of the equation. Deformations are a category and the construction is functorial (cf. [6, Section 4]).

- In this way we get for any algebroid curve C a functor

$$\chi_{es} : \text{equisingular–deformations of the parametrization of C} \\ \to \text{(usual) deformations of the equation of C.}$$

We call the image of χ_{es} (full subcategory) **equisingular deformations of the equation** or just **es–deformations** of C over Spec(A).
- The construction can be generalized to reducible C and a set of HN-expansions over A of the branches of C (with certain properties).

More generally, any deformation $\Phi : X \to T$, equisingular or not, of the parametrization of a plane curve singularity C induces a deformation of the equation by eliminating the uniformising variable. Without going to details, we mention that es-deformations of the parametrization are deformations with a given section, which map to deformations of the equation with section. However, as for es-deformations of the equation the equisingular section is unique [6, Theorem 4.2], they are a subcategory of the usual deformations (without section).

Question: Does the base space T of an arbitrary deformation of C admit a unique maximal subspace over which the deformation is equisingular? In other words, does there exist a unique **equisingularity stratum** of Φ in T?

The answer is well–known for $K = \mathbb{C}$. Recall that for any K and $f \in K[[x, y]]$

- $\mu(f) := \dim_K K[[x, y]]/\langle f_x, f_y\rangle$ is the **Milnor number** of f.

If $p = \text{char}(K) = 0$ $\mu(f)$ depends only on the ideal $\langle f\rangle$ and not on the choosen generator and then we write also $\mu(C)$ for $C = \text{Spec } K[[x, y]]/\langle f\rangle$.

For $K = \mathbb{C}$ or, more generally, if char$(K) = 0$, the equisingularity stratum of any deformation $\Phi : X \to T$ of C exists and is the **μ–constant stratum** of Φ, i.e. the set of points $t \in T$ such that Milnor number of the fibres X_t is constant along some section $\sigma : T \to X$ of Φ. If $\Phi : X \to T$ is the semiuniversal deformation of C, the μ–constant stratum is denoted by Δ_μ,

- $\Delta_\mu = \{t \in T : \exists \text{ a section } \sigma : T \to X \text{ of } \Phi \text{ s.t. } \mu(X_t, \sigma(t)) = \mu(C)\}$.

The restriction of Φ to Δ_μ may be considered as the **semiuniversal es–deformation** of C in the sense that any es–deformation of C over some base space T can be induced by a morphism $\varphi : T \to \Delta_\mu$, such that the tangent map of φ is unique. Moreover Δ_μ is known to be smooth (cf. e.g. [16, Theorem II.2.61]).

If K has positive characteristic, the situation is more complicated. A semiuniversal es–deformation of C exists, but an equisingularity stratum does not always exist. We have to distinguish between good and bad characteristic p. The **characteristic is good**, if either $p = 0$ or if $p > 0$ does not divide the multiplicity of any branch of C. The situation is described in the following theorem and in the next subsection.

Theorem 2.3 (Campillo, Greuel, Lossen; [6, Theorem 4.2, Corollary 9.6])

(1) *The functor χ_{es} defined above is smooth (unobstructed). In particular, the base spaces of the semiuniversal es-deformations of the pararmetrisation and of the equation differ by a smooth factor. They are isomorphic if the characteristic is good.*

(2) *The functor $\underline{\mathrm{Def}}_C^{es}$ of (isomorphism classes) of es–deformations of the equation has a semiuniversal deformation with smooth base space B_C^{es}.*

(3) *If char$(K) = 0$ then B_C^{es} coincides with the μ–constant stratum Δ_μ in the base space of the (usual) semiuniversal deformation of C.*

(4) *In good characteristic there exists for any deformation of C over some T a unique maximal equisingularity stratum $T_{es} \subset T$ (generalizing the μ–constant stratum).*

For an algorithm to compute the es-stratum and its tangent space (isomorphic to the equisingularity ideal of Wahl [20]) see [7, Algorithms 2 and 3]. It is implemented in SINGULAR [18].

2.3 Pathologies and Open Problems

If the characteristic is bad, i.e. $p > 0$ divides the multiplicity of some branch of C, we have the following **pathologies**:

(1) There exist deformations of C which are not equisingular but become equisingular after a finite base change. We call such deformations (with section) **weakly equisingular**.

(2) Let $\Phi_C : X_C \to B_C$ denote the semiuniversal deformation of C. In general no unique es–stratum in B_C exists. For example, let $p = 2$ and $f = x^4 + y^6 + y^7$. Then the following holds.

There exist infinitely many smooth subgerms $B_\alpha \subset B_C$ s.t.

- $B_\alpha \cong B_C^{es}$.
- All B_α have the same tangent space.
- The restriction of Φ_C to B_α is equisingular for any α.
- The restriction of Φ_C to $B_{\alpha_1} \cup B_{\alpha_2}$ is not equisingular if $\alpha_1 \neq \alpha_2$. Hence, a unique maximal equisingularity stratum of Φ_C does not exist.

(3) Let B_C^{wes} denote the Zariski–closure of the union of all subgerms $B' \subset B_C$ with the property that $\Phi_C|B'$ is equisingular. Then $\Phi_C|B_C^{wes}$ is the **semiuniversal weakly equisingular deformation** of C, i.e. it has the versality property for weakly es–deformations (as above for es–deformations). We have

- B_C^{wes} is irreducible but in general not smooth
- B_C^{wes} becomes smooth after a finite (purely inseparable) base change.

For a proof of these facts see [6].

Let us mention the following **Open Problems**:

- In [6] we develop a theory of weakly equisingular deformations with section. Since the sections for such deformations may not be unique, it would be interesting to develop a theory of weakly equisingular deformations without section and to see if for such deformations an equisingularity stratum exists.
- It follows from Theorem 2.3 that "p = good" is a sufficient condition for $B_C^{es} = B_C^{wes}$ (and hence that B_C^{wes} is smooth) but it is not a necessary condition. The problem is to find a necessary and sufficient condition for $B_C^{es} = B_C^{wes}$ (which does not only depend on p).

We do not yet fully understand the relation between weak and strong equisingularity (even between weak and strong triviality).

3 Classification of Singularities

In this section we consider hypersurface singularities $f \in K[[x]] := K[[x_1, \dots, x_n]]$, again with K algebraically closed and $\text{char}(K) = p \geq 0$. We recall the classical results for the classification of singularities in characteristic zero and present some more recent results in positive characteristic.

The two most important equivalence relations for power series are right equivalence and contact equivalence. The two equivalence relations lead of course to different classification results. It turns out that the classification of so called "simple singularities" w.r.t. contact equivalence is rather similar for $p = 0$ and for $p > 0$. However, for right equivalence the classification of simple singularities in positive characteristic is surprisingly different from that in characteristic zero.

Let $f, g \in K[[x]]$. Recall that

- f is **right equivalent** to g ($f \overset{r}{\sim} g$) $:\Leftrightarrow \exists\, \Phi \in \text{Aut}(K[[x]])$ such that $f = \Phi(g)$, i.e. f and g differ by an analytic coordinate change Φ with $\Phi(x_i) = \Phi_i$ and $f = g(\Phi_1, \cdots, \Phi_n)$.
- f is **contact equivalent** to g ($f \overset{c}{\sim} g$) $:\Leftrightarrow \exists\, \Phi \in \text{Aut}(K[[x]])$ and $u \in K[[x]]^*$ such that $f = u \cdot \Phi(g)$, i.e. the analytic K–algebras $K[[x]]/\langle f \rangle$ and $K[[x]]/\langle g \rangle$ are isomorphic.
- f is called **right–simple** (resp. **contact–simple**): $\Leftrightarrow \exists$ finite set $\{g_1, \dots, g_l\} \subset K[[x]]$ of power series such that for any deformation of f,

$$F_t(x) = F(x,t) = f(x) + \sum_{i=1}^{k} t_i h_i(x,t),$$

there exists a neighborhood $U = U(0) \subset K^k$ such that $\forall\, t \in U\ \exists\, 1 \leq j \leq l$ with $F_t \overset{r}{\sim} g_j$ (resp. $F_t \overset{c}{\sim} g_j$); in other words, f has **no moduli** or f is of **finite deformation type**.

3.1 Classification in Characteristic Zero

The most important classification result for hypersurface singularities in characteristic zero is the following result by V. Arnold:

Theorem 3.1 (Arnold; [1]) *Let $f \in \mathbb{C}\{x_1, \ldots, x_n\}$. Then f is right–simple $\Leftrightarrow$ f is right equivalent to an ADE singularity from the following list:*

$$\begin{array}{llll}
A_k: x_1^{k+1} + x_2^2 & +q, & k \geq 1, & q := x_3^2 + \cdots + x_n^2 \\
D_k: x_2(x_1^2 + x_2^{k-2}) & +q, & k \geq 4 & \\
E_6: x_1^3 + x_2^4 & +q & & \\
E_7: x_1(x_1^2 + x_2^3) & +q & & \\
E_8: x_1^3 + x_2^5 & +q & &
\end{array}$$

Later it was proved that $f \in \mathbb{C}\{x_1, \ldots, x_n\}$ is right–simple $\Leftrightarrow$ f is contact–simple.

Arnold's classification has numerous applications. The list of simple or ADE singularities appears in many other contexts of mathematics and is obtained also by classifications using a completely different equivalence relation (cf. [9, 10]). One such classification result is the following.

Theorem 3.2 (Buchweitz, Greuel, Schreyer; Knörrer, [3, Theorem A], [17, Theorem])
$f \in \mathbb{C}\{x_1, \ldots, x_r\}$ is simple $\Leftrightarrow$ f is of ***finite CM–type****, (i.e. there are only finitely many isomorphism classes of maximal indecomposable Cohen–Macaulay modules over the local ring $\mathbb{C}\{x\}/\langle f \rangle$).*

3.2 Classification in Positive Characteristic

The classification of hypersurface singularities in positive characteristic started with the following result. Let K be an algebraically closed field of characteristic $p > 0$.

Theorem 3.3 (Greuel, Kröning [11, Teorem 1.4])
The following are equivalent for $f \in K[[x_1, \cdots, x_n]]$:

(1) *f is contact–simple,*
(2) *f is of finite CM–type.*
(3) *f is an ADE–singularity, i.e. f is contact equivalent to a power series form Arnold's list, but with few extra normal forms in small characteristic ($p \leq 5$). E.g. for $p = 3$ $\exists$ two normal forms for E_6: $E_6^0 = x^3 + y^4$ and $E_6^1 = x^3 + x^2y^2 + y^5$.*

The classification of right–simple singularities in positive characteristic remained open for many years. It turned out that the result differs substantially from the characteristic zero case. For example, in characteristic zero there exist two infinite series A_k and D_k of right–simple singularities, but for any $p > 0$ there are only finitely many right–simple singularities.

Theorem 3.4 (Greuel, Nguyen [12, Theorem 3.2, 3.3])
$f \in K[[x_1, \cdots, x_n]]$ is right–simple $\Leftrightarrow$ f is right equivalent to one of the following:

(1) **n = 1** : $A_k : x^{k+1}$ $\quad 1 \le k < p-1$
(2) **n = 2, p > 2**:

$$\begin{array}{ll} A_k : x^{k+1} + y^2, & 1 \le k < p-1 \\ D_k : x(y^2 + x^{k-2}), & 4 \le k < p \\ E_6 : x^3 + y^4, & p > 3 \\ E_7 : x(x^2 + y^3), & p > 3 \\ E_8 : x^3 + y^5 & p > 5 \end{array}$$

(3) **n > 2, p > 2**

$$g(x_1, x_2) + x_3^2 + \cdots + x_n^2 \text{ with } g \text{ from (2)}$$

(4) **n ≥ 2, p = 2**

$$A_1 : x_1x_2 + x_3x_4 + \cdots + x_{n-1}x_n \text{ , } n \text{ even}$$

Note that Arnold proved a "singularity determinator" and accomplished the complete classification of unimodal and bimodal hypersurfaces w.r.t. right equivalence in characteristic zero with tables of normal forms (cf. [1]). Such a singularity determinator and a classification of unimodal and bimodal singularities in positive characteristic was achieved by Nguyen Hong Duc in [19].

3.3 *Pathologies and a Conjecture*

We comment on some differences of the classification in positive and zero characteristic and propose a conjecture.

For $f \in K[[x_1, \cdots, x_n]]$ the **Milnor number** of f is

$$\mu(f) = \dim_K K[[x]]/\langle f_{x_1}, \ldots, f_{x_n}\rangle.$$

If $f \in \mathfrak{m}^2, \mathfrak{m} = \langle x_1 \ldots, x_n\rangle$, and $\mu(f) < \infty$, we have seen the following pathologies:

- For all $p > 0$ there exist only finitely many right–simple singularities, in particular $\mu(f)$ is bounded by a function of p.
- For $p = 2$ and n odd there exists no right–simple singularity.

The reason why there are so few right–simple singularities in characteristic $p > 0$ can be seen from the following example: the dimension of the group $Aut(K[[x]])$ and hence the right orbit of f is too small (due to $(x + y)^p = x^p + y^p$).

Example We show that $f = x^p + x^{p+1}$ is not right–simple in characteristic p by showing that if $f_t = x^p + tx^{p+1}$ is right equivalent to $f_{t'} = x^p + t'x^{p+1}$ then $t = t'$. This shows that f can be deformed into infinitely many different normal forms.

To see this, assume $\Phi(f_t) = f_{t'}$ for $\Phi(x) = u_1x + u_2x^2 + \cdots \in \mathrm{Aut}(K[[x]])$. Then $(u_1x + u_2x^2 + \cdots)^p + t(u_1x + u_2x^2 + \cdots)^p(u_1x + \cdots) = x^p + t'x^{p+1}$ and hence $u_1^p - 1 - (u_1 - 1)^p = 0$. This implies $u_1 = 1$, $tu_1^{p+1} = t'$ and $t = t'$.

The classification suggests the following conjecture that is in strict contradiction to the characteristic 0 case:

Conjecture Let $\mathrm{char}(K) = p > 0$ and $(f_k)_k \in K[[x_1, \ldots, x_n]]$ a sequence of isolated singularities with Milnor number $\mu(f_k) < \infty$ going to infinity if $k \to \infty$. Then the right modality of f_k goes to infinity too.

The conjecture was proved by Nguyen H. D. for unimodal and bimodal singularities and follows from the classification. He shows in [19] that $\mu(f) \leq 4p$ if the right modality of f is less or equal to 2.

4 Finite Determinacy and Tangent Image

A power series is finitely determined (for a given equivalence relation) if it is equivalent to its truncation up to some finite order. For the classification of singularities the property of being finitely determined is indispensable. In this section we give a survey on finite determinacy in characteristic $p \geq 0$, not only for power series but also for ideals and matrices of power series. We consider algebraic group actions and their tangent maps where new phenomena appear for $p > 0$, which lead to interesting open problems.

4.1 Finite Determinacy for Hypersurfaces

A power series $f \in K[[x]] = K[[x_1, \cdots, x_n]]$ is called right (resp. contact) **k–determined** if for all $g \in K[[x]]$ such that $j_k(f) = j_k(g)$ we have $f \overset{r}{\sim} g$ (resp. $f \overset{c}{\sim} g$). f is called **finitely determined**, if it is k–determined for some $k < \infty$.

- Here

$$j_k : K[[x]] \to J^{(k)} := K[[x]]/\mathfrak{m}^{k+1}$$

denotes the canonical projection, called the k–jet. Usually we identify $j_k(f)$ with the power series expansion of f up to and including order k.

- We use ord(f), the **order** of f (the maximum k with $f \in \mathfrak{m}^k$) and the **Tjurina number** of f,

$$\tau(f) := \dim_K K[[x]]/\langle f, f_{x_1}, \ldots, f_{x_n}\rangle.$$

Theorem 4.1 (Boubakri, Greuel, Markwig, Pham [2, Therem 5], [14, Theorem 4.13])
For $f \in \mathfrak{m}$ the following holds.

(1) *f is finitely right determined $\Leftrightarrow \mu(f) < \infty$.*
If this holds, then f is right $(2\mu(f) - ord(f) + 2)$–determined.
(2) *f is finitely contact determined $\Leftrightarrow \tau(f) < \infty$.*
If this holds, then f is contact $(2\tau(f) - ordf(f) + 2)$–determined.

If the characteristic is 0 we have better bounds: f is right $(\mu(f)+1)$–determined (resp. $(\tau(f) + 1)$–determined) if $\mu(f) < \infty$ (which is equivalent to $\tau(f) < \infty$ for $p = 0$), see [16, Corollary I.2.24] for $K = \mathbb{C}$ and use the Lefschetz principle for arbitrary K with $p = 0$. The proof in characteristic $p > 0$ is more difficult than for $p = 0$, due to a pathology of algebriac group actions in positive characteristic, which we address in Sect. 4.3.

- We can express right (resp. contact) equivalence by actions of algebraic groups. We have

$$f \overset{r}{\sim} g \text{ (resp.} f \overset{c}{\sim} g) \Leftrightarrow f \in \text{orbit } G \cdot g \text{ of } g \text{ with groups}$$

$$G = \mathscr{R} := \text{Aut}(K[[x]]) \textbf{ right group} \text{ (for } \overset{r}{\sim}),$$

$$G = \mathscr{K} := K[[x]]^* \rtimes \mathscr{R} \textbf{ contact group} \text{ (for } \overset{c}{\sim}),$$

where G acts as $(\Psi = \Phi, f) \mapsto \Phi(f)$ (resp. $(\Psi = (u, \Phi), f) \mapsto u \cdot \Phi(f)$).
- G is not an algebraic group, but the k–jet $G^{(k)}$ is algebraic and the induced action on k–jets

$$G^{(k)} \times J^{(k)} \to J^{(k)}, \ (\Psi, f) \mapsto j_k(\Psi \cdot f)$$

is an algebraic action. If f is finitely determined then, for sufficiently large k, f is (right resp. contact) equivalent to g iff $j_k(f)$ is in the $G^{(k)}$–orbit of $j_k(g)$.

The determination of the tangent space $T_f(G^{(k)}f)$ of the orbit of $G^{(k)}$ at f is important, but there is a big difference for $p = 0$ and $p > 0$.

Consider the tangent map To_k to the orbit map

$$o_k : G^{(k)} \to G^{(k)}f \subset K[[x]], \ \Psi \mapsto j_k(\Psi \cdot f)$$

and denote the image of $To_k : T_e(G^{(k)}) \to T_f(G^{(k)}f)$ by $\widetilde{T}_f(G^{(k)}f)$. We have

$$\widetilde{T}_f(G^{(k)}f) \subset T_f(G^{(k)}f)$$

with equality if $p = 0$, but strict inclusion may happen if $p > 0$ as we shall see below. $\widetilde{T}_f(G^{(k)}f)$ and $T_f(G^{(k)}f)$ are inverse systems and we define the inverse limits as

$$T_f(Gf) := \lim_{\substack{\longleftarrow \\ k\geq 0}} T_f(G^{(k)}f)) \subset K[[x]], \textbf{ tangent space}, \text{ and}$$

$$\widetilde{T}_f(Gf) := \lim_{\substack{\longleftarrow \\ k\geq 0}} \widetilde{T}_{jf}(G^{(k)}f)) \subset K[[x]], \textbf{ tangent image}$$

to the orbit Gf of G.

The tangent images for $G = \mathscr{R}$ and $G = \mathscr{K}$ can be easily identified:

$$\widetilde{T}_f(\mathscr{R}f) = \mathfrak{m}\langle\frac{\partial f}{\partial x_1}, \ldots, \frac{\partial f}{\partial x_n}\rangle,$$

$$\widetilde{T}_f(\mathscr{K}f) = \langle f\rangle + \mathfrak{m}\langle\frac{\partial f}{\partial x_1}, \ldots, \frac{\partial f}{\partial x_n}\rangle.$$

If $\mathrm{char}(K) = 0$ then the orbit map o_k is separable, which implies $T_f(Gf) = \widetilde{T}_f(Gf)$. Moreover, in any characteristic we have:

- If the tangent space and the tangent image to Gf coincide (e.g. if $\mathrm{char}(K) = 0$), then f is finitely determined if and only if

$$\dim_K K[[x]]/\widetilde{T}_f(Gf) < \infty.$$

4.2 *Finite Determinacy for Ideals and Matrices*

We generalize the results of the previous section to ideals and matrices. Consider matrices

$$A = [a_{ij}] \in M_{r,s} := \mathrm{Mat}(r, s, K[[x]]) \text{ with } r \geq s,$$

$K[[x]] = K[[x_1, \cdots, x_n]]$, and the group

$$G = \mathrm{GL}(r, K[[x]]) \times \mathrm{GL}(s, K[[x]]) \rtimes \mathrm{Aut}(K[[x]])$$

acting on $M_{r,s}$ in the obvious way:

$$(U, V, \Phi, A) \mapsto U \cdot \Phi(A) \cdot V = U \cdot [a_{ij}(\Phi(x))] \cdot V.$$

If $r = s = 1$ and $A = [f]$ then $GA = \mathscr{K} f$ and the considerations of this section generalize contact equivalence for power series.

A is called G **k–determined** if for all $B \in M_{r,s}$ with $j_k(A) = j_k(B)$ we have $B \subset G \cdot A$. A is **finitely G–determined**, if A is G k–determined for some $k < \infty$.

As in the case of one power series we have:

- the induced action of G on jet–spaces gives algebraic group actions,
- the tangent image to the orbit of G is contained in the tangent space

$$\widetilde{T}_A(GA) \subset T_A(GA) \text{ with } \text{“} ='' \text{ if } p = \text{char}(K) = 0,$$

- In any characteristic $\widetilde{T}_A(GA)$ can be computed in terms of the entries of A and their partials, but $T_A(GA)$ in general not if $p > 0$.

Theorem 4.2 (Greuel, Pham [13, Theorem 3.2, Lemma 2.11])

(1) *If* $\dim_K(M_{r,s}/\widetilde{T}_A(GA)) < \infty \Rightarrow$ *A is finitely G–determined (in particular, the orbit GA contains a matrix with polynomial entries).*
(2) *If A is finitely G–determined* $\Rightarrow \dim_K(M_{r,s}/T_A(GA)) < \infty$

In general we do not know whether $\dim_K(M_{r,s}/\widetilde{T}_A(GA)) < \infty$ is necessary for finite G–determinacy of A for $p > 0$, except for the case of 1–column matrices:

Theorem 4.3 (Greuel, Pham [14, Theorem 3.8]) *Let $p \geq 0$. If $A \in M_{r,1}$, then A is finitely G–determined* $\Leftrightarrow \dim_K M_{r,1}/\widetilde{T}(GA) < \infty$.

Two ideals $I, J \subset K[[x_1, \ldots, x_n]]$ are called **contact equivalent** $\Leftrightarrow K[[x]]/I \cong K[[x]]/J$. It is not difficult to see that two ideals, which are given by the same number of generators, are contact equivalent iff the corresponding 1–column matrices are G-equivalent (cf. [14, Proposition 4.3]).

Let I be minimally generated by $f_1, \ldots, f_r$. Then I is called **finitely contact determined**, if for every ideal J of $K[[x]]$ that can be generated by r elements $b_1, \ldots, b_r$ with $b_i - a_i \in \mathfrak{m}^{k+1}$ for $i = 1, \ldots, m$ and some $k > 0$, we have that I and J are contact equivalent. It can be shown this notion is independent of the minimal set of generators. We get

Corollary 4.4 *Let $I = \langle f_1, \ldots, f_r \rangle \subset \mathfrak{m}$ an ideal with r the minimal number of generators of I and let I_r be the ideal generated by $r \times n$–minors of the Jacobian matrix* $[\frac{\partial f_i}{\partial x_j}]$.

(1) $r \geq n$: *I is finitely contact determined*
$\Leftrightarrow \dim_K(K[[x]]/I) < \infty$

(2) $r \leq n$: *I is finitely contact determined*
$$\Leftrightarrow \dim_K(K[[x]]/I + I_r) < \infty$$

The equivalence of $\dim_K(K[[x]]/I) < \infty$ and $\dim_K(K[[x]]/I + I_r) < \infty$ for $r = n$, which follows from the proof, can also easily be seen directly.

Theorem 4.5 (Greuel, Pham [14, Theorem 4.11]) *For an ideal* $I = \langle f_1, \ldots, f_r \rangle \subset \mathfrak{m}K[[x_1, \ldots, x_n]]$ *with r the minimal number of generators of I,* $r < n$*, the following are equivalent in any characteristic:*

(1) *I is finitely contact determined,*
(2) *The Tjurina number* $\tau(I) := \dim_K K[[x]]^r/(I \cdot K[[x]]^r + \langle \frac{\partial f}{\partial x_1}, \cdots, \frac{\partial f}{\partial x_n} \rangle) < \infty$*, with* $\frac{\partial f}{\partial x_i} = (\frac{\partial f_1}{\partial x_i}, \cdots, \frac{\partial f_r}{\partial x_i}) \in K[[x]]^r$,
(3) *I is an isolated complete intersection singularity.*

For the proof of (1) ⇒ (2) we need a result about Fitting ideals, which is of independent interest.

Proposition 4.6 (Greuel, Pham [14, Theorem 2.7]) *Let* $A \in M_{r,s}$ *be finitely G–determined and let* $I_t \subset K[[x_1, \cdots, x_n]]$ *be the Fitting ideal generated by the* $t \times t$ *minors of A. Then* I_t *has maximal height, i.e.*

$$ht(I_t) = \min\{s, (r - t + 1)(s - t + 1)\},\ t = 1, \cdots, n.$$

4.3 Pathology and a Problem

We show that $\widetilde{T}_f(Gf) \subsetneq T_f(Gf)$ may happen in positive characteristic:

Example 4.7 Let $G = \mathscr{K}$, $f = x^3 + y^4$, $\mathrm{char}(K) = 3$. We compute (using SINGULAR, see [15]):

- f is contact 5–determined
- $\dim_K \widetilde{T}_f(\mathscr{K}^{(5)} f) = 11$
- $\dim_K T_f(\mathscr{K}^{(5)} f) = 12$

For the computation of $\widetilde{T}_f$ we use the formula from section 4.1 but since the tangent space T_f has no description in terms of f and $\frac{\partial f}{\partial x_i}$ if char $(K) > 0$, we compute the stabilzer of G and its dimension with the help of Gröbner bases.

Problem Does finite determinacy of $A \in M_{r,s}$ always imply finite codimension of $\widetilde{T}_A(GA)$ if $p > 0$?

We may conjecture that this is not the case for arbitrary r, s, n.

References

1. Arnol'd, V.I., Gusein-Zade, S.M., Varchenko, A.N.: Singularities of Differentiable Maps. Volume I: The Classification of Critical Points, Caustics and Wave Fronts. Monographs in Mathematics, vol. 82, X, 382p. Birkhäuser, Boston/Basel/Stuttgart (1985). Zbl 0554.58001
2. Boubakri, Y., Greuel, G.-M., Markwig, T.: Invariants of hypersurface singularities in positive characteristic. Rev. Mat. Complut. **25**, 61–85 (2012). Zbl 1279.14004
3. Buchweitz, R.-O., Greuel, G.-M., Schreyer, F.-O.: Cohen-Macaulay modules on hypersurface singularities. II. Invent. Math. **88**, 165–182 (1987). Zbl 0617.14034
4. Campillo, A.: Algebroid Curves in Positive Characteristic. Lecture Notes in Mathematics, vol. 813, 168p. Springer, Berlin/Heidelberg/New York (1980). Zbl 0451.14010
5. Campillo, A.: Hamburger–Noether expansions over rings. Trans. Am. Math. Soc. **279**, 377–388 (1983). Zbl 0559.14020
6. Campillo, A., Greuel, G.-M., Lossen, C.: Equisingular deformations of plane curves in arbitrary characteristic. Compos. Math. **143**, 829–882 (2007). Zbl 1121.14003
7. Campillo, A., Greuel, G.-M., Lossen, C.: Equisingular calculations for plane curve singularities. J. Symb. Comput. **42**(1–2), 89–114 (2007). Zbl 1128.14003
8. Decker, W., Greuel, G.-M., Pfister, G., Schönemann, H.: Singular 4-1-0 – A computer algebra system for polynomial computations. http://www.singular.uni-kl.de (2016)
9. Durfee, A.H.: Fifteen characterizations of rational double points and simple critical points. Enseign. Math. II. Sér. **25**, 132–163 (1979). Zbl 0418.14020
10. Greuel, G.-M.: Deformation and classification of singularities and modules. (Deformation und Klassifikation von Singularitäten und Moduln.) Jahresbericht der DMV, Jubiläumstag., 100 Jahre DMV, Bremen/Dtschl. 1990, 177–238 (1992). Zbl 0767.32017
11. Greuel, G.-M., Kröning, H.: Simple singularities in positive characteristic. Math. Z. **203**(2), 339–354 (1990). Zbl 0715.14001
12. Greuel, G.-M., Nguyen, H.D.: Right simple singularities in positive characteristic. J. Reine Angew. Math. **712**, 81–106 (2016). Zbl 1342.14006
13. Greuel, G.-M., Pham, T.H.: On finite determinacy for matrices of power series. Math. Zeitschrift. (2016, to appear). arXiv:1609.05133
14. Greuel, G.-M., Pham, T.H.: Finite determinacy of matrices and ideals in arbitrary characteristic (2017). arXiv:1708.02442v1
15. Greuel, G.-M., Pham, T.H.: Algorithms for group actions in arbitrary characteristic and a problem in singularity theory. arXiv:1709.08592v1
16. Greuel, G.-M., Lossen, C., Shustin, E.: Introduction to Singularities and Deformations. Springer Monographs in Mathematics. Springer, Berlin (2007). Zbl 1125.32013
17. Knörrer, H.: Cohen-Macaulay modules on hypersurface singularities. I. Invent. Math. **88**, 153–164 (1987). Zbl 0617.14033
18. Lossen, C., Mindnich, A.: `equisingl.lib`. A Singular 4-1-0 library for computing the equisingularity stratum of a family of plane curves (2016)
19. Nguyen, H.D.: Right unimodal and bimodal singularities in positive characteristic (2017). arXiv:1507.03554. To be published in International Mathematics Research Notices (IMRN), https://doi.org/10.1093/imrn/rnx165
20. Wahl, J.M.: Equisingular deformations of plane algebroid curves. Trans. Am. Math. Soc. **193**, 143–170 (1974). Zbl 0294.14007

Ultrametric Spaces of Branches on Arborescent Singularities

Evelia R. García Barroso, Pedro D. González Pérez, and Patrick Popescu-Pampu

This paper is dedicated to Antonio Campillo and Arkadiusz Płoski.

Abstract Let S be a normal complex analytic surface singularity. We say that S is *arborescent* if the dual graph of any *good* resolution of it is a tree. Whenever A, B are distinct branches on S, we denote by $A \cdot B$ their intersection number in the sense of Mumford. If L is a fixed branch, we define $U_L(A, B) = (L \cdot A)(L \cdot B)(A \cdot B)^{-1}$ when $A \neq B$ and $U_L(A, A) = 0$ otherwise. We generalize a theorem of Płoski concerning smooth germs of surfaces, by proving that whenever S is arborescent, then U_L is an ultrametric on the set of branches of S different from L. We compute the maximum of U_L, which gives an analog of a theorem of Teissier. We show that U_L encodes topological information about the structure of the embedded resolutions of any finite set of branches. This generalizes a theorem of Favre and Jonsson concerning the case when both S and L are smooth. We generalize also from smooth germs to arbitrary arborescent ones their valuative interpretation of the dual trees of the resolutions of S. Our proofs are based in an essential way on a determinantal identity of Eisenbud and Neumann.

E. R. García Barroso
Departamento de Matemáticas, Estadística e I.O. Sección de Matemáticas, Universidad de La Laguna, La Laguna, Tenerife, España
e-mail: ergarcia@ull.es

P. D. González Pérez
Instituto de Ciencias Matemáticas (CSIC-UAM-UC3M-UCM), Madrid, Spain

Facultad de Ciencias Matemáticas, Departamento de Álgebra, Geometría y Topología, Universidad Complutense de Madrid, Madrid, España
e-mail: pgonzalez@mat.ucm.es

P. Popescu-Pampu (✉)
Département de Mathématiques, Université de Lille, Villeneuve d'Ascq Cedex, France
e-mail: patrick.popescu@math.univ-lille1.fr

G.-M. Greuel et al. (eds.), *Singularities, Algebraic Geometry, Commutative Algebra, and Related Topics*, https://doi.org/10.1007/978-3-319-96827-8_3

1 Introduction

Płoski proved the following theorem in his paper [42]:

> Let $f, g, h \in \mathbb{C}\{x, y\}$ be irreducible power series. Then in the sequence
>
> $$\frac{m_O(f, g)}{(\text{ord } f)\,(\text{ord } g)}, \quad \frac{m_O(f, h)}{(\text{ord } f)\,(\text{ord } h)}, \quad \frac{m_O(g, h)}{(\text{ord } g)\,(\text{ord } h)}$$
>
> there are two terms equal and the third is not less than the equal terms. Here $m_O(f, g)$ denotes the intersection multiplicity of the branches $f = 0$, $g = 0$ and ord f stands for the order of f.

Denote by $Z_f \subset (\mathbb{C}^2, 0)$ the *branch* (that is, the germ of irreducible curve) defined by the equation $f = 0$. One has analogously the branches Z_g, Z_h. Looking at the inverses of the previous quotients:

$$U(Z_f, Z_g) := \begin{cases} \dfrac{(\text{ord } f)\,(\text{ord } g)}{m_O(f, g)}, & \text{if } Z_f \neq Z_g \\ 0, & \text{if } Z_f = Z_g \end{cases},$$

one may express Płoski's theorem in the following equivalent way:

> U is an ultrametric distance on the set of branches of $(\mathbb{C}^2, 0)$.

Note that, if Z_f and Z_g are two different branches and if $l \in \mathbb{C}\{x, y\}$ defines a smooth branch transversal to Z_f and Z_g (that is, if one has ord $f = m_O(l, f)$ and ord $g = m_O(l, g)$), then:

$$U(Z_f, Z_g) = \frac{m_O(l, f)\, m_O(l, g)}{m_O(f, g)}.$$

This is the view-point we take in our paper. *Instead of working with multiplicities, we work with intersection multiplicities (also called intersection numbers) with a fixed branch*. More precisely, we study the properties of the quotients:

$$U_L(A, B) := \begin{cases} \dfrac{(L \cdot A)\,(L \cdot B)}{A \cdot B}, & \text{if } A \neq B, \\ 0, \ \text{if } A = B, \end{cases}$$

when A and B vary in the set of branches of a normal surface singularity S which are different from a fixed branch L. In the previous formula, $A \cdot B \in \mathbb{Q}_+^*$ denotes *Mumford's intersection number* of [36, II.b] (see Definition 17).

We focus on the germs of normal surfaces which have in common with $(\mathbb{C}^2, 0)$ the following crucial property: *the dual graphs of their resolutions with simple normal crossings are trees*. We call *arborescent* the normal surface singularities with this property. Note that in this definition we impose no conditions on the *genera* of the irreducible components of exceptional divisors (see also Remark 65).

We prove that (see Theorem 85):

Let S be an arborescent singularity and let L be a fixed branch on it. Then, U_L is an ultrametric distance on the set of branches of S different from L.

Given a finite set $\mathcal{F}$ of branches on S, one gets in this way an infinite family of ultrametric distances on it, parametrized by the branches L which do not belong to $\mathcal{F}$. But, whenever one has an ultrametric on a finite set $\mathcal{F}$, there is a canonically associated rooted tree whose set of leaves is $\mathcal{F}$ (the interior-rooted tree of Definition 45). In our context, we show that the previous infinite family of ultrametrics define all the same *unrooted* tree. This tree may be interpreted in the following way using the resolutions of S (see Theorem 87 and Remark 101):

Let S be an arborescent singularity and let $\mathcal{F}$ be a finite set of branches on it. Let L be another branch, which does not belong to $\mathcal{F}$. Then the interior-rooted tree associated to U_L is homeomorphic to the union of the geodesics joining the representative points of the branches of $\mathcal{F}$ in the dual graph of any embedded resolution of the reduced Weil divisor whose branches are the elements of $\mathcal{F}$.

Both theorems are based on the following result (reformulated differently in Proposition 69):

Let S be an arborescent singularity. Consider a resolution of it with simple normal crossing divisor E. Denote by $(E_v)_{v \in \mathcal{V}}$ the irreducible components of E and by $(\check{E}_v)_{v \in \mathcal{V}}$ their duals, that is, the divisors supported by E such that $\check{E}_v \cdot E_w = \delta_{vw}$, for any $v, w \in \mathcal{V}$ and δ_{vw} denotes the Kronecker delta. Then, for any $u, v, w \in \mathcal{V}$ such that v belongs to the geodesic of the dual tree of E which joins u and w, one has:

$$(-\check{E}_u \cdot \check{E}_v)(-\check{E}_v \cdot \check{E}_w) = (-\check{E}_v \cdot \check{E}_v)(-\check{E}_u \cdot \check{E}_w).$$

In turn, this last result is obtained from an identity between determinants of weighted trees proved by Eisenbud and Neumann [18, Lemma 20.2] (see Proposition 70 below). Therefore, our proof is completely different in spirit from Płoski's original proof, which used computations with Newton-Puiseux series. Instead, we work exclusively with the numbers $-\check{E}_u \cdot \check{E}_v$, which are positive and birationally invariant, in the sense that they are unchanged if one replaces E_u and E_v by their strict transforms on a higher resolution (see Corollary 20).

We were also inspired by an inequality of Teissier [45, Page 40]:

Let S be a normal surface singularity with marked point O. If A, B are two distinct branches on it and m_O denotes the multiplicity function at O, then one has the inequality:

$$\frac{m_O(A) \cdot m_O(B)}{A \cdot B} \leq m_O(S).$$

We prove the following analog of it in the setting of arborescent singularities (see Corollary 84, in which we describe also the case of equality):

Whenever L, A, B are three pairwise distinct branches on the arborescent singularity S and E_l is the unique component of the exceptional divisor of an embedded resolution of L which intersects the strict transform of L, one has:

$$U_L(A, B) \leq -\check{E}_l \cdot \check{E}_l.$$

Then, in Theorem 92, we prove the following analog of the fact that U_L is an ultrametric:

> Under the hypothesis that the generic hyperplane section of an arborescent singularity S is irreducible, the function U_O defined by the left-hand side of Teissier's inequality is also an ultrametric.

Our approach allows us also to extend from smooth germs to arbitrary arborescent singularities S a valuative intepretation given by Favre and Jonsson [20, Theorem 6.50] of the natural partial order on the rooted tree defined by U_L.

Namely, consider a branch A on S different from L, and an embedded resolution of $L + A$. As before, we denote by $(E_v)_{v \in \mathcal{V}}$ the irreducible components of its exceptional divisor. Look at the dual graph of the total transform of $L + A$ as a tree rooted at the strict transform of L, and denote by $\preceq_L$ the associated partial order on its set of vertices, identified with $\mathcal{V} \cup \{L, A\}$. To each $v \in \mathcal{V}$ is associated a valuation ord_v^L of the local ring $\mathcal{O}$ of S, proportional to the divisorial valuation of E_v and normalized relative to L. Similarly, to the branch A is associated a semivaluation int_A^L of $\mathcal{O}$, which is also normalized relative to L, defined by $\text{int}_A^L(h) = (A \cdot Z_h) \cdot (A \cdot L)^{-1}$ (see Definition 116). Given two semivaluations ν_1 and ν_2, say that $\nu_1 \leq_{val} \nu_2$ if $\nu_1(h) \leq \nu_2(h)$ for all $h \in \mathcal{O}$. This is obviously a partial order on the set of semivaluations of $\mathcal{O}$. We prove (see Theorem 119):

> For an arborescent singularity, the inequality $\text{ord}_u^L \leq_{val} \text{ord}_v^L$ is equivalent to the inequality $u \preceq_L v$. Similarly, the inequality $\text{ord}_u^L \leq_{val} \text{int}_A^L$ is equivalent to the inequality $u \preceq_L A$.

Even when S is smooth, this result is stronger than the result of Favre and Jonsson, which concerns only the case where the branch L *is also smooth*. In this last case, our Theorem 87 specializes to Lemma 3.69 of [20], in which $U_L(A, B)$ is expressed in terms of what they call the *relative skewness function* α_x on a tree of conveniently normalized valuations (we give more explanations in Remark 118).

As was the case for Płoski's treatment in [42], Favre and Jonsson's study in [20] is based in an important way on Newton-Puiseux series. We avoid completely the use of such series and we extend their results to all arborescent singularities, by using instead the dual divisors $\check{E}_u$ defined above. Our treatment in terms of the divisors $\check{E}_u$ and the numbers $-\check{E}_u \cdot \check{E}_v$ was inspired by the alternative presentation of the theory of [20] given by Jonsson in [32, Section 7.3], again in the smooth surface case.

As in this last paper, our study could be continued by looking at the projective system of embedded resolutions of divisors of the form $L + C$, for varying reduced Weil divisors C, and by gluing accordingly the corresponding ultrametric spaces and rooted trees (see Remark 121). One would get at the limit a description of a quotient of the Berkovich space of the arborescent singularity S. We decided not to do this in this paper, in order to isolate what we believe are the most elementary ingredients of such a construction, which do not depend in any way on Berkovich theory.

Let us mention also another difference with the treatments of smooth germs S in [20] and [32]. In both references, the authors treat simultaneously the relations between triples of functions on their trees (called *skewness*, *thinness* and *multiplicity*). Our paper shows that an important part of their theory (for instance, the reconstruction of the shape of the trees from valuation theory) may be done by looking at only one function (the one they call the skewness).

In the whole paper, we work for simplicity with *complex* normal surface singularities. Note that, in fact, our techniques make nearly everything work for singularities which are spectra of normal 2 dimensional local rings defined over algebraically closed fields of arbitrary characteristic. Indeed, our treatment is based on the fact that the intersection matrix of a resolution of the singularity is negative definite (see Theorem 3 below), a theorem which is true in this greater generality (see Lipman [35, Lemma 14.1]). The only exception to this possibility of extending our results to positive characteristic is Sect. 4.3, as it uses Newton-Puiseux series, which behave differently in positive characteristic (see Remark 114).

The paper is structured as follows. In Sect. 2 we recall standard facts about Mumford's intersection theory on normal surface singularities. In Sect. 3 we present basic relations between *ultrametrics*, *arborescent posets*, *hierarchies* on finite sets, *trees*, *rooted trees*, *height* and *depth* functions on rooted trees, and *additive distances* on unrooted ones. Even if those relations are standard, we could not find them formulated in a way adapted to our purposes. For this reason we present them carefully. The next two sections contain our results. Namely, the ultrametric spaces of branches of arborescent singularities are studied in Sect. 4 and the valuative interpretations are developed in Sect. 5. We dedicate a special Sect. 4.3 to the original case considered by Płoski, where both S and L are smooth, by giving an alternative proof of his theorem using so-called *Eggers-Wall trees*. Finally, Sect. 6 contains examples and a list of open problems which turn around the following question: *is it possible to extend at least partially our results to some singularities which are not arborescent*? Our examples show that there exist normal surface singularities and branches L on them for which U_L is not even a metric.

Since the first version of this paper, we have greatly extended its results, in collaboration with Matteo Ruggiero (see [25]), modifiying also substantially our approach: instead of working with rooted trees associated with ultrametrics, we work with unrooted trees associated with metrics satisfying the so-called *four point condition*. Nevertheless, we feel that this first approach remains interesting and potentially useful in other contexts.

2 A Reminder on Intersection Theory for Normal Surface Singularities

In this section we introduce the basic vocabulary and properties needed in the sequel about complex normal surface singularities S and about the branches on them. In particular, we recall the notions of *good resolution*, *associated dual graph*, natural pair of dual lattices and *intersection form* on S. We explain the notion of *determinant* of S and the way to define, following Mumford, a rational *intersection number of effective divisors without common components* on S. This definition is based in turn on the definition of the *exceptional transform* of such a divisor on any resolution of S. The exceptional transform belongs to the *nef cone* of the resolution. We show that the exceptional divisors belonging to the interior of the nef cone are proportional to exceptional transforms of principal divisors on S, a fact which we use later in the proof of Theorem 119.

2.1 The Determinant of a Normal Surface Singularity

In the whole paper, (S, O) denotes a **complex analytic normal surface singularity**, that is, a germ of complex analytic normal surface. The germ is allowed to be smooth, in which case it will still be called a singularity. This is a common abuse of language. Most of the time we will write simply S instead of (S, O).

Definition 1 A **resolution** of S is a proper bimeromorphic morphism $\pi : \tilde{S} \to S$ with total space $\tilde{S}$ smooth. By abuse of language, we will also say in this case that $\tilde{S}$ is a resolution of S. The **exceptional divisor** $E := \pi^{-1}(O)$ of the resolution is considered as a reduced curve. A resolution of S is **good** if its exceptional divisor has simple normal crossings, that is, if all its components are smooth and its singularities are ordinary double points.

A special case of the so-called *Zariski main theorem* (see [29, Cor. 11.4]) implies that E is connected, hence the associated *weighted dual graph* is also connected:

Definition 2 Let $\pi : \tilde{S} \to S$ be a good resolution of S. We denote the irreducible components of its exceptional divisor E by $(E_u)_{u \in \mathcal{V}}$. The **weighted dual graph** Γ of the resolution has $\mathcal{V}$ as vertex set. There are no loops, but as many edges between the distinct vertices u, v as the intersection number $E_u \cdot E_v$. Moreover, each vertex $v \in \mathcal{V}$ is weighted by the self-intersection number $E_v \cdot E_v$ of E_v on $\tilde{S}$.

Usually one decorates each vertex u also by the genus of the corresponding component E_u. *We do not do this, because those genera play no role in our study.*

The set $\mathcal{V}$ may be seen not only as the vertex set of the dual graph Γ, but also as a set of parameters for canonical bases of the two following dual lattices, introduced by Lipman in [35, Section 18]:

$$\Lambda := \bigoplus_{u \in \mathcal{V}} \mathbb{Z}\, E_u, \quad \check{\Lambda} := \mathrm{Hom}_{\mathbb{Z}}(\Lambda, \mathbb{Z}).$$

The **intersection form**:

$$\begin{array}{rccc} I: & \Lambda \times \Lambda & \to & \mathbb{Z} \\ & (D_1, D_2) & \to & D_1 \cdot D_2 \end{array}$$

is the symmetric bilinear form on Λ which computes the intersection number of compact divisors on $\tilde{S}$. The following fundamental theorem was proved by Du Val [16, Section 4] and Mumford [36, Page 6]. It is also a consequence of Zariski [48, Lemma 7.4]. See also a proof in Lipman [35, Lemma 14.1], where it is explained that the theorem remains true for normal surface singularities defined over arbitrary algebraically closed fields, possibly of positive characteristic:

Theorem 3 *The intersection form of any resolution of S is negative definite.*

As a consequence, the map:

$$\tilde{I} : \Lambda \to \check{\Lambda}$$

induced by the intersection form I is an embedding of lattices, which allows to see Λ as a sublattice of finite index of $\check{\Lambda}$, and $\check{\Lambda}$ as a lattice of the $\mathbb{Q}$-vector space $\Lambda_{\mathbb{Q}} := \Lambda \otimes_{\mathbb{Z}} \mathbb{Q}$. In particular, the real vector space $\check{\Lambda}_{\mathbb{R}}$ gets identified with $\Lambda_{\mathbb{R}}$. As the intersection form I extends canonically to $\Lambda_{\mathbb{Q}}$, one may restrict it to $\check{\Lambda}$. We will denote those extensions by the same symbol I. The following lemma is immediate:

Lemma 4 *Seen as a subset of $\Lambda_{\mathbb{Q}}$, the lattice $\check{\Lambda}$ may be characterized in the following way:*

$$\check{\Lambda} = \{D \in \Lambda_{\mathbb{Q}} \mid D \cdot E_u \in \mathbb{Z}, \textit{ for all } u \in \mathcal{V}\}.$$

Denote by δ_{uv} the Kronecker delta. The basis of $\check{\Lambda}$ which is dual to the basis $(E_u)_{u \in \mathcal{V}}$ of Λ may be characterized in the following way as a subset of $\Lambda_{\mathbb{Q}}$:

Definition 5 The **Lipman basis** of π is the set $(\check{E}_u)_{u \in \mathcal{V}}$ of elements of the lattice $\check{\Lambda}$ defined by:

$$\check{E}_u \cdot E_v = \delta_{uv}, \text{ for all } v \in \mathcal{V}.$$

Definition 5 implies immediately that:

$$E_u = \sum_{w \in \mathcal{V}} (E_w \cdot E_u) \check{E}_w, \quad \text{for all } u \in \mathcal{V}, \tag{1}$$

and that, conversely:

$$\check{E}_u = \sum_{w \in \mathcal{V}} (\check{E}_w \cdot \check{E}_u) E_w, \quad \text{for all } u \in \mathcal{V}. \tag{2}$$

More generally, whenever $D \in \Lambda_{\mathbb{R}} = \check{\Lambda}_{\mathbb{R}}$, one has:

$$D = \sum_{w \in \mathscr{V}} (D \cdot \check{E}_w) E_w = \sum_{w \in \mathscr{V}} (D \cdot E_w) \check{E}_w. \tag{3}$$

The following result is well-known and goes back at least to Zariski [48, Lemma 7.1] (see it also creeping inside Artin's proof of [2, Prop. 2 (i)]):

Proposition 6 *If $D \in \Lambda_{\mathbb{Q}} \setminus \{0\}$ is such that $D \cdot E_v \geq 0$ for all $v \in \mathscr{V}$, then all the coefficients $D \cdot \check{E}_v$ of D in the basis $(E_v)_{v \in \mathscr{V}}$ of $\Lambda_{\mathbb{Q}}$ are negative.*

Combining the equations (1) and (2), one gets:

Proposition 7 *Once one chooses a total order on $\mathscr{V}$, the matrices $(E_u \cdot E_v)_{(u,v) \in \mathscr{V}^2}$ and $(\check{E}_u \cdot \check{E}_v)_{(u,v) \in \mathscr{V}^2}$ are inverse of each other.*

By abuse of language, we say that the two previous functions from $\mathscr{V}^2$ to $\mathbb{Q}$ are *matrices*, even without a choice of total order on the index set $\mathscr{V}$. The intersection matrix $(E_u \cdot E_v)_{(u,v) \in \mathscr{V}^2}$ has negative entries on the diagonal (as a consequence of Theorem 3) and non-negative entries outside it. By contrast:

Proposition 8 *The inverse matrix $(\check{E}_u \cdot \check{E}_v)_{(u,v) \in \mathscr{V}^2}$ has all its entries negative.*

Proof By formula (2), the entries of this matrix are the coefficients of the various rational divisors $\check{E}_u$ in the basis $(E_v)_{v \in \mathscr{V}}$ of $\Lambda_{\mathbb{Q}}$. By Definition 5, we see that $\check{E}_u \cdot E_v \geq 0$ for all $v \in \mathscr{V}$. One may conclude using Proposition 6.

The fact that if the entries of a symmetric positive definite matrix are non-positive outside the diagonal, then the entries of the inverse matrix are non-negative was proved by Coxeter [11, Lemma 9.1] and differently by Du Val [15, Page 309], following a suggestion of Mahler. The stronger fact that in our case the entries of $(-\check{E}_u \cdot \check{E}_v)_{(u,v) \in \mathscr{V}^2}$ are positive comes from the fact that the dual graph of the initial matrix $(-E_u \cdot E_v)_{(u,v) \in \mathscr{V}^2}$ is connected. For historical details about this theme, see Coxeter [12, Section 10.9].

Proposition 6 may be reformulated as the fact that *the pointed nef cone of π is included in the interior of the opposite of the effective cone of π*, where we use the following terminology, which is standard for global algebraic varieties:

Definition 9 Let π be a resolution of S. The **effective cone** σ of π is the simplicial subcone of $\Lambda_{\mathbb{R}}$ consisting of those divisors with non-negative coefficients in the basis $(E_u)_{u \in \mathscr{V}}$. The **nef cone** $\check{\sigma}$ of π is the simplicial subcone of $\check{\Lambda}_{\mathbb{R}}$, identified to $\Lambda_{\mathbb{R}}$ through $\tilde{I}$, consisting of those divisors whose intersections with all effective divisors are non-negative. That is, σ is the convex cone generated by $(E_u)_{u \in \mathscr{V}}$ and $\check{\sigma}$ is the convex cone generated by $(\check{E}_u)_{u \in \mathscr{V}}$.

The determinant of a symmetric bilinear form on a lattice is well-defined: it is independent of the basis of the lattice used to compute it. When the bilinear form is positive definite, the determinant is positive. This motivates to look also at the opposite of the intersection form (see Neumann and Wahl [39, Sect. 12]). Up to the sign, the following notion was also studied in [18] and [38].

Definition 10 Let π be a good resolution of S with weighted dual graph Γ. The **determinant** $\det(\Gamma) \in \mathbb{N}^*$ of Γ is the determinant of the opposite $-I$ of the intersection form, that is, the determinant of the matrix $(-E_u \cdot E_v)_{(u,v)\in\mathscr{V}^2}$.

It is well-known (see for instance [13, Prop. 3.4 of Chap. 2]) that $\det(\Gamma)$ is equal to the cardinal of the torsion subgroup of the first integral homology group of the boundary (or link) of S. This shows that $\det(\Gamma)$ is independent of the choice of resolution of S. This fact could have been proved directly, by studying the effect of a blow-up of one point on the exceptional divisor of a given resolution and by using the fact that any two resolutions are related by a finite sequence of blow ups of points and of their inverse blow-downs. As a consequence, we define:

Definition 11 The **determinant** $\det(S)$ **of the singularity** S is the determinant of the weighted dual graph of any good resolution of it.

2.2 *Mumford's Rational Intersection Number of Branches*

As explained in the introduction, we will be mainly interested by the *branches* living on S:

Definition 12 A **branch** on S is a germ of irreducible formal curve on S. We denote by $\mathscr{B}(S)$ the set of branches on S. A **divisor** (respectively $\mathbb{Q}$**-divisor**) on S is an element of the free abelian group (respectively $\mathbb{Q}$**-vector space**) generated by the branches living on it. The divisor is **effective** if all its coefficients are non-negative. It is **principal** if it is the divisor of a germ of formal function on S.

Note that the divisors we consider are Weil divisors, as they are not necessarily principal.

We will study the divisors on S using their *embedded resolutions*, to which one extends the notion of *weighted dual graph*:

Definition 13 If D is a divisor on S, an **embedded resolution** of D is a resolution $\pi : \tilde{S} \to S$ of S such that the preimage $\pi^{-1}|D|$ of $|D|$ on $\tilde{S}$ has simple normal crossings (here $|D|$ denotes the support of D). The **weighted dual graph** Γ_D of D with respect to π is obtained from the weighted dual graph Γ of π by adding as new vertices the branches of D and by joining each such branch A to the unique vertex $u(A) \in \mathscr{V}$ such that $E_{u(A)}$ meets A.

The following construction of Mumford [36] of a canonical, possibly non-reduced structure of $\mathbb{Q}$-divisor on $\pi^{-1}|D|$, will be very important for us:

Definition 14 Let A be a divisor on S and $\pi : \tilde{S} \to S$ a resolution of S. The **total transform** of A on $\tilde{S}$ is the $\mathbb{Q}$-divisor $\pi^*A = \tilde{A} + (\pi^*A)_{ex}$ on $\tilde{S}$ such that:

1. $\tilde{A}$ is the **strict transform** of A on $\tilde{S}$ (that is, the sum of the closures inside $\tilde{S}$ of the branches of A, keeping unchanged the coefficient of each branch).

2. The support of the **exceptional transform** $(\pi^* A)_{ex}$ of A on $\tilde{S}$ (or by π) is included in the exceptional divisor E.
3. $\pi^* A \cdot E_u = 0$ for each irreducible component E_u of E.

Such a divisor $(\pi^* A)_{ex}$ exists and is unique:

Proposition 15 *Let A be a divisor on S and $\pi : \tilde{S} \to S$ a resolution of S. Then the exceptional transform $(\pi^* A)_{ex}$ of A is given by the following formula:*

$$(\pi^* A)_{ex} = - \sum_{u \in \mathscr{V}} (\tilde{A} \cdot E_u)\, \check{E}_u.$$

In particular, $(\pi^* A)_{ex}$ lies in the opposite of the nef cone.

Proof The third condition of Definition 14 implies that: $(\pi^* A)_{ex} \cdot E_u = -\tilde{A} \cdot E_u$, for all $u \in \mathscr{V}$. By combining this with Eq. (3), we get:

$$(\pi^* A)_{ex} = \sum_{u \in \mathscr{V}} ((\pi^* A)_{ex} \cdot E_u)\, \check{E}_u = - \sum_{u \in \mathscr{V}} (\tilde{A} \cdot E_u)\, \check{E}_u.$$

As $\tilde{A} \cdot E_u \geq 0$ for all $u \in \mathscr{V}$, we see that $-(\pi^* A)_{ex}$ lies in the cone generated by $(\check{E}_u)_{u \in \mathscr{V}}$ inside $\Lambda_{\mathbb{R}}$ which, by Definition 9, is the nef cone $\check{\sigma}$.

In the case in which A is principal, defined by a germ of holomorphic function f_A, then $(\pi^* A)_{ex}$ is simply the exceptional part of the principal divisor on $\tilde{S}$ defined by the pull-back function $\pi^* f_A$. By Proposition 15, in this case $(\pi^* A)_{ex}$ belongs to the semigroup $-\check{\sigma} \cap \Lambda$ of integral exceptional divisors whose opposites are nef. In general not all the elements of this semigroup consist in such exceptional transforms of principal divisors, but this is true for those lying in the interior of $-\check{\sigma}$:

Proposition 16 *Consider a resolution of S. Any element of the lattice Λ which lies in the interior of the cone $-\check{\sigma}$ has a multiple by a positive integer which is the exceptional transform of an effective principal divisor on S.*

Proof Denote by K a canonical divisor on the resolution $\tilde{S}$ and by $E = \sum_{v \in \mathscr{V}} E_v$ the reduced exceptional divisor. By [9, Theorem 4.1], any divisor $D \in \Lambda$ such that:

$$(D + E + K) \cdot E_u \leq -2 \text{ for all } u \in \mathscr{V} \tag{4}$$

is the exceptional transform of an effective principal divisor.

Assume now that $H \in \Lambda$ belongs to the *interior* of the opposite $-\check{\sigma}$ of the nef cone. This means that $H \cdot E_u < 0$ for any $u \in \mathscr{V}$. There exists therefore $n \in \mathbb{N}^*$ such that: $nH \cdot E_u < -(E + K) \cdot E_u - 2$, for all $u \in \mathscr{V}$. Equivalently, $D := nH$ satisfies the inequalities (4), therefore it is the exceptional transform of an effective principal divisor.

Definition 14 allowed Mumford to introduce a *rational intersection number* of any two divisors on S without common branches:

Definition 17 Let A, B be two divisors on S without common branches. Then their **intersection number** $A \cdot B \in \mathbb{Q}$ is defined by: $A \cdot B := \pi^*A \cdot \pi^*B$, for any embedded resolution π of $A + B$.

The fact that this definition is independent of the resolution was proved by Mumford [36, Section II (b)], by showing that it is unchanged if one blows up one point on E. As an immediate consequence we have:

Proposition 18 *Let A, B be two divisors on S without common branches and $(\pi^*A)_{ex}$, $(\pi^*A)_{ex}$ be their exceptional transforms on an embedded resolution $\tilde{S}$ of $A + B$. Then $A \cdot B = -(\pi^*A)_{ex} \cdot (\pi^*B)_{ex}$.*

Assume now that A is a branch on S. Consider an embedded resolution of it. Recall from Definition 13 that $u(A)$ denotes the unique index $u \in \mathcal{V}$ such that the strict transform $\tilde{A}$ meets E_u. Then one has another immediate consequence of the definitions:

Lemma 19 *If A is a branch on S and if $\tilde{S}$ is an embedded resolution of A, then $(\pi^*A)_{ex} = -\check{E}_{u(A)}$.*

By combining Proposition 18, Lemma 19 and Proposition 8, one gets:

Corollary 20

1. *Assume that A and B are distinct branches on S. If $\tilde{S}$ is an embedded resolution of $A + B$, then:*

$$A \cdot B = -\check{E}_{u(A)} \cdot \check{E}_{u(B)} > 0.$$

2. *If E_u, E_v are two components of the exceptional divisor of a resolution, then the intersection number $\check{E}_u \cdot \check{E}_v$ is independent of the resolution (that is, it will be the same if one replaces E_u, E_v by their strict transforms on another resolution).*

3 Generalities on Ultrametrics and Trees

In this section we explain basic relations between *ultrametrics, arborescent posets, hierarchies* on finite sets, *rooted* and *unrooted trees, height* and *depth* functions on rooted trees, and *additive distances* on unrooted ones. The fact that we couldn't find these relations described in the literature in a way adapted to our purposes, explains the level of detail of this section. The framework developed here allows us to formulate in the next two sections simple conceptual proofs of our main results: Theorems 85, 87, 92, 119 and Corollary 120.

We begin by explaining the relation between *rooted trees, arborescent posets* and *hierarchies* (see Sect. 3.1). In Sect. 3.2 we recall the notion of *ultrametric spaces* and we explain that ultrametrics on finite sets may be alternatively presented as rooted trees endowed with a *depth function*, the intermediate object in this structural metamorphosis being the *hierarchy* of closed balls of the metric, which is an *arborescent poset*. We also define a dual notion of *height function* on a tree. This allows us to relate in Sect. 3.3 ultrametrics with *additive distances* on unrooted trees: we explain that any choice of root allows to transform canonically such a distance into a height function. Finally, in Sect. 4.3 we apply the previous considerations by giving a new proof of Płoski's original theorem, using the notion of *Eggers-Wall tree*.

3.1 Trees, Rooted Trees, Arborescent Partial Orders and Hierarchies

If V is a set, we denote by $\binom{V}{2}$ the set of its subsets with 2 elements. There are two related notions of trees: combinatorial and topological.

Definition 21 A **combinatorial tree** is a finite combinatorial connected graph without cycles, that is, a pair $(\mathcal{V}, \mathcal{A})$ where $\mathcal{V}$ is a finite set of **vertices**, $\mathcal{A} \subset \binom{\mathcal{V}}{2}$ is the set of **edges** and for any two distinct vertices u, v, there is a unique sequence $\{v_0, v_1\}, \ldots, \{v_{k-1}, v_k\}$ of edges, with $v_0 = u$, $v_k = v$ and $v_0, \ldots, v_k$ pairwise distinct. Its **geometric realization** is the simplicial complex obtained by joining the vertices which are end-points of edges by segments in the real vector space $\mathbb{R}^{\mathcal{V}}$.

Definition 22 A **topological tree** T is a topological space homeomorphic to the geometric realization of a combinatorial tree. If u, v are two points of it, the unique embedded arc joining u and v is called the **geodesic joining** u **and** v and is denoted $[uv] = [vu]$. More generally, the **convex hull** $[V]$ of a finite subset V of T is the union of the geodesics joining pairwise the points of V. If $V = \{u, v, w, \ldots\}$, we denote also $[V] = [uvw\ldots]$. The **valency** of a point $u \in T$ is the number of connected components of its complement $T \setminus \{u\}$. The **ends** of the topological tree are its points of valency 1 and its **nodes** are its points of valency at least 3. An underlying combinatorial tree being fixed (which will always be the case in the sequel), its vertices are by definition the **vertices** of T. A point $u \in T$ is called **interior** if it is not an end. Denote:

- $\mathcal{A}(T) =$ the set of edges of T;
- $\mathcal{V}(T) =$ the set of vertices of T;
- $\mathcal{E}(T) =$ the set of ends of T;
- $\mathcal{N}(T) =$ the set of nodes of T;
- $\mathcal{I}(T) =$ the set of interior vertices of T.

One has the inclusion $\mathcal{N}(T) \cup \mathcal{E}(T) \subset \mathcal{V}(T)$, which is strict if and only if either there is at least one vertex of valency 2 or the tree is reduced to a point (that is, $\mathcal{V}(T)$ has only one element, hence $\mathcal{A}(T) = \emptyset$).

We will use also the following vocabulary about posets:

Definition 23 In a poset $(V, \preceq)$, we say that $u \in V$ is a **predecessor** of $v \in V$ or that v is a **successor** of u if $u \prec v$ (which means that the inequality is *strict*). We say that the predecessor u of v is a **direct predecessor** of v if it is not a predecessor of any other predecessor of v. Then we say also that v is a **direct successor** of u. Two elements of V are **comparable** if one is a predecessor of the other, and **directly comparable** if one is a direct predecessor of the other one.

In the sequel we will work also with *rooted trees*, where the root may be either a vertex or a point interior to some edge. A choice of root of a given tree endows it with a canonical partial order:

Definition 24 Let T be a topological tree. One says that it is **rooted** if a point $\rho \in T$ is chosen, called the **root**. Whenever we want to emphasize the root, we denote by T_ρ the tree T rooted at ρ. The **associated partial order** $\preceq_\rho$ on T_ρ is defined by:

$$u \preceq_\rho v \iff [\rho u] \subset [\rho v].$$

The maximal elements for this partial order are called the **leaves** of T_ρ. Their set is denoted by $\mathcal{L}(T_\rho)$. The **topological vertex set** $\mathcal{V}_{top}(T_\rho)$ of T_ρ consists of its leaves, its nodes and its root. That is, it is defined by:

$$\mathcal{V}_{top}(T_\rho) := \mathcal{L}(T_\rho) \cup \mathcal{N}(T) \cup \{\rho\}. \tag{1}$$

If V is a finite set, the restriction to V of a partial order $\preceq$ coming from a structure of rooted tree with vertex set *containing* V is called an **arborescent partial order**. In this case, $(V, \preceq)$ is called an **arborescent poset**.

It is immediate to see that for a rooted tree T_ρ, the root ρ is the absolute minimum of $(T_\rho, \preceq_\rho)$. In addition, the partial order $\preceq_\rho$ has well-defined infima of pairs of points. This motivates (see Fig. 1):

Notation 25 *Let T_ρ be a rooted tree. The **infimum** of a and b for the partial order $\preceq_\rho$, that is, the maximal element of $[\rho a] \cap [\rho b]$, is denoted $a \wedge_\rho b$.*

It will be important in the sequel to distinguish the trees in which the root is an end (which implies, by Definition 22, that the tree has at least two vertices):

Definition 26 An **end-rooted tree** is a rooted tree T_ρ whose root is an end. Then the root ρ has a unique direct successor ρ^+ and each leaf a has a unique direct

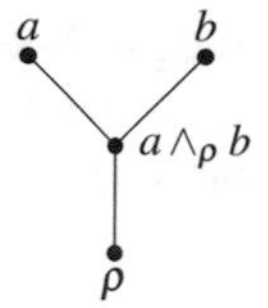

Fig. 1 The infimum of a and b for the partial order $\preceq_\rho$ (see Notation 25)

predecessor a^-. The **core** $\overset{\circ}{T}_\rho$ of the end-rooted tree T_ρ is the convex hull of $\{\rho^+\} \cup \{a^-,\ a \in \mathscr{L}(T_\rho)\}$, seen as a tree rooted at ρ^+.

A rooted tree which is not end-rooted, that is, such that the root is interior, is called **interior-rooted**.

Given an arbitrary rooted tree, there is a canonical way to embed it in an end-rooted tree:

Definition 27 Let T_ρ be a tree rooted at ρ. Its **extension** $\hat{T}_{\hat{\rho}}$ is the tree obtained from T_ρ by adding a new root $\hat{\rho}$, which is joined by an edge to ρ.

We defined arborescent posets starting from arbitrary rooted trees. Conversely, any arborescent poset $(V, \preceq)$ has a canonically associated rooted tree $T(V, \preceq)$ endowed with an underlying combinatorial tree. The root ρ may not belong to V, but the vertex set of $T(V, \preceq)$ is exactly $V \cup \{\rho\}$. More precisely:

Definition 28 Let $(V, \preceq)$ be an arborescent poset. Its **associated rooted tree** $T(V, \preceq)$ is defined by:

- If $(V, \preceq)$ has a unique minimal element, then the root coincides with it, and the edges are exactly the sets of the form $\{u, v\}$, where v is a direct predecessor of u. That is, $T(V, \preceq)$ is the underlying tree of the Hasse diagram of the poset.
- If $(V, \preceq)$ has several minimal elements, then one considers a new set $\hat{V} := V \sqcup \{m\}$ and one extends the order $\preceq$ to it by imposing that m is a predecessor of all the elements of V. Then one proceeds as in the previous case, working with $(\hat{V}, \preceq)$ instead of $(V, \preceq)$. In particular, the root is the new vertex m.

The **extended rooted tree** $\hat{T}(V, \preceq)$ **of** $(V, \preceq)$ is the extension of $T(V, \preceq)$ according to Definition 27.

The fact that the objects introduced in Definition 28 are always trees is a consequence of the following elementary proposition, whose proof we leave to the reader:

Proposition 29 *A partial order $\preceq$ on the finite set V is arborescent if and only if any element of it has at most one direct predecessor.*

The notion of *extended rooted tree* of an arborescent poset will play an important role in our context (see Remark 88).

Remark 30 We took the name of *arborescent partial order* from [19]. The characterization given in Proposition 29 was chosen in [19] as the definition of this notion.

One has the following fact, whose proof we leave to the reader:

Proposition 31 *Let T_ρ be a rooted tree. Then the rooted tree $T(\mathscr{V}_{top}(T_\rho), \preceq_\rho)$ associated to the arborescent poset $(\mathscr{V}_{top}(T_\rho), \preceq_\rho)$ is isomorphic to T_ρ by an isomorphism which fixes $\mathscr{V}_{top}(T_\rho)$.*

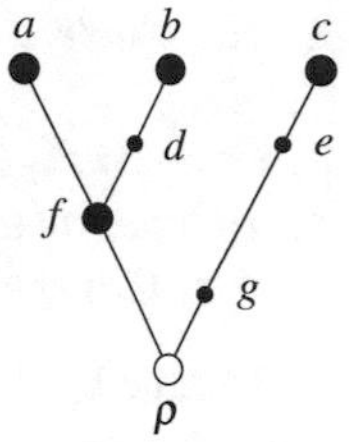

Fig. 2 The vertex set versus the topological vertex set in Example 33

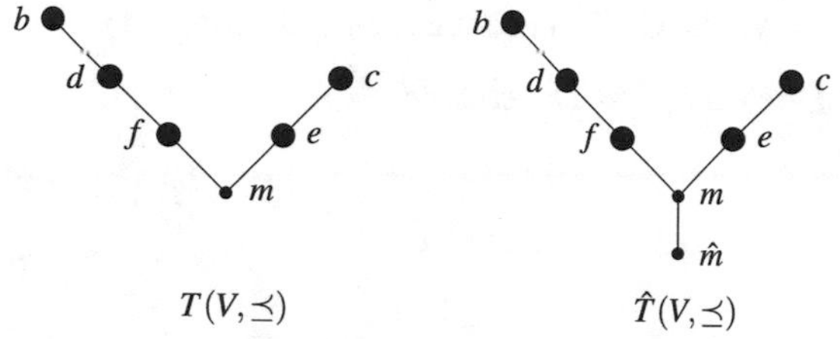

Fig. 3 The two trees associated canonically to the arborescent poset V of Example 33

A rooted tree may also be encoded as a supplementary structure on its set of leaves. Namely:

Definition 32 Let T_ρ be a rooted tree. To each point $v \in T_\rho$, associate its **cluster** $\mathbb{K}_\rho(v)$ as the set of leaves which have v as predecessor:

$$\mathbb{K}_\rho(v) := \{u \in \mathcal{L}(T_\rho),\ v \preceq_\rho u\}.$$

Example 33 Figure 2 shows a rooted tree T with vertex set $\{a, \dots, g\}$ indicated by black bullets. It is rooted at a point ρ which is not a vertex, indicated by a white bullet. Note that ρ lies in the interior of the edge $[fg]$. The topological vertex set of T, indicated with bigger bullets, is $\mathcal{V}_{top}(T_\rho) = \{a, b, c, f, \rho\}$. The arborescent poset $(V = \{b, c, d, e, f\}, \preceq_\rho)$ (which is taken distinct from the vertex set, see Definition 24), may be described by the strict inequalities, $f \prec_\rho d \prec_\rho b$ and $e \prec_\rho c$, between directly comparable elements of V. Notice that the poset V has two minimal elements e and f. The root m of the tree $T(V, \preceq)$ is a new point, since we are in the second case of Definition 28. The two rooted trees $T(V, \preceq)$ and $\hat{T}(V, \preceq)$ are drawn in Fig. 3. In both cases, the vertices corresponding to the elements of V are represented with bigger bullets. The clusters associated to the vertices and the root of T_ρ are:

$$\mathbb{K}_\rho(a) = \{a\},\ \mathbb{K}_\rho(b) = \mathbb{K}_\rho(d) = \{b\},\ \mathbb{K}_\rho(f) = \{a, b\},$$
$$\mathbb{K}_\rho(c) = \mathbb{K}_\rho(e) = \mathbb{K}_\rho(g) = \{c\},\ \mathbb{K}_\rho(\rho) = \{a, b, c\}.$$

One has the following direct consequences of Definition 32:

Proposition 34

1. *The cluster of a leaf u is $\{u\}$ and the cluster of the root is the entire set of leaves $\mathcal{L}(T_\rho)$.*

2. *The clusters* $\mathbb{K}_\rho(u)$ *and* $\mathbb{K}_\rho(v)$ *are disjoint if and only if* u *and* v *are incomparable.*
3. *One has* $\mathbb{K}_\rho(v) \subseteq \mathbb{K}_\rho(u)$ *if and only if* u *is a predecessor of* v.
4. *Two points* $u, v \in T$ *have the same cluster if and only if one is a predecessor of the other one and the geodesic* $[uv]$ *does not contain nodes of* T_ρ.

Denote by 2^W the *power set* of a set W, that is, its set of subsets. As an immediate consequence of Proposition 34 one has (recall that $\mathscr{V}_{top}(T_\rho)$ denotes the *topological vertex set* of T_ρ, introduced in Definition 24):

Corollary 35 *The* ***cluster map****:*

$$\begin{array}{rcl} \mathbb{K}_\rho : \mathscr{V}_{top}(T_\rho) & \to & 2^{\mathscr{L}(T_\rho)} \\ u & \to & \mathbb{K}_\rho(u) \end{array}$$

is decreasing from the poset $(\mathscr{V}_{top}(T_\rho), \preceq_\rho)$ *to the poset* $(2^{\mathscr{L}(T_\rho)}, \subseteq)$*. Moreover:*

1. If T_ρ is not end-rooted, *then* $\mathbb{K}_\rho$ *is injective.*
2. If T_ρ is end-rooted, *then* $\mathbb{K}_\rho$ *is injective in restriction to* $\mathscr{V}_{top}(T_\rho) \setminus \{\rho\}$ *and* $\mathbb{K}_\rho(\rho) = \mathbb{K}_\rho(\rho^+)$, *where* ρ^+ *is the unique direct successor of* ρ *in the poset* $(\mathscr{V}_{top}(T_\rho), \preceq_\rho)$.

Proposition 34 may be reformulated by saying that the image of the cluster map is a *hierarchy* in the following sense:

Definition 36 A **hierarchy** on the finite set X is a subset of $2^X \setminus \{\emptyset\}$ (whose elements are called **clusters**) satisfying the following properties:

1. All the one-element subsets of X as well as X itself are clusters.
2. Given any two clusters, they are either disjoint or one is included inside the other.

In fact, those properties characterize completely the images of cluster maps associated to rooted trees with given leaf set (folklore, see [4, Introduction]):

Proposition 37 *The images of the cluster maps associated to the rooted trees with finite sets of leaves* X *are exactly the* ***hierarchies*** *on* X.

Now, if one orders a hierarchy by *reverse inclusion*, one gets an arborescent poset:

Lemma 38 *Let* $\mathscr{H}$ *be a hierarchy on the finite set* X*. Define the partial order* $\preceq_{ri}$ *on* $\mathscr{H}$ *by:*

$$K_1 \preceq_{ri} K_2 \iff K_1 \supseteq K_2.$$

Then $(\mathscr{H}, \preceq_{ri})$ *is an arborescent poset.*

Proof By Proposition 29, it is enough to check that for any cluster $K_1 \neq X$, there exists a unique cluster K_2 strictly containing it and such that there are no other clusters between K_1 and K_2. But this comes from the fact that, by condition (2) of Definition 36, all the clusters containing K_1 form a chain (that is, a totally ordered set) under inclusion.

Conversely, one has the following characterization of arborescent posets coming from hierarchies, whose proof we leave to the reader:

Proposition 39 *An arborescent poset* $(V, \preceq)$ *is isomorphic to the poset defined by a hierarchy if and only if any non-maximal element has at least two direct successors. In particular, the associated rooted tree* $T(V, \preceq)$ *is never end-rooted.*

3.2 Ultrametric Spaces and Dated Rooted Trees

In this subsection we explain the relation between *finite ultrametric spaces* and rooted trees endowed with a *depth function*.

Let us first fix our notations and vocabulary about metric spaces:

Notation 40 *If* (X, d) *is a metric space, then a* **closed ball** *of it is a subset of the form:* $\mathbb{B}(a, r) := \{p \in X \mid d(a, p) \leq r\}$, *where a denotes any point of X and* $r \in [0, +\infty)$. *Each time a subset of X is presented in this way, one says that a is* **a center** *and r is* **a radius** *of it. The* **diameter** *of a subset* $Y \subset X$ *is diam*$(Y) := \sup\{d(x, y), x, y \in Y\} \in [0, +\infty]$.

In Euclidean geometry, a closed ball has a unique center and a unique radius. None of those two properties remains true in general. There is an extreme situation in which *any* point of a closed ball is a center of it:

Definition 41 Let (X, d) be a metric space. The distance function $d : X \times X \to [0, \infty)$ is called an **ultrametric** if one has the following strong form of the triangle inequality, called the **ultrametric inequality**:

$$d(a, b) \leq \max\{d(a, c), d(b, c)\}, \quad \text{for all } a, b, c \in X.$$

In this case, we say that (X, d) is **an ultrametric space**.

One has the following characterizations of ultrametricity, which result immediately from the definition:

Proposition 42 *Let* (X, d) *be a metric space. Then the following conditions are equivalent:*

1. *d is an ultrametric.*
2. *All the triangles are either equilateral or isosceles with the unequal side being the shortest.*
3. *For any closed ball, all its points are centers of it.*
4. *Given any two closed balls, they are either disjoint, or one of them is contained into the other one.*

As a consequence of Proposition 42, we have the following property:

Lemma 43 *Let (X, U) be a finite ultrametric space. If $\mathscr{D}$ is a closed ball, then its diameter* $\operatorname{diam}(\mathscr{D})$ *is the minimal radius r such that $\mathscr{D} = \mathbb{B}(a, r)$ for any $a \in \mathscr{D}$.*

The prototypical examples of ultrametric spaces are the fields of p-adic numbers or, more generally, all the fields endowed with a non-Archimedean norm. Our goal in the rest of this section is to describe the canonical presentation of finite ultrametric spaces as *sets of leaves of finite rooted trees* (see Proposition 50 below). A pleasant elementary introduction to this view-point is contained in Holly's paper [30].

The basic fact indicating that rooted trees are related with ultrametric spaces is the similarity of the conditions defining hierarchies (see Definition 36) with the characterization of ultrametrics given as point (4) of Proposition 42. This characterization, combined with the fact that the closed balls of radius 0 are exactly the subsets with one element, and the fact that on a finite metric space, the closed balls of sufficiently big radius are the whole set, shows that:

Lemma 44 *The set $\mathscr{B}(X, U)$ of closed balls of a finite ultrametric space (X, U) is a hierarchy on X.*

As a consequence of Lemmas 38 and 44, an ultrametric U on a finite set X defines canonically two rooted trees with leaf-set X and topological vertex set $\mathscr{B}(X, U)$:

Definition 45 Let U be an ultrametric on the finite set X. The **interior-rooted tree T^U associated to the ultrametric** U is the rooted tree $T(\mathscr{B}(X, U), \preceq_{ri})$ determined by the arborescent poset of closed balls of U. The **end-rooted tree $\hat{T}^U$ associated to the ultrametric** U is the extended rooted tree $\hat{T}(\mathscr{B}(X, U), \preceq_{ri})$.

The terminology is motivated by Proposition 39, which implies that T^U is indeed always interior-rooted and $\hat{T}^U$ always end-rooted. By Definition 28 and Lemma 38, the root ρ of T^U corresponds to the set X, while the root $\hat{\rho}$ of $\hat{T}^U$ is defined as the immediate predecessor of ρ in the tree $\hat{T}^U$.

One may encode also the *values* of the metric U on its end-rooted tree $\hat{T}^U$, as a decoration on its set $\mathscr{I}(\hat{T}^U)$ of interior vertices:

Definition 46 Let (X, U) be a finite ultrametric space. Its **diametral function**:

$$\delta^U : \mathscr{I}(\hat{T}^U) \to \mathbb{R}_+^*,$$

defined on the set of interior vertices of the end-rooted tree $\hat{T}^U$ of (X, U), associates to each vertex $u \in \mathscr{I}(\hat{T}^U)$ the diameter of its cluster $\mathbb{K}_{\hat{\rho}}(u)$.

The root ρ of T^U is always an interior vertex of $\hat{T}^U$, and its diameter $\delta^U(\rho)$ is equal to the diameter of X. Notice also that $\mathbb{K}_\rho(u) = \mathbb{K}_{\hat{\rho}}(u)$, for all $u \in \mathscr{I}(\hat{T}^U) = \mathscr{I}(T^U)$.

The diametral function of a finite ultrametric space is a *depth function on the associated end-rooted tree* in the following sense:

Definition 47 A **depth function** on a rooted tree T_ρ is a strictly decreasing function:

$$\delta : (\mathscr{I}(T_\rho), \preceq_\rho) \to (\mathbb{R}_+^*, \leq).$$

That is, $\delta(v) < \delta(u)$ whenever $u \prec_\rho v$. A pair (T_ρ, δ) of a rooted tree and a depth function on it is called a **depth-dated tree**.

Intuitively, such a function δ measures the depth of the interior vertices as seen from the leaves, if one imagines that the leaves are above the root, as modeled by the partial order $\preceq_\rho$.

We have explained how to pass from an ultrametric on a finite set to a depth-dated rooted tree (see Definitions 45 and 46). Conversely, given such a tree, one may construct an ultrametric on its set of leaves (recall that $a \wedge_\rho b$ is the infimum of a and b for the partial order $\preceq_\rho$, see Notation 25):

Lemma 48 *Let (T_ρ, δ) be a depth-dated rooted tree. Then the function U^δ : $\mathscr{L}(T_\rho) \times \mathscr{L}(T_\rho) \to \mathbb{R}_+$ defined by:*

$$U^\delta(a, b) := \begin{cases} \delta(a \wedge_\rho b) & \text{if } a \neq b, \\ 0 & \text{if } a = b, \end{cases}$$

is an ultrametric on $\mathscr{L}(T_\rho)$.

Proof Consider $a, b, c \in \mathscr{L}(T_\rho)$. The inequality:

$$U^\delta(a, b) \leq \max\{U^\delta(a, c), U^\delta(b, c)\}$$

is equivalent to:

$$\delta(a \wedge_\rho b) \leq \max\{\delta(a \wedge_\rho c), \delta(b \wedge_\rho c)\},$$

which is in turn a consequence of the facts that δ is a depth function and that:

$$a \wedge_\rho b \geq \min\{a \wedge_\rho c, b \wedge_\rho c\}.$$

The previous inequality, including the existence of this minimum (taken relative to the rooted tree partial order $\prec_\rho$) is a basic property of rooted trees.

Moreover, if the tree T_ρ is end-rooted, then one may reconstruct it as the end-rooted tree associated to the ultrametric space $(\mathscr{L}(T_\rho), U^\delta)$:

Proposition 49 *Let (T_ρ, δ) be a depth-dated end-rooted tree. There exists a unique isomorphism fixing the set of leaves between the combinatorial rooted trees* *underlying the depth-dated trees:*

- T_ρ *endowed with the topological vertex set* $\mathcal{V}_{top}(T_\rho)$ *and with the restriction to its set of interior vertices of the depth function* δ;
- *the depth-dated tree* $(\hat{T}^{U^\delta}, \delta^{U^\delta})$ *associated to the ultrametric* U^δ.

Taken together, the previous considerations prove the announced bijective correspondence between ultrametrics on a finite set X and a special type of depth-dated end-rooted trees with set of leaves X:

Proposition 50 *Let* X *be a finite set. The map which associates to an ultrametric* U *on* X *the diametral function* δ^U *on the end-rooted tree* $\hat{T}^U$ *realizes a bijective correspondence between ultrametrics on* X *and isomorphism classes of depth-dated end-rooted rooted trees* (T_ρ, δ) *with set of vertices equal to their topological vertex set and with set of leaves equal to* X.

The following notion is dual to that of *depth functions*:

Definition 51 A **height function** on a rooted tree T_ρ is a strictly increasing function:

$$h : (\mathcal{I}(T_\rho), \preceq_\rho) \to (\mathbb{R}_+, \leq).$$

That is, $h(u) < h(v)$ whenever $u \prec_\rho v$. A pair (T_ρ, h) of a rooted tree and a height function on it is called a **height-dated tree**.

Remark 52 Note the slight asymmetry of the two definitions: we impose that depth functions take positive values, but we allow a height function to vanish. This asymmetry is motivated by the fact that we use depth functions to define ultrametrics by Lemma 48. The condition of strict increase on a height function imposes that a vanishing may occur only at the minimal element of $\mathcal{I}(T_\rho)$, which is either the root ρ (if T_ρ is interior-rooted) or its immediate successor in $\mathcal{V}(T_\rho)$ (if T_ρ is end-rooted).

Any strictly decreasing function allows to transform height functions into depth functions:

Lemma 53 *Any strictly decreasing map:*

$$s : (\mathbb{R}_+, \leq) \to (\mathbb{R}_+^*, \leq)$$

transforms by left-composition all height functions on a rooted tree into depth functions.

Remark 54 In [4], Böcker and Dress defined more general *symbolically dating maps* on trees, taking values in arbitrary sets, and characterize the associated *symbolic ultrametrics* by a list of axioms. We don't use here that generalized setting. Nevertheless we mention it because that paper inspired us in our work. For instance, we introduced the names *depth-dated/height-dated tree* by following its "*dating*"

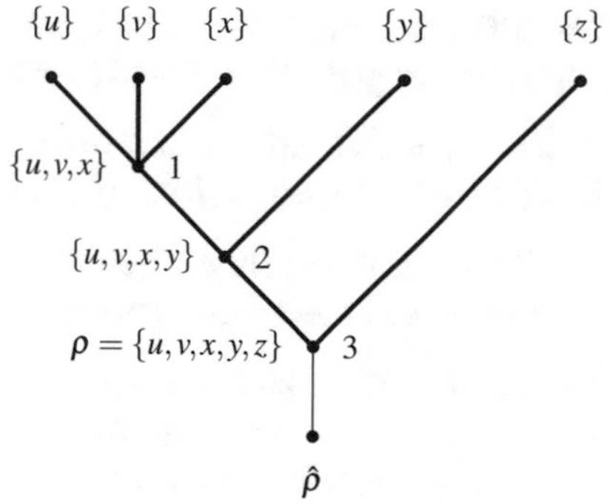

Fig. 4 The depth-dated tree associated to the ultrametric of Example 55

terminology (which seems standard in part of the mathematical literature concerned with problems of classifications, as mathematical phylogenetics).

Example 55 Consider a set $X = \{u, v, x, y, z\}$ and a function $U : X^2 \to \mathbb{R}_+$ whose matrix $(U(a, b))_{a,b \in X}$ is (for the order $u < \cdots < z$):

$$\begin{pmatrix} 0 & 1 & 1 & 2 & 3 \\ 1 & 0 & 1 & 2 & 3 \\ 1 & 1 & 0 & 2 & 3 \\ 2 & 2 & 2 & 0 & 3 \\ 3 & 3 & 3 & 3 & 0 \end{pmatrix}.$$

It is immediate to check that U is an ultrametric distance. The hierarchy of its closed balls is:

$$\{\,\{u\}, \ldots, \{z\}, \{u, v, x\}, \{u, v, x, y\}, \{u, v, x, y, z\}\,\}.$$

The associated rooted trees T^U and $\hat{T}^U$ are represented in Fig. 4, T^U being drawn with thicker segments. Near each vertex is written the associated cluster. Near each interior vertex u of $\hat{T}^U$ is also written the value of the diametral function $\delta^U(u)$, that is, the diameter of the cluster $\mathbb{K}_\rho(u)$.

3.3 Additive Distances on Trees

Let us pass now to the notion of *additive distance* on an unrooted tree. Our aim in this subsection is to explain in which way such a distance defines plenty of ultrametrics (see Proposition 62).

Definition 56 Let (V, A) be a combinatorial tree. An **additive distance** on it is a symmetric map $d : V \times V \to \mathbb{R}_+$ such that:

1. $d(u, v) = 0$ if and only if $u = v$;
2. $d(u, v) + d(v, w) = d(u, w)$ whenever $v \in [uw]$.

Of course, the additive distance d is a metric on V. Buneman [6] characterized such metrics in the following way:

Proposition 57 *A metric d on the finite set V comes from an additive distance function on a combinatorial tree with vertex set V if and only if it satisfies the following* ***four points condition****:*

$$d(l, u)+d(v, w) \leq \max\{d(l, v)+d(u, w), d(l, w)+d(u, v)\} \textit{ for any } l, u, v, w \in V.$$

In fact, one has the following more precise inequality, which we leave as an exercise to the reader:

Proposition 58 *Let d be an additive distance on the finite combinatorial tree (V, A). Consider also its geometric realization, inside which will be taken convex hulls. Then, for every $l, u, v, w \in V$, one has:*

$$d(l, v) + d(u, w) \geq d(l, u) + d(v, w) \iff [lv] \cap [uw] \neq \emptyset,$$

with equality if and only if one has moreover $[lu] \cap [vw] \neq \emptyset$, that is, if and only if the convex hull $[luvw]$ is like in the right-most tree of Fig. 5, or as in one of its degenerations.

More degenerate configurations are obtained by contracting to points one or more segments of Fig. 5. For instance, it is possible to have $m = n$ or $m = l$, etc.

Assume now that (T, d) is a topological tree endowed with an additive distance (on its underlying combinatorial tree). Choose a root $\rho \in \mathcal{V}(T)$. The distance function may now be encoded alternatively as a more general *height function*:

Definition 59 The **remoteness function** associated to the additive distance d on the rooted tree T_ρ is defined by:

$$\begin{aligned} h_{d,\rho} : \mathcal{V}(T_\rho) &\to \mathbb{R}_+ \\ u &\to d(u, \rho). \end{aligned}$$

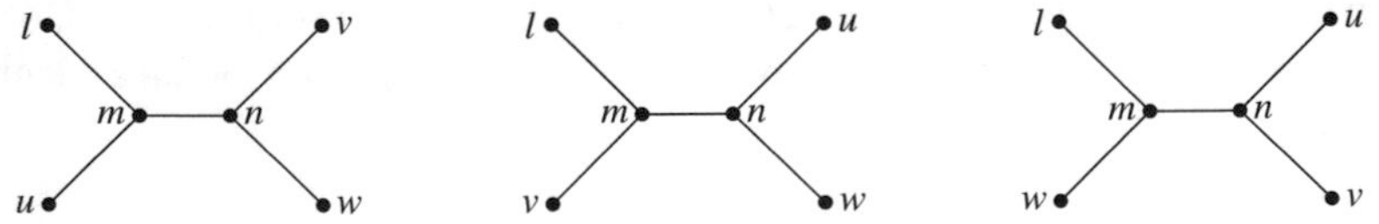

Fig. 5 The three possible generic trees with four leaves

Note that, as may be verified simply by looking at Fig. 1, the remoteness function allows to reconstruct the additive distance:

Lemma 60 *Assume that d is an additive distance on the rooted tree T_ρ. Then one has the following equivalent equalities:*

1. $d(a,b) = h_{d,\rho}(a) + h_{d,\rho}(b) - 2h_{d,\rho}(a \wedge_\rho b)$ *for all* $a, b \in \mathscr{V}(T_\rho)$.
2. $h_{d,\rho}(a \wedge_\rho b) = \frac{1}{2} \cdot (d(\rho,a) + d(\rho,b) - d(a,b))$ *for all* $a, b \in \mathscr{V}(T_\rho)$.

In the following results we consider an end-rooted tree $T_{\hat{\rho}}$ with non-empty core $\overset{\circ}{T}_\rho$ (see Definition 26). Notice that the set of vertices of the core $\overset{\circ}{T}_\rho$ is the set of interior vertices of $T_{\hat{\rho}}$ and that $\overset{\circ}{T}_\rho$ is considered as a rooted-tree, the root ρ being the immediate successor of $\hat{\rho}$ in $T_{\hat{\rho}}$. It is immediate to see that:

Lemma 61 *Consider an end-rooted tree $T_{\hat{\rho}}$ with non-empty core $\overset{\circ}{T}_\rho$. Then the map: $d \to h_{d,\rho}$, which associates to an additive distance function of $\overset{\circ}{T}_\rho$ its remoteness function, establishes a bijection from the set of additive distances on $\overset{\circ}{T}$ to the set of height functions of $T_{\hat{\rho}}$ which vanish at ρ.*

Combining Lemmas 61, 53 and 48, we get:

Proposition 62 *Assume that $T_{\hat{\rho}}$ is an end-rooted tree with non-empty core $\overset{\circ}{T}_\rho$, which is endowed with an additive distance function d. Then, for any strictly decreasing function $s : (\mathbb{R}_+, \leq) \to (\mathbb{R}_+^*, \leq)$, the function $U^{d,\rho,s} : \mathscr{L}(T_{\hat{\rho}}) \times \mathscr{L}(T_{\hat{\rho}}) \to \mathbb{R}_+$ defined by:*

$$U^{d,\rho,s}(a,b) := \begin{cases} s(h_{d,\rho}(a \wedge_\rho b)) & \text{if } a \neq b \\ 0 & \text{otherwise,} \end{cases}$$

is an ultrametric on the set of leaves $\mathscr{L}(T_{\hat{\rho}})$.

Therefore, starting from an end-rooted tree whose core is endowed with an additive distance, one gets plenty of ultrametric spaces, depending on the choice of map s. Each such ultrametric space has two associated rooted trees, as explained in Definition 45. By Proposition 50, the end-rooted one may be identified topologically with the initial end-rooted tree:

Proposition 63 *Let $T_{\hat{\rho}}$ be an end-rooted tree with non-empty core $\overset{\circ}{T}_\rho$. We assume that $\overset{\circ}{T}_\rho$ is endowed with an additive distance function d. Consider $T_{\hat{\rho}}$ as a combinatorial rooted tree with vertex set equal to its topological vertex set $\mathscr{V}_{top}(T_{\hat{\rho}})$, as defined by* (1). *Let $s : (\mathbb{R}_+, \leq) \to (\mathbb{R}_+^*, \leq)$ be a strictly decreasing map and let $U^{d,\rho,s} : \mathscr{L}(T_{\hat{\rho}}) \times \mathscr{L}(T_{\hat{\rho}}) \to \mathbb{R}_+$ be the ultrametric defined in Proposition* 62. *Then:*

1. *The end-rooted tree $\hat{T}^{U^{d,\rho,s}}$ is uniquely isomorphic with $T_{\hat{\rho}}$ as a combinatorial rooted tree with leaf-set $\mathscr{L}(T_{\hat{\rho}})$. In restriction to $\mathscr{I}(T_{\hat{\rho}})$, this isomorphism identifies the diametral function $\delta^{U^{d,\rho,s}}$ of Definition* 46 *with $s \circ h_{d,\rho}$.*

2. *The previous isomorphism identifies the interior-rooted tree* $T^{U^{d,\rho,s}}$ *with the convex hull* $[\mathscr{L}(T_{\hat{\rho}})]$, *as rooted trees with leaf-set* $\mathscr{L}(T_{\hat{\rho}})$.

Proposition 63 is a special case of the theory explained by Böcker and Dress in [4].

4 Arborescent Singularities and Their Ultrametric Spaces of Branches

In this section we introduce the notion of *arborescent* singularity and we prove that, given any good resolution of such a singularity, there is a natural additive metric on its dual tree, constructed from a determinantal identity of Eisenbud and Neumann. We deduce from this additivity the announced results about the fact that the functions U_L defined in the introduction are ultrametrics, and their relation with the dual trees of resolutions.

4.1 Determinant Products for Arborescent Singularities

In this subsection we explain the basic facts about arborescent singularities needed in the sequel.

Definition 64 A normal surface singularity is called **arborescent** if the dual graph of some good resolution of it is a tree.

It is immediate to see, using the fact that there exists a minimal good resolution, which any good resolution dominates by a sequence of blow-ups of points, that the dual graph of some good resolution is a tree if and only if this is the case for any good resolution.

Remark 65 All normal quasi-homogeneous, rational, minimally elliptic singularities which are not cusp singularities or Neumann and Wahl's [39] splice quotient singularities are arborescent. Note that the first three classes of singularities are special cases of splice quotients whenever their boundaries are rational homology spheres (which is always the case for rational ones, but means that the quotient by the $\mathbb{C}^*$-action is a rational curve in the quasihomogeneous case, and that one does not have a simply elliptic singularity – a special case of quasihomogeneous singularities, in which this quotient is elliptic and there are no special orbits – in the minimally elliptic case). This was proved by Neumann [37] for the quasi-homogeneous singularities and by Okuma [41] for the other classes (see also [39, Appendix]). Note that all splice quotients are special cases of normal surface singularities with rational homology sphere links, which in turn are all arborescent by [13, Proposition 3.4 of Chap. 2].

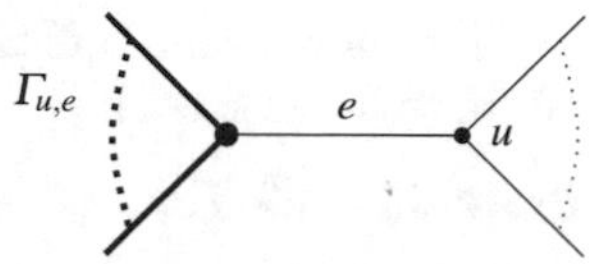

Fig. 6 The subtree $\Gamma_{u,e}$ of Definition 67 is sketched with thicker segments

Remark 66 José Seade told us that arborescent singularities had also appeared, but without receiving a special name, in Camacho's paper [7].

The notions explained in the following definition were introduced with slightly different terminology by Eisenbud and Neumann [18, Sect. 20, 21] (recall that the notion of *determinant* of a weighted graph was explained in Definition 10):

Definition 67 Let S be an arborescent singularity. Consider any good resolution of it. For each vertex u of its dual tree Γ and each edge e containing u, we say that the **subtree** $\Gamma_{u,e}$ **of** u **in the direction** e is the full subtree of Γ whose vertices are those vertices t of Γ distinct from u, which are seen from u in the direction of the edge e, that is, such that $e \subset [ut]$ (see Fig. 6). The **edge determinant** $\det_{u,e}(\Gamma)$ **at the vertex** u **in the direction** e is the determinant of $\Gamma_{u,e}$. For any $v, w \in \mathcal{V}$, define the **determinant product** $p(v, w)$ **of the pair** (v, w) as the product of the determinants at all the points of the geodesic $[vw]$ which connects v and w, in the directions of the edges which are *not* contained in that geodesic.

Note that the definition implies that $p(v, w) = p(w, v) \in \mathbb{N}^*$ for any $v, w \in \mathcal{V}$.

In order to compute determinant products in concrete examples, it is important to be able to compute rapidly determinants of weighted trees. One could use the general algorithms of linear algebra. Happily, there exists a special algorithm adapted to tree determinants, which was presented in Duchon's thesis [14, Sect. III.1] and studied in [18, Section 21]. We used it a lot for our experimentations. This algorithm may be formulated as follows:

Proposition 68 *Let Γ_u be the dual tree of a good resolution of an arborescent singularity, rooted at one of its vertices u. For any vertex v of Γ_u, denote by $-\alpha_v < 0$ the self-intersection of the component E_v. Denote also by $(e_j)_{j \in J(u)}$ the edges of Γ_u containing u. Each subtree Γ_{u,e_j} is considered to be rooted at the vertex of e_j different from u. Define recursively the* **continued fraction** *$cf(\Gamma_u)$ of the weighted rooted tree Γ_u by:*

$$cf(\Gamma_u) = \begin{cases} \alpha_u, & \text{if } \Gamma_u \text{ is reduced to the vertex } u, \\ \alpha_u - \sum_{j \in J(u)} (cf(\Gamma_{u,e_j}))^{-1} & \text{otherwise.} \end{cases}$$

Then:

$$\det(\Gamma) = cf(\Gamma_u) \cdot \prod_{j \in J(u)} \det(\Gamma_{u,e_j}).$$

The following multiplicative property of determinant products will be fundamental for us in the sequel:

Proposition 69 *For any three vertices $u, v, w \in \mathcal{V}$ such that $v \in [uw]$, one has:*

$$p(u, v) \cdot p(v, w) = p(v, v) \cdot p(u, w). \tag{1}$$

Equivalently:

$$\frac{p(u, v)}{\sqrt{p(u, u) \cdot p(v, v)}} \cdot \frac{p(v, w)}{\sqrt{p(v, v) \cdot p(w, w)}} = \frac{p(u, w)}{\sqrt{p(u, u) \cdot p(w, w)}}. \tag{2}$$

Proof The following proof is to be followed on Fig. 7. We define:

- $P(u)$ = the product of edge determinants at the vertex u, over the set of edges starting from u and not contained in $[uw]$. The products $P(v)$ and $P(w)$ are defined analogously.
- $P(uv)$ = the product of edge determinants at all vertices of Γ situated in the interior of the geodesic $[uv]$, over the edges not contained in $[uw]$. $P(vw)$ is defined analogously.
- M = the edge determinant at v in the direction of the unique edge starting from v and contained in $[uv]$. N is defined analogously.

Then one has the following formulae, clearly understandable on Fig. 7:

$$p(u, v) = P(u) \cdot P(uv) \cdot P(v) \cdot N, \qquad p(v, w) = M \cdot P(v) \cdot P(vw) \cdot P(w),$$
$$p(v, v) = M \cdot N \cdot P(v), \qquad p(u, w) = P(u) \cdot P(uv) \cdot P(v) \cdot P(vw) \cdot P(w).$$

The equality (1) is a direct consequence of the previous factorisations. Finally, it is immediate to see that (1) and (2) are equivalent.

The following proposition was proved by Eisenbud and Neumann [18, Lemma 20.2] (see also Neumann and Wahl [39, Theorem 12.2]) for trees corresponding to the singularities whose boundaries ∂M are rational homology spheres. Nevertheless, their proofs use only the fact that those are weighted trees appearing as dual graphs of singularities.

Fig. 7 Illustration for the proof of Proposition 69

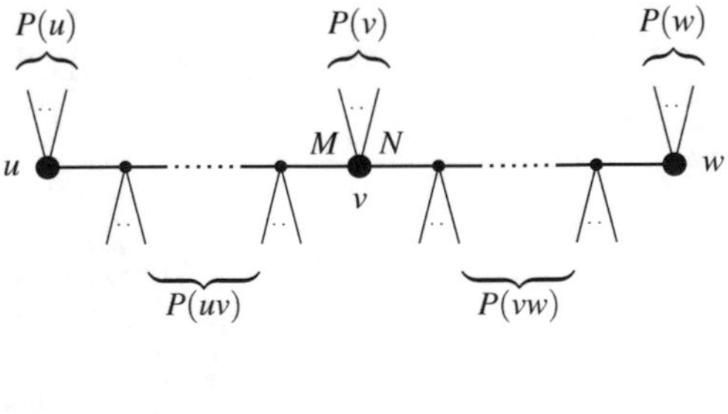

Proposition 70 *Let S be an arborescent singularity. Consider any good resolution of it. Then, for any $v, w \in \mathcal{V}$, one has:*

$$p(v, w) = (-\check{E}_v \cdot \check{E}_w) \cdot \det(S).$$

Remark 71 Using Proposition 70, formula (1) becomes the equality:

$$(-\check{E}_u \cdot \check{E}_v)(-\check{E}_v \cdot \check{E}_w) = (-\check{E}_v \cdot \check{E}_v)(-\check{E}_u \cdot \check{E}_w), \text{ for any } v \in [uw] \tag{3}$$

of the introduction. After having seen a previous version of this paper, Jonsson told us that he had proved this equality for dual trees of compactifying divisors of $\mathbb{C}^2$ and Némethi told us that the equality could also be proved using Lemma 4.0.1 from his paper [5] written with Braun.

Remark 72 Let us consider the real vector space $\Lambda_{\mathbb{R}} = \check{\Lambda}_{\mathbb{R}}$ endowed with the opposite of the intersection form. It is a Euclidean vector space. Measuring the angles with this Euclidean metric, the equality (3) becomes:

$$\cos(\angle \check{E}_u \check{E}_v) \cdot \cos(\angle \check{E}_v \check{E}_w) = \cos(\angle \check{E}_u \check{E}_w), \text{ for any } v \in [uw].$$

By the spherical pythagorean theorem, this means that the spherical triangle whose vertices are the unit vectors determined by $\check{E}_u, \check{E}_v, \check{E}_w$ is rectangle at the vertex corresponding to $\check{E}_v$. This interpretation was noticed by the third author in the announcement [44] of the main results of this paper.

Corollary 73 *For any $u, v \in \mathcal{V}$, one has the inequality:*

$$p(u, u) \cdot p(v, v) \geq p^2(u, v),$$

with equality if and only if $u = v$.

Proof Using Proposition 70, the desired inequality may be rewritten as:

$$(-\check{E}_u \cdot \check{E}_u) \cdot (-\check{E}_v \cdot \check{E}_v) \geq (-\check{E}_u \cdot \check{E}_v)^2$$

which is simply the Cauchy-Schwartz inequality implied by the positive definiteness of the intersection form $-I$ on $\check{\Lambda}_{\mathbb{R}}$. One has equality if and only if the vectors $\check{E}_u$ and $\check{E}_v$ are proportional, which is the case if and only if $u = v$.

Let us introduce a new function, which is well-defined thanks to Corollary 73:

Definition 74 The **determinant distance** $d : \mathcal{V} \times \mathcal{V} \to \mathbb{R}_+$ is defined by:

$$d(u, v) := -\log \frac{p(u, v)}{\sqrt{p(u, u) \cdot p(v, v)}}, \text{ for all } u, v \in \mathcal{V}.$$

Remark 75 In the paper [26], Gignac and Ruggiero defined the *angular distance* for any normal surface singularity, as:

$$\rho(u, v) := -\log \frac{(-\check{E}_u \cdot \check{E}_v)^2}{(-\check{E}_u \cdot \check{E}_u) \cdot (-\check{E}_v \cdot \check{E}_v)}, \text{ for all } u, v \in \mathscr{V}.$$

By Proposition 70, on arborescent singularities one gets $\rho = 2d$. In the sequel [25] to the present paper, concerning arbitrary normal surface singularities, we also work with the angular distance ρ as a replacement of what we call here the *determinant distance*.

Reformulated in terms of the determinant distance, Eq. (2) provides:

Proposition 76 *For any three vertices $u, v, w \in \mathscr{V}$ such that $v \in [uw]$, one has:*

$$d(u, v) + d(v, w) = d(u, w).$$

Moreover d is symmetric and $d(u, v) \geq 0$, with equality if and only if $u = v$. That is, the determinant distance d is an additive distance on the tree Γ, in the sense of Definition 56.

Therefore, Proposition 57 implies that:

Corollary 77 *Let S be an arborescent singularity. Consider any good resolution of it. Then, for any vertices $u, v, w, l \in \mathscr{V}$, one has:*

$$d(l, u) + d(v, w) \leq \max\{d(l, v) + d(u, w), d(l, w) + d(u, v)\}. \tag{4}$$

Equivalently:

$$p(l, u) \cdot p(v, w) \geq \min\{p(l, v) \cdot p(u, w), p(l, w) \cdot p(u, v)\}. \tag{5}$$

Remark 78 We could have worked instead with the function e^{-d}, which is a *multiplicative distance function*, that is, a distance with values in the cancellative abelian monoid $((0, +\infty), \cdot)$, in the sense of Bandelt and Steel [3]. We prefer to work instead with a classical additive distance, in order not to complicate the understanding of the reader who is not accustomed with this more general setting, which is generalized even more by Böcker and Dress [4]. Note also that, as a consequence of Proposition 70, one has:

$$e^{-d(u,v)} = \frac{-(\check{E}_u \cdot \check{E}_v)}{\sqrt{(-(\check{E}_u \cdot \check{E}_u)) \cdot (-(\check{E}_v \cdot \check{E}_v))}},$$

which is the cosine of the angle formed by the vectors $\check{E}_u$ and $\check{E}_v$ with respect to the euclidean scalar product $-I$ on $\check{\Lambda}_{\mathbb{R}}$.

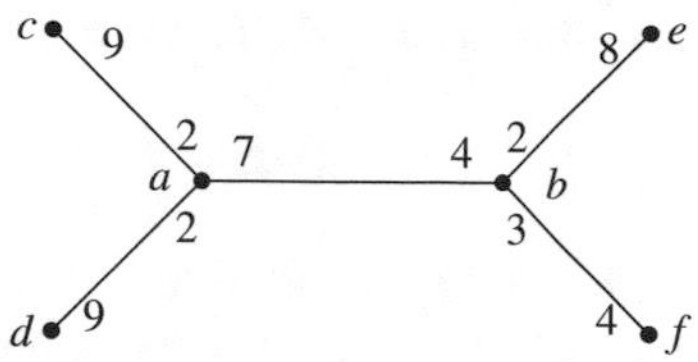

Fig. 8 The edge determinants in Example 80

Note the following consequence of Proposition 58, which refines inequality (5) from Corollary 77:

Proposition 79 *For any $u, v, w, l \in \mathcal{V}$, one has the equivalence:*

$$p(l, v) \cdot p(u, w) \leq p(l, u) \cdot p(v, w) \iff [lv] \cap [uw] \neq \emptyset$$

with equality if and only if one has moreover $[lu] \cap [vw] \neq \emptyset$.

This proposition or, alternatively, the weaker statement of Corollary 77 will imply in turn our main results announced in the introduction (that is, Corollary 84, Theorems 85 and 119).

Example 80 Let us consider a germ of arborescent singularity S which has a good resolution whose dual graph is indicated in Fig. 8. All self-intersections are equal to -2, with the exception of $E_f \cdot E_f = -3$. The genera are arbitrary. The edge determinants at each vertex are indicated in Fig. 8 near the corresponding edge. For instance, $\det_{a,[ab]}(\Gamma)$ is the determinant of the subtree $\Gamma_{a,[ab]}$, which is the full subtree with vertices e, b, f. This allows to compute the determinant product of any pair of vertices. For instance:

$$p(a, b) = \det_{a,[ac]}(\Gamma) \cdot \det_{a,[ad]}(\Gamma) \cdot \det_{b,[be]}(\Gamma) \cdot \det_{b,[bf]}(\Gamma) = 2 \cdot 2 \cdot 2 \cdot 3 = 24.$$

The matrix of determinant products $(p(u, v))_{u,v}$ is the following one, for the ordering $a < \cdots < f$ of the vertices of Γ:

$$\begin{pmatrix} 28 & 24 & 14 & 14 & 12 & 8 \\ 24 & 24 & 12 & 12 & 12 & 8 \\ 14 & 12 & 9 & 7 & 6 & 4 \\ 14 & 12 & 7 & 9 & 6 & 4 \\ 12 & 12 & 6 & 6 & 8 & 4 \\ 8 & 8 & 4 & 4 & 4 & 4 \end{pmatrix}.$$

Moreover, we have $\det(S) = 4$, as may be computed easily using Proposition 68.

Remark 81 Specialize Proposition 79 by putting $v = w$. Then one has automatically $[lv] \cap [uw] \neq \emptyset$, which implies:

$$p(l, v) \cdot p(u, v) \leq p(l, u) \cdot p(v, v)$$

for any $l, u, v \in \mathcal{V}$. This inequality may be rewritten as:

$$(-\check{E}_l \cdot \check{E}_v)(-\check{E}_u \cdot \check{E}_v) \leq (-\check{E}_l \cdot \check{E}_u)(-\check{E}_v \cdot \check{E}_v), \tag{6}$$

with equality if and only if $v \in [lu]$. In [26], Gignac and Ruggiero proved that the inequality (6) is valid for *all* normal surface singularities, and moreover that the equality in (6) holds if and only if v *separates* u *from* w *in the dual graph*. This condition is a generalization of the condition $v \in [lu]$ when S is arborescent. Their result is the central ingredient of the sequel [25] of the present paper, written in collaboration with Ruggiero.

4.2 The Ultrametric Associated to a Branch on an Arborescent Singularity

The main results of this subsection are our generalization of Płoski's theorem to arbitrary arborescent singularities (Theorem 85) and the interpretation of the associated rooted trees as convex hulls in dual graphs of resolutions (Theorem 87).

Recall the notation U_L explained in the introduction:

Notation 82 *Let S be a normal surface singularity and let L be a fixed branch on it. If A, B denote two branches on S different from L, define:*

$$U_L(A, B) := \begin{cases} \dfrac{(L \cdot A)\,(L \cdot B)}{A \cdot B}, & \text{if } A \neq B \\ 0, & \text{if } A = B. \end{cases}$$

The following proposition explains several ways to compute or to think about U_L in the case of *arborescent* singularities (recall that the notation $u(A)$ was introduced in Definition 13):

Proposition 83 *Let S be an* arborescent *singularity and let L be a fixed branch on it. Assume moreover that A, B are two distinct branches on S different from L and that $\tilde{S}$ is an embedded resolution of $A + B + L$, with dual tree Γ. Denote $a = u(A), b = u(B), l = u(L)$. Then:*

1. $U_L(A, B) = \det(S)^{-1} \cdot \dfrac{p(l, a) \cdot p(l, b)}{p(a, b)}$.
2. $U_L(A, B) = \det(S)^{-1} \cdot \dfrac{p^2(l, a \wedge_l b)}{p(a \wedge_l b, a \wedge_l b)}$.
3. $U_L(A, B) = \det(S)^{-1} \cdot p(l, l) \cdot e^{-2h_l(a \wedge_l b)}$, *where h_l is the remoteness function on Γ associated to the determinant distance d and the root l, as explained in* Definition 59.

Proof **Let us prove point 1.** By Corollary 20, $A \cdot B = -\check{E}_a \cdot \check{E}_b$. By Proposition 70, $-\check{E}_a \cdot \check{E}_b = \det(S)^{-1} \cdot p(a, b)$. Using the analogous formulae in order to transform the intersection numbers $L \cdot A$ and $L \cdot B$, we get the desired equality.

We prove now point 2. Given the equality of the previous point, the second equality is equivalent to:

$$p(l, a) \cdot p(l, b) \cdot p(a \wedge_l b, a \wedge_l b) = p(a, b) \cdot p^2(l, a \wedge_l b).$$

But this may be obtained by multiplying termwise the following special cases of formula (1) stated in Proposition 69 (in which, for simplicity, we have denoted $c := a \wedge_l b$):

$$\begin{aligned} p(l, a) \cdot p(c, c) &= p(l, c) \cdot p(c, a), \\ p(l, b) \cdot p(c, c) &= p(l, c) \cdot p(c, b), \\ p(a, c) \cdot p(c, b) &= p(a, b) \cdot p(c, c). \end{aligned}$$

Finally, let us prove point 3. Using Definition 74, the equality (83) may be rewritten as:

$$U_L(A, B) = \frac{p(l, l)}{\det(S)} \cdot \frac{p^2(l, a \wedge_l b)}{p(l, l) \cdot p(a \wedge_l b, a \wedge_l b)} = \frac{p(l, l)}{\det(S)} \cdot e^{-2d(l, a \wedge_l b)}.$$

But, by Definition 59, $d(l, a \wedge_l b) = h_l(a \wedge_l b)$, and the formula is proved.

The first equality stated in Proposition 83 allows to compute the maximum of U_L:

Corollary 84 *Whenever* L, A, B *are three pairwise distinct branches on the arborescent singularity* S*, one has:*

$$U_L(A, B) \leq -\check{E}_l \cdot \check{E}_l,$$

with equality if and only if $l \in [ab]$.

Proof As $[la] \cap [lb] \neq \emptyset$, Proposition 79 implies that: $p(l, a) \cdot p(l, b) \leq p(l, l) \cdot p(a, b)$. Combining this with the first equality stated in Proposition 83, we get:

$$U_L(A, B) \leq p(l, l) \cdot \det(S)^{-1} = -\check{E}_l \cdot \check{E}_l,$$

where the last equality is a consequence of Proposition 70. The fact that one has equality if an only if $l \in [ab]$ follows from Proposition 79.

Recall that $\mathcal{B}(S)$ denotes the set of branches on S. The following is our generalization of Płoski's theorem recalled at the beginning of the introduction:

Theorem 85 *For any four pairwise distinct branches* (L, A, B, C) *on the arborescent singularity* S*, one has:*

$$U_L(A, B) \leq \max\{U_L(A, C), U_L(B, C)\}.$$

Therefore, the function U_L is an ultrametric on $\mathscr{B}(S) \setminus \{L\}$.

Proof We will give two different proofs of this theorem.

The first proof Let $\pi : \tilde{S} \to S$ be an embedded resolution of $A + B + C + L$. By Proposition 18, we know that the pairwise intersection numbers on S of the four branches are the opposites of the intersection numbers on $\tilde{S}$ of their exceptional transforms by the morphism π. By Lemma 19, there exist four (possibly coinciding) indices $l, a, b, c \in \mathscr{V}$ such that $(\pi^* A)_{ex} = -\check{E}_a$, $(\pi^* B)_{ex} = -\check{E}_b$, $(\pi^* C)_{ex} = -\check{E}_c$, $(\pi^* L)_{ex} = -\check{E}_l$. Using the first equality of Proposition 83, the stated inequality is equivalent to:

$$\frac{p(l, a) \cdot p(l, b)}{p(a, b)} \leq \max \left\{ \frac{p(l, a) \cdot p(l, c)}{p(a, c)}, \frac{p(l, b) \cdot p(l, c)}{p(b, c)} \right\}.$$

By taking the inverses of the three fractions and multiplying then all of them by $p(l, a) \cdot p(l, b) \cdot p(l, c)$, we see that the previous inequality is equivalent to:

$$p(l, c) \cdot p(a, b) \geq \min\{p(l, b) \cdot p(a, c),\ p(l, a) \cdot p(b, c)\}.$$

But this inequality is true by Corollary 77.

The second proof We could have argued also by combining the third equality of Proposition 83 with Proposition 62. Indeed, the function $s : (\mathbb{R}_+, \leq) \to (\mathbb{R}_+^*, \leq)$ defined by: $s(x) := \frac{p(l,l)}{\det(S)} \cdot e^{-2x}$, is strictly decreasing. This line of reasoning may be easily followed on Fig. 9. Up to permuting a, b, c, it represents the generic tree $[labc]$. All other topological possibilities are degenerations of it. Using the third equality of Proposition 83, we have:

$$U_L(A, B) = \det(S)^{-1} \cdot p(l, l) \cdot e^{-2h_l(a \wedge_l b)},$$
$$U_L(A, C) = U_L(B, C) = \det(S)^{-1} \cdot p(l, l) \cdot e^{-2h_l(b \wedge_l c)}.$$

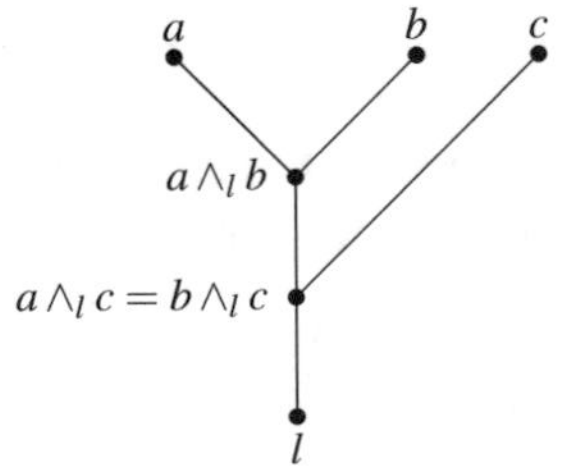

Fig. 9 A generic position of a, b, c and l

But the inequality $b \wedge_l c \preceq_l a \wedge_l b$ implies that: $h_l(b \wedge_l c) \leq h_l(a \wedge_l b)$. Therefore:

$$U_L(A, C) = U_L(B, C) \geq U_L(A, B),$$

which is the ultrametric inequality (recall Proposition 42 (2)).

Remark 86 The previous theorem was proved in this form for *smooth* complex germs S and L by Chądzyński and Płoski [10, Section 4] and again by Favre and Jonsson in [20, Lemma 3.56] for *smooth* germs S and L. Abío, Alberich-Carraminana and González-Alonso later explored in [1] the values taken by this ultrametric, also in the case of smooth germs. Their results were extended recently by the first author and Płoski [23] to smooth surfaces defined over algebraically closed fields of positive characteristic. Favre and Jonsson were not conscious about the result of Chądzyński and Płoski, and they attributed the theorem to the first author's thesis [21, Cor. 1.2.3]. This last reference states in fact that a related function is an ultrametric, a result which combined with [21, Prop. 1.2.2] implies indeed that U_L is an ultrametric. Note that at the time of writing [21], the first author did not know the papers [42] and [10].

As a consequence of Proposition 63, one gets also the announced topological interpretation of the two rooted trees associated to the ultrametric U_L (see Definition 45):

Theorem 87 *Let S be an arborescent singularity and L a fixed branch on it. Assume that $\mathscr{F}$ is a finite set of branches on S, all of them different from L. Consider an embedded resolution of the sum of L and of the branches in $\mathscr{F}$. Let $\Gamma_L(\mathscr{F})$ be the dual tree of the total transform of this divisor. Then we have:*

1. *the end-rooted tree $\hat{T}^{U_L}$ associated to the ultrametric space $(\mathscr{F}, U^L)$ is isomorphic as a rooted tree with leaf set $\mathscr{F}$ with the convex hull of $\{L\} \cup \mathscr{F}$ in $\Gamma_L(\mathscr{F})$, endowed with its topological vertex set and with root at L;*
2. *the previous isomorphism sends the interior-rooted tree T^{U_L} associated to the ultrametric space $(\mathscr{F}, U^L)$ onto the convex hull of $\mathscr{F}$ in $\Gamma_L(\mathscr{F})$.*

Remark 88 Note that the root L and the branches in $\mathscr{F}$ are always ends of $\Gamma_L(\mathscr{F})$. This is the reason why we have decided to associate systematically to an ultrametric a rooted tree in which the root is an end, its *end-rooted tree* (see Definition 45). Note also that the convex hull $[\{u(L)\} \cup \{u(A),\ A \in \mathscr{F}\}]$ inside $\Gamma_L(\mathscr{F})$, which is equal to the core of the end-rooted tree $\Gamma_L(\mathscr{F})$, is equipped with the additive distance d of Definition 74. This fact has to be used when one deduces Theorem 87 from Proposition 63.

Example 89 Let us consider an arborescent singularity S as in Example 80. That is, we assume that it admits a good resolution $\tilde{S}$ with weighted dual graph Γ as in Fig. 8. Consider branches L, A, B, C, D, E, F on S such that the total transform of $L + A + \cdots + F$ on $\tilde{S}$ is a normal crossings divisor. Moreover, we assume that the strict transforms of $A, \ldots, F$ intersect $a, \ldots, f$ respectively, and that the strict transform of L intersects E_a. Therefore, with the notations of Proposition 83, we have $l = a$. We see on Fig. 8 that when x, y vary among $\{a, \ldots, f\}$ and remain

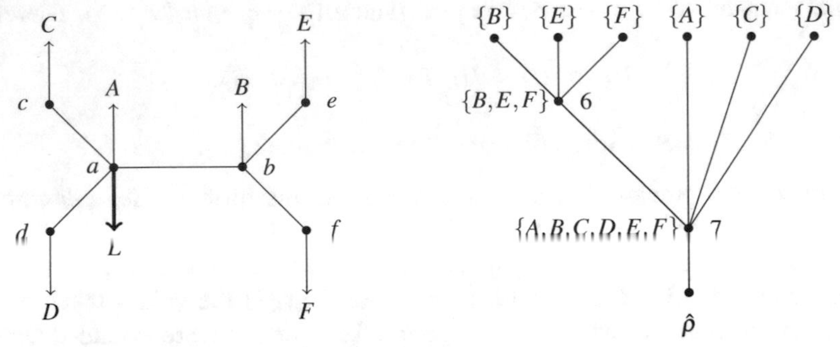

Fig. 10 An illustration of Theorem 87: Example 89

distinct, their infimum $x \wedge_l y$ relative to the root $l = a$ of Γ is either a or b. By the second equality of Proposition 83, we deduce that the only possible values of $\det(S) \cdot U_L(X, Y)$, when $X \neq Y$ vary among $\{A, \ldots, F\}$, are:

$$\frac{p^2(a,a)}{p(a,a)} = p(a,a) = 28, \quad \frac{p^2(a,b)}{p(b,b)} = \frac{24^2}{24} = 24.$$

Continuing to use the second equality of Proposition 83, we get the following values for the entries of the matrix $(U_L(X, Y))_{X,Y}$ (recall from Example 80 that $\det(S) = 4$):

$$\begin{pmatrix} 0&7&7&7&7&7 \\ 7&0&7&7&6&6 \\ 7&7&0&7&7&7 \\ 7&7&7&0&7&7 \\ 7&6&7&7&0&6 \\ 7&6&7&7&6&0 \end{pmatrix}.$$

One may check immediately on this matrix that U_L is an ultrametric on the set $\{A, \ldots, F\}$. In Fig. 10 we represent both the dual tree $\hat{\Gamma}$ of the total transform of $L + A + \cdots + F$ and the end-rooted tree $\hat{T}^{U_L}$ associated to the ultrametric U_L. Near the two nodes of $\hat{T}^{U_L}$ we indicate both the corresponding clusters and their diameters (as in Fig. 4).

In the introduction we recalled the following result of Teissier [45, page 40], which inspired us to formulate Corollary 84:

Proposition 90 *If S is a normal surface singularity, A, B are two divisors without common branches on it and m_O denotes the function which gives the multiplicity at O, then one has the inequality:*

$$\frac{m_O(A) \cdot m_O(B)}{A \cdot B} \leq m_O(S).$$

This result suggests to consider the following analog of the function U_L introduced in Notation 82:

Notation 91 *Let S be a normal surface singularity. If A, B denote two branches on S, define:*

$$U_O(A, B) := \begin{cases} \dfrac{m_O(A) \cdot m_O(B)}{A \cdot B}, & \text{if } A \neq B, \\ 0, & \text{if } A = B. \end{cases}$$

An immediate consequence of the previous results is the following analog of Theorem 85 (which holds for a restricted class of arborescent singularities):

Theorem 92 *For any three pairwise distinct branches (A, B, C) on the arborescent singularity S* with irreducible generic hyperplane section, *one has:*

$$U_O(A, B) \leq \max\{U_O(A, C), U_O(B, C)\}.$$

Therefore, the function U_O is an ultrametric on the set $\mathcal{B}(S)$ of branches of S.

Proof By definition, a *generic hyperplane section* of a normal surface singularity S is the divisor defined by a generic element of the maximal ideal of the local ring of S. For instance, if S is smooth, the generic hyperplane sections are smooth branches on S.

Fix an embedding of S in a smooth space $(\mathbb{C}^n, 0)$. Choose a generic hyperplane H in this space which is transversal to the three branches A, B, C. Therefore, its intersection numbers with the branches are equal to their multiplicities. Denote by L the intersection of S and H, which is a branch by hypothesis. Since the intersection number of a branch with H in the ambient smooth space $(\mathbb{C}^n, 0)$ is equal to the intersection number of the branch with L on S, we get:

$$m_O(A) = L \cdot A, \quad m_O(B) = L \cdot B, \quad m_O(C) = L \cdot C. \tag{7}$$

Therefore:

$$U_O(A, B) = U_L(A, B), \quad U_O(A, C) = U_L(A, C), \quad U_O(B, C) = U_L(B, C).$$

We conclude using Theorem 85.

Remark 93 Assume that S is a *rational* surface singularity and that $\tilde{S}$ is a resolution of it. Then the exceptional transform on $\tilde{S}$ of a generic hyperplane section L of S is the *fundamental cycle* Z_f of the resolution, defined by Artin [2, Page 132] (see also Ishii [31, Definition 7.2.3]). The total transform of L is in this case a normal crossings divisor. The number of branches of L whose strict transforms intersect a component E_u of E is equal to the intersection number $-Z_f \cdot E_u$. Therefore, the generic hyperplane section is irreducible if and only if all these numbers vanish, with the exception of one of them, which is equal to 1 (that is, if and only if there exists

$u \in \mathcal{V}$ such that $Z_f = -\check{E}_u$). This may be easily checked. For instance, starting from the list of rational surface singularities of multiplicity 2 or 3 given at the end of Artin's paper [2], one sees that among the rational double points A_n, D_n, E_n, only those of type A_n do not have irreducible generic hyperplane sections. And among rational triple points, those which do not have irreducible generic hyperplane sections are the first three of the left column and the first one of the right column of that paper.

Remark 94 Under the hypothesis of Theorem 92, Teissier's inequality stated in Proposition 90 may be proved as a particular case of the inequality stated in Corollary 84. Indeed, arguing as in the proof of Theorem 92, we may assume that we work with an irreducible hyperplane section L such that the equalities (7) hold. Let L' be a second hyperplane section, whose strict transform to the resolution with which we work is disjoint from the strict transform of L, but intersects the same component E_l. Moreover, we may assume that both are transversal to E_l. By Corollary 20, we have $L \cdot L' = -E_l^2$. But $L \cdot L' = m_O(S)$. This shows, as announced, that our inequality becomes Teissier's one.

4.3 Płoski's Theorem and the Ultrametric Nature of Eggers-Wall Trees

In this subsection we assume that both the surface S and the branch L are smooth. Consider a finite set $\mathcal{F}$ of branches on S, distinct from L. We explain first how to associate to them a rooted tree $\Theta_L(\mathcal{F})$, whose set of leaves is labeled by the elements of $\mathcal{F}$ and whose root is labeled by L: the *Eggers-Wall tree of $\mathcal{F}$ relative to L*. Then we prove that the function U_L is an ultrametric on $\mathcal{F}$ with associated end-rooted tree isomorphic to $\Theta_L(\mathcal{F})$, by showing that in restriction to $\mathcal{F}$, the function U_L corresponds to a depth function on $\Theta_L(\mathcal{F})$. Note that the content of this subsection cannot be extended to algebraically closed fields of positive characteristic, because in this setting there are Weiertrass polynomials whose roots are not expressible as Newton-Puiseux series (see the related Remark 114).

Choose a coordinate system (x, y) on S such that L is defined by $x = 0$.

The following considerations on characteristic exponents are classical. One may find information about their historical evolution in [24, Section 2].

Let A be a branch on S different from L. Relative to the coordinate system (x, y), it may be defined by a Weierstrass polynomial $f_A \in \mathbb{C}[[x]][y]$, which is unitary, irreducible and of degree $d_A = L \cdot A$. For simplicity, we mention only the dependency on A, not on the coordinate system (x, y).

By the Newton-Puiseux theorem, f_A has d_A roots inside $\mathbb{C}[[x^{1/d_A}]]$ (the **Newton-Puiseux roots** of A in the coordinate system (x, y)), which may be obtained from a fixed one of them $\xi(x^{1/d_A})$ by replacing x^{1/d_A} with the other d_A-th roots of x (here $\xi \in \mathbb{C}[[t]]$ denotes a formal power series with non-negative integral exponents).

Therefore, all the Newton-Puiseux roots have the same exponents. Some of those exponents may be distinguished by looking at the differences of roots:

Definition 95 The **characteristic exponents** of A relative to L are the x-orders $\nu_x(\eta - \eta')$ of the differences between Newton-Puiseux roots η, η' of A in the coordinate system (x, y).

The characteristic exponents may be read from a given Newton-Puiseux root $\eta \in \mathbb{C}[[x^{1/d_A}]]$ of f_A by looking at the increasing sequence of exponents appearing in η and by keeping those which cannot be written as a quotient of integers with the same smallest common denominator as the previous ones. In this sequence, one starts from the first exponent which is not an integer.

The *Eggers-Wall tree of* A *relative to* L is a geometrical way of encoding the sequences of characteristic exponents and of their successive common denominators:

Definition 96 The **Eggers-Wall tree** $\Theta_L(A)$ relative to L is a compact oriented segment endowed with the following supplementary structures:

- an increasing homeomorphism $\mathbf{e}_{L,A} : \Theta_L(A) \to [0, \infty]$, the **exponent function**;
- marked points, which are by definition the points whose exponents are the characteristic exponents of A relative to L, as well as the smallest end of $\Theta_L(A)$, labeled by L, and the greatest point, labeled by A.
- an **index function** $\mathbf{i}_{L,A} : \Theta_L(A) \to \mathbb{N}$, which associates to each point $P \in \Theta_L(A)$ the index of $(\mathbb{Z}, +)$ in the subgroup of $(\mathbb{Q}, +)$ generated by the characteristic exponents of A which are strictly smaller than $\mathbf{e}_{L,A}(P)$.

The index $\mathbf{i}_{L,A}(P)$ may be also seen as the smallest common denominator of the exponents of a Newton-Puiseux root of f_A which are strictly less than $\mathbf{e}_{L,A}(P)$.

Example 97 Assume that A has as Newton-Puiseux root $x + x^{5/2} + x^{8/3} + x^{17/6}$. The set of characteristic exponents of A relative to the branch L defined by $x = 0$ is $\mathscr{E}(A) = \{5/2, 8/3\}$. The Eggers-Wall tree $\Theta_L(A)$ is drawn in Fig. 11. We wrote the

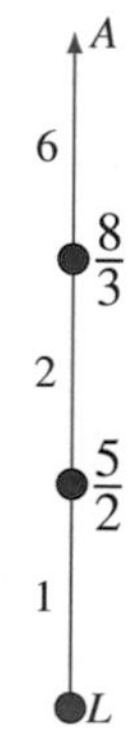

Fig. 11 The Eggers-Wall tree of the series of Example 97

value of the exponent function near each vertex and of the index function near each edge on which it is constant.

We give now the definition of the Eggers-Wall tree associated to several branches. In addition to the characteristic exponents of the individual branches, we need to know the *orders of coincidence* of the pairs of branches:

Definition 98 If A and B are two distinct branches, which are also distinct from L, then their **order of coincidence** relative to L is defined by:

$$k_L(A, B) := \max\{\nu_x(\eta_A - \eta_B) \mid \eta_A \in Z(f_A),\ \ \eta_B \in Z(f_B)\} \in \mathbb{Q}_+^*.$$

Informally speaking, the order of coincidence is the greatest rational number k for which one may find Newton-Puiseux roots of the two branches coinciding up to that number (k excluded).

Notice that the order of coincidence is symmetric: $k_L(A, B) = k_L(B, A)$.

Definition 99 Let $\mathscr{F}$ be a finite set of branches on S, different from L. The **Eggers-Wall tree** $\Theta_L(\mathscr{F})$ of $\mathscr{F}$ relative to L is the rooted tree obtained as the quotient of the disjoint union of the individual Eggers-Wall trees $\Theta_L(A)$, for $A \in \mathscr{F}$, by the equivalence relation generated by the following natural gluing of $\Theta_L(A)$ with $\Theta_L(B)$ along the initial segments $\mathbf{e}_{L,A}^{-1}[0, k_L(A, B)]$ and $\mathbf{e}_{L,B}^{-1}[0, k_L(A, B)]$:

$$\mathbf{e}_{L,A}^{-1}(\alpha) \sim \mathbf{e}_{L,B}^{-1}(\alpha), \quad \text{for all } \alpha \in [0, k_L(A, B)].$$

One endows $\Theta_L(\mathscr{F})$ with the **exponent function** $\mathbf{e}_L : \Theta_L(\mathscr{F}) \to [0, \infty]$ and the **index function** $\mathbf{i}_L : \Theta_L(\mathscr{F}) \to \mathbb{N}$ obtained by gluing the initial functions $\mathbf{e}_{L,A}$ and $\mathbf{i}_{L,A}$ respectively, for A varying among the elements of $\mathscr{F}$.

It is an instructive exercise to prove that indeed the index functions of the branches of $\mathscr{F}$ get glued into a single index function on $\Theta_L(\mathscr{F})$. Note that, by construction, $k_L(A, B) = e_L(A \wedge_L B)$ for any pair (A, B) of distinct branches of $\mathscr{F}$. Here $A \wedge_L B$ denotes the infimum of the leaves of $\Theta_L(\mathscr{F})$ labeled by A and B, relative to the partial order on the vertices of $\Theta_L(\mathscr{F})$ defined by the root L (see Notation 25).

Example 100 Consider a set $\mathscr{F}$ of branches in $(\mathbb{C}^2, 0)$, whose elements C_i (where $i \in \{1, \ldots, 5\}$) have the following Newton-Puiseux roots η_i:

$$\begin{aligned}
\eta_1 &= x^2 \\
\eta_2 &= x^{5/2} + x^{8/3} \\
\eta_3 &= x^{5/2} + x^{11/4} \\
\eta_4 &= x^{7/2} + x^{17/4} \\
\eta_5 &= x^{7/2} + 2x^{17/4} + x^{14/3}.
\end{aligned}$$

As before, we assume that L is defined by $x = 0$. Then $k_L(C_1, C_2) = k_L(C_1, C_3) = k_L(C_1, C_4) = k_L(C_1, C_5) = 2$, $k_L(C_2, C_4) = k_L(C_2, C_5) = 5/2$, $k_L(C_2, C_3) =$

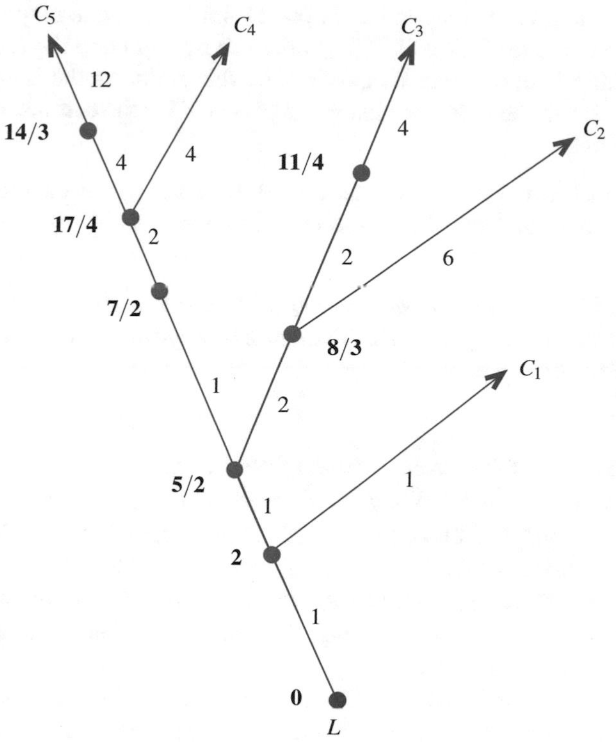

Fig. 12 The Eggers-Wall tree of Example 100

$8/3$, $k_L(C_3, C_4) = k_L(C_3, C_5) = 5/2$, $k_L(C_4, C_5) = 17/4$ and the Eggers-Wall tree of $\mathscr{F}$ relative to L is as drawn in Fig. 12.

Remark 101 Eggers [17] had constructed a tree which is nearly homeomorphic to the tree of Definition 45 (there may be a difference in the neighborhood of their roots, which the interested reader may find without difficulties). He did not consider on it the index function. Instead, he considered two types of edges which, given the exponent function, allowed to encode the same information as the index function. Another difference with Definition 99 is that Eggers assumed that the smooth branch L is transversal to the tangents of all the branches of $\mathscr{F}$. What we call *Eggers-Wall tree* was introduced by Wall [46] in his reinterpretation of Eggers' work using computations of 0-chains and 1-chains supported on the tree.

Remark 102 Before Eggers' work, Kuo and Lu [33] had introduced a related tree associated to a finite set $\mathscr{F}$ of branches on $(\mathbb{C}^2, 0)$, different from the y-axis. Namely, they represented by a segment each *Newton-Puiseux root* of the branches of $\mathscr{F}$, exactly as in Definition 96. Then the construction proceeded exactly like for Eggers-Wall trees, with a slight variant. Namely, they used a general graphical convention for building *dendrograms* in genetics, using only horizontal and vertical segments. If one proceeds instead as in our Definition 99, one may prove that the Galois group of the extension of $\mathbb{C}[[x]]$ obtained by adjoining the roots of $\mathscr{F}$ acts on the resulting tree, and that its quotient by this action is the Eggers-Wall tree $\Theta_L(\mathscr{F})$. Moreover, the index of each point of $\Theta_L(\mathscr{F})$ is the cardinal of the fiber of this quotient map.

The fact that in the previous notations $\Theta_L(\mathscr{F}), \mathbf{e}_L, \mathbf{i}_L$ we mentioned only the dependency on L, and not the whole coordinate system (x, y), comes from the following fact:

Proposition 103 *The Eggers-Wall tree $\Theta_L(\mathscr{F})$, seen as a rooted tree endowed with the exponent function $\mathbf{e}_L$ and the index function $\mathbf{i}_L$, depends only on the pair $(\mathscr{F}, L)$, where L is defined by $x = 0$.*

Proof

- *Assume first that A is a branch distinct from L.*
  Choose some $p \in \mathbb{N}^*$. Let $\phi_p : \tilde{S} \to S$ be the cyclic cover of S of degree p, ramified along L. Denote by $\tilde{O} \in \tilde{S}$ the preimage of $O \in S$. Consider then the (total) pullback $\phi_p^* A$. By computing in the coordinates (x, y), with respect to which the morphism ϕ_p is simply given by $x = u^p$, $y = v$ for suitably chosen coordinates (u, v) on $\tilde{S}$, one sees that this pullback has only smooth branches if and only if p is divisible by $A \cdot L$.
  Suppose that this is the case. Then again by computing in local coordinates, one sees that the set $\mathscr{E}(A)$ of characteristic exponents of A with respect to (x, y) is equal to the set of rationals of the form $\frac{1}{p} A_i \cdot_{\tilde{O}} A_j$, where (A_i, A_j) varies among the set of couples of distinct branches of $\phi_p^* A$ and the intersection numbers are computed at $\tilde{O}$. This shows that $\mathscr{E}(A)$ depends only on the pair (A, L), and not on the coordinate system (x, y) chosen such that L is defined by $x = 0$.
- *Assume now that A and B are two different branches distinct from L.* Take $p \in \mathbb{N}^*$ divisible by both $A \cdot L$ and $B \cdot L$. Again by computing in the coordinates (x, y), we see that $k_L(A, B)$ is the maximal value of the rational numbers of the form $\frac{1}{p} A_i \cdot_{\tilde{O}} B_j$, where A_i varies among the branches of $\phi_p^* A$ and B_j among those of $\phi_p^* B$. This shows that the order of coincidence of A and B with respect to (x, y) depends also only on L.
  By combining the two invariance properties we deduce the proposition.

Let us introduce a third function defined on the Eggers-Wall tree:

Definition 104 Let A be a branch on S. The **contact complexity** $\mathbf{c}_L(P)$ of a point $P \in \Theta_L(A)$ is defined by:

$$\mathbf{c}_L(P) := \left(\sum_{j=1}^{l} \frac{\alpha_j - \alpha_{j-1}}{\mathbf{i}_{j-1}} \right) + \frac{\mathbf{e}_L(P) - \alpha_l}{\mathbf{i}_l}.$$

where $\alpha_0 := 0, \alpha_1 < \ldots < \alpha_g$ are the characteristic exponents of A relative to L, that $\mathbf{i}_j := \mathbf{i}(\alpha_0, \ldots, \alpha_j)$ is the value of the index function $\mathbf{i}_L$ in restriction to the half-open interval $(P_j = \mathbf{e}_L^{-1}(\alpha_j), P_{j+1} = \mathbf{e}_L^{-1}(\alpha_{j+1})]$ and $\mathbf{e}_L(P) \in [\alpha_l, \alpha_{l+1}]$ is the value of the exponent function at the point P. The possibility $\alpha_l = \alpha_0 = 0$ is allowed.

Remark 105 The previous definition gives the same value to $\mathbf{c}_L(P)$ when $\mathbf{e}_L(P) = \alpha_l$, if we compute it by looking at α_l either as an element of $[\alpha_{l-1}, \alpha_l]$ or as an element of $[\alpha_l, \alpha_{l+1}]$.

Note that the right-hand side of the formula defining $\mathbf{c}_L(P)$ may be reinterpreted as an integral of the piecewise constant function $1/\mathbf{i}_L$ along the segment $[LP]$ of $\Theta_L(A)$, the measure being the one determined by the exponent function:

$$\mathbf{c}_L(P) = \int_L^P \frac{d\, \mathbf{e}_L}{\mathbf{i}_L}. \tag{8}$$

This allows to express conversely $\mathbf{e}_L$ in terms of $\mathbf{c}_L(P)$ and $\mathbf{i}_L$:

$$\mathbf{e}_L(P) = \int_L^P \mathbf{i}_L \, d\, \mathbf{c}_L. \tag{9}$$

Remark 106 Formulae (8) and (9) are inspired by the formulae (3.7) and (3.9) of Favre and Jonsson's book [20], relating *thinness* and *skewness* as functions on the valuative tree.

As the function $\mathbf{i}_L : \Theta_L(A) \to \mathbb{N}^*$ is increasing along the segment $\Theta_L(A)$, formulae (8) and (9) imply:

Corollary 107 *The contact complexity $\mathbf{c}_L$ is an increasing homeomorphism from $\Theta_L(A)$ to $[0, \infty]$. Moreover, it is piecewise affine and concave in terms of the parameter $\mathbf{e}_L$. Conversely, the function $\mathbf{e}_L$ is continuous piecewise affine and convex in terms of the parameter $\mathbf{c}_L$.*

Let us consider now the case of a finite set $\mathscr{F}$ of branches. As an easy consequence of Definition 104, we get:

Lemma 108 *The contact functions of the branches of $\mathscr{F}$ glue into a continuous strictly increasing surjection $\mathbf{c}_L : \Theta_L(\mathscr{F}) \to [0, \infty]$.*

This allows us to formulate the following definition:

Definition 109 Assume that both S and L are smooth. Let $\mathscr{F}$ be a finite set of branches on S. The **contact complexity** $\mathbf{c}_L : \Theta_L(\mathscr{F}) \to [0, \infty]$ is the function obtained by gluing the contact complexities of the individual branches of $\mathscr{F}$.

We chose the name of this function motivated by the following theorem, which is a reformulation of a result of Max Noether's paper [40] (see also [47, Theorem 4.1.6]):

Theorem 110 *Let A and B be two distinct branches of $\mathscr{F}$. Then, in terms of the partial order defined by the root L on the set of vertices of $\Theta_L(\mathscr{F})$, one has:*

$$U_L(A, B) = \mathbf{c}_L(A \wedge_L B)^{-1}.$$

As a consequence, we get the following strengthening of Płoski's theorem (what is stronger is the fact that the end-rooted tree associated to the ultrametric U_L is isomorphic to the Eggers-Wall tree $\Theta_L(\mathscr{F})$).

Theorem 111 *Let L be a smooth branch on a smooth germ of surface S. Consider a finite set $\mathscr{F}$ of branches on S, distinct from L. Then U_L is an ultrametric on $\mathscr{F}$ and the associated end-rooted tree is isomorphic to the Eggers-Wall tree $\Theta_L(\mathscr{F})$ of $\mathscr{F}$ relative to L.*

Proof By Lemma 108, $\mathbf{c}_L$ restricts to a height function on the rooted tree $\Theta_L(\mathscr{F})$. Therefore, its inverse $\mathbf{c}_L^{-1}$ is a depth function. By Lemma 48, we deduce that U_L is an ultrametric. Using then Proposition 49, applied to the end-rooted tree $\Theta_L(\mathscr{F})$ depth-dated by $\mathbf{c}_L^{-1}$, we deduce that the end-rooted tree associated to the ultrametric U_L on $\mathscr{F}$ is indeed isomorphic to $\Theta_L(\mathscr{F})$.

As a consequence of the previous theorem and of Theorem 87, we get:

Theorem 112 *Let L be a smooth branch on a smooth germ of surface S. Consider a finite set $\mathscr{F}$ of branches on S, distinct from L. Then the Eggers-Wall tree $\Theta_L(\mathscr{F})$ is isomorphic as a rooted tree with the convex hull of $\{L\} \cup \mathscr{F}$ in the dual graph of an embedded resolution of their sum, rooted at the strict transform of L.*

Remark 113 It was the third author who proved first an isomorphism theorem of this kind in [43, Theorem 4.4.1]. There the isomorphism was proved by embedding the two trees in a common space and proving that the images coincided. In that work a slightly different notion of Eggers-Wall tree was used, coinciding topologically with Eggers' original definition from [17]. Later, [43, Theorem 4.4.1] was refined by Wall [47, Sect. 9.4] (see also [47, Sect. 9.10, Rem. on Sect. 9.4]). Let us mention also that with the hypothesis of Theorem 112, the Eggers-Wall tree $\Theta_L(\mathscr{F})$ is combinatorially isomorphic to the dual graph of a *partial* embedded resolution of $L + \mathscr{F}$ (see [27, Section 3.4]).

Remark 114 When both S and L are smooth and $\mathscr{F}$ is a finite set of branches on S different from L, the fact that U_L is an ultrametric distance on $\mathscr{F}$ even when the base field has positive characteristic was proved before by the first author and Płoski in [22, Theorem 2.8]. The associated end-rooted tree provides, in a way, a generalization of the notion of characteristic exponents in positive characteristic introduced in Campillo's book (see [8, Chapter III]). As noted in the introduction, our approach also works for arborescent surface singularities S defined over algebraically closed fields in positive characteristic, therefore U_L is an ultrametric also in this generality, for any branch L on S.

5 Valuative Considerations

In this section we recall first the notions of *valuation* and *semivaluation* on the local ring of S and the natural partial order on the set of all semivaluations. We introduce the *order valuations* defined by irreducible exceptional divisors and the *intersection semivaluations* defined by the branches lying on S. The choice of a fixed branch L on S allows to define versions of the previous (semi)valuations which are *normalized relative to* L. We prove then that, for arborescent singularities, two such normalized (semi)valuations are in the same order relation as their representative points in the dual tree of an embedded resolution of them and of the branch L, seen as a tree rooted at L.

5.1 Basic Types of Valuations and Semivaluations

In this subsection we define the types of valuations and semivaluations considered in the sequel. We do not assume here that the normal surface singularity S is arborescent.

Denote by $\mathcal{O}$ the local ring of S and by $\mathfrak{m}$ its maximal ideal. Denote also:

$$\overline{\mathbb{R}}_+ := \mathbb{R}_+ \cup \{+\infty\} = [0, +\infty]$$

endowed with the usual total order.

In full generality, a *valuation* or a *semivaluation* takes its values in an arbitrary totally ordered abelian group enriched with a symbol $+\infty$ which is greater than any element of the group. Here we will restrict to the special case where the totally ordered abelian group is $(\mathbb{R}, +)$:

Definition 115 A **semivaluation** on $\mathcal{O}$ is a map $\nu : \mathcal{O} \to \overline{\mathbb{R}}_+$ such that:

- $\nu(gh) = \nu(g) + \nu(h)$ for all $g, h \in \mathcal{O}$.
- $\nu(g + h) \geq \min\{\nu(g), \nu(h)\}$ for all $g, h \in \mathcal{O}$.
- $\nu(1) = 0$ and $\nu(0) = +\infty$.

If $\nu^{-1}(+\infty) = \{0\}$, then one says that ν is a **valuation**. Denote by $\mathrm{Val}(S)$ the set of valuations of $\mathcal{O}$ and by $\overline{\mathrm{Val}}(S)$ the set of semivaluations. There is a natural partial order $\leq_{val}$ on $\overline{\mathrm{Val}}(S)$, defined by:

$$\nu_1 \leq_{val} \nu_2 \iff \nu_1(h) \leq \nu_2(h), \ \forall\, h \in \mathcal{O}.$$

In the sequel we will consider the following special types of valuations and semivaluations:

Definition 116 Let L be a branch on S and π be an embedded resolution of it. As usual, $(E_v)_{v \in \mathcal{V}}$ denote the irreducible components of its exceptional divisor. Let A be a branch different from L.

1. If $v \in \mathcal{V}$, the v**-order**, denoted by $\mathrm{ord}_v : \mathcal{O} \to \overline{\mathbb{R}}_+$, is defined by:

$$\mathrm{ord}_v(h) := \text{ order of vanishing of } \pi^*(h) \text{ along } E_v.$$

2. If $v \in \mathcal{V}$, the v**-order relative to** L, denoted by $\mathrm{ord}_v^L : \mathcal{O} \to \overline{\mathbb{R}}_+$, is defined by:

$$\mathrm{ord}_v^L(h) := \frac{\mathrm{ord}_v(h)}{-\check{E}_v \cdot \check{E}_l}.$$

3. The A**-intersection order**, denoted by $\mathrm{int}_A : \mathcal{O} \to \overline{\mathbb{R}}_+$, is defined by:

$$\mathrm{int}_A(h) := \begin{cases} A \cdot Z_h & \text{if } A \text{ is not a branch of } Z_h \\ +\infty & \text{otherwise.} \end{cases}$$

4. The A**-intersection order relative to** L, denoted by $\mathrm{int}_A^L : \mathcal{O} \to \overline{\mathbb{R}}_+$, is defined by:

$$\mathrm{int}_A^L(h) := \frac{\mathrm{int}_A(h)}{A \cdot L}.$$

Note that the functions ord_v and ord_v^L are valuations, but that the functions int_A and int_A^L are semivaluations which are not valuations. Indeed, they take the value $+\infty$ on all the elements of $\mathcal{O}$ which vanish on the branch A.

Remark 117 In [20], a semivaluation ν of the local ring $\mathbb{C}[[x, y]]$ is called *normalized* relative to the variable x if $\nu(x) = 1$. If L denotes the branch Z_x then, with our notations, ord^L and int_A^L are normalized relative to x. In our more general context of arbitrary normal surface singularities, the branch L is not necessarily a principal divisor.

Remark 118 In [32, Sect. 7.4.8], in which S is considered to be *smooth*, Jonsson defines a function α on the set of valuations of the local ring $\mathcal{O}$ which are proportional to the divisorial valuations ord_u, by the following formula: $\alpha(t \cdot \mathrm{ord}_u) = t^2(\check{E}_u \cdot \check{E}_u)$. That is, α is homogeneous of degree 2 and takes the value $\check{E}_u \cdot \check{E}_u$ on the valuation ord_u. In [32, Sect. 7.6.2, Note 13], he remarks that this function α is the *opposite* of the *skewness* function denoted with the same symbol α in [20]. A smooth germ S is arborescent and verifies $\det(S) = 1$. Therefore, by Proposition 70, his definition may be reexpressed in the following way in the same case of smooth germs S:

$$\alpha(t \cdot \mathrm{ord}_u) = -t^2 p(u, u) = -t^2 \det(S)^{-1} \cdot p(u, u).$$

This indicates two possible generalizations of the function α to arbitrary arborescent singularities, depending on which of the two previous equalities is taken as a definition.

5.2 *The Valuative Partial Order for Arborescent Singularities*

The following theorem extends Lemma 3.69 of Favre and Jonsson [20] from a smooth germ S of surface and a smooth branch L on it, to arborescent singularities and arbitrary branches on them:

Theorem 119 *Let L, A be two distinct branches on the arborescent singularity S. Let π be an embedded resolution of $L + A$ and let $\Gamma_L(A)$ be the dual tree of the total transform of $L + A$. Consider $\Gamma_L(A)$ as a combinatorial tree rooted at L and let $\preceq_L$ be the corresponding partial order. Assume that $u, v \in \mathcal{V}$. Then:*

1. $\mathrm{ord}_u^L \leq_{val} \mathrm{ord}_v^L$ *if and only if* $u \preceq_L v$.
2. $\mathrm{ord}_u^L \leq_{val} \mathrm{int}_A^L$ *if and only if* $u \preceq_L A$.

Proof The proof of this theorem is strongly based on the determinantal formula of Eisenbud and Neumann stated in Proposition 70.

The proof of the implication $u \preceq_L v \Longrightarrow \mathrm{ord}_u^L \leq_{val} \mathrm{ord}_v^L$ in point 1 Consider an arbitrary germ of function $h \in \mathfrak{m}$. We want to prove that $\mathrm{ord}_u^L(h) \leq \mathrm{ord}_v^L(h)$. Let us work with an embedded resolution of $L + Z_h$. This is no reduction of generality, as the truth of the relation $u \preceq_L v$ does not depend on the resolution on which E_u and E_v appear as irreducible components of the exceptional divisor. As $\mathrm{ord}_u(h)$ is the coefficient of E_u in the exceptional transform $(\pi^* Z_h)_{ex}$ of Z_h, the expansion (3) shows that:

$$\mathrm{ord}_u(h) = \check{E}_u \cdot (\pi^* Z_h)_{ex}. \tag{1}$$

The desired inequality $\mathrm{ord}_u^L(h) \leq \mathrm{ord}_v^L(h)$ becomes:

$$\frac{\check{E}_u \cdot (\pi^* Z_h)_{ex}}{-\check{E}_u \cdot \check{E}_l} \leq \frac{\check{E}_v \cdot (\pi^* Z_h)_{ex}}{-\check{E}_v \cdot \check{E}_l}.$$

The divisor $(\pi^* Z_h)_{ex}$ being a linear combination with non-negative coefficients of the divisors $(-\check{E}_w)_{w \in \mathcal{V}}$ (see Proposition 15), it is enough to prove that:

$$\frac{-\check{E}_u \cdot \check{E}_w}{-\check{E}_u \cdot \check{E}_l} \leq \frac{-\check{E}_v \cdot \check{E}_w}{-\check{E}_v \cdot \check{E}_l} \text{ for all } w \in \mathcal{V}.$$

By Proposition 70, the previous inequality is equivalent to:

$$p(l, v) \cdot p(u, w) \leq p(l, u) \cdot p(v, w).$$

But this last inequality is true, as a consequence of Proposition 79. Indeed, the inequality $u \preceq_L v$ implies that $u \in [lv] \cap [uw]$, therefore $[lv] \cap [uw] \neq \emptyset$.

The proof of the implication $\mathbf{ord}_u^L \preceq_{val} \mathbf{ord}_v^L \Longrightarrow u \preceq_L v$ in point 1 Assume by contradiction that the inequality $u \preceq_L v$ is not true. This means that $[lv] \cap [uu] = \emptyset$. As $[lu] \cap [vu] \neq \emptyset$ (because u belongs to this intersection), Proposition 79 implies the inequality:

$$p(l, v) \cdot p(u, u) > p(l, u) \cdot p(v, u),$$

which may be rewritten, using Proposition 70, as:

$$\frac{-\check{E}_u \cdot \check{E}_u}{-\check{E}_u \cdot \check{E}_l} > \frac{-\check{E}_v \cdot \check{E}_u}{-\check{E}_v \cdot \check{E}_l}.$$

Therefore, whenever the positive rational numbers $(\epsilon_w)_{w \in \mathscr{V} \setminus \{u\}}$ are small enough, one has also the strict inequality:

$$\frac{-\check{E}_u \cdot H}{-\check{E}_u \cdot \check{E}_l} > \frac{-\check{E}_v \cdot H}{-\check{E}_v \cdot \check{E}_l}, \tag{2}$$

where $H \in \Lambda_{\mathbb{Q}}$ is defined by: $H := \check{E}_u + \sum_{w \in \mathscr{V} \setminus \{u\}} \epsilon_w \check{E}_w$. As $H \in \check{\sigma}$, Proposition 16 shows that there exists $n \in \mathbb{N}^*$ such that $-nH$ is the exceptional transform of a principal divisor. Denote by $h \in \mathfrak{m}$ a defining function of such a divisor. Therefore, $-nH = (\pi^* Z_h)_{ex}$, and the inequality (2) implies:

$$\frac{\check{E}_u \cdot (\pi^* Z_h)_{ex}}{-\check{E}_u \cdot \check{E}_l} > \frac{\check{E}_v \cdot (\pi^* Z_h)_{ex}}{-\check{E}_v \cdot \check{E}_l}. \tag{3}$$

Using formula (1), this inequality may be rewritten as: $\mathrm{ord}_u^L(h) > \mathrm{ord}_v^L(h)$. But this contradicts the hypothesis $\mathrm{ord}_u^L \preceq_{val} \mathrm{ord}_v^L$.

The proof of the implication $u \preceq_L A \Longrightarrow \mathbf{ord}_u^L \preceq_{val} \mathbf{int}_A^L$ in point 2 Consider an arbitrary germ of function $h \in \mathfrak{m}$. We want to prove that $\mathrm{ord}_u^L(h) \leq \mathrm{int}_A^L(h)$. We may assume that we work with a resolution of $L + A + Z_h$. By Proposition 18, we have that $\mathrm{int}_A(h) = -(\pi^* A)_{ex} \cdot (\pi^* Z_h)_{ex}$. Using Lemma 19, we deduce the equality:

$$\mathrm{int}_A(h) = \check{E}_a \cdot (\pi^* Z_h)_{ex}, \tag{4}$$

where E_a denotes the unique component of the exceptional divisor E which intersects the strict transform of A. By Corollary 20, $A \cdot L = -\check{E}_a \cdot \check{E}_l$. Therefore, the desired inequality $\mathrm{ord}_u^L(h) \leq \mathrm{int}_A^L(h)$ becomes:

$$\frac{\check{E}_u \cdot (\pi^* Z_h)_{ex}}{-\check{E}_u \cdot \check{E}_l} \leq \frac{\check{E}_a \cdot (\pi^* Z_h)_{ex}}{-\check{E}_a \cdot \check{E}_l}.$$

As before, it is enough to prove that:

$$\frac{-\check{E}_u \cdot \check{E}_w}{-\check{E}_u \cdot \check{E}_l} \leq \frac{-\check{E}_a \cdot \check{E}_w}{-\check{E}_a \cdot \check{E}_l} \text{ for all } w \in \mathcal{V}.$$

By Proposition 70, the previous inequality is equivalent to:

$$p(l, a) \cdot p(u, w) \leq p(l, u) \cdot p(a, w).$$

But this last inequality is true, as a consequence of Proposition 79. Indeed, the inequality $u \leq_L A$ implies that $u \in [la] \cap [uw]$, therefore $[lv] \cap [uw] \neq \emptyset$.

The proof of the implication $\mathbf{ord}_u^L \leq_{val} \mathbf{int}_A^L \Longrightarrow u \preceq_L A$ in point 2 We reason again by contradiction, assuming that the inequality $u \preceq_L A$ is not true. This means that $[la] \cap [uu] = \emptyset$. As $[lu] \cap [au] \neq \emptyset$ (because u belongs to this intersection), Proposition 79 implies that:

$$p(l, a) \cdot p(u, u) > p(l, u) \cdot p(a, u).$$

Replacing v by a in the reasoning done in the proof of formula (3) above, we arrive at the following inequality:

$$\frac{\check{E}_u \cdot (\pi^* Z_h)_{ex}}{-\check{E}_u \cdot \check{E}_l} > \frac{\check{E}_a \cdot (\pi^* Z_h)_{ex}}{\check{E}_a \cdot \check{E}_l}.$$

Combining it with formulae (1) and (4), as well as Proposition 18, it becomes: $\text{ord}_u^L(h) > \text{int}_A^L(h)$. But this contradicts the hypothesis $\text{ord}_u^L \leq_{val} \text{int}_A^L$.

By combining Theorem 119 with Theorem 87, we get:

Corollary 120 *Let S be an arborescent singularity and $\mathcal{F}$ a finite set of branches on it. Let L be a branch not belonging to $\mathcal{F}$. Consider any embedded resolution π of the sum D of L with the elements of $\mathcal{F}$. Let $(E_u)_{u \in \mathcal{V}}$ be the components of the exceptional divisor of π. Then the partial order $\preceq_{val}$ is arborescent in restriction to the set:*

$$\{\text{ord}_u^L \mid u \in \mathcal{V}\} \cup \{\text{int}_A^L \mid A \in \mathcal{F}\}$$

and the associated extended rooted tree (in the sense of Definition 28) is isomorphic with the convex hull of $\{L\} \cup \mathcal{F}$ in the dual tree of the total transform of D by π, rooted at the strict transform of L.

Remark 121 Till now we worked with fixed embedded resolutions of the various reduced divisors considered on S. But, given a fixed divisor, one could consider the projective system of all its resolutions. One gets an associated direct system of embeddings of the dual graphs of total transforms. The associated ultrametric spaces are instead the same. Consider the set of all reduced divisors on S, directed by inclusion. One gets an associated direct system of isometric embeddings of ultrametric spaces, therefore of isometric embeddings of associated trees. One could prove then an analog of Jonsson's [32, Theorem 7.9] (which concerns only *smooth* germs S), which presents a valuative tree associated to the singularity S (that is, a quotient of a Berkovich space) as a projective limit of dual trees. We prefer not to do this here, in order to restrict to phenomena visible on fixed resolutions of S and which may be described by elementary combinatorial means, without any appeal to Berkovich geometry.

Remark 122 After having seen a previous version of this paper, Ruggiero sent us the preliminary version [26] of the paper he writes with Gignac. In that paper they extend to the spaces $\overline{\mathrm{Val}}(S)$ of semivaluations of normal surface singularities S, part of the theory described in [20] and [32]. This started our collaboration with Ruggiero leading to the sequel [25] of the present paper.

6 Perspectives on Non-arborescent Singularities

In this section we give two examples, showing that for singularities which are not necessarily arborescent, U_L is not necessarily an ultrametric or even a metric on the set of branches distinct from L. Then we state some open problems related with this phenomenon.

6.1 Non-arborescent Examples

Example 123 Consider the weighted dual graph Γ represented in Fig. 13 (the self-intersections being indicated between brackets and the genera being arbitrary). Denote by I the associated intersection form. Consider the matrix of $-I$, obtained after having ordered the vertices a, b, c, d. By computing its principal minors, one sees that this symmetric matrix is positive definite, which shows that I is negative definite. By a theorem of Grauert [28, Page 367] (see also Laufer [34, Theorem 4.9]),

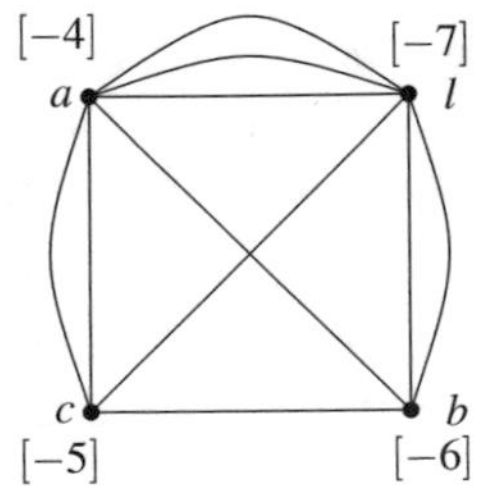

Fig. 13 The dual graph of the singularity in Example 123

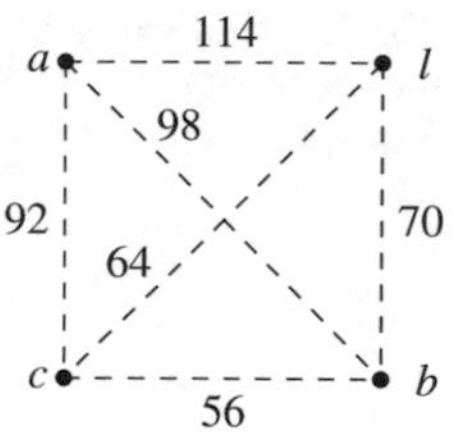

Fig. 14 The label on the edge $[uv]$ is the number $-(\check{E}_u \cdot \check{E}_v) \cdot \det(S)$ in Example 123

any divisor with normal crossings in a smooth complex surface which admits this weighted dual graph may be contracted to a normal surface singularity S. The graph Γ admitting cycles, the singularity is not arborescent. Denote by π the resolution of S whose dual graph is Γ.

Let us consider branches A, B, C, L on S whose strict transforms by π are smooth and intersect transversally E_a, E_b, E_c, E_l at smooth points of the total exceptional divisor E. Therefore $(\pi^* A)_{ex} = -\check{E}_a$, $(\pi^* B)_{ex} = -\check{E}_b$, $(\pi^* C)_{ex} = -\check{E}_c$, and $(\pi^* L)_{ex} = -\check{E}_l$.

The entries $-(\check{E}_u \cdot \check{E}_v) \cdot \det(S)$ of the adjoint matrix of $(-E_u \cdot E_v)_{uv}$ are as indicated in Fig. 14. Using Corollary 20, one may compute then the values of U_L, getting:

$$\det(S) \cdot U_L(A, B) = \frac{64 \cdot 70 \cdot 114}{6272},$$

$$\det(S) \cdot U_L(A, C) = \frac{64 \cdot 70 \cdot 114}{6440},$$

$$\det(S) \cdot U_L(B, C) = \frac{64 \cdot 70 \cdot 114}{6384}.$$

As the three values are pairwise distinct, we see that U_L is not an ultrametric on the set $\{A, B, C\}$. Therefore it is nor an ultrametric on the set of branches $\mathcal{B}(S)\setminus\{L\}$. Let us mention that $\det(S) = 56$, even if one does not need this in order to do the previous computations. One may check immediately on the above values that U_L is nevertheless a metric on the set $\{A, B, C\}$. We do not know if it is also a metric on $\mathcal{B}(S) \setminus \{L\}$.

Example 124 Consider the weighted graph Γ represented in Fig. 15 (the self-intersections being indicated between brackets and the genera being arbitrary). As in the previous example, we see that there exist normal surface singularities with such weighted dual graphs, and that they are not arborescent.

Let us consider branches A, B, C, L on S with the same properties as in the previous example.

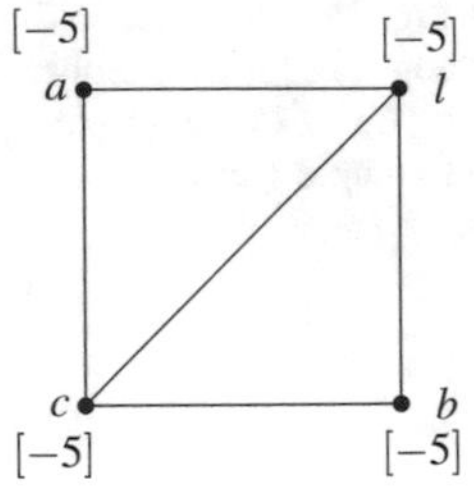

Fig. 15 The dual graph of the singularity in Example 124

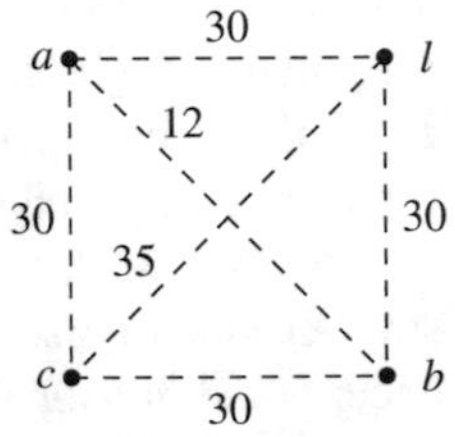

Fig. 16 The label on the edge $[uv]$ is the number $-(\check{E}_u \cdot \check{E}_v) \cdot \det(S)$ in Example 124

The entries $-(\check{E}_u \cdot \check{E}_v) \cdot \det(S)$ of the adjoint matrix of $(-E_u \cdot E_v)_{uv}$ are as indicated in Fig. 16. Using Corollary 20, one may compute then the values of U_L, getting:

$$\det(S) \cdot U_L(A, B) = 75, \quad \det(S) \cdot U_L(A, C) = 35, \quad \det(S) \cdot U_L(B, C) = 35.$$

One sees that in this case U_L is not even a metric on the set $\{A, B, C\}$.

6.2 Some Open Problems

Let us end this paper with some open problems:

1. *Characterize the normal surface singularities for which U_L is a metric (compare with Examples* 123 *and* 124*).*
2. *Characterize the normal surface singularities whose generic hyperplane section is irreducible (compare with Theorem* 92*).*
3. *Characterize the normal surface singularities for which U_O is an ultrametric (compare with Theorem* 92*).*
4. *Characterize the normal surface singularities for which U_O is a metric.*

Acknowledgements This research was partially supported by the French grant ANR-12-JS01-0002-01 SUSI and Labex CEMPI (ANR-11-LABX-0007-01), and also by the Spanish Projects MTM2016-80659-P, MTM2016-76868-C2-1-P. The third author is grateful to María Angelica Cueto, András Némethi and Dmitry Stepanov for inspiring conversations. We are also grateful to Nicholas Duchon for having sent us his thesis, and to Charles Favre, Mattias Jonsson, András Némethi, Walter Neumann and Matteo Ruggiero for their comments on a previous version of this paper.

References

1. Abío, I., Alberich-Carramiñana, M., González-Alonso, V.: The ultrametric space of plane branches. Commun. Algebra **39**(11), 4206–4220 (2011)
2. Artin, M.: On isolated rational singularities of surfaces. Am. J. Math. **83**, 129–136 (1966)
3. Bandelt, H.-J., Steel, M.A.: Symmetric matrices representable by weighted trees over a cancellative abelian monoid. SIAM J. Discrete Math. **8**(4), 517–525 (1995)
4. Böcker, S., Dress, S.: Recovering symbolically dated, rooted trees from symbolic ultrametrics. Adv. Math. **138**, 105–125 (1998)
5. Braun, G., Némethi, A.: Surgery formula for Seiberg-Witten invariants of negative definite plumbed 3-manifolds. J. Reine Angew. Math. **638**, 189–208 (2010)
6. Buneman, P.: A note on the metric properties of trees. J. Comb. Theory **17**, 48–50 (1974)
7. Camacho, C.: Quadratic forms and holomorphic foliations on singular surfaces. Math. Ann. **282**, 177–184 (1988)
8. Campillo, A.: Algebroid Curves in Positive Characteristic. Lecture Notes in Mathematics, vol. 813. Springer, Berlin (1980)
9. Caubel, C., Némethi, A., Popescu-Pampu, P.: Milnor open books and Milnor fillable contact 3-manifolds. Topology **45**, 673–689 (2006)
10. Chądzyński, J., Płoski, A.: An inequality for the intersection multiplicity of analytic curves. Bull. Pol. Acad. Sci. Math. **36**(3–4), 113–117 (1988)
11. Coxeter, H.S.M.: Discrete groups generated by reflections. Ann. Math. **35**(3), 588–621 (1934)
12. Coxeter, H.S.M.: Regular Polytopes. Dover Publications Inc., New York (1973)
13. Dimca, A.: Singularities and Topology of Hypersurfaces. Universitext. Springer, New York (1992)
14. Duchon, N.: Involutions of plumbed manifolds. Thesis, University of Maryland, College Park (1982). Available at http://sandsduchon.org/duchon/DuchonThesis.zip
15. Du Val, P.: The unloading problem for plane curves. Am. J. Math. **62**(1), 307–311 (1940)
16. Du Val, P.: On absolute and non-absolute singularities of algebraic surfaces. Revue de la Faculté des Sciences de l'Univ. d'Istanbul (A) **91**, 159–215 (1944)
17. Eggers, H.: Polarinvarianten und die Topologie von Kurvensingularitaten. Bonner Math. Schriften **147** (1983)
18. Eisenbud, D., Neumann, W.: Three-Dimensional Link Theory and Invariants of Plane Curve Singularities. Princeton University Press, Princeton (1985)
19. Fauvet, F., Menous, F., Sauzin, D.: Explicit linearization of one-dimensional germs through tree-expansions. https://hal.archives-ouvertes.fr/hal-01053805v2. Submitted on 22 Jan 2015
20. Favre, C., Jonsson, M.: The Valuative Tree. Lecture Notes in Mathematics, vol. 1853. Springer, Berlin (2004)
21. García Barroso, E.: Invariants des singularités de courbes planes et courbure des fibres de Milnor. PhD thesis, University of La Laguna, Tenerife (1996). Available at http://ergarcia.webs.ull.es/tesis.pdf
22. García Barroso, E., Płoski, A.: An approach to plane algebroid branches. Rev. Mat. Complut. **28**(1), 227–252 (2015)
23. García Barroso, E., Płoski, A.: On the intersection multiplicity of plane branches. ArXiv:1710.05346
24. García Barroso, E., González Pérez, P.D., Popescu-Pampu, P.: Variations on inversion theorems for Newton-Puiseux series. Math. Annalen **368**, 1359–1397 (2017)
25. García Barroso, E., González Pérez, P.D., Popescu-Pampu, P., Ruggiero, M.: Ultrametric distances on valuative spaces. ArXiv:1802.01165
26. Gignac, W., Ruggiero, M.: Local dynamics of non-invertible maps near normal surface singularities. Arxiv:1704.04726
27. González Pérez, P.D.: Toric embedded resolutions of quasi-ordinary hypersurface singularities. Ann. Inst. Fourier Grenoble **53**(6), 1819–1881 (2003)

28. Grauert, H.: Über Modifikationen und exzeptionnelle analytische Mengen. Math. Ann. **146**, 331–368 (1962)
29. Hartshorne, R.: Algebraic Geometry. Springer, New York (1977)
30. Holly, J.E.: Pictures of ultrametric spaces, the p-adic numbers, and valued fields. Am. Math. Mon. **108**, 721–728 (2001)
31. Ishii, S.: Introduction to Singularities. Springer, Tokyo (2014)
32. Jonsson, M.: Dynamics on Berkovich spaces in low dimensions. In: Berkovich Spaces and Applications. Lecture Notes in Mathematics, vol. 2119, pp. 205–366. Springer, Cham (2015)
33. Kuo, T.C., Lu, Y.C.: On analytic function germs of two complex variables. Topology **16**(4), 299–310 (1977)
34. Laufer, H.: Normal Two-Dimensional Singularities. Annals of Mathematics Studies, vol. 71. Princeton University Press, Princeton (1971)
35. Lipman, J.: Rational singularities with applications to algebraic surfaces and unique factorization. Inst. Hautes Études Sci. Publ. Math. No. **36**, 195–279 (1969)
36. Mumford, D.: The topology of normal singularities of an algebraic surface and a criterion for simplicity. Inst. Hautes Études Sci. Publ. Math. No. **9**, 5–22 (1961)
37. Neumann, W.: Abelian covers of quasihomogeneous surface singularities. In: Singularities, Arcata 1981. Proceedings of Symposia in Pure Mathematics, vol. 40, pp. 233–243. American Mathematical Society (1983)
38. Neumann, W.: On bilinear forms represented by trees. Bull. Aust. Math. Soc. **40**, 303–321 (1989)
39. Neumann, W., Wahl, J.: Complete intersection singularities of splice type as universal abelian covers. Geom. Topol. **9**, 699–755 (2005)
40. Noether, M.: Les combinaisons caractéristiques dans la transformation d'un point singulier. Rend. Circ. Mat. Palermo **IV**, 89–108; 300–301 (1890)
41. Okuma, T.: Universal abelian covers of certain surface singularities. Math. Ann. **334**, 753–773 (2006)
42. Płoski, A.: Remarque sur la multiplicité d'intersection des branches planes. Bull. Pol. Acad. Sci. Math. **33**, 601–605 (1985)
43. Popescu-Pampu, P.: Arbres de contact des singularités quasi-ordinaires et graphes d'adjacence pour les 3-variétés réelles. PhD thesis, University of Paris 7 (2001). Available at https://tel.archives-ouvertes.fr/tel-00002800v1
44. Popescu-Pampu, P.: Ultrametric spaces of branches on arborescent singularities. Math. Forsch. Oberwolfach Rep. **46**, 2655–2658 (2016)
45. Teissier, B.: Sur une inégalité à la Minkowski pour les multiplicités. Appendix to Eisenbud, D., Levine, H.: An algebraic formula for the degree of a C^∞-map germ. Ann. Math. **106**(38–44), 19–44 (1977)
46. Wall, C.T.C.: Chains on the Eggers tree and polar curves. In: Proceedings of the International Conference on Algebraic Geometry and Singularities, Sevilla 2001. Rev. Mat. Iberoamericana **19**(2), 745–754 (2003)
47. Wall, C.T.C.: Singular Points of Plane Curves. London Mathematical Society Student Texts, vol. 63. Cambridge University Press, Cambridge (2004)
48. Zariski, O.: The theorem of Riemann-Roch for high multiples of an effective divisor on an algebraic surface. Ann. Math. **76**(3), 560–615 (1962)

Two Points of the Boundary of Toric Geometry

Bernard Teissier

Abstract This note presents two observations which have in common that they lie at the boundary of toric geometry. The first one because it concerns the deformation of affine toric varieties into non toric germs in order to understand how to avoid some ramification problems arising in the study of local uniformization in positive characteristic, and the second one because it uses limits of projective systems of equivariant birational maps of toric varieties to study the space of additive preorders on $\mathbf{Z}^r$ for $r \geq 2$.

1 Using Toric Degeneration to Avoid Wild Ramification

In his book [1], Campillo introduces and studies a notion of characteristic exponents for plane branches over an algebraically closed field of positive characteristic. One of the definitions he gives is that the characteristic exponents are those of a plane branch in characteristic zero having the same process of embedded resolution of singularities. His basic definition is given in terms of Hamburger-Noether expansion[1] but we shall not go into this here. He then gives an example to show that even if you do have a Puiseux-type parametrization for a branch in positive characteristic, which is not always the case, the exponents you see in the parametrization are *not* in general the characteristic exponents. His example is this (see [1, Chap. III, §5, Example 3.5.4]):

Let p be a prime number. Let us choose a field k of characteristic p and consider the plane branch defined parametrically by $x = t^{p^3}$, $y = t^{p^3+p^2} + t^{p^3+p^2+p+1}$, and implicitly by a unitary polynomial of degree p^3 in y with coefficients in $k[[x]]$.

[1]The Hamburger-Noether expansion is an algorithm extracting in any characteristic a description of the resolution process by point blowing-ups from a parametric representation $x(t)$, $y(t)$.

B. Teissier (✉)
CNRS, Institut Mathématique de Jussieu-Paris Rive Gauche, Paris, France
e-mail: bernard.teissier@imj-prg.fr

G.-M. Greuel et al. (eds.), *Singularities, Algebraic Geometry, Commutative Algebra, and Related Topics*, https://doi.org/10.1007/978-3-319-96827-8_4

Campillo computes by a Hamburger-Noether expansion the characteristic exponents and finds $\beta_0 = p^3, \beta_1 = p^3 + p^2, \beta_2 = p^3 + 2p^2 + p, \beta_3 = p^3 + 2p^2 + 2p + 1$. Note that there are four characteristic exponents while the parametrization exhibits three exponents in all. In [13, Remark 7.19], the author had computed directly from the parametrization the generators of the semigroup of values of the t-adic valuation of $k[[t]]$ on the subring $k[[t^{p^3}, t^{p^3+p^2} + t^{p^3+p^2+p+1}]]$ and found the numerical semigroup with minimal system of generators:

$$\Gamma = \langle p^3,\ p^3 + p^2,\ p^4 + p^3 + p^2 + p,\ p^5 + p^4 + p^3 + p^2 + p + 1\rangle.$$

We can verify that Campillo's characteristic exponents and the generators of the semigroup satisfy the classical relations of Zariski, in accordance with [1, Proposition 4.3.5].

This example is very interesting because it shows that in positive characteristic, even if a Puiseux-type expansion exists, its exponents do not determine the resolution process. In thinking about resolution in positive characteristic, one should keep away from ideas inspired by Puiseux exponents. The semigroup, however, does determine the resolution process in all characteristics. It is shown in [13, 14] that in the case of analytically irreducible curves one can obtain embedded resolution by studying the embedded resolution of the monomial curve corresponding to the minimal system of generators of the numerical semigroup Γ of the values taken on the algebra of the curve by its unique valuation. Another, more classical, reason is that the semigroup determines the Puiseux exponents of the curve in characteristic zero having the same resolution process, and therefore this resolution process (see [1, Chap. IV, §3]).

The polynomial $f(x, y) \in k[[x]][y]$ defining our plane branch can be obtained by eliminating u_2, u_3 between three equations which are:

$$y^p - x^{p+1} - u_2 = 0,\ u_2^p - x^{p(p+1)}y - u_3 = 0,\ u_3^p - x^{p^2(p+1)}u_2 = 0.$$

This makes it apparent that our plane branch is a flat deformation of the curve defined by $y^p - x^{p+1} = 0,\ u_2^p - x^{p(p+1)}y = 0,\ u_3^p - x^{p^2(p+1)}u_2 = 0$, which is the monomial curve C^Γ in $\mathbf{A}^4(k)$ given parametrically by $x = t^{p^3}, y = t^{p^3+p^2}, u_2 = t^{p^4+p^3+p^2+p}, u_3 = t^{p^5+p^4+p^3+p^2+p+1}$. That is, the monomial curve whose affine algebra is the semigroup algebra $k[t^\Gamma]$. The binomial equations correspond to a system of generators of the arithmetical relations between the generators of the semigroup. Compare with [14, Remark 7.19]; we have chosen here the canonical system of relations between the generators of the semigroup, where all the exponents in the second monomial of the equations except the exponent of x are $< p$.

Moreover, if we give to each of the variables x, y, u_2, u_3 a weight equal to the exponent of t for this variable in the parametrization of C^Γ, we see that to each binomial is added a term of higher weight. This is an *overweight deformation* of a prime binomial ideal in the sense of [14, §3].

Eliminating u_2 and u_3 between the three equations gives the equation of our plane curve with semigroup Γ:

$$(y^p - x^{p+1})^{p^2} - 2x^{p^2(p+1)}y^p + x^{(p^2+1)(p+1)} = 0.$$

If $p = 2$ this reduces to $(y^2 - x^3)^4 - x^{15} = 0$, which looks like it should be the overweight deformation $y^2 - x^3 - u_2 = 0, u_2^4 - x^{15} = 0$ of the monomial curve with equations $y^2 - x^3 = 0, u_2^4 - x^{15} = 0$. However, such is not the case because the two binomials $y^2 - x^3, u_2^4 - x^{15}$ do not generate a prime ideal in $k[[x, y, u_2]]$. One verifies that $u_2^2 - x^6 y$ is not in the ideal but its square is. The binomials do not define an integral monomial curve, but in fact the non reduced curve given parametrically by $x = t^8, y = t^{12}, u_2 = t^{30}$ with non coprime exponents, while we know that the semigroup of our curve is $\Gamma = \langle 8, 12, 30, 63\rangle$. The equation of our plane curve is indeed irreducible, but it is *not* an overweight deformation of an integral monomial curve in $\mathbf{A}^3(k)$. One has to embed our plane curve in $\mathbf{A}^4(k)$ to view it as an overweight deformation of a monomial curve.

The reader can verify that the same phenomenon occurs in any positive characteristic. For $p \neq 2$ our curve looks like an overweight deformation of the curve in $\mathbf{A}^3(k)$ defined by the binomials $y^p - x^{p+1}, u_2^{p^2} - 2x^{p^2(p+1)}y^p$ but these binomials do not generate a prime ideal: the binomial $u_2^p - 2x^{p(p+1)}y$ is not in the ideal, but its p-th power is by Fermat's little theorem. So appearances can be deceptive also from the equational viewpoint.

We expect the same result in characteristic zero, and this is true as soon as the field k contains the p-th roots of 2. In this case, by (the proof of) [4, Theorem 2.1] the ideal I generated by $y^p - x^{p+1}, u_2^{p^2} - 2x^{p^2(p+1)}y^p$ is not prime because the lattice $\mathscr{L}$ in $\mathbf{Z}^3$ generated by $(-(p+1), p, 0)$ and $(-p^2(p+1), -p, p^2)$ is not saturated; the vector $w = (-p(p+1), -1, p)$ is not in $\mathscr{L}$, but pw is. Here the argument is that if I was a prime ideal, at least one of the factors of the product $\prod_{\zeta^p=2}(u_2^p - \zeta x^{p(p+1)}y) = u_2^{p^2} - 2x^{p^2(p+1)}y^p$ should be in I, which is clearly impossible.

This raises the following question: Given an algebraically closed field k, assuming that we have a Puiseux expansion $x = t^n, y = \sum_j a_j t^j,\ a_j \in k$, can one predict, from the set $\{j/a_j \neq 0\}$, useful information about the semigroup of the corresponding plane curve, even only a bound on the number of generators? More generally, denoting by $k[[t^{\mathbf{Q}_{\geq 0}}]]$ the Hahn ring of series whose exponents are non negative rational numbers forming a well-ordered set, and given a series $y(t) \in k[[t^{\mathbf{Q}_{\geq 0}}]]$ which is integral over $k[[t]]$, can one deduce from the knowledge about the exponents and coefficients of $y(t)$ provided by the work of Kedlaya in [7] useful information about the semigroup of the plane branch whose equation is the integral dependence relation?

This example has another interesting feature. We note that no linear projection of our plane curve to a line in the plane can be tamely ramified. However, our curve is a deformation of the monomial curve, for which the projection to the u_3-axis is tamely ramified. More precisely, the inclusion of k-algebras $k[u_3] \subset k[t^\Gamma]$ corresponding to $u_3 \mapsto t^{p^5+p^4+p^3+p^2+p+1}$ gives rise to an extension $\mathbf{Z} \subset \mathbf{Z}$ of value groups, for the u_3-adic and t-adic valuations respectively, whose index is $p^5 + p^4 + p^3 + p^2 + p + 1$ and hence prime to p, while there is no residue field extension. This is related to what we saw in the first part of this section; it is precisely the coordinate u_3 missing in $\mathbf{A}^3(k)$ which provides the tame projection.

It is a general fact that if $\Gamma \subset \mathbf{Z}^r$ is a finitely generated semigroup generating $\mathbf{Z}^r$ as a group, given a system of generators $\Gamma = \langle\gamma_1, \ldots, \gamma_N\rangle$ and an algebraically closed field k, there always exist r of the generators, say $\gamma_{i_1}, \ldots, \gamma_{i_r}$ such that the inclusion of k-algebras $k[u_{i_1}, \ldots, u_{i_r}] \subset k[t^\Gamma]$ defined by $u_{i_\ell} \mapsto t^{\gamma_{i_\ell}}$ defines a tame extension of the fraction fields. In fact, the corresponding map $\pi\colon \mathrm{Spec}k[t^\Gamma] \to \mathbf{A}^r(k)$ is étale on the torus of $\mathrm{Spec}k[t^\Gamma]$. Note that the subset $\{i_1, \ldots, i_r\} \subset \{1, \ldots, N\}$ depends in general on the characteristic of the field k. This is immediately visible in the case of a monomial curve, where the result follows directly from the fact that the generators of the semigroup are coprime since the semigroup generates $\mathbf{Z}$ as a group.

In the general case, as explained in [14, Proof of Proposition 3.20 and Proof of Proposition 7.4], modulo a Gale-type duality this fact is an avatar of the fact that the *relative torus* $\mathrm{Spec}\mathbf{Z}[t^{\mathbf{Z}^r}]$ of $\mathrm{Spec}\mathbf{Z}[t^\Gamma]$ is smooth over $\mathrm{Spec}\mathbf{Z}$ while of course the whole toric variety is not. This implies, for each prime number p, the non-vanishing on the torus of certain jacobian determinants which are exactly those whose non vanishing is needed to ensure the étaleness of the map π. We note that the map π is not finite in general; it is finite if and only if the vectors $\gamma_{i_1}, \ldots, \gamma_{i_r}$ generate the cone of $\mathbf{R}^r$ generated by Γ.

More precisely, recall the description found in [13], before Prop. 6.2, of the jacobian ideal of an r-dimensional affine toric variety defined by a prime binomial ideal $P \subset k[U_1, \ldots, U_N]$. The jacobian determinant $J_{G,\mathbf{L}'}$ of rank $c = N - r$ of the generators $(U^{m^\ell} - \lambda_\ell U^{n^\ell})_{\ell\in\{1,\ldots,L\}}$ of P, associated to a sequence $G = (k_1, \ldots, k_c)$ of distinct elements of $\{1, \ldots, N\}$ and a subset $\mathbf{L}' \subseteq \{1, \ldots, L\}$ of cardinality c, satisfies the congruence

$$U_{k_1} \ldots U_{k_c}.J_{G,\mathbf{L}'} \equiv \Big(\prod_{\ell\in\mathbf{L}'} U^{m^\ell}\Big)\mathrm{Det}_{G,\mathbf{L}'}(\langle m - n\rangle) \ \ \mathrm{mod}.P,$$

where $\big(\langle m - n\rangle\big)$ is the matrix of the vectors $(m^\ell - n^\ell)_{\ell\in\{1,\ldots,L\}}$, and $\mathrm{Det}_{G,\mathbf{L}'}$ indicates the minor in question. If the field k is of characteristic p, choosing a minor which is not divisible by p amounts to choosing r of the coordinates such that the corresponding projection to $\mathbf{A}^r(k)$ of a certain binomial variety containing the toric variety as one of its irreducible components (see [4, Corollary 2.3]) is étale outside of the coordinate hyperplanes.

It is shown in [14, Proof of Proposition 3.20] that for any prime p there exist minors $\mathrm{Det}_{G,\mathbf{L}'}(\langle m - n\rangle)$ which are not divisible by p. As mentioned above, this is the equational aspect of the smoothness over $\mathrm{Spec}\mathbf{Z}$ of the torus $\mathrm{Spec}\mathbf{Z}[t^{\mathbf{Z}^r}]$ of the affine toric variety over $\mathbf{Z}$ corresponding to the subsemigroup Γ of $\mathbf{Z}^r$.

The next step is to realize that an overweight deformation preserves the non vanishing. Let us illustrate this on our example:

The jacobian matrix of our three binomial equations over a field of characteristic p is

$$\begin{pmatrix} -x^p & 0 & 0 & 0 \\ 0 & -x^{p(p+1)} & 0 & 0 \\ 0 & 0 & -x^{p^2(p+1)} & 0 \end{pmatrix}$$

We see that there is only one minor which is non zero on the torus, where $x \neq 0$, corresponding to the inclusion $k[u_3] \subset k[t^\Gamma]$. After the overweight deformation which produces our plane branch, the jacobian matrix becomes

$$\begin{pmatrix} -x^p & 0 & -1 & 0 \\ 0 & -x^{p(p+1)} & 0 & -1 \\ 0 & 0 & -x^{p^2(p+1)} & 0 \end{pmatrix}$$

and we see that the same minor is $\neq 0$. The facts we have just mentioned imply that for each characteristic p, some of these minors are non zero modulo p. But the duality implies that if $u_{i_1}, \ldots, u_{i_r}$ are the variables not involved in the derivations producing one such minor, then the absolute value of this minor, which is not divisible by p, is the index of the extension of groups $\mathbf{Z}\gamma_{i_1} + \cdots + \mathbf{Z}\gamma_{i_r} \subset \mathbf{Z}^r$. This explains the tameness result we have just seen, and also why it is preserved by overweight deformation. In our example the matrix $(\langle m - n \rangle)$ after reduction modulo p is

$$\begin{pmatrix} -1 & 0 & 0 & 0 \\ 0 & -1 & 0 & 0 \\ 0 & 0 & -1 & 0 \end{pmatrix}$$

In conclusion the method which we can apply to our curve, which is that if we embed it in the affine space spanned by the monomial curve associated to its semigroup, in any characteristic we are certain to find a tame projection to a coordinate axis, will work for any finitely generated semigroup Γ in $\mathbf{Z}^r$ generating $\mathbf{Z}^r$ as a group and provide tame projections to $\mathbf{A}^r(k)$ of the affine toric variety Spec$k[t^\Gamma]$, which will remain tame after an overweight deformation. In order to obtain tame projections for the space whose valuation we want to uniformize, we have to first re-embed it in the space where the associated toric variety lives so that it can appear as an overweight deformation. This is an important element in the proof of local uniformization for rational Abhyankar valuations (those with value group $\mathbf{Z}^{\dim R}$) on equicharacteristic excellent local domains R with an algebraically closed residue field given in [14].

The reason is that the local uniformization of Abhyankar valuations on an equicharacteristic excellent local domain with algebraically closed residue field can be reduced to that of rational valuations of complete local domains whose semigroup is a finitely generated subsemigroup of $\mathbf{Z}^r$ and then the complete local domain is an overweight deformation of an affine toric variety by [14, Proposition 5.1].

2 Additive Preorders and Orders on $\mathbf{Z}^r$ and Projective Limits of Toric Varieties

In this section the space of additive preorders on $\mathbf{Z}^r$ with a topology defined by Kuroda and Sikora (see [8, 12]) is presented as homeomorphic to the projective limit of finite topological spaces which are spaces of orbits on toric varieties with the topology induced by the Zariski topology, and the space of additive orders as the closed subspace corresponding to the projective limit of the subsets of closed points of the preceding finite topological spaces. We use this to give a toric proof of a theorem of Sikora in [12] showing that the space of additive orders on $\mathbf{Z}^r$ with $r \geq 2$ is homeomorphic to the Cantor set. The relation between toric geometry and preorders on $\mathbf{Z}^r$ is due to Ewald-Ishida in [3] where they build the analogue in toric geometry of the Zariski-Riemann space of an algebraic variety. This relation was later developed by Pedro González Pérez and the author in [5], which contains in particular the results used here. In that text we quoted Sikora's result without realizing that we could give a direct proof in our framework. We begin with the:

Proposition 1 *Let I be a directed partially ordered set and $(b_{\iota',\iota}\colon X_{\iota'} \to X_\iota$ for $\iota' > \iota)$ a projective system indexed by I of surjective continuous maps between finite topological spaces. Then:*

(a) *The projective limit $\mathscr{X}$ of the projective system, endowed with the projective limit topology, is a quasi-compact space.*
(b) *In each X_ι consider the subset D_ι of closed points, on which the induced topology is discrete. If we assume that:*

1. *The maps $b_{\iota',\iota}$ map each $D_{\iota'}$ onto D_ι in such a way that the inverse image in the projective limit $\mathscr{D}$ of the D_ι by the canonical map $b_{\infty,\iota}\colon \mathscr{D} \to D_\iota$ of any element of a D_ι is infinite.*
2. *There exists a map $h\colon I \to \mathbf{N} \setminus \{0\}$ such that $h(\iota') \geq h(\iota)$ if $\iota' > \iota$ and $h^{-1}([1, m])$ is finite for all $m \in \mathbf{N}$, where $[1, m] = \{1, \ldots, m\}$.*

Then $\mathscr{D}$ is closed in $\mathscr{X}$ and homeomorphic to the Cantor set.

Proof Statement (a) is classical, see [6, 2–14] and [12]. The compactness comes from Tychonoff's theorem and the definition of the projective limit topology. To prove (b) we begin by showing that $\mathscr{D}$ can be endowed with a metric compatible with the projective limit topology. Given $w, w' \in \mathscr{D}$ with $w \neq w'$, define $r(w, w')$ to be the smallest integer n such that $b_{\infty,\iota}(w) \neq b_{\infty,\iota}(w')$ for some $\iota \in h^{-1}([1, n])$. This is the smallest element in a non-empty set of integers since if $w \neq w'$, by definition of the projective limit there is an index ι_0 such that $b_{\infty,\iota_0}(w) \neq b_{\infty,\iota_0}(w')$. Thus, our set contains $h(\iota_0)$. It follows from the definition that given three different

w, w', w'' we have that $r(w, w'') \geq \min(r(w, w'), r(w', w''))$, and that $r(w, w') = r(w', w)$. We can define a distance on $\mathscr{D}$ by setting $d(w, w) = 0$ and $d(w, w') = r(w, w')^{-1}$ for $w' \neq w$. By the definition of the projective limit topology on $\mathscr{D}$ we see that it is totally discontinuous because the D_ι are and that every ball $\mathbf{B}(w, \eta) = \{w'/d(w', w) \leq \eta\}$ is the intersection $\bigcap_{\iota \in h^{-1}([1, \lfloor \eta^{-1} \rfloor])} b_{\infty,\iota}^{-1}(w)$ of finitely many open sets . The first assumption implies that every ball of positive radius centered in a point of $\mathscr{D}$ is infinite, so that $\mathscr{D}$ is perfect. Finally, our space $\mathscr{D}$ is a perfect compact and totally disconnected metric space and thus homeomorphic to the Cantor set by [6, Corollary 2-98]. The fact that $\mathscr{D}$ is closed in $\mathscr{X}$ follows from the fact that each D_ι is closed in X_ι.

Remark 1 In [6, Theorem 2–95] it is shown that a compact totally disconnected metric space is homeomorphic to the projective limit of a projective system of finite discrete spaces. The authors then show that if two such spaces are perfect, the projective systems can be chosen so that their projective limits are homeomorphic.

We recall some definitions and facts of toric geometry that are needed for our purpose, referring to [2] for proofs. Let $M = \mathbf{Z}^r$ be the lattice of integral points in $\mathbf{R}^r$ and $N = \mathrm{Hom}_{\mathbf{Z}}(M, \mathbf{Z})$ its dual, a lattice in $N_{\mathbf{R}} = \check{\mathbf{R}}^r$. We assume that $r \geq 2$.

A (finite) fan is a finite collection $\Sigma = (\sigma_\alpha)_{\alpha \in A}$ of rational polyhedral strictly convex cones in $N_{\mathbf{R}}$ such that if τ is a face of a $\sigma_\alpha \in \Sigma$, then τ is a cone of the fan, and the intersection $\sigma_\alpha \cap \sigma_\beta$ of two cones of the fan is a face of each. A rational polyhedral convex cone is by definition the cone positively generated by finitely many vectors of the lattice N, called integral vectors. It is strictly convex if it does not contain any non zero vector spacc. Given a rational polyhedral cone σ, its convex dual $\check{\sigma} = \{u \in \mathbf{R}^r / < u, v > \geq 0\ \forall v \in \sigma\}$, where $< u, v >= v(u) \in \mathbf{R}$, is again a rational polyhedral convex cone, which is strictly convex if and only if the dimension of σ, that is, the dimension of the smallest vector subspace of $\check{\mathbf{R}}^r$ containing σ, is r. A refinement Σ' of a fan Σ is a fan such that every cone of Σ' is contained in a cone of Σ and the union of the cones of Σ' is the same as that of Σ. We denote this relation by $\Sigma' \prec \Sigma$.

By a theorem of Gordan (see [2, Chap. V, §3, Lemma 3.4]), for every rational strictly convex cone in $N_{\mathbf{R}}$ the semigroup $\check{\sigma} \cap M$ is a finitely generated semigroup generating M as a group. If we fix a field k the semigroup algebra $k[t^{\check{\sigma} \cap M}]$ is finitely generated and corresponds to an affine algebraic variety T_σ over k, which may be singular but is normal because the semigroup $\check{\sigma} \cap M$ is saturated in the sense that if for some $k \in \mathbf{N}_{>0}$ and $m \in M$ we have $km \in \check{\sigma} \cap M$, then $m \in \check{\sigma} \cap M$. To each fan Σ is associated a normal algebraic variety T_Σ obtained by glueing up the affine toric varieties $\mathrm{Spec}k[t^{\check{\sigma} \cap M}]$, $\sigma \in \Sigma$, along the affine varieties corresponding to faces that are the intersections of two cones of the fan. A refinement Σ' of a fan Σ gives rise to a proper birational map $T_{\Sigma'} \to T_\Sigma$.

Each toric variety admits a natural action of the torus $T_{\{0\}} = (k^*)^r = (k-$ points of) $\mathrm{Spec}k[t^M]$ which has a dense orbit corresponding to the cone $\sigma = \{0\}$ of the fan. There is an inclusion reversing bijection between the cones of the fan and the orbits of the action of the torus on T_Σ. In an affine chart T_σ, the traces of the orbits of the torus action correspond to *faces* F_τ of the semigroup $\check{\sigma} \cap M$:

they are the intersections of the semigroup with the linear duals $\tau^\perp$ of the faces $\tau \subset \sigma$. The monomial ideal of $k[t^{\check{\sigma}\cap M}]$ generated by the monomials t^δ; $\delta \notin F_\tau$ is prime and defines the closure of the orbit corresponding to τ. When it is of maximal dimension the cone σ itself corresponds to the zero dimensional orbit in T_σ and all zero dimensional orbits are obtained in this way. A prime monomial ideal of $k[t^{\check{\sigma}\cap M}]$ defines the intersection with T_σ of an irreducible subvariety invariant by the torus action.

A preorder on M is a binary relation $\preceq$ with the properties that for any $m, n, o \in M$ either $m \preceq n$ or $n \preceq m$ and $m \preceq n \preceq o$ implies $m \preceq o$.

An additive preorder on M is a preorder $\preceq$ such that if $m \preceq m'$ then $m + n \preceq m' + n$ for all $n \in M$. It is a fact (see [3, 8, 10]) that given any additive preorder $\preceq$ on M there exist an integer s, $1 \leq s \leq r$ and s vectors $v_1, \ldots, v_s$ in $N_{\mathbf{R}}$ such that

$$m \preceq n \text{ if and only if } (< m, v_1 >, \ldots, < m, v_s >) \leq_{lex} (< n, v_1 >, \ldots, < n, v_s >),$$

where lex means the lexicographic order. An additive order is an additive preorder which is an order. This means that the vector subspace of $N_{\mathbf{R}}$ generated by $\nu_1, \ldots, \nu_s$ is not contained in any rational hyperplane.

In accordance with the notations of [5] we denote by w a typical element of $ZR(\Sigma)$ and by $m \preceq_w n$ the corresponding binary relation on $\mathbf{Z}^r$. Following Ewald-Ishida in [3] we define a topology on the set of additive preorders as follows:

Definition 1 Let σ be a rational polyhedral cone in $N_{\mathbf{R}}$. Define $\mathcal{U}_\sigma$ to be the set of additive preorders w of M such that $0 \preceq_w \check{\sigma} \cap M$. The $\mathcal{U}_\sigma$ are a basis of open sets for a topology on the set $ZR(M)$ of additive preorders on M. Given a fan Σ the union $ZR(\Sigma) = \bigcup_{\sigma\in\Sigma} \mathcal{U}_\sigma \subset ZR(M)$ endowed with the induced topology is defined by Ewald-Ishida as the Zariski-Riemann manifold of the fan Σ. By [3, Proposition 2.10], it depends only on the support $|\Sigma| = \bigcup_{\sigma\in\Sigma} \sigma$ and if we assume $|\Sigma| = N_{\mathbf{R}}$, it is equal to $ZR(M)$.

This topology is the same as the topology defined in [8] and [12], where a pre-basis of open sets is as follows: given two elements $a, b \in M$ a set of the pre-basis is the set $\mathcal{U}_{a,b}$ of preorders w for which $a \preceq_w b$. Indeed, to say that a preorder w is in the intersection $\bigcap_i \mathcal{U}_{a_i,b_i}$ of finitely many such sets is the same as saying that $\check{\sigma} \cap M \subset \{m \in M/0 \preceq_w m\}$ where $\sigma \subset N_{\mathbf{R}}$ is the rational polyhedral cone dual to the cone $\check{\sigma}$ in $M_{\mathbf{R}}$ generated by the vectors $b_i - a_i$. If $\check{\sigma} = \mathbf{R}^r$, the intersection is the trivial preorder, where all elements are equivalent.

Theorem 1 (Ewald-Ishida in [3, Theorem 2.4]) *The space $ZR(M)$ is quasi-compact, and for any finite fan Σ the space $ZR(\Sigma)$ is quasi-compact.*

It is shown in [3, Proposition 2.6] that to any fan Σ and any preorder $w \in ZR(\Sigma)$ we can associate a cone $\sigma \in \Sigma$. It is the unique cone with the properties $0 \preceq_w \check{\sigma} \cap M$ and $\sigma^\perp \cap M = \{m \in \check{\sigma} \cap M/m \preceq_w 0 \text{ and } m \succeq_w 0\}$. Following Ewald-Ishida, we say that w dominates σ. If w is an order, σ is of maximal dimension r.

It is not difficult to verify that if Σ' is a refinement of Σ we have $ZR(\Sigma') = ZR(\Sigma)$ and moreover, given $w \in ZR(\Sigma)$ the corresponding cones σ', σ verify $\sigma' \subset \sigma$, so that we have a torus-equivariant map $T_{\sigma'} \to T_\sigma$ of the corresponding toric affine varieties.

Given a fan Σ its finite refinements, typically denoted by Σ', form a directed partially ordered set. It is partially ordered by the refinement relation $\Sigma' \prec \Sigma$. It is a directed set because any two finite fans with the same support have a common finite refinement (see [9, Chap. III] or [2, Chap. VI]). We are going to study three projective systems of sets indexed by it. The set $\{T_{\Sigma'}\}$ of torus-invariant irreducible subvarieties of $T_{\Sigma'}$, with the topology induced by the Zariski topology, the set $O_{\Sigma'}$ of the torus orbits of $T_{\Sigma'}$, again endowed with the Zariski topology, and finally the set of 0-dimensional orbits, which is the set of closed points of $O_{\Sigma'}$. The first two sets are in fact equal because a torus invariant irreducible subvariety of T_Σ is the closure of an orbit. This is the meaning of [5, Lemma 3.3]: a prime monomial ideal of $k[t^{\check{\sigma}\cap M}]$ is generated by the monomials that are not in a *face* of the semigroup, and the Lemma states that faces are the $\tau^\perp \cap \check{\sigma} \cap M$, where τ is a face of σ and so corresponds to an orbit.

The sets $\{T_{\Sigma'}\} = O_{\Sigma'}$ are finite since our collection of cones in each Σ' is finite. The topology induced by the Zariski topology of $T_{\Sigma'}$ on $\{T_{\Sigma'}\}$ is such that the closure of an element of $\{T_{\Sigma'}\}$ is the set of orbits contained in the closure of the corresponding orbit of the torus action.

Thus, the closed points of $\{T_{\Sigma'}\}$ are the zero dimensional orbits of the torus action on $T_{\Sigma'}$ and are in bijective correspondence with the cones of maximal dimension of Σ'.

Given a refinement $\Sigma' \prec \Sigma$, the corresponding proper birational equivariant map $T_{\Sigma'} \to T_\Sigma$ maps surjectively $\{T_{\Sigma'}\}$ to $\{T_\Sigma\}$ and zero dimensional orbits to zero dimensional orbits. The map induced on zero dimensional orbits is surjective because every orbit contains zero dimensional orbits in its closure.

Given an additive preorder $w \in ZR(\Sigma)$ it dominates a unique cone σ' in each refinement Σ' of Σ and defines a unique torus-invariant irreducible subvariety of $T_{\Sigma'}$ corresponding to the prime ideal of $k[t^{\check{\sigma}'\cap M}]$ generated by the monomials whose exponents are $\succeq_w 0$. This defines a map

$$Z : ZR(\Sigma) \longrightarrow \varprojlim_{\Sigma' \prec \Sigma} \{T_{\Sigma'}\}.$$

We now quote two results from [5]:

Theorem 2 (González Pérez-Teissier in [5, Proposition 14.8]) *For any finite fan Σ, the map Z is a homeomorphism between $ZR(\Sigma)$ with its Ewald-Ishida topology and $\varprojlim_{\Sigma' \prec \Sigma} \{T_{\Sigma'}\}$ with the projective limit of the topologies induced by the Zariski topology.*

Corollary 1 (González Pérez-Teissier in [5, Section 13]) *If $|\Sigma| = N_{\mathbf{R}}$ the map Z is a homeomorphism between the space $ZR(M)$ of additive preorders on $\mathbf{Z}^r$ and*

$\lim_{\substack{\longleftarrow \\ \Sigma' \prec \Sigma}} \{T_{\Sigma'}\}$ *and induces a homeomorphism between the space of additive orders on* M *and the projective limit of the discrete sets* $\{T_{\Sigma'}\}_0$ *of* 0*-dimensional orbits.*

Definition 2 The height $h(\Sigma)$ of a finite fan Σ in $N_{\mathbf{R}}$ is the maximum absolute value of the coordinates of the primitive vectors in N generating the one-dimensional cones of Σ. There are only finitely many fans of height bounded by a given integer.

We note that the fan consisting of the 2^r quadrants of $N_{\mathbf{R}}$ and their faces has height one.

Let us show that we can apply Proposition 1 to obtain the conjunction of Sikora's result in [12, proposition 1.7] and Ewald-Ishida's in [3, Proposition 2.3]:

Proposition 2 *For* $r \geq 2$ *the space of additive preorders on* Z^r *is quasi-compact. It contains as a closed subset the space of additive orders, which is homeomorphic to the Cantor set.*

Proof To apply Proposition 1, our partially ordered directed set is the set of finite fans in $N_{\mathbf{R}}$ with support $N_{\mathbf{R}}$. The set of additive preorders is quasi-compact by Proposition 1. Concerning the set of orders, we apply the second part of the proposition. Our discrete sets are the sets $\{T_{\Sigma'}\}_0$ of zero dimensional torus orbits of the toric varieties $T_{\Sigma'}$. Each refinement of a cone of maximal dimension contains cones of maximal dimension and since $r \geq 2$ each cone $\sigma' \in \Sigma'$ can be refined into infinitely many fans with support σ' which produce as many refinements of Σ'. To see this, one may use the fact that each cone of maximal dimension, given any integral vector in its interior, can be refined into a fan whose cones are all regular and which contains the cone generated by the given integral vector (see [9, Chap. III] or [2, Chap. VI]). Our function $h(\Sigma')$ is the height of Definition 2 above, which has the required properties since $h(\Sigma') \geq h(\Sigma)$ if $\Sigma' \prec \Sigma$ because the one dimensional cones of Σ are among those of Σ'.

Remark 2 There are of course many metrics compatible with the topology of the space of orders induced by that of $ZR(M)$ and another one is provided following [8, §1] and [12, Definition 1.2]: Let $\mathbf{B}(0, D)$ be the ball centered at 0 and with radius D in $\mathbf{R}^r$. Define the distance $\tilde{d}(w, w')$ to be 0 if $w = w'$ and otherwise $\frac{1}{D}$, where D is the largest integer such that w and w' induce the same order on $\mathbf{Z}^r \cap \mathbf{B}(0, D)$. It would be interesting to verify directly, perhaps using Siegel's Lemma (see [11]), that the distances $d(w, w')$ and $\tilde{d}(w, w')$ define the same topology.

Remark 3 The homeomorphism Z of Theorem 2 is the toric avatar of Zariski's homeomorphism between the space of valuations of a field of algebraic functions and the projective limit of the proper birational models of this field. As explained in [5, §13] the space of orders is the analogue for the theory of preorders of zero dimensional valuations in the theory of valuations.

Acknowledgements I am grateful to Hussein Mourtada for interesting discussions of the first topic of this note and for calling my attention to the phenomenon described in the case $p = 2$.

References

1. Campillo, A.: Algebroid Curves in Positive Characteristic. Springer Lecture Notes in Mathematics, vol. 813. Springer, Berlin/Heidelberg (1980)
2. Ewald, G.: Combinatorial Convexity and Algebraic Geometry. Graduate Texts in Mathematics, vol. 168. Springer, New York (1996)
3. Ewald, G., Ishida, M.-N.: Completion of real fans and Zariski-Riemann spaces. Tohôku Math. J. **58**(2), 189–218 (2006)
4. Eisenbud, D., Sturmfels, B.: Binomial ideals. Duke Math. J. **84**, 1–45 (1996)
5. González Pérez, P., Teissier, B.: Toric geometry and the semple-nash modification. Revista de la Real Academia de Ciencias Exactas, Físicas y Naturales, Seria A, Matemáticas **108**(1), 1–46 (2014)
6. Hocking, J.G., Young, G.S.: Topology. Dover (1988). ISBN:0-486-65676-4
7. Kedlaya, K.: On the algebraicity of generalized power series. Beitr. Algebra Geom. **58**(3), 499–527 (2017)
8. Kuroda, O.: The infiniteness of the SAGBI bases for certain invariant rings. Osaka J. Math. **39**, 665–680 (2002)
9. Kempf, G., Knudsen, F., Mumford, D., Saint-Donat, B.: Toroidal Embeddings I. Springer Lecture Notes in Mathematics, vol. 339. Springer, New York (1973)
10. Robbiano, L.: On the theory of graded structures. J. Symb. Comput. **2**, 139–170 (1986)
11. Schmidt, W.M.: Siegel's lemma and heights, chapter I. In: Masser, D., Nesterenko, Y.V., Schlickewei, H.P., Schmidt, W.M., Waldschmidt, M. (eds.) Diophantine Approximation. Lectures from the C.I.M.E. Summer School held in Cetraro, 28 June–6 July 2000. Edited by Amoroso, F., Zannier, U. (eds.) Fondazione CIME/CIME Foundation Subseries. Springer Lecture Notes in Mathematics, vol. 1819, pp. 1–33. Springer, Vienna (2000)
12. Sikora, A.: Topology on the spaces of orderings of groups. Bull. Lond. Math. Soc. **36**(4), 519–526 (2004)
13. Teissier, B.: Valuations, deformations, and toric geometry. In: Valuation Theory and Its Applications, vol. II, pp. 361–459. Fields Institute Communications 33. AMS, Providence (2003). Available at http://webusers.imj-prg.fr/~bernard.teissier/
14. Teissier, B.: Overweight deformations of affine toric varieties and local uniformization. In: Campillo, A., Kuhlmann, F.-V., Teissier, B. (eds.) Valuation Theory in Interaction. Proceedings of the Second International Conference on Valuation Theory, Segovia–El Escorial, 2011. European Mathematical Society Publishing House, Congress Reports Series, Sept 2014, pp. 474–565

On the Milnor Formula in Arbitrary Characteristic

Evelia R. García Barroso and Arkadiusz Płoski

Dedicated to Antonio Campillo on the occasion of his 65th birthday

Abstract The Milnor formula $\mu = 2\delta - r + 1$ relates the Milnor number μ, the double point number δ and the number r of branches of a plane curve singularity. It holds over the fields of characteristic zero. Melle and Wall based on a result by Deligne proved the inequality $\mu \geq 2\delta - r + 1$ in arbitrary characteristic and showed that the equality $\mu = 2\delta - r + 1$ characterizes the singularities with no wild vanishing cycles. In this note we give an account of results on the Milnor formula in characteristic p. It holds if the plane singularity is Newton non-degenerate (Boubakri et al. Rev. Mat. Complut. 25:61–85, 2010) or if p is greater than the intersection number of the singularity with its generic polar (Nguyen Annales de l'Institut Fourier, Tome 66(5):2047–2066, 2016). Then we improve our result on the Milnor number of irreducible singularities (Bull. Lond. Math. Soc. 48:94–98, 2016). Our considerations are based on the properties of polars of plane singularities in characteristic p.

2010 ***Mathematics Subject Classification*** Primary 14H20; Secondary 32S05.

E. R. García Barroso (✉)
Departamento de Matemáticas, Estadística e I.O. Sección de Matemáticas, Universidad de La Laguna, La Laguna, España
e-mail: ergarcia@ull.es

A. Płoski
Department of Mathematics and Physics, Kielce University of Technology, Kielce, Poland
e-mail: matap@tu.kielce.pl

G.-M. Greuel et al. (eds.), *Singularities, Algebraic Geometry, Commutative Algebra, and Related Topics*, https://doi.org/10.1007/978-3-319-96827-8_5

1 Introduction

John Milnor proved in his celebrated book [17] the formula

$$\mu = 2\delta - r + 1, \tag{M}$$

where μ is the Milnor number, δ the double point number and r the number of branches of a plane curve singularity. The Milnor's proof of (M) is based on topological considerations. A proof given by Risler [21] is algebraic and shows that (M) holds in characteristic zero.

On the other hand Melle and Wall based on a result by Deligne [5] proved the inequality $\mu \geq 2\delta - r + 1$ in arbitrary characteristic and showed that the Milnor formula holds if and only if the singularity has not wild *vanishing cycles* [16]. In the sequel we will call a *tame singularity* any plane curve singularity verifying (M).

Recently some papers on the singularities satisfying (M) in characteristic p appeared. In [1] the authors showed that planar Newton non-degenerate singularities are tame. Different notions of non-degeneracy for plane curve singularities are discussed in [10]. In [18] the author proved that if the characteristic p is greater than the kappa invariant then the singularity is tame. In [7] and [11] the case of irreducible singularities is investigated. Our aim is to give an account of the above-mentioned results.

In Sect. 2 we prove that any semi-quasihomogeneous singularity is tame. Our proof is different from that given in [1] and can be extended to the case of Kouchnirenko nondegenerate singularities ([1, Theorem 9]). In Sects. 3 and 4 we generalize Teissier's lemma ([22, Chap. II, Proposition 1.2]) relating the intersection number of the singularity with its polar and the Minor number to the case of arbitrary characteristic and reprove the result due to H.D. Nguyen [18, Corollary 3.2] in the following form: if $p > \mu(f) + \mathrm{ord}(f) - 1$ then the singularity is tame.

Section 5 is devoted to the strengthened version of our result on the Milnor number of irreducible singularities.

2 Semi-quasihomogeneous Singularities

Let $\mathbf{K}$ be an algebraically closed field of characteristic $p \geq 0$. For any formal power series $f \in \mathbf{K}[[x, y]]$ we denote by $\mathrm{ord}(f)$ (resp. $\mathrm{in}(f)$) the *order* (resp. the *initial form* of f). A power series $l \in \mathbf{K}[[x, y]]$ is called a *regular parameter* if $\mathrm{ord}(l) = 1$. A *plane curve singularity* (in short: *a singularity*) is a nonzero power series f of order greater than one. For any power series $f, g \in \mathbf{K}[[x, y]]$ we put $i_0(f, g) := \dim_{\mathbf{K}} \mathbf{K}[[x, y]]/(f, g)$ and called it the *intersection number* of f and g. The *Milnor number* of f is

$$\mu(f) := \dim_{\mathbf{K}} \mathbf{K}[[x, y]]/\left(\frac{\partial f}{\partial x}, \frac{\partial f}{\partial y}\right).$$

If Φ is an automorphism of $\mathbf{K}[[x, y]]$ then $\mu(f) = \mu(\Phi(f))$ (see [1, p. 62]). If the characteristic of $\mathbf{K}$ is $p = \operatorname{char}\mathbf{K} > 0$ then we can have $\mu(f) = +\infty$ and $\mu(uf) < +\infty$ for a unit $u \in \mathbf{K}[[x, y]]$ (take $f = x^p + y^{p-1}$ and $u = 1 + x$).

Let $f \in \mathbf{K}[[x, y]]$ be a reduced (without multiple factors) power series and consider a regular parameter $l \in \mathbf{K}[[x, y]]$. Assume that l does not divide f. We call the *polar of f with respect to l* the power series

$$\mathscr{P}_l(f) = \frac{\partial(f, l)}{\partial(x, y)} = \frac{\partial f}{\partial x}\frac{\partial l}{\partial y} - \frac{\partial f}{\partial y}\frac{\partial l}{\partial x}.$$

If $l = -bx + ay$ for $(a, b) \neq (0, 0)$ then $\mathscr{P}_l(f) = a\frac{\partial f}{\partial x} + b\frac{\partial f}{\partial y}$.

For any reduced power series f we put $\mathscr{O}_f = \mathbf{K}[[x, y]]/(f)$, $\overline{\mathscr{O}_f}$ the integral closure of $\mathscr{O}_f$ in the total quotient ring of $\mathscr{O}_f$ and $\delta(f) = \dim_{\mathbf{K}} \overline{\mathscr{O}_f}/\mathscr{O}_f$ (the double point number). Let $\mathscr{C}$ be the *conductor* of $\mathscr{O}_f$, that is the largest ideal in $\mathscr{O}_f$ which remains an ideal in $\overline{\mathscr{O}_f}$. We define $c(f) = \dim_{\mathbf{K}} \overline{\mathscr{O}_f}/\mathscr{C}$ (the *degree of conductor*) and $r(f)$ the number of irreducible factors of f. The *semigroup $\Gamma(f)$* associated with the irreducible power series f is defined as the set of intersection numbers $i_0(f, h)$, where h runs over power series such that $h \not\equiv 0 \pmod f$.

The degree of conductor $c(f)$ is equal to the smallest element c of $\Gamma(f)$ such that $c + N \in \Gamma(f)$ for all integers $N \geq 0$ (see [2, 9]).

For any reduced power series f we define

$$\overline{\mu}(f) := c(f) - r(f) + 1.$$

In particular, if f is irreducible then $\overline{\mu}(f) = c(f)$.

Proposition 2.1 *Let $f = f_1 \cdots f_r \in \mathbf{K}[[x, y]]$ be a reduced power series, where f_i is irreducible for $i = 1, \ldots, r$. Then*

(i) $\overline{\mu}(f) = \overline{\mu}(uf)$ *for any unit u of* $\mathbf{K}[[x, y]]$.

(ii)

$$\overline{\mu}(f) + r - 1 = \sum_{i=1}^{r} \overline{\mu}(f_i) + 2 \sum_{1 \leq i < j \leq r} i_0(f_i, f_j).$$

(iii) *Let l be a regular parameter such that $i_0(f_i, l) \not\equiv 0 \pmod p$ for $i = 1, \ldots, r$. Then*

$$i_0(f, \mathscr{P}_l(f)) = \overline{\mu}(f) + i_0(f, l) - 1.$$

(iv) $\overline{\mu}(f) = \mu(f)$ *if and only if* $\mu(f) = 2\delta(f) - r(f) + 1$.

(v) $\overline{\mu}(f) \geq 0$ *and* $\overline{\mu}(f) = 0$ *if and only if* $\operatorname{ord}(f) = 1$.

Proof Property (i) is obvious. To check (ii) observe that

$$\sum_{i=1}^{r}\overline{\mu}(f_i)+2\sum_{1\le i<j\le r} i_0(f_i,f_j)=\sum_{i=1}^{r}c(f_i)+2\sum_{1\le i<j\le r} i_0(f_i,f_j)=c(f)=\overline{\mu}(f)+r-1,$$

by [3, Lemma 2.1, p. 381]. Property (iii) in the case $r = 1$ reduces to the Dedekind formula $i_0(f, \mathscr{P}_l(f)) = c(f) + i_0(f, l) - 1$ provided that $i_0(f, l) \not\equiv 0 \pmod p$ [7, Lemma 3.1]. To check the general case we apply the Dedekind formula to the irreducible factors f_i of f and we get

$$\begin{aligned}
i_0(f,\mathscr{P}_l(f)) &= \sum_{i=1}^{r} i_0(f_i,\mathscr{P}_l(f)) = \sum_{i=1}^{r} i_0\left(f_i,\mathscr{P}_l(f_i)\frac{f}{f_i}\right)\\
&= \sum_{i=1}^{r}\left(i_0(f_i,\mathscr{P}_l(f_i)) + \sum_{j\neq i} i_0(f_i,f_j)\right)\\
&= \sum_{i=1}^{r}\left(\overline{\mu}(f_i) + i_0(f_i,l) - 1 + \sum_{j\neq i} i_0(f_i,f_j)\right)\\
&= \sum_{i=1}^{r}\overline{\mu}(f_i) + 2\sum_{1\le i<j\le r} i_0(f_i,f_j) + i_0(f,l) - r\\
&= \overline{\mu}(f) + r - 1 + i_0(f,l) - r = \overline{\mu}(f) + i_0(f,l) - 1.
\end{aligned}$$

Property (iv) follows since $c(f) = 2\delta(f)$ for any reduced power series f by the Gorenstein theorem (see for example [20, Section 5]).

Now we prove Property (v). If f is irreducible then $\overline{\mu}(f) = c(f) \ge 0$ with equality if and only if $\operatorname{ord}(f) = 1$. Suppose that $r > 1$. Then by (ii) we get

$$\overline{\mu}(f) + r - 1 \ge 2\sum_{1\le i<j\le r} i_0(f_i,f_j) \ge r(r-1)$$

and $\overline{\mu}(f) \ge (r-1)^2 > 0$, which proves (v).

Remark 2.2 Using Proposition 2.1(ii) we check the following property:

Let $f = g_1 \cdots g_s \in \mathbf{K}[[x, y]]$ be a reduced power series, where the power series g_i for $i = 1, \ldots, s$ are pairwise coprime. Then

$$\overline{\mu}(f) + s - 1 = \sum_{i=1}^{s}\overline{\mu}(g_i) + 2\sum_{1\le i<j\le s} i_0(g_i,g_j).$$

Let $\overrightarrow{w} = (n, m) \in (\mathbf{N}_+)^2$ be a pair of strictly positive integers. In the sequel we call $\overrightarrow{w}$ a *weight*.

Let $f = \sum c_{\alpha\beta} x^\alpha y^\beta \in \mathbf{K}[[x, y]]$ be a power series. Then

- the $\overrightarrow{w}$ *-order* of f is $\mathrm{ord}_{\overrightarrow{w}}(f) = \inf\{\alpha n + \beta m \ : \ c_{\alpha\beta} \neq 0\}$,
- the $\overrightarrow{w}$*-initial form* of f is $\mathrm{in}_{\overrightarrow{w}}(f) = \sum_{\alpha n + \beta m = w} c_{\alpha\beta} x^\alpha y^\beta$, where $w = \mathrm{ord}_{\overrightarrow{w}}(f)$,
- $R_{\overrightarrow{w}}(f) = f - \mathrm{in}_{\overrightarrow{w}}(f)$.

Thus $R_{\overrightarrow{w}}(f)$ is a power series of $\overrightarrow{w}$-order greater than $\mathrm{ord}_{\overrightarrow{w}}(f)$.

Note that $\mathrm{ord}_{\overrightarrow{w}}(x) = n$ and $\mathrm{ord}_{\overrightarrow{w}}(y) = m$.

A power series f is *semi-quasihomogeneous* (with respect to $\overrightarrow{w}$) if the system of equations

$$\begin{cases} \frac{\partial}{\partial x} \mathrm{in}_{\overrightarrow{w}}(f) = 0, \\ \\ \frac{\partial}{\partial y} \mathrm{in}_{\overrightarrow{w}}(f) = 0 \end{cases}$$

has the only solution $(x, y) = (0, 0)$.

A power series f is *convenient* if $f(x, 0) \cdot f(0, y) \neq 0$.

Suppose that $\mathrm{in}_{\overrightarrow{w}}(f)$ is convenient and the line $\alpha n + \beta m = \mathrm{ord}_{\overrightarrow{w}}(f)$ intersects the axes in points $(m, 0)$ and $(0, n)$. Let $d = \gcd(m, n)$. Then $\mathrm{in}_{\overrightarrow{w}}(f) = F(x^{m/d}, y^{n/d})$, where $F(u, v) \in \mathbf{K}[u, v]$ is a homogeneous polynomial of degree d.

Proposition 2.3 *Suppose that* $\mathrm{in}_{\overrightarrow{w}}(f)$ *has no multiple factors. Then*

$$\overline{\mu}(f) = \left(\frac{\mathrm{ord}_{\overrightarrow{w}}(f)}{n} - 1\right) \cdot \left(\frac{\mathrm{ord}_{\overrightarrow{w}}(f)}{m} - 1\right).$$

Proof In the proof we will use lemmas collected in the Appendix.

Observe that if $\mathrm{in}_{\overrightarrow{w}}(f)$ has no multiple factors then $\mathrm{in}_{\overrightarrow{w}}(f) = m_{\overrightarrow{w}}(f) \left(\mathrm{in}_{\overrightarrow{w}}(f)\right)^o$, where $m_{\overrightarrow{w}}(f) \in \{1, x, y, xy\}$ and $\left(\mathrm{in}_{\overrightarrow{w}}(f)\right)^o$ is a convenient power series or a constant. To prove the proposition we will use Hensel's Lemma (see Lemma A.3) and Remark 2.2. We have to consider several cases.

Case 1: $\mathrm{in}_{\overrightarrow{w}}(f) = (\mathrm{const}) \cdot x$ or $\mathrm{in}_{\overrightarrow{w}}(f) = (\mathrm{const}) \cdot y$.

In this case $\mathrm{ord}(f) = 1$ and by Proposition 2.1*(v)* $\overline{\mu}(f) = 0$. If $\mathrm{in}_{\overrightarrow{w}}(f) = (\mathrm{const}) \cdot x$ (resp. $\mathrm{in}_{\overrightarrow{w}}(f) = (\mathrm{const}) \cdot y$) then $\mathrm{ord}_{\overrightarrow{w}}(f) = n$ (resp. $\mathrm{ord}_{\overrightarrow{w}}(f) = m$) and

$$\left(\frac{\mathrm{ord}_{\overrightarrow{w}}(f)}{n} - 1\right)\left(\frac{\mathrm{ord}_{\overrightarrow{w}}(f)}{m} - 1\right) = 0.$$

Case 2: $\quad \mathrm{in}_{\vec{w}}(f) = (\mathrm{const}) \cdot xy.$

By Hensel's Lemma (see Lemma A.3) $f = f_1 f_2$, where $\mathrm{in}_{\vec{w}}(f_1) = c_1 x$, $\mathrm{in}_{\vec{w}}(f_2) = c_2 y$ with constants $c_1, c_2 \neq 0$. Using Remark 2.2 and Lemma A.1 we get

$$\overline{\mu}(f) + 1 = \overline{\mu}(f_1 f_2) + 1 = \overline{\mu}(f_1) + \overline{\mu}(f_2) + 2i_0(f_1, f_2) = 0 + 0 + 2.1$$

and $\overline{\mu}(f)$ 1. On the other hand $\mathrm{ord}_{\vec{w}}(f) = n + m$ and

$$\left(\frac{\mathrm{ord}_{\vec{w}}(f)}{n} - 1\right)\left(\frac{\mathrm{ord}_{\vec{w}}(f)}{m} - 1\right) = 1.$$

Case 3: $\quad$ The power series $\mathrm{in}_{\vec{w}} f$ is convenient.

Assume additionally that the line $n\alpha + m\beta = \mathrm{ord}_{\vec{w}}(f)$ intersects the axes in points $(m, 0)$ and $(0, n)$. Let $d = \gcd(n, m)$. Then the $\vec{w}$-initial form of f is

$$\mathrm{in}_{\vec{w}} f = \prod_{i=1}^{d} \left(a_i x^{m/d} + b_i y^{n/d}\right),$$

where $a_i x^{m/d} + b_i y^{n/d}$ are pairwise coprime. By Hensel's Lemma (see Lemma A.3) we get a factorization $f = \prod_{i=1}^{d} f_i$, where $\mathrm{in}_{\vec{w}} f_i = a_i x^{m/d} + b_i y^{n/d}$ for $i = 1, \dots, d$. The factors f_i are irreducible with semigroup $\Gamma(f_i) = \frac{m}{d}\mathbf{N} + \frac{n}{d}\mathbf{N}$ and

$$\overline{\mu}(f_i) = c(f_i) = \left(\frac{m}{d} - 1\right)\left(\frac{n}{d} - 1\right)$$

(see, for example [6]). Moreover by Lemma A.1 we have

$$i_0(f_i, f_j) = \frac{\mathrm{ord}_{\vec{w}} f_i \, \mathrm{ord}_{\vec{w}} f_j}{mn} = \frac{mn}{d^2}, \text{ for } i \neq j$$

and we get by Proposition 2.1(ii)

$$\overline{\mu}(f) + d - 1 = \sum_{i=1}^{d} \overline{\mu}(f_i) + 2 \sum_{1 \le i < j \le d} i_0(f_i, f_j) = d\left(\frac{m}{d} - 1\right)\left(\frac{n}{d} - 1\right) + 2\frac{d(d-1)}{2}\frac{mn}{d^2},$$

which implies $\overline{\mu}(f) = (m-1)(n-1) = \left(\frac{\mathrm{ord}_{\vec{w}} f}{n} - 1\right)\left(\frac{\mathrm{ord}_{\vec{w}} f}{m} - 1\right)$, since the weighted order of f is $\mathrm{ord}_{\vec{w}} f = mn$.

Now consider the general case, that is when the line $n\alpha + m\beta = \mathrm{ord}_{\vec{w}}(f)$ intersects the axes in points $(m_1, 0) = \left(\frac{\mathrm{ord}_{\vec{w}} f}{n}, 0\right)$ and $(0, n_1) = \left(0, \frac{\mathrm{ord}_{\vec{w}}(f)}{m}\right)$. Then f is semi-quasihomogeneous with respect to $\vec{w_1} = (n_1, m_1)$ and the line $n_1\alpha + m_1\beta = \mathrm{ord}_{\vec{w_1}}(f)$ intersects the axes in points $(m_1, 0)$ and $(0, n_1)$. By the first part of the proof we get

$$\overline{\mu}(f) = (m_1 - 1)(n_1 - 1) = \left(\frac{\mathrm{ord}_{\overrightarrow{w}}(f)}{n} - 1\right)\left(\frac{\mathrm{ord}_{\overrightarrow{w}}(f)}{m} - 1\right),$$

which proves the proposition in Case 3.

Case 4: $\mathrm{in}_{\overrightarrow{w}}(f) = x\left(\mathrm{in}_{\overrightarrow{w}}(f)\right)^o$ or $\mathrm{in}_{\overrightarrow{w}}(f) = y\left(\mathrm{in}_{\overrightarrow{w}}(f)\right)^o$, where $\left(\mathrm{in}_{\overrightarrow{w}}(f)\right)^o$ is convenient.

This case follows from Hensel's Lemma (Lemma A.3), Cases 1 and 3.

Case 5: $\mathrm{in}_{\overrightarrow{w}}(f) = xy\left(\mathrm{in}_{\overrightarrow{w}}(f)\right)^o$, where $\left(\mathrm{in}_{\overrightarrow{w}}(f)\right)^o$ is convenient.

This case follows from Hensel's Lemma (Lemma A.3), Cases 2 and 3.

Theorem 2.4 *Suppose that* $\mathrm{in}_{\overrightarrow{w}}(f)$ *has no multiple factors. Then* f *is tame if and only if* f *is a semi-quasihomogeneous singularity with respect to* $\overrightarrow{w}$.

Proof We have $\overline{\mu}(f) = \left(\frac{\mathrm{ord}_{\overrightarrow{w}}(f)}{n} - 1\right)\left(\frac{\mathrm{ord}_{\overrightarrow{w}}(f)}{m} - 1\right)$ by Proposition 2.3. On the other hand, by Lemma A.2, we get that $\mu(f) = \left(\frac{\mathrm{ord}_{\overrightarrow{w}}(f)}{n} - 1\right)\left(\frac{\mathrm{ord}_{\overrightarrow{w}}(f)}{m} - 1\right)$ if and only if the system of equations

$$\begin{cases} \frac{\partial}{\partial x}\mathrm{in}_{\overrightarrow{w}}(f) = 0, \\ \frac{\partial}{\partial y}\mathrm{in}_{\overrightarrow{w}}(f) = 0 \end{cases}$$

has the only solution $(x, y) = (0, 0)$. The theorem follows from Proposition 2.1(iv).

Example 2.5 Let $f(x, y) = x^m + y^n + \sum_{\alpha n + \beta m > nm} c_{\alpha\beta} x^\alpha y^\beta$ and let $d = \gcd(m, n)$. Then $\mathrm{in}_{\overrightarrow{w}}(f) = x^m + y^n$ has no multiple factors if and only if $d \not\equiv 0$ (mod p). If $d \not\equiv 0$ (mod p) then f is tame if and only if $m \not\equiv 0$ (mod p) and $n \not\equiv 0$ (mod p).

Corollary 2.6 *The semi-quasihomogeneous singularities are tame.*

Corollary 2.6 is a particular case of the following

Theorem 2.7 (Boubakri, Greuel, Markwig [1, Theorem 9].) *The planar Newton non-degenerate singularities are tame.*

3 Teissier's Lemma in Characteristic $p \geq 0$

The intersection theoretical approach to the Milnor number in characteristic zero [4] is based on a lemma due to Teissier who proved a more general result (the case of hypersurfaces) in [22, Chapter II, Proposition 1.2]. A general formula on isolated complete intersection singularity is due to Greuel [8] and Lê [14]. In this section we study Teissier's Lemma in arbitrary characteristic $p \geq 0$.

Let $f \in \mathbf{K}[[x, y]]$ be a reduced power series and $l \in \mathbf{K}[[x, y]]$ be a regular parameter. Assume that l does not divide f and consider the polar $\mathscr{P}_l(f) = \frac{\partial f}{\partial x}\frac{\partial l}{\partial y} - \frac{\partial f}{\partial y}\frac{\partial l}{\partial x}$ of f with respect to l. In this section we assume, without loss of generality, that $\mathrm{ord}(l(0, y)) = 1$.

Lemma 3.1 *Let $f \in \mathbf{K}[[x, y]]$ be a reduced power series and $l \in \mathbf{K}[[x, y]]$ be a regular parameter. Then $i_0(l, \mathscr{P}_l(f)) \geq i_0(f, l) - 1$ with equality if and only if $i_0(f, l) \not\equiv 0 \pmod{p}$.*

Proof Recall that $\mathrm{ord}(l(0, y)) = 1$. Let $\phi(t) = (\phi_1(t), \phi_2(t))$ be a good parametrization of the curve $l(x, y) = 0$ (see [19, Section 2]). In particular $0 = l(\phi(t))$ so $\frac{d}{dt}l(\phi(t)) = 0$. On the other hand we have $\mathrm{ord}(\phi_1(t)) = i_0(x, l) = \mathrm{ord}(l(0, y)) = 1$ and $\phi_1'(0) \neq 0$. Differentiating $f(\phi(t))$ and $l(\phi(t))$ we get

$$\frac{d}{dt}f(\phi(t)) = \frac{\partial f}{\partial x}(\phi(t))\phi_1'(t) + \frac{\partial f}{\partial y}(\phi(t))\phi_2'(t) \tag{1}$$

and

$$0 = \frac{d}{dt}l(\phi(t)) = \frac{\partial l}{\partial x}(\phi(t))\phi_1'(t) + \frac{\partial l}{\partial y}(\phi(t))\phi_2'(t). \tag{2}$$

From (2) we have $\frac{\partial l}{\partial x}(\phi(t))\phi_1'(t) = -\frac{\partial l}{\partial y}(\phi(t))\phi_2'(t)$ and by (1) and the definition of $\mathscr{P}_l(f)$ we get

$$\mathscr{P}_l(f)(\phi(t))\phi_1'(t) = \frac{d}{dt}f(\phi(t))\frac{\partial l}{\partial y}(\phi(t)).$$

Since $\phi_1'(t)$ and $\frac{\partial l}{\partial y}(\phi(t))$ are units in $\mathbf{K}[[t]]$ we have

$$\mathrm{ord}(\mathscr{P}_l(f)(\phi(t))) = \mathrm{ord}\left(\frac{d}{dt}f(\phi(t))\right) \geq \mathrm{ord}(f(\phi(t))) - 1,$$

with equality if and only if $\mathrm{ord}(f(\phi(t))) \not\equiv 0 \pmod{p}$. Now the lemma follows from the formula $i_0(h, l) = \mathrm{ord}(h(\phi(t)))$ which holds for every power series $h \in \mathbf{K}[[x, y]]$.

Corollary 3.2 *Suppose that $i_0(f, l) = \mathrm{ord}(f) \not\equiv 0 \pmod{p}$ for a regular parameter $l \in \mathbf{K}[[x, y]]$. Then*

(a) $i_0(l, \mathscr{P}_l(f)) = \mathrm{ord}(f) - 1$,
(b) $\mathrm{ord}(\mathscr{P}_l(f)) = \mathrm{ord}(f) - 1$,
(c) *if h is an irreducible factor of $\mathscr{P}_l(f)$ then $i_0(l, h) = \mathrm{ord}(h)$.*

Proof Property (a) follows immediately from Lemma 3.1. To check (b) observe that we get $\mathrm{ord}(\mathscr{P}_l(f)) = \mathrm{ord}(\mathscr{P}_l(f)) \cdot \mathrm{ord}(l) \leq i_0(l, \mathscr{P}_l(f)) = \mathrm{ord}(f) - 1$, where the last equality follows from (a). The inequality $\mathrm{ord}(\mathscr{P}_l(f)) \geq \mathrm{ord}(f) - 1$ is obvious.

Let $\mathscr{P}_l(f)=\prod_{i=1}^{s} h_i$, where h_i is irreducible. From (a) and (b) we get

$$0=i_0(l,\mathscr{P}_l(f))-\mathrm{ord}(\mathscr{P}_l(f))=\sum_{i=1}^{s}(i_0(l,h_i)-\mathrm{ord}(h_i)).$$

Since $i_0(l,h_i)\geq \mathrm{ord}(h_i)$ we have $i_0(l,h_i)=\mathrm{ord}(h_i)$ for $i=1,\dots,s$ which proves (c).

Proposition 3.3 (Teissier's Lemma in characteristic p.**)** *Let* $f\in \mathbf{K}[[x,y]]$ *be a reduced power series. Suppose that*

(i) $i_0(f,l)\not\equiv 0 \pmod p$,
(ii) *for any irreducible factor* h *of* $\mathscr{P}_l(f)$ *we get* $i_0(l,h)\not\equiv 0 \pmod p$.

Then

$$i_0(f,\mathscr{P}_l(f))\leq \mu(f)+i_0(f,l)-1$$

with equality if and only if

(iii) *for any irreducible factor* h *of* $\mathscr{P}_l(f)$ *we get* $i_0(f,h)\not\equiv 0 \pmod p$.

Proof Fix an irreducible factor h of $\mathscr{P}_l(f)$ and let $\psi(t)=(\psi_1(t),\psi_2(t))$ be a good parametrization of the curve $h(x,y)=0$. Then $\mathrm{ord}(l(\psi(t)))=i_0(l,h)\not\equiv 0 \pmod p$ by (ii) and $\mathrm{ord}\left(\frac{d}{dt}l(\psi(t))\right)=\mathrm{ord}(l(\psi(t)))-1$. Differentiating $f(\psi(t))$ and $l(\psi(t))$ we get

$$\frac{d}{dt}f(\psi(t))=\frac{\partial f}{\partial x}(\psi(t))\psi_1'(t)+\frac{\partial f}{\partial y}(\psi(t))\psi_2'(t), \tag{3}$$

and

$$\frac{d}{dt}l(\psi(t))=\frac{\partial l}{\partial x}(\psi(t))\psi_1'(t)+\frac{\partial l}{\partial y}(\psi(t))\psi_2'(t). \tag{4}$$

Since $\mathscr{P}_l(f)(\psi(t))=0$, it follows from (3) and (4) that

$$\frac{d}{dt}f(\psi(t))\frac{\partial l}{\partial y}(\psi(t))=\frac{d}{dt}l(\psi(t))\frac{\partial f}{\partial y}(\psi(t)). \tag{5}$$

Since $\frac{\partial l}{\partial y}(\psi(t))$ is a unit in $\mathbf{K}[[t]]$, taking orders in (5) we have

$$\begin{aligned}\mathrm{ord}(f(\psi(t)))-1\leq \mathrm{ord}\left(\frac{d}{dt}f(\psi(t))\right)&=\mathrm{ord}\left(\frac{d}{dt}l(\psi(t))\right)+\mathrm{ord}\left(\frac{\partial f}{\partial y}(\psi(t))\right)\\&=\mathrm{ord}(l(\psi(t)))-1+\mathrm{ord}\left(\frac{\partial f}{\partial y}(\psi(t))\right),\end{aligned}$$

where the last equality follows from $\operatorname{ord}(l(\psi(t))) \not\equiv 0 \pmod p$.

Hence $i_0(f,h) \le i_0(l,h) + i_0\left(\frac{\partial f}{\partial y}, h\right)$.

Summing up over all h counted with multiplicities as factors of $\mathscr{P}_l(f)$ we obtain

$$i_0(f, \mathscr{P}_l(f)) \le i_0(l, \mathscr{P}_l(f)) + i_0\left(\frac{\partial f}{\partial y}, \mathscr{P}_l(f)\right). \tag{6}$$

By Lemma 3.1 and assumption (i) we have $i_0(l, \mathscr{P}_l(f)) = i_0(f,l) - 1$. Moreover $i_0\left(\frac{\partial f}{\partial y}, \mathscr{P}_l(f)\right) = \mu(f)$ since $\operatorname{ord}(l(0,y)) = 1$ and we get from the equality (6)

$$i_0(f, \mathscr{P}_l(f)) \le \mu(f) + i_0(f,l) - 1.$$

The equality holds if and only if $i_0(f,h) = i_0(l,h) + i_0\left(\frac{\partial f}{\partial y}, h\right)$ for every h, which is equivalent to the condition $i_0(f,h) \not\equiv 0 \pmod p$, since $i_0(f,h) \not\equiv 0 \pmod p$ if and only if $\operatorname{ord}\left(\frac{d}{dt} f(\psi(t))\right) = \operatorname{ord}(f(\psi(t))) - 1$.

Corollary 3.4 (Teissier [22, Chapter II, Proposition 1.2]) *If* $\operatorname{char} \mathbf{K} = 0$ *then*

$$i_0(f, \mathscr{P}_l(f)) = \mu(f) + i_0(f,l) - 1.$$

Corollary 3.5 *Suppose that* $p = \operatorname{char} \mathbf{K} > \operatorname{ord}(f)$ *and let* $i_0(f,l) = \operatorname{ord}(f)$. *Then*

$$i_0(\mathscr{P}_l(f), f) \le \mu(f) + i_0(f,l) - 1.$$

The equality holds if and only if for any irreducible factor h *of* $\mathscr{P}_l(f)$ *we get* $i_0(f,h) \not\equiv 0 \pmod p$.

Proof If $\operatorname{ord}(f) < p$ then $i_0(f,l) = \operatorname{ord}(f) \not\equiv 0 \pmod p$ and by Corollary 3.2 for any irreducible factor h of $\mathscr{P}_l(f)$ we get

$$i_0(l,h) = \operatorname{ord}(h) \le \operatorname{ord}(\mathscr{P}_l(f)) = \operatorname{ord}(f) - 1 < p.$$

Hence $i_0(l,h) \not\equiv 0 \pmod p$ and the corollary follows from Proposition 3.3.

Example 3.6 Let $f = x^{p+2} + y^{p+1} + x^{p+1}y$, where $p = \operatorname{char} K > 2$. Take $l = y$. Then $i_0(f,l) = p+2 \not\equiv 0 \pmod p$, $\mathscr{P}_l(f) = \frac{\partial f}{\partial x} = x^p(2x+y)$ and the irreducible factors of $\mathscr{P}_l(f)$ are $h_1 = x$ and $h_2 = 2x+y$. Clearly $i_0(l,h_1) = i_0(l,h_2) = 1 \not\equiv 0 \pmod p$. Moreover $i_0(f,h_1) = i_0(f,h_2) = p+1$ and all assumptions of Proposition 3.3 are satisfied.

Hence $i_0(f, \mathscr{P}_l(f)) = \mu(f) + i_0(f,l) - 1$ and $\mu(f) = i_0(f, \mathscr{P}_l(f)) - i_0(f,l) + 1 = p(p+1)$. Note that $l = 0$ is a curve of maximal contact with $f = 0$. Let $l_1 = x$. Then $i_0(f,l_1) = \operatorname{ord}(f) = p+1$, $\mathscr{P}_{l_1}(f) = -(y^p + x^{p+1})$ and $h = y^p + x^{p+1}$ is the only irreducible factor of the polar $\mathscr{P}_{l_1}(f)$. Since $i_0(l_1,h) = p$, the condition (ii) of Proposition 3.3 is not satisfied. However, $i_0(f, \mathscr{P}_{l_1}(f)) = \mu(f) + i_0(f,l_1) - 1$, which we check directly.

4 Tame Singularities

Assume that f is a plane curve singularity.

Proposition 4.1 *Let $f = f_1 \cdots f_r \in \mathbf{K}[[x, y]]$ be a reduced power series, where f_i is irreducible for $i = 1, \ldots, r$. Suppose that there exists a regular parameter l such that $i_0(f_i, l) \not\equiv 0 \pmod p$ for $i = 1, \ldots, r$. Then f is tame if and only if Teissier's lemma holds, that is if $i_0(f, \mathscr{P}_l(f)) = \mu(f) + i_0(f, l) - 1$.*

Proof By Proposition 2.1(iii) we have that $i_0(f, \mathscr{P}_l(f)) = \overline{\mu}(f) + i_0(f, l) - 1$. Thus $i_0(f, \mathscr{P}_l(f)) = \mu(f) + i_0(f, l) - 1$ if and only if $\mu(f) = \overline{\mu}(f)$. We finish the proof using Proposition 2.1(iv).

Proposition 4.2 (Milnor [17], Risler [21]) *If* $\operatorname{char} \mathbf{K} = 0$ *then any plane singularity is tame.*

Proof Teissier's Lemma holds by Corollary 3.4. Use Proposition 4.1.

Proposition 4.3 *Let $p = \operatorname{char} \mathbf{K} > 0$. Suppose that $p > \operatorname{ord}(f)$. Let l be a regular parameter such that $i_0(f, l) = \operatorname{ord}(f)$. Then f is tame if and only if for any irreducible factor h of $\mathscr{P}_l(f)$ we get $i_0(f, h) \not\equiv 0 \pmod p$.*

Proof Take a regular parameter l such that $i_0(f, l) = \operatorname{ord}(f)$. By hypothesis we get $i_0(f, l) < p$ so $i_0(f, l) \not\equiv 0 \pmod p$. By Corollary 3.2 the assumption (ii) of Proposition 3.3 is satisfied.

Hence $i_0(f, \mathscr{P}_l(f)) \leq \mu(f) + i_0(f, l) - 1$ with equality if and only if $i_0(f, h) \not\equiv 0 \pmod p$ for any irreducible factor h of $\mathscr{P}_l(f)$. Use Proposition 4.1.

Proposition 4.4 (Nguyen [18]) *Let $p = \operatorname{char} \mathbf{K} > 0$. Suppose that there exists a regular parameter l such that $i_0(f, l) = \operatorname{ord}(f)$ and $i_0(f, \mathscr{P}_l(f)) < p$. Then f is tame.*

Proof We have $p > i_0(f, \mathscr{P}_l(f)) \geq \operatorname{ord}(f) \cdot \operatorname{ord}(\mathscr{P}_l(f))$. Hence $p > \operatorname{ord}(f)$ and we may apply Proposition 4.3. Since $i_0(f, \mathscr{P}_l(f)) < p$ for any irreducible factor h of $\mathscr{P}_l(f)$ we have that $i_0(f, h) < p$ and obviously $i_0(f, h) \not\equiv 0 \pmod p$. The proposition follows from Proposition 4.3.

Theorem 4.5 (Nguyen [18]) *If $p > \mu(f) + \operatorname{ord}(f) - 1$ then f is tame.*

Proof Since f is a singularity we get $\mu(f) > 0$ and by hypothesis the characteristic of the field verifies $p > \mu(f) - 1 + \operatorname{ord}(f) \geq \operatorname{ord}(f)$. By the first part of the proof of Proposition 4.3 we have $i_0(f, \mathscr{P}_l(f)) \leq \mu(f) + \operatorname{ord}(f) - 1$, where l is a regular parameter such that $i_0(f, l) = \operatorname{ord}(f)$. Hence $i_0(f, \mathscr{P}_l(f)) < p$ and the theorem follows from Proposition 4.4.

5 The Milnor Number of Plane Irreducible Singularities

Let $f \in \mathbf{K}[[x, y]]$ be an irreducible power series of order $n = \mathrm{ord}(f)$ and let $\Gamma(f)$ be the semigroup associated with $f = 0$.

Let $\overline{\beta_0}, \ldots, \overline{\beta_g}$ be the minimal sequence of generators of $\Gamma(f)$ defined by the conditions

- $\overline{\beta_0} = \min(\Gamma(f)\backslash\{0\}) = \mathrm{ord}(f) = n$,
- $\overline{\beta_k} = \min(\Gamma(f)\backslash\mathbf{N}\overline{\beta_0} + \cdots + \mathbf{N}\overline{\beta_{k-1}})$ for $k \in \{1, \ldots, g\}$,
- $\Gamma(f) = \mathbf{N}\overline{\beta_0} + \cdots + \mathbf{N}\overline{\beta_g}$.

Let $e_k = \gcd(\overline{\beta_0}, \ldots, \overline{\beta_k})$ for $k \in \{1, \ldots, g\}$. Then $n = e_0 > e_1 > \cdots e_{g-1} > e_g = 1$. Let $n_k = e_{k-1}/e_k$ for $k \in \{1, \ldots, g\}$. We have $n_k > 1$ for $k \in \{1, \ldots, g\}$ and $n = n_1 \cdots n_g$. Let $n^* = \max(n_1, \ldots, n_g)$. Then $n^* \leq n$ with equality if and only if $g = 1$.

The following theorem is a sharpened version of the main result of [7].

Theorem 5.1 *Let* $f \in \mathbf{K}[[x, y]]$ *be an irreducible power series of order* $n > 1$ *and let* $\overline{\beta_0}, \ldots, \overline{\beta_g}$ *be the minimal system of generators of* $\Gamma(f)$. *Suppose that* $p = \mathrm{char}\,\mathbf{K} > n^*$. *Then the following two conditions are equivalent:*

(i) $\overline{\beta_k} \not\equiv 0 \pmod p$ *for* $k \in \{1, \ldots, g\}$,
(ii) f *is tame.*

In [7] the equivalence of (i) and (ii) is proved under the assumption that $p > n$.

If $f \in \mathbf{K}[[x, y]]$ is an irreducible power series then we get $\mathrm{ord}(f(x, 0)) = \mathrm{ord}(f)$ or $\mathrm{ord}(f(0, y)) = \mathrm{ord}(f)$. In the sequel we assume that $\mathrm{ord}(f(0, y)) = \mathrm{ord}(f) = n$. The proof of Theorem 5.1 is based on Merle's factorization theorem:

Theorem 5.2 (Merle [15], García Barroso-Płoski [7]) *Suppose that* $\mathrm{ord}(f(0, y)) = \mathrm{ord}(f) = n \not\equiv 0 \pmod p$. *Then* $\frac{\partial f}{\partial y} = h_1 \cdots h_g$ *in* $\mathbf{K}[[x, y]]$, *where*

(a) $\mathrm{ord}(h_k) = \frac{n}{e_k} - \frac{n}{e_{k-1}}$ *for* $k \in \{1, \ldots, g\}$.
(b) *If* $h \in \mathbf{K}[[x, y]]$ *is an irreducible factor of* h_k, $k \in \{1, \ldots, g\}$, *then*

 (b1) $\frac{i_0(f,h)}{\mathrm{ord}(h)} = \frac{e_{k-1}\overline{\beta_k}}{n}$, *and*
 (b2) $\mathrm{ord}(h) \equiv 0 \left(mod\ \frac{n}{e_{k-1}}\right)$.

Lemma 5.3 *Suppose that* $p > n^*$. *Then* $i_0\left(f, \frac{\partial f}{\partial y}\right) \leq \mu(f) + \mathrm{ord}(f) - 1$ *with equality if and only if* $\overline{\beta_k} \not\equiv 0 \pmod p$ *for* $k \in \{0, \ldots, g\}$.

Proof Obviously $n_k \not\equiv 0 \pmod p$ for $k = 1, \ldots, g$ and $n = n_1 \cdots n_g \not\equiv 0 \pmod p$. Let h be an irreducible factor of $\frac{\partial f}{\partial y}$. Then, by Corollary 3.2(c) $i_0(h, x) = \mathrm{ord}(h)$. By Theorem 5.2 (b2) $\mathrm{ord}(h) = m_k \frac{n}{e_{k-1}}$, for an index $k \in \{1, \ldots, g\}$, where $m_k \geq 1$ is an integer. Hence $m_k \frac{n}{e_{k-1}} = \mathrm{ord}(h) \leq \mathrm{ord}(h_k) = \frac{n}{e_{k-1}}(n_k - 1)$ and $m_k \leq n_k - 1 < n_k < p$, which implies $m_k \not\equiv 0 \pmod p$ and $\mathrm{ord}(h) \not\equiv 0 \pmod p$. By Proposition 3.3 we get $i_0\left(f, \frac{\partial f}{\partial y}\right) \le \mu(f) + \mathrm{ord}(f) - 1$. By Theorem 5.2 (b1) we have the equalities $i_0(f, h) = \left(\frac{e_{k-1}\overline{\beta_k}}{n}\right) \mathrm{ord}(h) = m_k\overline{\beta_k}$ and we get $i_0(f, h) \not\equiv 0 \pmod{p}$ if and only if $\overline{\beta_k} \not\equiv 0 \pmod{p}$, which proves the second part of Lemma 5.3.

Proof of Theorem 5.1 Use Lemma 5.3 and Proposition 4.1. ■

Example 5.4 Let $f(x, y) = (y^2 + x^3)^2 + x^5 y$. Then f is irreducible and its semigroup is $\Gamma(f) = 4\mathbf{N} + 6\mathbf{N} + 13\mathbf{N}$. Here $e_0 = 4$, $e_1 = 2$, $e_2 = 1$ and $n_1 = n_2 = 2$. Hence $n^* = 2$.

Let $p > n^* = 2$. If $p = \mathrm{char}\,\mathbf{K} \ne 3, 13$ then f is tame. On the other hand if $p = 2$ then $\mu(f) = +\infty$ since x is a common factor of $\frac{\partial f}{\partial y}$ and $\frac{\partial f}{\partial x}$. Hence f is tame if and only if $p \ne 2, 3, 13$. Note that for any l with $\mathrm{ord}(l) = 1$ we have $i_0(f, l) \equiv 0 \pmod{2}$.

Proposition 5.5 *If* $\Gamma(f) = \overline{\beta_0}\mathbf{N} + \overline{\beta_1}\mathbf{N}$ *then* f *is tame if and only if* $\overline{\beta_0} \not\equiv 0 \pmod{p}$ *and* $\overline{\beta_1} \not\equiv 0 \pmod{p}$.

Proof Let $\overrightarrow{w} = (\overline{\beta_0}, \overline{\beta_1})$. There exists a system of coordinates x, y such that we can write $f = y^{\overline{\beta_0}} + x^{\overline{\beta_1}} +$ terms of weight greater than $\overline{\beta_0}\,\overline{\beta_1}$. The proposition follows from Theorem 2.4 (see also [7, Example 2]).

In [11] the authors proved, without any restriction on $p = \mathrm{char}\,\mathbf{K}$, the following profound result:

Theorem 5.6 (Hefez, Rodrigues, Salomão [11, 12]) *Let* $\Gamma(f) = \overline{\beta_0}\mathbf{N} + \cdots + \overline{\beta_g}\mathbf{N}$. *If* $\overline{\beta_k} \not\equiv 0 \pmod{p}$ *for* $k = 0, \ldots, g$ *then* f *is tame.*

The question as to whether the converse of Theorem 5.6 is true remains open.

Acknowledgements The first-named author was partially supported by the Spanish Project MTM 2016-80659-P.

Appendix

Let $\overrightarrow{w} = (n, m) \in (\mathbf{N}_+)^2$ be a weight.

Lemma A.1 *Let* $f, g \in \mathbf{K}[[x, y]]$ *be power series without constant term. Then*

$$i_0(f, g) \ge \frac{\left(\mathrm{ord}_{\overrightarrow{w}}(f)\right)\left(ord_{\overrightarrow{w}}(g)\right)}{mn},$$

with equality if and only if the system of equations

$$\begin{cases} \mathrm{in}_{\overrightarrow{w}}(f) = 0, \\ \\ \mathrm{in}_{\overrightarrow{w}}(g) = 0 \end{cases}$$

has the only solution $(x, y) = (0, 0)$.

Proof By a basic property of the intersection multiplicity (see for example [19, Proposition 3.8 (v)]) we have that for any nonzero power series $\tilde{f}$, $\tilde{g}$

$$i_0(\tilde{f}, \tilde{g}) \geq \mathrm{ord}(\tilde{f})\mathrm{ord}(\tilde{g}), \tag{7}$$

with equality if and only if the system of equations $\mathrm{in}(\tilde{f}) = 0$, $\mathrm{in}(\tilde{g}) = 0$ has the only solution $(0, 0)$. Consider the power series $\tilde{f}(u, v) = f(u^n, v^m)$ and $\tilde{g}(u, v) = g(u^n, v^m)$. Then $i_0(\tilde{f}, \tilde{g}) = i_0(f, g)i_0(u^n, v^m) = i_0(f, g)nm$, $\mathrm{ord}(\tilde{f}) = \mathrm{ord}_{\overrightarrow{w}}(f)$, $\mathrm{ord}(\tilde{g}) = \mathrm{ord}_{\overrightarrow{w}}(g)$ and the lemma follows from (7).

Lemma A.2 *Let* $f \in \mathbf{K}[[x, y]]$ *be a non-zero power series. Then*

$$i_0\left(\frac{\partial f}{\partial x}, \frac{\partial f}{\partial y}\right) \geq \left(\frac{\mathrm{ord}_{\overrightarrow{w}}(f)}{n} - 1\right)\left(\frac{\mathrm{ord}_{\overrightarrow{w}}(f)}{m} - 1\right)$$

with equality if and only if f *is a semi-quasihomogeneous singularity with respect to* $\overrightarrow{w}$.

Proof The following two properties are useful:

$$\mathrm{ord}_{\overrightarrow{w}}\left(\frac{\partial f}{\partial x}\right) \geq \mathrm{ord}_{\overrightarrow{w}}(f) - n \text{ with equality if and only if } \frac{\partial}{\partial x}\mathrm{in}_{\overrightarrow{w}}(f) \neq 0, \tag{8}$$

$$\text{if } \frac{\partial}{\partial x}\mathrm{in}_{\overrightarrow{w}}(f) \neq 0 \text{ then } \mathrm{in}_{\overrightarrow{w}}\left(\frac{\partial f}{\partial x}\right) = \frac{\partial}{\partial x}\mathrm{in}_{\overrightarrow{w}}(f). \tag{9}$$

By the first part of Lemma A.1 and Property (8) we get

$$\begin{aligned} i_0\left(\frac{\partial f}{\partial x}, \frac{\partial f}{\partial y}\right) &\geq \frac{\left(\mathrm{ord}_{\overrightarrow{w}}\left(\frac{\partial f}{\partial x}\right)\right)\left(ord_{\overrightarrow{w}}\left(\frac{\partial f}{\partial y}\right)\right)}{nm} \geq \frac{\left(\mathrm{ord}_{\overrightarrow{w}}(f) - n\right)\left(ord_{\overrightarrow{w}}(f) - m\right)}{nm} \\ &= \left(\frac{\mathrm{ord}_{\overrightarrow{w}}(f)}{n} - 1\right)\left(\frac{\mathrm{ord}_{\overrightarrow{w}}(f)}{m} - 1\right). \end{aligned}$$

Using the second part of Lemma A.1 and Properties (8) and (9) we check that $i_0\left(\frac{\partial f}{\partial x}, \frac{\partial f}{\partial y}\right) = \left(\frac{\mathrm{ord}_{\overrightarrow{w}}(f)}{n} - 1\right)\left(\frac{\mathrm{ord}_{\overrightarrow{w}}(f)}{m} - 1\right)$ if and only if f is a semi-quasihomogeneous singularity with respect to $\overrightarrow{w}$.

Lemma A.3 (Hensel's Lemma [13, Theorem 16.6]) *Suppose that* $\mathrm{in}_{\vec{w}}(f) = \psi_1 \cdots \psi_s$ *with pairwise coprime* ψ_i. *Then* $f = g_1 \cdots g_s \in \mathbf{K}[[x, y]]$ *with* $\mathrm{in}_{\vec{w}}(g_i) = \psi_i$ *for* $i = 1, \ldots, s$.

References

1. Boubakri, Y., Greuel G.-M., Markwig, T.: Invariants of hypersurface singularities in positive characteristic. Rev. Mat. Complut. **25**, 61–85 (2010)
2. Campillo, A.: Algebroid Curves in Positive Characteristic. Lecture Notes in Mathematics, vol. 813. Springer, Berlin (1980)
3. Campillo, A.: Hamburger-Noether expansions over rings. Trans. Am. Math. Soc. **279**, 377–388 (1983)
4. Cassou-Noguès, P., Płoski, A.: Invariants of plane curve singularities and Newton diagrams. Univ. Iag. Acta Mathematica **49**, 9–34 (2011)
5. Deligne, P.: La formule de Milnor. Sem. Geom. algébrique, Bois-Marie 1967–1969, SGA 7 II. Lecture Notes in Mathematics, 340, Exposé XVI, 197–211 (1973)
6. García Barroso, E., Płoski, A.: An approach to plane algebroid branches. Rev. Mat. Complut. **28**, 227–252 (2015)
7. García Barroso, E., Płoski, A.: The Milnor number of plane irreducible singularities in positive characteristic. Bull. Lond. Math. Soc. **48**(1), 94–98 (2016). https://doi.org/10.1112/blms/bdv095
8. Greuel, G.-M.: Der Gauss-Manin-Zusammenhang isolierter Singularitäten von vollständigen Durchschnitten. Math. Ann. **214**, 235–266 (1975)
9. Greuel, G.-M., Lossen, C., Shustin, E.: Introduction to Singularities and Deformations. Springer, Berlin (2007)
10. Greuel, G.-M., Nguyen, H.D.: Some remarks on the planar Kouchnirenko's theorem. Rev. Mat. Complut. **25**, 557–579 (2012)
11. Hefez, A, Rodrigues, J.H.O., Salomão, R.: The Milnor number of a hypersurface singularity in arbitrary characteristic. arXiv: 1507.03179v1 (2015)
12. Hefez, A., Rodrigues, J.H.O., Salomão, R.: The Milnor number of plane branches with tame semigroup of values. arXiv: 1708.07412 (2017)
13. Kunz, E.: Introduction to Plane Algebraic Curves. Translated from the 1991 German edition by Richard G. Belshoff. Birkhäuser, Boston (2005)
14. Lê, D.T.: Calculation of Milnor number of isolated singularity of complete intersection. Funct. Anal. Appl. **8**, 127–131 (1974)
15. Merle, M.: Invariants polaires des courbes planes. Invent. Math. **41**(2), 103–111 (1977)
16. Melle-Hernández, A., Wall, C.T.C.: Pencils of curves on smooth surfaces. Proc. Lond. Math. Soc. **83**(2), 257–278 (2001)
17. Milnor, J.W.: Singular Points of Complex Hypersurfaces. Princeton University Press, Princeton (1968)
18. Nguyen, H.D.: Invariants of plane curve singularities, and Plücker formulas in positive characteristic. Annal. Inst. Fourier **66**, 2047–2066 (2016)
19. Płoski, A.: Introduction to the local theory of plane algebraic curves. In: Krasiński, T., Spodzieja, S. (eds.) Analytic and Algebraic Geometry, pp. 115–134. Łódź University Press, Łódź (2013)
20. Płoski, A.: Plane algebroid branches after R. Apéry. Łódź 35–44 (2014). Accesible in konfrogi.math.uni.lodz.pl
21. Risler, J.J.: Sur l'idéal jacobien d'une courbe plane. Bull. Soc. Math. Fr. **99**(4), 305–311 (1971)
22. Teissier, B.: Cycles évanescents, section planes et condition de Whitney. Astérisque **7–8**, 285–362 (1973)

Foliations in the Plane Uniquely Determined by Minimal Subschemes of its Singularities

Jorge Olivares

Abstract Let $\mathbb{P}^n$ be the projective space over an algebraically closed ground field $\mathbf{K}$. In a previous paper, we have shown that the space of foliations by curves of degree greater than or equal to two which are uniquely determined by a subscheme of minimal degree of its scheme of singularities, contains a nonempty Zariski-open subset and hence, that the set of non-degenerate foliations with this property contains a Zariski-open subset. Moreover, we posed the question whether every non-degenerate foliation in $\mathbb{P}^2$ has this property. In this paper, we prove that this is true, in $\mathbb{P}^2$, in degrees 4 and 5.

1 Introduction and Statement of the Results

Let $\mathbb{P}^n = \mathrm{Proj}\,\mathbf{K}[x_0, \ldots, x_n]$ be the projective space of dimension $n \geq 2$ over an algebraically closed ground field $\mathbf{K}$ and let $\mathscr{O}_{\mathbb{P}^n}$, $\Theta_{\mathbb{P}^n}$, $\Omega^1_{\mathbb{P}^n}$ and $\mathscr{H}$ denote its structure, tangent, cotangent and hyperplane sheaves. We shall denote $\bigwedge^p \Omega^1_{\mathbb{P}^n}$ by $\Omega^p_{\mathbb{P}^n}$, for $1 < p \leq n$. For an $\mathscr{O}_{\mathbb{P}^n}$-sheaf $\mathscr{E}$, we will write $\mathscr{E}(d)$ for $\mathscr{E} \otimes \mathscr{H}^{\otimes d}$, if $d \geq 0$ and $\mathscr{E} \otimes (\mathscr{H}^*)^{\otimes |d|}$, if $d < 0$.

Let

$$\mathbf{E} = \mathbf{E}(n, r-1) = \mathrm{H}^0(\mathbb{P}^n, \Theta_{\mathbb{P}^n}(r-1)), \text{ and } \mathbf{e} = \mathbf{e}(n, r-1) = \dim_{\mathbf{K}} \mathbf{E}. \tag{1}$$

A foliation by curves with singularities (or simply a *foliation* in the sequel) of degree r on $\mathbb{P}^n$ is the class $[s] \in \mathbb{P}\mathbf{E}$ of a global section $s \in \mathbf{E}$. We denote the scheme of zeroes of $[s]$ by $[s]_0$. We say that $[s]$ has isolated singularities if $\dim([s]_0) = 0$ and we say that it is non-degenerate if it has isolated singularities and $[s]_0$ is reduced.

For a closed subscheme $Z \subset \mathbb{P}^n$ with sheaf of ideals I_Z, we have the short exact sequence

J. Olivares (✉)
Centro de Investigación en Matemáticas, A.C., Guanajuato, México
e-mail: olivares@cimat.mx

G.-M. Greuel et al. (eds.), *Singularities, Algebraic Geometry, Commutative Algebra, and Related Topics*, https://doi.org/10.1007/978-3-319-96827-8_6

$$0 \longrightarrow I_Z \longrightarrow \mathcal{O}_{\mathbb{P}^n} \longrightarrow \mathcal{O}_Z \longrightarrow 0. \tag{2}$$

For a zero-dimensional subscheme $Y \subset \mathbb{P}^n$, the space $\mathrm{H}^0(\mathbb{P}^n, \Theta_{\mathbb{P}^n} \otimes I_Y(r-1))$ of sections that vanish on Y will be denoted by

$$\mathbf{E}_Y = \mathrm{H}^0(\mathbb{P}^n, \Theta_{\mathbb{P}^n} \otimes I_Y(r-1)), \text{ with } \mathbf{e}_Y = \dim_{\mathbf{K}} \mathbf{E}_Y. \tag{3}$$

Let Y be a reduced zero-dimensional closed subscheme of $\mathbb{P}^n$ of degree y. With a little abuse of notation, let $I = I_Y = \bigoplus_{s \geq d+1} \mathrm{H}^0(\mathbb{P}^n, I_Y(s)) = \bigoplus_{s \geq d+1} I_s$ be the homogeneous ideal of forms that vanish on Y, let A be its homogeneous coordinate ring: $A = \mathbf{K}[x_0, \ldots, x_n]/I = \bigoplus_s (A_s = \mathbf{K}[x_0, \ldots, x_n]_s / I_s)$, and $\mathrm{H}_A(s) = \dim_{\mathbf{K}} A_s$, its Hilbert function. Following [2], we say that Y is *generic* if $\mathrm{H}_A(s) = \min\{y, \binom{n+s}{s}\}$. It is easy to see that this is equivalent to

$$\mathrm{H}_A(s) = \begin{cases} \binom{n+s}{s} & s < b \\ y & s \geq b. \end{cases} \Leftrightarrow \begin{cases} \mathrm{h}^0(\mathbb{P}^n, I_Y(s)) = 0 & s < b \\ \mathrm{h}^1(\mathbb{P}^n, I_Y(s)) = 0 & s \geq b, \end{cases} \tag{4}$$

$$\text{where } b \in \mathbb{N} \text{satisfies } \binom{n+b-1}{b-1} \leq y < \binom{n+b}{b}.$$

For $b = r+1$ and generic Y, $I = \langle I_{r+1}, I_{r+2} \rangle$: Its homogeneous ideal is generated in degrees $r+1$ and $r+2$ [8, Proposition 1.4]. In this case, we follow [1] and say that a generic Y is *general* if

$$\iota_R^0 : I_{r+1}^{\oplus(n+1)} \longrightarrow I_{r+2};\ \iota_R^0(F_0, \ldots, F_n) = \sum_{i=0}^{n} x_i F_i \tag{5}$$

has maximal rank. If Y is considered as a point in the symmetric product $S^y\mathbb{P}^n$, it turns out that the set $N_y = \{Y \in S^y\mathbb{P}^n \mid Y \text{is general}\} \subset S^y\mathbb{P}^n$ is open [8] and nonempty (this last statement is the affirmative solution to the so-called Cohen-Macaulay Type Conjecture: see [1] for references). This, together with [2, Proposition 1.2] (which is crystal-clear in dimension $n = 2$ if one works with 1-forms, see (12) below), gives [1, Theorem 0.1]: Under our conventions (1) and (3), it states that

$$\begin{aligned} &\text{For a general } Y \in S^y\mathbb{P}^n, \text{ with } y > n \text{ and } n \geq 2 \text{ and } r \geq 2, \\ &\mathbf{e}_Y = \max\{0, \mathbf{e} - n \cdot y\}, \text{ so that } \mathbf{e}_Y \neq 0 \Leftrightarrow y \leq \omega = \omega(n, r-1), \end{aligned} \tag{6}$$

where $\boldsymbol{\omega} = \omega(n, r-1) = [\frac{\mathbf{e}-1}{n}]$ denotes the integral part of the number between brackets.

It follows from equation (2.2) in [5] that $Y \in S^y\mathbb{P}^n$ is general if and only if $\mathrm{h}^1(\mathbb{P}^n, \Theta_{\mathbb{P}^n} \otimes I_Y(r-1)) = 0$, that is, if and only if Y imposes exactly $n \cdot y$ independent conditions on $\mathbf{E}$.

Let $0 \leq \rho \leq n-1$ be the unique integer such that $\mathbf{e} - 1 = n \cdot \omega + \rho$. It is immediate from (6) that

$$m(n, r-1) = \begin{cases} \omega & \text{if } \rho = 0, \text{and} \\ \omega + 1 & \text{if } 1 \leq \rho \leq n-1. \end{cases} \tag{7}$$

is the minimal possible degree of a zero-dimensional reduced subscheme $Y \subset [s]_0$ such that $\mathbf{e}_Y = 1$ (that is, such that $\mathbb{P}\mathbf{E}_Y = \{[s]\}$).

Definition 1 Let $[s] \in \mathbb{P}\mathbf{E}$ be a foliation with isolated singularities of degree r in $\mathbb{P}^n$, with $S = [s]_0$. We say that $Y \subset S$ is a minimal subscheme for $[s]$ if $\deg Y = m(n, r-1)$ and $\mathbb{P}\mathbf{E}_Y = \{[s]\}$. Hence, a minimal subscheme for $[s]$ is a subscheme of S which uniquely determines $[s]$ and has minimal degree.

It follows, moreover (from [1], for $\rho = 0$ and from [5], for $1 \leq \rho \leq n-1$), that the set of those $[s] \in \mathbb{P}\mathbf{E}$ having a minimal subscheme $Y \subset [s]_0$ contains a non-empty Zariski-open subset of $\mathbb{P}\mathbf{E}$. Hence, the set of non-degenerate sections with this property contains a non-empty Zariski-open subset of $\mathbb{P}\mathbf{E}$.

In case $n = 2$, we have shown at the end of [5] that *every* non-degenerate foliation of degrees $r = 2$ and $r = 3$ has a minimal subscheme and announced that the same is true, for degrees $r = 4$ and $r = 5$: This improvement of [4, Theorem 3.5] is our main result:

Theorem 1 *Let* $[s] \in \mathbb{P}\mathbf{E}(2, r-1)$ *be a non-degenerate foliation of degree* $r \geq 4$, *and let* $S = [s]_0$. *Then, there exists a subscheme* $Z^1 \subset S$ *of degree* $M_r - 1 = r(r+5)/2 - 1$ *such that* $[s]$ *is uniquely determined by* Z^1. *It follows in particular that every non-degenerate foliation* $[s]$ *of degrees* $r = 4$ *and* 5 *in* $\mathbb{P}^2$ *has a minimal subscheme* $Z^1 \subset [s]_0$.

Still in dimension $n = 2$, we further study the question of whether *every* non-degenerate foliation $[s]$ of degree $r \geq 6$, with $S = [s]_0$ has or not a minimal subscheme $Y \subset S$.

Let N_j denote $h^0(\mathbb{P}^2, \mathcal{O}_{\mathbb{P}^2}(j)) = \binom{j+2}{2}$. We have

Corollary 1 *Let* $[s] \in \mathbb{P}\mathbf{E}(2, r-1)$ *be a non-degenerate foliation of degree* $r \geq 4$ *and let* $S = [s]_0$. *Then, there exists a general subscheme* $\hat{Z}_{12} \subset S$ *of degree* $N_r + 2$.

2 Foliations in the Projective Plane

To fix our notation, we give a quick review of foliations in $\mathbb{P}^2 = \operatorname{Proj} \mathbf{K}[x_0, x_1, x_2]$. Recall that $\mathbf{e}(2, r-1) = (r+1)(r+3)$ (see [1]) and that, counting multiplicities, $\deg [s]_0 = r^2 + r + 1$, for $s \in \mathbf{E}(2, r-1)$ with isolated singularities (see [4]). The foliation defined by an $s \in \mathbf{E}(2, r-1)$ may be also defined through the projective 1-form Ω that annihilates s (as described in [4, §2]). Indeed, let $R = \sum_{k=0}^{2} x_k \frac{\partial}{\partial x_k}$ be the radial vector field and let ι_R denote the *contraction by* R operator in differential forms. The dual of the twisted Euler sequence ([10, Example 8.20.1], or [9, p. 409], in the complex case) is

$$0 \longrightarrow \Omega^1_{\mathbb{P}^2}(r+2) \longrightarrow \mathcal{O}_{\mathbb{P}^2}(r+1)^{\oplus 3} \xrightarrow{\iota_R} \mathcal{O}_{\mathbb{P}^2}(r+2) \longrightarrow 0. \tag{8}$$

This correspondence induces a **K**-vector space isomorphism between $\mathbf{E}(2, r-1)$ and $H^0(\mathbb{P}^2, \Omega^1_{\mathbb{P}^2}(r+2))$, and the subspace $\mathbf{E}_Y$ from (3) corresponds to $H^0(\mathbb{P}^2, \Omega^1_{\mathbb{P}^2}(r+2) \otimes I_Y)$.

We also have the short exact sequence

$$0 \longrightarrow \Omega^2_{\mathbb{P}^2}(r+2) \longrightarrow \mathcal{O}_{\mathbb{P}^2}(r)^{\oplus 3} \xrightarrow{\iota_R} \Omega^1_{\mathbb{P}^2}(r+2) \longrightarrow 0, \tag{9}$$

where $\Omega^2_{\mathbb{P}^2}(r+2) \cong \mathcal{O}_{\mathbb{P}^2}(r-1)$.

For a zero-dimensional subscheme $Y \subset \mathbb{P}^2$ with sheaf of ideals I_Y, these give rise, respectively, to the exact sequences

$$0 \longrightarrow \Omega^1_{\mathbb{P}^2}(r+2) \otimes I_Y \xrightarrow{j} I_Y(r+1)^{\oplus 3} \xrightarrow{\iota_R} I_Y(r+2) \longrightarrow 0, \tag{10}$$

$$0 \longrightarrow I_Y(r-1) \longrightarrow I_Y(r)^{\oplus 3} \xrightarrow{\iota_R} \Omega^1_{\mathbb{P}^2}(r+2) \otimes I_Y \longrightarrow 0. \tag{11}$$

Although it is not relevant for what follows, it is not hard to see that $H^2(\mathbb{P}^2, \Omega^1_{\mathbb{P}^2}(r+2) \otimes I_Y) = 0$. Hence, the long exact cohomology sequence associated to (10) is

$$\begin{aligned} 0 \to H^0(\mathbb{P}^2, \Omega^1_{\mathbb{P}^2}(r+2) \otimes I_Y) &\overset{j^0}{\to} H^0(\mathbb{P}^2, I_Y(r+1))^{\oplus 3} \overset{\iota_R^0}{\to} H^0(\mathbb{P}^2, I_Y(r+2)) \to \\ \to H^1(\mathbb{P}^2, \Omega^1_{\mathbb{P}^2}(r+2) \otimes I_Y) &\overset{j^1}{\to} H^1(\mathbb{P}^2, I_Y(r+1))^{\oplus 3} \overset{\iota_R^1}{\to} H^1(\mathbb{P}^2, I_Y(r+2)) \to 0. \end{aligned} \tag{12}$$

2.1 *Special Subschemes*

Let $r \geq 2$ be an integer, let $[s] \in \mathbb{P}H^0(\mathbb{P}^2, \Omega_{\mathbb{P}^2}(r+2))$ be a non-degenerate foliation of degree r, represented by a 1-form $\Omega = A dx_0 + B dx_1 + C dx_2$, and let $S = [s]_0$. A closed, reduced, zero-dimensional subscheme $Z \subset \mathbb{P}^2$ will be said to be of **type C** if $Z \subset [s]_0$, for some $[s]$, $\deg(Z) = M_r = r(r+5)/2$, $h^0(\mathbb{P}^2, I_Z(r+1)) = 3$ and $h^1(\mathbb{P}^2, I_Z(r+1)) = 0$. Type C subschemes were called *special* in [4]. Special subschemes Z exist for every nondegenerate $[s]$, $\mathbb{P}H^0(\mathbb{P}^2, \Omega^1_{\mathbb{P}^2}(r+2) \otimes I_Z) = \{[s]\}$ [4, Lemma 3.4 and Theorem 3.5] and $Z \neq S$ if and only if $r \geq 3$. It follows from (4) with $b = r+1$ that Z is generic, hence the homogeneous ideal of Z is generated in degrees $r+1$ and $r+2$.

From the long exact cohomology sequence associated with (2), it follows easily that $\mathrm{h}^0(\mathbb{P}^2, I_Z(r+2)) = r+6$. Since the rank of ι_R^0 in the sequence (12) associated with Z is equal to 8, it follows that $\mathrm{h}^1(\mathbb{P}^2, \Omega^1_{\mathbb{P}^2}(r+2)\otimes I_Z) = r-2$.

Since $N_{r+1} = M_r + 3$, it is a standard fact that there exists a basis $\{A, B, C, D_p : p \in Z\}$ of $\mathrm{H}^0(\mathbb{P}^2, \mathscr{O}_{\mathbb{P}^2}(r+1))$ such that

$$D_p(p') \begin{cases} \neq 0 & \text{if } p' = p, \text{ and} \\ = 0 & \text{if } p' \neq p, p' \in Z. \end{cases}$$

These polynomials will be referred to as *test polynomials* for Z.

In a similar vein, a reduced, closed, zero-dimensional subscheme $\hat{Z} \subset \mathbb{P}^2$ will be said to be of **type B** if $\deg(\hat{Z}) = N_r$ and $\mathrm{h}^q(\mathbb{P}^2, I_{\hat{Z}}(r)) = 0$, for $q = 0, 1$. It is not hard to see that any two of these three conditions imply the third one. It follows moreover from [8, Proposition 1.4] that the homogeneous ideal of $\hat{Z}$ is generated in degree $r+1$ and that $\mathrm{h}^0(\mathbb{P}^2, I_{\hat{Z}}(r+1)) = r+2$, hence the map ι_R^0 in the sequence (12) associated with $\hat{Z}$ is surjective (so that $\mathrm{h}^1(\mathbb{P}^2, \Omega^1_{\mathbb{P}^2}(r+2)\otimes I_{\hat{Z}}) = 0$) and $\mathrm{h}^0(\mathbb{P}^2, \Omega^1_{\mathbb{P}^2}(r+2)\otimes I_{\hat{Z}}) = r+1$, as one easily computes. In sum, every type B subscheme $\hat{Z}$ is general (of degree N_r).

This can also be seen from the long exact cohomology sequence associated with (11) for $Y = \hat{Z}$: The fact that $\mathrm{h}^2(\mathbb{P}^2, I_{\hat{Z}}(r-1)) = 0$ (see [7]) proves that $\mathrm{h}^1(\mathbb{P}^2, \Omega^1_{\mathbb{P}^2}(r+2)\otimes I_{\hat{Z}}) = 0$ and hence, that ι_R^0 is surjective. This approach shows that ι_R^0 is surjective for every Y satisfying $\mathrm{h}^1(\mathbb{P}^2, I_Y(r)) = 0$: Type B subschemes are those with maximal degree that satisfy this property.

Let $Z \subset S$ be of type C, let $\{p_1, \ldots, p_m\}$ be closed points in the support of Z, with $1 \leq m \leq r-1$, and consider $Z^m = Z \setminus \{p_1, \ldots, p_m\}$. It is clear that $\mathrm{h}^1(\mathbb{P}^2, I_{Z^m}(r+1)) = 0$ and hence, again by [8, Corollary 1.6], that the homogeneous ideal of Z^m is generated in degrees $r+1$ and $r+2$. Moreover, $\mathrm{h}^0(\mathbb{P}^2, I_{Z^m}(r+1)) = 3+m$ and $\mathrm{h}^0(\mathbb{P}^2, I_{Z^m}(r+2)) = r+6+m$. A basis of the former can be given in terms of test polynomials, namely $\{A, B, C, D_{p_i} : i = 1, \ldots, m\}$.

For $m = r-1$, the degree of the subscheme $Z^{r-1} \subset Z$ is equal to N_r and satisfies $\mathrm{h}^1(\mathbb{P}^2, I_{Z^{r-1}}(r+1)) = 0$. If $r \geq 4$, we can always choose the points so that $Z^{r-1} \subset Z$ is of type B:

Lemma 1 *Let $[s] \in \mathbb{P}H^0(\mathbb{P}^2, \Omega^1_{\mathbb{P}^2}(r+2))$ be a non-degenerate foliation of degree $r \geq 4$ in $\mathbb{P}^2$ with $S = [s]_0$, and let $Z \subset S$ be a subscheme of type C. Then there exist collections $B = \{p_1, \ldots, p_{r-1}\}$ of closed points in the support of Z such that the subscheme $Z^{r-1} = Z \setminus \{p_1, \ldots, p_{r-1}\} \subset Z$ is of type B.*

Proof For $Y \subset S$, consider the residual subscheme Y' of Y in S (see [3] and [4]). It follows from [3, Proposition 3.1] that

$$\mathrm{h}^1(\mathbb{P}^2, I_Y(m)) = \mathrm{h}^0(\mathbb{P}^2, I_{Y'}(2r-2-m)), \quad \text{for } r \leq m \leq 2r-2, \tag{13}$$

so that $0 = \mathrm{h}^1(\mathbb{P}^2, I_Z(r+1)) = \mathrm{h}^0(\mathbb{P}^2, I_{Z'}(r-3))$, $\deg(Z') = N_{r-3}$, $\mathrm{h}^1(\mathbb{P}^2, I_{Z'}(r-3)) = 0$ and $\mathrm{h}^1(\mathbb{P}^2, I_Z(r)) = \mathrm{h}^0(\mathbb{P}^2, I_{Z'}(r-2)) = r-1$. By [8, Proposition 1.4] again, the homogeneous ideal of Z' is generated in degree $r-2$. Let $\mathrm{H}^0(\mathbb{P}^2, I_{Z'}(r-2)) = \langle F_1, \ldots, F_{r-1}\rangle_{\mathbf{K}}$.

If $Y = Z^{r-1}$ for some $B = \{p_1, \ldots, p_{r-1}\} \subset Z$, then $Y' = Z' \cup B$ and (13) gives $\mathrm{h}^1(\mathbb{P}^2, I_{Z^{r-1}}(r)) = \mathrm{h}^0(\mathbb{P}^2, I_{Z' \cup B}(r-2))$: These are equal to zero if and only if the points in B impose exactly $r-1$ independent conditions on $\mathrm{H}^0(\mathbb{P}^2, I_{Z'}(r-2))$ and we can choose them to do so, through the following procedure: Since $\mathrm{H}^0(\mathbb{P}^2, I_S(r-2)) = 0$ (by [3, Theorem 3.2]), there exists a $p = p_1 \in Z$ such that $F_{r-1}(p_1) \neq 0$ and hence, the subspace $\mathrm{H}^0(\mathbb{P}^2, I_{Z' \cup p_1}(r-2)) = \langle F_1^{(1)}, \ldots, F_{r-2}^{(1)}\rangle_{\mathbf{K}} \subset \mathrm{H}^0(\mathbb{P}^2, I_{Z'}(r-2))$ has codimension equal to 1. For the same reason as before, there exists a $p = p_2 \in Z \setminus p_1$ such that $F_{r-2}^{(1)}(p_2) \neq 0$ and hence, the subspace $\mathrm{H}^0(\mathbb{P}^2, I_{Z' \cup \{p_1, p_2\}}(r-2)) = \langle F_1^{(2)}, \ldots, F_{r-3}^{(2)}\rangle_{\mathbf{K}} \subset \mathrm{H}^0(\mathbb{P}^2, I_{Z' \cup p_1}(r-2))$ has codimension equal to 1. We iterate $r-1$ times to get a $B = \{p_1, \ldots, p_{r-1}\} \subset Z$ such that $\mathrm{h}^0(\mathbb{P}^2, I_{Z' \cup B}(r-2)) = 0$. □

3 The Proofs

Proof of Theorem 1 The second part of the statement follows from the first one, together with the relations (15) below. For the first part, let $\Omega = A dx_0 + B dx_1 + C dx_2$ be a 1-form representing $[s]$, and let $S = [s]_0$. Let $Z \subset S$ be a subscheme of type C and let Z' be the residual subscheme of Z in S. Consider $p \in Z$ and $D = D_p$.

We will show, by contradiction, that $\mathrm{h}^0(\mathbb{P}^2, \Omega^1_{\mathbb{P}^2} \otimes I_{Z^1}(r+2)) = 1$, for some $Z^1 = Z \setminus p$:

Consider the sequence (12) with $Y = Z \setminus p$. Then, $\mathrm{h}^0(\mathbb{P}^2, \Omega^1_{\mathbb{P}^2} \otimes I_{Z \setminus p}(r+2)) \geq 2$ if and only if there exist linear forms L_A, L_B, L_C and L_D, with $L_D \neq 0$ such that

$$L_A \cdot A + L_B \cdot B + L_C \cdot C + L_D \cdot D = 0. \tag{14}$$

Let $S[p] = \{q \in S : D_p(q) \neq 0\}$, so that $Z'[p] = \{q \in Z' : D_p(q) \neq 0\}$. Then $\deg Z'[p] \geq r-2$ (otherwise, the foliations associated with Ω and another linearly independent section share at least r^2+3 singularities, in contradiction with [4, Corollary 3.3]). Evaluating (14) in $q \in S[p]$, we get $L_D(q) \cdot D(q) = 0$ and hence, all points $q \in S[p]$ lie in the line $L_D = L_{D_p} = 0$. In particular, every $q \in Z'[p]$, together with p, lie in this line. We claim, moreover, that $\deg Z'[p] = r-2$ (otherwise, since $\deg Z' = N_{r-3} = r-2+N_{r-4}$, then $\deg\,(Z' \setminus Z'[p]) = \deg Z' - \deg Z'[p] < N_{r-4}$ and $Z' \setminus Z'[p]$ lies in at least one curve Q of degree $r-4$. Hence, Z' lies in the curve of degree $r-3$ given by $Q \cdot L_D = 0$, in contradiction with [4, Lemma 3.4]).

The first conclusion is that the line $L_{D_p} = 0$ contains $r - 1$ points of S: the point $p \in Z$ itself and exactly other $r - 2$ points in Z'.

Our second claim is that in the support of Z', there exist at most $r - 1$ subsets consisting of $r - 2$ aligned points: Indeed, $N_{r-3} = \sum_{k=1}^{r-2} k = (r - 2) + (r - 3) + \cdots + 1$: In the first line, we have $r - 2$ points; in the second, one point from the first and $r - 3$ other points; in the third, one from the first, another from the second, and $r - 4$ other points. Inductively we see that in the $(r - 2) -$ th line, there lies at most $r - (r - 1) = 1$ points not already counted, and at this step we have already counted all the points in Z', so that in the $(r - 1) -$ th line, all its $r - 2$ points have been already counted.

The third claim is that $3(r - 1) < \deg Z = r(r + 5)/2$. Indeed, this inequality is equivalent to $r^2 - r + 6 > 0$ and this is obvious, as actually $r^2 - r > 0$.

Our fourth (and final) claim is the well-known fact (see [6], for instance) that the maximal number of points of S that lie in a line is $r + 1$ (in which case, the line is invariant by the foliation) and it is r, if the line is not invariant.

Now we are in position to finish the proof: If (14) holds for every $p \in Z$, then there exists a line $L_{D_p} = 0$ that contains p and $r-2$ points of Z', there exist at most $r-1$ of such lines $L_{D_p} = 0$, and (by the fourth claim) each of these lines contains at most two other points (say) p_1 and p_2 in S, that actually belong to Z (because those in Z' have been already counted). The conclusion is that there exist *at most* three points $\{p, p_1, p_2\}$ in Z in each line $L_{D_p} = 0$ and $L_{D_p} = L_{D_{p_1}} = L_{D_{p_2}}$ because, as $r \geq 4$, the $r - 2 \geq 2$ points from Z' determine the line L_{D_p}. Hence, $3(r - 1)$ should be $\geq \deg Z = r(r+5)/2$, in contradiction with the third claim. Hence, there exists a line $L_{D_p} = 0$ having at least four points in Z, and this line contains at least $4+r-2 = r+2$ singularities of $[s]$, in contradiction with the assumption that S is reduced. □

Corollary 2 *Let $r \geq 4$ be an integer, let $[s] \in \mathbb{P}H^0(\mathbb{P}^2, \Omega_{\mathbb{P}^2}(r + 2))$ be a non-degenerate foliation of degree r in $\mathbb{P}^2$, let $S = [s]_0$, let $Z \subset S$ be a subscheme of type C, with test polynomials $\{D_p : p \in Z\}$ and let $S[p] = \{q \in S : D_p(q) \neq 0\}$. Then $[s]$ is uniquely determined by $Z^1 = Z \setminus p$ if the (at least) $r - 1$ points in $S[p]$ are not aligned.*

Proof It is clear form the proof of Theorem 1 that $S[p]$ consists of at least $r - 1$ points, which include p itself. Consider (14) with $D = D_p$ and denote L_{D_p} by L_p. Evaluate (14) at all points $q \in S[p]$ to conclude that $L_p(q) = 0$. Our hypothesis on $S[p]$ imply that, necessarily, L_p is identically zero. □

Proof of Corollary 1 Denote s by s_0 and S by S_0. Consider a $Z^1 \subset S_0$ as in Theorem 1 and, as in Lemma 1, drop $r - 2$ points $D = \{p_1, \ldots, p_{r-2}\}$ from the support of Z^1 to obtain a type B subscheme $\hat{Z} = Z^1 \setminus D \subset Z^1$, so that $\mathbf{e}_{\hat{Z}} = r + 1$. Since s_0 is non-degenerate, we can assume that $\mathbf{E}_{\hat{Z}}$ has a $\mathbf{K}-$basis $\{s_0, s_1, \ldots, s_r\}$ with all s_i non-degenerate. Let $S_i = [s_i]_0$. There exists a $p \in D$ such that $p \notin S_i$, for some $i \in \{1, \ldots, r\}$ (otherwise, $Z^1 \subset S_i$, for every $i \in \{1, \ldots, r\}$, in contradiction with Theorem 1). Rename this point p as $\hat{p}_1$ and let $\hat{Z}_1 = \hat{Z} \cup \hat{p}_1$: Then $\mathbf{E}_{\hat{Z}} \supset \mathbf{E}_{\hat{Z}_1}$ has codimension at least equal to one and we can still assume that $\mathbf{E}_{\hat{Z}_1}$ has a

$\mathbf{K}$–basis $\{s_0, s_1^1, \ldots, s_{r-1}^1\}$ of non-degenerate sections. We can repeat the argument above to prove that there exists a $\hat{p}_2 \in D \setminus \hat{p}_1$ such that $\hat{p}_2 \notin S_j^1 = [s_j^1]_0$, for some $j \in \{1, \ldots, r-1\}$. For $\hat{Z}_2 = \hat{Z}_1 \cup \hat{p}_2$, we see that $\mathbf{E}_{\hat{Z}_1} \supset \mathbf{E}_{\hat{Z}_2}$ has codimension at least equal to one. We iterate to get, for $k = 1, \ldots, r-2$, a $\hat{Z}_k = \hat{Z} \cup \{\hat{p}_1, \ldots, \hat{p}_k\}$ such that

$$\mathbf{E}_{\hat{Z}} \supset \mathbf{E}_{\hat{Z}_1} \supset \mathbf{E}_{\hat{Z}_2} \supset \cdots \supset \mathbf{E}_{\hat{Z}_{r-2}} = \mathbf{E}_{Z^1}.$$

Two links of this chain must have codimension equal to 2 (otherwise, $\mathbf{e}_{\mathbf{Z}^1} = 3$) and if these occur at the points $\hat{p}_{k_1}, \hat{p}_{k_2}$, it is not hard to see that $\mathbf{E}_{\hat{Z}} \supset \mathbf{E}_{\hat{Z} \cup \{p_{k_1}, p_{k_2}\}}$ has codimension equal to 4, so that $\hat{Z}_{12} = \hat{Z} \cup \{p_{k_1}, p_{k_2}\} \subset S$ is general of degree $N_r + 2$. □

4 Work in Progress

For $n = 2$, an easy computation shows that the minimal degree (7) is

$$m(2, r-1) = \begin{cases} \omega \quad = \frac{\mathbf{e}-1}{2} = \frac{(r+1)(r+3)-1}{2} & \text{if } r = 2t, \text{ and} \\ \omega + 1 = \frac{\mathbf{e}}{2} \quad = \frac{(r+1)(r+3)}{2} & \text{if } r = 2t+1, \end{cases}$$

and satisfies the relations

$$m(2, r-1) = M_r - (t-1) = \begin{cases} N_r + t & \text{if } r = 2t, \text{ and} \\ N_r + t + 1 & \text{if } r = 2t+1. \end{cases} \tag{15}$$

Let $[s]$ be a non-degenerate foliation of degree $r \geq 6$ with $S = [s]_0$. Roughly speaking, Lemma 1 and relations (15) suggest two possible strategies to decide the existence or not of a minimal subscheme Y for $[s]$, namely:

4.1 Drop Points from a Type C

Given a $Z \subset S$ of Type C, find conditions on a collection of $t-1$ points $D_{t-1} = \{p_1, \ldots, p_{t-1}\}$ in the support of Z, such that the scheme with support at $Z \setminus D_{t-1} \subset S$ is a minimal Y for $[s]$. Having these, try to decide if every such an $[s]$ satisfies these conditions.

A basic tool in this direction is Corollary 2, because every $p_j \in D_{t-1}$ must satisfy its conditions.

4.2 Add Points to a Type B

Given a $\hat{Z} \subset S$ of Type B (and its residual subscheme $\hat{Z}'$ in S), find conditions on a collection of t points $B_t = \{p_1, \dots, p_t\}$ in the support of $\hat{Z}'$, such that the scheme with support at $Z \cup B_t \subset S$ is a general $\hat{Y} \subset S$ (that is B_t should impose exactly $2t$ independent conditions on $\mathbf{E}_{\hat{Z}}$). If $r = 2t$, then $Y = \hat{Y}$ is minimal; if $r = 2t + 1$, then $\mathbf{E}_{\hat{Y}} = \langle s, s_1 \rangle_{\mathbf{K}}$. It follows from [5] that $Y = \hat{Y} \cup p$ is minimal for $[s]$, for every $p \in S \setminus [s_1]_0$.

Having these, again, try to decide if every such an $[s]$ satisfies these conditions. It is in this direction that Corollary 1 becomes relevant.

References

1. Ballico, E.: On meromorphic vector fields on projective spaces. Am. J. Math. **115**(5), 1135–1138 (1993)
2. Ballico, E., Geramita, A.V.: The minimal free resolution of the ideal of s general points in $\mathbb{P}^3$. In: Proceedings of the 1984 Vancouver Conference in Algebraic Geometry, pp. 1–10. CMS Conference Proceeding, vol. 6. American Mathematical Society, Providence (1986)
3. Campillo, A., Olivares, J.: Polarity with respect to a foliation and Cayley-Bacharach Theorems. J. Reine Angew. Math. **534**, 95–118 (2001)
4. Campillo, A., Olivares, J.: Special subschemes of the scheme of singularities of a plane foliation. C. R. Math. Acad. Sci. Paris **344**(9), 581–585 (2007)
5. Campillo, A., Olivares, J.: Foliations by curves uniquely determined by minimal subschemes of its singularities. To appear in Greuel, G.-M., Hamm, H., Trang, L.-D. (eds.) Special Volume Dedicated to the Memory of Egbert Brieskorn. Journal of Singularities (2018)
6. Campillo, A., Farrán, J.I., Pizabarro, M.J.: Evaluation codes at singular points of algebraic differential equations. Appl. Algebra Eng. Commun. Comput. **18**(1–2), 191–203 (2007)
7. Esteves, E., Kleiman, S.: Bounds on leaves of foliations in the plane. In: Gaffney, T. (ed.) Real and Complex Singularities. Contemporary Mathematics, vol. 354, pp. 57–67 (2004)
8. Geramita, A.V., Maroscia, P.: The ideal of forms vanishing at a finite set of points in $\mathbb{P}^n$. J. Algebra **90**(2), 528–555 (1984)
9. Griffiths, Ph., Harris, J.: Principles of Algebraic Geometry. Pure and Applied Mathematics. Wiley, New York (1978)
10. Hartshorne, R.: Algebraic Geometry. Graduate Texts in Mathematics, vol. 52. Springer, New York/Heidelberg (1977)

Newton Transformations and the Motivic Milnor Fiber of a Plane Curve

Pierrette Cassou-Noguès and Michel Raibaut

Pour Antonio, en témoignage de notre amitié

Abstract In this article we give an expression of the motivic Milnor fiber at the origin of a polynomial in two variables with coefficients in an algebraically closed field. The expression is given in terms of some motives associated to the faces of the Newton polygons appearing in the Newton algorithm. In the complex setting, we deduce a computation of the Euler characteristic of the Milnor fiber in terms of the area of the surfaces under the Newton polygons encountered in the Newton algorithm which generalizes the Milnor number computation by Kouchnirenko in the isolated case.

1 Introduction

Let $\mathbf{k}$ be an algebraically closed field of characteristic zero. Let f be a regular map defined on a smooth $\mathbf{k}$-variety X. Using the motivic integration theory introduced by Kontsevich in [20], Denef and Loeser defined in [8, 11], the motivic Milnor fiber of f at a point x of X, denoted by $S_{f,x}$ as an element of $\mathscr{M}_{\mathbf{k}}^{\hat{\mu}}$, a modified Grothendieck ring of varieties over $\mathbf{k}$ endowed with an action of the group of roots of unity $\hat{\mu}$. They proved that the motive $S_{f,x}$ is a "motivic" incarnation of the topological Milnor fiber of f at x denoted by F_x and endowed with its monodromy action T_x. For instance, when $\mathbf{k}$ is the field of complex numbers, they proved in [8, Theorem 4.2.1,

P. Cassou-Noguès (✉)
Institut de Mathématiques de Bordeaux, (UMR 5251), Université de Bordeaux, Talence Cedex, France
e-mail: Pierrette.Cassou-Nogues@math.u-bordeaux.fr

M. Raibaut
Université Grenoble Alpes, Université Savoie Mont Blanc, CNRS, LAMA, 73000, Chambéry, France
e-mail: Michel.Raibaut@univ-smb.fr

G.-M. Greuel et al. (eds.), *Singularities, Algebraic Geometry, Commutative Algebra, and Related Topics*, https://doi.org/10.1007/978-3-319-96827-8_7

Corollay 4.3.1] that the motive $S_{f,x}$ realizes on usual invariants of (F_x, T_x) as the Euler characteristic, the monodromy zeta function or Steenbrink's spectrum.

In [15] Guibert computed the motivic Milnor fiber at the origin of a polynomial which is non degenerate with respect to its Newton polyhedron. The computation is given in terms of the faces of the Newton polyhedron. For more general point of views we refer to the memoir of Artal-Bartolo, Cassou-Nogues, Luengo and Melle-Hernandez [2], the composition of a non-degenerate polynomial with regular functions with separated variables of Guibert, Loeser and Merle in [16] and the logarithmic approach of Bultot and Nicaise in [3, 4].

In [15] Guibert computed also the motivic fiber at the origin of a function in two variables in terms of the Puiseux pairs of its branches and their orders of contact. This approach is generalized in [18], where the authors compute the motivic Milnor fiber at the origin of a composition of a polynomial with two functions with the same variables but transversality assumptions. More recently using resolution of singularites Lê Quy Thuong computed in [24] the motivic Milnor fiber in an inductive and combinatoric way using the extended simplified resolution graph. An other recursive approach is given by González-Villa, Kennedy and McEwan in [14]. The quasi-ordinary case is treated in the article of González-Villa and González-Pérez [13].

Inspired by the works of the first author and Veys in the case of an ideal of $\mathbf{k}[[x, y]]$ in [5, 6], the aim of this article is to give in Theorem 2 an expression of the motivic Milnor fiber at the origin of a polynomial in $\mathbf{k}[x, y]$ in terms of some motives associated to the faces of the Newton polygons appearing in the Newton algorithm. The Newton algorithm is an algorithm which allows to compute the tree of the resolution of a germ at the origin in terms of Newton polygons.

We present the strategy of the proof of the theorem and the structure of the article. Let $\mathbb{G}_m$ be the multiplicative group of $\mathbf{k}$. We will work with a Grothendieck ring $\mathscr{M}_{\mathbb{G}_m}^{\mathbb{G}_m}$ of the category $\mathrm{Var}_{\mathbb{G}_m}^{\mathbb{G}_m}$ of $\mathbb{G}_m$-algebraic varieties endowed with a monomial $\mathbb{G}_m$-action (see Sect. 2 for details). This ring is isomorphic to the ring $\mathscr{M}_{\mathbf{k}}^{\hat{\mu}}$ (see [17, §2.4]).

We denote by $\mathscr{L}(\mathbb{A}_{\mathbf{k}}^2)$ the arc space of the affine plane $\mathbb{A}_{\mathbf{k}}^2$ whose $\mathbf{k}$-rational points are formal series in $\mathbf{k}[[t]]^2$. The multiplicative group $\mathbb{G}_m$ acts canonically on $\mathscr{L}(\mathbb{A}_{\mathbf{k}}^2)$ by $\lambda.\varphi(t)$ equal to $\varphi(\lambda t)$ for any λ in $\mathbb{G}_m$ and any arc φ in $\mathscr{L}(\mathbb{A}_{\mathbf{k}}^2)$. For a non zero element ψ in $\mathbf{k}[[t]]$, we denote by $\mathrm{ord}\,(\psi)$ the *order* of ψ and by $\overline{\mathrm{ac}}\,(\psi)$ its *angular component* equal to its first non-zero coefficient with the convention that $\overline{\mathrm{ac}}\,(0)$ is zero.

Let f be a polynomial in $\mathbf{k}[x, y]$ with $f(0, 0) = 0$. The *motivic zeta function* of f at the origin is the formal series

$$\left(Z_f(T)\right)_{(0,0)} = \sum_{n\geq 1} \mathrm{mes}\,(X_n(f))T^n$$

in $\mathscr{M}_{\mathbb{G}_m}^{\mathbb{G}_m}[[T]]$, where for any $n \geq 1$, mes $(X_n(f))$ is the *motivic measure* of the arc space

$$X_n(f) = \{\varphi \in \mathscr{L}(\mathbb{A}_{\mathbf{k}}^2) \mid \varphi(0) = (0,0),\ \operatorname{ord} f(\varphi) = n\}$$

endowed with the arrow "angular component" $\overline{\mathrm{ac}}\,(f)$ to $\mathbb{G}_m$ and the standard action of $\mathbb{G}_m$ on arcs. Denef and Loeser show in [8, 11] that this zeta function is rational by giving a formula of $\left(Z_f(T)\right)_{(0,0)}$ in terms of a resolution of f. It admits a limit when T goes to ∞ and the *motivic Milnor fiber* of f at the origin is the motive of $\mathscr{M}_{\mathbb{G}_m}^{\mathbb{G}_m}$ defined as

$$\left(S_f\right)_{(0,0)} = -\lim_{T\to\infty}\left(Z_f(T)\right)_{(0,0)}.$$

Following the strategy of [6] we describe now how to compute the motivic Milnor fiber $\left(S_f\right)_{(0,0)}$ using the Newton algorithm and taking account of $\mathbb{G}_m$-actions (see Remark 7). Assume that f is written as $f(x,y) = \sum_{(a,b)\in\mathbb{N}^2} c_{a,b}x^a y^b$. The Newton diagram of f at the origin, denoted by $\mathscr{N}_f$, is the convex hull of the set of points $\{(a,b) + \mathbb{R}_+^2 | c_{a,b} \neq 0\}$. We denote by $\mathfrak{m}$ the usual support function of $\mathscr{N}_f$ (see Proposition 2). Let γ be a face of $\mathscr{N}_f$ and f_γ be the *face polynomial* of f associated to γ and defined as

$$f_\gamma(x,y) = \sum_{(a,b)\in\gamma} c_{a,b}x^a y^b = cx^{a_\gamma}y^{b_\gamma}\prod_{1\leq i\leq r}(y^p - \mu_i x^q)^{\nu_i},$$

with c in $\mathbb{G}_m$, $(p,q) \in \mathbb{N}^2$ coprime, μ_i in $\mathbb{G}_m$ all different and called roots of f_γ and ν_i in $\mathbb{N}^*$. The set of roots μ_i is denoted by R_γ. The last equality needs the assumption $\mathbf{k}$ algebraically closed. The main remark is that for any arc $\varphi = (x(t), y(t))$ with $\operatorname{ord} x(t)$ and $\operatorname{ord} y(t)$ in $\mathbb{N}$ we have

$$\mathfrak{m}(\operatorname{ord} x(t), \operatorname{ord} y(t)) \leq \operatorname{ord} f(\varphi)$$

with equality if and only if $f_\gamma(\overline{\mathrm{ac}}\,x(t), \overline{\mathrm{ac}}\,y(t))$ is non zero, where γ is the face of $\mathscr{N}_f$ whose dual cone $\mathscr{C}_\gamma$ contains the couple $(\operatorname{ord} x(t), \operatorname{ord} y(t))$. This implies a decomposition of the motivic zeta function

$$\left(Z_f(T)\right)_{(0,0)} = \sum_{\gamma\in\mathscr{N}(f)} Z_\gamma(T) = \sum_{\gamma\in\mathscr{N}(f)} Z_\gamma^{=}(T) + Z_\gamma^{<}(T)$$

in terms of faces of the Newton polygon and a dichotomy based on the fact that the face polynomial vanishes or doesn't vanish (see Sect. 4.1 and Proposition 4). The case of arcs such that the face polynomial evaluated on their angular components does not vanish is done in Sect. 4.3.1. The other case, necessarily for one dimensional faces, is treated in Remark 12. The main tool is the Newton map. For instance, consider γ a one dimensional face of $\mathcal{N}(f)$ and μ one of the roots of the face polynomial f_γ. The face γ is supported by a line with normal vector (p, q) in $\mathbb{N}^2$ with $\gcd(p, q) = 1$. Let (p', q') be in $\mathbb{N}^2$ such that $pp' - qq' = 1$. A Newton map of γ and μ is the map

$$\begin{array}{rcl} \sigma_{(p,q,\mu)} : \mathbf{k}[x, y] & \longrightarrow & \mathbf{k}[x_1, y_1] \\ f(x, y) & \mapsto & f(\mu^{q'} x_1^p, x_1^q(y_1 + \mu^{p'})) = f_\sigma(x_1, y_1) \end{array}.$$

Then in Proposition 7 we obtained the refined decomposition

$$Z_\gamma^<(T) = \sum_{\mu \in R_\gamma} \left(Z_{f_{\sigma(p,q,\mu)}, \omega_{p,q}}(T)\right)_{(0,0)}$$

in terms of roots and Newton transforms. Here $\omega_{p,q}$ is the differential form induced by the Newton map and motivic zeta functions with differential form are recalled in Sect. 2. The main step is Proposition 6 comparing a measure of arcs for f and for f_σ. Finally, the motivic zeta function can be computed inductively using the Newton algorithm which consists in considering the Newton polygon of f_σ and applying the same process (see Lemma 2). It is proven (see Lemma 4) that after a finite number of steps, we end up with a monomial or $x^k(y + \mu x^q + g(x, y))^\nu$ multiplied by a unit in $\mathbf{k}[[x, y]]$ with k in $\mathbb{N}$, $g(x, y) = \sum_{a+bq>q} c_{a,b} x^a y^b$ in $\mathbf{k}[x, y]$. These base cases are studied in Sect. 4.3.3. After the proof of Theorem 2, we conclude Sect. 4 by two interesting examples. Cauwergs' example in [7] shows that two functions which have the same topological type can have different motivic Milnor fibers. Schrauwen, Steenbrink and Stevens example in [26] shows, on the other hand, that two functions with different topological type can have same motivic Milnor fiber.

In the last sections, we apply the previous results to the computation of invariants of f using what is known for quasihomogeneous polynomials. We use in particular computations of the monodromy zeta function by Martin-Moralés in [23]. In Sect. 5 we deduce from Theorem 2 a generalization of Kouchnirenko's theorem which computes the Milnor number, in the isolated case, in terms of the area of the surfaces under the Newton polygons appearing in the Newton algorithm. Furthermore, in Sect. 3 we recover the formula for the monodromy Zeta function given by Eisenbud and Neumann in [12] and that of Varchenko in [27].

2 Motivic Milnor Fibers

Below we explain some definitions and properties that will be used throughout the paper. We refer to [10, 17, 21, 22] and [16] for further discussion.

2.1 Motivic Setting

2.1.1 Grothendieck Rings

Let $\mathbf{k}$ be a field of characteristic 0 and denote by $\mathbb{G}_m$ its multiplicative group. We call $\mathbf{k}$-*variety*, a separated reduced scheme of finite type over $\mathbf{k}$. We denote by $Var_{\mathbf{k}}$ the category of $\mathbf{k}$-varieties and by $Var_{\mathbb{G}_m}$ the category of $\mathbb{G}_m$-varieties, where objects are morphisms $X \to \mathbb{G}_m$ in $Var_{\mathbf{k}}$. As usual we denote by $\mathscr{M}_{\mathbb{G}_m}$ the localization of the Grothendieck ring of varieties over $\mathbb{G}_m$ with respect to the relative line. We will use also the $\mathbb{G}_m$-equivariant variant $\mathscr{M}_{\mathbb{G}_m}^{\mathbb{G}_m}$ introduced in [17, §2] or [16, §2], which is generated by classes of objects $Y \to \mathbb{G}_m$ endowed with a monomial $\mathbb{G}_m$-action. In this context the class of the projection from $\mathbb{A}^1_{\mathbf{k}} \times \mathbb{G}_m$ to $\mathbb{G}_m$ endowed with the trivial action is denoted by $\mathbb{L}$.

2.1.2 Rational Series

Let A be one of the rings $\mathbb{Z}[\mathbb{L}, \mathbb{L}^{-1}]$ or $\mathscr{M}_{\mathbb{G}_m}^{\mathbb{G}_m}$. We denote by $A[[T]]_{sr}$ the A-submodule of $A[[T]]$ generated by 1 and finite products of terms $p_{e,i}(T)$ defined as $\mathbb{L}^e T^i/(1 - \mathbb{L}^e T^i)$ with e in $\mathbb{Z}$ and i in $\mathbb{N}_{>0}$. There is a unique A-linear morphism $\lim_{T\to\infty} : A[[T]]_{sr} \to A$ such that for any subset $(e_i, j_i)_{i\in I}$ of $\mathbb{Z} \times \mathbb{N}_{>0}$ with I finite or empty, $\lim_{T\to\infty}(\prod_{i\in I} p_{e_i,j_i}(T))$ is equal to $(-1)^{|I|}$.

We will use the following lemma similar to [5, §3] or [15, 2.1.5] and [17, 2.9].

Lemma 1 *Let φ and η be two $\mathbb{Z}$-linear forms defined on $\mathbb{Z}^2$ with $\varphi(\mathbb{N}^2)$ and $\eta(\mathbb{N}^2)$ included in $\mathbb{N}$. Let C be a rational polyhedral convex cone of $\mathbb{R}^2 \setminus \{(0,0)\}$. We assume that for any $n \geq 1$, the set C_n defined as $\varphi^{-1}(n) \cap C \cap \mathbb{N}^2$ is finite. We consider the formal series in $\mathbb{Z}\left[\mathbb{L}, \mathbb{L}^{-1}\right][[T]]$.*

$$S_{\varphi,\eta,C}(T) = \sum_{n\geq 1} \sum_{(k,l)\in C_n} \mathbb{L}^{-\eta(k,l)} T^n.$$

- *If C is equal to $\mathbb{R}_{>0}\omega_1 + \mathbb{R}_{>0}\omega_2$ where ω_1 and ω_2 are two non colinear primitive vectors in $\mathbb{N}^2$ with $\varphi(\omega_1) \neq 0$ and $\varphi(\omega_2) \neq 0$ then, if*

$$\mathscr{P} = (]0,1]\omega_1 +]0,1]\omega_2) \cap \mathbb{N}^2$$

we have

$$S_{\varphi,\eta,C}(T) = \sum_{(k_0,l_0)\in\mathscr{P}} \frac{\mathbb{L}^{-\eta(k_0,l_0)} T^{\varphi(k_0,l_0)}}{(1 - \mathbb{L}^{-\eta(\omega_1)} T^{\varphi(\omega_1)})(1 - \mathbb{L}^{-\eta(\omega_2)} T^{\varphi(\omega_2)})} \tag{1}$$

and $\lim_{T\to\infty} S_{\varphi,\eta,C}(T) = 1$.

- *If C is equal to $\mathbb{R}_{>0}\omega$ where ω is a primitive vector in $\mathbb{N}^2$ with $\varphi(\omega) \neq 0$ then we have*

$$S_{\varphi,\eta,C}(T) = \frac{\mathbb{L}^{-\eta(\omega)}T^{\varphi(\omega)}}{1-\mathbb{L}^{-\eta(\omega)}T^{\varphi(\omega)}} \text{ and } \lim_{T\to\infty} S_{\varphi,\eta,C}(T) = -1. \tag{2}$$

Proof Assume that C is equal to $\mathbb{R}_{>0}\omega_1 + \mathbb{R}_{>0}\omega_2$ where ω_1 and ω_2 are two non colinear primitive vectors in $\mathbb{N}^2$ with $\varphi(\omega_1) \neq 0$ and $\varphi(\omega_2) \neq 0$. Let $\mathscr{P}$ be defined as $(]0,1]\omega_1+]0,1]\omega_2)\cap\mathbb{N}^2$. For any $(k,l) \in C$ there is a unique $(k_0,l_0) \in \mathscr{P}$, there is a unique (α,β) in $\mathbb{N}^2$ such that

$$(k,l) = (k_0,l_0) + \alpha\omega_1 + \beta\omega_2.$$

As the set $\mathscr{P}$ is finite, we have

$$S_{\varphi,\eta,C}(T) = \sum_{(k_0,l_0)\in\mathscr{P}} \mathbb{L}^{-\eta(k_0,l_0)}T^{\varphi(k_0,l_0)}S_{k_0,l_0}(T)$$

where

$$\begin{aligned} S_{k_0,l_0}(T) &= \sum\nolimits_{n\geq 1} \sum_{\substack{\alpha\geq 0,\beta\geq 0\\ \alpha\varphi(\omega_1)+\beta\varphi(\omega_2)=n-\varphi(k_0,l_0)}} \left(\mathbb{L}^{-\eta(\omega_1)}T^{\varphi(\omega_1)}\right)^{\alpha}\left(\mathbb{L}^{-\eta(\omega_2)}T^{\varphi(\omega_2)}\right)^{\beta} \\ &= \sum\nolimits_{n\geq \varphi(k_0,l_0)} \sum_{\substack{\alpha\geq 0,\beta\geq 0\\ \alpha\varphi(\omega_1)+\beta\varphi(\omega_2)=n-\varphi(k_0,l_0)}} \left(\mathbb{L}^{-\eta(\omega_1)}T^{\varphi(\omega_1)}\right)^{\alpha}\left(\mathbb{L}^{-\eta(\omega_2)}T^{\varphi(\omega_2)}\right)^{\beta} \\ &= \sum\nolimits_{m\geq 0} \sum_{\substack{\alpha\geq 0,\beta\geq 0\\ \alpha\varphi(\omega_1)+\beta\varphi(\omega_2)=m}} \left(\mathbb{L}^{-\eta(\omega_1)}T^{\varphi(\omega_1)}\right)^{\alpha}\left(\mathbb{L}^{-\eta(\omega_2)}T^{\varphi(\omega_2)}\right)^{\beta} \end{aligned}$$

which implies the equality (1). As $\varphi(1,0)$ and $\varphi(0,1)$ are non negative integers, for any (k_0,l_0) in $\mathscr{P}$, $\varphi(k_0,l_0) \leq \varphi(\omega_1) + \varphi(\omega_2)$ with equality only in the case

$$(k_0,l_0) = \omega_1 + \omega_2$$

which implies the result on the limit in (1). The proof of (2) is similar.

2.2 *Arcs*

2.2.1 Arc Spaces

Let X be a $\mathbf{k}$-variety. We denote by $\mathscr{L}_n(X)$ the *space of n-jets* of X. This set is a $\mathbf{k}$-scheme of finite type and its K-rational points are morphisms from $Spec\ K[t]/t^{n+1}$ to X, for any extension K of $\mathbf{k}$. There are canonical morphisms form $\mathscr{L}_{n+1}(X)$ to $\mathscr{L}_n(X)$ induced by the truncation modulo t^{n+1}. These morphisms are $\mathbb{A}^d_{\mathbf{k}}$-bundles when X is smooth with pure dimension d. The *arc space* of X, denoted by $\mathscr{L}(X)$, is the projective limit of this system. This set is a $\mathbf{k}$-scheme and we denote by π_n from $\mathscr{L}(X)$ to $\mathscr{L}_n(X)$ the canonical morphisms. For more details we refer for instance to [9, 22].

2.2.2 Origin, Order, Angular Component and Action

For a non zero element φ in $K[[t]]$ or in $K[t]/t^{n+1}$, we denote by ord (φ) the valuation of φ and by $\overline{\text{ac}}\ (\varphi)$ its first non-zero coefficient. By convention $\overline{\text{ac}}\ (0)$ is zero. The scalar $\overline{\text{ac}}\ (\varphi)$ is called the *angular component* of φ. The multiplicative group $\mathbb{G}_m$ acts canonically on $\mathscr{L}_n(X)$ and on $\mathscr{L}(X)$ by $\lambda.\varphi(t) := \varphi(\lambda t)$. We consider the application *origin* which maps an arc φ to $\varphi(0) = \varphi \mod t$.

2.3 *The Motivic Milnor Fiber*

Let f be a polynomial in $\mathbf{k}[x, y]$, with $f(0, 0) = 0$. For any $n \geq 1$,

$$X_n(f) = \left\{\varphi \in \mathscr{L}(\mathbb{A}^2_{\mathbf{k}}) \mid \varphi(0) = (0, 0),\ \text{ord}\, f(\varphi) = n\right\}$$

is a scheme endowed with the arrow "angular component" $\overline{\text{ac}}\ (f)$ to $\mathbb{G}_m$ and the standard action of $\mathbb{G}_m$ on arcs. In particular, for any $m \geq n$ the image of $X_n(f)$ by the truncation map π_m is a variety in $\mathscr{M}^{\mathbb{G}_m}_{\mathbb{G}_m}$ denoted by $X_n^{(m)}(f)$. In particular, by smoothness of $\mathbb{A}^2_{\mathbf{k}}$ we have the equality

$$\left[X_n^{(m)}(f)\right]\mathbb{L}^{-2m} = \left[X_n^{(n)}(f)\right]\mathbb{L}^{-2n} \in \mathscr{M}^{\mathbb{G}_m}_{\mathbb{G}_m}$$

this element is the *motivic measure* of $X_n(f)$ denoted by mes $(X_n(f))$ and introduced by Kontsevich in [20].

Denef and Loeser defined in [8] the *motivic zeta function* of f at the origin as

$$\left(Z_f(T)\right)_{(0,0)} = \sum_{n\geq 1} \text{mes}\,(X_n(f))T^n$$

in $\mathscr{M}^{\mathbb{G}_m}_{\mathbb{G}_m}[[T]]$. They show that the motivic zeta function is rational by giving a formula for $Z_f(T)$ in terms of a resolution of f, see for instance [8, 11]. It admits a limit when T goes to ∞ and by definition the *motivic Milnor fiber* of f at $(0, 0)$ is

$$\left(S_f\right)_{(0,0)} = -\lim_{T\to\infty}\left(Z_f(T)\right)_{(0,0)} \in \mathscr{M}^{\mathbb{G}_m}_{\mathbb{G}_m}.$$

Denef and Loeser proved that this motive contains usual invariants of the Milnor fiber of f at $(0, 0)$ see for instance [8, Theorem 4.2.1, Corollary 4.3.1] or [11, 19]. We will apply these theorems in Sects. 5 and 6 to compute the Euler characteristic and the monodromy zeta function of the Milnor fiber in terms of the Newton polygons of f.

2.4 *Motivic Zeta Function and Differential Form*

Let f be a polynomial in $\mathbf{k}[x, y]$ with $f(0, 0) = 0$. Let ω be the differential form $\omega = x^{\nu-1} dx \wedge dy$ with $\nu \geq 1$. If x does not divide f then we assume $\nu = 1$. For any (n, l) in $(\mathbb{N}^*)^2$, we define

$$X_{n,l}(f, \omega) = \left\{\varphi \in \mathscr{L}(\mathbb{A}^2_{\mathbf{k}}) \mid \varphi(0) = (0, 0),\ \text{ord}\, f(\varphi(t)) = n,\ \text{ord}\, \omega(\varphi(t)) = l\right\}$$

endowed with the arrow "angular component" $\overline{\text{ac}}(f)$ to $\mathbb{G}_m$ and the standard action of $\mathbb{G}_m$ on arcs. We consider the following motivic zeta function

$$\left(Z_{f,\omega}(T)\right)_{(0,0)} = \sum_{n\geq 1} \left(\sum_{l\geq 1} \text{mes}\ \left(X_{n,l}(f, \omega)\right) \mathbb{L}^{-l}\right) T^n \in \mathscr{M}_{\mathbb{G}_m}^{\mathbb{G}_m}[[T]]. \tag{3}$$

The assumption on ω ensures that the sum over l is finite. This zeta function is for instance studied in [2, 5, 7, 28]. It is rational by Denef-Loeser standard arguments, and admits a limit when T goes to ∞, and we denote

$$\left(S_{f,\omega}\right)_{(0,0)} = -\lim_{T\to\infty} \left(Z_{f,\omega}(T)\right)_{(0,0)}.$$

Proposition 1 *The motives* $\left(S_f\right)_{(0,0)}$ *and* $\left(S_{f,\omega}\right)_{(0,0)}$ *are equal.*

Proof By taking the limit when T goes to infinity, this equality follows immediately from the expression of the rational form of the motivic zeta functions $\left(Z_{f,\omega}(T)\right)_{(0,0)}$ and $\left(Z_f(T)\right)_{(0,0)}$ on a log-resolution of $(\mathbb{A}^2, f^{-1}(0) \cup \{x = 0\})$ adapted to $x = 0$. See for instance [7, §1.4].

3 Newton Algorithm

Let $\mathbf{k}$ be an algebraically closed field of characteristic zero, with multiplicative group denoted by $\mathbb{G}_m$.

Definition 1 Let f be an element of $\mathbf{k}[x, y]$ written as $f(x, y) = \sum_{(a,b)\in\mathbb{Z}^2} c_{a,b} x^a y^b$.

We define the *support* of f by

$$\operatorname{Supp} f = \left\{(a,b) \in \mathbb{Z}^2 | c_{a,b} \neq 0\right\}.$$

In this section (Lemma 2), we define the local Newton algorithm for a polynomial f in $\mathbf{k}[x, y]$.

3.1 Newton Polygons

Definition 2 For any set E in $\mathbb{N} \times \mathbb{N}$, denote by $\Delta(E)$ the smallest convex set containing

$$E + (\mathbb{R}_{>0})^2 = \left\{a + b, a \in E, b \in (\mathbb{R}_{>0})^2\right\}.$$

A subset Δ of $\mathbb{R}^2$ is called *Newton diagram* if there exists a set E in $\mathbb{N} \times \mathbb{N}$, such that Δ is equal to $\Delta(E)$. The smallest set E_0 of $\mathbb{N} \times \mathbb{N}$ such that Δ is equal to $\Delta(E_0)$ is called the *set of vertices* of Δ.

Remark 1 The set of vertices of a Newton diagram is finite.

Definition 3 Let E_0 be the set $\{v_0, \cdots, v_d\}$ with $v_i = (a_i, b_i)$ in $\mathbb{N} \times \mathbb{N}$ satisfying $a_{i-1} < a_i$ and $b_{i-1} > b_i$ for any i in $\{1, \ldots, d\}$. For such i, we denote by S_i the segment $[v_{i-1}, v_i]$ and by l_{S_i} the line supporting the segment S_i. We call the set

$$\mathcal{N}(\Delta) = \{S_i\}_{i \in \{1,\ldots,d\}} \cup \{v_i\}_{i \in \{0,\ldots,d\}}$$

the *Newton polygon* of Δ. The integer $h(\Delta) = b_0 - b_d$ is called the *height* of Δ.

Let f be an element of $k[x, y]$ equal to

$$f(x, y) = \sum_{(a,b) \in \mathbb{N} \times \mathbb{N}} c_{a,b} x^a y^b.$$

Definition 4 The *Newton diagram* of f is the Newton diagram $\Delta(f)$ equal to $\Delta(\operatorname{Supp} f)$. The *Newton polygon at the origin* of f is the Newton polygon $\mathcal{N}(f)$ equal to $\mathcal{N}(\Delta(f))$. The *height of f* denoted by $h(f)$ is the height $h(\Delta(f))$.

Definition 5 Let l be a line in $\mathbb{R}^2$. The *initial part* of f with respect to l is the quasi-homogeneous polynomial

$$\operatorname{in}(f, l) = \sum_{(a,b) \in l \cap \mathcal{N}(f)} c_{a,b} x^a y^b.$$

Notation 1 *In the following, for a face γ of $\mathcal{N}(f)$ contained in a line l, instead of writing* $\mathrm{in}(f, l)$ *we will simply write* f_γ.

Remark 2 Let γ be a one-dimensional face of $\mathcal{N}(f)$. We have the equality

$$f_\gamma = x^{a_\gamma} y^{b_\gamma} F_\gamma(x^q, y^p),$$

where (a_γ, b_γ) belongs to $\mathbb{N} \times \mathbb{N}$, p and q are coprime positive integers and

$$F_\gamma(x, y) = c \prod_{1 \leq i \leq r} (y - \mu_i x)^{\nu_i},$$

with c in $\mathbf{k}^*$, μ_i in $\mathbf{k}^*$ (all different) and ν_i in $\mathbb{N}^*$. We call f_γ the *face polynomial* of f associated to γ.

Definition 6 We say that f is *non degenerate* with respect to its Newton polygon $\mathcal{N}(f)$ if for each one dimensional face γ in $\mathcal{N}(f)$, the polynomial F_γ has simple roots.

Remark 3 The polynomial f is non degenerate with respect to $\mathcal{N}(f)$ if and only if for any one-dimensional face γ, the initial part f_γ has no critical point on the torus $(\mathbf{k}^*)^2$.

3.2 Newton Algorithm

Definition 7 (Newton map) Let (p, q) be in $\mathbb{N}^2$ with $\gcd(p, q) = 1$. Let (p', q') be in $\mathbb{N}^2$ such that $pp' - qq' = 1$. Let μ be in $\mathbb{G}_m$. We define the application

$$\begin{array}{rcl} \sigma_{(p,q,\mu)} : \mathbf{k}[x, y] & \longrightarrow & \mathbf{k}[x_1, y_1] \\ f(x, y) & \mapsto & f\left(\mu^{q'} x_1^p, x_1^q (y_1 + \mu^{p'})\right) \end{array}$$

We say that $\sigma_{(p,q,\mu)}$ is a *Newton map*.

Remark 4 The Newton map $\sigma_{(p,q,\mu)}$ depends on (p', q') but this doesn't affect our results. Indeed, if $(p' + lq, q' + lp)$ is another pair, then

$$f\left(\mu^{q'+lp} x_1^p, x_1^q (y_1 + \mu^{p'+lq})\right) = f\left(\mu^{q'} (x_1 \mu^l)^p, (x_1 \mu^l)^q (y_1 \mu^{-lq} + \mu^{p'})\right).$$

Furthermore, there is exactly one choice of (p', q') satisfying $pp' - qq' = 1$ and moreover $p' \leq q$ and $q' < p$. In the sequel we will always assume these inequalities. This will make procedures canonical.

In the well-known following proposition we introduce notations used in the rest of the paper.

Proposition 2 *Let E be a set of some (m, n) for n and m in $\mathbb{N}$. Let (p, q) be in $\mathbb{N}^2$ with $\gcd(p, q) = 1$. We consider the linear form $l_{(p,q)}$ which maps (a, b) on $ap + bq$.*

1. *The minimum of $l_{(p,q)|\Delta(E)}$, denoted by $\mathfrak{m}(p, q)$, is obtained on a face of $\Delta(E)$ denoted by $\gamma(p, q)$. In particular, $l_{(p,q)}$ is constant on $\gamma(p, q)$.*
2. *For any face γ of $\Delta(E)$, we denote by C_γ the interior, in its own generated vector space in $\mathbb{R}^2$, of the positive cone generated by the set $\{(p, q) \in \mathbb{N}^2 \mid \gamma(p, q) = \gamma\}$. This set is a rational polyhedral cone, convex and relatively open.*

For a one dimensional face γ, we denote by $\mathbf{n}_\gamma$ the normal vector to γ with integral non negative coordinates and the smallest norm. With that notation we have :

3. *For a one dimensional face γ, $C_\gamma = \mathbb{R}_{>0}\mathbf{n}_\gamma$.*
4. *A zero dimensional face γ is an intersection of two one dimensional faces γ_1 and γ_2, and $C_\gamma = \mathbb{R}_{>0}\mathbf{n}_{\gamma_1} + \mathbb{R}_{>0}\mathbf{n}_{\gamma_2}$.*
5. *The set of cones C_γ is a fan, called the dual fan of $\Delta(E)$.*

Lemma 2 (Newton algorithm) *Let (p, q) be in $\mathbb{N}^2$ with $\gcd(p, q) = 1$. Let μ be in $\mathbb{G}_m$.*

Let f be a non zero element in $\mathbf{k}[x, y]$ and f_1 be its Newton transform $\sigma_{(p,q,\mu)}(f)$ in $\mathbf{k}[x_1, y_1]$.

1. *If there does not exist a face S of $\mathcal{N}(f)$ of dimension 1 whose supporting line has equation $pa + qb = N$, for some N, then*

$$f_1(x_1, y_1) = x_1^{\mathfrak{m}(p,q)} u(x_1, y_1)$$

with $u(x_1, y_1)$ in $\mathbf{k}[x_1, y_1]$ and $u(0, 0) \neq 0$.

2. *If there exists a face S of $\mathcal{N}(f)$ of dimension 1 whose supporting line has equation $pa + qb = N$, and if $F_S(1, \mu) \neq 0$, then $\mathfrak{m}(p, q) = N$ and*

$$f_1(x_1, y_1) = x_1^N u(x_1, y_1)$$

with $u(x_1, y_1)$ in $\mathbf{k}[x_1, y_1]$ and $u(0, 0) \neq 0$.

3. *If there exists a face S of $\mathcal{N}(f)$ of dimension 1 whose supporting line has equation $pa + qb = N$, and if $F_S(1, \mu) = 0$, then $\mathfrak{m}(p, q) = N$ and*

$$f_1(x_1, y_1) = x_1^N g_1(x_1, y_1)$$

with $g_1(x_1, y_1)$ in $\mathbf{k}[x_1, y_1]$ and $g_1(0, 0) = 0$, $g_1(0, y_1) = dy_1^\nu + \cdots$, where $d \neq 0$ and ν is the multiplicity of μ as root of $F_S(1, Y) = 0$. In particular in that case, $\nu \geq h(\sigma_{(p,q,\mu)}(f))$.

Proof We consider the linear form $l_{(p,q)}$ which maps (a,b) on $ap+bq$. Writing $f(x,y)=\sum_{(a,b)\in \mathrm{supp}(f)} c_{a,b}x^a y^b$ we obtain

$$f_1(x_1,y_1)=\sum_{(a,b)\in \mathrm{supp}(f)}\left(c_{a,b}\mu^{q'a}(y_1+\mu^{p'})^b\right)x_1^{l_{(p,q)}(a,b)}.$$

Let $\mathfrak{m}(p,q)$ be the minimum $\min_{(a,b)\in \mathrm{supp}(f)} l_{(p,q)}(a,b)$. By convexity and definition of the Newton diagram of f, this minimum is reached on a zero dimensional or one dimensional face γ. Then, we can write

$$\begin{aligned} f_1(x_1,y_1)&=x_1^{\mathfrak{m}(p,q)}\sum\nolimits_{(a,b)\in\gamma} c_{a,b}\mu^{q'a}\left(y_1+\mu^{p'}\right)^b \\ &+x_1^{\mathfrak{m}(p,q)}\sum\nolimits_{(a,b)\notin\gamma}\left(c_{a,b}\mu^{q'a}(y_1+\mu^{p'})^b\right)x_1^{l_{(p,q)}(a,b)-\mathfrak{m}(p,q)}. \end{aligned}$$

Remark that for any (a,b) not in γ, $l_{(p,q)}(a,b)-\mathfrak{m}(p,q)>0$.

1. If there does not exist a face S of $\mathcal{N}(f)$ of dimension 1 whose supporting line has direction $pa+qb=0$, then the face γ is a vertex and the results follows.
2. If there exists a one dimensional face S of $\mathcal{N}(f)$ whose supporting line has equation $pa+qb=N$, then γ is equal to S. In particular we have $N=\mathfrak{m}(p,q)$ and

$$x_1^{\mathfrak{m}(p,q)}\sum_{(a,b)\in\gamma} c_{a,b}\mu^{q'a}(y_1+\mu^{p'})^b=\mathrm{in}(f,\gamma)\left(\mu^{q'}x_1^p, x_1^q(y_1+\mu^{p'})\right),$$

and the result follows by computations.

Remark 5 If $f_1(x_1,y_1)$ is equal to $x_1^{n_1}y_1^{m_1}u(x_1,y_1)$, where (n_1,m_1) belongs to $\mathbb{N}^2$ and $u\in k[x_1,y_1]$ is a unit in $\mathbf{k}[[x_1,y_1]]$, we say for short that f_1 is a *monomial times a unit.*

Remark 6 From this lemma, we see that there are a finite number of (p,q,μ) such that $\sigma_{(p,q,\mu)}(f)$ is eventually not a monomial times a unit in $\mathbf{k}[[x_1,y_1]]$. These triples are given by the equations of the faces of the Newton polygon and the roots of the corresponding face polynomials.

Notation 2 *If* $\sigma=\sigma_{(p,q,\mu)}$ *is a Newton map, we denote by* $f_\sigma=\sigma_{(p,q,\mu)}(f)$.

Lemma 3 (Lemma 2.11 [5]) *If the height of* f_σ *is equal to the height of* f*, then the Newton polygon of* f *has a unique face* S *with* $f_S=x^k y^l (y-\mu x^q)^\nu$*, with* (k,l,ν) *in* $\mathbb{N}\times\mathbb{N}^2$ *and* q *in* $\mathbb{N}^*$.

Let $\Sigma_n=(\sigma_1,\cdots,\sigma_n)$ where σ_i is a Newton map for all i, we define f_{Σ_n} by induction: $f_{\Sigma_1}=f_{\sigma_1}$, $f_{\Sigma_i}=(f_{\Sigma_{i-1}})_{\sigma_i}$.

Definition 8 If the height in the Newton process remains constant, we say that the Newton algorithm stabilizes.

Lemma 4 *Let f be in $\mathbf{k}[x, y]$. If the Newton algorithm of f stabilizes with height ν then*

$$f(x, y) = U(x, y)x^k(y + x^q + g(x, y))^\nu,$$

with $(k, q) \in \mathbb{N}^2$, $g(x, y) \in \mathbf{k}[x, y]$, $g(x, y) = \sum_{a+bq>q} c_{a,b}x^a y^b$, $U(x, y) \in \mathbf{k}[[x, y]]$ and $U(0, 0) \neq 0$.

Proof Assume that the Newton algorithm of f is stabilized in height ν. We can write

$$f(x, y) = \sum_{i>\nu} y^i a_i(x) + y^\nu a_\nu(x) + \sum_{j<\nu} y^j a_j(x)$$

with $a_\nu(x) = 1 + \cdots$, for $j < \nu$, $a_j(x) = \frac{\nu!}{j!(\nu-j)!}\mu^{\nu-j}x^{q(\nu-j)} + \cdots$ where $\cdots$ means higher order terms. Let $f_1(x, y)$ be the $(\nu - 1)$th derivative of f with respect to y.

$$f_1(x, y) = \sum_{i>\nu} \frac{i!}{(i-\nu)!} y^{i-\nu+1} a_i(x) + \nu! y a_\nu(x) + (\nu - 1)! a_{\nu-1}(x)$$

$$f_1(x, y) = \nu!(y + \mu x^q + g(x, y))$$

with $g(x, y) \subset \mathbf{k}[x, y]$, $g(x, y) = \sum_{a+bq>q} c_{a,b}x^a y^b$. The hypothesis that the Newton algorithm is stabilized in height ν, implies that at each step the face polynomial is of the form $x^l(y + \tilde{\mu}x^r)^\nu$ for some $l, \tilde{\mu}, r$ and it implies that f has a Puiseux series of the form $y_0 = \sum c_i x^i$ with $i > 0$ as its unique root of multiplicity ν. This point follows from Puiseux theory, see for instance [29, 2.1]. Then, there is a formal series $u(x, y)$ in $\mathbf{k}[[x, y]]$ with $u(0, 0) \neq 0$ such that

$$f(x, y) = u(x, y)(y - y_0)^\nu.$$

Then y_0 is also the unique root of the polynomial f_1, and

$$f_1(x, y) = \nu!(y + \mu x^q + g(x, y)) = u_1(x, y)(y - y_0)$$

with $u_1(x, y) \in \mathbf{k}[[x, y]]$, $u_1(0, 0) \neq 0$. Then we obtain

$$f(x, y) = (\nu!)^\nu \frac{u(x, y)}{u_1(x, y)^\nu}(y + \mu x^q + g(x, y))^\nu = U(x, y)(y + \mu x^q + g(x, y))^\nu$$

with $U(x, y) \in \mathbf{k}[[x, y]]$, $U(0, 0) \neq 0$.

Theorem 1 *For all $f(x, y)$ in $\mathbf{k}[x, y]$, there exists a natural integer n_0, such that for any sequence of Newton maps $\Sigma_n = (\sigma_1, \cdots, \sigma_n)$ with $n \geq n_0$, f_{Σ_n} is either a* *monomial times a unit or there exists an integer* ν *in* $\mathbb{N}$ *such that* f_{Σ_n} *is of the form* $U(x, y)x^k(y + \mu x^q + g(x, y))^\nu$ *where* k *is an integer in* $\mathbb{N}$, $g(x, y)$ *is a polynomial in* $\mathbf{k}[x, y]$ *equal to* $\sum_{a+bq>q} c_{a,b}x^a y^b$, μ *belongs to* $\mathbb{C}^*$ *and* $U(x, y)$ *is in* $\mathbf{k}[[x, y]]$ *with* $U(0, 0) \neq 0$.

Proof From Lemma 2, we first observe that the number of Newton maps σ, such that f_σ is not a monomial times a unit is finite, bounded by the sum on all faces S of the number of roots of F_S.

We argue by induction on the height of f. If $h(f) = 0$, f is a monomial times a unit and $n_0 = 0$. Consider the case where $h(f) > 0$. In that case, $\mathcal{N}(f)$ has at least one face of dimension 1. Choose one, S, and a root of F_S with multiplicity $\nu \geq 1$. Let $ap + bq = N$ be the equation of the supporting line of S. Then

$$f_\sigma(x_1, y_1) = x_1^N g_1(x_1, y_1),$$

with N in $\mathbb{N}$, $g_1(x_1, y_1)$ in $\mathbf{k}[x_1, y_1]$ and the height of $\Delta(f_1)$ is equal to $\nu \leq h(f)$. Either we have $h(f_\sigma) = 0$, or $h(f) \geq h(f_\sigma) \geq 1$. We continue the process and either the height vanishes or it stabilizes at a positive value.

Example 1 Let f be the polynomial

$$f(x, y) = (x^3 - y^2)^2 x^2 + y^7.$$

The Newton polygon has two compact faces of dimension one. Then, we consider $v_0 = (0, 7), v_1 = (2, 4), v_2 = (8, 0)$. The face $\gamma_1 = [v_0, v_1]$ has supporting line with equation $3\alpha + 2\beta = 14$ and polynomial face $y^4(x^2 + y^3)$. The face $\gamma_2 = [v_1, v_2]$ has supporting line $2\alpha + 3\beta = 16$ and face polynomial $x^2(x^3 - y^2)^2$. Relatively to the face γ_1, the Newton map gives $x = x_1^3, y = x_1^2(y_1 - 1)$ and the Newton transform is

$$f_1^1(x_1, y_1) = x_1^{14}(3y_1 - 2x_1^5 + g(x_1, y_1)).$$

The Newton algorithm stabilizes in height 1. Relatively to the face γ_2, the Newton map gives $x = x_1^2, y = x_1^3(y_1 + 1)$ and the Newton transform is

$$f_1^2(x_1, y_1) = x_1^{16}(4y_1^2 + x_1^5 + g_1(x_1, y_1)).$$

There is a second step $x_1 = -x_2^2, y = x_2^5(y_2 + 1/2)$ and the Newton transform is

$$f_2^2(x_2, y_2) = x_2^{42}(-512y_2 + 664x_2^{10} + g_2(x_2, y_2)).$$

The Newton algorithm stabilizes in height 1.

4 Motivic Milnor Fibers and Newton Algorithm

Notation 1 *Let* $\mathbf{k}$ *be an algebraically closed field of characteristic zero. Let* f *be a polynomial in* $\mathbf{k}[x, y]$ *with* $f(0,0) = 0$*. Let* ω *be the differential form* $\omega = x^{\nu-1} dx \wedge dy$ *with* $\nu \geq 1$*. If* x *does not divide* f *then we assume* $\nu = 1$*.*

In this section, using the Newton algorithm Sect. 3.1 and the strategy of the first author and Veys in [6] we express the motivic zeta function $(Z_{f,\omega}(T))_{(0,0)}$ and the motivic Milnor fiber $(S_f)_{(0,0)}$ in terms of the Newton polygons of f and its Newton transforms.

Remark 7 Compared to [6], we work in this article relatively to the standard action on arcs, varieties involved are, then, endowed with an action of the multiplicative group $\mathbb{G}_m$. Applying the Newton algorithm as in [6] we have to take into account these actions and use the construction of the Grothendieck ring $\mathscr{M}_{\mathbb{G}_m}^{\mathbb{G}_m}$ to identify classes $[X \to \mathbb{G}_m, \sigma]$ and $[X \to \mathbb{G}_m, \sigma']$ for different actions σ and σ' of $\mathbb{G}_m$. This identification allows factorization in zeta functions, see for instance Proposition 5 and its proof.

Theorem 2 *Let* f *be a polynomial in* $\mathbf{k}[x, y]$ *with* $f(0,0) = 0$*, let* ω *be the differential form* $x^{\nu-1} dx \wedge dy$*, with* $\nu = 1$ *if* x *does not divide* f*. Denote by* $\gamma_h = (a_h, b_h)$ *and* $\gamma_v = (a_v, b_v)$ *the zero dimensional faces contained respectively in the horizontal and vertical faces of* $\mathscr{N}(f)$*. The motivic zeta function of* f *in* $(0,0)$ *relatively to the differential form* ω *is equal to*

$$\begin{aligned}(Z_{f,\omega}(T))_{(0,0)} &= \textstyle\sum_{(a,b)\in\{\gamma_h,\gamma_v\}} [x^a y^b : \mathbb{G}_m^r \to \mathbb{G}_m, \sigma_{\mathbb{G}_m}] R_{x^a y^b,\omega}(T) \\ &+ \textstyle\sum_{\gamma\in\mathscr{N}(f)\setminus\{\gamma_h,\gamma_v\}} \left[f_\gamma : \mathbb{G}_m^2 \setminus (f_\gamma = 0) \to \mathbb{G}_m, \sigma_\gamma\right] R_{\gamma,\omega}(T) \\ &+ \textstyle\sum_{\gamma\in\mathscr{N}(f), \dim\gamma=1} \sum_{\mu\in R_\gamma} \left(Z_{f_{\sigma(p,q,\mu)},\omega_{p,q}}(T)\right)_{(0,0)},\end{aligned} \tag{4}$$

where

- $r = 1$ *if* $a = 0$ *or* $b = 0$ *and* $r = 2$ *if* a *and* b *are non zero,*
- $\sigma(p, q, \mu)$ *are the associated Newton transforms and* $\omega_{p,q}$ *is the differential form* $\omega_{p,q}(v, w) = v^{(\nu p+q-1)} dv \wedge dw$*,*
- *for a face* γ*,* $R_{\gamma,\omega}$ *are rational fonctions in formulas* (12)*,* (13)*,* (14) *and* (15) *of Proposition* 3 *and formulas* (19) *and* (20) *of Proposition* 5*,*
- *For a face* (a, b) *in* $\{\gamma_h, \gamma_v\}$ *the action* $\sigma_{\mathbb{G}_m}$ *of* $\mathbb{G}_m$ *is defined in Proposition* 3*. For a face* γ *not contained in* $\{\gamma_h, \gamma_v\}$ *the action* σ_γ *of* $\mathbb{G}_m$ *is defined in Proposition* 5*.*

Furthermore, the motivic Milnor fiber of f *in* $(0,0)$ *is :*

$$\begin{aligned}(S_f)_{(0,0)} &= \textstyle\sum_{(a,b)\in\{\gamma_h,\gamma_v\}} (-1)^{r+1} [x^a y^b : \mathbb{G}_m^r \to \mathbb{G}_m, \sigma_{\mathbb{G}_m}] \\ &+ \textstyle\sum_{\gamma\in\mathscr{N}(f)\setminus\{\gamma_h,\gamma_v\}} (-1)^{\dim(\gamma)-1} \left[f_\gamma : \mathbb{G}_m^2 \setminus (f_\gamma = 0) \to \mathbb{G}_m, \sigma_\gamma\right] \\ &+ \textstyle\sum_{\gamma\in\mathscr{N}(f), \dim\gamma=1} \sum_{\mu\in R_\gamma} \left(S_{f_{\sigma(p,q,\mu)}}\right)_{(0,0)}.\end{aligned} \tag{5}$$

Proof In Sect. 4.1, formula (10), we consider the decomposition

$$\left(Z_{f,\omega}(T)\right)_{(0,0)} = \sum_{\gamma \in \mathcal{N}(f)} Z_{\gamma}(T).$$

In Sect. 4.2, Proposition 3, we show the rationality and compute the limit of $Z_{\gamma}(T)$ for γ a face contained in a non compact face of $\mathcal{N}(f)$. This limit does not depend on ω. In Sect. 4.3 we consider the case of a face γ not contained in a non compact face of $\mathcal{N}(f)$. Depending on the fact that the face polynomial f_{γ} of the face γ vanishes or does not vanish on the angular components of the coordinates of an arc (Remark 10), we decompose in formula (17) the zeta function $Z_{\gamma}(T)$ as a sum of $Z_{\gamma}^{=}(T)$ and $Z_{\gamma}^{<}(T)$. In particular if the face γ is zero dimensional then $Z_{\gamma}^{<}(T)$ is zero. In Proposition 5 we show the rationality and compute the limit of the zeta function $Z_{\gamma}^{=}(T)$. This limit does not depend on ω. In Sect. 4.3.2, Proposition 7 we prove the decomposition

$$Z_{\gamma}^{<}(T) = \sum_{\mu \in R_{\gamma}} \left(Z_{f_{\sigma(p,q,\mu)},\omega_{p,q}}(T)\right)_{(0,0)}.$$

Applying the Newton algorithm (Theorem 1) inductively, using the base cases given in the Examples 2 and 3, we obtain the rationality of the motivic zeta function $Z_{f,\omega}(T)$, we recover that its limit does not depend on ω and by Proposition 1 we obtain the decomposition of the motivic Milnor fiber $\left(S_f\right)_{(0,0)}$ in (5).

Remark 8 (Newton algorithm) In the formula (5) each of the Newton transforms $f_{\sigma(p,q,\mu)}$ is a polynomial in $\mathbf{k}[x, y]$. Applying the Newton algorithm (Theorem 1) inductively we can compute the motivic Milnor fiber $\left(S_{f,\omega}\right)_{(0,0)}$ using the base cases given in the Examples 2 and 3.

Applying the Theorem 2 in the quasi-homogeneous case we obtain

Corollary 1 *Let f be a quasi-homogeneous polynomial in $\mathbf{k}[x, y]$. We assume its Newton polygon contains a one-dimensional face γ supported by a line of equation $\alpha p + \beta q = N$,* gcd$(p, q) = 1$. *The polynomial f can be written as*

$$f(x, y) = c x^a y^b \prod_{i=1}^{r} (x^q - \mu_i y^p)^{\nu_i},$$

with $r \geq 1$. We denote by (a_v, b_v) and (a_h, b_h) the faces γ_v and γ_h contained respectively in the vertical face and the horizontal face. The motivic Milnor fiber of f at the origin is

$$\begin{aligned}\left(S_f\right)_{(0,0)} &= (-1)^{r_v+1}[x^{a_v} y^{b_v} : \mathbb{G}_m^{r_v} \to \mathbb{G}_m, \sigma_{\mathbb{G}_m}] + (-1)^{r_h+1}[x^{a_h} y^{b_h} : \mathbb{G}_m^{r_h} \to \mathbb{G}_m, \sigma_{\mathbb{G}_m}] \\ &+ \left[f_{\gamma} : \mathbb{G}_m^2 \setminus (f_{\gamma} = 0) \to \mathbb{G}_m, \sigma_{\gamma}\right] + \textstyle\sum_{i=1}^{r} \left(S_{f_{\sigma(p,q,\mu_i)}}\right)_{(0,0)},\end{aligned} \tag{6}$$

with

- *for any i in $\{1,\dots,r\}$,*

$$\left(S_{f_{\sigma(p,q,\mu_i)}}\right)_{(0,0)} = \left(S_{x^N y^{\nu_i}}\right)_{(0,0)} \tag{7}$$

- $r_v = 1$ *(resp* $r_h = 1$*) if* $a_v = 0$ *or* $b_v = 0$ *(resp* $a_h = 0$ *or* $b_h = 0$*) and* $r_v = 2$ *(resp* $r_h = 1$*) if* $a_v \neq 0$ *and* $b_v \neq 0$ *(resp* $a_h \neq 0$ *and* $b_h \neq 0$*).*

Proof The proof follows from Theorem 2 applied in the quasi-homogeneous case. Furthermore, for a root μ_i with multiplicity ν_i, by the proof of Lemma 2 the corresponding Newton transform is

$$f_{\sigma(p,q,\mu_i)}(x_1, y_1) = x_1^N y_1^{\nu_i} U_i(x_1, y_1)$$

where U_i is a unit on $\mathbf{k}[x_1, y_1]$ and we have

$$\left(S_{f_\sigma(p,q,\mu_i)}\right)_{(0,0)} = \left(S_{x_1^N y_1^{\nu_i}}\right)_{(0,0)}.$$

Theorem 3 *Let f be a polynomial in $\mathbf{k}[x, y]$ with $f(0,0) = 0$. Denote by γ_h and γ_v the zero dimensional faces contained respectively in the horizontal and vertical faces of $\mathcal{N}(f)$.*

$$\begin{aligned}\left(S_f\right)_{(0,0)} = {} & \textstyle\sum_{\gamma\in\mathcal{N}(f),\,\dim\gamma=1} \left[\left(S_{f_\gamma}\right)_{(0,0)} + \sum_{\mu\in R_\gamma}\left(S_{f_{\sigma(p,q,\mu)}}\right)_{(0,0)} - \left(S_{(f_\gamma)_{\sigma(p,q,\mu)}}\right)_{(0,0)}\right] \\ & - \textstyle\sum_{\gamma\in\mathcal{N}(f)\setminus\{\gamma_h,\gamma_v\},\dim\gamma=0}\left(S_{f_\gamma}\right)_{(0,0)}.\end{aligned} \tag{8}$$

Proof Remark first that a zero-dimensional face non contained in the horizontal or vertical face, belongs to two one-dimensional faces. Then, applying Corollary 1 for each face polynomial f_γ with γ of dimension one, and summing, we obtain

$$\textstyle\sum_{\gamma\in\mathcal{N}(f),\dim\gamma=1}\left(S_\gamma\right)_{(0,0)} =$$

$$\textstyle\sum_{(a,b)\in\{\gamma_h,\gamma_v\}}(-1)^{r+1}[x^a y^b : \mathbb{G}_m^r \to \mathbb{G}_m, \sigma_{\mathbb{G}_m}]$$

$$+2\textstyle\sum_{\gamma\in\mathcal{N}(f)\setminus\{\gamma_h,\gamma_v\},\dim\gamma=0}(-1)^{\dim(\gamma)-1}\left[f_\gamma : \mathbb{G}_m^2\setminus(f_\gamma=0)\to\mathbb{G}_m, \sigma_\gamma\right]$$

$$+\textstyle\sum_{\gamma\in\mathcal{N}(f)\setminus\{\gamma_h,\gamma_v\},\dim\gamma=1}(-1)^{\dim(\gamma)-1}\left[f_\gamma : \mathbb{G}_m^2\setminus(f_\gamma=0)\to\mathbb{G}_m, \sigma_\gamma\right]$$

$$+\textstyle\sum_{\gamma\in\mathcal{N}(f),\dim\gamma=1}\sum_{\mu\in R_\gamma}\left(S_{(f_\gamma)_{\sigma(p,q,\mu)}}\right)_{(0,0)},$$

where $r = 1$ if $a = 0$ or $b = 0$ and $r = 2$ if a and b are non zero, Then, formula (8) follows from formula (5) of Theorem 2 and the fact that the motivic Milnor fiber at the origin of a monomial $x^a y^b$ with $a \geq 1$ and $b \geq 1$ is

$$\left(S_{x^a y^b}\right)_{(0,0)} = -\left[x^a y^b : \mathbb{G}_m^2 \to \mathbb{G}_m, \sigma_{\mathbb{G}_m}\right].$$

Remark 9 The motivic vanishing cycles version of formula (8) is

$$\begin{aligned}\left(S_f\right)^{(1),\phi}_{(0,0)} = & \sum_{\gamma\in\mathcal{N}(f),\,\dim\gamma=1} \left(S_{f_\gamma}\right)^{(1),\phi}_{(0,0)} + \sum_{\mu\in\Omega_\gamma} \left(S_{f_{\sigma(p,q,\mu)}}\right)^{(1),\phi}_{(0,0)} - \left(S_{(f_\gamma)_{\sigma(p,q,\mu)}}\right)^{(1),\phi}_{(0,0)} \\ & - \sum_{\gamma\in\mathcal{N}(f)\setminus\{\gamma_h,\gamma_v\},\dim\gamma=0} \left(S_{f_\gamma}\right)^{(1),\phi}_{(0,0)},\end{aligned} \tag{9}$$

where $\left(S_f\right)^{(1)}_{(0,0)}$ is the fiber in 1 of $\left(S_f\right)_{(0,0)}$ and $\left(S_f\right)^{(1),\phi}_{(0,0)} = \left(S_f\right)^{(1)}_{(0,0)} - 1$.

Proof Indeed, the number of zero-dimensional faces of $\mathcal{N}(f)$ non contained in the vertical or horizontal face is equal to the number of one-dimensional faces of $\mathcal{N}(f)$ minus one.

4.1 Decomposition of the Zeta Function Along $\mathcal{N}(f)$

We decompose the motivic zeta function $\left(Z_{f,\omega}(T)\right)_{(0,0)}$ defined in Sect. 2.4 along the Newton polygon $\mathcal{N}(f)$ of f : for any face γ of $\mathcal{N}(f)$, for any $k \geq 1$ and $n \geq 1$, we consider

$$X_{n,k,\gamma} = \left\{ (x(t), y(t)) \in \mathcal{L}(\mathbb{A}^2_{\mathbf{k}}) \;\middle|\; \begin{array}{l} \operatorname{ord} x(t) = k \\ (\operatorname{ord} x(t), \operatorname{ord} y(t)) \in C_\gamma \\ \operatorname{ord} f((x(t), y(t))) = n \end{array} \right\}$$

where C_γ is the dual cone of the face γ (see Proposition 2). Then, by additivity of the measure we have

$$\left(Z_{f,\omega}(T)\right)_{(0,0)} = \sum_{\gamma\in\mathcal{N}(f)} Z_\gamma(T) \tag{10}$$

with for any face γ in $\mathcal{N}(f)$

$$Z_\gamma(T) = \sum_{n\geq 1} \left(\sum_{k\geq 1} \mathbb{L}^{-(\nu-1)k} \operatorname{mes}\left(X_{n,k,\gamma}\right) \right) T^n. \tag{11}$$

We argue differently if the face γ is contained or not in a non compact face.

4.2 Rationality and Limit of $Z_\gamma(T)$, for a Face γ of $\mathcal{N}(f)$ Contained in a Non Compact Face

Proposition 3 *Let γ be a zero dimensional face of $\mathcal{N}(f)$ contained in a non compact face of $\mathcal{N}(f)$. Then, the motivic zeta function $Z_\gamma(T)$ is rational and more precisely,*

- *if γ is equal to $(a, 0)$, then γ is contained in the horizontal face, its dual cone $C_\gamma = \mathbb{R}_{>0}(0, 1) + \mathbb{R}_{>0}(p, q)$ and we have*

$$Z_\gamma(T) = [x^a : \mathbb{G}_m \to \mathbb{G}_m, \sigma_{\mathbb{G}_m}]\left(-1 + \frac{1}{1 - \mathbb{L}^{-\nu p - q}T^{pa}} \sum_{r=0}^{p-1} \mathbb{L}^{-\nu r - [qr/p]}T^{ra}\right), \tag{12}$$

furthermore

$$-\lim Z_\gamma(T) = [x^a : \mathbb{G}_m \to \mathbb{G}_m, \sigma_{\mathbb{G}_m}]$$

with $\sigma_{\mathbb{G}_m}(\lambda, x) = \lambda.x$ for any (λ, x) in $\mathbb{G}_m^2$, and for any r in $\{0, \ldots, p-1\}$, $[qr/p]$ is the integral part of qr/p,

- *if γ is equal to (a, b) with $a \neq 0, b \neq 0$ and contained in the horizontal face, then $C_\gamma = \mathbb{R}_{>0}(0, 1) + \mathbb{R}_{>0}(p, q)$ and*

$$Z_\gamma(T) = [x^a y^b : \mathbb{G}_m^2 \to \mathbb{G}_m, \sigma_{\mathbb{G}_m}] \sum_{(k_0, l_0) \in \mathcal{P}_\gamma} \frac{\mathbb{L}^{-\nu k_0 - l_0}T^{ak_0 + bl_0}}{(1 - \mathbb{L}^{-(\nu p + q)}T^{pa + qb})(1 - \mathbb{L}^{-1}T^b)}, \tag{13}$$

furthermore we have

$$-\lim Z_\gamma(T) = -[x^a y^b : \mathbb{G}_m^2 \to \mathbb{G}_m, \sigma_{\mathbb{G}_m}]$$

with $\mathcal{P}_\gamma = (]0, 1](0, 1) +]0, 1](p, q)) \cap \mathbb{N}^2$ and $\sigma_{\mathbb{G}_m}(\lambda, (x, y)) = (\lambda.x, \lambda.y)$ for any (λ, x, y) in $\mathbb{G}_m^3$,

- *if γ is equal to $(0, b)$, then γ is contained in the vertical face then C_γ is the cone $\mathbb{R}_{>0}(1, 0) + \mathbb{R}_{>0}(p, q)$ and we have*

$$Z_\gamma(T) = [y^b : \mathbb{G}_m \to \mathbb{G}_m, \sigma_{\mathbb{G}_m}]\left(-1 + \frac{1}{1 - \mathbb{L}^{-p-q}T^{bq}} \sum_{r=0}^{q-1} \mathbb{L}^{-[pr/q] - r}T^{rb}\right), \tag{14}$$

furthermore

$$-\lim Z_\gamma(T) = [y^b : \mathbb{G}_m \to \mathbb{G}_m, \sigma_{\mathbb{G}_m}]$$

with $\sigma_{\mathbb{G}_m}(\lambda, y) = \lambda.y$ *for any* (λ, y) *in* $\mathbb{G}_m^2$ *and for any* r *in* $\{0, \ldots, q-1\}$, $[pr/q]$ *is the integral part of* $\frac{pr}{q}$,

- *if* γ *is equal to* (a, b) *with* $a \neq 0, b \neq 0$, *and contained in the vertical face, then* $C_\gamma = \mathbb{R}_{>0}(1,0) + \mathbb{R}_{>0}(p,q)$ *and we have*

$$Z_\gamma(T) = [x^a y^b : \mathbb{G}_m^2 \to \mathbb{G}_m, \sigma_{\mathbb{G}_m}] \sum_{(k_0,l_0)\in\mathscr{P}_\gamma} \frac{\mathbb{L}^{-\nu k_0 - l_0} T^{ak_0+bl_0}}{(1-\mathbb{L}^{-(\nu p+q)} T^{pa+qb})(1-\mathbb{L}^{-1}T^a)}, \tag{15}$$

furthermore

$$-\lim Z_\gamma(T) = -[x^a y^b : \mathbb{G}_m^2 \to \mathbb{G}_m, \sigma_{\mathbb{G}_m}]$$

with $\mathscr{P}_\gamma = (]0,1](1,0)+]0,1](p,q)) \cap \mathbb{N}^2$ *and* $\sigma_{\mathbb{G}_m}(\lambda, (x, y)) = (\lambda.x, \lambda.y)$ *for any* (λ, x, y) *in* $\mathbb{G}_m^3$.

Proof Assume that γ is contained in the horizontal face then its dual cone C_γ is equal to $\mathbb{R}_{>0}(0,1) + \mathbb{R}_{>0}(p,q)$.

- Assume $\gamma = (a, 0)$ with $a \neq 0$. For any (n, k) in $(\mathbb{N}^*)^2$, for any $(x(t), y(t))$ in $X_{n,k,\gamma}$, the order of $x(t)$ is k and the order ord $f(x(t), y(t))$ is equal to ka. Then

$$Z_\gamma(T) = \sum_{k\geq 1} \mathbb{L}^{-(\nu-1)k} \text{mes}\,(X_{ka,k,\gamma}) T^{ka}.$$

A couple (k, l) in $(\mathbb{N}^*)^2$ belongs to C_γ if and only if $pl > qk$ then, for any $k \geq 1$,

$$X_{ka,k,\gamma} = \{(x(t), y(t)) \in \mathscr{L}(\mathbb{A}^2_{\mathbf{k}}) \mid \text{ord}\, x(t) = k,\ p.\text{ord}\, y(t) > qk\}$$

Remark that the condition $p.\text{ord}\, y(t) > qk$ is equivalent to $\text{ord}\, y(t) \geq [qk/p]+1$ where $[qk/p]$ denotes the integral part of $\frac{qk}{p}$. Then we deduce, that for any $k \geq 1$,

$$\text{mes}\,(X_{ka,k,\gamma}) = [x^a : \mathbb{G}_m \to \mathbb{G}_m, \sigma_{\mathbb{G}_m}].\mathbb{L}^{-[qk/p]-k}.$$

In particular

$$Z_\gamma(T) = [x^a : \mathbb{G}_m \to \mathbb{G}_m, \sigma_{\mathbb{G}_m}] \left(\sum_{k\geq 1} \mathbb{L}^{-\nu k-[qk/p]} T^{ka} \right).$$

Let $k \geq 1$, there is a unique integer $r \in \{0, \cdots, p-1\}$ such that $k = [k/p]p + r$. There is also a unique integer β_r in $\{0, \cdots, p-1\}$ such that $qr = [qr/p]p + \beta_r$. Then, $[qk/p] = [k/p]q + [qr/p]$. Writing k as $lp + r$ with $l \in \mathbb{N}$ and $r \in$

$\{0, \cdots, p-1\}$ we decompose the zeta function $Z_\gamma(T)$ as the sum of p formal series in $\mathscr{M}_{\mathbb{G}_m}^{\mathbb{G}_m}[[T]]$

$$Z_\gamma(T) = [x^a : \mathbb{G}_m \to \mathbb{G}_m, \sigma_{\mathbb{G}_m}] \left(-1 + \sum_{r=0}^{p-1} \left(\sum_{l \geq 0} \mathbb{L}^{-\nu(lp+r)-lq-[qr/p]} T^{lpa+ra} \right) \right)$$

which implies (12).

- Assume $\gamma = (a, b)$ with $b \neq 0$. For any arc $(x(t), y(t))$ in $X_{n,k,\gamma}$, we have the equality ord $f(x(t), y(t)) = ka + b.\text{ord } y$. In particular we have

$$Z_\gamma(T) = \sum_{n \geq 1} \sum_{\substack{(k,l) \in C_\gamma \cap (\mathbb{N}^*)^2 \\ n = ka+lb}} \mathbb{L}^{-(\nu-1)k} \text{mes}(X_{k,l}) T^n$$

with for any (k, l) in $C_\gamma \cap (\mathbb{N}^*)^2$

$$X_{k,l} = \{(x(t), y(t)) \in \mathscr{L}(\mathbb{A}^2_{\mathbf{k}}) \mid \text{ord } x = k, \text{ord } y = l\}$$

and mes $(X_{k,l}) = [x^a y^b : \mathbb{G}_m^2 \to \mathbb{G}_m, \mathbb{G}_m^2]\mathbb{L}^{-k-l}$. Then formula (13) follows from Lemma 1 applied to C_γ with $\eta : \mathbb{Z}^2 \to \mathbb{Z}$ mapping (k, l) to $-\nu k - l$ and $\varphi : \mathbb{Z}^2 \to \mathbb{Z}$ mapping (k, l) to $ak + bl$.

The case where γ is a zero dimensional face contained in the vertical face is similar. Note in particular that in the case $(0, b)$, by assumption $\nu = 1$.

4.3 *Rationality and Limit of $Z_\gamma(T)$ for a Face γ of $\mathscr{N}(f)$ Not Contained in a Non Compact Face*

We consider the case where γ is a face not contained in a non compact face of $\mathscr{N}(f)$. We use the function $\mathfrak{m}$ defined in Proposition 2. Let f_γ be the initial part of f corresponding to the face γ.

Remark 10 For any arc $(x(t), y(t))$ in $\mathscr{L}(\mathbb{A}^2_{\mathbf{k}})$ with (ord $x(t)$, ord $y(t)$) in C_γ, the order ord $f(x(t), y(t))$ is at least equal to $\mathfrak{m}(\text{ord } x(t), \text{ord } y(t))$ with equality if and only if $f_\gamma(\overline{\text{ac}}\, x(t), \overline{\text{ac}}\, y(t)) \neq 0$.

Proposition 4 *For any $n \geq 1$, for any (α, β) in C_γ we denote by*

$$X_{n,(\alpha,\beta)} = \{(x(t), y(t)) \in \mathscr{L}(\mathbb{A}^2_{\mathbf{k}}) \mid ordx(t) = \alpha, ordy(t) = \beta, ord f(x(t), y(t)) = n\}. \quad (16)$$

The motivic zeta function $Z_\gamma(T)$ can be decomposed as

$$Z_\gamma(T) = Z_\gamma^{=}(T) + Z_\gamma^{<}(T), \quad (17)$$

with

$$Z_\gamma^{=}(T) = \sum_{n\geq 1} \sum_{\substack{(\alpha,\beta)\in C_\gamma \\ \mathfrak{m}(\alpha,\beta)=n}} \mathbb{L}^{-(\nu-1)\alpha} mes\left(X_{n,(\alpha,\beta)}\right) T^n$$

and

$$Z_\gamma^{<}(T) = \sum_{n\geq 1} \sum_{\substack{(\alpha,\beta)\in C_\gamma \\ \mathfrak{m}(\alpha,\beta)<n}} \mathbb{L}^{-(\nu-1)\alpha} mes\left(X_{n,(\alpha,\beta)}\right) T^n.$$

We will simply write

$$Z_\gamma^{<}(T) = \sum_{(n,\alpha,\beta)\in C_\gamma^{<}\cap\mathbb{N}^3} \mathbb{L}^{-(\nu-1)\alpha} mes\left(X_{n,(\alpha,\beta)}\right) T^n, \tag{18}$$

where for any compact face γ in the Newton polytope $\mathcal{N}(f)$ of f, we consider the following cone

$$C_\gamma^{<} = \left\{ (n,\alpha,\beta)\in\mathbb{R}_{>0}\times C_\gamma \,\middle|\, \mathfrak{m}(\alpha,\beta)<n \right\}.$$

Remark 11 If $\dim\gamma = 0$ then $Z_\gamma^{<}(T) = 0$ as an empty sum.

4.3.1 Rationality and Limit of $Z_\gamma^{=}(T)$

Notation 3 *Let (α,β) be in $C_\gamma\cap(\mathbb{N}^*)^2$. By construction of the Grothendieck ring of varieties in* [16, §2] *and* [17, §2] *(see* [25, Proposition 3.13] *for details) The class $\left[f_\gamma(x,y) : \mathbb{G}_m^2\setminus(f_\gamma=0)\to\mathbb{G}_m, \sigma_{\alpha,\beta}\right]$ with $\sigma_{\alpha,\beta}$ the action of $\mathbb{G}_m$ defined by $\sigma_{\alpha,\beta}(\lambda,(x,y)) = (\lambda^\alpha x, \lambda^\beta y)$ does not depend on (α,β) and we replace $\sigma_{\alpha,\beta}$ by σ_γ.*

Similarly to [15] or [16] we have

Proposition 5 *The motivic zeta function $Z_\gamma^{=}(T)$ is rational. In particular,*

- *If γ is a one dimensional face supported by a line of equation $ap+bq=N$ with dual cone $C_\gamma=\mathbb{R}_{>0}(p,q)$ then,*

$$Z_\gamma^{=}(T) = \left[f_\gamma : \mathbb{G}_m^2\setminus(f_\gamma=0)\to\mathbb{G}_m, \sigma_\gamma\right] \frac{\mathbb{L}^{-\nu p-q}T^N}{1-\mathbb{L}^{-\nu p-q}T^N}. \tag{19}$$

- *If γ is a zero dimensional face (a,b) with dual cone $C_\gamma=\mathbb{R}_{>0}u_1+\mathbb{R}_{>0}u_2$ then*

$$Z_\gamma^{=}(T) = \left[f_\gamma : \mathbb{G}_m^2\setminus(f_\gamma=0)\to\mathbb{G}_m, \sigma_\gamma\right] R_\gamma(T) \tag{20}$$

where

$$R_\gamma(T) = \sum_{(\alpha_0,\beta_0)\in P_\gamma} \frac{\mathbb{L}^{-\nu\alpha_0-\beta_0}T^{((a,b)|(\alpha_0,\beta_0))}}{(1-\mathbb{L}^{-((\nu,1)|u_1)}T^{((a,b)|u_1)})(1-\mathbb{L}^{-((\nu,1)|u_2)}T^{((a,b)|u_2)})},$$

and $P_\gamma = (]0,1]u_1+]0,1]u_2) \cap \mathbb{N}^2$.

Furthermore, the limit of the zeta function $Z_\gamma^{=}(T)$ *is*

$$\lim_{T\to\infty} Z_\gamma^{=}(T) = (-1)^{\dim\gamma}\left[f_\gamma : \mathbb{G}_m^2 \setminus (f_\gamma = 0) \to \mathbb{G}_m, \sigma_\gamma\right] \in \mathscr{M}_{\mathbb{G}_m}^{\mathbb{G}_m}.$$

Proof Indeed, for any face γ in $\mathscr{N}(f)$, for any (α,β) in $C_\gamma \cap (\mathbb{N}^*)^2$, for any integer $m \geq \mathfrak{m}(\alpha,\beta)$, the m-jet space $\pi_m\left(X_{\mathfrak{m}(\alpha,\beta),(\alpha,\beta)}\right)$ with the canonical $\mathbb{G}_m$-action is a bundle over $\left(f_\gamma(x,y), \mathbb{G}_m^2 \setminus (f_\gamma=0), \sigma_{\alpha,\beta}\right)$ with fiber $\mathbb{A}^{2m-\alpha-\beta}$ and for any (λ,x,y) in $\mathbb{G}_m^3$, $\sigma_{\alpha,\beta}(\lambda,(x,y)) = (\lambda^\alpha x, \lambda^\beta y)$. Then by definition of the motivic measure we get

$$\text{mes}\left(X_{\mathfrak{m}(\alpha,\beta),(\alpha,\beta)}\right) = \mathbb{L}^{-\alpha-\beta}\left[f_\gamma(x,y) : \mathbb{G}_m^2 \setminus (f_\gamma=0) \to \mathbb{G}_m, \sigma_{\alpha,\beta}\right] \in \mathscr{M}_{\mathbb{G}_m}^{\mathbb{G}_m}.$$

Then, using Notation 3 we have

$$Z_\gamma^{=}(T) = \left[f_\gamma(x,y) : \mathbb{G}_m^{*2} \setminus (f_\gamma=0) \to \mathbb{G}_m, \sigma_\gamma\right] \sum_{n\geq 1} \sum_{\substack{(\alpha,\beta)\in C_\gamma\cap\mathbb{N}^{*2} \\ \mathfrak{m}(\alpha,\beta)=n}} \mathbb{L}^{-(\nu-1)\alpha-\alpha-\beta}T^n.$$

and, formulas (19) and (20) follow by Lemma 1.

4.3.2 Rationality and Limit of $Z_\gamma^{<}(T)$

Remark 12 Let γ be a one dimensional face of $\mathscr{N}(f)$. The face γ is supported by a line with equation $mp + nq = N$ with $gcd(p,q) = 1$ and $N > 0$. Following Remark 2 the initial part f_γ of f can be factored as

$$f_\gamma(x,y) = cx^a y^b \prod_{\mu_i \in R_\gamma} (y^p - \mu_i x^q)^{\nu_i}$$

with c in $\mathbf{k}$, a and b in $\mathbb{N}$ and N equal to $pa + qb + (\sum_i \nu_i)pq$.

Remark 13 By additivity of the measure, for any one dimensional face γ in $\mathscr{N}(f)$, and using the Notation 18 the motivic zeta function $Z_\gamma^{<}(T)$ has the following decomposition

$$Z_\gamma^{<}(T) = \sum_{\mu\in R_\gamma} \sum_{(n,\alpha,\beta)\in C_\gamma^{<}\cap\mathbb{N}^3} \mathbb{L}^{-(\nu-1)\alpha}\text{mes}\,(X_{(n,\alpha,\beta),\mu})T^n$$

where

$$X_{(n,\alpha,\beta),\mu} = \left\{ \left(x(t) = t^{\alpha}\tilde{x}(t),\, y(t) = t^{\beta}\tilde{y}(t)\right) \in \mathscr{L}(\mathbb{A}^2_{\mathbf{k}}) \;\middle|\; \begin{array}{l} \tilde{y}(0)^p = \mu\tilde{x}(0)^q, \\ \text{ord}\, x(t) = \alpha,\ \text{ord}\, y(t) = \beta \\ \text{ord}\, f(x(t), y(t)) = n \end{array} \right\}.$$

Proposition 6 *Let γ be a one dimensional face of $\mathscr{N}(f)$, let μ be a root of f_γ and $\sigma(p, q, \mu)$ the induced Newton transform defined in Definition 7. For any $k > 0$, denoting $\alpha = pk$ and $\beta = qk$, we have*

$$mes\left(X_{(n,\alpha,\beta),\mu}\right) = \mathbb{L}^{-(p+q-1)k} mes\left(Y^{\sigma(p,q,\mu)}_{(n,k)}\right)$$

with

$$Y^{\sigma(p,q,\mu)}_{(n,k)} = \left\{ (v(t), w(t)) \in \mathscr{L}(\mathbb{A}^2_k) \;\middle|\; \begin{array}{l} ord\, v(t) = k,\ ord\, w(t) > 0 \\ ord\, f_{\sigma(p,q,\mu)}(v(t), w(t)) = n \end{array} \right\}.$$

Proof The proof is inspired from [5, Lemma 3.3]. Let L be an integer bigger than α, β and n. We consider the jet spaces,

$$X^{(L)}_{(n,\alpha,\beta),\mu} = \left\{ (t^{\alpha}\tilde{x}(t), t^{\beta}\tilde{y}(t)) \in \mathscr{L}_L(\mathbb{A}^2_{\mathbf{k}}) \;\middle|\; \begin{array}{l} \tilde{x}(0) \neq 0,\ \tilde{y}(0) \neq 0,\ \tilde{y}(0)^p = \mu\tilde{x}(0)^q \\ \tilde{x}(t) \in \mathscr{L}_{L-\alpha}(\mathbb{A}^1_{\mathbf{k}}),\ \tilde{y}(t) \in \mathscr{L}_{L-\beta}(\mathbb{A}^1_{\mathbf{k}}) \\ \text{ord}\, f(t^{\alpha}\tilde{x}(t), t^{\beta}\tilde{y}(t)) = n \end{array} \right\},$$

which are the jet-spaces $\pi_L\left(X_{(n,\alpha,\beta),\mu}\right)$,

$$\overline{X}^{(L)}_{(n,\alpha,\beta),\mu} = \left\{ (\varphi_1(t), \varphi_2(t)) \in \mathscr{L}_L(\mathbb{A}^2_{\mathbf{k}}) \;\middle|\; \begin{array}{l} \varphi_1(0) \neq 0,\ \varphi_2(0) \neq 0,\ \varphi_2(0)^p = \mu\varphi_1(0)^q \\ \text{ord}\, f(t^{\alpha}\varphi_1, t^{\beta}\varphi_2) = n \end{array} \right\},$$

$$\overline{Y}^{(L)}_{(n,k)} = \left\{ (\psi_1(t), \psi_2(t)) \in \mathscr{L}_L(\mathbb{A}^2_{\mathbf{k}}) \;\middle|\; \begin{array}{l} \text{ord}\, f_{\sigma(p,q,\mu)}(t^k\psi_1(t), \psi_2(t)) = n \\ \text{ord}\, \psi_1(t) = 0,\ \text{ord}\, \psi_2(t) \geq 1 \end{array} \right\},$$

$$Y^{(L)}_{(n,k)} = \left\{ (v'(t), w'(t)) \in \mathscr{L}_L(\mathbb{A}^2_{\mathbf{k}}) \;\middle|\; \begin{array}{l} \text{ord}\, f_{\sigma(p,q,\mu)}(v'(t), w'(t)) = n \\ \text{ord}\, v'(t) = k,\ \text{ord}\, w'(t) \geq 1 \end{array} \right\},$$

which are the jet spaces $\pi_L\left(Y^{\sigma(p,q,\mu)}_{(n,k)}\right)$.

The application $(\varphi_1, \varphi_2) \mapsto \left(t^{\alpha}\varphi_1 \mod t^{L+1}, t^{\beta}\varphi_2 \mod t^{L+1}\right)$ induces a structure of bundle on $\overline{X}^{(L)}_{(n,\alpha,\beta),\mu}$ over $X^{(L)}_{(n,\alpha,\beta),\mu}$ with fiber $\mathbb{A}^{\alpha+\beta}$. Also, the application $(\psi_1, \psi_2) \mapsto \left(t^k\psi_1 \mod t^{L+1}, \psi_2 \mod t^{L+1}\right)$ induces a structure of bundle on $\overline{Y}^{(L)}_{(n,k)}$ over $Y^{(L)}_{(n,k)}$ with fiber $\mathbb{A}^k$. We deduce the equalities $\left[X^{(L)}_{(n,\alpha,\beta),\mu}\right] = \mathbb{L}^{-\alpha-\beta}\left[\overline{X}^{(L)}_{(n,\alpha,\beta),\mu}\right]$ and $\left[\overline{Y}^{(L)}_{(n,k)}\right] = \mathbb{L}^k\left[Y^{(L)}_{(n,k)}\right]$.

We consider the application

$$\begin{aligned} \Phi^{\sigma(p,q,\mu)}: \ \overline{Y}^{(L)}_{(n,k)} &\to \overline{X}^{(L)}_{(n,\alpha,\beta),\mu} \\ (\psi_1,\psi_2) &\mapsto \left(\mu^{q'}\psi_1^p, \psi_1^q(\psi_2+\mu^{p'})\right). \end{aligned}$$

Using the relation $pp' - qq' = 1$, we can check that $\Phi^{\sigma(p,q,\mu)}\left(\overline{Y}^{(L)}_{(n,k)}\right) \subset \overline{X}^{(L)}_{(n,\alpha,\beta),\mu}$. Indeed, if $(\varphi_1(t), \varphi_2(t))$ is equal to $\Phi^{\sigma(p,q,\mu)}(\psi_1(t), \psi_2(t))$ then

$$\varphi_2(0)^p - \mu\varphi_1(0)^q = \left(\mu^{p'}\psi_1(0)^q\right)^p - \mu\left(\mu^{q'}\psi_1(0)^p\right)^q = 0$$

and using the relations $\alpha = pk$, $\beta = qk$, and the definitions we deduce the equality

$$f\left(t^\alpha\varphi_1(t), t^\beta\varphi_2(t)\right) = f_{\sigma(p,q,\mu)}\left(t^k\psi_1(t), \psi_2(t)\right).$$

We prove that $\Phi^{\sigma(p,q,\mu)}$ is an isomorphism building the inverse application. Consider $\varphi(t) = (\varphi_1(t), \varphi_2(t))$ in $\overline{X}^{(L)}_{(n,\alpha,\beta),\mu}$. Remark that if there is $\psi(t) = (\psi_1(t), \psi_2(t))$ in $\overline{Y}^{(L)}_{(n,k)}$ such that $\varphi(t) = \Phi^{\sigma(p,q,\mu)}(\psi_1(t), \psi_2(t))$ then we have

$$\frac{\varphi_1(0)^{p'}}{\varphi_2(0)^{q'}} = \psi_1(0).$$

Furthermore, denoting $\varphi_1(t) = \varphi_1(0)\tilde{\varphi_1}(t)$, by Hensel lemma, there is a unique formal series $a(t)$ such that

$$a(0) = 1 \text{ and } a(t)^p = \tilde{\varphi_1}(t).$$

The formal series $a(t)$ is denoted by $\tilde{\varphi_1}(t)^{1/p}$. Hence, the inverse map, maps en arc $(\varphi_1(t), \varphi_2(t))$ to the arc

$$\left(\frac{\varphi_1(0)^{p'}}{\varphi_2(0)^{q'}}\tilde{\varphi_1}(t)^{1/p} \mod t^{L+1}, -\mu^{p'} + \left(\frac{\varphi_1(0)^{p'}}{\varphi_2(0)^{q'}}\tilde{\varphi_1}(t)^{1/p}\right)^{-q}\varphi_2(t) \mod t^{L+1}\right).$$

This proves that $\Phi^{\sigma(p,q,\mu)}$ is an isomorphism and we deduce the equality

$$\left[X^{(L)}_{(n,\alpha,\beta),\mu}\right] = \mathbb{L}^{-\alpha-\beta+k}\left[Y^{(L)}_{(n,k)}\right].$$

Finally, the result follows by definition of the motivic measure.

Remark 14 By construction for any n and k we have $\mathfrak{m}(pk, qk) \leq n$. And we are studying the case $\mathfrak{m}(pk, qk) < n$. In particular the arc space $X_{(n,\alpha,\beta),\mu}$ is empty if $\mathfrak{m}(pk, qk) \geq n$ and it follows from the proof that $Y_{(n,k)}^{\sigma(p,q,\mu)}$ is also empty.

Proposition 7 *Let γ be a one dimensional face of $\mathcal{N}(f)$. The motivic zeta function $Z_\gamma^<(T)$ can be decomposed as*

$$Z_\gamma^<(T) = \sum_{\mu \in R_\gamma} Z_{f_{\sigma(p,q,\mu)},\omega_{p,q}}(T)$$

where R_γ is the set of roots μ of f_γ, $\sigma(p, q, \mu)$ are the associated Newton transforms and $\omega_{p,q}$ is the differential form

$$\omega_{p,q}(v, w) = v^{(\nu p+q-1)} dv \wedge dw.$$

Remark 15 Let γ be a one dimensional face of $\mathcal{N}(f)$. This face is supported by a line of equation $ap + bq = N > 0$. Let μ be a root of f_γ and $\sigma(p, q, \mu)$ the induced Newton transform. By Lemma 2, there is a polynomial $\tilde{f}_{\sigma(p,q,\mu)}$ in $\mathbf{k}[v, w]$ such that

$$f_{\sigma(p,q,\mu)}(v, w) = v^N \tilde{f}_{\sigma(p,q,\mu)}(v, w).$$

In particular the Newton transform $f_{\sigma(p,q,\mu)}$ is a polynomial in $\mathbf{k}[v, w]$ and vanishes at (0, 0).

Proof Let γ be a one dimensional face of $\mathcal{N}(f)$. For any element (α, β) in C_γ there is $k > 0$ such that $\alpha = pk$ and $\beta = qk$. The set of points with integral coefficients in the cone $C_\gamma^<$,

$$C_\gamma^< \cap \mathbb{N}^3 = \left\{ (n, \alpha, \beta) \in \mathbb{N}^* \times C_\gamma \,\middle|\, \mathfrak{m}(\alpha, \beta) < n \right\},$$

is in bijection with the cone

$$\overline{C}_\gamma^< = \left\{ (n, k) \in \mathbb{N}^{*2} \,\middle|\, \mathfrak{m}(pk, qk) < n \right\}.$$

Using this notation we prove the equality

$$Z_\gamma^<(T) = \sum_{\mu \in R_\gamma} \left(Z_{f_{\sigma(p,q,\mu)},\omega_{p,q}}(T) \right)_{(0,0)}.$$

Indeed, we have

$$Z^{<}_{\gamma}(T) = \sum_{\mu\in R_\gamma}\sum_{(n,\alpha,\beta)\in C^{<}_{\gamma}\cap\mathbb{N}^3} \mathbb{L}^{-(\nu-1)pk}\mathrm{mes}\,(X_{(n,\alpha,\beta),\mu})T^n$$

$$= \sum_{\mu\in R_\gamma}\sum_{(n,k)\in\overline{C}^{<}_{\gamma}} \mathbb{L}^{-(\nu-1)pk}\mathbb{L}^{-(p+q-1)k}\mathrm{mes}\left(Y^{\sigma(p,q,\mu)}_{n,k}\right)T^n,$$

but using the definition of the zeta function and the Remark 14 we have

$$\left(Z_{f_{\sigma(p,q,\mu)},\omega_{p,q}}(T)\right)_{(0,0)} = \sum_{n\geq 1}\left(\sum_{k\geq 1}\mathbb{L}^{-(\nu p+q-1)k}\mathrm{mes}\left(Y^{\sigma(p,q,\mu)}_{n,k}\right)\right)T^n$$
$$= \sum_{(n,k)\in\overline{C}^{<}_{\gamma}}\mathbb{L}^{-(\nu p+q-1)k}\mathrm{mes}\left(Y^{\sigma(p,q,\mu)}_{n,k}\right)T^n$$

with

$$Y^{\sigma(p,q,\mu)}_{n,k} = \left\{(v(t),w(t))\in\mathscr{L}(\mathbb{A}^2_k)\;\middle|\;\begin{array}{l}(v(0),w(0))=0,\ \mathrm{ord}\,v(t)=k\\ \mathrm{ord}\,f_{\sigma(p,q,\mu)}(v(t),w(t))=n\\ \mathrm{ord}\,\omega_{p,q}(v(t),w(t))=(\nu p+q-1)k\end{array}\right\}.$$

4.3.3 Base Cases

Example 2 In this example we consider $f(x,y)=x^a y^b u(x,y)$ with $a\geq 0, b\geq 0$ and $u(0,0)\neq 0$. The following computations are standard.

- If $a>0$ and $b>0$ then

$$\left(Z_{f,\omega}(T)\right)_{(0,0)} = [x^a y^b : \mathbb{G}_m^2\to\mathbb{G}_m]\frac{\mathbb{L}^{-\nu}T^a}{1-\mathbb{L}^{-\nu}T^a}\frac{\mathbb{L}^{-1}T^b}{1-\mathbb{L}^{-1}T^b}$$

  and

$$\left(S_f\right)_{(0,0)} = -[x^a y^b : \mathbb{G}_m^2\to\mathbb{G}_m,\sigma_{\mathbb{G}_m}].$$

- If $b>0$ and $a=0$ (remark that by assumption on ν, $\nu=1$) and we have

$$\left(Z_{f,\omega}(T)\right)_{(0,0)} = [y^b : \mathbb{G}_m\to\mathbb{G}_m,\sigma_{\mathbb{G}_m}]\frac{\mathbb{L}^{-1}T^b}{1-\mathbb{L}^{-1}T^b}$$

  and

$$\left(S_f\right)_{(0,0)} = [y^b : \mathbb{G}_m\to\mathbb{G}_m,\sigma_{\mathbb{G}_m}].$$

- If $a>0$ and $b=0$ then we have

$$\left(Z_{f,\omega}(T)\right)_{(0,0)} = [x^a : \mathbb{G}_m\to\mathbb{G}_m,\sigma_{\mathbb{G}_m}]\frac{\mathbb{L}^{-\nu}T^a}{1-\mathbb{L}^{-\nu}T^a}$$

and

$$\left(S_f\right)_{(0,0)} = [x^a : \mathbb{G}_m \to \mathbb{G}_m, \sigma_{\mathbb{G}_m}].$$

- If $a = b = 0$ then $\left(Z_{f,\omega}(T)\right)_{(0,0)} = 0$ and $\left(S_f\right)_{(0,0)} = 0$.

The following example corresponds to the base case of Lemma 4

Example 3 Let f be in $\mathbf{k}[x, y]$ which is not a monomial. If the Newton algorithm of f stabilizes with height m then by Lemma 4

$$f(x, y) = U(x, y)x^M(y - \mu x^q + g(x, y))^m,$$

with μ in $\mathbf{k}^*$, $(M, q) \in \mathbb{N} \times \mathbb{N}_{>0}$, $g(x, y) \in \mathbf{k}[x, y]$, $g(x, y) = \sum_{a+bq>q} c_{a,b}x^a y^b$ and $U(x, y) \in \mathbf{k}[[x, y]]$ with $U(0, 0) \neq 0$. As U is a unit, as all the arcs $(x(t), y(t))$ used in the computation of the motivic Milnor fiber at the origin satisfy $(x(0), y(0)) = (0, 0)$, we can assume in the following $U(x, y) = 1$. We denote by $h(x, y)$ the polynomial $y - \mu x^q + g(x, y)$. We denote by γ the compact one dimensional face of the Newton polygon of f with face polynomial $x^M(y - \mu x^q)^m$. The motivic zeta function $\left(Z_{f,\omega}(T)\right)_{(0,0)}$ is rational and

- if $M = 0$, then

$$\left(S_f\right)_{(0,0)} = [z^m : \mathbb{G}_m \to \mathbb{G}_m, \sigma_{\mathbb{G}_m}] \tag{21}$$

  with $\sigma_{\mathbb{G}_m}(\lambda, x) = \lambda.x$,
- If $M > 0$ then

$$\left(S_f\right)_{(0,0)} = -[x^M y^m : \mathbb{G}_m^2 \to \mathbb{G}_m, \sigma_{\mathbb{G}_m^2}] \tag{22}$$

  with $\sigma_{\mathbb{G}_m^2}(\lambda, (x, y)) = (\lambda x, \lambda y)$.

Proof The Newton polygon of f has three face polynomials x^{M+qm}, $x^M y^m$ and $f_\gamma(x, y) = x^M(y - \mu x^q)^m$. Applying the decomposition formula (10) we get

$$\left(Z_{f,\omega}(T)\right)_{(0,0)} = Z_{x^{qm+M}}(T) + Z_{x^M y^m}(T) + Z_\gamma^=(T) + Z_\gamma^<(T).$$

The rationality form and the limit of $Z_{x^{qm+M}}(T)$, $Z_{x^M y^m}(T)$ and $Z_\gamma^=(T)$ are given in Propositions 3 and 5. We study now the zeta function $Z_\gamma^<(T)$. We treat simultaneously the cases $M = 0$ (with $\nu = 1$) and $M > 0$. Recall that

$$Z_\gamma^<(T) = \sum_{n\geq 1} \sum_{\substack{(\alpha,\beta) \in C_\gamma \\ \mathrm{m}(\alpha,\beta) < n}} \mathbb{L}^{-(\nu-1)\alpha} \mathrm{mes}\left(X_{n,(\alpha,\beta)}\right) T^n.$$

As the cone $C_\gamma = \mathbb{R}_{>0}(1, q)$, we get

$$Z_\gamma^<(T) = \sum_{n\geq 1} \sum_{\substack{k \geq 1 \\ Mk + qkm < n}} \mathbb{L}^{-(\nu-1)k} \mathrm{mes}\left(X_{n,(k,kq)}\right) T^n.$$

Using the formula (16) defining $X_{n,(\alpha,\beta)}$, we have for any $n \geq 1$ and $k \geq 1$

$$X_{n,(q,qk)} = \left\{ \varphi(t) = (x(t), y(t)) \in \mathscr{L}(\mathbb{A}_{\mathbf{k}}^2) \;\middle|\; \begin{array}{l} \mathrm{ord}\, x(t) = k,\ \mathrm{ord}\, y(t) = qk \\ \overline{\mathrm{ac}}\, y(t) - \mu \overline{\mathrm{ac}}\, x(t)^q = 0, \\ \mathrm{ord}\, h(\varphi(t)) = \frac{n-kM}{m} \end{array} \right\}.$$

For any $(k, l) \in (\mathbb{N}^*)^2$ we introduced

$$X_{l,k}^<(h) = \left\{ (x(t), y(t)) \in \mathscr{L}(\mathbb{A}_{\mathbf{k}}^2) \;\middle|\; \begin{array}{l} \mathrm{ord}\, x(t) = k,\ \mathrm{ord}\, y(t) = qk \\ \overline{\mathrm{ac}}\, y(t) = \mu \overline{\mathrm{ac}}\, x(t)^q,\ \mathrm{ord}\, h(x(t), y(t)) = l \end{array} \right\}$$

endowed with the stardard action on arcs $\lambda.\varphi(t) = \varphi(\lambda t)$ and the structural map to $\mathbb{G}_m$:

$$\varphi(t) = (x(t), y(t)) \mapsto \overline{\mathrm{ac}}\, f(\varphi(t)) = \overline{\mathrm{ac}}\, x(t)^M \overline{\mathrm{ac}}\, h(\varphi(t))^m.$$

Remark in particular that if $X_{l,k}(h)^<$ is non empty then $l > qk$. We consider the cone $C = \{(k, l) \in \mathbb{R}_{>0}^2 \mid l > kq\}$ and we have

$$Z_\gamma^<(T) = \sum_{n\geq 1} \sum_{\substack{(k,l) \in C \\ Mk + lm = n}} \mathbb{L}^{-(\nu-1)k} \mathrm{mes}\left(X_{l,k}^<(h)\right) T^n.$$

As h is a non degenerate polynomial, then using arguments of [15, Lemme 2.1.1], [16] or [25, Lemme 3.17], we obtain for any (k, l) in C

$$\mathrm{mes}\left(X_{l,k}^<(h)\right) = \left[\begin{array}{lll} (y = \mu x^q) \cap \mathbb{G}_m^2 \times \mathbb{G}_m & \to & \mathbb{G}_m \\ ((x, y), \xi) & \mapsto & x^M \xi^m \end{array}, \sigma_{k,l} \right] \mathbb{L}^{-k-l},$$

where $\sigma_{k,l}(\lambda, (x, y, \xi)) = (\lambda^k x, \lambda^{kq} y, \lambda^l \xi)$. We observe that

$$\mathrm{mes}\left(X_{l,k}^<(h)\right) = \left[x^M \xi^m : \mathbb{G}_m^2 \to \mathbb{G}_m, \sigma_{k,l}\right] \mathbb{L}^{-k-l},$$

with $\sigma_{k,l}(\lambda, (x, \xi)) = (\lambda^k x, \lambda^l \xi)$. We prove now the equality

$$\left[x^M \xi^m : \mathbb{G}_m^2 \to \mathbb{G}_m, \sigma_{k,l}\right] = \left[x^M \xi^m : \mathbb{G}_m^2 \to \mathbb{G}_m, \sigma_{1,1}\right]. \tag{23}$$

We use [16, §2.2] and [17, §2.4, (2.5.1)]. Indeed by the colimit construction of the Grothendieck ring $\mathcal{M}_{\mathbb{G}_m^2}^{\mathbb{G}_m^2}$ we have the equality

$$[id : \mathbb{G}_m^2 \to \mathbb{G}_m^2, \tau_{k,l}] = [id : \mathbb{G}_m^2 \to \mathbb{G}_m^2, \tau_{1,1}]$$

with for any $k \geq 1$ and $l \geq 1$, $\tau_{k,l}$ is the action of $\mathbb{G}_m^2$

$$\tau_{k,l}((\lambda, \mu), (x, \xi)) = (\lambda^k x, \mu^l \xi).$$

Using the diagonal morphism $(id, id) : \mathbb{G}_m \to \mathbb{G}_m^2$ we consider the canonical morphism

$$\Delta : \mathcal{M}_{\mathbb{G}_m^2}^{\mathbb{G}_m^2} \to \mathcal{M}_{\mathbb{G}_m^2}^{\mathbb{G}_m}$$

which maps $[X \to \mathbb{G}_m^2, \sigma_{X,\mathbb{G}_m^2}]$ to $[X \to \mathbb{G}_m^2, \sigma_{X,\mathbb{G}_m}]$, with $\sigma_{X,\mathbb{G}_m}$ the action of $\mathbb{G}_m$ on X defined for (λ, x) in $X \times \mathbb{G}_m$ by

$$\sigma_{X,\mathbb{G}_m}(\lambda, x) = \sigma_{X,\mathbb{G}_m^2}((\lambda, \lambda), x).$$

Applying Δ we obtain the equality in $\mathcal{M}_{\mathbb{G}_m^2}^{\mathbb{G}_m}$

$$[id : \mathbb{G}_m^2 \to \mathbb{G}_m^2, \sigma_{k,l}] = [id : \mathbb{G}_m^2 \to \mathbb{G}_m^2, \sigma_{1,1}].$$

We consider the map $x^M \xi^m : \mathbb{G}_m^2 \to \mathbb{G}_m$ and we obtain the equality in $\mathcal{M}_{\mathbb{G}_m}^{\mathbb{G}_m}$

$$(x^M \xi^m)_![id : \mathbb{G}_m^2 \to \mathbb{G}_m^2, \sigma_{k,l}] = (x^M \xi^m)_![id : \mathbb{G}_m^2 \to \mathbb{G}_m^2, \sigma_{1,1}]$$

this gives equality (23). We use the notation $\sigma_{\mathbb{G}_m^2}$ instead of $\sigma_{1,1}$. Then, we obtain

$$Z_\gamma^<(T) = \left[x^M \xi^m : \mathbb{G}_m^2 \to \mathbb{G}_m, \sigma_{\mathbb{G}_m^2}\right] \sum_{n \geq 1} \sum_{\substack{(k,l) \in C \\ Mk + lm = n}} \mathbb{L}^{-(\nu-1)k-k-l} T^{ml+kM}.$$

Writing $C = \mathbb{R}_{>0}(0, 1) + \mathbb{R}_{>0}(1, q)$ and using Lemma 1 we obtain

$$Z_\gamma^<(T) = \left[x^M \xi^m : \mathbb{G}_m^2 \to \mathbb{G}_m, \sigma_{\mathbb{G}_m^2}\right] \sum_{(k_0,l_0) \in P_\gamma} \frac{\mathbb{L}^{-\nu k_0 - l_0} T^{Mk_0+ml_0}}{(1 - \mathbb{L}^{-\nu-q} T^{M+mq})(1 - \mathbb{L}^{-1} T^m)}$$

and

$$\lim_{T \to \infty} Z_\gamma^<(T) = \left[x^M \xi^m : \mathbb{G}_m^2 \to \mathbb{G}_m, \sigma_{\mathbb{G}_m^2}\right]. \tag{24}$$

Taking the limit when T goes to infinity, we obtain the formula (21) and (22) in the following way.

- If $M = 0$ then

$$\begin{aligned}(S_f)_{(0,0)} &= [y^m : \mathbb{G}_m \to \mathbb{G}_m, \sigma_{\mathbb{G}_m}] + [x^{qm} : \mathbb{G}_m \to \mathbb{G}_m, \sigma_{\mathbb{G}_m}] \\ &+[(y-\mu x^q)^m : \mathbb{G}_m^2 \setminus (y = \mu x^q) \to \mathbb{G}_m, \sigma_\gamma] - [\xi^m : \mathbb{G}_m^2 \to \mathbb{G}_m, \sigma_{\mathbb{G}_m^2}].\end{aligned}$$

By additivity in the Grothendieck ring we have

$$(S_f)_{(0,0)} = [(y-\mu x^q)^m : \mathbb{A}_{\mathbf{k}}^2 \setminus (y = \mu x^q) \to \mathbb{G}_m, \sigma_\gamma] - [\xi^m : \mathbb{G}_m^2 \to \mathbb{G}_m, \sigma_{\mathbb{G}_m^2}],$$

where for any (λ, x, y) in $\mathbb{G}_m^3$, $\sigma_{\mathbb{G}_m}(\lambda, x) = \lambda.x$, $\sigma_{\mathbb{G}_m^2}(\lambda, (x, \xi)) = (\lambda.x, \lambda.\xi)$ and $\sigma_\gamma(\lambda, (x, y)) = (\lambda x, \lambda^q y)$.

By the equivariant isomorphism

$$\begin{array}{ccc}\mathbb{A}_{\mathbf{k}}^2 \setminus \{y = \mu x^q\} & \to & \mathbb{A}_{\mathbf{k}}^1 \times \mathbb{G}_m \\ (x, y) & \mapsto & (x, z = y - \mu x^q)\end{array}$$

we obtain (using the above identification of actions in the construction of the Grothendieck ring of varieties)

$$[(y-\mu x^q)^m : \mathbb{A}_{\mathbf{k}}^2 \setminus (y = \mu x^q) \to \mathbb{G}_m, \sigma_\gamma] = [z^m : \mathbb{A}_{\mathbf{k}}^1 \times \mathbb{G}_m \to \mathbb{G}_m, \sigma_{\mathbb{G}_m}].$$

In particular, we obtain

$$\begin{aligned}(S_f)_{(0,0)} &= [\xi^m : \mathbb{A}_{\mathbf{k}}^1 \times \mathbb{G}_m \to \mathbb{G}_m, \sigma_{\mathbb{G}_m^2}] - [\xi^m : \mathbb{G}_m^2 \to \mathbb{G}_m, \sigma_{\mathbb{G}_m^2}] \\ &= \mathbb{L}[\xi^m : \mathbb{G}_m \to \mathbb{G}_m, \sigma_{\mathbb{G}_m}] - (\mathbb{L}-1)[\xi^m : \mathbb{G}_m \to \mathbb{G}_m, \sigma_{\mathbb{G}_m}] \\ &= [\xi^m : \mathbb{G}_m \to \mathbb{G}_m, \sigma_{\mathbb{G}_m}].\end{aligned}$$

- As well, if $M > 0$ then

$$\begin{aligned}(S_f)_{(0,0)} &= -[x^M y^m : \mathbb{G}_m^2 \to \mathbb{G}_m, \sigma_{\mathbb{G}_m^2}] + [x^{M+qm} : \mathbb{G}_m \to \mathbb{G}_m, \sigma_{\mathbb{G}_m}] \\ &+[x^M(y-\mu x^q)^m : \mathbb{G}_m^2 \setminus (y = \mu x^q) \to \mathbb{G}_m, \sigma_\gamma] \\ &-[x^M \xi^m : \mathbb{G}_m^2 \to \mathbb{G}_m, \sigma_{\mathbb{G}_m^2}]\end{aligned}$$

by additivity in the Grothendieck ring of varieties

$$\begin{aligned}(S_f)_{(0,0)} &= -2[x^M y^m : \mathbb{G}_m^2 \to \mathbb{G}_m, \sigma_{\mathbb{G}_m^2}] \\ &+[x^M(y-\mu x^q)^m : \mathbb{G}_m \times \mathbb{A}_{\mathbf{k}}^1 \setminus (y = \mu x^q) \to \mathbb{G}_m, \sigma_\gamma]^{\cdot}\end{aligned}$$

Then, by the equivariant isomorphism

$$\begin{array}{ccc}(\mathbb{G}_m \times \mathbb{A}^1_{\mathbf{k}}) \setminus \{y = \mu x^q\} & \to & \mathbb{G}_m^2 \\ (x, y) & \mapsto & (x, z = y - \mu x^q)\end{array}$$

we obtain (using the above identification of actions in the construction of the Grothendieck ring of varieties)

$$[x^M (y - \mu x^q)^m : \mathbb{G}_m \times \mathbb{A}^1_{\mathbf{k}} \setminus (y = \mu x^q) \to \mathbb{G}_m, \sigma_\gamma] = [x^M z^m : \mathbb{G}_m^2 \to \mathbb{G}_m, \sigma_{\mathbb{G}_m^2}].$$

In particular we obtain,

$$\left(S_f\right)_{(0,0)} = -2[x^M y^m : \mathbb{G}_m^2 \to \mathbb{G}_m, \sigma_{\mathbb{G}_m^2}] + [x^M z^m : \mathbb{G}_m^2 \to \mathbb{G}_m, \sigma_{\mathbb{G}_m^2}]$$

and then

$$\left(S_f\right)_{(0,0)} = -[x^M \xi^m : \mathbb{G}_m^2 \to \mathbb{G}_m, \sigma_{\mathbb{G}_m^2}].$$

Remark 16 In the proof of this example if we had used the Newton algorithm then we would have

$$-\lim_{T \to \infty} Z_\gamma^<(T) = \left(S_{f_{\sigma(1,q,\mu)}}\right)_{(0,0)}$$

where the Newton transform polynomial $f_{\sigma(1,q,\mu)}$ is of the form

$$f_{\sigma(1,q,\mu)} = x_1^{M+mq} (y_1 + \lambda x_1^{q_1} + g_1(x_1, y_1))^m \tag{25}$$

with

$$g_1(x_1, y_1) = \sum_{(a_1,b_1), a_1+bq_1>q_1} c_{a,b}^{(1)} x_1^a y_1^b$$

and we use $x_1^{\nu+q-1} dx_1 \wedge y_1$ the induced differential form. Using the isomorphism

$$\begin{array}{ccc}\mathbb{G}_m^2 & \to & \mathbb{G}_m^2 \\ (x_1, y_1) & \mapsto & (x = x_1, y = x_1^q y_1),\end{array}$$

we observe that

- if $M = 0$, then by (25) and formula (22) we have

$$S_{f_{\sigma(1,q,\mu)}} = -[x_1^{mq} y_1^m : \mathbb{G}_m^2 \to \mathbb{G}_m, \sigma_{\mathbb{G}_m^2}] = -[y^m : \mathbb{G}_m^2 \to \mathbb{G}_m, \sigma_{\mathbb{G}_m^2}],$$

- if $M > 0$, then by (25) and formula (22)

$$S_{f_{\sigma(1,q,\mu)}} = -[x_1^{M+mq} y_1^m : \mathbb{G}_m^2 \to \mathbb{G}_m, \sigma_{\mathbb{G}_m^2}] = -[x^M y^m : \mathbb{G}_m^2 \to \mathbb{G}_m, \sigma_{\mathbb{G}_m^2}].$$

This is compatible with the formula (24).

Remark 17 If f is non degenerate with respect to its Newton polygon, we have two formulae to compute $(S_f)_{(0,0)}$, the formula given by Guibert in [15] (see articles of Guibert-Loeser-Merle [16] and Bultot-Nicaise [3, 4] for generalizations and a version of the formula without misprints) and the formula given in this article. Let's prove they give the same answer. Using Guibert, we have

$$\begin{aligned}(S_f)_{(0,0)} &= \textstyle\sum_{x^a y^b \in \{\gamma_h,\gamma_v\}} (-1)^{r+1} [x^a y^b : \mathbb{G}_m^r \to \mathbb{G}_m, \sigma_{\mathbb{G}_m}] \\ &\quad - \textstyle\sum_{\gamma \in \mathcal{N}(f)\setminus\{\gamma_h,\gamma_v\}} (-1)^{\dim(\gamma)} \left[f_\gamma : \mathbb{G}_m^2 \setminus (f_\gamma = 0) \to \mathbb{G}_m, \sigma_\gamma\right] \\ &\quad - \textstyle\sum_{\gamma \in \mathcal{N}(f), \dim \gamma = 1} \left[\xi : (f_\gamma = 0 \cap \mathbb{G}_m^2) \times \mathbb{G}_m \to \mathbb{G}_m, \sigma_\gamma\right],\end{aligned} \tag{26}$$

with $r = 1$ if $b_h = 0$ (resp $a_v = 0$) and $r = 2$ if $b_h \neq 0$ (resp $a_v \neq 0$).

Using the formula in our paper

$$\begin{aligned}(S_f)_{(0,0)} &= \textstyle\sum_{x^a y^b \in \{\gamma_h,\gamma_v\}} (-1)^{r+1} [x^a y^b : \mathbb{G}_m^r \to \mathbb{G}_m, \sigma_{\mathbb{G}_m}] \\ &\quad - \textstyle\sum_{\gamma \in \mathcal{N}(f)\setminus\{\gamma_h,\gamma_v\}} (-1)^{\dim(\gamma)} \left[f_\gamma : \mathbb{G}_m^2 \setminus (f_\gamma = 0) \to \mathbb{G}_m, \sigma_\gamma\right] \\ &\quad + \textstyle\sum_{\gamma \in \mathcal{N}(f), \dim \gamma = 1} \sum_{\mu \in R_\gamma} \left(S_{f_{\sigma(p,q,\mu)}}\right)_{(0,0)},\end{aligned} \tag{27}$$

with $r = 1$ if $b_h = 0$ (resp $a_v = 0$) and $r = 2$ if $b_h \neq 0$ (resp $a_v \neq 0$).

Then we have to prove that

$$-\sum_{\substack{\gamma \in \mathcal{N}(f) \\ \dim \gamma = 1}} \left[\xi : (f_\gamma = 0 \cap \mathbb{G}_m^2) \times \mathbb{G}_m \to \mathbb{G}_m, \sigma_\gamma\right] = \sum_{\substack{\gamma \in \mathcal{N}(f) \\ \dim \gamma = 1}} \sum_{\mu \in R_\gamma} \left(S_{f_{\sigma(p,q,\mu)}}\right)_{(0,0)}.$$

In particular, we prove that for any one dimensional face γ we have the equality

$$-\left[\xi : ((f_\gamma = 0) \cap \mathbb{G}_m^2) \times \mathbb{G}_m \to \mathbb{G}_m, \sigma_\gamma\right] = \sum_{\mu \in R_\gamma} \left(S_{f_{\sigma(p,q,\mu)}}\right)_{(0,0)}.$$

As f is non degenerate, we can assume

$$f_\gamma(x, y) = x^a y^b \prod_{\mu \in R_\gamma} (y^p - \mu x^q)$$

with $\gcd(p, q) = 1$. Then

$$\left[((f_\gamma = 0) \cap \mathbb{G}_m^2) \times \mathbb{G}_m \xrightarrow{\xi} \mathbb{G}_m, \sigma_\gamma\right] = \sum_{\mu \in R_\gamma} \left[((y^p = \mu x^q) \cap \mathbb{G}_m^2) \times \mathbb{G}_m \xrightarrow{\xi} \mathbb{G}_m, \sigma_\gamma\right].$$

We have the equality

$$\left[\xi : ((y^p = \mu x^q) \cap \mathbb{G}_m^2) \to \mathbb{G}_m, \sigma_\gamma\right] = \left(S_{f_{\sigma(p,q,\mu)}}\right)_{(0,0)}.$$

Indeed, we have $f_{\sigma(p,q,\mu)}(x,y) = x^N(y + \rho x^{q_1} + g(x,y))$ Then by example 3.15, we have

$$\begin{aligned}\left(S_{f_{\sigma(p,q,\mu)}}\right)_{(0,0)} &= -[x^N z : \mathbb{G}_m^2 \to \mathbb{G}_m, \sigma_{\mathbb{G}_m^2}] = [\xi : \mathbb{G}_m^2 \to \mathbb{G}_m, \sigma_{\mathbb{G}_m^2}] \\ &= -(\mathbb{L}-1)[\xi : \mathbb{G}_m \to \mathbb{G}_m, \sigma_{\mathbb{G}_m}]\end{aligned}$$

using the equivariant isomorphism

$$\begin{aligned}\mathbb{G}_m^2 &\to \mathbb{G}_m^2 \\ (x,z) &\mapsto (x, x^N z)\end{aligned}$$

and the identification of action in $\mathscr{M}_{\mathbb{G}_m}^{\mathbb{G}_m}$. As well, we have

$$\begin{aligned}\left[\xi : ((y^p = \mu x^q) \cap \mathbb{G}_m^2) \times \mathbb{G}_m \to \mathbb{G}_m, \sigma_\gamma\right] &= [\xi : \mathbb{G}_m^2 \to \mathbb{G}_m, \sigma_{\mathbb{G}_m^2}] \\ &= (\mathbb{L}-1)[\xi : \mathbb{G}_m \to \mathbb{G}_m, \sigma_{\mathbb{G}_m}].\end{aligned}$$

using the equivariant isomorphism

$$\begin{aligned}\mathbb{G}_m^2 &\to ((y^p = \mu x^q) \cap \mathbb{G}_m^2) \times \mathbb{G}_m \\ (\tau, \xi) &\mapsto \quad (\mu^{q'} \tau^p, \mu^{p'} \tau^q, \xi)\end{aligned}$$

with p' and q' integers satisfying the equality $pp' - qq' = 1$.

Example 4 We go on with Example 1. Using Theorem 2, we have

$$\begin{aligned}\left(S_f\right)_{(0,0)} &= [y^7 : \mathbb{G}_m \to \mathbb{G}_m, \sigma_{\mathbb{G}_m}] + [x^8 : \mathbb{G}_m \to \mathbb{G}_m, \sigma_{\mathbb{G}_m}] \\ &+ \left[y^4(x^2+y^3) : \mathbb{G}_m^2 \setminus (x^2+y^3=0) \to \mathbb{G}_m, \sigma_{\gamma_1}\right] \\ &+ \left[x^2(x^3-y^2)^2 : \mathbb{G}_m^2 \setminus (x^3-y^2=0) \to \mathbb{G}_m, \sigma_{\gamma_2}\right] - \left[x^2y^4 : \mathbb{G}_m^2 \setminus \to \mathbb{G}_m, \sigma_\gamma\right] \\ &+ \left(S_{f_1^1}\right)_{(0,0)} + \left(S_{f_1^2}\right)_{(0,0)}.\end{aligned} \tag{28}$$

Using Example 3, we have

$$\left(S_{f_1^1}\right)_{(0,0)} = -[x^{14} y : \mathbb{G}_m^2 \to \mathbb{G}_m, \sigma_{\mathbb{G}_m^2}].$$

Using again Theorem 2,

$$\begin{aligned}\left(S_{f_1^2}\right)_{(0,0)} &= [x^{21} : \mathbb{G}_m \to \mathbb{G}_m, \sigma_{\mathbb{G}_m}] - \left[x^{16}y^2 : \mathbb{G}_m^2 \setminus \to \mathbb{G}_m, \sigma_\gamma\right] + \left(S_{f_2^2}\right)_{(0,0)} \\ &+ \left[x^{16}(4y^2 + x^5) : \mathbb{G}_m^2 \setminus (4y^2 + x^5 = 0) \to \mathbb{G}_m, \sigma_\gamma\right].\end{aligned} \tag{29}$$

Finally, using Example 3, we have

$$\left(S_{f_2^2}\right)_{(0,0)} = -[x^{42}y : \mathbb{G}_m^2 \to \mathbb{G}_m, \sigma_{\mathbb{G}_m^2}].$$

By Theorem 3, we have

$$\begin{aligned}(S_f)_{(0,0)} = \left(S_{y^4(x^2+y^3)}\right)_{(0,0)} + \left(S_{x^2(x^3-y^2)}\right)_{(0,0)} + \left(S_{x^{16}(4y^2+x^5)}\right)_{(0,0)} \\ - \left(S_{x^{16}y^2}\right)_{(0,0)} - \left(S_{x^2y^4}\right)_{(0,0)}.\end{aligned}$$

4.4 Cauwbergs' Example

In [7] Cauwbergs considers the example

$$f_\lambda(x, y) = xy^2(x - y)(x - \lambda y)$$

with $\lambda \in \mathbb{C} \setminus \{0, 1\}$. He proved in [7, Proposition 4.1], that there exists λ and λ' in $\mathbb{C} \setminus \{0, 1\}$ such that

$$\left(Z_{f_\lambda}(T)\right)_{(0,0)} \neq \left(Z_{f_{\lambda'}}(T)\right)_{(0,0)}.$$

As a corollary we deduce from that proposition that the motivic Milnor fibers $\left(S_{f_\lambda,\omega}\right)_{(0,0)}$ and $\left(S_{f_{\lambda'},\omega}\right)_{(0,0)}$ are also different. Indeed, using results of this section we can express the motivic zeta function $(Z_{f_\lambda}(T))_{(0,0)}$ as

$$\begin{aligned}Z_\lambda(T) &= [xy^4 : \mathbb{G}_m^2 \to \mathbb{G}_m, \sigma_{\mathbb{G}_m^2}] \textstyle\sum_{(k,l)\in C_{xy^4}} \mathbb{L}^{-k-l}T^{4k+l} \\ &+ [x^3y^2 : \mathbb{G}_m^2 \to \mathbb{G}_m, \sigma_{\mathbb{G}_m^2}] \textstyle\sum_{(k,l)\in C_{x^3y^2}} \mathbb{L}^{-k-l}T^{3k+2l} \\ &+ [f_\lambda : \mathbb{G}_m^2 \setminus (f_\lambda = 0) \to \mathbb{G}_m, \sigma_{\mathbb{G}_m^2}] \textstyle\sum_{(k,k)\in C_\gamma^{\leq}} \mathbb{L}^{-2k}T^{5k} \\ &+ [\xi : ((f_\lambda = 0) \cap \mathbb{G}_m^2) \times \mathbb{G}_m \to \mathbb{G}_m, \sigma_{\mathbb{G}_m}] \textstyle\sum_{((k,k),n)\in C_\gamma^{<}} \mathbb{L}^{-2k}T^n\end{aligned}$$

as well for $Z_{\lambda'}(T)$. We remark also that we can write

$$[((f_\lambda = 0) \cap \mathbb{G}_m^2) \times \mathbb{G}_m \xrightarrow{\xi} \mathbb{G}_m, \sigma_{\mathbb{G}_m}] = [(x-y)(x-\lambda y) = 0 \cap \mathbb{G}_m^2][\mathbb{G}_m \xrightarrow{\xi} \mathbb{G}_m, \tau_{\mathbb{G}_m}]$$
$$= (2\mathbb{L} - 1)[\mathbb{G}_m \xrightarrow{\xi} \mathbb{G}_m, \tau_{\mathbb{G}_m}].$$

Then, from the difference between $Z_\lambda(T)$ and $Z_{\lambda'}(T)$ and the previous expressions we deduce

$$[f_\lambda : \mathbb{G}_m^2 \setminus (f_\lambda = 0) \to \mathbb{G}_m, \sigma_{\mathbb{G}_m^2}] \neq [f_{\lambda'} : \mathbb{G}_m^2 \setminus (f_{\lambda'} = 0) \to \mathbb{G}_m, \sigma_{\mathbb{G}_m^2}],$$

and we conclude passing to the limit in $Z_\lambda(T)$ or $Z_{\lambda'}(T)$ or also by Remark 17 that

$$\left(S_{f_\lambda,\omega}\right)_{(0,0)} \neq \left(S_{f_{\lambda'},\omega}\right)_{(0,0)}$$

whereas the topological Milnor fibers of f_λ and $f_{\lambda'}$ have the same topological type.

4.5 Schrauwen, Steenbrink and Stevens Example

In [26], the authors give examples of plane curve singularities with different topological types but the same spectral pairs. We shall use these examples to show that we can get the same motivic Milnor fiber at the origin with different topological types.

Let

$$f_1(x,y) = f_{11,00}(x,y) = ((y-x^2)^2-x^6)((y+x^2)^2-x^6)((x-y^2)^2-y^5)((x+y^2)^2-y^5)$$

$$f_2(x,y) = f_{10,10}(x,y) = ((y-x^2)^2-x^6)((y+x^2)^2-x^5)((x-y^2)^2-y^6)((x+y^2)^2-y^5).$$

The two functions have the same Newton polygon. There are two one dimensional compact faces with face polynomial $x^4(y^2 - x^4)^2$ and $y^4(x^2 - y^4)^2$ and 3 zero dimensional faces with face polynomial x^{12}, x^4y^4 and y^{12}. We shall apply the same Newton maps. In the following formula we write f for f_1 or f_2.

$$\begin{aligned}\left(S_f\right)_{(0,0)} &= [x^{12} : \mathbb{G}_m \to \mathbb{G}_m, \sigma_{\mathbb{G}_m}] + [y^{12} : \mathbb{G}_m \to \mathbb{G}_m, \sigma_{\mathbb{G}_m}] - [x^4y^4 : \mathbb{G}_m^2 \to \mathbb{G}_m, \sigma_{\mathbb{G}_m}] \\ &+ \left[(y^2 - x^4)^2x^4 : \mathbb{G}_m^2 \setminus (y^2 - x^4 = 0) \to \mathbb{G}_m, \sigma_\gamma\right] \\ &+ \left[(x^2 - y^4)^2y^4 : \mathbb{G}_m^2 \setminus (x^2 - y^4 = 0) \to \mathbb{G}_m, \sigma_\gamma\right] \\ &+ \sum_{\gamma \in \mathcal{N}(f), \dim \gamma = 1} \sum_{\mu \in R_\gamma} \left(S_{f_{\sigma(p,q,\mu)}}\right)_{(0,0)} .\end{aligned} \tag{30}$$

We use the following Newton transformations

- Let $\sigma_1(x,y) = (x, x^2(y+1))$. Then f_{1,σ_1} has one one dimensional compact face with face polynomial $4x^{12}(y^2 - x^2)$ which is non degenerate. The same for f_{2,σ_1}.

- Let $\sigma_2(x, y) = (x, x^2(y-1))$. Then f_{1,σ_2} has one one dimensional compact face with face polynomial $4x^{12}(y^2 - x^2)$ which is non degenerate, and f_{2,σ_2} has one one dimensional face with face polynomial $4x^{12}(y^2 - x)$ which is also non degenerate.
- Let $\sigma_3(x, y) = (x^2, x(y+1))$. Then f_{1,σ_3} has one one dimensional compact face with face polynomial $4x^{12}(4y^2 - x)$ which is non degenerate, and f_{2,σ_3} has one one dimensional face with face polynomial $4x^{12}(4y^2 - x^2)$ which is also non degenerate.
- Finally, let $\sigma_4(x, y) = (-x^2, x(y-1))$. Then f_{1,σ_4} has onc one dimensional compact face with face polynomial $4x^{12}(4y^2 + x)$ which is non degenerate, and f_{2,σ_4} has one one dimensional face with face polynomial $4x^{12}(4y^2 + x)$ which is also non degenerate. Then, expressing the formula of the Theorem 2, we deduce after computations that

$$\left(S_{f_1}\right)_{(0,0)} = \left(S_{f_2}\right)_{(0,0)},$$

thanks to the following equalities:

$$[x^{12}(y^2 - x^2) : \mathbb{G}_m^2 \setminus (y^2 = x^2) \to \mathbb{G}_m, \sigma_{\mathbb{G}_m^2}]$$
$$= [x^{12}(4y^2 - x^2) : \mathbb{G}_m^2 \setminus (4y^2 = x^2) \to \mathbb{G}_m, \sigma_{\mathbb{G}_m^2}],$$

$$[((x^{12}(y^2 - x^2) = 0) \cap \mathbb{G}_m^2) \times \mathbb{G}_m \xrightarrow{\xi} \mathbb{G}_m, \sigma_{\mathbb{G}_m}]$$
$$= [((x^{12}(4y^2 - x^2) = 0) \cap \mathbb{G}_m^2) \times \mathbb{G}_m \xrightarrow{\xi} \mathbb{G}_m, \sigma_{\mathbb{G}_m}],$$

thanks to the isomorphim of $\mathbb{G}_m^2$, $(x, y) \mapsto (x, 2y)$, and also

$$[x^{12}(y^2 - x) : \mathbb{G}_m^2 \setminus (y^2 = x) \to \mathbb{G}_m, \sigma_{\mathbb{G}_m^2}]$$
$$= [x^{12}(4y^2 + x) : \mathbb{G}_m^2 \setminus (4y^2 = -x) \to \mathbb{G}_m, \sigma_{\mathbb{G}_m^2}],$$

$$[((x^{12}(y^2 - x) = 0) \cap \mathbb{G}_m^2) \times \mathbb{G}_m \xrightarrow{\xi} \mathbb{G}_m, , \sigma_{\mathbb{G}_m}]$$
$$= [((x^{12}(4y^2 + x) = 0) \cap \mathbb{G}_m^2) \times \mathbb{G}_m \xrightarrow{\xi} \mathbb{G}_m, , \sigma_{\mathbb{G}_m}],$$

thanks to the isomorphim of $\mathbb{G}_m^2$, $(x, y) \mapsto (-x, 2y)$.

5 Generalized Kouchnirenko's Formula

5.1 Euler Characteristic of the Milnor Fiber of a Quasi Homogeneous Polynomial

Proposition 8 *Let*

$$F(x, y) = cx^a y^b \prod_{1\leq i\leq r} (y^p - \mu_i x^q)^{\nu_i}$$

be a quasihomogeneous polynomial. We denote by $F_{(0,0)}$ the Milnor fiber of F at the origin. We have

$$\chi_c(F_{(0,0)}) = -Nr + \epsilon_b N/p + \epsilon_a N/q \tag{31}$$

where $\epsilon_b = 0$ if $b \neq 0$ and $\epsilon_b = 1$ if $b = 0$, and $\epsilon_a = 0$ if $a \neq 0$ and $\epsilon_a = 1$ if $a = 0$, and $N = ap + bq + pq \sum \nu_i$.

Proof The Euler characteristic $\chi_c(F_{(0,0)})$ is computed in [23, Theorem 2.8]. Indeed, the weighted blow-up at the origin with weights (p, q) is a $\mathbb{Q}$-resolution $\pi : X \to \mathbb{C}^2$ of $H = F^{-1}(0)$ where

$$X = \{(x, y), [u : v]_{(p,q)} \in \mathbb{C}^2 \times \mathbb{P}^1_{(p,q)} \mid \exists\lambda \in \mathbb{C},\ x = \lambda^p u,\ y = \lambda^q v\}.$$

The exceptional divisor E has one irreducible component. Let $\hat{H}$ be the strict transform of H. We denote $\check{E} = E \setminus (E \cap \hat{H})$.

X is the union of two charts $X_{u\neq 0}$ and $X_{v\neq 0}$.

- A parametrization of the chart $X_{u\neq 0}$ is given by

$$\begin{array}{rccc} \Phi: & \mathbb{C}^2 & \to & X_{u\neq 0} \\ & (x_1, y_1) & \mapsto & \left((x_1^p, x_1^q y_1), [1 : y_1]_{(p,q)}\right) \end{array}$$

which implies that the chart $X_{u\neq 0}$ is isomorphic to $\mathbb{C}^2/\mu_p$ (denoted by $X(p, (-1, q))$ in [23]) the quotient of $\mathbb{C}^2$ by the action of the group of p-roots of unity

$$\begin{array}{ccc} \mu_p \times \mathbb{C}^2 & \to & \mathbb{C}^2 \\ (\xi, (x, y)) & \mapsto & (\xi^{-1}x, \xi^q y). \end{array}$$

In particular the singular locus of the chart $X_{u\neq 0}$ is the quotient singularity at the origin denoted by P. The total transform of F has equation

$$g(x_1, y_1) = f(x_1^p, x_1^q y_1) = C x_1^N y_1^b \prod_{i=1}^{r} (y_1^p - \mu_i)^{\nu_i} = 0.$$

- A parametrization of the chart $X_{v\neq 0}$ is given by

$$\begin{array}{rccc} \Phi: & \mathbb{C}^2 & \to & X_{v\neq 0} \\ & (x_1, y_1) & \mapsto & \left((x_1 y_1^p, y_1^q), [x_1 : 1]_{(p,q)}\right) \end{array}$$

which implies that the chart $X_{v\neq 0}$ is isomorphic to $\mathbb{C}^2/\mu_q$ (denoted by $X(q, (p, -1))$ in [23]) the quotient of $\mathbb{C}^2$ by the action of the group of q-roots of unity

$$\begin{array}{rcl} \mu_q \times \mathbb{C}^2 & \to & \mathbb{C}^2 \\ (\xi, (x, y)) & \mapsto & (\xi^p x, \xi^{-1} y). \end{array}$$

In particular the singular locus of the chart $X_{u\neq 0}$ is the quotient singularity at the origin denoted by Q. The total transform of F has equation

$$g(x_1, y_1) = f(x_1^p, x_1^q y_1) = C y_1^N x_1^a \prod_{i=1}^{r} (1 - \mu_i x_i^q)^{\nu_i} = 0.$$

Hence the variety X has two singular points P and Q and we consider the stratification

$$X = \{P\} \cup \{Q\} \cup X \setminus \{P, Q\}.$$

By application of Theorem 2.8 in [23] we have

$$\chi_c(F_{(0,0)}) = m_{\check{E}\cap X\setminus\{P,Q\}} \chi_c(\check{E}\cap X\setminus\{P, Q\}) + m_{\check{E}\cap\{P\}} \chi_c(\check{E}\cap\{P\}) + m_{\check{E}\cap\{Q\}} \chi_c(\check{E}\cap\{Q\})$$

where the multiplicities are defined in [23]. Formula (31) follows in that way.

- If $a \neq 0$ and $b \neq 0$, then the strict transform $\hat{H}$ has $r+2$ components and contains the points P and Q. Then, $\check{E}$ is contained in the stratum $X \setminus \{P, Q\}$ which is smooth, its multiplicity $m_{\check{E}\cap X\setminus\{P,Q\}}$ is equal to N and its Euler characteristic is equal to $-r$. Then

$$\chi_c(F_{(0,0)}) = -Nr.$$

- If $a = 0$ and $b \neq 0$ then the strict transform has $r + 1$ components with one containing the point P. The point Q does not belong to $\hat{H}$. The multiplicities are

$$m_{\check{E}\cap X\setminus\{P,Q\}} = N,\ m_{\check{E}\cap\{Q\}} = \frac{N}{q},$$

and we obtain

$$\chi_c(F_{(0,0)}) = \chi_c(\{Q\})\frac{N}{q} + \chi_c(\check{E} \cap X \setminus \{P, Q\})N = \frac{N}{q} - Nr.$$

- The cases ($a \neq 0, b = 0$) and ($a = 0, b = 0$) are similar.

Remark 18 Using above notations, let $\mathscr{S}$ be the area of the triangle with vertices $(0, 0)$, $v_0 = (a, b + p\sum_{i=1}^{r} \nu_i)$ and $v_d = (a + q\sum_{i=1}^{r} \nu_i, b)$. We have $N = 2\mathscr{S}/\sum_{i=1}^{r} \nu_i$. Indeed, we have

$$2\mathscr{S} = \left|\det\begin{pmatrix} a & a + q\sum \nu_i \\ b + p\sum \nu_i & b \end{pmatrix}\right| = |N|\sum_{i=1}^{r} \nu_i.$$

Then we have another expression for the Euler characteristic:

$$\chi_c(F_{(0,0)}) = -\frac{(r - \epsilon_b/p - \epsilon_a/q)2\mathscr{S}}{\sum_{i=1}^{r} \nu_i}.$$

Definition 9 Let Δ be a Newton diagram, with one dimensional faces S_i with $i \in \{1, \cdots, d\}$. Let f be a polynomial such that $\Delta(f) = \Delta$. For each face S_i, we denote by r_i the number of roots of F_{S_i}, we denote by s_i the number of points with integer coordinates on the face S_i without its vertices and by $\mathscr{S}_i$ its area defined in Remark 18. We define the *area of Δ with respect to f* as

$$\mathscr{S}_{\Delta,f} = \sum_{i=1}^{d} r_i\mathscr{S}_i/(s_i + 1).$$

Remark 19 Note that if f is non degenerate, then $\mathscr{S}_{\Delta,f} = \mathscr{S}_\Delta$ where S_Δ is the area of Δ.

Using Definition 9 and Proposition 8 we have

Proposition 9 *If F is a quasi-homogeneous polynomial, its Newton diagram has one face $[v_0, v_d]$. We have*

$$-\chi_c(F_{(0,0)}) = 2\mathscr{S}_{\Delta,f} - \epsilon_0 b_0 - \epsilon_d a_d$$

where $v_0 = (a_0, b_0)$ and $\epsilon_0 = 1$ if $a_0 = 0$ and 0 otherwise, and $v_d = (a_d, b_d)$ and $\epsilon_d = 1$ if $b_d = 0$ and 0 otherwise.

5.2 Generalized Kouchnirenko's Formula

Notation 4 *In the following, we denote by $\tilde{\chi}_c : \mathscr{M} \to \mathbb{Z}$ the realisation which factorizes the Euler characteristic invariant χ_c through the motive map from $Var_{\mathbf{k}}$ to $\mathscr{M}$.*

The following result is a well-known corollary of Denef-Loeser' Theorem 4.2.1 in [8]. In the isolated case, it follows from the A'Campo formula [1].

Theorem 4 *Let f be a polynomial in $\mathbf{k}[x, y]$ with $f(0, 0) = 0$, we denote by $F_{(0,0)}$ its Milnor fiber at the origin. Then,*

$$\chi_c(F_{(0,0)}) = \tilde{\chi}_c\left(\left(S_f\right)^{(1)}_{(0,0)}\right)$$

where $\left(S_f\right)^{(1)}_{(0,0)}$ is the fiber in 1 of the motivic Milnor fiber $\left(S_f\right)_{(0,0)}$.

Then, we obtain the generalized Kouchnirenko's formula

Theorem 5 *Let f be a polynomial in $\mathbf{k}[x, y]$ with $f(0, 0) = 0$. We denote by $F_{(0,0)}$ its Milnor fiber at the origin. Then,*

$$-\chi_c\left(F_{(0,0)}\right) = -\tilde{\chi}_c\left(\left(S_f\right)^{(1)}_{(0,0)}\right) = 2\mathscr{S}_{\Delta(f),f} - \varepsilon_{b_h} a_h - \varepsilon_{a_v} b_v + \sum_{\sigma} \tilde{\chi}_c\left(\left(S_{f_\sigma}\right)^{(1)}_{(0,0)}\right)$$

where the summation is taken over all possible Newton maps associated to the Newton polygon of f.

Remark 20 If f is non degenerate, we get Kouchnirenko's theorem

$$-\chi_c\left(F_{(0,0)}\right) = 2\mathscr{S}_{\Delta(f),f} - \varepsilon_{b_h} a_h - \varepsilon_{a_v} b_v.$$

Otherwise, the Euler characteristic is expressed in terms of the areas of the Newton diagrams of f and the Euler characteristic of the Newton transforms of f, which can be computed by induction in terms of area of Newton polygons.

Proof Recall first that for any integers $a \geq 1$ and $b \geq 1$, the Euler characteristic $\tilde{\chi}_c([x^a y^b = 1])$ is equal to zero. Then taking the fiber in 1, and applying the realization $\tilde{\chi}_c$, it follows from Eqs. (8), (7) and (4) that

$$\tilde{\chi}_c\left(\left(S_f\right)^{(1)}_{(0,0)}\right) = \sum_{\gamma \in \mathscr{N}(f),\, \dim \gamma = 1} \left[\tilde{\chi}_c\left(\left(S^{(1)}_{f_\gamma}\right)_{(0,0)}\right) + \sum_{\mu \in R_\gamma} \tilde{\chi}_c\left(\left(S_{f_{\sigma(p,q,\mu)}}\right)^{(1)}_{(0,0)}\right)\right], \tag{32}$$

and the formula follows from the quasi-homogeneous case in Proposition 9 and Definition 9.

Example 5 We come back to the Example 1. We have shown

$$(S_f)_{(0,0)} = \left(S_{y^4(x^2+y^3)}\right)_{(0,0)} + \left(S_{x^2(x^3-y^2)}\right)_{(0,0)} + \left(S_{x^{16}(4y^2+x^5)}\right)_{(0,0)} \\ - \left(S_{x^{16}y^2}\right)_{(0,0)} - \left(S_{x^2y^4}\right)_{(0,0)}.$$

Classically we have

$$\chi_c\left(\left(S_{x^{16}y^2}\right)^{(1)}_{(0,0)}\right) = \chi_c\left(\left(S_{x^2y^4}\right)^{(1)}_{(0,0)}\right) = 0$$

and using Proposition 8 we deduce,

$$-\chi_c\left(\left(S_{y^4(x^2+y^3)}\right)^{(1)}_{(0,0)}\right) = 14 - 7, \; -\chi_c\left(\left(S_{x^2(x^3-y^2)^2}\right)^{(1)}_{(0,0)}\right) = 32/2 - 8,$$

$$-\chi_c\left(\left(S_{x^{16}(4y^2+x^5)}\right)^{(1)}_{(0,0)}\right) = 42 - 21,$$

and we get

$$-\chi_c\left(\left(S_f\right)^{(1)}_{(0,0)}\right) = 7 + 8 + 21 = 36.$$

6 Monodromy Zeta Function

6.1 The Invariant Monodromy Zeta Function

In this section we assume $\mathbf{k} = \mathbb{C}$. Let $(X \to \mathbb{G}_m, \sigma_{\mathbb{G}_m})$ be an element of $\mathrm{Var}_{\mathbb{G}_m}^{\mathbb{G}_m}$ with dimension d. The fiber $X^{(1)}$ is a variety endowed with a monomial action $\sigma_{\hat{\mu}}$ of the group of roots of unity. For any i, the cohomology group $H^i_c(X^{(1)}, \mathbb{Q})$ of $X^{(1)}$ is endowed with a quasi-unipotent operator T_i. The *monodromy zeta function* of $(X \to \mathbb{G}_m, \sigma_{\mathbb{G}_m})$ is defined as

$$Z_X^{\mathrm{mon}}(t) = \prod_{i=0}^{\dim X} P_{T_i}(t)^{(-1)^i}$$

where for any i, $P_{T_i}(t)$ is the characteristic polynomial of T_i.

The invariant

$$\begin{array}{rcl} Z^{\mathrm{mon}} : & \mathrm{Var}_{\mathbb{G}_m}^{\mathbb{G}_m} & \to (\mathbb{C}(t)^*, \times) \\ & (X \to \mathbb{G}_m, \sigma_{\mathbb{G}_m}) & \mapsto Z_X^{\mathrm{mon}}(t) \end{array}$$

factorizes through the Grothendieck group $K_0(\mathrm{Var}_{\mathbb{G}_m}^{\mathbb{G}_m})$ and we denote the associated realization by $\tilde{Z}^{\mathrm{mon}}$

$$\begin{array}{l} \tilde{Z}^{\mathrm{mon}} : \left(K_0(\mathrm{Var}_{\mathbb{G}_m}^{\mathbb{G}_m}), +\right) \to (\mathbb{C}(t)^*, \times) \\ \quad [X \to \mathbb{G}_m, \sigma_{\mathbb{G}_m}] \mapsto \tilde{Z}^{\mathrm{mon}}([X \to \mathbb{G}_m, \sigma_{\mathbb{G}_m}]) := Z_X^{\mathrm{mon}}(t) \end{array}.$$

It is a group morphism which extends to $(\mathscr{M}_{\mathbb{G}_m}^{\mathbb{G}_m}, +)$.

6.2 The Monodromy Zeta Function of the Milnor Fiber

Let f be a polynomial in $\mathbb{C}[x, y]$ with $f(0,0) = 0$. We denote by $F_{(0,0)}$ its Milnor fiber at the origin and for i in $\{0, 1, 2\}$, $T_i : H_c^i(F_{(0,0)}, \mathbb{Q}) \to H_c^i(F_{(0,0)}, \mathbb{Q})$ the monodomy operators on the cohomology groups of the Milnor fiber with $P_{T_i}(t)$ their characteristic polynomial. The *monodromy zeta function* of f at the origin is

$$Z_f^{\mathrm{mon}}(t) = \frac{P_{T_0}(t) P_{T_2}(t)}{P_{T_1}(t)}.$$

As a corollary of Denef-Loeser's Theorem 4.2.1 in [8] or [11], or the A'Campo formula [1] in the isolated case we have

Theorem 6 *Let f be a polynomial in $\mathbf{k}[x, y]$ with $f(0,0) = 0$, we denote by $F_{(0,0)}$ its Milnor fiber at the origin. Then,*

$$Z_f^{mon}(t) = \tilde{Z}^{mon}\left(\left(S_f\right)_{(0,0)}^{(1)}\right),$$

where $\left(S_f\right)_{(0,0)}^{(1)}$ is the fiber in 1 of the motivic Milnor fiber $\left(S_f\right)_{(0,0)}$.

With a similar proof to that of Proposition 8, the computation of the monodromy zeta function at the origin for a quasi-homogenenous polynomial follows from [23, Theorem 2.8] and [23, Lemma 2.5]

Proposition 10 *Let*

$$F(x, y) = cx^a y^b \prod_{1 \leq i \leq r} (y^p - \mu_i x^q)^{\nu_i},$$

be a quasihomogeneous polynomial. The monodromy zeta function of F at the origin is equal to

$$Z_F(t) = (1 - t^N)^r (1 - t^{N/p})^{-\epsilon_b} (1 - t^{N/q})^{-\epsilon_a}$$

where $\epsilon_b = 0$ *if* $b \neq 0$ *and* $\epsilon_b = 1$ *if* $b = 0$, *and* $\epsilon_a = 0$ *if* $a \neq 0$ *and* $\epsilon_a = 1$ *if* $a = 0$, *and* $N = ap + bq + pq \sum \nu_i$.

If F *is a monomial,* $F(x, y) = cx^a y^b$ *with* $a \geq 1$ *and* $b \geq 1$ *then* $Z_F(t) = 1$.

Then using this Proposition and Eqs. (8), (7) and (4) we deduce

Proposition 11 *Let* f *be a polynomial in* $\mathbb{C}[x, y]$ *with* $f(0, 0) = 0$. *The monodromy zeta function at the origin is equal*

$$Z_f(t) = \prod_S (1 - t^{N_S})^{r_S} (1 - t^{a_h})^{-\varepsilon_{b_h}} (1 - t^{b_v})^{-\varepsilon_{a_v}} \prod_\sigma Z_{f_\sigma}(t)$$

where the first product is taken over all faces S *of the Newton polygon of* f, *and the second product is taken over all Newton maps associated to the Newton polygon of* f.

Remark 21 This proposition allows to recover Theorem 12.1 of Eisenbud-Neumann in [12]. Furthermore, in the case of a non-degenerate polynomial, we recover Varchenko's formula for the monodromy zeta function in [27].

Example 6 By application of Proposition 11 to Example 1, we get

$$Z(t) = \frac{(1 - t^{16})(1 - t^{14})(1 - t^{42})}{(1 - t^8)(1 - t^7)(1 - t^{21})}.$$

Acknowledgements The first author is partially supported by the projects MTM2016-76868-C2-1-P (UCM Madrid) and MTM2016-76868-C2-2-P (Zaragoza). The second author is partially supported by the project ANR-15-CE40-0008 (Dfigo). The authors thank the referees for theirs suggestions.

References

1. A'Campo, N.: La fonction zêta d'une monodromie. Comment. Math. Helv. **50**, 233–248 (1975)
2. Artal Bartolo, E., Cassou-Noguès, P., Luengo, I., Melle Hernández, A.: Quasi-ordinary power series and their zeta functions. Mem. Am. Math. Soc. **178**(841), vi+85 (2005)
3. Bultot, E.: Computing zeta functions on log smooth models. C. R. Math. Acad. Sci. Paris **353**(3), 261–264 (2015)
4. Bultot, E., Nicaise, J.: Computing zeta functions on log smooth models (2016). arXiv:1610.00742
5. Cassou-Noguès, P., Veys, W.: Newton trees for ideals in two variables and applications. Proc. Lond. Math. Soc. (3) **108**(4), 869–910 (2014)
6. Cassou-Noguès, P., Veys, W.: The Newton tree: geometric interpretation and applications to the motivic zeta function and the log canonical threshold. Math. Proc. Camb. Philos. Soc. **159**(3), 481–515 (2015)
7. Cauwbergs, T.: Splicing motivic zeta functions. Rev. Mat. Complut. **29**(2), 455–483 (2016)
8. Denef, J., Loeser, F.: Motivic Igusa zeta functions. J. Algebraic Geom. **7**(3), 505–537 (1998)
9. Denef, J., Loeser, F.: Germs of arcs on singular algebraic varieties and motivic integration. Invent. Math. **135**(1), 201–232 (1999)

10. Denef, J., Loeser, F.: Geometry on arc spaces of algebraic varieties. In: European Congress of Mathematics, vol. I, Barcelona, 2000. Volume 201 of Progress in Mathematics, pp. 327–348. Birkhäuser, Basel (2001)
11. Denef, J., Loeser, F.: Lefschetz numbers of iterates of the monodromy and truncated arcs. Topology **41**(5), 1031–1040 (2002)
12. Eisenbud, D., Neumann, W.: Three-Dimensional Link Theory and Invariants of Plane Curve Singularities. Volume 110 of Annals of Mathematics Studies. Princeton University Press, Princeton (1985)
13. González Pérez, P.D., González Villa, M.: Motivic Milnor fiber of a quasi-ordinary hypersurface. J. Reine Angew. Math. **687**, 159–205 (2014)
14. Gonzalez Villa, M., Kenned, G., McEwan, L.J.: A recursive formula for the motivic milnor fiber of a plane curve (2016). arXiv:1610.08487
15. Guibert, G.: Espaces d'arcs et invariants d'Alexander. Comment. Math. Helv. **77**(4), 783–820 (2002)
16. Guibert, G., Loeser, F., Merle, M.: Nearby cycles and composition with a nondegenerate polynomial. Int. Math. Res. Not. **31**, 1873–1888 (2005)
17. Guibert, G., Loeser, F., Merle, M.: Iterated vanishing cycles, convolution, and a motivic analogue of a conjecture of Steenbrink. Duke Math. J. **132**(3), 409–457 (2006)
18. Guibert, G., Loeser, F., Merle, M.: Composition with a two variable function. Math. Res. Lett. **16**(3), 439–448 (2009)
19. Hrushovski, E., Loeser, F.: Monodromy and the Lefschetz fixed point formula. Ann. Sci. Éc. Norm. Supér. (4) **48**(2), 313–349 (2015)
20. Kontsevich, M.: Lecture at Orsay. Décembre 7 (1995)
21. Loeser, F.: Seattle lectures on motivic integration. In: Algebraic Geometry—Seattle 2005. Part 2. Volume 80 of Proceedings of Symposia in Pure Mathematics, pp. 745–784. American Mathematical Society, Providence (2009)
22. Looijenga, E.: Motivic measures. Astérisque **276**, 267–297 (2002). Séminaire Bourbaki, vol. 1999/2000
23. Martín-Morales, J.: Monodromy zeta function formula for embedded **Q**-resolutions. Rev. Mat. Iberoam. **29**(3), 939–967 (2013)
24. Quy-Thuong Lł.: Motivic milnor fibers of plane curve singularities (2017). arXiv:1703.04820
25. Raibaut, M.: Singularités à l'infini et intégration motivique. Bull. Soc. Math. Fr. **140**(1), 51–100 (2012)
26. Schrauwen, R., Steenbrink, J., Stevens, J.: Spectral pairs and the topology of curve singularities. In: Complex Geometry and Lie Theory (Sundance, UT, 1989). Volume 53 of Proceedings of Symposia in Pure Mathematics, pp. 305–328. American Mathematical Society, Providence (1991).
27. Varchenko, A.N.: Zeta-function of monodromy and Newton's diagram. Invent. Math. **37**(3), 253–262 (1976)
28. Veys, W.: Zeta functions and "Kontsevich invariants" on singular varieties. Can. J. Math. **53**(4), 834–865 (2001)
29. Wall, C.T.C.: Singular Points of Plane Curves. Volume 63 of London Mathematical Society Student Texts. Cambridge University Press, Cambridge (2004)

Nash Modification on Toric Curves

Daniel Duarte and Daniel Green Tripp

Abstract We revisit the problem of resolution of singularities of toric curves by iterating the Nash modification. We give a bound on the number of iterations required to obtain the resolution. We also introduce a different approach on counting iterations by dividing the combinatorial algorithm of the Nash modification of toric curves into several division algorithms.

1 Introduction

The Nash modification of an equidimensional algebraic variety is a modification that replaces singular points by limits of tangent spaces. It has been proposed to iterate this construction to obtain resolution of singularities [14, 16]. So far, only in very few cases it is known that this method works: the case of curves [14] and the family $\{z^p + x^q y^r = 0\} \subset \mathbb{C}^3$ for positive integers p, q, r [15]. In [14] it is also proved that Nash modifications do not resolve singularities in positive characteristic.

There is a variant of this problem which consists in iterating the Nash modification followed by normalization. This variant was studied in[7, 8, 12, 17], culminating with the theorem by M. Spivakovsky that the normalized Nash modification solves singularities of complex surfaces.

Several years later, the original question received a new wave of attention with the appearance of several papers exploring the case of toric varieties [1, 4, 9, 10]. It was proved in [9, 10] that the iteration of the Nash modification for toric varieties corresponds to a purely combinatorial algorithm on the semigroups defining the toric variety. Using this combinatorial description, some new partial results were

D. Duarte (✉)
CONACYT, Universidad Autónoma de Zacatecas, Zacatecas, Mexico
e-mail: aduarte@uaz.edu.mx

D. G. Tripp
Universidad Nacional Autónoma de México, México City, México
e-mail: dangreen@ciencias.unam.mx

G.-M. Greuel et al. (eds.), *Singularities, Algebraic Geometry, Commutative Algebra, and Related Topics*, https://doi.org/10.1007/978-3-319-96827-8_8

obtained in the mentioned papers. Nevertheless, as far as we know, the question for toric varieties has not been completely solved so far.

In this note, we take a step backward hoping that a better understanding of the simplest case (that of toric curves) might throw some light on the problem for higher-dimensional toric varieties.

As we mentioned before, it is already known that the Nash modification solves singularities of curves. In this note we revisit this problem by giving a combinatorial proof of that fact for toric curves. In addition, we give a bound on the number of iterations required to obtain the desingularization. We also introduce a different approach on counting iterations. As we will see, the combinatorial algorithm for toric curves can be divided into several division algorithms that record the significant improvements that take place during the algorithm. We will give an effective bound for the number of such division algorithms.

We conclude with a brief discussion on some special features of the Nash modification of toric curves. These features include comments on Hilbert-Samuel multiplicity, embedding dimension, the support of an ideal defining the Nash modification, and a possible generalization of our results for toric surfaces.

2 Nash Modification of a Toric Variety

Let us start by recalling the definition of the Nash modification of an equidimensional algebraic variety. We consider complex algebraic varieties even though the results of this paper also hold for algebraically closed fields of characteristic zero.

Definition 1 Let $X \subset \mathbb{C}^n$ be an equidimensional algebraic variety of dimension d. Consider the Gauss map:

$$\begin{aligned} G : X \setminus \mathrm{Sing}(X) &\to G(d, n) \\ x &\mapsto T_x X, \end{aligned}$$

where $G(d, n)$ is the Grassmanian of d-dimensional vector spaces in $\mathbb{C}^n$, and $T_x X$ is the tangent space to X at x. Denote by X^* the Zariski closure of the graph of G. Call ν the restriction to X^* of the projection of $X \times G(d, n)$ to X. The pair (X^*, ν) is called the *Nash modification* of X.

We are interested in studying this construction in the case of toric curves. Let us recall the definition of an affine toric variety (see, for instance, [3, Section 1.1] or [18, Chapter 4]).

Let $\mathscr{A} = \{a_1, \ldots, a_n\} \subset \mathbb{Z}^d$ be a finite set satisfying $\mathbb{Z}\mathscr{A} = \{\sum_i \lambda_i a_i | \lambda_i \in \mathbb{Z}\} = \mathbb{Z}^d$. The set $\mathscr{A}$ induces a homomorphism of semigroups

$$\pi_{\mathscr{A}} : \mathbb{N}^n \to \mathbb{Z}^d, \quad \alpha = (\alpha_1, \ldots, \alpha_n) \mapsto \alpha_1 a_1 + \cdots + \alpha_n a_n.$$

Consider the ideal

$$I_{\mathscr{A}} := \langle x^{\alpha} - x^{\beta} | \alpha, \beta \in \mathbb{N}^n, \ \pi_{\mathscr{A}}(\alpha) = \pi_{\mathscr{A}}(\beta) \rangle \subset \mathbb{C}[x_1, \ldots, x_n].$$

Definition 2 We call $Y_{\mathscr{A}} := \mathbf{V}(I_{\mathscr{A}}) \subset \mathbb{C}^n$ the toric variety defined by $\mathscr{A}$.

It is well known that a variety obtained in this way is irreducible, contains a dense open set isomorphic to $(\mathbb{C}^*)^d$ and such that the natural action of $(\mathbb{C}^*)^d$ on itself extends to an action on $Y_{\mathscr{A}}$.

2.1 Combinatorial Algorithm for Affine Toric Varieties

It was recently proved that the iteration of the Nash modification of a toric variety $Y_{\mathscr{A}}$ corresponds to a combinatorial algorithm involving the elements in $\mathscr{A}$ (see [9, 10]). Here we follow the results as presented in [10, Section 4].

Let $\mathscr{A} = \{a_1, \ldots, a_n\} \subset \mathbb{Z}^d$ be a finite set such that $\mathbb{Z}\mathscr{A} = \mathbb{Z}^d$ and $0 \notin \mathrm{Conv}(\mathscr{A})$, where $\mathrm{Conv}(\mathscr{A})$ is the convex hull of $\mathscr{A}$ in $\mathbb{R}^d$. Let $Y_{\mathscr{A}} \subset \mathbb{C}^n$ be the corresponding affine toric variety. The Nash modification $Y^*_{\mathscr{A}}$ of $Y_{\mathscr{A}}$ can be described as follows.

For a given $n \in \mathbb{N}$, denote $[n] := \{1, \ldots, n\}$. Let $J = (j_1, \ldots, j_d) \in [n]^d$, where $1 \leq j_1 < j_2 < \cdots < j_d \leq n$. Denote $\det(a_J) := \det(a_{j_1}, \ldots, a_{j_d})$. Now fix $J_0 \in [n]^d$ such that $\det(a_{J_0}) \neq 0$ and for any other $J \in [n]^d$ denote $\Delta_J^{J_0} := \sum_{j \in J} a_j - \sum_{k \in J_0} a_k$.

Theorem 1 ([10, Section 4]; [9, Proposition 60]) *$Y^*_{\mathscr{A}}$ is covered by affine toric varieties $Y_{\mathscr{A}_{J_0}}$, where $\mathscr{A}_{J_0} = \{a_j\}_{j \in J_0} \cup \{\Delta_J^{J_0} | J \in [n]^d, \#(J \setminus J_0) = 1, \det(a_J) \neq 0\}$ and $0 \notin \mathrm{Conv}(\mathscr{A}_{J_0})$.*

In addition, we consider the following criterion of non-singularity for toric varieties.

Proposition 1 ([10, Criterion 3.16]) *Let $\mathscr{A} \subset \mathbb{Z}^d$ be as before. The affine toric variety $Y_{\mathscr{A}}$ is non-singular if and only if there are d elements in $\mathbb{N}\mathscr{A}$ such that $\mathbb{N}\mathscr{A}$ is generated by those elements as a semigroup.*

It follows from the theorem and proposition that iteration of the Nash modification of toric varieties coincides with the following combinatorial algorithm:

Algorithm 2.1

1. *Let $\mathscr{A} = \{a_1, \ldots, a_n\} \subset \mathbb{Z}^d$ be a finite set such that $\mathbb{Z}\mathscr{A} = \mathbb{Z}^d$ and $0 \notin \mathrm{Conv}(\mathscr{A})$.*
2. *For every $J_0 \in [n]^d$ such that $\det(a_{J_0}) \neq 0$, consider those sets $\mathscr{A}_{J_0}$ with the property $0 \notin \mathrm{Conv}(\mathscr{A}_{J_0})$.*
3. *If every $\mathbb{N}\mathscr{A}_{J_0}$ from step 2 is generated by d elements then stop. Otherwise replace $\mathscr{A}$ by each $\mathscr{A}_{J_0}$ that is not generated by d elements and return to step 1.*

2.2 Combinatorial Algorithm for Affine Toric Curves

We are interested in applying the previous algorithm for toric curves. Let us describe how the algorithm looks like in this special case.

First, we can assume that $\mathscr{A} = \{a_1, \ldots, a_n\} \subset \mathbb{Z}$ is such that $0 < a_1 < \ldots < a_n$. This follows from the fact that $0 \notin \text{Conv}(\mathscr{A})$ if and only if $\mathscr{A} \subset \mathbb{N}$ or $\mathscr{A} \subset \mathbb{Z} \backslash \mathbb{N}$ and $Y_{\mathscr{A}} \cong Y_{-\mathscr{A}}$. In addition, the condition $\mathbb{Z}\mathscr{A} = \mathbb{Z}$ is equivalent to $\gcd(\mathscr{A}) = 1$. For these reasons, from now on, $\mathscr{A} \subset \mathbb{Z}$ will denote a finite set $\{a_1, \ldots, a_n\}$ such that

$$0 < a_1 < \ldots < a_n \text{ and } \gcd(\mathscr{A}) = 1. \qquad (*)$$

Finally, Proposition 1 is equivalent to ask that $1 \in \mathscr{A}$ for toric curves.

Now we can describe Algorithm 2.1 in this case. Observe that in this case only for $J_0 = (1) \in [n]$ it happens that $0 \notin \text{Conv}(\mathscr{A}_{J_0})$ (see step 2 of Algorithm 2.1). In other words, the Nash modification of $Y_{\mathscr{A}}$ is contained in a single affine chart according to Theorem 1.

Algorithm 2.2 1. *Let* $\mathscr{A} = \{a_1, \ldots, a_n\} \subset \mathbb{Z}$ *be a finite set satisfying* (∗).
2. *Let* $\mathscr{A}' := \{a_1\} \cup \{a_2 - a_1, \ldots, a_n - a_1\}$.
3. *If* $1 \in \mathscr{A}'$ *the algorithm stops. Otherwise replace* $\mathscr{A}$ *with* $\mathscr{A}'$ *and go to step* 1.

Example 1 Let $\mathscr{A} = \{12, 28, 33\}$. Then:

$$\begin{aligned} \mathscr{A}' &= \{12\} \cup \{28 - 12, 33 - 12\} = \{12, 16, 21\}, \\ \mathscr{A}'' &= \{12\} \cup \{16 - 12, 21 - 12\} = \{4, 9, 12\}, \\ \mathscr{A}''' &= \{4\} \cup \{9 - 4, 12 - 4\} = \{4, 5, 8\}, \\ \mathscr{A}'''' &= \{4\} \cup \{5 - 4, 8 - 4\} = \{1, 4\}. \end{aligned}$$

Since $1 \in \mathscr{A}''''$, the algorithm stops, that is, $Y_{\mathscr{A}''''}$ is a non-singular curve.

Remark 1 Notice the resemblance of this algorithm with Euclid's algorithm (this is why in [10] Algorithm 2.1 is called *multidimensional Euclidean algorithm*). Actually, for $n = 2$, Algorithm 2.2 is exactly Euclid's algorithm for a_1 and a_2.

Remark 2 Let $\mathscr{A}$ be as before. If $a_i = qa_1$ for some $q \in \mathbb{N}$ and $i > 1$ it follows that $\mathbb{N}\mathscr{A} = \mathbb{N}(\mathscr{A} \setminus \{a_i\})$. In particular, $Y_{\mathscr{A}} \cong Y_{\mathscr{A} \setminus \{a_i\}}$. Therefore, some times we will assume that no multiples of the minimum of $\mathscr{A}$ other than itself appear in $\mathscr{A}$.

3 Resolution of Toric Curves and Number of Iterations

It is well known that the iteration of the Nash modification resolves singularities of curves [14, Corollary 1]. In this section we revisit this problem for toric curves giving a combinatorial proof in this special case. We also consider the problem of finding a bound for the number of iterations required to obtain a non-singular curve.

3.1 Resolution of Toric Curves

Given a finite set $\mathscr{A} \subset \mathbb{Z}$ satisfying (∗), we denote $\mathscr{A}^1 := \mathscr{A}'$, where $\mathscr{A}'$ is the set obtained after applying once Algorithm 2.2 to $\mathscr{A}$. Similarly, $\mathscr{A}^k := (\mathscr{A}^{k-1})'$, for $k \geq 2$.

Lemma 1 *Let $\mathscr{A} = \{a_1, \ldots, a_n\} \subset \mathbb{Z}$ be a finite set satisfying (∗) and $1 < a_1$. Then $\min(\mathscr{A}^1) \leq \min(\mathscr{A})$. In addition, there is $k \in \mathbb{N}$ such that $\min(\mathscr{A}^k) < \min(\mathscr{A})$.*

Proof From Remark 2, we can assume that $\mathscr{A}$ contains no multiples of $\min(\mathscr{A})$. From Algorithm 2.2 we know that $\mathscr{A}^1 = \{a_1, a_2 - a_1, \ldots, a_n - a_1\}$. From the assumption on $\mathscr{A}$, $a_1 \neq a_2 - a_1$. If $a_1 < a_2 - a_1$ then $\min(\mathscr{A}^1) = a_1 = \min(\mathscr{A})$. If $a_2 - a_1 < a_1$ then $\min(\mathscr{A}^1) < \min(\mathscr{A})$. This proves that $\min(\mathscr{A}^1) \leq \min(\mathscr{A})$.

Now we perform division algorithm for a_1 and a_2 (since $1 < a_1$ then $|\mathscr{A}| \geq 2$ according to (∗)): there exist $q, r \in \mathbb{N}$ such that $a_2 = a_1 q + r$, where $0 < r < a_1$. Notice that, for any $0 < l < q$, $\mathscr{A}^l = \{a_1, a_2 - la_1, \ldots, a_n - la_1\}$, where $a_1 < a_j - la_1$ for $j > 1$, and $\mathscr{A}^q = \{a_1, a_2 - qa_1, \ldots, a_n - qa_1\}$, where $a_2 - qa_1 = r < a_1$ and $a_2 - qa_1 < a_j - qa_1$ for $j > 2$. Hence, $\min(\mathscr{A}^q) = a_2 - qa_1 = r < a_1 = \min(\mathscr{A})$.

Remark 3 The proof of the previous lemma shows that the smallest k such that $\min(\mathscr{A}^k) < \min(\mathscr{A})$ can be obtained from the division algorithm applied to the two smallest elements of $\mathscr{A}$. In particular, Algorithm 2.2 can be divided into several division algorithms that record the significant improvements of the algorithm (see Sect. 4).

Remark 4 It is also important to observe (following the notation of the proof of the lemma) that $\min(\mathscr{A}^q \setminus \{r\})$ is not necessarily a_1 (for example if $\mathscr{A} = \{7, 17, 19\}$).

Proposition 2 *A finite iteration of Nash modifications resolves singularities of toric curves.*

Proof Let $Y_{\mathscr{A}}$ be a toric curve defined by $\mathscr{A} = \{a_1, \ldots, a_n\} \subset \mathbb{Z}$ satisfying (∗) and such that $1 < a_1$. Assume that $\mathscr{A}$ contains no multiples of $\min(\mathscr{A})$. By the division algorithm, $a_2 = a_1 q + r$, where $0 < r < a_1$. From the proof of Lemma 1, it follows that $1 \leq \min(\mathscr{A}^q) < a_1$. If $1 = \min(\mathscr{A}_q)$ then $Y_{\mathscr{A}^q}$ is non-singular. Otherwise repeat the process for $\mathscr{A}^q$. Continuing like this we obtain a sequence

$$a_1 > \min(\mathscr{A}^q) > \min(\mathscr{A}^{q_1}) > \ldots \geq 1.$$

This decreasing sequence cannot be infinite, so $1 \in \mathscr{A}^k$, for some k.

Remark 5 There is a geometric interpretation of the number $\min(\mathscr{A})$ in terms of Hilbert-Samuel multiplicity (see Sect. 5).

3.2 Number of Iterations

Now we consider the problem of giving a bound to the number of iterations required to resolve a toric curve using Nash modifications.

Definition 3 Let $\mathscr{A} \subset \mathbb{Z}$ be a finite set satisfying $(*)$. We denote as $\eta(\mathscr{A})$ the number of iterations required to solve the singularities of $Y_{\mathscr{A}}$ using the Nash modification. In other words, $\eta(\mathscr{A})$ is the minimum $k \in \mathbb{N}$ such that $1 \in \mathscr{A}^k$.

In general, to estimate $\eta(\mathscr{A})$ is not a simple task. Let $\mathscr{A} := \{a_1^{(0)}, \ldots, a_{n_0}^{(0)}\}$ be such that no multiple of $\min(\mathscr{A})$ (other than itself) is contained in $\mathscr{A}$. Following the proof of Lemma 1, Algorithm 2.2 can be summarized as follows (at every row we assume that $\mathscr{A}_i^q$ contains no multiples of $\min(\mathscr{A}_i^q)$ and its elements are ordered increasingly):

$$
\begin{aligned}
\mathscr{A} &= \{a_1^{(0)}, a_2^{(0)}, \ldots, a_{n_0}^{(0)}\}, & a_2^{(0)} &= a_1^{(0)} q_1 + r_1, \\
\mathscr{A}^{q_1} &= \{a_1^{(1)}, a_2^{(1)}, \ldots, a_{n_1}^{(1)}\}, & a_1^{(1)} &= r_1,\ a_2^{(1)} = a_1^{(1)} q_2 + r_2, \\
\mathscr{A}^{q_1+q_2} &= \{a_1^{(2)}, a_2^{(2)}, \ldots, a_{n_2}^{(2)}\}, & a_1^{(2)} &= r_2,\ a_2^{(2)} = a_1^{(2)} q_3 + r_3, \\
&\vdots & &\vdots \\
\mathscr{A}^{q_1+\cdots+q_{s-1}} &= \{a_1^{(s-1)}, a_2^{(s-1)}, \ldots, a_{n_{s-1}}^{(s-1)}\}, & a_1^{(s-1)} &= r_{s-1},\ a_2^{(s-1)} = a_1^{(s-1)} q_s + 1. \\
\mathscr{A}^{q_1+\ldots+q_s} &= \{1\}.
\end{aligned}
\tag{1}
$$

Summarized like this, $\eta(\mathscr{A}) = q_1 + q_2 + \cdots + q_s$, by definition. The input of the algorithm is $\mathscr{A}$, thus we can compute q_1 directly from it. Unfortunately, for $i \geq 2$, it is not clear how to bound the numbers q_i since, to begin with, there is no control over the values of $a_1^{(i)}$ and $a_2^{(i)}$. Those values depend on how $a_2^{(0)}, \ldots, a_{n_0}^{(0)}$ are distributed.

There is, however, another way to bound $\eta(\mathscr{A})$. The following proposition was adapted from [4, Lemma 5.5].

Proposition 3 *Let $\mathscr{A} = \{a_1, \ldots, a_n\} \subset \mathbb{Z}$ be such that $(*)$ holds. Let*

$$v(\mathscr{A}) := \min\{a_i \,|\, \gcd(a_1, \ldots, a_i) = 1\}.$$

Then $v(\mathscr{A}^1) \leq v(\mathscr{A}) - 2$. Therefore, if $a_1 > 1$, after at most $\lfloor \frac{v(\mathscr{A})}{2} \rfloor$ iterations of the algorithm, the resulting set contains 1. In other words, $\eta(\mathscr{A}) \leq \lfloor \frac{v(\mathscr{A})}{2} \rfloor$.

Proof Assume that $1 < a_1$. Let $a_k = v(\mathscr{A})$, where $2 \leq k \leq n$. By applying once the algorithm we obtain, in particular, $\{a_1, a_2 - a_1, \ldots, a_k - a_1\} \subset \mathscr{A}^1$. Now denote $N = \max\{a_1, a_k - a_1\}$. Since $\gcd(a_1, a_2 - a_1, \ldots, a_k - a_1) = 1$ we have $v(\mathscr{A}^1) \leq N$. If $N = a_k - a_1$ then, since $a_1 \geq 2$ we have $v(\mathscr{A}^1) \leq a_k - a_1 \leq v(\mathscr{A}) - 2$. Suppose now that $N = a_1$. If $a_k = a_1 + 1$ then $a_k \geq 3$ and since $1 = a_k - a_1 \in \mathscr{A}^1$

we obtain $v(\mathscr{A}^1) = 1 < 3 \leq a_k = v(\mathscr{A})$, so $v(\mathscr{A}^1) \leq v(\mathscr{A}) - 2$. Otherwise $a_k > a_1 + 1$ which implies $v(\mathscr{A}^1) \leq a_1 \leq a_k - 2 = v(\mathscr{A}) - 2$. This proves the proposition.

Unfortunately, the bound given in the previous proposition for $\eta(\mathscr{A})$ is not sharp in general, as the following example shows.

Example 2 Let $\mathscr{A} = \{10^{10}, 2 \cdot 10^{10} + 1\}$. It follows that $\mathscr{A}^1 = \{10^{10}, 10^{10} + 1\}$ and $\mathscr{A}^2 = \{1, 10^{10}\}$. Thus, it took two iterations to terminate the algorithm but the bound is $\lfloor \frac{v(\mathscr{A})}{2} \rfloor = 10^{10}$.

Even though we did not succeed in giving a more precise bound for $\eta(\mathscr{A})$, the summary of Algorithm 2.2 given in (1) shows that the algorithm can be divided into several division algorithms. In the next section we give a bound for the number of those division algorithms (the number s in the notation of (1)). As we will see, this is a simpler task.

4 Counting Division Algorithms

Let us start with an example.

Example 3 In this example, it takes twelve iterations of Algorithm 2.2 to get the element 1 which can be summarized in three division algorithms.

$$\mathscr{A} = \{20, 165, 172\},$$
$$\mathscr{A}^8 = \{20, 165 - 20 \cdot 8, 172 - 20 \cdot 8\} = \{5, 12, 20\},$$
$$\mathscr{A}^{10} = \{5, 12 - 5 \cdot 2, 20 - 5 \cdot 2\} = \{2, 5, 10\}$$
$$\mathscr{A}^{12} = \{2, 5 - 2 \cdot 2, 10 - 2 \cdot 2\} = \{1, 2, 6\}.$$

Definition 4 Let $\mathscr{A} = \{a_1, \ldots, a_n\} \subset \mathbb{Z}$ be such that $(*)$ holds. Assume that $1 < a_1$. We denote as $\delta(\mathscr{A})$ the number of division algorithms required for Algorithm 2.2 to stop (this corresponds to number s in (1)).

Remark 6 Notice that $\delta(\mathscr{A}) \leq \eta(\mathscr{A})$.

The goal of this section is to give a bound for $\delta(\mathscr{A})$. As we will see, the bound depends only on the first two elements of $\mathscr{A}$. First we need to introduce the Fibonacci sequence.

Definition 5 The *Fibonacci sequence* is defined as

$$F_1 := 1,\ F_2 := 2 \text{ and } F_{m+2} := F_{m+1} + F_m \text{ for each } m \geq 1.$$

The following theorem is a direct generalization of the analogous result for $n = 2$, which seems to be well known (see, for instance, [11]).

Theorem 2 *Let* $\mathscr{A} = \{a_1, \ldots, a_n\} \subset \mathbb{Z}$ *be such that* $(*)$ *holds and* $1 < a_1$. *Then* $a_1 \geq F_{\delta(\mathscr{A})+1}$ *and* $a_2 \geq F_{\delta(\mathscr{A})+2}$.

Proof We proceed by induction on $\delta(\mathscr{A})$. For $\delta(\mathscr{A}) = 1$ the result is true since, by hypothesis, $a_1 \geq 2 = F_{1+1}$ and $a_2 \geq a_1 + 1 \geq 2 + 1 = 3 = F_{1+2}$.

Assume that the statement is true for all $\mathscr{A}$ such that $\delta(\mathscr{A}) = m$. Suppose that $\delta(\mathscr{A}) = m + 1$. Assume that $\mathscr{A}$ does not contain multiples of $\min(\mathscr{A})$. Applying the division algorithm to a_1 and a_2 we obtain $a_2 = a_1 q + r$, where $0 < r < a_1$. Then $\delta(\mathscr{A}^q) = m$ and $\min(\mathscr{A}^q) = r$. By induction, $r \geq F_{m+1}$. Let us prove that $a_1 \geq F_{m+2}$. Since $a_1 > r$ then $a_1 \geq \min(\mathscr{A}^q \setminus \{r\})$. On the other hand, by induction we know that $\min(\mathscr{A}^q \setminus \{r\}) \geq F_{m+2}$. Consequently, $a_1 \geq F_{m+2}$.

Thus, $r \geq F_{m+1}$ and $a_1 \geq F_{m+2}$. Finally, since $q \geq 1$ it follows

$$a_2 = a_1 q + r \geq a_1 + r \geq F_{m+2} + F_{m+1} = F_{m+3}.$$

Therefore, $a_1 \geq F_{m+2} = F_{\delta(\mathscr{A})+1}$ and $a_2 \geq F_{m+3} = F_{\delta(\mathscr{A})+2}$.

Corollary 1 *Let* $\mathscr{A} = \{a_1, \ldots, a_n\} \subset \mathbb{Z}$ *such that* $(*)$ *holds. Assume* $1 < a_1$. *Let* $m \in \mathbb{N}$ *be such that* $a_1 < F_{m+1}$ *or* $a_2 < F_{m+2}$. *Then* $\delta(\mathscr{A}) < m$.

Proof Suppose that $\delta(\mathscr{A}) \geq m$. By the previous theorem we have $a_1 \geq F_{\delta(\mathscr{A})+1}$ and $a_2 \geq F_{\delta(\mathscr{A})+2}$. On the other hand, $F_{\delta(\mathscr{A})+1} \geq F_{m+1}$ and $F_{\delta(\mathscr{A})+2} \geq F_{m+2}$ implying $a_1 \geq F_{m+1}$ and $a_2 \geq F_{m+2}$.

Even though the previous corollary gives a good bound for $\delta(\mathscr{A})$, in practice it may not be easy to find the smallest m such that $a_1 < F_{m+1}$ or $a_2 < F_{m+2}$ (especially for large numbers). Now we give a simpler bound using the number of digits in a_1 and a_2 (written in base 10). We use the following well-known fact (see, for instance, [11]).

Lemma 2 *Let* $k \geq 1$. *If* $m \geq 5k$ *then* F_{m+1} *has at least* $k + 1$ *digits.*

Corollary 2 *If* a_1 *and* a_2 *have* k_1 *and* k_2 *digits, respectively, then*

$$\delta(\mathscr{A}) < \min\{5k_1, 5k_2 - 1\}.$$

Proof Let $m := \delta(\mathscr{A})$ and assume that $m \geq 5k_1$. By Theorem 2, $a_1 \geq F_{m+1}$ and by Lemma 2, F_{m+1} has at least $k_1 + 1$ digits. Consequently, a_1 has at least $k_1 + 1$ digits. This shows that if a_1 has k_1 digits then $\delta(\mathscr{A}) < 5k_1$. Similarly, if we assume that $m \geq 5k_2 - 1$ then $m + 1 \geq 5k_2$. Theorem 2 and Lemma 2 imply that a_2 has at least $k_2 + 1$ digits. Therefore, if a_2 has k_2 digits then $\delta(\mathscr{A}) < 5k_2 - 1$.

5 Some Other Features

In this last section we collect some other features regarding the Nash modification of toric curves. We will see that there are some nice properties of the Nash modification of toric varieties that hold only in dimension 1.

5.1 *Hilbert-Samuel Multiplicity*

It is well known that Hilbert-Samuel multiplicity plays a fundamental role in theorems of resolution of singularities. This invariant has been used to measure improvements on singularities after a suitable blowup. What about for the Nash modification?

The proof we gave for the resolution of singularities of toric curves iterating Nash modifications was a purely combinatorial one and the improvements in the algorithm were measured by looking at the semigroup itself. Because of the following result (which seems to be well known, see for instance [6, Theorem 3.14, Chapter 5]) the improvements on the singularities of toric curves after a Nash modification can also be measured using Hilbert-Samuel multiplicity.

Proposition 4 *Let* $\mathscr{A} = \{a_1, \dots, a_n\} \subset \mathbb{Z}$ *be a finite set satisfying* $(*)$. *Let* R *be the localization of* $\mathbb{C}[Y_{\mathscr{A}}]$ *at the maximal ideal corresponding to* $0 \in Y_{\mathscr{A}}$. *Then* $mult(Y_{\mathscr{A}}, 0) = a_1$, *where* $mult(Y_{\mathscr{A}}, 0)$ *is the Hilbert-Samuel multiplicity of* R.

Now Lemma 1 can be restated as follows.

Corollary 3 *Let* $Y_{\mathscr{A}}$ *be a singular toric curve defined by a set* $\mathscr{A} \subset \mathbb{Z}$ *satisfying* $(*)$. *Then* $mult(Y_{\mathscr{A}^1}, 0) \leq mult(Y_{\mathscr{A}}, 0)$. *In addition, the Hilbert-Samuel multiplicity drops after a finite number of iterations of the Nash modification.*

Unfortunately, for higher-dimensional toric varieties, the previous corollary is false in general.

Example 4 Let $a_1 = (1, 0)$, $a_2 = (1, 1)$, and $a_3 = (3, 4)$. The set $\mathscr{A} = \{a_1, a_2, a_3\}$ defines the normal toric surface $Y_{\mathscr{A}} = \mathbf{V}(xz - y^4) \subset \mathbb{C}^3$. Since $\mathscr{A} \subset \mathbb{Z}^2$ satisfies $\mathbb{Z}\mathscr{A} = \mathbb{Z}^2$ and $(0, 0) \notin \mathrm{Conv}(\mathscr{A})$, we can apply Algorithm 2.1 to compute the Nash modification of $Y_{\mathscr{A}}$.

Step 2 for $J_0 = (1, 2) \in \{1, 2, 3\}^2$ gives $\mathscr{A}_{J_0} = \{a_1, a_2\} \cup \{(2, 3), (2, 4)\}$. The toric ideal associated to $\mathscr{A}_{J_0}$ is $I_{\mathscr{A}_{J_0}} = \langle y^2w - z^2, xw - yz, xz - y^3 \rangle$. A direct computation shows that $(0, 0, 0, 0) \in Y_{\mathscr{A}_{J_0}} \subset \mathbb{C}^4$ is a singular point. Now notice that $mult(Y_{\mathscr{A}}, (0, 0, 0)) = 2$. Using `SINGULAR 4-0-2` [5], we computed the Hilbert-Samuel multiplicity of $Y_{\mathscr{A}_{J_0}}$ to find $mult(Y_{\mathscr{A}_{J_0}}, (0, 0, 0, 0)) = 3$. Thus, there is a point in the preimage of the Nash modification of $(0, 0, 0) \in Y_{\mathscr{A}}$ whose Hilbert-Samuel multiplicity increases.

5.2 Embedding Dimension

It is well known that the Nash modification of an algebraic variety may not preserve embedding dimension. A. Nobile illustrated this fact with a plane curve $X \subset \mathbb{C}^2$ given by the parametrization $x = t^4$, $y = t^{11} + t^{13}$. Nobile proved that the embedding dimension of the only point of $\nu^{-1}((0,0))$ is 3 (see [14, Example 3]). In the following proposition we see that for toric curves the embedding dimension does not increase after applying a Nash modification.

Proposition 5 *Let $\mathscr{A} = \{a_1, \ldots, a_n\} \subset \mathbb{Z}$ be a finite set satisfying (∗). Let $\mathscr{A}' \subset \mathbb{Z}$ be the set obtained from Algorithm 2.2. Then the embedding dimension of $Y_{\mathscr{A}'}$ is less or equal than the embedding dimension of $Y_{\mathscr{A}}$.*

Proof Suppose that the semigroup $\mathbb{N}\mathscr{A}$ is minimally generated by $\mathscr{A}$. Then $Y_{\mathscr{A}}$ has embedding dimension n. According to Algorithm 2.2, $|\mathscr{A}^1| \leq n$. In particular, the embedding dimension of $Y^*_{\mathscr{A}} = Y_{\mathscr{A}^1}$ is less or equal than n.

The previous proposition is not true for general toric varieties as the following example shows.

Example 5 Let $\mathscr{A} = \{(1,0), (1,1), (2,3)\} \subset \mathbb{Z}^2$. Let $Y_{\mathscr{A}} = \mathbf{V}(xz - y^3) \subset \mathbb{C}^3$ be the corresponding toric surface. The embedding dimension of $Y_{\mathscr{A}}$ is 3. Applying the combinatorial Algorithm 2.1 to $\mathscr{A}$ we obtain the following two sets in $\mathbb{Z}^2$, which determine the affine charts covering $Y^*_{\mathscr{A}}$: $\mathscr{A}_1 = \{(1,0), (1,1), (1,2), (1,3)\}$ and $\mathscr{A}_2 = \{(-1,-2), (0,-1), (1,1), (2,3)\}$. A direct computation shows that the semigroups $\mathbb{N}\mathscr{A}_1$ and $\mathbb{N}\mathscr{A}_2$ are minimally generated by $\mathscr{A}_1$ and $\mathscr{A}_2$, respectively. Thus, the embedding dimension of $Y_{\mathscr{A}_1} \subset \mathbb{C}^4$ or $Y_{\mathscr{A}_2} \subset \mathbb{C}^4$ at the origin is 4. We conclude that there are points in $Y^*_{\mathscr{A}}$ on which the embedding dimension increases.

5.3 Zero Locus of an Ideal Defining the Nash Modification

Let $\mathscr{A} = \{a_1, \ldots, a_n\} \subset \mathbb{Z}^d$ be a set of vectors defining a toric variety $Y_{\mathscr{A}} \subset \mathbb{C}^n$. It was proved in [7, 9, 13] that an ideal whose blowup defines the Nash modification of a toric variety can be described in the following combinatorial way (this ideal is usually called *logarithmic jacobian ideal*).

Theorem 3 *Let $\mathscr{J}$ be the ideal of the coordinate ring $\mathbb{C}[u_1, \ldots, u_n]/I_{\mathscr{A}}$ of $Y_{\mathscr{A}}$ generated by the products $u_{i_1} \cdots u_{i_d}$ such that $\det(a_{i_1}, \ldots, a_{i_d}) \neq 0$. Then the Nash modification of $Y_{\mathscr{A}}$ is isomorphic to the blowup of $Y_{\mathscr{A}}$ along the ideal $\mathscr{J}$.*

Notice that in the case of curves, where $\mathscr{A} \subset \mathbb{Z}$ satisfies (∗), $\mathscr{J} = \langle u_1, \ldots, u_n \rangle$.

Corollary 4 *The Nash modification of a singular toric curve $Y_{\mathscr{A}}$ is isomorphic to blowing up an ideal whose zero locus is the origin in $Y_{\mathscr{A}}$.*

In other words, the zero locus of the ideal $\mathcal{J}$, whose blowup defines the Nash modification of the toric curve $Y_{\mathscr{A}}$, satisfies $\mathrm{Sing}(Y_{\mathscr{A}}) = \mathbf{V}(\mathcal{J})$. For toric surfaces this is not necessarily true.

Example 6 Let $\mathscr{A} = \{(1, 0), (1, 1), (1, 2)\} \subset \mathbb{Z}^2$. Let $Y_{\mathscr{A}} = \mathbf{V}(xz - y^2) \subset \mathbb{C}^3$ be the corresponding toric surface. In this case $\mathcal{J} = \langle xy, xz, yz\rangle$. Then

$$\mathrm{Sing}(Y_{\mathscr{A}}) = \{(0, 0, 0)\} \subsetneq \{(x, 0, z) \in Y_{\mathscr{A}} | x = 0 \text{ or } z = 0\} = \mathbf{V}(\mathcal{J}).$$

Actually, it can be proved that, under some hypothesis, for toric surfaces $Y_{\mathscr{A}}$ with isolated singular point, there is no choice of an ideal $\mathcal{J}$ defining the Nash modification of $Y_{\mathscr{A}}$ such that $\mathrm{Sing}(Y_{\mathscr{A}}) = \mathbf{V}(\mathcal{J})$ (see [2, Theorem 3.6]).

5.4 *Possible Generalizations for Toric Surfaces*

The results presented in this note were motivated by the same questions for toric surfaces: does the Nash modification resolve singularities of toric surfaces? and, if so, how many iterations does it take?

In [4] both questions were explored and some partial answers were given (in [9] the first question was also partially answered for toric varieties of any dimension). Unfortunately, the bounds given in [4] are not at all sharp. On the other hand, the cases considered in that paper were studied by separating Algorithm 2.1 into two independent algorithms: one algorithm in $\{0\} \times \mathbb{Z}$ followed from an algorithm in $\mathbb{Z} \times \{0\}$. These two algorithms are quite similar to the one studied in this note.

Because of this resemblance, we strongly believe that the approach of counting division algorithms that we developed in Sect. 4 can be applied in dimension 2, at least in the cases considered in [4]. This way of counting iterations might result in a better understanding of those cases for which we still do not know whether Nash modifications solve singularities.

We hope to report in a future paper on this approach for toric surfaces.

Acknowledgements We would like to thank the referees for their careful reading and their comments. The second author would also like to thank José Seade for financial support through the project FORDECyT-265667.

References

1. Atanasov, A., Lopez, C., Perry, A., Proudfoot, N., Thaddeus, M.: Resolving toric varieties with Nash blow-ups. Exp. Math. **20**(3), 288–303 (2011)
2. Chávez Martínez, E., Duarte, D.: On the zero locus of ideals defining the Nash blowup of toric surfaces. Commun. Algebra (2017). https://doi.org/10.1080/00927872.2017.1335740

3. Cox, D., Little, J., Schenck, H.: Toric Varieties. Graduate Studies in Mathematics, vol. 124. AMS, Providence (2011)
4. Duarte, D.: Nash modification on toric surfaces. Revista de la Real Academia de Ciencias Exactas, Físicas y Naturales, Serie A Matemáticas **108**(1), 153–171 (2014)
5. Decker, W., Greuel, G.-M., Pfister, G., Schönemann, H.: Singular 4-0-2—A computer algebra system for polynomial computations (2015). http://www.singular.uni-kl.de
6. Gel'fand, I.M., Kapranov, M.M., Zelevinsky, A.V.: Discriminants, Resultants, and Multidimensional Determinants. Mathematics: Theory and Applications. Birkhauser, Boston (1994)
7. Gonzalez-Sprinberg, G.: Eventails en dimension 2 et transformé de Nash. Publ. de l'E.N.S Paris, pp. 1–68 (1977)
8. Gonzalez-Sprinberg, G.: Résolution de Nash des points doubles rationnels. Ann. Inst. Fourier Grenoble **32**(2), 111–178 (1982)
9. González, P.D., Teissier, B.: Toric geometry and the Semple-Nash modification. Revista de la Real Academia de Ciencias Exactas, Físicas y Naturales, Serie A Matemáticas **108**(1), 1–48 (2014)
10. Grigoriev, D., Milman, P.: Nash desingularization for binomial varieties as Euclidean division. A priori termination bound, polynomial complexity in essential dim 2. Adv. Math. **231**(6), 3389–3428 (2012)
11. Grossman, H.: On the number of divisions in finding a G.C.D. Am. Math. Mon. **31**(9), 443 (1924)
12. Hironaka, H.: On Nash blowing-up. In: Artin, M. (ed.) Arithmetic and Geometry II. Progress in Mathematics, vol. 36. Birkhauser, Boston, pp. 103–111 (1983)
13. Lejeune-Jalabert, M., Reguera, A.: The Denef-Loefer series for toric surfaces singularities. In: Proceedings of the International Conference on Algebraic Geometry and Singularities (Spanish), Sevilla (2001); Rev. Mat. Iberoamericana **19**, 581–612 (2003)
14. Nobile, A.: Some properties of the Nash blowing-up. Pac. J. Math. **60**, 297–305 (1975)
15. Rebassoo, V.: Desingularisation properties of the Nash blowing-up process. Thesis, University of Washington (1977)
16. Semple, J.G.: Some investigations in the geometry of curve and surface elements. Proc. Lond. Math. Soc. (3) **4**, 24–49 (1954)
17. Spivakovsky, M.: Sandwiched singularities and desingularisation of surfaces by normalized Nash transformations. Ann. Math. (2) **131**(3), 411–491 (1990)
18. Sturmfels, B.: Gröbner Bases and Convex Polytopes. University Lecture Series, vol. 8. American Mathematical Society, Providence (1996)

A Recursive Formula for the Motivic Milnor Fiber of a Plane Curve

Manuel González Villa, Gary Kennedy, and Lee J. McEwan

Dedicated to Antonio Campillo on the occasion of his 65th birthday

Abstract We find a recursive formula for the motivic Milnor fiber of an irreducible plane curve, using the notions of a truncation and derived curve. We then apply natural transformations to obtain a similar recursion for the Hodge-theoretic spectrum.

1 Introduction

In this paper we develop a recursive formula for the motivic Milnor fiber of a singular point of a plane curve, expressed in terms of the essential exponents in its Puiseux expansion. This subject has been treated in [5], and later in [4], but the novel feature here is that we develop a recursion at the level of virtual motives, adapting the topological techniques of [6]. Alternative approaches to the computation of the motivic Milnor fibre of a singular point of a plane curve proposed in the literature are to use an embedded or log resolution of the singular point [3], the Newton process [1], or splicing [2].

After developing the recursion, we apply natural transformations to obtain similar recursions for the Hodge-theoretic spectrum and the topological monodromy. We conclude with an extended example in which, using our recursion, we write out the virtual motive of a specific curve and compute its spectrum; for comparison

M. González Villa (✉)
Centro de Investigación en Matemáticas, A.C., Guanajuato, Gto, México
e-mail: manuel.gonzalez@cimat.mx

G. Kennedy · L. J. McEwan
Ohio State University at Mansfield, Mansfield, OH, USA
e-mail: kennedy@math.ohio-state.edu; mcewan@math.ohio-state.edu

G.-M. Greuel et al. (eds.), *Singularities, Algebraic Geometry, Commutative Algebra, and Related Topics*, https://doi.org/10.1007/978-3-319-96827-8_9

we compute the same virtual motive using the original construction of Denef and Loeser [3].

2 The Theorem

Consider a irreducible germ at the origin of a complex plane curve C with Puiseux expansion

$$\zeta = \sum c_\lambda x^\lambda$$

in which the essential exponents are $\lambda_1 < \lambda_2 < \cdots < \lambda_e$. Let $\lambda_1 = n/m$, where m and n are relatively prime. If necessary we will change coordinates so that none of the exponents are integers and so that $c_{\lambda_1} = 1$. Then C is defined by the vanishing of a function which can be written both as a product and a sum:

$$f(x, y) = \prod (y - \zeta) = (y^m - x^n)^{d'} + \cdots . \tag{1}$$

Here d' is some positive integer, and the unnamed terms have higher order with respect to the weighting $\mathrm{wt}(x) = m$, $\mathrm{wt}(y) = n$; the product is taken over all possible conjugates, the number of which is $d = md'$.

As in [6], we define two associated curves. The *truncation* C_1 is the curve with Puiseux expansion $\zeta_1 = x^{\lambda_1}$ and defining function $f_1(x, y) = y^m - x^n$; it has m conjugates. The *derived curve* C' has Puiseux expansion

$$\zeta' = \sum_{\lambda > \lambda_1} c_\lambda x^{\lambda'}, \tag{2}$$

where $\lambda' = m(\lambda - \lambda_1 + n)$. It has d' conjugates, and its defining function is

$$f'(x, y) = \prod (y - \zeta') = y^{d'} + \ldots, \tag{3}$$

a product using each conjugate once. Its essential exponents are

$$\lambda'_1, \lambda'_2, \ldots, \lambda'_{e-1}$$

where

$$\lambda'_i = m(\lambda_{i+1} - \lambda_1 + n).$$

Denote by μ_N the group of roots of the unity of order N and by $\hat{\mu}$ the inverse limit of the groups of roots of unity (with respect to the morphisms $\mu_M \to \mu_N$ given by $x \mapsto x^{M/N}$, whenever M is a multiple of N). An action of the group μ_N

in a variety is called good if each orbit is contained in an affine subvariety. A good action of $\hat{\mu}$ on a variety is an action of $\hat{\mu}$ which factors through a good μ_N-action for some N. Following Denef and Loeser [3, Subsection 2.4], let $K_0^{\hat{\mu}}(\mathrm{Var}_{\mathbb{C}})$ denote the monodromic Grothendieck ring of varieties with good $\hat{\mu}$-action, and let $\mathcal{M}_{\mathbb{C}}^{\hat{\mu}}$ be the localization obtained by inverting $\mathbb{L}$, the class of the affine line. The class of a variety X in this ring will be denoted $[X]$ and will be called a *virtual motive*.

In the space $\mathcal{L}_N(\mathbb{C}^2_{x,y})_{\mathbf{0}}$ of N-jets based at the origin, let $\mathcal{X}_{N,1}(f)$ be the subspace consisting of jets φ for which

$$(f \circ \varphi)(t) = t^N + \text{ higher order terms.}$$

Notice that the group μ_N acts on $\mathcal{X}_{N,1}(f)$ by $a \cdot \varphi(t) \mapsto \varphi(at)$.

The *local motivic zeta function* of f at the origin is

$$Z_{\mathbf{0}}(f) = \sum_{N=1}^{\infty} [\mathcal{X}_{N,1}(f)]\mathbb{L}^{-2N} T^N.$$

The *motivic Milnor fiber* of f at the origin is the limit

$$S_{\mathbf{0}}(f) = - \lim_{T \to \infty} Z_{\mathbf{0}}(f).$$

(In fact $Z_{\mathbf{0}}(f)$ is known to be a rational function of degree ≤ 0, so that it makes sense to consider its limit at $T = +\infty$.)

In Sect. 3 we prove the following recursive formula.

Theorem 1 $S_{\mathbf{0}}(f) = S_{\mathbf{0}}((f_1)^{d'}) + S_{\mathbf{0}}(f') - [\mu_{d'}]$.

In this formula $[\mu_{d'}]$ denotes the virtual motive of roots of unity of order d'. The complex multiplication in $\mu_{d'}$ induces a good $\hat{\mu}$-action on $\mu_{d'}$.

To make the base case of this recursion more explicit, we introduce the following notation: $[(f_1)^{d'} - 1]$ represents the virtual motive associated to the variety $\{(x, y) \in \mathbb{C}^2 | (f_1(x, y))^{d'} = 1\}$.

Theorem 2 $S_{\mathbf{0}}((f_1)^{d'}) = [(f_1)^{d'} - 1] - [\mu_{d'}](\mathbb{L} - 1)$.

This formula is implicit in the results of Section 3 of Guibert [5], but we prefer to give a short self-contained proof in Sect. 4 below. Combining Theorems 1 and 2, we obtain the following formula:

$$S_{\mathbf{0}}(f) = [(f_1)^{d'} - 1] - [\mu_{d'}]\mathbb{L} + S_{\mathbf{0}}(f'). \tag{4}$$

As a consequence of (4), the motivic Milnor fiber of f at the origin $S_{\mathbf{0}}(f)$ is determined by the essential exponents of the complex plane curve C.

3 Proof of Theorem 1

We begin with two basic properties of the motivic Milnor fiber.

Lemma 1 *For the coordinate function y, we have* $S_{\mathbf{0}}(y^{d'}) = [\mu_{d'}]$.

Proof This follows from a direct calculation.

Lemma 2 *If the quotient* g/h *is a unit in the local ring of the complex plane at the origin, then* $[\mathcal{X}_{N,1}(g)] = [\mathcal{X}_{N,1}(h)]$ *for each N.*

Proof Suppose that $g = uh$. Choose λ to be an Nth root of $1/u(0,0)$. Given an arc φ, map it to the arc ψ, where

$$\psi(t) = \varphi(\lambda t).$$

This sets up an equivariant identification between $\mathcal{X}_{N,1}(g)$ and $\mathcal{X}_{N,1}(h)$.

To prove Theorem 1, we will use a pair of maps $\mathbb{C}^2_{v,w} \to \mathbb{C}^2_{x,y}$ defined as follows:

$$\pi(v,w) = (v^m, v^n(1+w)),$$
$$\pi'(v,w) = (v, v^{mn}w),$$

and we will work with jets in $\mathcal{L}_N(\mathbb{C}^2_{v,w})_{\mathbf{0}}$.

Lemma 3 *The quotient* $\frac{f'\circ\pi'}{f\circ\pi}$ *is a unit in the local ring of the complex plane at the origin. Thus* $[\mathcal{X}_{N,1}(f\circ\pi)] = [\mathcal{X}_{N,1}(f'\circ\pi')]$ *for each N.*

Proof Looking at the product in (1) and composing with π, we observe that we may write

$$\begin{aligned} y - \zeta &= v^n(1+w) - \left((v^m)^{n/m} + \sum_{\lambda>\lambda_1} c_\lambda (v^m)^\lambda \right) \\ &= v^n \left(1 + w - 1^{n/m} - \sum_{\lambda>\lambda_1} c_\lambda v^{m\lambda - n} \right). \end{aligned}$$

The interpretation of $1^{n/m}$ depends on the choice of conjugate; each possible mth root of unity occurs equally often. Thus in d' of the factors the interpretation is that $1^{n/m} = 1$. In the remaining factors $1^{n/m}$ is some other mth root of unity, and within each of these factors the subfactor within parentheses is a unit. The full product

$f \circ \pi(v, w)$ is thus a unit times

$$v^{nd'} \prod (w - \sum_{\lambda > \lambda_1} c_\lambda v^{m\lambda - n}),$$

where the product is taken over all d' conjugates.

Now looking at the product in (3) and composing with π', we see that

$$\begin{aligned} f' \circ \pi'(v, w) &= \prod (v^{mn} w - \sum_{\lambda > \lambda_1} c_\lambda v^{\lambda'}) \\ &= v^{mnd'} \prod (w - \sum_{\lambda > \lambda_1} c_\lambda v^{\lambda' - mn}), \end{aligned}$$

precisely the same product over all d' conjugates.

Given a jet $\rho \in \mathcal{L}_N(\mathbb{C}^2_{v,w})_{\mathbf{0}}$, its *pushforward* by π is the jet $\pi \circ \rho \in \mathcal{L}_N(\mathbb{C}^2_{x,y})_{\mathbf{0}}$. Note that the pushforward of a jet in $\mathcal{X}_{N,1}(f \circ \pi)$ is a jet in $\mathcal{X}_{N,1}(f)$. The pushforward jets form a subspace which we denote by $\mathcal{X}^{\pi}_{N,1}(f)$. In the same way, we define three similar subspaces:

- $\mathcal{X}^{\pi}_{N,1}((f_1)^{d'})$ inside $\mathcal{X}_{N,1}((f_1)^{d'})$,
- $\mathcal{X}^{\pi'}_{N,1}(f')$ inside $\mathcal{X}_{N,1}(f')$,
- $\mathcal{X}^{\pi'}_{N,1}(y^{d'})$ inside $\mathcal{X}_{N,1}(y^{d'})$.

Lemma 4 *Each jet in $\mathcal{X}^{\pi}_{N,1}(f)$ is the pushforward of a unique jet in $\mathcal{X}_{N,1}(f \circ \pi)$. Thus the map π identifies the space $\mathcal{X}_{N,1}(f \circ \pi)$ with the space $\mathcal{X}^{\pi}_{N,1}(f)$. It likewise identifies $\mathcal{X}_{N,1}((f_1)^{d'} \circ \pi)$ with $\mathcal{X}^{\pi}_{N,1}((f_1)^{d'})$. The jet spaces $\mathcal{X}_{N,1}(f) \setminus \mathcal{X}^{\pi}_{N,1}(f)$ and $\mathcal{X}_{N,1}((f_1)^{d'}) \setminus \mathcal{X}^{\pi}_{N,1}((f_1)^{d'})$ obtained by set-theoretic difference are equal.*

Proof We claim that $\mathcal{X}^{\pi}_{N,1}(f)$ consists of all jets in $\mathcal{X}_{N,1}(f)$ of the form

$$\varphi(t) = (x_0 t^{mp} + \ldots, y_0 t^{np} + \ldots), \tag{5}$$

where p is a positive integer, with x_0 and y_0 being nonzero numbers satisfying

$$(y_0)^m = (x_0)^n. \tag{6}$$

Indeed, consider a jet ρ in $\mathcal{X}_{N,1}(f \circ \pi)$; suppose that $\rho(t) = (v(t), w(t))$, where $v(t) = v_0 t^p + \ldots$, with $v_0 \neq 0$. Then

$$\pi \circ \rho(t) = ((v_0)^m t^{mp} + \ldots, (v_0)^n t^{np} + \ldots).$$

Conversely, given

$$\varphi(t) = (x(t), y(t)) = (x_0 t^{mp} + \ldots, y_0 t^{np} + \ldots)$$

with $(y_0)^m = (x_0)^n$, note that there is a unique number v_0 for which $x_0 = (v_0)^m$ and $y_0 = (v_0)^n$. Explicitly, letting r and s be the smallest positive integers for which

$$\det \begin{bmatrix} m & n \\ r & s \end{bmatrix} = 1,$$

we have $v_0 = (x_0)^s/(y_0)^r$. Let $v(t)$ be the mth root of $x(t)$ with this leading coefficient, i.e.,

$$v(t) = v_0 t^p + \ldots \quad \text{and} \quad (v(t))^m = x(t).$$

Let $w(t) = y(t)/(v(t))^n - 1$. Then $\rho(t) = (v(t), w(t))$ gives the unique jet whose pushforward is φ.

The same argument applies with $(f_1)^{d'}$ in place of f, showing that $\mathcal{X}^{\pi}_{N,1}((f_1)^{d'})$ consists of all jets in $\mathcal{X}_{N,1}((f_1)^{d'})$ of the form specified by by equations (5) and (6), and that each such jet is a pushforward in a unique way.

Again consider a jet $\varphi \in \mathcal{X}^{\pi}_{N,1}(f)$. Equations (5) and (6) imply a cancellation: there is no term of degree $mnpd'$, and thus we must have $N > mnpd'$. The same remark applies to a jet in $\mathcal{X}^{\pi}_{N,1}((f_1)^{d'})$. For a jet $\varphi \in \mathcal{X}_{N,1}(f) \setminus \mathcal{X}^{\pi}_{N,1}(f)$, however, the cancellation just described does not occur. This means that $(f \circ \varphi)(t)$ and $((f_1)^{d'} \circ \varphi)(t)$ have the same lowest-order term. Thus the jet spaces $\mathcal{X}_{N,1}(f) \setminus \mathcal{X}^{\pi}_{N,1}(f)$ and $\mathcal{X}_{N,1}((f_1)^{d'}) \setminus \mathcal{X}^{\pi}_{N,1}((f_1)^{d'})$ coincide.

Lemma 5 *Each jet in $\mathcal{X}^{\pi'}_{N,1}(f')$ is the pushforward of a unique jet in $\mathcal{X}_{N,1}(f' \circ \pi')$. Thus the map π' identifies the space $\mathcal{X}_{N,1}(f' \circ \pi')$ with the space $\mathcal{X}^{\pi'}_{N,1}(f')$. It likewise identifies $\mathcal{X}_{N,1}(y^{d'} \circ \pi')$ with $\mathcal{X}^{\pi'}_{N,1}(y^{d'})$. The $\mathcal{X}_{N,1}(f') \setminus \mathcal{X}^{\pi'}_{N,1}(f')$ and $\mathcal{X}_{N,1}(y^{d'}) \setminus \mathcal{X}^{\pi'}_{N,1}(y^{d'})$ are equal.*

Proof We claim that $\mathcal{X}^{\pi'}_{N,1}(f')$ consists of all jets in $\mathcal{X}_{N,1}(f')$ of the form

$$\varphi(t) = (x(t), y(t)) = (x_0 t^p + \ldots, y_0 t^q + \ldots), \tag{7}$$

(with $x_0 \neq 0$ and $y_0 \neq 0$), where $q > mnp$, together with all jets in which $y(t)$ is identically zero. Indeed, consider a jet $\rho(t) = (v(t), w(t))$ in $\mathcal{X}_{N,1}(f' \circ \pi')$. Note that $v(t)$ can't be identically zero, and write it as $v(t) = x_0 t^p + \ldots$. Then its pushforward has the required form. Conversely, given a jet $\varphi(t) = (x(t), y(t))$ of this form, compose with the birational inverse of π', i.e., let $v(t) = x(t)$ and $w(t) = y(t)/(x(t))^{mn}$, to obtain the unique jet ρ whose pushforward is φ. The same argument applies with $y^{d'}$ in place of f'.

Thus a jet φ in $\mathcal{X}_{N,1}(f') \setminus \mathcal{X}^{\pi'}_{N,1}(f')$ has the form shown in equation (7), with $q \leq mnp$, and likewise for a jet in $\mathcal{X}_{N,1}(y^{d'}) \setminus \mathcal{X}^{\pi'}_{N,1}(y^{d'})$ If we expand the product in (3) and then compose each term with such a jet, the leading term $y^{d'}$ has order $d'q$. Observe that each exponent λ' appearing in (2) is greater than mn. Thus each other term in the expansion will have order greater than $d'q$. Thus $f \circ \varphi$ and $y^{d'} \circ \varphi$ have the same leading term.

Recall that Lemma 3 states this equation of virtual motives:

- $[\mathcal{X}_{N,1}(f \circ \pi)] = [\mathcal{X}_{N,1}(f' \circ \pi')]$

Lemma 4 implies the following equations:

- $[\mathcal{X}^{\pi}_{N,1}(f)] = [\mathcal{X}_{N,1}(f \circ \pi)]$
- $[\mathcal{X}^{\pi}_{N,1}((f_1)^{d'})] = [\mathcal{X}_{N,1}((f_1)^{d'} \circ \pi)]$
- $[\mathcal{X}_{N,1}(f)] - [\mathcal{X}^{\pi}_{N,1}(f)] = [\mathcal{X}_{N,1}((f_1)^{d'})] - [\mathcal{X}^{\pi}_{N,1}((f_1)^{d'})]$

Lemma 5 gives us the following equations:

- $[\mathcal{X}^{\pi'}_{N,1}(f')] = [\mathcal{X}_{N,1}(f' \circ \pi')]$
- $[\mathcal{X}^{\pi'}_{N,1}(y^{d'})] = [\mathcal{X}_{N,1}(y^{d'} \circ \pi')]$
- $[\mathcal{X}_{N,1}(f')] - [\mathcal{X}^{\pi'}_{N,1}(f')] = [\mathcal{X}_{N,1}(y^{d'})] - [\mathcal{X}^{\pi'}_{N,1}(y^{d'})]$

Thus

$$\begin{aligned} -[\mathcal{X}_{N,1}(f)] = &- [\mathcal{X}_{N,1}((f_1)^{d'})] - [\mathcal{X}_{N,1}(f')] \\ &+ [\mathcal{X}_{N,1}((f_1)^{d'} \circ \pi)] + [\mathcal{X}_{N,1}(y^{d'})] - [\mathcal{X}_{N,1}(y^{d'} \circ \pi')]. \end{aligned}$$

Using each term as coefficient in a power series $\sum a_N \mathbb{L}^{-2N} T^N$ and taking the limit as T approaches infinity, we obtain this equation:

$$\begin{aligned} S_0(f) = &S_0((f_1)^{d'}) + S_0(f') \\ &- S_0((f_1)^{d'} \circ \pi) - S_0(y^{d'}) + S_0(y^{d'} \circ \pi') \end{aligned} \tag{8}$$

Both $(f_1)^{d'} \circ \pi(v, w)$ and $y^{d'} \circ \pi'(v, w)$ are $v^{mnd'} w$ times a unit; Lemma 2 tells us that $S_0((f_1)^{d'} \circ \pi) = S_0(y^{d'} \circ \pi)$. Thus by applying Lemma 1, we obtain the statement of Theorem 1.

4 Proof of Theorem 2

To prove Theorem 2, we will again use the decomposition of the jet space $\mathcal{X}_{N,1}((f_1)^{d'})$ into $\mathcal{X}^{\pi}_{N,1}((f_1)^{d'})$ plus its complement. The theorem is an immediate consequence of these two equations:

$$-\lim_{T\to\infty}\sum_{N=1}^{\infty}[\mathcal{X}^{\pi}_{N,1}((f_1)^{d'})]\mathbb{L}^{-2N}(f))T^N = -[\mu_{d'}](\mathbb{L}-1) \tag{9}$$

$$-\lim_{T\to\infty}\sum_{N=1}^{\infty}[\mathcal{X}_{N,1}((f_1)^{d'})\setminus\mathcal{X}^{\pi}_{N,1}((f_1)^{d'})]\mathbb{L}^{-2N}(f))T^N = [(f_1)^{d'}-1] \tag{10}$$

To prove equation (9), we first invoke Lemma 4 to replace $\mathcal{X}^{\pi}_{N,1}((f_1)^{d'})$ by $\mathcal{X}_{N,1}((f_1)^{d'}\circ\pi)$, and remark that

$$(f_1)^{d'}\circ\pi(u,v) = (v^{mn}(mw+\ldots))^{d'}.$$

Consider a jet ρ in $[\mathcal{X}_{N,1}((f_1)^{d'}\circ\pi)]$, and write it as

$$\rho(t) = (v_p t^p+\ldots, w_q t^q+\ldots),$$

with nonzero v_p and w_q. Then $N = d'(mnp+q)$ and $(mv_pw_q)^{d'} = 1$. Thus

$$\begin{aligned}\sum_{N=1}^{\infty}[\mathcal{X}_{N,1}((f_1)^{d'}\circ\pi)]\mathbb{L}^{-2N}T^N &= \sum_{p=1}^{\infty}\sum_{q=1}^{\infty}[\mu'_d](\mathbb{L}-1)\mathbb{L}^{N-p}\mathbb{L}^{N-q}\mathbb{L}^{-2N}T^N\\ &= [\mu'_d](\mathbb{L}-1)\sum_{p=1}^{\infty}\mathbb{L}^{-p}T^{d'mnp}\sum_{q=1}^{\infty}\mathbb{L}^{-q}T^{d'q}\\ &= [\mu'_d](\mathbb{L}-1)\cdot\frac{\mathbb{L}^{-1}T^{d'mn}}{1-\mathbb{L}^{-1}T^{d'mn}}\cdot\frac{\mathbb{L}^{-1}T^{d'}}{1-\mathbb{L}^{-1}T^{d'}},\end{aligned}$$

which yields the desired formula.

To prove equation (10), we examine a jet

$$\varphi\in\mathcal{X}_{N,1}((f_1)^{d'})\setminus\mathcal{X}^{\pi}_{N,1}((f_1)^{d'}).$$

As we remarked in the proof of Lemma 4, for such a jet we do not see the cancellation of terms implied by equations (5) and (6).

Suppose that the order of vanishing of $(f_1)^{d'} \circ \varphi$ is divisible by md' but not mnd'. Then φ must have the following form:

$$\varphi(t) = (x(t), y(t)) = (x_0 t^p + \ldots, y_0 t^q + \ldots),$$

where the following conditions are satisfied:

- q is not a multiple of n,
- $(y_0)^{md'} = 1$,
- $p = \lceil \frac{mq}{n} \rceil$.

Note that x_0 may vanish; in fact $x(t)$ may even be identically zero. The order of vanishing of $(f_1)^{d'} \circ \varphi$ is $N = mqd'$. The virtual motive associated to all such jets is

$$[\mu_{md'}]\mathbb{L}^{2mqd'-p-q+1},$$

and thus its contribution to the sum in (10) is

$$[\mu_{md'}]\mathbb{L}^{-p-q+1}T^{mqd'}.$$

Observe that if we increase q by n, then p increases by m and the order of vanishing increases by mnd'. Thus the contributions of these jet spaces to the sum in (10) may be packaged as $n-1$ geometric series with common ratio $r = \mathbb{L}^{-(m+n)}T^{mnd'}$. For each series, the leading term is determined by a value of q that is less than n, and thus the order of vanishing is less than mnd'. Hence the contribution of this series to the limit as $T \to \infty$ is zero. A similar argument shows that there is no contribution from jets for which the order of vanishing of $(f_1)^{d'} \circ \varphi$ is divisible by nd' but not mnd'.

It remains to analyze those jets for which the order of vanishing N of $(f_1)^{d'} \circ \varphi$ is divisible by mnd'. Such a jet has the form given in equation (5), where

$$N = mnpd'$$

$$\text{and } ((y_0)^m - (x_0)^n)^{d'} = 1;$$

note that x_0 or y_0 may vanish. The virtual motive associated to such jets is

$$[(f_1)^{d'} - 1]\mathbb{L}^{2mnpd'-(m+n)p},$$

and thus the total contribution of these jets to the sum in equation (10) is

$$[(f_1)^{d'} - 1]\sum_{p=1}^{\infty} \mathbb{L}^{-(m+n)p}T^{mnpd'} = [(f_1)^{d'} - 1]\frac{\mathbb{L}^{-(m+n)}T^{mnd'}}{1 - \mathbb{L}^{-(m+n)}T^{mnd'}}.$$

Taking the limit as T goes to infinity gives us $-[(f_1)^{d'} - 1]$.

5 A Recursive Formula for the Spectrum

Steenbrink [9, (2.1)] and [10, Section 1] and Saito [7, Section 2] associate to a plane curve singularity a multiset of rational numbers called its *spectrum* — it can be regarded as an element of the group ring $\mathbb{Z}[t^{\mathbb{Q}}]$. As Steenbrink [9] says, it "gathers the information about the eigenvalues of the monodromy operator . . . and about the Hodge filtration on the vanishing cohomology." We briefly explain this construction.

We have defined the motivic Milnor fiber $S_{\mathbf{0}}(f)$ as an element of $\mathcal{M}_{\mathbb{C}}^{\hat{\mu}}$; the class $1 - S_{\mathbf{0}}(f)$ is called the *motivic vanishing cycle*. There is a natural map χ_h^{mon}, called the *Hodge characteristic*, from $\mathcal{M}_{\mathbb{C}}^{\hat{\mu}}$ to $K_0(HS^{\text{mon}})$, the Grothendieck ring of mixed Hodge structures with a finite automorphism. For a variety X with good $\hat{\mu}$-action, we associate the sum

$$\sum_i (-1)^i [H_c^i(X, \mathbb{Q})] \in K_0(HS^{\text{mon}}).$$

Here $H_c^i(X, \mathbb{Q})$ indicates the ith cohomology group with compact supports, which is naturally endowed with a mixed Hodge structure and a finite automorphism. This map passes to the localization; thus we have a linear map

$$\chi_h^{\text{mon}} : \mathcal{M}_{\mathbb{C}}^{\hat{\mu}} \to K_0(HS^{\text{mon}}).$$

Steenbrink [8] and Varchenko [11] showed that one can define a natural mixed Hodge structure $\text{MHS}(f)$ on the cohomology of the classical Milnor fiber, which in turn determines an element of $K_0(HS^{\text{mon}})$. In Theorem 3.5.5 of [3], Denef and Loeser prove that the Hodge characteristic of the motivic Milnor fiber $S_{\mathbf{0}}(f)$ is $\text{MHS}(f)$.

There is also a natural linear map from $K_0(HS^{\text{mon}})$ to $\mathbb{Z}[t^{\mathbb{Q}}]$, called the *Hodge spectrum*. For a mixed Hodge structure H endowed with a finite automorphism, let $H_\alpha^{p,q}$ denote the eigenspace of $H^{p,q}$ associated to the eigenvalue $\exp(2\pi i \alpha)$. We define

$$\text{hsp}(H) = \sum_{\alpha \in \mathbb{Q} \cap [0,1)} t^\alpha \sum t^p \dim(H_\alpha^{p,q}).$$

Composing with the Hodge characteristic, we obtain a map

$$\text{Sp} : \mathcal{M}_{\mathbb{C}}^{\hat{\mu}} \to \mathbb{Z}[t^{\mathbb{Q}}].$$

Denef and Loeser [3] show that the spectrum of f (as defined by Steenbrink [10] and Saito [7]) can be computed using

$$\text{spectrum}(f) = \text{Sp}(1 - S_{\mathbf{0}}(f)).$$

Applying Sp to the formula of Theorem 1, we obtain this recursive formula for the spectrum:

$$\mathrm{spectrum}(f) = \mathrm{spectrum}((f_1)^{d'}) + \mathrm{spectrum}(f') + \mathrm{Sp}([\mu_{d'}] - 1). \tag{11}$$

As an alternative, here is the result of applying Sp to formula (4):

$$\mathrm{spectrum}(f) = -\,\mathrm{Sp}([(f_1)^{d'} - 1]) + \mathrm{Sp}([\mu_{d'}]\mathbb{L}) + \mathrm{spectrum}(f'). \tag{12}$$

To make this a usable recursion, we supplement formula (12) with results from Guibert [5]. In Lemme 3.4.2(ii), he tells us that

$$\mathrm{Sp}([f_1 - 1]) = t - \frac{t^{1/m} - t}{1 - t^{1/m}} \cdot \frac{t^{1/n} - t}{1 - t^{1/n}}.$$

Thus, as a multiset, it consists of $(m-1)(n-1)/2$ numbers in the interval $(0, 1)$ and an equal number in the interval $(1, 2)$ (all of these counting negatively), together with the number 1. Furthermore we remark that the multiset is symmetric with respect to reflection across the number 1. We may write $\mathrm{Sp}([f_1 - 1])$ as a sum $\mathrm{Sp}^{(0)}([f_1 - 1]) + \mathrm{Sp}^{(1)}([f_1 - 1])$, using first the terms with exponents between 0 and 1, and then those with exponents greater than or equal to 1. Guibert's Lemme 3.4.3 says that

$$\begin{aligned}\mathrm{Sp}([(f_1)^{d'} - 1])(t) = {} & \frac{1-t}{1-t^{1/d'}} \cdot \\ & \left(\mathrm{Sp}^{(0)}([f_1 - 1])(t^{1/d'}) + t^{1-1/d'}\,\mathrm{Sp}^{(1)}([f_1 - 1])(t^{1/d'})\right).\end{aligned} \tag{13}$$

We claim that the term $\mathrm{Sp}([\mu_{d'}]\mathbb{L})$ in formula (12) is

$$\frac{1-t}{1-t^{1/d}} \cdot t.$$

Indeed, the action on the product $\mu_{d'} \times \mathbb{C}$ has as eigenvalues the d'-th roots of unity, and the cohomology lives in $H^{1,1}$. Now note that this term cancels some of the contributions to formula (13); this has the same effect as if the initial term t had been omitted from $\mathrm{Sp}^{(1)}([f_1 - 1])$. Thus again the spectrum has reflectional symmetry. The upshot is that the contribution of the first two terms on the left of (12) to the spectrum of f is obtained from the spectrum of f_1 by the following process:

- Discard the spectral numbers greater than 1.
- Compress by a factor of $1/d'$, i.e., multiply each spectral number by this value.
- Take the union of d' copies: use the first copy unaltered, add $1/d'$ to each spectral number to get the second copy, etc.
- To obtain the remaining contributions in the interval $(1, 2)$, reflect across 1.

Finally we remark that formula (11) and the auxiliary formulas imply formula (6) of Theorem 2.3 in [6], a recursion for the monodromy:

$$\mathbf{H}(t) = \frac{\mathbf{H}_1(t^{d'}) \cdot \mathbf{H}'(t)}{t^{d'} - 1}. \tag{14}$$

To see this, observe that formula (13) implies that the eigenvalues of the monodromy for $(f_1)^{d'}$ are the d'th roots of the eigenvalues of the monodromy for f_1. Combining this with formula (11) yields the result.

6 An Example

Let us consider Example 2.2 from [6]. We begin with the curve whose Puiseux expansion is

$$\zeta = x^{3/2} + x^{7/4} + x^{11/6}. \tag{15}$$

Then its truncation $f_1 = y^2 - x^3$ is parametrized by $\zeta_1 = x^{3/2}$, and its derived curve is parametrized by

$$\zeta' = x^{13/2} + x^{20/3}.$$

Repeating this process, we obtain a truncation $f_1' = y^2 - x^{13}$, parametrized by $\zeta_1' = x^{13/2}$ and a second derived curve $f'' = y^3 - x^{79}$ with parametrization

$$\zeta'' = x^{79/3}.$$

The number of conjugates of the curve, its truncation, the derived curve, the second truncation, and the second derived curve are $d = 1$, $d_1 = 2$, $d' = 6$, $d_1' = 2$, and $d'' = 3$.

Applying our recursive formula (4) repeatedly, the motivic Milnor fiber $S_{\mathbf{0}}(f)$ equals

$$[(y^2 - x^3)^6 - 1] + [(y^2 - x^{13})^3 - 1] + [y^3 - x^{79} - 1] - [\mu_6]\mathbb{L} - [\mu_3]\mathbb{L} - \mathbb{L} + 1.$$

Before computing the spectrum, we remark that formula (14) implies a recursion for the Euler characteristic of the Milnor fiber (formula (4) of the same cited theorem). Applying that formula here, we infer that the Milnor number is 204; thus our recursion will yield 204 spectral numbers.

To compute the spectrum of f, we first apply the process described in the previous section to the two spectral numbers $\frac{5}{6}$ and $\frac{7}{6}$ of $f_1 = y^2 - x^3$. Using just the spectral number $\frac{5}{6}$, compression and copying gives these six spectral numbers in the interval (0, 1):

$$\frac{5}{36}, \frac{11}{36}, \frac{17}{36}, \frac{23}{36}, \frac{29}{36}, \frac{35}{36}.$$

Similarly, the spectral numbers $\frac{15}{26}, \frac{17}{26}, \frac{19}{26}, \frac{21}{26}, \frac{23}{26}, \frac{25}{26}$ of $f_1' = y^2 - x^{13}$ give us eighteen additional spectral numbers of f:

$$\frac{15}{78}, \dots \frac{25}{78}, \quad \frac{41}{78}, \dots \frac{51}{78}, \quad \frac{67}{78}, \dots \frac{77}{78}.$$

Likewise there are contributions coming from the 78 spectral numbers of $y^3 - x^{79}$ which lie in the interval (0, 1); these contributions are

$$\frac{82}{237}, \frac{85}{237}, \frac{88}{237}, \frac{91}{237}, \dots, \frac{235}{237}$$

and

$$\frac{161}{237}, \frac{164}{237}, \frac{167}{237}, \frac{170}{237}, \dots, \frac{236}{237}.$$

In total this gives 102 spectral numbers less than 1; after reflection we obtain an additional 102 values, which together account for the Milnor number.

We now compare our computation with Denef and Loeser's computation in terms of an embedded resolution [3], continuing to work with the function f whose Puiseux expansion is specified in (15). Let $h : Y \to \mathbb{C}^2$ be a function achieving this embedded resolution. Using index set J, denote by E_i the irreducible components of $h^{-1}(f^{-1}(0))$. For each $i \in J$, denote by N_i the multiplicity of E_i in the divisor of $f \circ h$ on Y. For $i, j \in J$, we consider the non singular varieties $E_i^\circ := E_i \setminus \cup_{j \neq i} E_j$ and $E_{i,j}^\circ := E_i \cap E_j$. We let $m_{i,j} := \gcd(N_i, N_j)$.

Denef and Loeser proved the following formula:

$$S_{\mathbf{0}}(f) = \sum_{i \in J} -[\tilde{E}_I^\circ \cap h^{-1}(0)] + \sum_{i,j \in J,\, E_i \cap E_j \neq \emptyset} [\tilde{E}_{i,j}^\circ \cap h^{-1}(0)](\mathbb{L} - 1),$$

where $E_{i,j}^\circ = \mu_{m_{i,j}}$ and $\tilde{E}_i^\circ$ is an unramified Galois cover of E_i° with Galois group μ_{N_i} defined as follows. Let U be a Zariski open subset of Y such that $f \circ h = uv^{N_i}$, with u a unit on U and v a morphism from U to $\mathbb{C}$. The restriction of $\tilde{E}_i^\circ$ above the set $E^\circ \cap U$ is defined by

$$\{(z, y) \in \mathbb{C} \times (E_i^\circ \cap U) \,:\, z^{N_i} = u^{-1}\}.$$

Considering an open covering of E° and gluing the corresponding restrictions we get the cover $\tilde{E}_i^\circ$ which has a natural μ_{N_i}-action obtained by multiplying the z-coordinate by the elements of μ_{N_i}.

To obtain the minimal embedded resolution of the curve whose Puiseux expansion is shown in (15), we perform 8 point blow-ups. The corresponding irreducible components E_i, which have been labeled according to their order of appearance, intersect according to the following dual resolution graph:

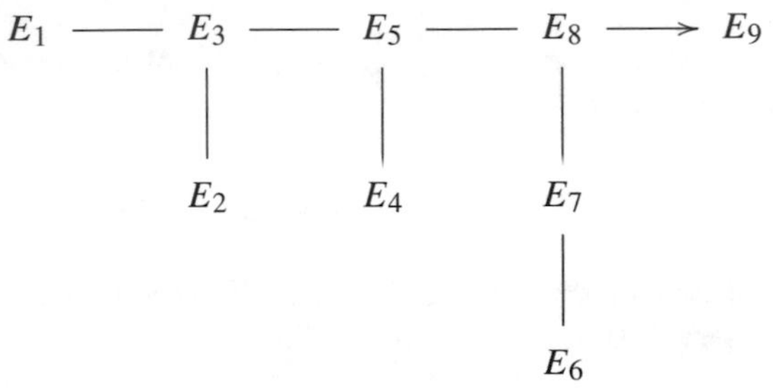

The irreducible component E_9 corresponds to the strict transform of the curve. The multiplicities N_i are given in the following table

	E_1	E_2	E_3	E_4	E_5	E_6	E_7	E_8	E_9
N_i	12	18	36	39	78	79	158	237	1

We now use this resolution to compute $S_0(f)$ in $K_0^{\hat{\mu}}(\mathrm{Var}_{\mathbb{C}})$. First, observe that $E_i^\circ \cap h^{-1}(0) = E_i^\circ$ except when $i = 9$; in the latter case, $E_9^\circ \cap h^{-1}(0)$ is the point $E_8 \cap E_9$. Because E_1°, E_2°, E_4°, and E_6° are each isomorphic to $\mathbb{C}$, we have

$$[\tilde{E}_1^\circ] = [\mu_{12}]\mathbb{L}, \quad [\tilde{E}_2^\circ] = [\mu_{18}]\mathbb{L}, \quad [\tilde{E}_4^\circ] = [\mu_{39}]\mathbb{L}, \quad [\tilde{E}_6^\circ] = [\mu_{78}]\mathbb{L}.$$

Each $E_{i,j}^\circ$ is a point; thus

$$[\tilde{E}_{1,3}^\circ] = [\mu_{12}], [\tilde{E}_{2,3}^\circ] = [\mu_{18}], [\tilde{E}_{3,5}^\circ] = [\mu_6], [\tilde{E}_{4,5}^\circ] = [\mu_{39}],$$
$$[\tilde{E}_{5,8}^\circ] = [\mu_3], [\tilde{E}_{6,7}^\circ] = [\mu_{79}], [\tilde{E}_{7,8}^\circ] = [\mu_{79}], [\tilde{E}_{8,9}^\circ] = 1,$$

where we have used the relevant $m_{i,j}$'s (for example $m_{1,3} = 12$). Since the greatest common divisor of N_6, N_7, and N_8 is 79, and E_7° is isomorphic to $\mathbb{C}^*$, it follows that

$$[\tilde{E}_7^\circ] = [\mu_{79}](\mathbb{L} - 1).$$

To analyze E_3, we choose an affine coordinate w so that the intersection with E_5 is the point at infinity, and so that the intersections with E_1 and E_2 are at 0 and 1, respectively. Thus we must consider

$$\{(z, w) \in \mathbb{C} \times E_3^\circ : z^{N_3} = w^{m_{1,3}}(1 - w)^{m_{3,5}}, \text{ i.e., } z^{36} = w^{12}(1 - w)^6\}.$$

There is an isomorphism between this locus and

$$\{(x, y) \in (\mathbb{C}^*)^2 : (y^2 - x^3)^6 = 1\}$$

given by

$$z = \frac{x}{y}, w = \frac{x^3}{y^2} \qquad x = \frac{w}{z^2}, y = \frac{w}{z^3}.$$

It identifies $[\tilde{E}_3^\circ]$ with $[\{(x, y) \in (\mathbb{C}^*)^2 : (y^2 - x^3)^6 = 1\}]$ (together with the usual quasi-homogeneous action) in $K_0^{\hat{\mu}}(\mathrm{Var}_\mathbb{C})$.

Similarly, the isomorphism between $\{(z, w) \in \mathbb{C} \times E_3^\circ : z^{78} = w^{36}(1-w)^3\}$ and $\{(x, y) \in (\mathbb{C}^*)^2 : (y^2 - x^{13})^3 = 1\}$, given by

$$z = \frac{x^6}{y}, w = \frac{x^{13}}{y^2} \qquad x = \frac{w}{z^2}, y = \frac{w^6}{z^{13}},$$

identifies $[\tilde{E}_5^\circ]$ with $[\{(x, y) \in (\mathbb{C}^*)^2 : (y^2 - x^{13})^3 = 1\}]$. In the same way, the isomorphism between $\{(z, w) \in \mathbb{C} \times E_8^\circ : z^{237} = w^{79}(1-w)\}$ and $\{(x, y) \in (\mathbb{C}^*)^2 : (y^2 - x^{13})^3 = 1\}$, given by

$$z = \frac{x^{26}}{y}, w = \frac{x^{79}}{y^3} \qquad x = \frac{w}{z^3}, y = \frac{w^{26}}{z^{79}},$$

identifies $[\tilde{E}_8^\circ]$ with $[\{(x, y) \in (\mathbb{C}^*)^2 : y^3 - x^{79} = 1\}]$.

Assembling these equations, we see that

- $[\tilde{E}_1^\circ] + [\tilde{E}_2^\circ] + [\tilde{E}_3^\circ] - [\tilde{E}_{1,3}^\circ](\mathbb{L} - 1) - [\tilde{E}_{2,3}^\circ](\mathbb{L} - 1) = [(y^2 - x^3)^6 - 1]$;
- $[\tilde{E}_4^\circ] + [\tilde{E}_5^\circ] - [\tilde{E}_{3,5}^\circ](\mathbb{L} - 1) - [\tilde{E}_{4,5}^\circ](\mathbb{L} - 1) = [(y^2 - x^{13})^3 - 1)] - [\mu_6]\mathbb{L}$;
- $[\tilde{E}_6^\circ] + [\tilde{E}_7^\circ] + [\tilde{E}_8^\circ] - [\tilde{E}_{5,8}^\circ](\mathbb{L} - 1) - [\tilde{E}_{6,7}^\circ](\mathbb{L} - 1) - [\tilde{E}_{7,8}^\circ](\mathbb{L} - 1) = [y^3 - x^{79} - 1)] - [\mu_3]\mathbb{L}$;

in $K_0^{\hat{\mu}}(\mathrm{Var}_\mathbb{C})$. These equalities, together with $[\tilde{E}_{8,9}^\circ](\mathbb{L} - 1) = (\mathbb{L} - 1)$, give us the expression for $S_\mathbf{0}(f)$ computed with our recursive formula.

Acknowledgements This paper benefited greatly from extensive conversations with Mirel Caibăr, and Alejandro Melle Hernández; we also received helpful advice from François Loeser. The first author was supported by Spanish Ministerio de Ciencia y Tecnología Grant no. MTM2016-76868-C2-1-P. The second author was supported by a Collaboration Grant from the Simons Foundation, and worked on this project while in residence at the Fields Institute.

References

1. Artal Bartolo, E., Cassou-Noguès, P., Luengo, I., Melle Hernández, A.: Quasi-ordinary power series and their zeta functions. Mem. Am. Math. Soc. **178**(841), vi+85 (2005)
2. Cauwbergs, T.: Splicing motivic zeta functions. Rev. Mat. Complut. **29**(2), 455–483 (2016)
3. Denef, J., Loeser, F.: Geometry on ARC Spaces of Algebraic Varieties. European Congress of Mathematics, vol. I (Barcelona, 2000). Progress in Mathematics, vol. 201, pp. 327–348. Birkhäuser, Basel (2001)
4. González Pérez, P.D., González Villa, M.: Motivic Milnor fiber of a quasi-ordinary hypersurface. J. Reine Angew. Math. **687**, 159–205 (2014)
5. Guibert, G.: Espaces d'arcs et invariants d'Alexander. Comment. Math. Helv. **77**(4), 783–820 (2002)
6. Kennedy, G., McEwan, L.: Monodromy of plane curves and quasi-ordinary surfaces. J. Singul. **1**, 146–168 (2010)
7. Saito, M.: On Steenbrink's conjecture. Math. Ann. **289**(4), 703–716 (1991)
8. Steenbrink, J.H.M.: Mixed Hodge structure on the vanishing cohomology. Real and complex singularities. Sijthoff and Noordhoff, Alphen aan den Rijn, pp. 525–563 (1977)
9. Steenbrink, J.H.M.: Semicontinuity of the singularity spectrum. Invent. Math. **79**, 557–565 (1985)
10. Steenbrink, J.H.M.: The spectrum of hypersurface singularities. Astérisque **179–180**, 163–184 (1989)
11. Varchenko, A.N.: Asymptotic Hodge structure on vanishing cohomology. Izv. Akad. Nauk SSSR Ser. Mat. **45**(3), 540–591, 688 (1981)

On Hasse–Schmidt Derivations: The Action of Substitution Maps

Luis Narváez Macarro

Dedicated to Antonio Campillo on the ocassion of his 65th birthday

Abstract We study the action of substitution maps between power series rings as an additional algebraic structure on the groups of Hasse–Schmidt derivations. This structure appears as a counterpart of the module structure on classical derivations.

1 Introduction

For any commutative algebra A over a commutative ring k, the set $\operatorname{Der}_k(A)$ of k-derivations of A is an ubiquous object in Commutative Algebra and Algebraic Geometry. It carries an A-module structure and a k-Lie algebra structure. Both structures give rise to a *Lie-Rinehart algebra* structure over (k, A). The k-derivations of A are contained in the filtered ring of k-linear differential operators $\mathscr{D}_{A/k}$, whose graded ring is commutative and we obtain a canonical map of graded A-algebras

$$\tau : \operatorname{Sym}_A \operatorname{Der}_k(A) \longrightarrow \operatorname{gr} \mathscr{D}_{A/k}.$$

If $\mathbb{Q} \subset k$ and $\operatorname{Der}_k(A)$ is a finitely generated projective A-module, the map τ is an isomorphism ([9, Corollary 2.17]) and we can deduce that the ring $\mathscr{D}_{A/k}$ is the enveloping algebra of the Lie-Rinehart algebra $\operatorname{Der}_k(A)$ (cf. [11, Proposition 2.1.2.11]).

Partially supported by MTM2016-75027-P, P12-FQM-2696 and FEDER.

L. Narváez Macarro (✉)
Departamento de Álgebra & Instituto de Matemáticas (IMUS), Facultad de Matemáticas, Universidad de Sevilla, Sevilla, Spain
e-mail: narvaez@us.es

G.-M. Greuel et al. (eds.), *Singularities, Algebraic Geometry, Commutative Algebra, and Related Topics*, https://doi.org/10.1007/978-3-319-96827-8_10

If we are not in characteristic 0, even if A is "smooth" (in some sense) over k, e.g. A is a polynomial or a power series ring with coefficients in k, the map τ has no chance to be an isomorphism.

In [9] we have proved that, if we denote by $\operatorname{Ider}_k(A) \subset \operatorname{Der}_k(A)$ the A-module of *integrable derivations* in the sense of Hasse–Schmidt (see Definition 11), then there is a canonical map of graded A-algebras

$$\vartheta : \Gamma_A \operatorname{Ider}_k(A) \longrightarrow \operatorname{gr} \mathscr{D}_{A/k},$$

where $\Gamma_A(-)$ denotes the *divided power algebra* functor, such that:

(i) $\tau = \vartheta$ when $\mathbb{Q} \subset k$ (in that case $\operatorname{Ider}_k(A) = \operatorname{Der}_k(A)$ and $\Gamma_A = \operatorname{Sym}_A$).
(ii) ϑ is an isomorphism whenever $\operatorname{Ider}_k(A) = \operatorname{Der}_k(A)$ and $\operatorname{Der}_k(A)$ is a finitely generated projective A-module.

The above result suggests an idea: under the "smoothness" hypothesis (ii), can be the ring $\mathscr{D}_{A/k}$ and their modules functorially reconstructed from Hasse–Schmidt derivations? To tackle it, we first need to explore the algebraic structure of Hasse–Schmidt derivations.

Hasse–Schmidt derivations of length $m \geq 1$ form a group, non-abelian for $m \geq 2$, which coincides with the (abelian) additive group of usual derivations $\operatorname{Der}_k(A)$ for $m = 1$. But $\operatorname{Der}_k(A)$ has also an A-module structure and a natural questions arises: Do Hasse–Schmidt derivations of any length have some natural structure extending the A-module structure of $\operatorname{Der}_k(A)$ for length $= 1$?

This paper is devoted to study the action of *substitution maps* (between power series rings) on Hasse–Schmidt derivations as an answer to the above question. This action plays a key role in [12].

Now let us comment on the content of the paper.

In Sect. 2 we have gathered, due to the lack of convenient references, some basic facts and constructions about rings of formal power series in an arbitrary number of variables with coefficients in a non-necessarily commutative ring. In the case of a finite number of variables many results and proofs become simpler, but we need the infinite case in order to study ∞-variate Hasse-Schmidt derivations later.

Sections 3 and 4 are devoted to the study of substitution maps between power series rings and their action on power series rings with coefficients on a (bi)module.

In Sect. 5 we study multivariate (possibly ∞-variate) Hasse–Schmidt derivations. They are a natural generalization of usual Hasse–Schmidt derivations and they provide a convenient framework to deal with Hasse–Schmidt derivations.

In Sect. 6 we see how substitution maps act on Hasse–Schmidt derivations and we study some compatibilities on this action with respect to the group structure.

In Sect. 7 we show how the action of substitution maps allows us to express any HS-derivation in terms of a fixed one under some natural hypotheses. This result generalizes Theorem 2.8 in [3] and provides a conceptual proof of it.

2 Rings and (Bi)modules of Formal Power Series

From now on R will be a ring, k will be a commutative ring and A a commutative k-algebra. A general reference for some of the constructions and results of this section is [2, §4].

Let $\mathbf{s}$ be a set and consider the free commutative monoid $\mathbb{N}^{(\mathbf{s})}$ of maps $\alpha : \mathbf{s} \to \mathbb{N}$ such that the set $\operatorname{supp} \alpha := \{s \in \mathbf{s} \mid \alpha(s) \neq 0\}$ is finite. If $\alpha \in \mathbb{N}^{(\mathbf{s})}$ and $s \in \mathbf{s}$ we will write α_s instead of $\alpha(s)$. The elements of the canonical basis of $\mathbb{N}^{(\mathbf{s})}$ will be denoted by $\mathbf{s}^t$, $t \in \mathbf{s}$: $\mathbf{s}^t_u = \delta_{tu}$ for $t, u \in \mathbf{s}$. For each $\alpha \in \mathbb{N}^{(\mathbf{s})}$ we have $\alpha = \sum_{t\in\mathbf{s}} \alpha_t \mathbf{s}^t$.

The monoid $\mathbb{N}^{(\mathbf{s})}$ is endowed with a natural partial ordering. Namely, for $\alpha, \beta \in \mathbb{N}^{(\mathbf{s})}$, we define

$$\alpha \leq \beta \quad \stackrel{\text{def.}}{\Longleftrightarrow} \quad \exists \gamma \in \mathbb{N}^{(\mathbf{s})} \text{ such that } \beta = \alpha + \gamma \quad \Leftrightarrow \quad \alpha_s \leq \beta_s \quad \forall s \in \mathbf{s}.$$

Clearly, $t \in \operatorname{supp} \alpha \Leftrightarrow \mathbf{s}^t \leq \alpha$. The partial ordered set $(\mathbb{N}^{(\mathbf{s})}, \leq)$ is a directed ordered set: for any $\alpha, \beta \in \mathbb{N}^{(\mathbf{s})}$, $\alpha, \beta \leq \alpha \vee \beta$ where $(\alpha \vee \beta)_t := \max\{\alpha_t, \beta_t\}$ for all $t \in \mathbf{s}$. We will write $\alpha < \beta$ when $\alpha \leq \beta$ and $\alpha \neq \beta$.

For a given $\beta \in \mathbb{N}^{(\mathbf{s})}$ the set of $\alpha \in \mathbb{N}^{(\mathbf{s})}$ such that $\alpha \leq \beta$ is finite. We define $|\alpha| := \sum_{s\in\mathbf{s}} \alpha_s = \sum_{s\in\operatorname{supp}\alpha} \alpha_s \in \mathbb{N}$. If $\alpha \leq \beta$ then $|\alpha| \leq |\beta|$. Moreover, if $\alpha \leq \beta$ and $|\alpha| = |\beta|$, then $\alpha = \beta$. The $\alpha \in \mathbb{N}^{(\mathbf{s})}$ with $|\alpha| = 1$ are exactly the elements $\mathbf{s}^t$, $t \in \mathbf{s}$, of the canonical basis.

A *formal power series* in $\mathbf{s}$ with coefficients in R is a formal expression $\sum_{\alpha\in\mathbb{N}^{(\mathbf{s})}} r_\alpha \mathbf{s}^\alpha$ with $r_\alpha \in R$ and $\mathbf{s}^\alpha = \prod_{s\in\mathbf{s}} s^{\alpha_s} = \prod_{s\in\operatorname{supp}\alpha} s^{\alpha_s}$. Such a formal expression is uniquely determined by the family of coefficients a_α, $\alpha \in \mathbb{N}^{(\mathbf{s})}$.

If $r = \sum_{\alpha\in\mathbb{N}^{(\mathbf{s})}} r_\alpha \mathbf{s}^\alpha$ and $r' = \sum_{\alpha\in\mathbb{N}^{(\mathbf{s})}} r'_\alpha \mathbf{s}^\alpha$ are two formal power series in $\mathbf{s}$ with coefficients in R, their sum and their product are defined in the usual way

$$r + r' := \sum_{\alpha\in\mathbb{N}^{(\mathbf{s})}} S_\alpha \mathbf{s}^\alpha, \quad S_\alpha := r_\alpha + r'_\alpha,$$

$$rr' := \sum_{\alpha\in\mathbb{N}^{(\mathbf{s})}} P_\alpha \mathbf{s}^\alpha, \; P_\alpha := \sum_{\beta+\gamma=\alpha} r_\beta r'_\gamma.$$

The set of formal power series in $\mathbf{s}$ with coefficients in R endowed with the above internal operations is a ring called the *ring of formal power series in* $\mathbf{s}$ *with coefficients in* R and is denoted by $R[[\mathbf{s}]]$. It contains the polynomial ring $R[\mathbf{s}]$ (and so the ring R) and all the monomials $\mathbf{s}^\alpha$ are in the center of $R[[\mathbf{s}]]$. There is a natural ring epimorphism, that we call the *augmentation*, given by

$$\sum_{\alpha\in\mathbb{N}^{(\mathbf{s})}} r_\alpha \mathbf{s}^\alpha \in R[[\mathbf{s}]] \longmapsto r_0 \in R, \tag{1}$$

which is a retraction of the inclusion $R \subset R[[\mathbf{s}]]$. Clearly, the ring $R[[\mathbf{s}]]$ is commutative if and only if R is commutative and $R^{\text{opp}}[[\mathbf{s}]] = R[[\mathbf{s}]]^{\text{opp}}$.

Any ring homomorphism $f : R \to R'$ induces a ring homomorphism

$$\overline{f} : \sum_{\alpha \in \mathbb{N}^{(\mathbf{s})}} r_\alpha \mathbf{s}^\alpha \in R[[\mathbf{s}]] \longmapsto \sum_{\alpha \in \mathbb{N}^{(\mathbf{s})}} f(r_\alpha) \mathbf{s}^\alpha \in R'[[\mathbf{s}]], \tag{2}$$

and clearly the correspondences $R \mapsto R[[\mathbf{s}]]$ and $f \mapsto \overline{f}$ define a functor from the category of rings to itself. If $\mathbf{s} = \emptyset$, then $R[[\mathbf{s}]] = R$ and the above functor is the identity.

Definition 1 A *k-algebra over A* is a (non-necessarily commutative) k-algebra R endowed with a map of k-algebras $\iota : A \to R$. A map between two k-algebras $\iota : A \to R$ and $\iota' : A \to R'$ over A is a map $g : R \to R'$ of k-algebras such that $\iota' = g \circ \iota$.

If R is a k-algebra (over A), then $R[[\mathbf{s}]]$ is also a $k[[\mathbf{s}]]$-algebra (over $A[[\mathbf{s}]]$).

If M is an $(A; A)$-bimodule, we define in a completely similar way the set of formal power series in $\mathbf{s}$ with coefficients in M, denoted by $M[[\mathbf{s}]]$. It carries an addition $+$, for which it is an abelian group, and left and right products by elements of $A[[\mathbf{s}]]$. With these operations $M[[\mathbf{s}]]$ becomes an $(A[[\mathbf{s}]]; A[[\mathbf{s}]])$-bimodule containing the polynomial $(A[\mathbf{s}]; A[\mathbf{s}])$-bimodule $M[\mathbf{s}]$. There is also a natural *augmentation* $M[[\mathbf{s}]] \to M$ which is a section of the inclusion $M \subset M[\mathbf{s}]$ and $M^{\mathrm{opp}}[[\mathbf{s}]] = M[[\mathbf{s}]]^{\mathrm{opp}}$. If $\mathbf{s} = \emptyset$, then $M[[\mathbf{s}]] = M$.

The *support* of a series $m = \sum_\alpha m_\alpha \mathbf{s}^\alpha \in M[[\mathbf{s}]]$ is $\operatorname{supp}(x) := \{\alpha \in \mathbb{N}^{(\mathbf{s})} \mid m_\alpha \neq 0\} \subset \mathbb{N}^{(\mathbf{s})}$. It is clear that $m = 0 \Leftrightarrow \operatorname{supp}(m) = \emptyset$. The *order* of a non-zero series $m = \sum_\alpha m_\alpha \mathbf{s}^\alpha \in M[[\mathbf{s}]]$ is $\operatorname{ord}(m) := \min\{|\alpha| \mid \alpha \in \operatorname{supp}(m)\} \in \mathbb{N}$. If $m = 0$ we define $\operatorname{ord}(0) = \infty$. It is clear that for $a \in A[[\mathbf{s}]]$ and $m, m' \in M[[\mathbf{s}]]$ we have $\operatorname{supp}(m + m') \subset \operatorname{supp}(m) \cup \operatorname{supp}(m')$, $\operatorname{supp}(am), \operatorname{supp}(ma) \subset \operatorname{supp}(m) + \operatorname{supp}(a)$, $\operatorname{ord}(m + m') \geq \min\{\operatorname{ord}(m), \operatorname{ord}(m')\}$ and $\operatorname{ord}(am), \operatorname{ord}(ma) \geq \operatorname{ord}(a) + \operatorname{ord}(m)$. Moreover, if $\operatorname{ord}(m') > \operatorname{ord}(m)$, then $\operatorname{ord}(m + m') = \operatorname{ord}(m)$.

Any $(A; A)$-linear map $h : M \to M'$ between two $(A; A)$-bimodules induces in an obvious way and $(A[[\mathbf{s}]]; A[[\mathbf{s}]])$-linear map

$$\overline{h} : \sum_{\alpha \in \mathbb{N}^{(\mathbf{s})}} m_\alpha \mathbf{s}^\alpha \in M[[\mathbf{s}]] \longmapsto \sum_{\alpha \in \mathbb{N}^{(\mathbf{s})}} h(m_\alpha) \mathbf{s}^\alpha \in M'[[\mathbf{s}]], \tag{3}$$

and clearly the correspondences $M \mapsto M[[\mathbf{s}]]$ and $h \mapsto \overline{h}$ define a functor from the category of $(A; A)$-bimodules to the category $(A[[\mathbf{s}]]; A[[\mathbf{s}]])$-bimodules.

For each $\beta \in M^{(\mathbf{s})}$, let us denote by $\mathfrak{n}_\beta^M(\mathbf{s})$ the subset of $M[[\mathbf{s}]]$ whose elements are the formal power series $\sum m_\alpha \mathbf{s}^\alpha$ with $m_\alpha = 0$ for all $\alpha \leq \beta$. One has $\mathfrak{n}_\beta^M(\mathbf{s}) \subset \mathfrak{n}_\gamma^M(\mathbf{s})$ whenever $\gamma \leq \beta$, and $\mathfrak{n}_{\alpha \vee \beta}^M(\mathbf{s}) \subset \mathfrak{n}_\alpha^M(\mathbf{s}) \cap \mathfrak{n}_\beta^M(\mathbf{s})$.

It is clear that the $\mathfrak{n}_\beta^M(\mathbf{s})$ are sub-bimodules of $M[[\mathbf{s}]]$ and $\mathfrak{n}_\beta^A(\mathbf{s}) M[[\mathbf{s}]] \subset \mathfrak{n}_\beta^M(\mathbf{s})$ and $M[[\mathbf{s}]] \mathfrak{n}_\beta^A(\mathbf{s}) \subset \mathfrak{n}_\beta^M(\mathbf{s})$. For $\beta = 0$, $\mathfrak{n}_0^M(\mathbf{s})$ is the kernel of the augmentation $M[[\mathbf{s}]] \to M$.

In the case of a ring R, the $\mathfrak{n}^R_\beta(\mathbf{s})$ are two-sided ideals of $R[[\mathbf{s}]]$, and $\mathfrak{n}^R_0(\mathbf{s})$ is the kernel of the augmentation $R[[\mathbf{s}]] \to R$.

We will consider $R[[\mathbf{s}]]$ as a topological ring with $\{\mathfrak{n}^R_\beta(\mathbf{s}), \beta \in \mathbb{N}^{(\mathbf{s})}\}$ as a fundamental system of neighborhoods of 0. We will also consider $M[[\mathbf{s}]]$ as a topological $(A[[\mathbf{s}]]; A[[\mathbf{s}]])$-bimodule with $\{\mathfrak{n}^M_\beta(\mathbf{s}), \beta \in \mathbb{N}^{(\mathbf{s})}\}$ as a fundamental system of neighborhoods of 0 for both, a topological left $A[[\mathbf{s}]]$-module structure and a topological right $A[[\mathbf{s}]]$-module structure. If $\mathbf{s}$ is finite, then $\mathfrak{n}^M_\beta(\mathbf{s}) = \sum_{s\in\mathbf{s}} s^{\beta_s+1}M[[\mathbf{s}]] = \sum_{s\in\mathbf{s}} M[[\mathbf{s}]]s^{\beta_s+1}$ and so the above topologies on $R[[\mathbf{s}]]$, and so on $A[[\mathbf{s}]]$, and on $M[[\mathbf{s}]]$ coincide with the $\langle\mathbf{s}\rangle$-adic topologies.

Let us denote by $\mathfrak{n}^M_\beta(\mathbf{s})^{\mathbf{c}} \subset M[\mathbf{s}]$ the intersection of $\mathfrak{n}^M_\beta(\mathbf{s})$ with $M[\mathbf{s}]$, i.e. the subset of $M[\mathbf{s}]$ whose elements are the finite sums $\sum m_\alpha \mathbf{s}^\alpha$ with $m_\alpha = 0$ for all $\alpha \leq \beta$. It is clear that the natural map $R[\mathbf{s}]/\mathfrak{n}^R_\beta(\mathbf{s})^{\mathbf{c}} \longrightarrow R[[\mathbf{s}]]/\mathfrak{n}^R_\beta(\mathbf{s})$ is an isomorphism of rings and the quotient $R[[\mathbf{s}]]/\mathfrak{n}^R_\beta(\mathbf{s})$ is a finitely generated free left (and right) R-module with basis the set of the classes of monomials $\mathbf{s}^\alpha$, $\alpha \leq \beta$.

In the same vein, the $\mathfrak{n}^M_\beta(\mathbf{s})^{\mathbf{c}}$ are sub-$(A[\mathbf{s}]; A[\mathbf{s}])$-bimodules of $M[\mathbf{s}]$ and the natural map $M[\mathbf{s}]/\mathfrak{n}^M_\beta(\mathbf{s})^{\mathbf{c}} \longrightarrow M[[\mathbf{s}]]/\mathfrak{n}^M_\beta(\mathbf{s})$ is an isomorphism of $(A[\mathbf{s}]/\mathfrak{n}^A_\beta(\mathbf{s})^{\mathbf{c}}; A[\mathbf{s}]/\mathfrak{n}^A_\beta(\mathbf{s})^{\mathbf{c}})$-bimodules. Moreover, we have a commutative diagram of natural $\mathbb{Z}$-linear isomorphisms

$$
\begin{array}{ccccc}
A[\mathbf{s}]/\mathfrak{n}^A_\beta(\mathbf{s})^{\mathbf{c}} \otimes_A M & \xrightarrow[\simeq]{\varrho} & M[\mathbf{s}]/\mathfrak{n}^M_\beta(\mathbf{s})^{\mathbf{c}} & \xleftarrow[\simeq]{\lambda} & M \otimes_A A[\mathbf{s}]/\mathfrak{n}^A_\beta(\mathbf{s})^{\mathbf{c}} \\
{\scriptstyle \text{nat.}\otimes\text{Id}}\downarrow\simeq & & \downarrow\simeq & & \simeq\downarrow{\scriptstyle \text{Id}\otimes\text{nat.}} \\
A[[\mathbf{s}]]/\mathfrak{n}^A_\beta(\mathbf{s}) \otimes_A M & \xrightarrow[\simeq]{\varrho'} & M[[\mathbf{s}]]/\mathfrak{n}^M_\beta(\mathbf{s}) & \xleftarrow[\simeq]{\lambda'} & M \otimes_A A[[\mathbf{s}]]/\mathfrak{n}^A_\beta(\mathbf{s})
\end{array}
\quad (4)
$$

where ϱ (resp. ϱ') is an isomorphism of $(A[\mathbf{s}]/\mathfrak{n}^A_\beta(\mathbf{s})^{\mathbf{c}}; A)$-bimodules (resp. of $(A[[\mathbf{s}]]/\mathfrak{n}^A_\beta(\mathbf{s}); A)$-bimodules) and λ (resp. λ') is an isomorphism of bimodules over $(A; A[\mathbf{s}]/\mathfrak{n}^A_\beta(\mathbf{s})^{\mathbf{c}})$(resp. over $(A; A[[\mathbf{s}]]/\mathfrak{n}^A_\beta(\mathbf{s}))$.

It is clear that the natural map

$$R[[\mathbf{s}]] \longrightarrow \varprojlim_{\beta\in\mathbb{N}^{(\mathbf{s})}} R[[\mathbf{s}]]/\mathfrak{n}^R_\beta(\mathbf{s}) \equiv \varprojlim_{\beta\in\mathbb{N}^{(\mathbf{s})}} R[\mathbf{s}]/\mathfrak{n}^R_\beta(\mathbf{s})^{\mathbf{c}}$$

is an isomorphism of rings and so $R[[\mathbf{s}]]$ is complete (hence, separated). Moreover, $R[[\mathbf{s}]]$ appears as the completion of the polynomial ring $R[\mathbf{s}]$ endowed with the topology with $\{\mathfrak{n}^R_\beta(\mathbf{s})^{\mathbf{c}}, \beta \in \mathbb{N}^{(\mathbf{s})}\}$ as a fundamental system of neighborhoods of 0.

Similarly, the natural map

$$M[[\mathbf{s}]] \longrightarrow \varprojlim_{\beta\in\mathbb{N}^{(\mathbf{s})}} M[[\mathbf{s}]]/\mathfrak{n}^M_\beta(\mathbf{s}) \equiv \varprojlim_{\beta\in\mathbb{N}^{(\mathbf{s})}} M[\mathbf{s}]/\mathfrak{n}^M_\beta(\mathbf{s})^{\mathbf{c}}$$

is an isomorphism of $(A[[\mathbf{s}]]; A[[\mathbf{s}]])$-bimodules, and so $M[[\mathbf{s}]]$ is complete (hence, separated). Moreover, $M[[\mathbf{s}]]$ appears as the completion of the bimodule $M[\mathbf{s}]$ over

$(A[\mathbf{s}]; A[\mathbf{s}])$ endowed with the topology with $\{\mathfrak{n}_\beta^M(\mathbf{s})^{\mathbf{c}}, \beta \in \mathbb{N}^{(\mathbf{s})}\}$ as a fundamental system of neighborhoods of 0.

Since the subsets $\{\alpha \in \mathbb{N}^{(\mathbf{s})} \mid \alpha \leq \beta\}$, $\beta \in \mathbb{N}^{(\mathbf{s})}$, are cofinal among the finite subsets of $\mathbb{N}^{(\mathbf{s})}$, the additive isomorphism

$$\sum_{\alpha\in\mathbb{N}^{(\mathbf{s})}} m_\alpha \mathbf{s}^\alpha \in M[[\mathbf{s}]] \mapsto \{m_\alpha\}_{\alpha\in\mathbb{N}^{(\mathbf{s})}} \in M^{\mathbb{N}^{(\mathbf{s})}}$$

is a homeomorphism, where $M^{\mathbb{N}^{(\mathbf{s})}}$ is endowed with the product of discrete topologies on each copy of M. In particular, any formal power series $\sum m_\alpha \mathbf{s}^\alpha$ is the limit of its finite partial sums $\sum_{\alpha\in F} m_\alpha \mathbf{s}^\alpha$, over the filter of finite subsets $F \subset \mathbb{N}^{(\mathbf{s})}$.

Since the quotients $A[[\mathbf{s}]]/\mathfrak{n}_\beta^A(\mathbf{s})$ are free A-modules, we have exact sequences

$$0 \longrightarrow \mathfrak{n}_\beta^A(\mathbf{s}) \otimes_A M \longrightarrow A[[\mathbf{s}]] \otimes_A M \longrightarrow \frac{A[[\mathbf{s}]]}{\mathfrak{n}_\beta^A(\mathbf{s})} \otimes_A M \longrightarrow 0$$

and the tensor product $A[[\mathbf{s}]] \otimes_A M$ is a topological left $A[[\mathbf{s}]]$-module with $\{\mathfrak{n}_\beta^A(\mathbf{s}) \otimes_A M, \beta \in \mathbb{N}^{(\mathbf{s})}\}$ as a fundamental system of neighborhoods of 0. The natural $(A[[\mathbf{s}]]; A)$-linear map

$$A[[\mathbf{s}]] \otimes_A M \longrightarrow M[[\mathbf{s}]]$$

is continuous and, if we denote by $A[[\mathbf{s}]]\widehat{\otimes}_A M$ the completion of $A[[\mathbf{s}]] \otimes_A M$, the induced map $A[[\mathbf{s}]]\widehat{\otimes}_A M \longrightarrow M[[\mathbf{s}]]$ is an isomorphism of $(A[[\mathbf{s}]]; A)$-bimodules, since we have natural $(A[[\mathbf{s}]]; A)$-linear isomorphisms

$$(A[[\mathbf{s}]] \otimes_A M) \,/\, \left(\mathfrak{n}_\beta^A(\mathbf{s}) \otimes_A M\right) \simeq \left(A[[\mathbf{s}]]/\mathfrak{n}_\beta^A(\mathbf{s})\right) \otimes_A M \simeq M[[\mathbf{s}]]/\mathfrak{n}_\beta^M(\mathbf{s})$$

for $\beta \in \mathbb{N}^{(\mathbf{s})}$, and so

$$A[[\mathbf{s}]]\widehat{\otimes}_A M = \lim_{\substack{\longleftarrow \\ \beta\in\mathbb{N}^{(\mathbf{s})}}} \left(\frac{A[[\mathbf{s}]] \otimes_A M}{\mathfrak{n}_\beta^A(\mathbf{s}) \otimes_A M}\right) \simeq \lim_{\substack{\longleftarrow \\ \beta\in\mathbb{N}^{(\mathbf{s})}}} \left(\frac{M[[\mathbf{s}]]}{\mathfrak{n}_\beta^M(\mathbf{s})}\right) \simeq M[[\mathbf{s}]]. \tag{5}$$

Similarly, the natural $(A; A[[\mathbf{s}]])$-linear map $M \otimes_A A[[\mathbf{s}]] \to M[[\mathbf{s}]]$ induces an isomorphism $M\widehat{\otimes}_A A[[\mathbf{s}]] \xrightarrow{\sim} M[[\mathbf{s}]]$ of $(A; A[[\mathbf{s}]])$-bimodules.

If $h : M \to M'$ is an $(A; A)$-linear map between two $(A; A)$-bimodules, the induced map $\overline{h} : M[[\mathbf{s}] \to M'[[\mathbf{s}]$ (see (3)) is clearly continuous and there is a commutative diagram

$$\begin{array}{ccccc}
A[[\mathbf{s}]]\widehat{\otimes}_A M & \xrightarrow{\simeq} & M[[\mathbf{s}]] & \xleftarrow{\simeq} & M\widehat{\otimes}_A A[[\mathbf{s}]] \\
{\scriptstyle \mathrm{Id}\widehat{\otimes} h}\downarrow & & {\scriptstyle \overline{h}}\downarrow & & {\scriptstyle h\widehat{\otimes}\mathrm{Id}}\downarrow \\
A[[\mathbf{s}]]\widehat{\otimes}_A M' & \xrightarrow{\simeq} & M'[[\mathbf{s}]] & \xleftarrow{\simeq} & M'\widehat{\otimes}_A A[[\mathbf{s}]].
\end{array}$$

Similarly, for any ring homomorphism $f : R \to R'$, the induced ring homomorphism $\overline{f} : R[[\mathbf{s}]] \to R'[[\mathbf{s}]]$ is also continuous.

Definition 2 We say that a subset $\Delta \subset \mathbb{N}^{(\mathbf{s})}$ is an *ideal* of $\mathbb{N}^{(\mathbf{s})}$ (resp. a *co-ideal* of $\mathbb{N}^{(\mathbf{s})}$) if whenever $\alpha \in \Delta$ and $\alpha \leq \alpha'$ (resp. $\alpha' \leq \alpha$), then $\alpha' \in \Delta$.

It is clear that Δ is an ideal if and only if its complement Δ^c is a co-ideal, and that the union and the intersection of any family of ideals (resp. of co-ideals) of $\mathbb{N}^{(\mathbf{s})}$ is again an ideal (resp. a co-ideal) of $\mathbb{N}^{(\mathbf{s})}$. Examples of ideals (resp. of co-ideals) of $\mathbb{N}^{(\mathbf{s})}$ are the $\beta + \mathbb{N}^{(\mathbf{s})}$ (resp. the $\mathfrak{n}_\beta(\mathbf{s}) := \{\alpha \in \mathbb{N}^{(\mathbf{s})} \mid \alpha \leq \beta\}$) with $\beta \in \mathbb{N}^{(\mathbf{s})}$. The $\mathfrak{t}_m(\mathbf{s}) := \{\alpha \in \mathbb{N}^{(\mathbf{s})} \mid |\alpha| \leq m\}$ with $m \geq 0$ are also co-ideals. Actually, a subset $\Delta \subset \mathbb{N}^{(\mathbf{s})}$ is an ideal (resp. a co-ideal) if and only if $\Delta = \cup_{\beta\in\Delta}\left(\beta + \mathbb{N}^{(\mathbf{s})}\right) = \Delta + \mathbb{N}^{(\mathbf{s})}$ (resp. $\Delta = \cup_{\beta\in\Delta}\mathfrak{n}_\beta(\mathbf{s})$).

We say that a co-ideal $\Delta \subset \mathbb{N}^{(\mathbf{s})}$ is bounded if there is an integer $m \geq 0$ such that $|\alpha| \leq m$ for all $\alpha \in \Delta$. In other words, a co-ideal $\Delta \subset \mathbb{N}^{(\mathbf{s})}$ is bounded if and only if there is an integer $m \geq 0$ such that $\Delta \subset \mathfrak{t}_m(\mathbf{s})$. Also, a co-ideal $\Delta \subset \mathbb{N}^{(\mathbf{s})}$ is non-empty if and only if $\mathfrak{t}_0(\mathbf{s}) = \mathfrak{n}_0(\mathbf{s}) = \{0\} \subset \Delta$.

For a co-ideal $\Delta \subset \mathbb{N}^{(\mathbf{s})}$ and an integer $m \geq 0$, we denote $\Delta^m := \Delta \cap \mathfrak{t}_m(\mathbf{s})$.

For each co-ideal $\Delta \subset \mathbb{N}^{(\mathbf{s})}$, we denote by Δ_M the sub-$(A[[\mathbf{s}]; A[[\mathbf{s}]])$-bimodule of $M[[\mathbf{s}]]$ whose elements are the formal power series $\sum_{\alpha\in\mathbb{N}^{(\mathbf{s})}} m_\alpha \mathbf{s}^\alpha$ such that $m_\alpha = 0$ whenever $\alpha \in \Delta$. One has

$$\Delta_M = \cdots = \left\{ m \in M[[\mathbf{s}]] \mid \operatorname{supp}(m) \subset \bigcap_{\beta\in\Delta} \mathfrak{n}_\beta(\mathbf{s})^c \right\} =$$

$$\bigcap_{\beta\in\Delta} \left\{ m \in M[[\mathbf{s}]] \mid \operatorname{supp}(m) \subset \mathfrak{n}_\beta(\mathbf{s})^c \right\} = \bigcap_{\beta\in\Delta} \mathfrak{n}_\beta^M(\mathbf{s}),$$

and so Δ_M is closed in $M[[\mathbf{s}]]$. Let $\Delta' \subset \mathbb{N}^{(\mathbf{s})}$ be another co-ideal. We have

$$\Delta_M + \Delta'_M = (\Delta \cap \Delta')_M.$$

If $\Delta \subset \Delta'$, then $\Delta'_M \subset \Delta_M$, and if $a \in \Delta'_A$, $m \in \Delta_M$ we have

$$\operatorname{supp}(am) \subset \operatorname{supp}(a) + \operatorname{supp}(m) \subset \left(\Delta'\right)^c + \Delta^c \subset \left(\Delta'\right)^c \cap \Delta^c = \left(\Delta' \cup \Delta\right)^c,$$

and so $\Delta'_A \Delta_M \subset (\Delta' \cup \Delta)_M$. Is a similar way we obtain $\Delta_M \Delta'_A \subset (\Delta' \cup \Delta)_M$.

Let us denote by $M[[\mathbf{s}]]_\Delta := M[[\mathbf{s}]]/\Delta_M$ endowed with the quotient topology. The elements in $M[[\mathbf{s}]]_\Delta$ are power series of the form

$$\sum_{\alpha\in\Delta} m_\alpha \mathbf{s}^\alpha, \quad m_\alpha \in M.$$

It is clear that $M[[\mathbf{s}]]_\Delta$ is a topological $(A[[\mathbf{s}]]_\Delta; A[[\mathbf{s}]]_\Delta)$-bimodule. A fundamental system of neighborhoods of 0 in $M[[\mathbf{s}]]_\Delta$ consist of

$$\frac{\mathfrak{n}_\beta^M(\mathbf{s}) + \Delta_M}{\Delta_M} = \frac{(\mathfrak{n}_\beta(\mathbf{s}) \cap \Delta)_M}{\Delta_M}, \quad \beta \in \mathbb{N}^{(\mathbf{s})},$$

and since the subsets $\mathfrak{n}_\beta(\mathbf{s}) \cap \Delta$, $\beta \in \mathbb{N}^{(\mathbf{s})}$, are cofinal among the finite subsets of Δ, we conclude that the additive isomorphism

$$\sum_{\alpha\in\Delta} m_\alpha \mathbf{s}^\alpha \in M[[\mathbf{s}]]_\Delta \mapsto \{m_\alpha\}_{\alpha\in\Delta} \in M^\Delta$$

is a homeomorphism, where M^Δ is endowed with the product of discrete topologies on each copy of M.

For $\Delta \subset \Delta'$ co-ideals of $\mathbb{N}^{(\mathbf{s})}$, we have natural continuous $(A[[\mathbf{s}]]_{\Delta'}; A[[\mathbf{s}]]_{\Delta'})$-linear projections $\tau_{\Delta'\Delta} : M[[\mathbf{s}]]_{\Delta'} \longrightarrow M[[\mathbf{s}]]_\Delta$, that we also call *truncations*,

$$\tau_{\Delta'\Delta} : \sum_{\alpha\in\Delta'} m_\alpha \mathbf{s}^\alpha \in M[[\mathbf{s}]]_{\Delta'} \longmapsto \sum_{\alpha\in\Delta} m_\alpha \mathbf{s}^\alpha \in M[[\mathbf{s}]]_\Delta,$$

and continuous $(A; A)$-linear scissions

$$\sum_{\alpha\in\Delta} m_\alpha \mathbf{s}^\alpha \in M[[\mathbf{s}]]_\Delta \longmapsto \sum_{\alpha\in\Delta} m_\alpha \mathbf{s}^\alpha \in M[[\mathbf{s}]]_{\Delta'}.$$

which are topological immersions.

In particular we have natural continuous $(A; A)$-linear topological embeddings $M[[\mathbf{s}]]_\Delta \hookrightarrow M[[\mathbf{s}]]$ and we define the *support* (resp. the *order*) of any element in $M[[\mathbf{s}]]_\Delta$ as its support (resp. its order) as element of $M[[\mathbf{s}]]$.

We have a bicontinuous isomorphism of $(A[[\mathbf{s}]]_\Delta; A[[\mathbf{s}]]_\Delta)$-bimodules

$$M[[\mathbf{s}]]_\Delta = \varprojlim_{m\in\mathbb{N}} M[[\mathbf{s}]]_{\Delta^m}.$$

For a ring R, the Δ_R are two-sided closed ideals of $R[[\mathbf{s}]]$, $\Delta_R \Delta'_R \subset (\Delta \cup \Delta')_R$ and we have a bicontinuous ring isomorphism

$$R[[\mathbf{s}]]_\Delta = \varprojlim_{m\in\mathbb{N}} R[[\mathbf{s}]]_{\Delta^m}.$$

When $\mathbf{s}$ is finite, $\mathfrak{t}_m(\mathbf{s})_R$ coincides with the $(m+1)$-power of the two-sided ideal generated by all the variables $s \in \mathbf{s}$.

As in (5) one proves that $A[[\mathbf{s}]]_\Delta \otimes_A M$ (resp. $M \otimes_A A[[\mathbf{s}]]_\Delta$) is endowed with a natural topology in such a way that the natural map $A[[\mathbf{s}]]_\Delta \otimes_A M \to M[[\mathbf{s}]]_\Delta$

(resp. $M \otimes_A A[[\mathbf{s}]]_\Delta \to M[[\mathbf{s}]]_\Delta$) is continuous and gives rise to a $(A[[\mathbf{s}]]_\Delta; A)$-linear (resp. to a $(A; A[[\mathbf{s}]]_\Delta)$-linear) isomorphism

$$A[[\mathbf{s}]]_\Delta \widehat{\otimes}_A M \xrightarrow{\sim} M[[\mathbf{s}]]_\Delta \qquad (\text{resp. } M \widehat{\otimes}_A A[[\mathbf{s}]]_\Delta \xrightarrow{\sim} M[[\mathbf{s}]]_\Delta).$$

If $h : M \to M'$ is an $(A; A)$-linear map between two $(A; A)$-bimodules, the map $\overline{h} : M[[\mathbf{s}]] \to M'[[\mathbf{s}]]$ (see (3)) obviously satisfies $\overline{h}(\Delta_M) \subset \Delta_{M'}$, and so induces another natural $(A[[\mathbf{s}]]_\Delta; A[[\mathbf{s}]]_\Delta)$-linear continuous map $M[[\mathbf{s}]]_\Delta \to M'[[\mathbf{s}]]_\Delta$, that will be still denoted by $\overline{h}$. We have a commutative diagram

$$\begin{array}{ccccc}
A[[\mathbf{s}]]_\Delta \widehat{\otimes}_A M & \xrightarrow{\simeq} & M[[\mathbf{s}]]_\Delta & \xleftarrow{\simeq} & M \widehat{\otimes}_A A[[\mathbf{s}]]_\Delta \\
{\scriptstyle \mathrm{Id}\widehat{\otimes} h}\big\downarrow & & {\scriptstyle \overline{h}}\big\downarrow & & {\scriptstyle h\widehat{\otimes}\mathrm{Id}}\big\downarrow \\
A[[\mathbf{s}]]_\Delta \widehat{\otimes}_A M' & \xrightarrow{\simeq} & M'[[\mathbf{s}]]_\Delta & \xleftarrow{\simeq} & M' \widehat{\otimes}_A A[[\mathbf{s}]]_\Delta .
\end{array}$$

Remark 1 In the same way that the correspondences $M \mapsto M[[\mathbf{s}]]$ and $h \mapsto \overline{h}$ define a functor from the category of $(A; A)$-bimodules to the category of $(A[[\mathbf{s}]]; A[[\mathbf{s}]])$-bimodules, we may consider functors $M \mapsto M[[\mathbf{s}]]_\Delta$ and $h \mapsto \overline{h}$ from the category of $(A; A)$-bimodules to the category of $(A[[\mathbf{s}]]_\Delta; A[[\mathbf{s}]]_\Delta)$-bimodules. We may also consider functors $R \mapsto R[[\mathbf{s}]]_\Delta$ and $f \mapsto \overline{f}$ from the category of rings to itself. Moreover, if R is a k-algebra (over A), then $R[[\mathbf{s}]]_\Delta$ is a $k[[\mathbf{s}]]_\Delta$-algebra (over $A[[\mathbf{s}]]_\Delta$).

Lemma 1 *Under the above hypotheses, Δ_M is the closure of $\Delta_\mathbb{Z} M[[\mathbf{s}]]$.*

Proof Any element in Δ_M is of the form $\sum_{\alpha\in\Delta} m_\alpha \mathbf{s}^\alpha$, but $\mathbf{s}^\alpha m_\alpha \in \Delta_\mathbb{Z} M[[\mathbf{s}]]$ whenever $\alpha \in \Delta$ and so it belongs to the closure of $\Delta_\mathbb{Z} M[[\mathbf{s}]]$.

Lemma 2 *Let R be a ring, $\mathbf{s}$ a set and $\Delta \subset \mathbb{N}^{(\mathbf{s})}$ a non-empty co-ideal. The units in $R[[\mathbf{s}]]_\Delta$ are those power series $r = \sum r_\alpha \mathbf{s}^\alpha$ such that r_0 is a unit in R. Moreover, in the special case where $r_0 = 1$, the inverse $r^* = \sum r^*_\alpha \mathbf{s}^\alpha$ of r is given by $r^*_0 = 1$ and*

$$r^*_\alpha = \sum_{d=1}^{|\alpha|} (-1)^d \sum_{\alpha^\bullet \in \mathscr{P}(\alpha,d)} r_{\alpha^1} \cdots r_{\alpha^d} \quad \textit{for } \alpha \neq 0,$$

where $\mathscr{P}(\alpha, d)$ is the set of d-uples $\alpha^\bullet = (\alpha^1, \dots, \alpha^d)$ with $\alpha^i \in \mathbb{N}^{(\mathbf{s})}$, $\alpha^i \neq 0$, and $\alpha^1 + \cdots + \alpha^d = \alpha$.

Proof The proof is standard and it is left to the reader.

Notation 1 *Let R be a ring, $\mathbf{s}$ a set and $\Delta \subset \mathbb{N}^{(\mathbf{s})}$ a non-empty co-ideal. We denote by $\mathscr{U}^\mathbf{s}(R; \Delta)$ the multiplicative sub-group of the units of $R[[\mathbf{s}]]_\Delta$ whose 0-degree coefficient is 1. Clearly, $\mathscr{U}^\mathbf{s}(R; \Delta)^{\mathrm{opp}} = \mathscr{U}^\mathbf{s}(R^{\mathrm{opp}}; \Delta)$. For $\Delta \subset \Delta'$ co-ideals we have $\tau_{\Delta'\Delta}\left(\mathscr{U}^\mathbf{s}(R; \Delta')\right) \subset \mathscr{U}^\mathbf{s}(R; \Delta)$ and the truncation map $\tau_{\Delta'\Delta} : \mathscr{U}^\mathbf{s}(R; \Delta') \to \mathscr{U}^\mathbf{s}(R; \Delta)$ is a group homomorphisms. Clearly, we have*

$$\mathscr{U}^\mathbf{s}(R; \Delta) = \varprojlim_{m\in\mathbb{N}} \mathscr{U}^\mathbf{s}(R; \Delta^m).$$

For any ring homomorphism $f : R \to R'$, the induced ring homomorphism $\overline{f}$: $R[[\mathbf{s}]]_\Delta \to R'[[\mathbf{s}]]_\Delta$ sends $\mathscr{U}^{\mathbf{s}}(R;\Delta)$ into $\mathscr{U}^{\mathbf{s}}(R';\Delta)$ and so it induces natural group homomorphisms $\mathscr{U}^{\mathbf{s}}(R;\Delta) \to \mathscr{U}^{\mathbf{s}}(R';\Delta)$.

Definition 3 Let R be a ring, $\mathbf{s}, \mathbf{t}$ sets and $\nabla \subset \mathbb{N}^{(\mathbf{s})}$, $\Delta \subset \mathbb{N}^{(\mathbf{t})}$ non-empty co-ideals. For each $r \in R[[\mathbf{s}]]_\nabla$, $r' \in R[[\mathbf{t}]]_\Delta$, the *external product* $r \boxtimes r' \in R[[\mathbf{s} \sqcup \mathbf{t}]]_{\nabla\times\Delta}$ is defined as

$$r \boxtimes r' := \sum_{(\alpha,\beta)\in\nabla\times\Delta} r_\alpha r'_\beta \mathbf{s}^\alpha \mathbf{t}^\beta.$$

Let us notice that the above definition is consistent with the existence of natural isomorphism of $(R;R)$-bimodules $R[[\mathbf{s}]]_\nabla \widehat{\otimes}_R R[[\mathbf{t}]]_\Delta \simeq R[[\mathbf{s}\sqcup\mathbf{t}]]_{\nabla\times\Delta} \simeq R[[\mathbf{t}\sqcup\mathbf{s}]]_{\Delta\times\nabla} \simeq R[[\mathbf{t}]]_\Delta \widehat{\otimes}_R R[[\mathbf{s}]]_\nabla$. Let us also notice that $1\boxtimes 1 = 1$ and $r\boxtimes r' = (r\boxtimes 1)(1\boxtimes r')$. Moreover, if $r \in \mathscr{U}^{\mathbf{s}}(R;\nabla)$, $r' \in \mathscr{U}^{\mathbf{t}}(R;\Delta)$, then $r\boxtimes r' \in \mathscr{U}^{\mathbf{s}\sqcup\mathbf{t}}(R;\nabla\times\Delta)$ and $(r\boxtimes r')^* = r'^* \boxtimes r^*$.

Let $k \to A$ be a ring homomorphism between commutative rings, E, F two A-modules, $\mathbf{s}$ a set and $\Delta \subset \mathbb{N}^{(\mathbf{s})}$ a non-empty co-ideal, i.e $\mathfrak{n}_0(\mathbf{s}) = \{0\} \subset \Delta$.

Proposition 1 *Under the above hypotheses, let $f : E[[\mathbf{s}]]_\Delta \to F[[\mathbf{s}]]_\Delta$ be a continuous $k[[\mathbf{s}]]_\Delta$-linear map. Then, for any co-ideal $\Delta' \subset \mathbb{N}^{(\mathbf{s})}$ with $\Delta' \subset \Delta$ we have*

$$f\left(\Delta'_E/\Delta_E\right) \subset \Delta'_F/\Delta_F$$

and so there is a unique continuous $k[[\mathbf{s}]]_{\Delta'}$-linear map $\overline{f} : E[[\mathbf{s}]]_{\Delta'} \to F[[\mathbf{s}]]_{\Delta'}$ such that the following diagram is commutative

$$\begin{array}{ccc}
E[[\mathbf{s}]]_\Delta & \xrightarrow{f} & F[[\mathbf{s}]]_\Delta \\
{\scriptstyle nat.}\downarrow & & \downarrow{\scriptstyle nat.} \\
E[[\mathbf{s}]]_{\Delta'} & \xrightarrow{\overline{f}} & F[[\mathbf{s}]]_{\Delta'}.
\end{array}$$

Proof It is a straightforward consequence of Lemma 1.

Notation 2 *Under the above hypotheses, the set of all continuous $k[[\mathbf{s}]]_\Delta$-linear maps from $E[[\mathbf{s}]]_\Delta$ to $F[[\mathbf{s}]]_\Delta$ will be denoted by*

$$\operatorname{Hom}^{\mathrm{top}}_{k[[\mathbf{s}]]_\Delta}(E[[\mathbf{s}]]_\Delta, F[[\mathbf{s}]]_\Delta).$$

It is an $(A[[\mathbf{s}]]_\Delta; A[[\mathbf{s}]]_\Delta)$-bimodule central over $k[[\mathbf{s}]]_\Delta$. For any co-ideals $\Delta' \subset \Delta \subset \mathbb{N}^{(\mathbf{s})}$, Proposition 1 provides a natural $(A[[\mathbf{s}]]_\Delta; A[[\mathbf{s}]]_\Delta)$-linear map

$$\operatorname{Hom}^{\mathrm{top}}_{k[[\mathbf{s}]]_\Delta}(E[[\mathbf{s}]]_\Delta, F[[\mathbf{s}]]_\Delta) \longrightarrow \operatorname{Hom}^{\mathrm{top}}_{k[[\mathbf{s}]]_{\Delta'}}(E[[\mathbf{s}]]_{\Delta'}, F[[\mathbf{s}]]_{\Delta'}).$$

For $E = F$, $\operatorname{End}^{\text{top}}_{k[[\mathbf{s}]]_\Delta}(E[[\mathbf{s}]]_\Delta)$ *is a* $k[[\mathbf{s}]]_\Delta$*-algebra over* $A[[\mathbf{s}]]_\Delta$.

1. For each $r = \sum_\beta r_\beta \mathbf{s}^\beta \in \operatorname{Hom}_k(E, F)[[\mathbf{s}]]_\Delta$ we define $\widetilde{r} : E[[\mathbf{s}]]_\Delta \to F[[\mathbf{s}]]_\Delta$ by

$$\widetilde{r}\left(\sum_{\alpha\in\Delta} e_\alpha \mathbf{s}^\alpha\right) := \sum_{\alpha\in\Delta}\left(\sum_{\beta+\gamma=\alpha} r_\beta(e_\gamma)\right)\mathbf{s}^\alpha,$$

which is obviously a continuous $k[[\mathbf{s}]]_\Delta$-linear map.

Let us notice that $\widetilde{r} = \sum_\beta \mathbf{s}^\beta \widetilde{r_\beta}$. It is clear that the map

$$r \in \operatorname{Hom}_k(E, F)[[\mathbf{s}]]_\Delta \longmapsto \widetilde{r} \in \operatorname{Hom}^{\text{top}}_{k[[\mathbf{s}]]_\Delta}(E[[\mathbf{s}]]_\Delta, F[[\mathbf{s}]]_\Delta) \tag{6}$$

is $(A[[\mathbf{s}]]_\Delta; A[[\mathbf{s}]]_\Delta)$-linear.

If $f : E[[\mathbf{s}]]_\Delta \to F[[\mathbf{s}]]_\Delta$ is a continuous $k[[\mathbf{s}]]_\Delta$-linear map, let us denote by $f_\alpha : E \to F$, $\alpha \in \Delta$, the k-linear maps defined by

$$f(e) = \sum_{\alpha\in\Delta} f_\alpha(e)\mathbf{s}^\alpha, \quad \forall e \in E.$$

If $g : E \to F[[\mathbf{s}]]_\Delta$ is a k-linear map, we denote by $g^e : E[[\mathbf{s}]]_\Delta \to F[[\mathbf{s}]]_\Delta$ the unique continuous $k[[\mathbf{s}]]_\Delta$-linear map extending g to $E[[\mathbf{s}]]_\Delta = k[[\mathbf{s}]]_\Delta \widehat{\otimes}_k E$. It is given by

$$g^e\left(\sum_\alpha e_\alpha \mathbf{s}^\alpha\right) := \sum_\alpha g(e_\alpha)\mathbf{s}^\alpha.$$

We have a $k[[\mathbf{s}]]_\Delta$-bilinear and $A[[\mathbf{s}]]_\Delta$-balanced map

$$\langle -, - \rangle : (r, e) \in \operatorname{Hom}_k(E, F)[[\mathbf{s}]]_\Delta \times E[[\mathbf{s}]]_\Delta \longmapsto \langle r, e\rangle := \widetilde{r}(e) \in F[[\mathbf{s}]]_\Delta.$$

Lemma 3 *With the above hypotheses, the following properties hold:*

(1) *The map (6) is an isomorphism of* $(A[[\mathbf{s}]]_\Delta; A[[\mathbf{s}]]_\Delta)$*-bimodules. When* $E = F$ *it is an isomorphism of* $k[[\mathbf{s}]]_\Delta$*-algebras over* $A[[\mathbf{s}]]_\Delta$.
(2) *The restriction map*

$$f \in \operatorname{Hom}^{\text{top}}_{k[[\mathbf{s}]]_\Delta}(E[[\mathbf{s}]]_\Delta, F[[\mathbf{s}]]_\Delta) \mapsto f|_E \in \operatorname{Hom}_k(E, F[[\mathbf{s}]]_\Delta)$$

is an isomorphism of $(A[[\mathbf{s}]]_\Delta; A)$*-bimodules.*

Proof

(1) One easily sees that the inverse map of $r \mapsto \widetilde{r}$ is $f \mapsto \sum_\alpha f_\alpha \mathbf{s}^\alpha$.
(2) One easily sees that the inverse map of the restriction map $f \mapsto f|_E$ is $g \mapsto g^e$.

Let us call $R = \mathrm{End}_k(E)$. As a consequence of the above lemma, the composition of the maps

$$R[[\mathbf{s}]]_\Delta \xrightarrow{r \mapsto \widetilde{r}} \mathrm{End}^{\mathrm{top}}_{k[[\mathbf{s}]]_\Delta}(E[[\mathbf{s}]]_\Delta) \xrightarrow{f \mapsto f|_E} \mathrm{Hom}_k(E, E[[\mathbf{s}]]_\Delta) \tag{7}$$

is an isomorphism of $(A[[\mathbf{s}]]_\Delta; A)$-bimodules, and so $\mathrm{Hom}_k(E, E[[\mathbf{s}]]_\Delta)$ inherits a natural structure of $k[[\mathbf{s}]]_\Delta$-algebra over $A[[\mathbf{s}]]_\Delta$. Namely, if $g, h \in \mathrm{Hom}_k(E, E[[\mathbf{s}]]_\Delta)$ with

$$g(e) = \sum_{\alpha \in \Delta} g_\alpha(e)\mathbf{s}^\alpha, \ h(e) = \sum_{\alpha \in \Delta} h_\alpha(e)\mathbf{s}^\alpha, \quad \forall e \in E, \quad g_\alpha, h_\alpha \in \mathrm{Hom}_k(E, E),$$

then the product $hg \in \mathrm{Hom}_k(E, E[[\mathbf{s}]]_\Delta)$ is given by

$$(hg)(e) = \sum_{\alpha \in \Delta} \left(\sum_{\beta+\gamma=\alpha} (h_\beta \circ g_\gamma)(e) \right) \mathbf{s}^\alpha. \tag{8}$$

Definition 4 Let $\mathbf{s}, \mathbf{t}$ be sets and $\Delta \subset \mathbb{N}^{(\mathbf{s})}$, $\nabla \subset \mathbb{N}^{(\mathbf{t})}$ non-empty co-ideals. For each $f \in \mathrm{End}^{\mathrm{top}}_{k[[\mathbf{s}]]_\Delta}(E[[\mathbf{s}]]_\Delta)$ and each $g \in \mathrm{End}^{\mathrm{top}}_{k[[\mathbf{t}]]_\nabla}(E[[\mathbf{t}]]_\nabla)$, with

$$f(e) = \sum_{\alpha \in \Delta} f_\alpha(e)\mathbf{s}^\alpha, \quad g(e) = \sum_{\beta \in \nabla} g_\beta(e)\mathbf{t}^\beta \quad \forall e \in E,$$

we define $f \boxtimes g \in \mathrm{End}^{\mathrm{top}}_{k[[\mathbf{s}\sqcup\mathbf{t}]]_{\Delta\times\nabla}}(E[[\mathbf{s} \sqcup \mathbf{t}]]_{\Delta\times\nabla})$ as $f \boxtimes g := h^e$, with:

$$h(x) := \sum_{(\alpha,\beta)\in\Delta\times\nabla} (f_\alpha \circ g_\beta)(x)\mathbf{s}^\alpha\mathbf{t}^\beta \quad \forall x \in E.$$

The proof of the following lemma is clear and it is left to the reader.

Lemma 4 *With the above hypotheses, or each $r \in R[[\mathbf{s}]]_\Delta, r' \in R[[\mathbf{t}]]_\nabla$, we have $\widetilde{r \boxtimes r'} = \widetilde{r} \boxtimes \widetilde{r'}$ (see Definition 3).*

Lemma 5 *Let us call $R = \mathrm{End}_k(E)$. For any $r \in R[[\mathbf{s}]]_\Delta$, the following properties are equivalent:*

(a) $r_0 = \mathrm{Id}$.
(b) *The endomorphism $\widetilde{r}$ is compatible with the natural augmentation $E[[\mathbf{s}]]_\Delta \to E$, i.e. $\widetilde{r}(e) \equiv e \mod \mathfrak{n}_0^E(\mathbf{s})/\Delta_E$ for all $e \in E[[\mathbf{s}]]_\Delta$.*

Moreover, if the above properties hold, then $\widetilde{r} : E[[\mathbf{s}]]_\Delta \to E[[\mathbf{s}]]_\Delta$ is a bi-continuous $k[[\mathbf{s}]]_\Delta$-linear automorphism.

Proof The equivalence of (a) and (b) is clear. For the second part, r is invertible since $r_0 = \mathrm{Id}$. So $\widetilde{r}$ is invertible too and $\widetilde{r}^{-1} = \widetilde{r^{-1}}$ is also continuous.

Notation 3 *We denote:*

$$\mathrm{Hom}_k^\circ(E, E[[\mathbf{s}]]_\Delta) :=$$
$$\left\{f \in \mathrm{Hom}_k(E, E[[\mathbf{s}]]_\Delta) \mid f(e) \equiv e \bmod \mathfrak{n}_0^E(\mathbf{s})/\Delta_E \quad \forall e \in E\right\},$$

$$\mathrm{Aut}^\circ_{k[[\mathbf{s}]]_\Delta}(E[[\mathbf{s}]]_\Delta) :=$$
$$\left\{f \in \mathrm{Aut}^{\mathrm{top}}_{k[[\mathbf{s}]]_\Delta}(E[[\mathbf{s}]]_\Delta) \mid f(e) \equiv e_0 \bmod \mathfrak{n}_0^E(\mathbf{s})/\Delta_E \quad \forall e \in E[[\mathbf{s}]]_\Delta\right\}.$$

Let us notice that a $f \in \mathrm{Hom}_k(E, E[[\mathbf{s}]]_\Delta)$*, given by* $f(e) = \sum_{\alpha\in\Delta} f_\alpha(e)\mathbf{s}^\alpha$*, belongs to* $\mathrm{Hom}_k^\circ(E, E[[\mathbf{s}]]_\Delta)$ *if and only if* $f_0 = \mathrm{Id}_E$.

The isomorphism in (7) gives rise to a group isomorphism

$$r \in \mathscr{U}^{\mathbf{s}}(\mathrm{End}_k(E); \Delta) \stackrel{\sim}{\longmapsto} \widetilde{r} \in \mathrm{Aut}^\circ_{k[[\mathbf{s}]]_\Delta}(E[[\mathbf{s}]]_\Delta) \tag{9}$$

and to a bijection

$$f \in \mathrm{Aut}^\circ_{k[[\mathbf{s}]]_\Delta}(E[[\mathbf{s}]]_\Delta) \stackrel{\sim}{\longmapsto} f|_E \in \mathrm{Hom}_k^\circ(E, E[[\mathbf{s}]]_\Delta). \tag{10}$$

So, $\mathrm{Hom}_k^\circ(E, E[[\mathbf{s}]]_\Delta)$ is naturally a group with the product described in (8).

3 Substitution Maps

In this section we will assume that k is a commutative ring and A a commutative k-algebra. The following notation will be used extensively.

Notation 4

(i) *For each integer* $r \geq 0$ *let us denote* $[r] := \{1, \ldots, r\}$ *if* $r > 0$ *and* $[0] = \emptyset$.
(ii) *Let* $\mathbf{s}$ *be a set. Maps from a set* Λ *to* $\mathbb{N}^{(\mathbf{s})}$ *will be usually denoted as* $\alpha^\bullet : l \in \Lambda \longmapsto \alpha^l \in \mathbb{N}^{(\mathbf{s})}$*, and its* support *is defined by* $\mathrm{supp}\,\alpha^\bullet := \{l \in \Lambda \mid \alpha^l \neq 0\}$.
(iii) *For each set* Λ *and for each map* $\alpha^\bullet : \Lambda \to \mathbb{N}^{(\mathbf{s})}$ *with finite support, its* norm *is defined by* $|\alpha^\bullet| := \sum_{l\in\mathrm{supp}\,\alpha^\bullet} \alpha^l = \sum_{l\in\Lambda} \alpha^l$. *When* $\Lambda = \emptyset$*, the unique map* $\Lambda \to \mathbb{N}^{(\mathbf{s})}$ *is the inclusion* $\emptyset \hookrightarrow \mathbb{N}^{(\mathbf{s})}$ *and its norm is* $0 \in \mathbb{N}^{(\mathbf{s})}$.
(iv) *If* Λ *is a set and* $e \in \mathbb{N}^{(\mathbf{s})}$*, we define*

$$\mathscr{P}^\circ(e, \Lambda) := \{\alpha^\bullet : \Lambda \to \mathbb{N}^{(\mathbf{s})} \mid \#\,\mathrm{supp}\,\alpha^\bullet < +\infty, |\alpha^\bullet| = e\}.$$

If F *is a finite set and* $e \in \mathbb{N}^{(\mathbf{s})}$*, we define*

$$\mathscr{P}(e, F) := \{\alpha : F \to \mathbb{N}^{(\mathbf{s})}_* \mid |\alpha| = e\} \subset \mathscr{P}^\circ(e, F).$$

It is clear that $\mathscr{P}(e,F)=\emptyset$ *whenever* $\#F>|e|$, $\mathscr{P}^\circ(e,\emptyset)=\emptyset$ *if* $e\neq 0$, $\mathscr{P}^\circ(0,\Lambda)$ *consists of only the constant map* 0 *and that* $\mathscr{P}(0,\emptyset)=\mathscr{P}^\circ(0,\emptyset)$ *consists of only the inclusion* $\emptyset\hookrightarrow\mathbb{N}_*^{(\mathbf{s})}$. *If* $\#F=1$ *and* $e\neq 0$, *then* $\mathscr{P}(e,F)$ *also consists of only one map: the constant map with value* e.

The natural map $\coprod_{F\in\mathfrak{P}_f(\Lambda)}\mathscr{P}(e,F)\longrightarrow\mathscr{P}^\circ(e,\Lambda)$ *is obviously a bijection.*

If $r>0$ *is an integer we will denote* $\mathscr{P}(e,r):=\mathscr{P}(e,[r])$.

(v) *Assume that* Λ *is a finite set,* $\mathbf{t}$ *is an arbitrary set and* $\pi:\Lambda\to\mathbf{t}$ *is map. Then, there is a natural bijection*

$$\mathscr{P}^\circ(e,\Lambda)\leftrightarrow\coprod_{e^\bullet\in\mathscr{P}^\circ(e,\mathbf{t})}\prod_{t\in\mathbf{t}}\mathscr{P}^\circ(e^t,\pi^{-1}(t))=\coprod_{e^\bullet\in\mathscr{P}^\circ(e,\mathbf{t})}\prod_{t\in\operatorname{supp}e^\bullet}\mathscr{P}^\circ(e^t,\pi^{-1}(t)).$$

Namely, to each $\alpha^\bullet\in\mathscr{P}^\circ(e,\Lambda)$ *we associate* $e^\bullet\in\mathscr{P}^\circ(e,\mathbf{t})$ *defined by* $e^t=\sum_{\pi(l)=t}\alpha^l$, *and* $\{\alpha^{t\bullet}\}_{t\in\mathbf{t}}\in\prod_{t\in\mathbf{t}}\mathscr{P}^\circ(e^t,\pi^{-1}(t))$ *with* $\alpha^{t\bullet}=\alpha^\bullet|_{\pi^{-1}(t)}$. *Let us notice that if for some* $t_0\in\mathbf{t}$ *one has* $\pi^{-1}(t_0)=\emptyset$ *and* $e^{t_0}\neq 0$, *then* $\mathscr{P}^\circ(e^{t_0},\pi^{-1}(t_0))=\emptyset$ *and so* $\prod_{t\in\mathbf{t}}\mathscr{P}^\circ(e^t,\pi^{-1}(t))=\emptyset$. *Hence*

$$\coprod_{e^\bullet\in\mathscr{P}^\circ(e,\mathbf{t})}\prod_{t\in\mathbf{t}}\mathscr{P}^\circ(e^t,\Lambda_t)=\coprod_{e^\bullet\in\mathscr{P}^\circ_\pi(e,\mathbf{t})}\prod_{t\in\mathbf{t}}\mathscr{P}^\circ(e^t,\pi^{-1}(t))=$$
$$\coprod_{e^\bullet\in\mathscr{P}^\circ_\pi(e,\mathbf{t})}\prod_{t\in\operatorname{supp}e^\bullet}\mathscr{P}^\circ(e^t,\pi^{-1}(t)),$$

where $\mathscr{P}^\circ_\pi(e,\mathbf{t})$ *is the subset of* $\mathscr{P}^\circ(e,\mathbf{t})$ *whose elements are the* $e^\bullet\in\mathscr{P}^\circ(e,\mathbf{t})$ *such that* $e^t=0$ *whenever* $\pi^{-1}(t)=\emptyset$ *and* $|e^t|\geq\#\pi^{-1}(t)$ *otherwise.*

The preceding bijection induces a bijection

$$\mathscr{P}(e,\Lambda)\longleftrightarrow\coprod_{e^\bullet\in\mathscr{P}^\circ_\pi(e,\mathbf{t})}\prod_{t\in\mathbf{t}}\mathscr{P}(e^t,\pi^{-1}(t))=\coprod_{e^\bullet\in\mathscr{P}^\circ_\pi(e,\mathbf{t})}\prod_{t\in\operatorname{supp}e^\bullet}\mathscr{P}(e^t,\pi^{-1}(t)).\tag{11}$$

(vi) *If* $\alpha\in\mathbb{N}^{(\mathbf{t})}$, *we denote*

$$[\alpha]:=\{(t,r)\in\mathbf{t}\times\mathbb{N}_*\mid 1\leq r\leq\alpha_t\}$$

endowed with the projection $\pi:[\alpha]\to\mathbf{t}$. *It is clear that* $|\alpha|=\#[\alpha]$, *and so* $\alpha=0\Longleftrightarrow[\alpha]=\emptyset$. *We denote* $\mathscr{P}(e,\alpha):=\mathscr{P}(e,[\alpha])$. *Elements in* $\mathscr{P}(e,\alpha)$ *will be written as*

$$\mathscr{E}^{\bullet\bullet}:(t,r)\in[\alpha]\longmapsto\mathscr{E}^{tr}\in\mathbb{N}^{(\mathbf{s})},\quad\text{with}\ \sum_{(t,r)\in[\alpha]}\mathscr{E}^{tr}=e.$$

For each $\mathscr{E}^{\bullet\bullet} \in \mathscr{P}(e,\alpha)$ and each $t \in \mathbf{t}$, we denote

$$\mathscr{E}^{t\bullet} : r \in [\alpha_t] \longmapsto \mathscr{E}^{tr} \in \mathbb{N}^{(\mathbf{s})}, \quad [\mathscr{E}]^\bullet : t \in \mathbf{t} \longmapsto [\mathscr{E}]^t := |\mathscr{E}^{t\bullet}| = \sum_{r=1}^{\alpha_t} \mathscr{E}^{tr} \in \mathbb{N}^{(\mathbf{s})}.$$

Notice that $|[\mathscr{E}]^t| \geq \alpha_t$, $[\mathscr{E}]^t = 0$ whenever $\alpha_t = 0$ and $|[\mathscr{E}]^\bullet| = e$. The bijection (11) gives rise to a bijection

$$\mathscr{P}(e,\alpha) \longleftrightarrow \coprod_{e^\bullet \in \mathscr{P}^\circ_\alpha(e,\mathbf{t})} \prod_{t\in\mathbf{t}} \mathscr{P}(e^t,\alpha_t) = \coprod_{e^\bullet \in \mathscr{P}^\circ_\alpha(e,\mathbf{t})} \prod_{t\in\operatorname{supp} e^\bullet} \mathscr{P}(e^t,\alpha_t), \tag{12}$$

where $\mathscr{P}^\circ_\alpha(e,\mathbf{t})$ is the subset of $\mathscr{P}^\circ(e,\mathbf{t})$ whose elements are the $e^\bullet \in \mathscr{P}^\circ(e,\mathbf{t})$ such that $e^t = 0$ if $\alpha_t = 0$ and $|e^t| \geq \alpha_t$ otherwise.

2. Let $\mathbf{t}$, $\mathbf{u}$ be sets and $\Delta \subset \mathbb{N}^{(\mathbf{u})}$ a non-empty co-ideal. Let $\varphi_0 : A[\mathbf{t}] \to A[[\mathbf{u}]]_\Delta$ be an A-algebra map given by:

$$\varphi_0(t) =: c^t = \sum_{\substack{\beta\in\Delta\\ 0<|\beta|}} c^t_\beta \mathbf{u}^\beta \in \mathfrak{n}^A_0(\mathbf{u})/\Delta_A \subset A[[\mathbf{u}]]_\Delta, \quad t \in \mathbf{t}.$$

Let us write down the expression of the image $\varphi_0(a)$ of any $a \in A[\mathbf{t}]$ in terms of the coefficients of a and the c^t, $t \in \mathbf{t}$. First, for each $r \geq 0$ and for each $t \in \mathbf{t}$ we have

$$\varphi_0(t^r) = (c^t)^r = \cdots = \sum_{\substack{e\in\Delta\\ |e|\geq r}} \left(\sum_{\beta^\bullet \in \mathscr{P}(e,r)} \prod_{k=1}^r c^t_{\beta^k} \right) \mathbf{u}^e.$$

Observe that

$$\sum_{\beta^\bullet \in \mathscr{P}(e,r)} \prod_{k=1}^r c^t_{\beta^k} = \begin{cases} 1 \text{ if } |e| = r = 0 \\ 0 \text{ if } |e| > r = 0. \end{cases} \tag{13}$$

So, for each $\alpha \in \mathbb{N}^{(\mathbf{t})}$ we have

$$\varphi_0(\mathbf{t}^\alpha) = \prod_{t\in\mathbf{t}} (c^t)^{\alpha_t} = \prod_{t\in\operatorname{supp}\alpha} (c^t)^{\alpha_t} = \prod_{t\in\operatorname{supp}\alpha} \left(\sum_{\substack{e\in\Delta\\ |e|\geq\alpha_t}} \left(\sum_{\beta^\bullet \in \mathscr{P}(e,\alpha_t)} \prod_{k=1}^{\alpha_t} c^t_{\beta^k} \right) \mathbf{u}^e \right) =$$

$$\sum_{\substack{e^t\in\Delta, t\in\operatorname{supp}\alpha\\ |e^t|\geq\alpha_t}} \prod_{t\in\operatorname{supp}\alpha} \left(\left(\sum_{\beta^\bullet \in \mathscr{P}(e^t,\alpha_t)} \prod_{k=1}^{\alpha_t} c^t_{\beta^k} \right) \mathbf{u}^{e^t} \right) =$$

$$\sum_{\substack{e^t \in \Delta, t \in \operatorname{supp} \alpha \\ |e^t| \geq \alpha_t}} \left(\sum_{\substack{\beta^{t\bullet} \in \mathscr{P}(e^t, \alpha_t) \\ t \in \operatorname{supp} \alpha}} \left(\prod_{t \in \operatorname{supp} \alpha} \prod_{k=1}^{\alpha_t} c^t_{\beta^{tk}} \right) \right) \left(\prod_{t \in \operatorname{supp} \alpha} \mathbf{u}^{e^t} \right) =$$

$$\sum_{\substack{e \in \Delta \\ |e| \geq |\alpha|}} \left(\sum_{\substack{e^t \in \Delta, t \in \operatorname{supp} \alpha \\ |e^t| \geq \alpha_t \\ |e^\bullet| = e}} \left(\sum_{\substack{\beta^{t\bullet} \in \mathscr{P}(e^t, \alpha_t) \\ t \in \operatorname{supp} \alpha}} \left(\prod_{t \in \operatorname{supp} \alpha} \prod_{k=1}^{\alpha_t} c^t_{\beta^{tk}} \right) \right) \right) \mathbf{u}^e =$$

$$\sum_{\substack{e \in \Delta \\ |e| \geq |\alpha|}} \left(\sum_{e^\bullet \in \mathscr{P}^\circ_\alpha(e, \mathbf{t})} \left(\sum_{\substack{\beta^{t\bullet} \in \mathscr{P}(e^t, \alpha_t) \\ t \in \operatorname{supp} \alpha}} \left(\prod_{t \in \operatorname{supp} \alpha} \prod_{k=1}^{\alpha_t} c^t_{\beta^{tk}} \right) \right) \right) \mathbf{u}^e = \sum_{\substack{e \in \Delta \\ |e| \geq |\alpha|}} \mathbf{C}_e(\varphi_0, \alpha) \mathbf{u}^e,$$

with (see (12)):

$$\mathbf{C}_e(\varphi_0, \alpha) = \sum_{\beta^{\bullet\bullet} \in \mathscr{P}(e, \alpha)} C_{\beta^{\bullet\bullet}}, \quad C_{\beta^{\bullet\bullet}} = \prod_{t \in \operatorname{supp} \alpha} \prod_{r=1}^{\alpha_t} c^t_{\beta^{tr}}, \quad \text{for } |\alpha| \leq |e|. \tag{14}$$

We have $\mathbf{C}_0(\varphi_0, 0) = 1$ and $\mathbf{C}_e(\varphi_0, 0) = 0$ for $e \neq 0$. For a fixed $e \in \mathbb{N}^{(\mathbf{u})}$ the support of any $\alpha \in \mathbb{N}^{(\mathbf{t})}$ such that $|\alpha| \leq |e|$ and $\mathbf{C}_e(\varphi_0, \alpha) \neq 0$ is contained in the set

$$\bigcup_{\substack{\beta \in \Delta \\ \beta \leq e}} \{t \in \mathbf{t} \mid c^t_\beta \neq 0\}$$

and so the set of such α's is finite provided that property (17) holds. We conclude that

$$\varphi_0 \left(\sum_{\alpha \in \mathbb{N}^{(\mathbf{t})}} a_\alpha \mathbf{t}^\alpha \right) = \sum_{\alpha \in \mathbb{N}^{(\mathbf{t})}} a_\alpha c^\alpha = \sum_{e \in \Delta} \left(\sum_{\substack{\alpha \in \mathbb{N}^{(\mathbf{t})} \\ |\alpha| \leq |e|}} \mathbf{C}_e(\varphi_0, \alpha) a_\alpha \right) \mathbf{u}^e. \tag{15}$$

Observe that for each non-zero $\alpha \in \mathbb{N}^{(\mathbf{t})}$ we have:

$$\operatorname{supp}(\varphi_0(\mathbf{t}^\alpha)) = \operatorname{supp} \left(\prod_{t \in \operatorname{supp} \alpha} (c^t)^{\alpha_t} \right) \subset \sum_{t \in \operatorname{supp}(\alpha)} \alpha_t \cdot \operatorname{supp}(c^t). \tag{16}$$

Let us notice that if we assign the weight $|\beta|$ to c^t_β, then $\mathbf{C}_e(\varphi_0, \alpha)$ is a quasi-homogeneous polynomial in the variables c^t_β, $t \in \operatorname{supp} \alpha$, $|\beta| \leq |e|$, of weight $|e|$.

The proof of the following lemma is easy and it is left to the reader.

Lemma 6 *For each $e \in \Delta$ and for each $\alpha \in \mathbb{N}^{(\mathbf{t})}$ with $0 < |\alpha| \leq |e|$, the following properties hold:*

(1) *If $|\alpha| = 1$, then $\mathbf{C}_e(\varphi_0, \alpha) = c^s_e$, where $\operatorname{supp} \alpha = \{s\}$, i.e. $\alpha = \mathbf{t}^s$ ($\mathbf{t}^s_t = \delta_{st}$).*
(2) *If $|\alpha| = |e|$, then*

$$\mathbf{C}_e(\varphi_0, \alpha) = \sum_{\substack{e^t \in \Delta, t \in \operatorname{supp} \alpha \\ |e^t| = \alpha_t, |e^\bullet| = e}} \left(\prod_{t \in \operatorname{supp} \alpha} \prod_{v \in \operatorname{supp} e^t} \left(c^t_{\mathbf{u}^v}\right)^{e^t_v} \right).$$

Proposition 2 *Let $\mathbf{t}, \mathbf{u}$ be sets and $\Delta \subset \mathbb{N}^{(\mathbf{u})}$ a non-empty co-ideal. For each family*

$$c = \left\{ c^t = \sum_{\substack{\beta \in \Delta \\ \beta \neq 0}} c^t_\beta \mathbf{u}^\beta \in \mathfrak{n}^A_0(\mathbf{u})/\Delta_A \subset A[[\mathbf{u}]]_\Delta,\ t \in \mathbf{t} \right\}$$

(we are assuming that $c^t_0 = 0$) satisfying the following property

$$\#\{t \in \mathbf{t} \mid c^t_\beta \neq 0\} < \infty \qquad \text{for all } \beta \in \Delta, \tag{17}$$

there is a unique continuous A-algebra map $\varphi : A[[\mathbf{t}]] \to A[[\mathbf{u}]]_\Delta$ such that $\varphi(t) = c^t$ for all $t \in \mathbf{t}$. Moreover, if $\nabla \subset \mathbb{N}^{(\mathbf{t})}$ is a non-empty co-ideal such that $\varphi(\nabla_A) = 0$, then φ induces a unique continuous A-algebra map $A[[\mathbf{t}]]_\nabla \to A[[\mathbf{u}]]_\Delta$ sending (the class of) each $t \in \mathbf{t}$ to c^t.

Proof Let us consider the unique A-algebra map $\varphi_0 : A[\mathbf{t}] \to A[[\mathbf{u}]]_\Delta$ defined by $\varphi_0(t) = c^t$ for all $t \in \mathbf{t}$. From (14) and (15) in 2, we know that

$$\varphi_0 \left(\sum_{\substack{\alpha \in \mathbb{N}^{(\mathbf{t})} \\ \text{finite}}} a_\alpha \mathbf{t}^\alpha \right) = \sum_{e \in \Delta} \left(\sum_{\substack{\alpha \in \mathbb{N}^{(\mathbf{t})} \\ |\alpha| \leq |e|}} \mathbf{C}_e(\varphi_0, \alpha) a_\alpha \right) \mathbf{u}^e.$$

Since for a fixed $e \in \mathbb{N}^{(\mathbf{u})}$ the support of the $\alpha \in \mathbb{N}^{(\mathbf{t})}$ such that $|\alpha| \leq |e|$ and $\mathbf{C}_e(\varphi_0, \alpha) \neq 0$ is contained in the finite set

$$\bigcup_{\substack{\beta \in \Delta \\ \beta \leq e}} \{t \in \mathbf{t} \mid c^t_\beta \neq 0\},$$

the set of such α's is always finite and we deduce that φ_0 is continuous, and so there is a unique continuous extension $\varphi : A[[\mathbf{t}]] \to A[[\mathbf{u}]]_\Delta$ such that $\varphi(t) = c^t$ for all $t \in \mathbf{t}$.

The last part is clear.

Remark 2 Let us notice that, after (16), to get the equality $\varphi(\nabla_A) = 0$ in the above proposition it is enough to have for each $\alpha \in \nabla^c$ (actually, it will be enough to consider the $\alpha \in \nabla^c$ minimal with respect to the ordering $\leq$ in $\mathbb{N}^{(\mathbf{t})}$):

$$\sum_{t \in \operatorname{supp}(\alpha)} \alpha_t \cdot \operatorname{supp}(c^t) \subset \Delta^c.$$

Definition 5 Let $\nabla \subset \mathbb{N}^{(\mathbf{t})}$, $\Delta \subset \mathbb{N}^{(\mathbf{u})}$ be non-empty co-ideals. An A-algebra map $\varphi : A[[\mathbf{t}]]_\nabla \to A[[\mathbf{u}]]_\Delta$ will be called a *substitution map* if the following properties hold:

(1) φ is continuous.
(2) $\varphi(t) \in \mathfrak{n}_0^A(\mathbf{u})/\Delta_A$ for all $t \in \mathbf{t}$.
(3) The family $c = \{\varphi(t), t \in \mathbf{t}\}$ satisfies property (17).

The set of substitution maps $A[[\mathbf{t}]]_\nabla \to A[[\mathbf{u}]]_\Delta$ will be denoted by $\mathcal{S}_A(\mathbf{t}, \mathbf{u}; \nabla, \Delta)$. The *trivial* substitution map $A[[\mathbf{t}]]_\nabla \to A[[\mathbf{u}]]_\Delta$ is the one sending any $t \in \mathbf{t}$ to 0. It will be denoted by $\mathbf{0}$.

Remark 3 In the above definition, a such φ is uniquely determined by the family $c = \{\varphi(t), t \in \mathbf{t}\}$, and will be called the *substitution map associated* with c. Namely, the family c can be lifted to $A[[\mathbf{u}]]$ by means of the natural A-linear scission $A[[\mathbf{u}]]_\Delta \hookrightarrow A[[\mathbf{u}]]$ and we may consider the unique continuous A-algebra map $\psi : A[[\mathbf{t}]] \to A[[\mathbf{u}]]$ such that $\psi(s) = c^s$ for all $s \in \mathbf{s}$. Since φ is continuous, we have a commutative diagram

$$\begin{array}{ccc} A[[\mathbf{t}]] & \xrightarrow{\psi} & A[[\mathbf{u}]] \\ \text{proj.}\downarrow & & \downarrow\text{proj.} \\ A[[\mathbf{t}]]_\nabla & \xrightarrow{\varphi} & A[[\mathbf{u}]]_\Delta, \end{array}$$

and so $\psi(\nabla_A) \subset \Delta_A$. Then, we may identify

$$\mathcal{S}_A(\mathbf{t}, \mathbf{u}; \nabla, \Delta) \equiv \left\{ \overline{\psi} \in \mathcal{S}_A(\mathbf{t}, \mathbf{u}; \mathbb{N}^{(\mathbf{t})}, \Delta) \mid \overline{\psi}(\nabla_A) = 0 \right\}.$$

For $\alpha \in \nabla$ and $e \in \Delta$ with $|\alpha| \leq |e|$ we will write $\mathbf{C}_e(\varphi, \alpha) := \mathbf{C}_e(\varphi_0, \alpha)$, where $\varphi_0 : A[\mathbf{t}] \to A[[\mathbf{u}]]_\Delta$ is the A-algebra map given by $\varphi_0(t) = \varphi(t)$ for all $t \in \mathbf{t}$ (see (14) in 2).

Example 1 For any family of integers $\nu = \{\nu_t \geq 1, t \in \mathbf{t}\}$, we will denote $[\nu] : A[[\mathbf{t}]]_\nabla \to A[[\mathbf{t}]]_{\nu\nabla}$ the substitution map determined by $[\nu](t) = t^{\nu_t}$ for all $t \in \mathbf{t}$, where

$$\nu\nabla := \{\gamma \in \mathbb{N}^{(\mathbf{t})} \mid \exists \alpha \in \nabla, \gamma \leq \nu\alpha\}.$$

We obviously have $[\nu\nu'] = [\nu]\circ[\nu']$.

Lemma 7 *The composition of two substitution maps* $A[[\mathbf{t}]]_\nabla \xrightarrow{\varphi} A[[\mathbf{u}]]_\Delta \xrightarrow{\psi} A[[\mathbf{s}]]_\Omega$ *is a substitution map and we have*

$$\mathbf{C}_f(\psi\circ\varphi,\alpha) = \sum_{\substack{e\in\Delta\\ |f|\geq|e|\geq|\alpha|}} \mathbf{C}_e(\varphi,\alpha)\mathbf{C}_f(\psi,e), \quad \forall f\in\Omega, \forall\alpha\in\nabla, |\alpha|\leq|f|.$$

Moreover, if one of the substitution maps is trivial, then the composition is trivial too.

Proof Properties (1) and (2) in Definition 5 are clear. Let us see property (3). For each $t\in\mathbf{t}$ let us write:

$$\varphi(t) =: c^t = \sum_{\substack{\beta\in\Delta\\ 0<|\beta|}} c^t_\beta \mathbf{u}^\beta \in \mathfrak{n}_0^A(\mathbf{u})/\Delta_A \subset A[[\mathbf{u}]]_\Delta,$$

and so

$$(\psi\circ\varphi)(t) = \psi\left(\sum_{\substack{\beta\in\Delta\\ 0<|\beta|}} c^t_\beta \mathbf{u}^\beta\right) = \sum_{\substack{\beta\in\Delta\\ 0<|\beta|}} c^t_\beta \left(\sum_{\substack{f\in\Omega\\ |f|\geq|\beta|}} \mathbf{C}_f(\psi,\beta)\mathbf{s}^f\right) = \sum_{\substack{f\in\Omega\\ |f|>0}} d^t_f \mathbf{s}^f$$

with

$$d^t_f = \sum_{\substack{\beta\in\Delta\\ 0<|\beta|\leq|f|}} c^t_\beta \mathbf{C}_f(\psi,\beta)$$

and for a fixed $f\in\Omega$ the set

$$\{t\in\mathbf{t}\mid d^t_f\neq 0\} \subset \bigcup_{\substack{\beta\in\nabla,|\beta|\leq|f|\\ \mathbf{C}_f(\psi,\beta)\neq 0}} \{t\in\mathbf{t}\mid c^t_\beta\neq 0\}$$

is finite. On the other hand

$$(\psi\circ\varphi)(\mathbf{t}^\alpha) = \psi\left(\sum_{\substack{e\in\Delta\\ |e|\geq|\alpha|}} \mathbf{C}_e(\varphi,\alpha)\mathbf{u}^e\right) = \sum_{\substack{e\in\Delta\\ |e|\geq|\alpha|}} \mathbf{C}_e(\varphi,\alpha)\left(\sum_{\substack{f\in\Omega\\ |f|\geq|e|}} \mathbf{C}_f(\psi,e)\mathbf{s}^f\right) =$$

$$\sum_{\substack{f\in\Omega\\ |f|\geq|\alpha|}} \left(\sum_{\substack{e\in\Delta\\ |f|\geq|e|\geq|\alpha|}} \mathbf{C}_e(\varphi,\alpha)\mathbf{C}_f(\psi,e)\right)\mathbf{u}^f$$

and so

$$\mathbf{C}_f(\psi\circ\varphi,\alpha)=\sum_{\substack{e\in\Delta\\ |f|\geq|e|\geq|\alpha|}}\mathbf{C}_e(\varphi,\alpha)\mathbf{C}_f(\psi,e),\quad \forall f\in\Omega,\forall\alpha\in\nabla,|\alpha|\leq|f|.$$

If B is a commutative A-algebra, then any substitution map $\varphi : A[[\mathbf{s}]]_\nabla \to A[[\mathbf{t}]]_\Delta$ induces a natural substitution map $\varphi_B : B[[\mathbf{s}]]_\nabla \to B[[\mathbf{t}]]_\Delta$ making the following diagram commutative

$$\begin{array}{ccc}
B\widehat{\otimes}_A A[[\mathbf{s}]]_\nabla & \xrightarrow{\mathrm{Id}\widehat{\otimes}\varphi} & B\widehat{\otimes}_A A[[\mathbf{t}]]_\Delta\\
\text{nat.}\downarrow\simeq & & \simeq\downarrow\text{nat.}\\
B[[\mathbf{s}]]_\nabla & \xrightarrow{\varphi_B} & B[[\mathbf{t}]]_\Delta.
\end{array}$$

3. For any substitution map $\varphi : A[[\mathbf{s}]]_\nabla \to A[[\mathbf{t}]]_\Delta$ and for any integer $n\geq 0$ we have $\varphi(\nabla^n_A/\nabla_A)\subset\Delta^n_A/\Delta_A$ and so there are induced substitution maps $\tau_n(\varphi) : A[[\mathbf{s}]]_{\nabla^n} \to A[[\mathbf{t}]]_{\Delta^n}$ making commutative the following diagram

$$\begin{array}{ccc}
A[[\mathbf{s}]]_\nabla & \xrightarrow{\varphi} & A[[\mathbf{t}]]_\Delta\\
\text{nat.}\downarrow & & \downarrow\text{nat.}\\
A[[\mathbf{s}]]_{\nabla^n} & \xrightarrow{\tau_n(\varphi)} & A[[\mathbf{t}]]_{\Delta^n}.
\end{array}$$

Moreover, if φ is the substitution map associated with a family $c=\{c^s, s\in\mathbf{s}\}$,

$$c^s=\sum_{\beta\in\Delta}c^s_\beta\mathbf{t}^\beta\in\mathfrak{n}^A_0(\mathbf{t})/\Delta_A\subset A[[\mathbf{t}]]_\Delta,$$

then $\tau_n(\varphi)$ is the substitution map associated with the family $\tau_n(c)=\{\tau_n(c)^s, s\in\mathbf{s}\}$, with

$$\tau_n(c)^s:=\sum_{\substack{\beta\in\Delta\\ |\beta|\leq n}}c^s_\beta\mathbf{t}^\beta\in\mathfrak{n}^A_0(\mathbf{t})/\Delta^n_A\subset A[[\mathbf{t}]]_{\Delta^n}.$$

So, we have truncations $\tau_n : \mathcal{S}_A(\mathbf{s},\mathbf{t};\nabla,\Delta)\longrightarrow\mathcal{S}_A(\mathbf{s},\mathbf{t};\nabla^n,\Delta^n)$, for $n\geq 0$.

We may also add two substitution maps $\varphi,\varphi' : A[[\mathbf{s}]]\to A[[\mathbf{t}]]_\Delta$ to obtain a new substitution map $\varphi+\varphi' : A[[\mathbf{s}]]\to A[[\mathbf{t}]]_\Delta$ determined by[1]:

$$(\varphi+\varphi')(s)=\varphi(s)+\varphi'(s),\quad \text{for all } s\in\mathbf{s}.$$

[1]Pay attention that $(\varphi+\varphi')(r)\neq\varphi(r)+\varphi'(r)$ for arbitrary $r\in A[[\mathbf{s}]]_\nabla$.

It is clear that $\mathcal{S}_A(\mathbf{s}, \mathbf{t}; \mathbb{N}^{(\mathbf{s})}, \Delta)$ becomes an abelian group with the addition, the zero element being the trivial substitution map $\mathbf{0}$.

If $\psi : A[[\mathbf{t}]]_\Delta \to A[[\mathbf{u}]]_\Omega$ is another substitution map, we clearly have

$$\psi \circ (\varphi + \varphi') = \psi \circ \varphi + \psi \circ \varphi'.$$

However, if $\psi : A[[\mathbf{u}]] \to A[[\mathbf{s}]]$ is a substitution map, we have in general

$$(\varphi + \varphi') \circ \psi \neq \varphi \circ \psi + \varphi' \circ \psi.$$

Definition 6 We say that a substitution map $\varphi : A[[\mathbf{t}]]_\nabla \to A[[\mathbf{u}]]_\Delta$ has *constant coefficients* if $c^t_\beta \in k$ for all $t \in \mathbf{t}$ and all $\beta \in \Delta$, where

$$\varphi(t) = c^t = \sum_{\substack{\beta\in\Delta \\ 0<|\beta|}} c^t_\beta \mathbf{u}^\beta \in \mathfrak{n}^A_0(\mathbf{u})/\Delta_A \subset A[[\mathbf{u}]]_\Delta.$$

This is equivalent to saying that $\mathbf{C}_e(\varphi, \alpha) \in k$ for all $e \in \Delta$ and for all $\alpha \in \nabla$ with $0 < |\alpha| \leq |e|$. Substitution maps which constant coefficients are induced by substitution maps $k[[\mathbf{t}]]_\nabla \to k[[\mathbf{u}]]_\Delta$.

We say that a substitution map $\varphi : A[[\mathbf{t}]]_\nabla \to A[[\mathbf{u}]]_\Delta$ is *combinatorial* if $\varphi(t) \in \mathbf{u}$ for all $t \in \mathbf{t}$. A combinatorial substitution map has constant coefficients and is determined by (and determines) a map $\mathbf{t} \to \mathbf{u}$, necessarily with finite fibers. If $\iota : \mathbf{t} \to \mathbf{u}$ is such a map, we will also denote by $\iota : A[[\mathbf{t}]]_\nabla \to A[[\mathbf{u}]]_{\iota_*(\nabla)}$ the corresponding substitution map, with

$$\iota_*(\nabla) := \{\beta \in \mathbb{N}^{(\mathbf{u})} \mid \beta \circ \iota \in \nabla\}.$$

4. Let $\varphi : A[[\mathbf{s}]]_\nabla \to A[[\mathbf{t}]]_\Delta$ be a continuous A-linear map. It is determined by the family $K = \{K_{e,\alpha}, e \in \Delta, \alpha \in \nabla\} \subset A$, with $\varphi(\mathbf{s}^\alpha) = \sum_{e\in\Delta} K_{e,\alpha}\mathbf{t}^e$. We will assume that

- φ is compatible with the order filtration, i.e. $\varphi(\nabla^n_A/\nabla_A) \subset \Delta^n_A/\Delta_A$ for all $n \geq 0$.
- φ is compatible with the natural augmentations $A[[\mathbf{s}]]_\nabla \to A$ and $A[[\mathbf{t}]]_\Delta \to A$.

These properties are equivalent to the fact that $K_{e,\alpha} = 0$ whenever $|\alpha| > |e|$ and $K_{0,0} = 1$.

Let $K = \{K_{e,\alpha}, e \in \Delta, \alpha \in \nabla, |\alpha| \leq |e|\}$ be a family of elements of A with

$$\#\{\alpha \in \nabla \mid |\alpha| \leq |e|, K_{e,\alpha} \neq 0\} < +\infty, \quad \forall e \in \Delta,$$

and $K_{0,0} = 1$, and let $\varphi : A[[\mathbf{s}]]_\nabla \to A[[\mathbf{t}]]_\Delta$ be the A-linear map given by

$$\varphi\left(\sum_{\alpha\in\nabla} a_\alpha \mathbf{s}^\alpha\right) = \sum_{e\in\Delta}\left(\sum_{\substack{\alpha\in\nabla\\|\alpha|\le|e|}} K_{e,\alpha} a_\alpha\right)\mathbf{t}^e.$$

It is clearly continuous and since $\varphi(\mathbf{s}^\alpha) = \sum_{\substack{e\in\Delta\\|\alpha|\le|e|}} K_{e,\alpha}\mathbf{t}^e$, it determines the family K.

Proposition 3 *With the above notations, the following properties are equivalent:*

(a) *φ is a substitution map.*
(b) *For each $\mu, \nu \in \nabla$ and for each $e \in \Delta$ with $|\mu+\nu| \le |e|$, the following equality holds:*

$$K_{e,\mu+\nu} = \sum_{\substack{\beta+\gamma=e\\|\mu|\le|\beta|,|\nu|\le|\gamma|}} K_{\beta,\mu}K_{\gamma,\nu}.$$

Moreover, if the above equality holds, then $K_{e,0} = 0$ whenever $|e| > 0$ and φ is the substitution map determined by

$$\varphi(u) = \sum_{\substack{e\in\Delta\\0<|e|}} K_{e,\mathbf{s}^u}\mathbf{t}^e, \quad u \in \mathbf{s}.$$

Proof (a) $\Rightarrow$ (b) If φ is a substitution map, there is a family

$$c^s = \sum_{\beta\in\Delta} c^s_\beta \mathbf{t}^\beta \in A[[\mathbf{t}]]_\Delta, \quad s \in \mathbf{s},$$

such that $\varphi(s) = c^s$. So, from (15), we deduce

$$K_{e,\alpha} = \mathbf{C}_e(\varphi,\alpha) = \sum_{f^{\bullet\bullet}\in\mathscr{P}(e,\alpha)} C_{f^{\bullet\bullet}} \quad \text{for } |\alpha| \le |e|,$$

with $C_{f^{\bullet\bullet}} = \prod_{s\in\operatorname{supp}\alpha}\prod_{r=1}^{\alpha_s} c^s_{f^{sr}}$.

For each ordered pair (r, s) of non-negative integers there are natural injective maps

$$i \in [r] \mapsto i \in [r+s], \quad i \in [s] \mapsto r+i \in [r+s]$$

inducing a natural bijection $[r] \sqcup [s] \longleftrightarrow [r+s]$. Consequently, for $(\mu,\nu) \in \mathbb{N}^{(\mathbf{s})} \times \mathbb{N}^{(\mathbf{s})}$ there are natural injective maps $[\mu] \hookrightarrow [\mu+\nu] \hookleftarrow [\nu]$ inducing a natural bijection $[\mu] \sqcup [\nu] \longleftrightarrow [\mu + \nu]$. So, for each $e \in \mathbb{N}^{(\mathbf{t})}$ and each $\mathscr{f}^{\bullet\bullet} \in \mathscr{P}(e, \mu + \nu)$, we can consider the restrictions $\mathscr{g}^{\bullet\bullet} = \mathscr{f}^{\bullet\bullet}|_{[\mu]} \in \mathscr{P}(\beta, \mu)$, $\mathscr{h}^{\bullet\bullet} = \mathscr{f}^{\bullet\bullet}|_{[\nu]} \in \mathscr{P}(\gamma, \nu)$, with $\beta = |\mathscr{g}^{\bullet\bullet}|$ and $\gamma = |\mathscr{h}^{\bullet\bullet}|$, $\beta + \gamma = e$. The correspondence $\mathscr{f}^{\bullet\bullet} \longmapsto (\beta, \gamma, \mathscr{g}^{\bullet\bullet}, \mathscr{h}^{\bullet\bullet})$ establishes a bijection between $\mathscr{P}(e, \mu + \nu)$ and the set of $(\beta, \gamma, \mathscr{g}^{\bullet\bullet}, \mathscr{h}^{\bullet\bullet})$ with $\beta, \gamma \in \mathbb{N}^{(\mathbf{t})}$, $\mathscr{g}^{\bullet\bullet} \in \mathscr{P}(\beta, \mu)$, $\mathscr{h}^{\bullet\bullet} \in \mathscr{P}(\gamma, \nu)$ and $|\beta| \geq |\mu|$, $|\gamma| \geq |\nu|$, $\beta + \gamma = e$. Moreover, under this bijection we have $C_{\mathscr{f}^{\bullet\bullet}} = C_{\mathscr{g}^{\bullet\bullet}} C_{\mathscr{h}^{\bullet\bullet}}$ and we deduce

$$K_{e,\mu+\nu} = \mathbf{C}_e(\varphi, \mu + \nu) = \sum_{\mathscr{f}^{\bullet\bullet}} C_{\mathscr{f}^{\bullet\bullet}} = \sum_{\substack{\beta+\gamma=e \\ |\mu|\leq|\beta| \\ |\nu|\leq|\gamma|}} \sum_{\mathscr{g}^{\bullet\bullet},\mathscr{h}^{\bullet\bullet}} C_{\mathscr{g}^{\bullet\bullet}} C_{\mathscr{h}^{\bullet\bullet}} =$$

$$\sum_{\substack{\beta+\gamma=e \\ |\mu|\leq|\beta| \\ |\nu|\leq|\gamma|}} \left(\sum_{\mathscr{g}^{\bullet\bullet}} C_{\mathscr{g}^{\bullet\bullet}} \right) \left(\sum_{\mathscr{h}^{\bullet\bullet}} C_{\mathscr{h}^{\bullet\bullet}} \right) = \sum_{\substack{\beta+\gamma=e \\ |\mu|\leq|\beta| \\ |\nu|\leq|\gamma|}} \mathbf{C}_\beta(\varphi, \mu) \mathbf{C}_\gamma(\varphi, \nu) = \sum_{\substack{\beta+\gamma=e \\ |\mu|\leq|\beta| \\ |\nu|\leq|\gamma|}} K_{\beta,\mu} K_{\gamma,\nu}.$$

where $\mathscr{f}^{\bullet\bullet} \in \mathscr{P}(e, \mu + \nu)$, $\mathscr{g}^{\bullet\bullet} \in \mathscr{P}(\beta, \mu)$ and $\mathscr{h}^{\bullet\bullet} \in \mathscr{P}(\gamma, \nu)$.

(b) $\Rightarrow$ (a) First, one easily proves by induction on $|e|$ that $K_{e,0} = 0$ whenever $|e| > 0$, and so $\varphi(1) = \varphi(\mathbf{s}^0) = K_{0,0} = 1$. Let $a = \sum_\alpha a_\alpha \mathbf{s}^\alpha$, $b = \sum_\alpha b_\alpha \mathbf{s}^\alpha$ be elements in $A[[t]]_\Delta$, and $c = ab = \sum_\alpha c_\alpha \mathbf{s}^\alpha$ with $c_\alpha = \sum_{\mu+\nu=\alpha} a_\mu b_\nu$. We have:

$$\varphi(ab) = \varphi(c) = \sum_{e\in\Delta} \left(\sum_{\substack{\alpha\in\nabla \\ |\alpha|\leq|e|}} K_{e,\alpha} c_\alpha \right) \mathbf{t}^e = \sum_{e\in\Delta} \left(\sum_{\substack{\mu,\nu\in\nabla \\ |\mu+\nu|\leq|e|}} K_{e,\mu+\nu} a_\mu b_\nu \right) \mathbf{t}^e =$$

$$\sum_e \left(\sum_{|\mu+\nu|\leq|e|} \sum_{\substack{\beta+\gamma=e \\ |\mu|\leq|\beta|,|\nu|\leq|\gamma|}} K_{\beta,\mu} K_{\gamma,\nu} a_\mu b_\nu \right) \mathbf{t}^e = \cdots = \varphi(a)\varphi(b).$$

We conclude that φ is a (continuous) A-algebra map determined by the images

$$\varphi(u) = \varphi\left(\mathbf{s}^{\mathbf{s}^u}\right) = \sum_{\substack{e\in\Delta \\ 0<|e|}} K_{e,\mathbf{s}^u} \mathbf{t}^e, \quad u \in \mathbf{s},$$

(remember that $\{\mathbf{s}^u\}_{u\in\mathbf{s}}$ is the canonical basis of $\mathbb{N}^{(\mathbf{s})}$) and so it is a substitution map.

Definition 7 The *tensor product* of two substitution maps $\varphi : A[[\mathbf{s}]]_\nabla \to A[[\mathbf{t}]]_\Delta$, $\psi : A[[\mathbf{u}]]_{\nabla'} \to A[[\mathbf{v}]]_{\Delta'}$ is the unique substitution map

$$\varphi \otimes \psi : A[[\mathbf{s} \sqcup \mathbf{u}]]_{\nabla\times\nabla'} \longrightarrow A[[\mathbf{t} \sqcup \mathbf{v}]]_{\Delta\times\Delta'}$$

making commutative the following diagram

$$\begin{array}{ccccc} A[[\mathbf{s}]]_\nabla & \longrightarrow & A[[\mathbf{s}\sqcup\mathbf{u}]]_{\nabla\times\nabla'} & \longleftarrow & A[[\mathbf{u}]]_{\nabla'} \\ \downarrow{\scriptstyle\varphi} & & \downarrow{\scriptstyle\varphi\otimes\psi} & & \downarrow{\scriptstyle\psi} \\ A[[\mathbf{t}]]_\Delta & \longrightarrow & A[[\mathbf{t}\sqcup\mathbf{v}]]_{\Delta\times\Delta'} & \longleftarrow & A[[\mathbf{v}]]_{\Delta'}, \end{array}$$

where the horizontal arrows are the combinatorial substitution maps induced by the inclusions $\mathbf{s},\mathbf{u}\hookrightarrow\mathbf{s}\sqcup\mathbf{u}$, $\mathbf{t},\mathbf{v}\hookrightarrow\mathbf{t}\sqcup\mathbf{v}$[2].

For all $(\alpha,\beta)\in\nabla\times\nabla'\subset\mathbb{N}^{(\mathbf{s})}\times\mathbb{N}^{(\mathbf{u})}\equiv\mathbb{N}^{(\mathbf{s}\sqcup\mathbf{u})}$ we have

$$(\varphi\otimes\psi)(\mathbf{s}^\alpha\mathbf{u}^\beta)=\varphi(\mathbf{s}^\alpha)\psi(\mathbf{u}^\beta)=\cdots=\sum_{\substack{e\in\Delta,f\in\Delta'\\|e|\geq|\alpha|\\|f|\geq|\beta|}}\mathbf{C}_e(\varphi,\alpha)\mathbf{C}_f(\psi,\beta)\mathbf{t}^e\mathbf{v}^f$$

and so, for all $(e,f)\in\Delta\times\Delta'$ and all $(\alpha,\beta)\in\nabla\times\nabla'$ with $|e|+|f|=|(e,f)|\geq|(\alpha,\beta)|=|\alpha|+|\beta|$ we have

$$\mathbf{C}_{(e,f)}(\varphi\otimes\psi,(\alpha,\beta))=\begin{cases}\mathbf{C}_e(\varphi,\alpha)\mathbf{C}_f(\psi,\beta) & \text{if } |\alpha|\leq|e| \text{ and } |\beta|\leq|f|,\\ 0 & \text{otherwise.}\end{cases}$$

4 The Action of Substitution Maps

In this section k will be a commutative ring, A a commutative k-algebra, M an $(A;A)$-bimodule, $\mathbf{s}$ and $\mathbf{t}$ sets and $\nabla\subset\mathbb{N}^{(\mathbf{s})}$, $\Delta\subset\mathbb{N}^{(\mathbf{t})}$ non-empty co-ideals.

Any A-linear continuous map $\varphi: A[[\mathbf{s}]]_\nabla\to A[[\mathbf{t}]]_\Delta$ satisfying the assumptions in 4 induces $(A;A)$-linear maps

$$\varphi_M:=\varphi\widehat{\otimes}\mathrm{Id}_M: M[[\mathbf{s}]]_\nabla\equiv A[[\mathbf{s}]]_\Delta\widehat{\otimes}_A M\longrightarrow M[[\mathbf{t}]]_\Delta\equiv A[[\mathbf{t}]]_\Delta\widehat{\otimes}_A M$$

and

$${}_M\varphi:=\mathrm{Id}_M\widehat{\otimes}\varphi: M[[\mathbf{s}]]_\nabla\equiv M\widehat{\otimes}_A A[[\mathbf{s}]]_\nabla\longrightarrow M[[\mathbf{t}]]_\Delta\equiv M\widehat{\otimes}_A A[[\mathbf{t}]]_\Delta.$$

[2]Let us notice that there are canonical continuous isomorphisms of A-algebras $A[[\mathbf{s}\sqcup\mathbf{u}]]_{\nabla\times\nabla'}\simeq A[[\mathbf{s}]]_\nabla\widehat{\otimes}_A A[[\mathbf{u}]]_{\nabla'}$, $A[[\mathbf{s}\sqcup\mathbf{u}]]_{\Delta\times\Delta'}\simeq A[[\mathbf{s}]]_\Delta\widehat{\otimes}_A A[[\mathbf{u}]]_{\Delta'}$.

If φ is determined by the family $K = \{K_{e,\alpha}, e \in \nabla, \alpha \in \Delta, |\alpha| \le |e|\} \subset A$, with $\varphi(\mathbf{s}^\alpha) = \sum_{\substack{e\in\Delta\\|e|\ge|\alpha|}} K_{e,\alpha}\mathbf{t}^e$, then

$$\varphi_M\left(\sum_{\alpha\in\nabla} m_\alpha \mathbf{s}^\alpha\right) = \sum_{\alpha\in\nabla}\varphi(\mathbf{s}^\alpha)m_\alpha = \sum_{e\in\Delta}\left(\sum_{\substack{\alpha\in\nabla\\|\alpha|\le|e|}} K_{e,\alpha}m_\alpha\right)\mathbf{t}^e, \quad m \in M[[\mathbf{s}]]_\nabla,$$

$$_M\varphi\left(\sum_{\alpha\in\nabla} m_\alpha \mathbf{s}^\alpha\right) = \sum_{\alpha\in\nabla}m_\alpha\varphi(\mathbf{s}^\alpha) = \sum_{e\in\Delta}\left(\sum_{\substack{\alpha\in\nabla\\|\alpha|\le|e|}} m_\alpha K_{e,\alpha}\right)\mathbf{t}^e, \quad m \in M[[\mathbf{s}]]_\nabla.$$

If $\varphi' : A[[\mathbf{t}]]_\Delta \to A[[\mathbf{u}]]_\Omega$ is another A-linear continuous map satisfying the assumptions in 4 and $\varphi'' = \varphi \circ \varphi'$, we have $\varphi''_M = \varphi_M \circ \varphi'_M$, $_M\varphi'' = {}_M\varphi \circ {}_M\varphi'$.

If $\varphi : A[[\mathbf{s}]]_\nabla \to A[[\mathbf{t}]]_\Delta$ is a substitution map and $m \in M[[\mathbf{s}]]_\nabla$, $a \in A[[\mathbf{s}]]_\nabla$, we have

$$\varphi_M(am) = \varphi(a)\varphi_M(m), \quad {}_M\varphi(ma) = {}_M\varphi(m)\varphi(a),$$

i.e. φ_M is $(\varphi; A)$-linear and $_M\varphi$ is $(A;\varphi)$-linear. Moreover, φ_M and $_M\varphi$ are compatible with the augmentations, i.e.

$$\varphi_M(m) \equiv m_0, \; {}_M\varphi(m) \equiv m_0 \bmod \mathfrak{n}_0^M(\mathbf{t})/\Delta_M, \quad m \in M[[\mathbf{s}]]_\nabla. \tag{18}$$

If φ is the trivial substitution map (i.e. $\varphi(s) = 0$ for all $s \in \mathbf{s}$), then $\varphi_M : M[[\mathbf{s}]]_\nabla \to M[[\mathbf{t}]]_\Delta$ and $_M\varphi : M[[\mathbf{s}]]_\nabla \to M[[\mathbf{t}]]_\Delta$ are also trivial, i.e.

$$\varphi_M(m) = {}_M\varphi(m) = m_0, \; m \in M[[\mathbf{s}]]_\nabla.$$

5. The above constructions apply in particular to the case of any k-algebra R over A, for which we have two induced continuous maps, $\varphi_R = \varphi\widehat{\otimes}\mathrm{Id}_R : R[[\mathbf{s}]]_\nabla \to R[[\mathbf{t}]]_\Delta$, which is $(A; R)$-linear, and $_R\varphi = \mathrm{Id}_R\widehat{\otimes}\varphi : R[[\mathbf{s}]]_\nabla \to R[[\mathbf{t}]]_\Delta$, which is $(R; A)$-linear.

   For $r \in R[[\mathbf{s}]]_\nabla$ we will denote

$$\varphi\bullet r := \varphi_R(r), \quad r\bullet\varphi := {}_R\varphi(r).$$

Explicitly, if $r = \sum_\alpha r_\alpha \mathbf{s}^\alpha$ with $\alpha \in \nabla$, then

$$\varphi\bullet r = \sum_{e\in\Delta}\left(\sum_{\substack{\alpha\in\nabla\\|\alpha|\le|e|}} \mathbf{C}_e(\varphi,\alpha)r_\alpha\right)\mathbf{t}^e, \quad r\bullet\varphi = \sum_{e\in\Delta}\left(\sum_{\substack{\alpha\in\nabla\\|\alpha|\le|e|}} r_\alpha\mathbf{C}_e(\varphi,\alpha)\right)\mathbf{t}^e. \tag{19}$$

From (18), we deduce that $\varphi_R(\mathscr{U}^{\mathbf{s}}(R;\nabla)) \subset \mathscr{U}^{\mathbf{t}}(R;\Delta)$ and ${}_R\varphi(\mathscr{U}^{\mathbf{s}}(R;\nabla)) \subset \mathscr{U}^{\mathbf{t}}(R;\Delta)$. We also have $\varphi\bullet 1 = 1\bullet\varphi = 1$.

If φ is a substitution map with constant coefficients, then $\varphi_R = {}_R\varphi$ is a ring homomorphism over φ. In particular, $\varphi\bullet r = r\bullet\varphi$ and $\varphi\bullet(rr') = (\varphi\bullet r)(\varphi\bullet r')$.

If $\varphi = \mathbf{0} : A[[\mathbf{s}]]_\nabla \to A[[\mathbf{t}]]_\Delta$ is the trivial substitution map, then $\mathbf{0}\bullet r = r\bullet\mathbf{0} = r_0$ for all $r \in R[[\mathbf{s}]]_\nabla$. In particular, $\mathbf{0}\bullet r = r\bullet\mathbf{0} = 1$ for all $r \in \mathscr{U}^{\mathbf{s}}(R;\nabla)$.

If $\psi : R[[\mathbf{t}]]_\Delta \to R[[\mathbf{u}]]_\Omega$ is another substitution map, one has

$$\psi\bullet(\varphi\bullet r) = (\psi\circ\varphi)\bullet r, \quad (r\bullet\varphi)\bullet\psi = r\bullet(\psi\circ\varphi).$$

Since $(R[[\mathbf{s}]]_\nabla)^{\mathrm{opp}} = R^{\mathrm{opp}}[[\mathbf{s}]]_\nabla$, for any substitution map $\varphi : A[[\mathbf{s}]]_\nabla \to A[[\mathbf{t}]]_\Delta$ we have $(\varphi_R)^{\mathrm{opp}} = {}_{R^{\mathrm{opp}}}\varphi$ and $({}_R\varphi)^{\mathrm{opp}} = \varphi_{R^{\mathrm{opp}}}$.

The proof of the following lemma is straightforward and it is left to the reader.

Lemma 8 *If $\varphi : A[[\mathbf{s}]]_\nabla \to A[[\mathbf{t}]]_\Delta$ is a substitution map, then:*

(i) *φ_R is left φ-linear, i.e. $\varphi_R(ar) = \varphi(a)\varphi_R(r)$ for all $a \in A[[\mathbf{s}]]_\nabla$ and for all $r \in R[[\mathbf{s}]]_\nabla$.*

(ii) *${}_R\varphi$ is right φ-linear, i.e. ${}_R\varphi(ra) = {}_R\varphi(r)\varphi(a)$ for all $a \in A[[\mathbf{s}]]_\nabla$ and for all $r \in R[[\mathbf{s}]]_\nabla$.*

Let us assume again that $\varphi : A[[\mathbf{s}]]_\nabla \to A[[\mathbf{t}]]_\Delta$ is an A-linear continuous map satisfying the assumptions in 4. We define the $(A;A)$-linear map

$$\varphi_* : f \in \mathrm{Hom}_k(A, A[[\mathbf{s}]]_\nabla) \longmapsto \varphi_*(f) = \varphi\circ f \in \mathrm{Hom}_k(A, A[[\mathbf{t}]]_\Delta)$$

which induces another one $\overline{\varphi_*} : \mathrm{End}^{\mathrm{top}}_{k[[\mathbf{s}]]_\nabla}(A[[\mathbf{s}]]_\nabla) \longrightarrow \mathrm{End}^{\mathrm{top}}_{k[[\mathbf{t}]]_\Delta}(A[[\mathbf{t}]]_\Delta)$ defined by

$$\overline{\varphi_*}(f) := (\varphi_*(f|_A))^e = (\varphi\circ f|_A)^e, \quad f \in \mathrm{End}^{\mathrm{top}}_{k[[\mathbf{s}]]_\nabla}(A[[\mathbf{s}]]_\nabla).$$

More generally, for a given left A-module E (which will be considered as a trivial $(A;A)$-bimodule) we have $(A;A)$-linear maps

$$(\varphi_E)_* : f \in \mathrm{Hom}_k(E, E[[\mathbf{s}]]_\nabla) \mapsto (\varphi_E)_*(f) = \varphi_E\circ f \in \mathrm{Hom}_k(E, E[[\mathbf{t}]]_\Delta),$$
$$\overline{(\varphi_E)_*} : \mathrm{End}^{\mathrm{top}}_{k[[\mathbf{s}]]_\nabla}(E[[\mathbf{s}]]_\nabla) \to \mathrm{End}^{\mathrm{top}}_{k[[\mathbf{t}]]_\Delta}(E[[\mathbf{t}]]_\Delta), \quad \overline{(\varphi_E)_*}(f) := (\varphi_E\circ f|_A)^e.$$

Let us denote $R = \mathrm{End}_k(E)$. For each $r \in R[[\mathbf{s}]]_\nabla$ and for each $e \in E$ we have

$$\widetilde{\varphi_R(r)}(e) = \varphi_E\left(\widetilde{r}(e)\right),$$

or more graphically, the following diagram is commutative (see (7)):

$$\begin{array}{ccccc} R[[\mathbf{s}]]_\nabla & \xrightarrow[r\mapsto\widetilde{r}]{\sim} & \mathrm{End}^{\mathrm{top}}_{k[[\mathbf{s}]]_\nabla}(E[[\mathbf{s}]]_\nabla) & \xrightarrow[\text{rest.}]{\sim} & \mathrm{Hom}_k(E,E[[\mathbf{s}]]_\nabla) \\ {\scriptstyle \varphi_R}\downarrow & & \downarrow{\scriptstyle \overline{(\varphi_E)_*}} & & {\scriptstyle (\varphi_E)_*}\downarrow \\ R[[\mathbf{t}]]_\Delta & \xrightarrow[r\mapsto\widetilde{r}]{\sim} & \mathrm{End}^{\mathrm{top}}_{k[[\mathbf{t}]]_\Delta}(E[[\mathbf{t}]]_\Delta) & \xrightarrow[\text{rest.}]{\sim} & \mathrm{Hom}_k(E,E[[\mathbf{t}]]_\Delta). \end{array}$$

In order to simplify notations, we will also write

$$\varphi\bullet f := \overline{(\varphi_E)_*}(f) \quad \forall f\in \mathrm{End}^{\mathrm{top}}_{k[[\mathbf{s}]]_\nabla}(E[[\mathbf{s}]]_\nabla)$$

and so have $\widetilde{\varphi\bullet r} = \varphi\bullet\widetilde{r}$ for all $r\in R[[\mathbf{s}]]_\nabla$. Let us notice that $(\varphi\bullet f)(e) = (\varphi_E\circ f)(e)$ for all $e\in E$, i.e.

$$\boxed{(\varphi\bullet f)|_E = (\varphi_E\circ f)|_E,\ \text{but in general}\, \varphi\bullet f\neq \varphi_E\circ f.} \tag{20}$$

If φ is the trivial substitution map, then $(\varphi_E)_*$ (resp. $\overline{(\varphi_E)_*}$) is also trivial in the sense that if $f=\sum_\alpha f_\alpha \mathbf{s}^\alpha \in \mathrm{Hom}_k(E,E[[\mathbf{s}]]_\nabla)$ (resp. $f=\sum_\alpha f_\alpha \mathbf{s}^\alpha \in \mathrm{End}_k(E)[[\mathbf{s}]]_\nabla \equiv \mathrm{End}^{\mathrm{top}}_{k[[\mathbf{s}]]_\nabla}(E[[\mathbf{s}]]_\nabla)$), then $(\varphi_E)_*(f) = f_0 \in \mathrm{End}_k(E)\subset \mathrm{Hom}_k(E,E[[\mathbf{s}]]_\nabla)$ (resp. $\overline{(\varphi_E)_*}(f) = f_0^e \in \mathrm{End}^{\mathrm{top}}_{k[[\mathbf{s}]]_\nabla}(E[[\mathbf{s}]]_\nabla)$, with $f_0^e(\sum_\alpha e_\alpha \mathbf{s}^\alpha) = \sum_\alpha f_0(e_\alpha)\mathbf{s}^\alpha$).

If $\varphi: A[[\mathbf{s}]]_\nabla \to A[[\mathbf{t}]]_\nabla$ is a substitution map, we have

$$(\varphi_E)_*(af) = \varphi(a)(\varphi_E)_*(f) \quad \forall a\in A[[\mathbf{s}]]_\nabla, \forall f\in \mathrm{Hom}_k(E,E[[\mathbf{s}]]_\nabla)$$

and so

$$\overline{(\varphi_E)_*}(af) = \varphi(a)\overline{(\varphi_E)_*}(f) \quad \forall a\in A[[\mathbf{s}]]_\nabla, \forall f\in \mathrm{End}^{\mathrm{top}}_{k[[\mathbf{s}]]_\nabla}(E[[\mathbf{s}]]_\nabla).$$

Moreover, the following inclusions hold

$$(\varphi_E)_*(\mathrm{Hom}_k^\circ(E,M[[\mathbf{s}]]_\nabla)) \subset \mathrm{Hom}_k^\circ(E,E[[\mathbf{t}]]_\Delta),$$
$$\overline{(\varphi_E)_*}\left(\mathrm{Aut}^\circ_{k[[\mathbf{s}]]_\nabla}(E[[\mathbf{s}]]_\nabla)\right) \subset \mathrm{Aut}^\circ_{k[[\mathbf{t}]]_\Delta}(E[[\mathbf{t}]]_\Delta),$$

and so we have a commutative diagram:

$$\begin{array}{ccccc} \mathscr{U}^{\mathbf{s}}(R;\nabla) & \xrightarrow[r\mapsto\widetilde{r}]{\sim} & \mathrm{Aut}^\circ_{k[[\mathbf{s}]]_\nabla}(E[[\mathbf{s}]]_\nabla) & \xrightarrow[\text{rest.}]{\sim} & \mathrm{Hom}^\circ_k(E,E[[\mathbf{s}]]_\nabla) \\ {\scriptstyle \varphi_R}\downarrow & & \downarrow{\scriptstyle \overline{(\varphi_E)_*}} & & {\scriptstyle (\varphi_E)_*}\downarrow \\ \mathscr{U}^{\mathbf{t}}(R;\Delta) & \xrightarrow[r\mapsto\widetilde{r}]{\sim} & \mathrm{Aut}^\circ_{k[[\mathbf{t}]]_\Delta}(E[[\mathbf{t}]]_\Delta) & \xrightarrow[\text{rest.}]{\sim} & \mathrm{Hom}^\circ_k(E,E[[\mathbf{t}]]_\Delta). \end{array} \tag{21}$$

Lemma 9 *With the notations above, if $\varphi : k[[\mathbf{s}]]_\nabla \to k[[\mathbf{t}]]_\Delta$ is a substitution map with constant coefficients, then*

$$\langle \varphi \bullet r, \varphi_E(e)\rangle = \varphi_E(\langle r, e\rangle), \quad \forall r \in R[[\mathbf{s}]]_\nabla, \forall e \in E[[\mathbf{s}]]_\nabla.$$

Proof Let us write $r = \sum_\alpha r_\alpha \mathbf{s}^\alpha$, $r_\alpha \in R = \operatorname{End}_k(E)$ and $e = \sum_\alpha e_\alpha \mathbf{s}^\alpha$, $e_\alpha \in E$. We have

$$\langle \varphi \bullet r, \varphi_E(e)\rangle = (\widetilde{\varphi \bullet r})(\varphi_E(e)) = \left(\sum_\alpha \varphi(\mathbf{s}^\alpha)\widetilde{r_\alpha}\right)\left(\sum_\alpha \varphi(\mathbf{s}^\alpha)e_\alpha\right) =$$

$$\sum_{\alpha,\beta} \varphi(\mathbf{s}^\alpha)\widetilde{r_\alpha}\left(\varphi(\mathbf{s}^\beta)e_\beta\right) = \sum_{\alpha,\beta} \varphi(\mathbf{s}^\alpha)\varphi(\mathbf{s}^\beta)\widetilde{r_\alpha}\left(e_\beta\right) = \sum_{\alpha,\beta} \varphi(\mathbf{s}^{\alpha+\beta})\widetilde{r_\alpha}(e_\beta) =$$

$$\sum_\gamma \varphi(\mathbf{s}^\gamma)\left(\sum_{\alpha+\beta=\gamma} \widetilde{r_\alpha}(e_\beta)\right) = \varphi_E\left(\sum_\gamma \left(\sum_{\alpha+\beta=\gamma} \widetilde{r_\alpha}(e_\beta)\right)\mathbf{s}^\gamma\right)$$

$$= \varphi_E\left(\widetilde{r}(e)\right) = \varphi_E(\langle r, e\rangle).$$

Notice that if $\varphi : k[[\mathbf{s}]]_\nabla \to k[[\mathbf{t}]]_\Delta$ is a substitution map with constant coefficients, we already pointed out that ${}_R\varphi = \varphi_R$, and indeed, $\varphi \bullet r = r \bullet \varphi$ for all $r \in R[[\mathbf{s}]]_\nabla$.

6. Let us denote $\iota : A[[\mathbf{s}]]_\nabla \to A[[\mathbf{s} \sqcup \mathbf{t}]]_{\nabla\times\Delta}$, $\kappa : A[[\mathbf{t}]]_\Delta \to A[[\mathbf{s} \sqcup \mathbf{t}]]_{\nabla\times\Delta}$ the combinatorial substitution maps given by the inclusions $\mathbf{s} \hookrightarrow \mathbf{s} \sqcup \mathbf{t}$, $\mathbf{t} \hookrightarrow \mathbf{s} \sqcup \mathbf{t}$.

Let us notice that for $r \in R[[\mathbf{s}]]_\nabla$ and $r' \in R[[\mathbf{t}]]_\Delta$, we have (see Definition 3) $r \boxtimes r' = (\iota \bullet r)(\kappa \bullet r') \in R[[\mathbf{s} \sqcup \mathbf{t}]]_{\nabla\times\Delta}$.

If $\nabla' \subset \nabla \subset \mathbb{N}^{(\mathbf{s})}$, $\Delta' \subset \Delta \subset \mathbb{N}^{(\mathbf{t})}$ are non-empty co-ideals, we have

$$\tau_{\nabla\times\Delta,\nabla'\times\Delta'}(r \boxtimes r') = \tau_{\nabla,\nabla'}(r) \boxtimes \tau_{\Delta,\Delta'}(r').$$

If we denote by $\Sigma : R[[\mathbf{s} \sqcup \mathbf{s}]]_{\nabla\times\nabla} \to R[[\mathbf{s}]]_\nabla$ the combinatorial substitution map given by the co-diagonal map $\mathbf{s} \sqcup \mathbf{s} \to \mathbf{s}$, it is clear that for each $r, r' \in R[[\mathbf{s}]]_\nabla$ we have

$$rr' = \Sigma \bullet (r \boxtimes r'). \tag{22}$$

If $\varphi : A[[\mathbf{s}]]_\nabla \to A[[\mathbf{u}]]_\Omega$ and $\psi : A[[\mathbf{t}]]_\Delta \to A[[\mathbf{v}]]_{\Omega'}$ are substitution maps, we have new substitution maps $\varphi \otimes \operatorname{Id} : A[[\mathbf{s} \sqcup \mathbf{t}]]_{\nabla\times\Delta} \to A[[\mathbf{u} \sqcup \mathbf{t}]]_{\Omega\times\Delta}$ and

$\mathrm{Id} \otimes \psi : A[[\mathbf{s} \sqcup \mathbf{t}]]_{\nabla\times\Delta} \to A[[\mathbf{s} \sqcup \mathbf{v}]]_{\nabla\times\Omega'}$ (see Definition 7) taking part in the following commutative diagrams of $(A; A)$-bimodules

$$\begin{array}{ccc} R[[\mathbf{s}]]_\nabla \otimes_R R[[\mathbf{t}]]_\Delta & \xrightarrow{\varphi_R\otimes \mathrm{Id}} & R[[\mathbf{u}]]_\Omega \otimes_R R[[\mathbf{t}]]_\Delta \\ \text{can.}\downarrow & & \downarrow\text{can.} \\ R[[\mathbf{s}\sqcup\mathbf{t}]]_{\nabla\times\Delta} & \xrightarrow{(\varphi\otimes\mathrm{Id})_R} & R[[\mathbf{u}\sqcup\mathbf{t}]]_{\Omega\times\Delta} \end{array}$$

and

$$\begin{array}{ccc} R[[\mathbf{s}]]_\nabla \otimes_R R[[\mathbf{t}]]_\Delta & \xrightarrow{\mathrm{Id}\otimes\psi} & R[[\mathbf{s}]]_\nabla \otimes_R R[[\mathbf{v}]]_{\Omega'} \\ \text{can.}\downarrow & & \downarrow\text{can.} \\ R[[\mathbf{s}\sqcup\mathbf{t}]]_{\nabla\times\Delta} & \xrightarrow{(\mathrm{Id}\otimes\varphi)_R} & R[[\mathbf{s}\sqcup\mathbf{v}]]_{\nabla\times\Omega'}. \end{array}$$

So $(\varphi\bullet r)\boxtimes r' = (\varphi\otimes\mathrm{Id})\bullet(r\boxtimes r')$ and $r\boxtimes(r'\bullet\psi) = (r\boxtimes r')\bullet(\mathrm{Id}\otimes\psi)$.

5 Multivariate Hasse-Schmidt Derivations

In this section we study multivariate (possibly ∞-variate) Hasse–Schmidt derivations. The original reference for 1-variate Hasse–Schmidt derivations is [4]. This notion has been studied and developed in [8, §27] (see also [13] and [10]). In [6] the authors study "finite dimensional" Hasse–Schmidt derivations, which correspond in our terminology to p-variate Hasse–Schmidt derivations.

From now on k will be a commutative ring, A a commutative k-algebra, $\mathbf{s}$ a set and $\Delta \subset \mathbb{N}^{(\mathbf{s})}$ a non-empty co-ideal.

Definition 8 A $(\mathbf{s}, \Delta)$*-variate Hasse-Schmidt derivation*, or a $(\mathbf{s}, \Delta)$*-variate HS-derivation* for short, of A over k is a family $D = (D_\alpha)_{\alpha\in\Delta}$ of k-linear maps $D_\alpha : A \longrightarrow A$, satisfying the following Leibniz type identities:

$$D_0 = \mathrm{Id}_A, \quad D_\alpha(xy) = \sum_{\beta+\gamma=\alpha} D_\beta(x) D_\gamma(y)$$

for all $x, y \in A$ and for all $\alpha \in \Delta$. We denote by $\mathrm{HS}_k^{\mathbf{s}}(A; \Delta)$ the set of all $(\mathbf{s}, \Delta)$-variate HS-derivations of A over k and $\mathrm{HS}_k^{\mathbf{s}}(A) =$ for $\Delta = \mathbb{N}^{(\mathbf{s})}$. In the case where $\mathbf{s} = \{1, \ldots, p\}$, a $(\mathbf{s}, \Delta)$-variate HS-derivation will be simply called a (p, Δ)*-variate HS-derivation* and we denote $\mathrm{HS}_k^p(A; \Delta) := \mathrm{HS}_k^{\mathbf{s}}(A; \Delta)$ and $\mathrm{HS}_k^p(A) := \mathrm{HS}_k^{\mathbf{s}}(A)$. For $p = 1$, a 1-variate HS-derivation will be simply called a *Hasse–Schmidt derivation* (a HS-derivation for short), or a *higher derivation*[3], and we will simply write $\mathrm{HS}_k(A; m) := \mathrm{HS}^1_k(A; \Delta)$ for $\Delta = \{q \in \mathbb{N} \mid q \le m\}$[4] and $\mathrm{HS}_k(A) := \mathrm{HS}^1_k(A)$.

7. The above Leibniz identities for $D \in \mathrm{HS}^{\mathbf{s}}_k(A; \Delta)$ can be written as

$$D_\alpha x = \sum_{\beta+\gamma=\alpha} D_\beta(x) D_\gamma, \quad \forall x \in A, \forall \alpha \in \Delta. \tag{23}$$

Any $(\mathbf{s}, \Delta)$-variate HS-derivation D of A over k can be understood as a power series

$$\sum_{\alpha\in\Delta} D_\alpha \mathbf{s}^\alpha \in \mathrm{End}_k(A)[[\mathbf{s}]]_\Delta$$

and so we consider $\mathrm{HS}^{\mathbf{s}}_k(A; \Delta) \subset \mathrm{End}_k(A)[[\mathbf{s}]]_\Delta$.

Proposition 4 *Let $D \in \mathrm{HS}^{\mathbf{s}}_k(A; \Delta)$ be a HS-derivation. Then, for each $\alpha \in \Delta$, the component $D_\alpha : A \to A$ is a k-linear differential operator or order $\le |\alpha|$ vanishing on k. In particular, if $|\alpha| = 1$ then $D_\alpha : A \to A$ is a k-derivation.*

Proof The proof follows by induction on $|\alpha|$ from (23).

The map

$$D \in \mathrm{HS}^{\mathbf{s}}_k(A; \mathfrak{t}_1(\mathbf{s})) \mapsto \{D_\alpha\}_{|\alpha|=1} \in \mathrm{Der}_k(A)^{\mathbf{s}} \tag{24}$$

is clearly a bijection.

The proof of the following proposition is straightforward and it is left to the reader (see Notation 1 and 2).

Proposition 5 *Let us denote $R = \mathrm{End}_k(A)$ and let $D = \sum_\alpha D_\alpha \mathbf{s}^\alpha \in R[[\mathbf{s}]]_\Delta$ be a power series. The following properties are equivalent:*

(a) *D is a $(\mathbf{s}, \Delta)$-variate HS-derivation of A over k.*
(b) *The map $\widetilde{D} : A[[\mathbf{s}]]_\Delta \to A[[\mathbf{s}]]_\Delta$ is a (continuous) $k[[\mathbf{s}]]_\Delta$-algebra homomorphism compatible with the natural augmentation $A[[\mathbf{s}]]_\Delta \to A$.*
(c) *$D \in \mathcal{U}^{\mathbf{s}}(R; \Delta)$ and for all $a \in A[[\mathbf{s}]]_\Delta$ we have $Da = \widetilde{D}(a) D$.*
(d) *$D \in \mathcal{U}^{\mathbf{s}}(R; \Delta)$ and for all $a \in A$ we have $Da = \widetilde{D}(a) D$.*

Moreover, in such a case $\widetilde{D}$ is a bi-continuous $k[[\mathbf{s}]]_\Delta$-algebra automorphism of $A[[\mathbf{s}]]_\Delta$.

Corollary 1 *Under the above hypotheses, $\mathrm{HS}^{\mathbf{s}}_k(A; \Delta)$ is a (multiplicative) subgroup of $\mathcal{U}^{\mathbf{s}}(R; \Delta)$.*

[3]This terminology is used for instance in [8].

[4]These HS-derivations are called of length m in [10].

If $\Delta' \subset \Delta \subset \mathbb{N}^{(\mathbf{s})}$ are non-empty co-ideals, we obviously have group homomorphisms $\tau_{\Delta\Delta'} : \operatorname{HS}^{\mathbf{s}}_k(A;\Delta) \longrightarrow \operatorname{HS}^{\mathbf{s}}_k(A;\Delta')$. Since any $D \in \operatorname{HS}^{\mathbf{s}}_k(A;\Delta)$ is determined by its finite truncations, we have a natural group isomorphism

$$\operatorname{HS}^{\mathbf{s}}_k(A) = \varprojlim_{\substack{\Delta' \subset \Delta \\ \sharp\Delta' < \infty}} \operatorname{HS}^{\mathbf{s}}_k(A;\Delta').$$

In the case $\Delta' = \Delta^1 = \Delta \cap \mathrm{t}_1(\mathbf{s})$, since $\operatorname{HS}^{\mathbf{s}}_k(A;\Delta^1) \simeq \operatorname{Der}_k(A)^{\Delta^1}$, we can think on $\tau_{\Delta\Delta^1}$ as a group homomorphism $\tau_{\Delta\Delta^1} : \operatorname{HS}^{\mathbf{s}}_k(A;\Delta) \to \operatorname{Der}_k(A)^{\Delta^1}$ whose kernel is the normal subgroup of $\operatorname{HS}^{\mathbf{s}}_k(A;\Delta)$ consisting of HS-derivations D with $D_\alpha = 0$ whenever $|\alpha| = 1$.

In the case $\Delta' = \Delta^n = \Delta \cap \mathrm{t}_n(\mathbf{s})$, for $n \geq 1$, we will simply write $\tau_n = \tau_{\Delta,\Delta^n}$: $\operatorname{HS}^{\mathbf{s}}_k(A;\Delta) \longrightarrow \operatorname{HS}^{\mathbf{s}}_k(A;\Delta^n)$.

Remark 4 Since for any $D \in \operatorname{HS}^{\mathbf{s}}_k(A;\Delta)$ we have $D_\alpha \in \mathscr{D}\mathrm{iff}^{|\alpha|}_{A/k}(A)$, we may also think on D as an element in a generalized Rees ring of the ring of differential operators:

$$\widehat{\mathscr{R}}^{\mathbf{s}}\left(\mathscr{D}_{A/k}(A);\Delta\right) := \left\{\sum_{\alpha\in\Delta} r_\alpha \mathbf{s}^\alpha \in \mathscr{D}_{A/k}(A)[[\mathbf{s}]]_\Delta \mid r_\alpha \in \mathscr{D}\mathrm{iff}^{|\alpha|}_{A/k}(A)\right\}.$$

The group operation in $\operatorname{HS}^{\mathbf{s}}_k(A;\Delta)$ is explicitly given by

$$(D,E) \in \operatorname{HS}^{\mathbf{s}}_k(A;\Delta) \times \operatorname{HS}^{\mathbf{s}}_k(A;\Delta) \longmapsto D \circ E \in \operatorname{HS}^{\mathbf{s}}_k(A;\Delta)$$

with

$$(D \circ E)_\alpha = \sum_{\beta+\gamma=\alpha} D_\beta \circ E_\gamma,$$

and the identity element of $\operatorname{HS}^{\mathbf{s}}_k(A;\Delta)$ is $\mathbb{I}$ with $\mathbb{I}_0 = \mathrm{Id}$ and $\mathbb{I}_\alpha = 0$ for all $\alpha \neq 0$. The inverse of a $D \in \operatorname{HS}^{\mathbf{s}}_k(A;\Delta)$ will be denoted by D^*.

Proposition 6 *Let $D \in \operatorname{HS}^{\mathbf{s}}_k(A;\Delta)$, $E \in \operatorname{HS}^{\mathbf{t}}_k(A;\nabla)$ be HS-derivations. Then their external product $D \boxtimes E$ (see Definition 3) is a $(\mathbf{s} \sqcup \mathbf{t}, \nabla \times \Delta)$-variate HS-derivation.*

Proof From Lemma 4 we know that $\widetilde{D \boxtimes E} = \widetilde{D} \boxtimes \widetilde{E}$ and we conclude by Proposition 5.

Definition 9 For each $a \in A^{\mathbf{s}}$ and for each $D \in \operatorname{HS}^{\mathbf{s}}_k(A;\Delta)$, we define $a \bullet D$ as

$$(a \bullet D)_\alpha := a^\alpha D_\alpha, \quad \forall \alpha \in \Delta.$$

It is clear that $a \bullet D \in \operatorname{HS}^{\mathbf{s}}_k(A;\Delta)$, $a' \bullet (a \bullet D) = (a'a) \bullet D$, $1 \bullet D = D$ and $0 \bullet D = \mathbb{I}$.

If $\Delta' \subset \Delta \subset \mathbb{N}^{(\mathbf{s})}$ are non-empty co-ideals, we have $\tau_{\Delta\Delta'}(a \bullet D) = a \bullet \tau_{\Delta\Delta'}(D)$. Hence, in the case $\Delta' = \Delta^1 = \Delta \cap \mathrm{t}_1(\mathbf{s})$, since $\mathrm{HS}^{\mathbf{s}}_k(A; \Delta^1) \simeq \mathrm{Der}_k(A)^{\Delta^1}$, the image of $\tau_{\Delta\Delta^1} : \mathrm{HS}^{\mathbf{s}}_k(A; \Delta) \to \mathrm{Der}_k(A)^{\Delta^1}$ is an A-submodule.

The following lemma provides a dual way to express the Leibniz identity (23), 7.

Lemma 10 *For each $D \in \mathrm{HS}^{\mathbf{s}}_k(A; \Delta)$ and for each $\alpha \in \Delta$, we have*

$$x D_\alpha = \sum_{\beta+\gamma=\alpha} D_\beta\, D^*_\gamma(x), \quad \forall x \in A.$$

Proof We have

$$\sum_{\beta+\gamma=\alpha} D_\beta\, D^*_\gamma(x) = \sum_{\beta+\gamma=\alpha} \sum_{\mu+\nu=\beta} D_\mu(D^*_\gamma(x)) D_\nu =$$

$$\sum_{e+\nu=\alpha} \left(\sum_{\mu+\gamma=e} D_\mu(D^*_\gamma(x)) \right) D_\nu = x D_\alpha.$$

It is clear that the map (24) is an isomorphism of groups (with the addition on $\mathrm{Der}_k(A)$ as internal operation) and so $\mathrm{HS}^{\mathbf{s}}_k(A; \mathrm{t}_1(\mathbf{s}))$ is abelian.

Notation 5 *Let us denote*

$$\mathrm{Hom}^\circ_{k-\mathrm{alg}}(A, A[[\mathbf{s}]]_\Delta) :=$$

$$\left\{ f \in \mathrm{Hom}_{k-\mathrm{alg}}(A, A[[\mathbf{s}]]_\Delta) \mid f(a) \equiv a \bmod \mathfrak{n}^A_0(\mathbf{s})/\Delta_A \ \forall a \in A \right\},$$

$$\mathrm{Aut}^\circ_{k[[\mathbf{s}]]_\Delta-\mathrm{alg}}(A[[\mathbf{s}]]_\Delta) :=$$

$$\left\{ f \in \mathrm{Aut}^{\mathrm{top}}_{k[[\mathbf{s}]]_\Delta-\mathrm{alg}}(A[[\mathbf{s}]]_\Delta) \mid f(a) \equiv a_0 \bmod \mathfrak{n}^A_0(\mathbf{s})/\Delta_A \ \forall a \in A[[\mathbf{s}]]_\Delta \right\}.$$

It is clear that (see Notation 3) $\mathrm{Hom}^\circ_{k-\mathrm{alg}}(A, A[[\mathbf{s}]]_\Delta) \subset \mathrm{Hom}^\circ_k(A, A[[\mathbf{s}]]_\Delta)$ and $\mathrm{Aut}^\circ_{k[[\mathbf{s}]]_\Delta-\mathrm{alg}}(A[[\mathbf{s}]]_\Delta) \subset \mathrm{Aut}^\circ_{k[[\mathbf{s}]]_\Delta}(A[[\mathbf{s}]]_\Delta)$ are subgroups and we have group isomorphisms (see (10) and (9)):

$$\mathrm{HS}^{\mathbf{s}}_k(A; \Delta) \xrightarrow[\simeq]{D \mapsto \widetilde{D}} \mathrm{Aut}^\circ_{k[[\mathbf{s}]]_\Delta-\mathrm{alg}}(A[[\mathbf{s}]]_\Delta) \xrightarrow[\simeq]{\text{restriction}} \mathrm{Hom}^\circ_{k-\mathrm{alg}}(A, A[[\mathbf{s}]]_\Delta). \tag{25}$$

The composition of the above isomorphisms is given by

$$D \in \mathrm{HS}^{\mathbf{s}}_k(A; \Delta) \stackrel{\sim}{\longmapsto} \Phi_D := \left[a \in A \mapsto \sum_{\alpha\in\Delta} D_\alpha(a) \mathbf{s}^\alpha \right] \in \mathrm{Hom}^\circ_{k-\mathrm{alg}}(A, A[[\mathbf{s}]]_\Delta). \tag{26}$$

For each HS-derivation $D \in \mathrm{HS}^{\mathbf{s}}_k(A;\Delta)$ we have

$$\widetilde{D}\left(\sum_{\alpha\in\Delta} a_\alpha \mathbf{s}^\alpha\right) = \sum_{\alpha\in\Delta} \Phi_D(a_\alpha)\mathbf{s}^\alpha,$$

for all $\sum_\alpha a_\alpha \mathbf{s}^\alpha \in A[[\mathbf{s}]]_\Delta$, and for any $E \in \mathrm{HS}^{\mathbf{s}}_k(A;\Delta)$ we have $\Phi_{D\circ E} = \widetilde{D}\circ\Phi_E$. If $\Delta' \subset \Delta$ is another non-empty co-ideal and we denote by $\pi_{\Delta\Delta'} : A[[\mathbf{s}]]_\Delta \to A[[\mathbf{s}]]_{\Delta'}$ the projection, one has $\Phi_{\tau_{\Delta\Delta'}(D)} = \pi_{\Delta\Delta'}\circ\Phi_D$.

Definition 10 For each HS-derivation $E \in \mathrm{HS}^{\mathbf{s}}_k(A;\Delta)$, we denote

$$\ell(E) := \min\{r \geq 1 \mid \exists\alpha \in \Delta, |\alpha| = r, E_\alpha \neq 0\} \geq 1$$

if $E \neq \mathbb{I}$ and $\ell(E) = \infty$ if $E = \mathbb{I}$. In other words, $\ell(E) = \mathrm{ord}(E - \mathbb{I})$. Clearly, if Δ is bounded, then $\ell(E) > \max\{|\alpha| \mid \alpha \in \Delta\} \Longleftrightarrow \ell(E) = \infty \Longleftrightarrow E = \mathbb{I}$.

We obviously have $\ell(E\circ E') \geq \min\{\ell(E), \ell(E')\}$ and $\ell(E^*) = \ell(E)$. Moreover, if $\ell(E') > \ell(E)$, then $\ell(E\circ E') = \ell(E)$:

$$\ell(E\circ E') = \mathrm{ord}(E\circ E' - \mathbb{I}) = \mathrm{ord}(E\circ(E' - \mathbb{I}) + (E - \mathbb{I}))$$

and since $\mathrm{ord}(E\circ(E' - \mathbb{I})) \geq$[5]$\mathrm{ord}(E' - \mathbb{I}) = \ell(E') > \ell(E) = \mathrm{ord}(E - \mathbb{I})$ we obtain

$$\ell(E\circ E') = \cdots = \mathrm{ord}(E\circ(E' - \mathbb{I}) + (E - \mathbb{I})) = \mathrm{ord}(E - \mathbb{I}) = \ell(E).$$

Proposition 7 *For each $D \in \mathrm{HS}^{\mathbf{s}}_k(A;\Delta)$ we have that D_α is a k-linear differential operator or order $\leq \lfloor \frac{|\alpha|}{\ell(D)} \rfloor$ for all $\alpha \in \Delta$. In particular, D_α is a k-derivation if $|\alpha| = \ell(D)$, whenever $\ell(D) < \infty$ ($\Leftrightarrow D \neq \mathbb{I}$).*

Proof We may assume $D \neq \mathbb{I}$. Let us call $n := \ell(D) < \infty$ and, for each $\alpha \in \Delta$, $q_\alpha := \lfloor \frac{|\alpha|}{n} \rfloor$ and $r_\alpha := |\alpha| - q_\alpha n$, $0 \leq r_\alpha < n$. We proceed by induction on q_α. If $q_\alpha = 0$, then $|\alpha| < n$, $D_\alpha = 0$ and the result is clear. Assume that the order of D_β is less or equal than q_β whenever $0 \leq q_\beta \leq q$. Now take $\alpha \in \Delta$ with $q_\alpha = q + 1$. For any $a \in A$ we have

$$[D_\alpha, a] = \sum_{\substack{\gamma+\beta=\alpha\\ |\gamma|>0}} D_\gamma(a) D_\beta = \sum_{\substack{\gamma+\beta=\alpha\\ |\gamma|\geq n}} D_\gamma(a) D_\beta,$$

but any β in the index set of the above sum must have norm $\leq |\alpha| - n$ and so $q_\beta < q_\alpha = q + 1$ and D_β has order $\leq q_\beta$. Hence $[D_\alpha, a]$ has order $\leq q$ for any $a \in A$ and D_α has order $\leq q + 1 = q_\alpha$.

[5]Actually, here an equality holds since the 0-term of E (as a series) is 1.

The following example shows that the group structure on HS-derivations takes into account the Lie bracket on usual derivations.

Example 2 If $D, E \in \mathrm{HS}^{\mathbf{s}}_k(A;\Delta)$, then we may apply the above proposition to $[D,E] = D\circ E\circ D^*\circ E^*$ to deduce that $[D,E]_\alpha \in \mathrm{Der}_k(A)$ whenever $|\alpha| = 2$. Actually, for $|\alpha| = 2$ we have:

$$[D,E]_\alpha = \begin{cases} [D_{\mathbf{s}^t}, E_{\mathbf{s}^t}] & \text{if } \alpha = 2\mathbf{s}^t \\ [D_{\mathbf{s}^t}, E_{\mathbf{s}^u}] + [D_{\mathbf{s}^u}, E_{\mathbf{s}^t}] & \text{if } \alpha = \mathbf{s}^t + \mathbf{s}^u, \text{ with } t \neq u. \end{cases}$$

Proposition 8 *For any $D, E \in \mathrm{HS}^{\mathbf{s}}_k(A;\Delta)$ we have $\ell([D,E]) \geq \ell(D) + \ell(E)$.*

Proof We may assume $D, E \neq \mathbb{I}$. Let us write $m = \ell(D) = \ell(D^*)$, $n = \ell(E) = \ell(E^*)$. We have $D_\beta = D^*_\beta = 0$ whenever $0 < |\beta| < m$ and $E_\gamma = E^*_\gamma = 0$ whenever $0 < |\gamma| < n$.

Let $\alpha \in \Delta$ be with $0 < |\alpha| < m+n$. If $|\alpha| < m$ or $|\alpha| < n$ it is clear that $[D,E]_\alpha = 0$. Assume that $m, n \leq |\alpha| < m+n$:

$$[D,E]_\alpha = \sum_{\beta+\gamma+\lambda+\mu=\alpha} D_\beta \circ E_\gamma\, D^*_\lambda\, E^*_\mu = \sum_{\gamma+\mu=\alpha} E_\gamma\, E^*_\mu +$$

$$\sum_{\substack{\beta+\gamma+\lambda+\mu=\alpha\\ |\beta+\lambda|>0}} D_\beta\, E_\gamma\, D^*_\lambda\, E^*_\mu = 0 + \sum_{\substack{\gamma+\lambda+\mu=\alpha\\ |\lambda|>0}} E_\gamma\, D^*_\lambda\, E^*_\mu + \sum_{\substack{\beta+\gamma+\mu=\alpha\\ |\beta|>0}} D_\beta\, E_\gamma\, E^*_\mu +$$

$$\sum_{\substack{\beta+\gamma+\lambda+\mu=\alpha\\ |\beta|,|\lambda|>0}} D_\beta\, E_\gamma\, D^*_\lambda\, E^*_\mu = \sum_{\substack{\gamma+\lambda+\mu=\alpha\\ |\lambda|\geq m}} E_\gamma\, D^*_\lambda\, E^*_\mu + \sum_{\substack{\beta+\gamma+\mu=\alpha\\ |\beta|\geq m}} D_\beta\, E_\gamma\, E^*_\mu +$$

$$\sum_{\substack{\beta+\gamma+\lambda+\mu=\alpha\\ |\beta|,|\lambda|\geq m}} D_\beta\, E_\gamma\, D^*_\lambda\, E^*_\mu = D^*_\alpha + \sum_{\substack{\gamma+\lambda+\mu=\alpha\\ |\lambda|\geq m, |\gamma+\mu|>0}} E_\gamma\, D^*_\lambda\, E^*_\mu + D_\alpha +$$

$$\sum_{\substack{\beta+\mu=\alpha\\ |\beta|\geq m\\ |\gamma+\mu|>0}} D_\beta\, E_\gamma\, E^*_\mu + \sum_{\substack{\beta+\lambda=\alpha\\ |\beta|,|\lambda|\geq m}} D_\beta\, D^*_\lambda + \sum_{\substack{\beta+\gamma+\lambda+\mu=\alpha\\ |\beta|,|\lambda|\geq m\\ |\gamma+\mu|>0}} D_\beta\, E_\gamma\, D^*_\lambda\, E^*_\mu =$$

$$D^*_\alpha + 0 + D_\alpha + 0 + \sum_{\substack{\beta+\lambda=\alpha\\ |\beta|,|\lambda|>0}} D_\beta\, D^*_\lambda + 0 = \sum_{\beta+\lambda=\alpha} D_\beta\, D^*_\lambda = 0.$$

So, $\ell([D,E]) \geq \ell(D) + \ell(E)$.

Corollary 2 *Assume that Δ is bounded and let m be the* max *of $|\alpha|$ with $\alpha \in \Delta$. Then, the group $\mathrm{HS}^{\mathbf{s}}_k(A;\Delta)$ is nilpotent of nilpotent class $\leq m$, where a central series is*[6]

$$\{\mathbb{I}\} = \{E \mid \ell(E) > m\} \lhd \{E \mid \ell(E) \geq m\} \lhd \cdots \lhd \{E \mid \ell(E) \geq 1\} = \mathrm{HS}^{\mathbf{s}}_k(A;\Delta).$$

[6]Let us notice that $\{E \in \mathrm{HS}^{\mathbf{s}}_k(A;\Delta) \mid \ell(E) > r\} = \ker \tau_{\Delta,\Delta_r}$.

Proposition 9 *For each* $D \in \mathrm{HS}_k^{\mathbf{s}}(A;\Delta)$*, its inverse* D^* *is given by* $D_0^* = \mathrm{Id}$ *and*

$$D_\alpha^* = \sum_{d=1}^{|\alpha|} (-1)^d \sum_{\alpha^\bullet \in \mathscr{P}(\alpha,d)} D_{\alpha^1} \circ \cdots \circ D_{\alpha^d}, \quad \alpha \in \Delta.$$

Moreover, $\sigma_{|\alpha|}(D_\alpha^*) = (-1)^{|\alpha|}\sigma_{|\alpha|}(D_\alpha)$.

Proof The first assertion is a straightforward consequence of Lemma 2. For the second assertion, first we have $D_\alpha^* = -D_\alpha$ for all α with $|\alpha| = 1$, and if we denote by $-\mathbf{1} \in A^{\mathbf{s}}$ the constant family -1 and $E = D \circ ((-\mathbf{1}) \bullet D)$, we have $\ell(E) > 1$. So, $D^* = ((-\mathbf{1}) \bullet D) \circ E^*$ and

$$D_\alpha^* = \sum_{\beta+\gamma=\alpha} (-1)^{|\beta|} D_\beta E_\gamma^* = (-1)^{|\alpha|} D_\alpha + \sum_{\substack{\beta+\gamma=\alpha \\ |\gamma|>0}} (-1)^{|\beta|} D_\beta E_\gamma^*.$$

From Proposition 7, we know that E_γ^* is a differential operator of order strictly less than $|\gamma|$ and so $\sigma_{|\alpha|}(D_\alpha^*) = (-1)^{|\alpha|}\sigma_{|\alpha|}(D_\alpha)$.

6 The Action of Substitution Maps on HS-Derivations

In this section, k will be a commutative ring, A a commutative k-algebra, $R = \mathrm{End}_k(A)$, $\mathbf{s}, \mathbf{t}$ sets and $\Delta \subset \mathbb{N}^{(\mathbf{s})}$, $\nabla \subset \mathbb{N}^{(\mathbf{t})}$ non-empty co-ideals.

We are going to extend the operation $(a, D) \in A^{\mathbf{s}} \times \mathrm{HS}_k^{\mathbf{s}}(A;\Delta) \mapsto a \bullet D \in \mathrm{HS}_k^{\mathbf{s}}(A;\Delta)$ (see Definition 9) by means of the constructions in section 4.

Proposition 10 *For any substitution map* $\varphi : A[[\mathbf{s}]]_\Delta \to A[[\mathbf{t}]]_\nabla$*, we have:*

(1) $\varphi_*\left(\mathrm{Hom}_{k\text{-alg}}^\circ(A, A[[\mathbf{s}]]_\Delta)\right) \subset \mathrm{Hom}_{k\text{-alg}}^\circ(A, A[[\mathbf{t}]]_\nabla)$,
(2) $\varphi_R\left(\mathrm{HS}_k^{\mathbf{s}}(A;\Delta)\right) \subset \mathrm{HS}_k^{\mathbf{t}}(A;\nabla)$,
(3) $\overline{\varphi_*}\left(\mathrm{Aut}_{k[[\mathbf{s}]]_\Delta\text{-alg}}^\circ(A[[\mathbf{s}]]_\Delta)\right) \subset \mathrm{Aut}_{k[[\mathbf{t}]]_\nabla\text{-alg}}^\circ(A[[\mathbf{t}]]_\nabla)$.

Proof By using diagram (21) and (25), it is enough to prove the first inclusion, but if $f \in \mathrm{Hom}_{k\text{-alg}}^\circ(A, A[[\mathbf{s}]]_\Delta)$, it is clear that $\varphi_*(f) = \varphi \circ f : A \to A[[\mathbf{t}]]_\nabla$ is a k-algebra map. Moreover, since $\varphi(\mathfrak{t}_0^A(\mathbf{s})/\Delta_A) \subset \mathfrak{t}_0^A(\mathbf{t})/\nabla_A$ (see 3) and $f(a) \equiv a \mod \mathfrak{t}_0^A(\mathbf{s})/\Delta_A$ for all $a \in A$, we deduce that $\varphi(f(a)) \equiv \varphi(a) \mod \mathfrak{t}_0^A(\mathbf{t})/\nabla_A$ for all $a \in A$, but φ is an A-algebra map and $\varphi(a) = a$. So $\varphi_*(f) \in \mathrm{Hom}_{k\text{-alg}}^\circ(A, A[[\mathbf{t}]]_\nabla)$.

As a consequence of the above proposition and diagram (21) we have a commutative diagram:

$$
\begin{array}{ccccc}
\mathrm{Hom}^{\circ}_{k-\mathrm{alg}}(A, A[[\mathbf{s}]]_{\Delta}) & \xleftarrow[\Phi_D \leftarrow\!\shortmid D]{\sim} & \mathrm{HS}^{\mathbf{s}}_k(A;\Delta) & \xrightarrow{\sim} & \mathrm{Aut}^{\circ}_{k[[\mathbf{s}]]_{\Delta}-\mathrm{alg}}(A[[\mathbf{s}]]_{\Delta}) \\
\varphi_* \downarrow & & \varphi_R \downarrow & & \downarrow \overline{\varphi_*} \\
\mathrm{Hom}^{\circ}_{k-\mathrm{alg}}(A, A[[\mathbf{t}]]_{\nabla}) & \xleftarrow[\Phi_D \leftarrow\!\shortmid D]{\sim} & \mathrm{HS}^{\mathbf{t}}_k(A;\nabla) & \xrightarrow{\sim} & \mathrm{Aut}^{\circ}_{k[[\mathbf{t}]]_{\nabla}-\mathrm{alg}}(A[[\mathbf{t}]]_{\nabla}).
\end{array}
\tag{27}
$$

The inclusion (2) in Proposition 10 can be rephrased by saying that for any substitution map $\varphi : A[[\mathbf{s}]]_\Delta \to A[[\mathbf{t}]]_\nabla$ and for any HS-derivation $D \in \mathrm{HS}^{\mathbf{s}}_k(A;\Delta)$ we have $\varphi \bullet D \in \mathrm{HS}^{\mathbf{t}}_k(A;\nabla)$ (see 5). Moreover $\Phi_{\varphi \bullet D} = \varphi \circ \Phi_D$.

It is clear that for any co-ideals $\Delta' \subset \Delta$ and $\nabla' \subset \nabla$ with $\varphi\left(\Delta'_A/\Delta_A\right) \subset \nabla'_A/\nabla_A$ we have

$$
\tau_{\nabla\nabla'}(\varphi \bullet D) = \varphi' \bullet \tau_{\Delta\Delta'}(D), \tag{28}
$$

where $\varphi' : A[[\mathbf{s}]]_{\Delta'} \to A[[\mathbf{t}]]_{\nabla'}$ is the substitution map induced by φ.

Let us notice that any $a \in A^{\mathbf{s}}$ gives rise to a substitution map $\varphi : A[[\mathbf{s}]]_\Delta \to A[[\mathbf{s}]]_\Delta$ given by $\varphi(s) = a_s s$ for all $s \in \mathbf{s}$, and one has $a \bullet D = \varphi \bullet D$.

8. Let $\varphi \in \mathcal{S}_A(\mathbf{s},\mathbf{t};\nabla,\Delta)$, $\psi \in \mathcal{S}_A(\mathbf{t},\mathbf{u};\Delta,\Omega)$ be substitution maps and $D, D' \in \mathrm{HS}^{\mathbf{s}}_k(A;\nabla)$ HS-derivations. From 5 we deduce the following properties:

- If we denote $E := \varphi \bullet D \in \mathrm{HS}^{\mathbf{t}}_k(A;\Delta)$, we have

$$
E_0 = \mathrm{Id}, \qquad E_e = \sum_{\substack{\alpha \in \nabla \\ |\alpha| \le |e|}} \mathbf{C}_e(\varphi,\alpha) D_\alpha, \qquad \forall e \in \Delta. \tag{29}
$$

- If φ has <u>constant coefficients</u>, then $\varphi \bullet (D \circ D') = (\varphi \bullet D) \circ (\varphi \bullet D')$. The general case will be treated in Proposition 11.
- If $\varphi = \mathbf{0}$ is the trivial substitution map or if $D = \mathbb{I}$, then $\varphi \bullet D = \mathbb{I}$.
- $\psi \bullet (\varphi \bullet D) = (\psi \circ \varphi) \bullet D$.

Remark 5 We recall that a HS-derivation $D \in \mathrm{HS}_k(A)$ is called *iterative* (see [8, pg. 209]) if

$$
D_i \circ D_j = \binom{i+j}{i} D_{i+j} \quad \forall i, j \ge 0.
$$

This notion makes sense for $\mathbf{s}$-variate HS-derivations of any length. Actually, iterativity may be understood through the action of substitution maps. Namely, if we denote by $\iota, \iota' : s \hookrightarrow s \sqcup s$ the two canonical inclusions and $\iota + \iota' : A[[\mathbf{s}]] \to A[[\mathbf{s} \sqcup \mathbf{s}]]$ is the substitution map determined by

$$
(\iota + \iota')(s) = \iota(s) + \iota'(s), \quad \forall s \in \mathbf{s},
$$

then a HS-derivation $D \in \mathrm{HS}^{\mathbf{s}}_k(A)$ is iterative if and only if

$$(\iota + \iota')\bullet D = (\iota\bullet D)\circ(\iota'\bullet D).$$

A similar remark applies for any formal group law instead of $\iota + \iota'$ (cf. [5]).

Proposition 11 *Let $\varphi : A[[\mathbf{s}]]_\nabla \to A[[\mathbf{t}]]_\Delta$ be a substitution map. Then, the following assertions hold:*

(i) *For each $D \in \mathrm{HS}^{\mathbf{s}}_k(A;\nabla)$ there is a unique substitution map $\varphi^D : A[[\mathbf{s}]]_\nabla \to A[[\mathbf{t}]]_\Delta$ such that $\left(\widetilde{\varphi\bullet D}\right)\circ\varphi^D = \varphi\circ\widetilde{D}$. Moreover, $(\varphi\bullet D)^* = \varphi^D\bullet D^*$ and $\varphi^{\mathbb{I}} = \varphi$.*

(ii) *For each $D, E \in \mathrm{HS}^{\mathbf{s}}_k(A;\nabla)$, we have $\varphi\bullet(D\circ E) = (\varphi\bullet D)\circ(\varphi^D\bullet E)$ and $\left(\varphi^D\right)^E = \varphi^{D\circ E}$. In particular, $\left(\varphi^D\right)^{D^*} = \varphi$.*

(iii) *If ψ is another composable substitution map, then $(\varphi\circ\psi)^D = \varphi^{\psi\bullet D}\circ\psi^D$.*

(iv) *$\tau_n(\varphi^D) = \tau_n(\varphi)^{\tau_n(D)}$, for all $n \geq 1$.*

(v) *If φ has constant coefficients then $\varphi^D = \varphi$.*

Proof

(i) We know that

$$\widetilde{D} \in \mathrm{Aut}^{\circ}_{k[[\mathbf{s}]]_\nabla-\mathrm{alg}}(A[[\mathbf{s}]]_\nabla) \quad \text{and} \quad \widetilde{\varphi\bullet D} \in \mathrm{Aut}^{\circ}_{k[[\mathbf{t}]]_\Delta-\mathrm{alg}}(A[[\mathbf{t}]]_\Delta).$$

The only thing to prove is that

$$\varphi^D := \left(\widetilde{\varphi\bullet D}\right)^{-1}\circ\varphi\circ\widetilde{D}$$

is a substitution map $A[[\mathbf{s}]]_\nabla \to A[[\mathbf{t}]]_\Delta$ (see Definition 5). Let start by proving that φ^D is an A-algebra map. Let us write $E = \varphi\bullet D$. For each $a \in A$ we have

$$\varphi^D(a) = \widetilde{E}^{-1}\left(\varphi\left(\widetilde{D}(a)\right)\right) = \widetilde{E}^{-1}\left(\varphi\left(\Phi_D(a)\right)\right) =$$
$$\widetilde{E}^{-1}\left((\varphi\circ\Phi_D)(a)\right) = \widetilde{E}^{-1}\left(\Phi_{\varphi\bullet D}(a)\right) = \widetilde{E}^{-1}\left(\left(\widetilde{\varphi\bullet D}\right)(a)\right) = a,$$

and so φ^D is A-linear. The continuity of φ^D is clear, since it is the composition of continuous maps. For each $s \in \mathbf{s}$, let us write

$$\varphi(s) = \sum_{\substack{\beta\in\Delta\\ |\beta|>0}} c^s_\beta \mathbf{t}^\beta.$$

Since φ is a substitution map, property (17) holds:

$$\#\{s \in \mathbf{s} \mid c^s_\beta \neq 0\} < \infty \qquad \text{for all } \beta \in \Delta.$$

We have

$$\varphi^D(s) = \widetilde{E^*}\left(\varphi(\widetilde{D}(s))\right) = \widetilde{E^*}\left(\varphi(s)\right) = \sum_{\beta\in\Delta}\left(\sum_{\alpha+\gamma=\beta} E^*_\alpha(c^s_\gamma)\right)\mathbf{t}^\beta = \sum_{\beta\in\Delta} d^s_\beta \mathbf{t}^\beta$$

with $d^s_\beta = \sum_{\alpha+\gamma=\beta} E^*_\alpha(c^s_\gamma)$. So, for each $\beta \in \Delta$ we have

$$\{s \in \mathbf{s} \mid c^s_\beta \neq 0\} \subset \bigcup_{\gamma\leq\beta} \{s \in \mathbf{s} \mid c^s_\gamma \neq 0\}$$

and φ^D satisfies property (17) too. We conclude that φ^D is a substitution map, and obviously it is the only one such that $\left(\widetilde{\varphi\bullet D}\right) \circ \varphi^D = \varphi \circ \widetilde{D}$. From there, we have

$$\varphi^D \circ \widetilde{D^*} = \varphi^D \circ \widetilde{D}^{-1} = \left(\widetilde{\varphi\bullet D}\right)^{-1} \circ \varphi = \widetilde{(\varphi\bullet D)^*} \circ \varphi,$$

and taking restrictions to A we obtain $\varphi^D \circ \Phi_{D^*} = \Phi_{(\varphi\bullet D)^*}$ and so $\varphi^D \bullet D^* = (\varphi\bullet D)^*$.

On the other hand, it is clear that if $D = \mathbb{I}$, then $\varphi^{\mathbb{I}} = \varphi$ and if $\varphi = \mathbf{0}$, $\mathbf{0}^D = \mathbf{0}$.

(ii) In order to prove the first equality, we need to prove the equality $\widetilde{\varphi\bullet(D\circ E)} = \left(\widetilde{\varphi\bullet D}\right) \circ \left(\widetilde{\varphi^D\bullet E}\right)$. For this it is enough to prove the equality after restriction to A, but

$$\left(\widetilde{\varphi\bullet(D\circ E)}\right)|_A = \Phi_{\varphi\bullet(D\circ E)} = \varphi\circ\Phi_{D\circ E} = \varphi\circ\widetilde{D}\circ\Phi_E,$$

$$\left(\left(\widetilde{\varphi\bullet D}\right)\circ\left(\widetilde{\varphi^D\bullet E}\right)\right)|_A = \left(\widetilde{\varphi\bullet D}\right)\circ\Phi_{\varphi^D\bullet E} = \left(\widetilde{\varphi\bullet D}\right)\circ\varphi^D\circ\Phi_E$$

and both are equal by (i). For the second equality, we have $\left(\varphi^D\right)^{D^*} = \varphi^{\mathbb{I}} = \varphi$.

(iii) Since

$$\left(\widetilde{(\varphi\circ\psi)\bullet D}\right)\circ\left(\varphi^{\psi\bullet D}\circ\psi^D\right) = \left(\widetilde{\varphi\bullet(\psi\bullet D)}\right)\circ\varphi^{\psi\bullet D}\circ\psi^D =$$
$$\varphi\circ\left(\widetilde{\psi\bullet D}\right)\circ\psi^D = \varphi\circ\psi\circ\widetilde{D},$$

we deduce that $(\varphi\circ\psi)^D = \varphi^{\psi\bullet D}\circ\psi^D$ from the uniqueness in (i).

Part (iv) is also a consequence of the uniqueness property in (i).

(v) Let us assume that φ has constant coefficients. We know from Lemma 9 that $\langle\varphi\bullet D, \varphi(a)\rangle = \varphi\left(\langle D, a\rangle\right)$ for all $a \in A[[\mathbf{s}]]_\nabla$, and so $\left(\widetilde{\varphi\bullet D}\right)\circ\varphi = \varphi\circ\widetilde{D}$. Hence, by the uniqueness property in (i) we deduce that $\varphi^D = \varphi$.

The following proposition gives a recursive formula to obtain φ^D from φ.

Proposition 12 *With the notations of Proposition 11, we have*

$$\mathbf{C}_e(\varphi, f+\nu) = \sum_{\substack{\beta+\gamma=e \\ |f+g|\leq|\beta|, |\nu|\leq|\gamma|}} \mathbf{C}_\beta(\varphi, f+g) D_g(\mathbf{C}_\gamma(\varphi^D, \nu))$$

for all $e \in \Delta$ and for all $f, \nu \in \nabla$ with $|f+\nu| \leq |e|$. In particular, we have the following recursive formula

$$\mathbf{C}_e(\varphi^D, \nu) := \mathbf{C}_e(\varphi, \nu) - \sum_{\substack{\beta+\gamma=e \\ |g|\leq|\beta|, |\nu|\leq|\gamma|<|e|}} \mathbf{C}_\beta(\varphi, g) D_g(\mathbf{C}_\gamma(\varphi^D, \nu)).$$

for $e \in \Delta$, $\nu \in \nabla$ with $|e| \geq 1$ and $|\nu| \leq |e|$, starting with $\mathbf{C}_0(\varphi^D, 0) = 1$.

Proof First, the case $f = 0$ easily comes from the equality

$$\sum_{\substack{e\in\Delta \\ |\nu|\leq|e|}} \mathbf{C}_e(\varphi, \nu)\mathbf{t}^e = \varphi(\mathbf{s}^\nu) = (\varphi \circ \widetilde{D})(\mathbf{s}^\nu) = \left(\left(\widetilde{\varphi \bullet D}\right) \circ \varphi^D\right)(\mathbf{s}^\nu) \quad \forall \nu \in \nabla.$$

For arbitrary f one has to use Proposition 3. Details are left to the reader.

The proof of the following corollary is a consequence of Lemma 10.

Corollary 3 *Under the hypotheses of Proposition 11, the following identity holds for each $e \in \Delta$*

$$(\varphi \bullet D)^*_e = \sum_{|\mu+\nu|\leq|e|} D^*_\mu \cdot D_\nu \left(\mathbf{C}_e(\varphi^D, \mu+\nu)\right).$$

Proposition 13 *Let $D \in \mathrm{HS}^{\mathbf{t}}_k(A; \Delta)$ be a HS-derivation and $\varphi : A[[\mathbf{s}]]_\nabla \to A[[\mathbf{t}]]_\Delta$ a substitution map. Then, the following identity holds:*

$$\widetilde{D} \circ \varphi = (D(\varphi) \otimes \pi) \circ \left(\widetilde{\kappa \bullet D}\right) \circ \iota,$$

where:

- $D(\varphi) : A[[\mathbf{s}]]_\nabla \to A[[\mathbf{t}]]_\Delta$ *is the substitution map determined by $D(\varphi)(s) = \widetilde{D}(\varphi(s))$ for all $s \in \mathbf{s}$.*
- $\pi : A[[\mathbf{t}]]_\Delta \to A$ *is the augmentation, or equivalently, the substitution map*[7] *given by $\pi(t) = 0$ for all $t \in \mathbf{t}$.*

[7]The map π can be also understood as the truncation $\tau_{\Delta,\{0\}} : A[[\mathbf{t}]]_\Delta \to A[[\mathbf{t}]]_{\{0\}} = A$.

- *$\iota : A[[\mathbf{s}]]_\nabla \to A[[\mathbf{s} \sqcup \mathbf{t}]]_{\nabla\times\Delta}$ and $\kappa : A[[\mathbf{t}]]_\Delta \to A[[\mathbf{s} \sqcup \mathbf{t}]]_{\nabla\times\Delta}$ are the combinatorial substitution maps determined by the inclusions $\mathbf{s} \hookrightarrow \mathbf{s} \sqcup \mathbf{t}$ and $\mathbf{t} \hookrightarrow \mathbf{s} \sqcup \mathbf{t}$, respectively.*

Proof It is enough to check that both maps coincide on any $a \in A$ and on any $s \in \mathbf{s}$. Details are left to the reader.

Remark 6 Let us notice that with the notations of Propositions 11 and 13, we have $\psi^{D} = (\psi \bullet D)^{\Pi}(\varphi)$.

The following proposition will not be used in this paper and will be stated without proof.

Proposition 14 *For any HS-derivation $D \in \mathrm{HS}^{\mathbf{s}}_k(A; \nabla)$ and any substitution map $\varphi \in \mathcal{S}(\mathbf{t}, \mathbf{u}; \Delta, \Omega)$, there exists a substitution map $D\star\varphi \in \mathcal{S}(\mathbf{s}\sqcup\mathbf{t}, \mathbf{s}\sqcup\mathbf{u}; \nabla\times\Delta, \nabla\times\Omega)$ such that for each HS-derivation $E \in \mathrm{HS}^{\mathbf{t}}_k(A; \Delta)$ we have:*

$$D \boxtimes (\varphi \bullet E) = (D \star \varphi) \bullet (D \boxtimes E).$$

7 Generating HS-Derivations

In this section we show how the action of substitution maps allows us to express any HS-derivation in terms of a fixed one under some natural hypotheses. We will be concerned with $(\mathbf{s}, \mathfrak{t}_m(\mathbf{s}))$-variate HS-derivations, where $\mathfrak{t}_m(\mathbf{s}) = \{\alpha \in \mathbb{N}^{(\mathbf{s})} \mid |\alpha| \leq m\}$. To simplify we will write $A[[\mathbf{s}]]_m := A[[\mathbf{s}]]_{\mathfrak{t}_m(\mathbf{s})}$ and $\mathrm{HS}^{\mathbf{s}}_k(A; m) := \mathrm{HS}^{\mathbf{s}}_k(A; \mathfrak{t}_m(\mathbf{s}))$ for any integer $m \geq 1$, and $\mathrm{HS}^{\mathbf{s}}_k(A; \infty) := \mathrm{HS}^{\mathbf{s}}_k(A)$. For $m \geq n \geq 1$ we will denote $\tau_{mn} : \mathrm{HS}^{\mathbf{s}}_k(A; m) \to \mathrm{HS}^{\mathbf{s}}_k(A; n)$ the truncation map.

Assume that $m \geq 1$ is an integer and let $\varphi : A[[\mathbf{s}]]_m \to A[[\mathbf{t}]]_m$ be a substitution map. Let us write

$$\varphi(s) = c^s = \sum_{\substack{\beta\in\mathbb{N}^{(\mathbf{t})}\\ 0<|\beta|\leq m}} c^s_\beta \mathbf{t}^\beta \in \mathfrak{n}_0(\mathbf{t})/\mathfrak{t}_m(\mathbf{t}) \subset A[[\mathbf{t}]]_m, \quad s \in \mathbf{s}$$

and let us denote by $\varphi_m, \varphi_{<m} : A[[\mathbf{s}]]_m \to A[[\mathbf{t}]]_m$ the substitution maps determined by

$$\varphi_m(s) = c^s_m := \sum_{\substack{\beta\in\mathbb{N}^{(\mathbf{t})}\\ |\beta|=m}} c^s_\beta \mathbf{t}^\beta \in \mathfrak{n}_0(\mathbf{t})/\mathfrak{t}_m(\mathbf{t}) \in A[[\mathbf{t}]]_m, \quad s \in \mathbf{s},$$

$$\varphi_{<m}(s) = c^s_{<m} := \sum_{\substack{\beta\in\mathbb{N}^{(\mathbf{t})}\\ 0<|\beta|<m}} c^s_\beta \mathbf{t}^\beta \in \mathfrak{n}_0(\mathbf{t})/\mathfrak{t}_m(\mathbf{t}) \in A[[\mathbf{t}]]_m, \quad s \in \mathbf{s}.$$

We have $c^s = c^s_m + c^s_{<m}$ and so $\varphi = \varphi_m + \varphi_{<m}$ (see 3).

Proposition 15 *With the above notations, for any HS-derivation $D \in \mathrm{HS}^{\mathbf{s}}_k(A; m)$ the following properties hold:*

(1) $(\varphi_m \bullet D)_e = 0$ *for* $0 < |e| < m$ *and* $(\varphi_m \bullet D)_e = \sum_{t \in \mathbf{s}} c^t_e D_{\mathbf{s}^t}$ *for* $|e| = m$, *where the* $\mathbf{s}^t$ *are the elements of the canonical basis of* $\mathbb{N}^{(\mathbf{s})}$.
(2) $\varphi \bullet D = (\varphi_m \bullet D) \circ (\varphi_{<m} \bullet D) = (\varphi_{<m} \bullet D) \circ (\varphi_m \bullet D)$.

Proof

(1) Let us denote $E' = \varphi_m \bullet D$. Since $\tau_{m,m-1}(E')$ coincides with $\tau_{m,m-1}(\varphi_m) \bullet \tau_{m,m-1}(D)$ (see (28)) and $\tau_{m,m-1}(\varphi_m)$ is the trivial substitution map, we deduce that $\tau_{m,m-1}(E') = \mathbb{I}$, i.e. $E_e = 0$ whenever $0 < |e| < m$.

From (29) and (14), for $|e| > 0$ we have $E'_e = \sum_{0<|\alpha|\le|e|} \mathbf{C}_e(\varphi_m, \alpha) D_\alpha$, with

$$\mathbf{C}_e(\varphi_m, \alpha) = \sum_{\ell^{\bullet\bullet} \in \mathcal{P}(e,\alpha)} C_{\ell^{\bullet\bullet}} \quad \text{for } |\alpha| \le |e|, \quad C_{\ell^{\bullet\bullet}} = \prod_{s \in \operatorname{supp}\alpha} \prod_{r=1}^{\alpha_s} (c^s_m)_{\ell^{sr}}.$$

Assume now that $|e| = m$, $1 < |\alpha| \le m$ and let $\ell^{\bullet\bullet} \in \mathcal{P}(e, \alpha)$. Since

$$\sum_{s \in \operatorname{supp}\alpha} \sum_{r=1}^{\alpha_s} \ell^{sr} = e,$$

we deduce that $|\ell^{sr}| < |e| = m$ for all s, r and so $(c^s_m)_{\ell^{sr}} = 0$ and $C_{\ell^{\bullet\bullet}} = 0$. Consequently, $\mathbf{C}_e(\varphi_m, \alpha) = 0$.

If $|\alpha| = 1$, then α must be an element $\mathbf{s}^t$ of the canonical basis of $\mathbb{N}^{(\mathbf{s})}$ and from Lemma 6, (1), we know that $\mathbf{C}_e(\varphi_m, \mathbf{s}^t) = (c^t_m)_e$. We conclude that

$$E'_e = \cdots = \sum_{t \in \mathbf{s}} \mathbf{C}_e(\varphi_m, \mathbf{s}^t) D_{\mathbf{s}^t} = \sum_{t \in \mathbf{s}} (c^t_m)_e D_{\mathbf{s}^t} = \sum_{t \in \mathbf{s}} c^t_e D_{\mathbf{s}^t}.$$

(2) Let us write $E = \varphi \bullet D$, $E' = \varphi_m \bullet D$ and $E'' = \varphi_{<m} \bullet D$. We have

$$\tau_{m,m-1}(E) = \tau_{m,m-1}(\varphi) \bullet \tau_{m,m-1}(D) =$$
$$\tau_{m,m-1}(\varphi_{<m}) \bullet \tau_{m,m-1}(D) = \tau_{m,m-1}(E'').$$

By property (1), we know that $\tau_{m,m-1}(E')$ is the identity and we deduce that $\tau_{m,m-1}(E) = \tau_{m,m-1}(E' \circ E'') = \tau_{m,m-1}(E'' \circ E')$. So $E_e = (E' \circ E'')_e = (E'' \circ E')_e$ for $|e| < m$.

Now, let $e \in \mathbb{N}^{(\mathbf{t})}$ be with $|e| = m$. By using again that $\tau_{m,m-1}(E')$ is the identity, we have $(E' \circ E'')_e = \cdots = E'_e + E''_e = \cdots = (E'' \circ E')_e$, and we conclude that $E' \circ E'' = E'' \circ E'$.

On the other hand, from Lemma 6, (1), we have that $\mathbf{C}_e(\varphi_{<m},\alpha) = 0$ whenever $|\alpha| = 1$, and one can see that $\mathbf{C}_e(\varphi,\alpha) = \mathbf{C}_e(\varphi_{<m},\alpha)$ whenever that $2 \leq |\alpha| \leq |e|$. So:

$$E_e = \sum_{1\leq|\alpha|\leq m} \mathbf{C}_e(\varphi,\alpha) D_\alpha = \sum_{|\alpha|=1} \mathbf{C}_e(\varphi,\alpha) D_\alpha + \sum_{2\leq|\alpha|\leq m} \mathbf{C}_e(\varphi,\alpha) D_\alpha =$$

$$\sum_{t\in\mathbf{s}} c_e^t D_{\mathbf{s}^t} + \sum_{2\leq|\alpha|\leq m} \mathbf{C}_e(\varphi_{<m},\alpha) D_\alpha = E'_e + \sum_{1\leq|\alpha|\leq m} \mathbf{C}_e(\varphi_{<m},\alpha) D_\alpha = E'_e + E''_e$$

and $E = E' \circ E'' = E'' \circ E'$.

The following theorem generalizes Theorem 2.8 in [3] to the case where $\operatorname{Der}_k(A)$ is not necessarily a finitely generated A-module. The use of substitution maps makes its proof more conceptual.

Theorem 1 *Let $m \geq 1$ be an integer, or $m = \infty$, and $D \in \operatorname{HS}^{\mathbf{s}}_k(A; m)$ a $\mathbf{s}$-variate HS-derivation of length m such that $\{D_\alpha, |\alpha| = 1\}$ is a system of generators of the A-module $\operatorname{Der}_k(A)$. Then, for each set $\mathbf{t}$ and each HS-derivation $G \in \operatorname{HS}^{\mathbf{t}}_k(A; m)$ there is a substitution map $\varphi : A[[\mathbf{s}]]_m \to A[[\mathbf{t}]]_m$ such that $G = \varphi \bullet D$. Moreover, if $\{D_\alpha, |\alpha| = 1\}$ is a basis of $\operatorname{Der}_k(A)$, φ is uniquely determined.*

Proof For m finite, we will proceed by induction on m. For $m = 1$ the result is clear. Assume that the result is true for HS-derivations of length $m-1$ and consider a $D \in \operatorname{HS}^{\mathbf{s}}_k(A; m)$ such that $\{D_\alpha, |\alpha| = 1\}$ is a system of generators of the A-module $\operatorname{Der}_k(A)$ and a $G \in \operatorname{HS}^{\mathbf{t}}_k(A; m)$. By the induction hypothesis, there is a substitution map $\varphi' : A[[\mathbf{s}]]_{m-1} \to A[[\mathbf{t}]]_{m-1}$, given by $\varphi'(s) = \sum_{|\beta|\leq m-1} c^s_\beta \mathbf{t}^\beta$, $s \in \mathbf{s}$, and such that $\tau_{m,m-1}(G) = \varphi' \bullet \tau_{m,m-1}(D)$. Let $\varphi'' : A[[\mathbf{s}]]_m \to A[[\mathbf{u}]]_m$ be the substitution map lifting φ' (i.e. $\tau_{m,m-1}(\varphi'') = \varphi'$) given by $\varphi''(s) = \sum_{|\beta|\leq m-1} c^s_\beta \mathbf{t}^\beta \in A[[\mathbf{t}]]_m$, $s \in \mathbf{s}$, and consider $F = \varphi'' \bullet D$. We obviously have $\tau_{m,m-1}(F) = \tau_{m,m-1}(G)$ and so, for $H = G \circ F^*$, the truncation $\tau_{m,m-1}(H)$ is the identity and $H_e = 0$ for $0 < |e| < m$. We deduce that each component of H of highest order, H_e with $|e| = m$, must be a k-derivation of A and so there is a family $\{c^s_e, s \in \mathbf{s}\}$ of elements of A such that $c^s_e = 0$ for all s except a finite number of indices and $H_e = \sum_{s\in\mathbf{s}} c^{\mathbf{s}}_e D_{\mathbf{s}^s}$, where $\{\mathbf{s}^s, s \in \mathbf{s}\}$ is the canonical basis of $\mathbb{N}^{(\mathbf{s})}$. To finish, let us consider the substitution map $\varphi : A[[\mathbf{s}]]_m \to A[[\mathbf{t}]]_m$ given by $\varphi(s) = \sum_{|\beta|\leq m} c^s_\beta \mathbf{t}^\beta$, $s \in \mathbf{s}$. From Proposition 15 we have

$$\varphi \bullet D = (\varphi_m \bullet D) \circ (\varphi_{<m} \bullet D) = H \circ (\varphi'' \bullet D) = H \circ F = G.$$

For HS-derivations of infinite length, following the above procedure we can construct φ as a projective limit of substitution maps $A[[\mathbf{s}]]_m \to A[[\mathbf{t}]]_m$, $m \geq 1$.

Now assume that the set $\{D_\alpha, |\alpha| = 1\}$ is linearly independent over A and let us prove that

$$\varphi \bullet D = \psi \bullet D \quad \Longrightarrow \quad \varphi = \psi. \tag{30}$$

The infinite length case can be reduced to the finite case since $\varphi = \psi$ if and only if all their finite truncations are equal. For the finite length case, we proceed by induction on the length m. Assume that the substitution maps are given by

$$\varphi(s) = c^s := \sum_{\substack{\beta \in \mathbb{N}^{(\mathbf{t})} \\ 0<|\beta|\leq m}} c^s_\beta \mathbf{t}^\beta \in \mathfrak{n}_0(\mathbf{t})/\mathfrak{t}_m(\mathbf{t}) \subset A[[\mathbf{t}]]_m, \quad s \in \mathbf{s}$$

$$\psi(s) = d^s := \sum_{\substack{\beta \in \mathbb{N}^{(\mathbf{t})} \\ 0<|\beta|\leq m}} d^s_\beta \mathbf{t}^\beta \in \mathfrak{n}_0(\mathbf{t})/\mathfrak{t}_m(\mathbf{t}) \subset A[[\mathbf{t}]]_m, \quad s \in \mathbf{s}.$$

If $m = 1$, then $\varphi = \varphi_1$ and $\psi = \psi_1$ and for each $e \in \mathbb{N}^{(\mathbf{t})}$ with $|e| = 1$ we have from Proposition 15

$$\sum_{s \in \mathbf{s}} c^s_e D_{\mathbf{s}^s} = (\varphi_1 \bullet D)_e = (\varphi \bullet D)_e = (\psi \bullet D)_e = (\psi_1 \bullet D)_e = \sum_{s \in \mathbf{s}} d^s_e D_{\mathbf{s}^s}$$

and we deduce that $c^s_e = d^s_e$ for all $s \in \mathbf{s}$ and so $\varphi = \psi$.

Now assume that (30) is true whenever the length is $m - 1$ and take D, φ and ψ as before of length m with $\varphi \bullet D = \psi \bullet D$. By considering $(m - 1)$-truncations and using the induction hypothesis we deduce that $\tau_{m,m-1}(\varphi) = \tau_{m,m-1}(\psi)$, or equivalently $\varphi_{<m} = \psi_{<m}$.

From Proposition 15 we obtain first that $\varphi_m \bullet D = \psi_m \bullet D$ and second that for each $e \in \mathbb{N}^{(\mathbf{t})}$ with $|e| = m$

$$\sum_{s \in \mathbf{s}} c^s_e D_{\mathbf{s}^s} = \sum_{s \in \mathbf{s}} d^s_e D_{\mathbf{s}^s}.$$

We conclude that $\varphi_m = \psi_m$ and so $\varphi = \psi$.

Now we recall the definition of integrability.

Definition 11 (Cf. [1, 7]) Let $m \geq 1$ be an integer or $m = \infty$ and $\mathbf{s}$ a set.

(i) We say that a k-derivation $\delta : A \to A$ is *m-integrable* (over k) if there is a Hasse–Schmidt derivation $D \in \operatorname{HS}_k(A; m)$ such that $D_1 = \delta$. Any such D will be called an *m-integral* of δ. The set of m-integrable k-derivations of A is denoted by $\operatorname{Ider}_k(A; m)$. We simply say that δ is *integrable* if it is ∞-integrable and we denote $\operatorname{Ider}_k(A) := \operatorname{Ider}_k(A; \infty)$.

(ii) We say that a $\mathbf{s}$-variate HS-derivation $D' \in \operatorname{HS}^{\mathbf{s}}_k(A; n)$, with $1 \leq n < m$, is *m-integrable* (over k) if there is a $\mathbf{s}$-variate HS-derivation $D \in \operatorname{HS}^{\mathbf{s}}_k(A; m)$ such that $\tau_{mn} D = D'$. Any such D will be called an *m-integral* of D'. The set of m-integrable $\mathbf{s}$-variate HS-derivations of A over k of length n is denoted by $\operatorname{IHS}^{\mathbf{s}}_k(A; n; m)$. We simply say that D' is *integrable* if it is ∞-integrable and we denote $\operatorname{IHS}^{\mathbf{s}}_k(A; n) := \operatorname{IHS}^{\mathbf{s}}_k(A; n; \infty)$.

Corollary 4 *Let $m \geq 1$ be an integer or $m = \infty$. The following properties are equivalent:*

(1) $\mathrm{Ider}_k(A; m) = \mathrm{Der}_k(A)$.
(2) $\mathrm{IHS}^{\mathbf{s}}_k(A; n; m) = \mathrm{HS}^{\mathbf{s}}_k(A; n)$ *for all n with $1 \leq n < m$ and all sets* $\mathbf{s}$.

Proof We only have to prove (1) $\Longrightarrow$ (2). Let $\{\delta_t, t \in \mathbf{t}\}$ be a system of generators of the A-module $\mathrm{Der}_k(A)$, and for each $t \in \mathbf{t}$ let $D^t \in \mathrm{HS}_k(A; m)$ be an m-integral of δ_t. By considering some total ordering $\prec$ on $\mathbf{t}$, we can define $D \in \mathrm{HS}^{\mathbf{t}}_k(A; m)$ as the external product (see Definition 3) of the ordered family $\{D^t, t \in \mathbf{t}\}$, i.e. $D_0 = \mathrm{Id}$ and for each $\alpha \in \mathbb{N}^{(\mathbf{t})}$, $\alpha \neq 0$,

$$D_\alpha = D^{t_1}_{\alpha_{t_1}} \circ \cdots \circ D^{t_e}_{\alpha_{t_e}} \quad \text{with} \quad \operatorname{supp} \alpha = \{t_1 < \cdots < t_e\}.$$

Let n be an integer with $1 \leq n < m$, $\mathbf{s}$ a set and $E \in \mathrm{HS}^{\mathbf{s}}_k(A; n)$. After Theorem 1, there exists a substitution map $\varphi : A[[\mathbf{t}]]_n \to A[[\mathbf{s}]]_n$ such that $E = \varphi \bullet \tau_{mn}(D)$. By considering any substitution map $\varphi' : A[[\mathbf{t}]]_m \to A[[\mathbf{s}]]_m$ lifting φ we find that $\varphi' \bullet D$ is an m-integral of E and so $E \in \mathrm{IHS}^{\mathbf{s}}_k(A; n; m)$.

References

1. Brown, W.C.: On the embedding of derivations of finite rank into derivations of infinite rank. Osaka J. Math. **15**, 381–389 (1978).
2. Bourbaki, N.: Elements of Mathematics. Algebra II, chaps. 4–7. Springer, Berlin (2003)
3. Fernández-Lebrón, M., Narváez-Macarro, L.: Hasse-Schmidt derivations and coefficient fields in positive characteristics. J. Algebra **265**(1), 200–210 (2003). arXiv:math/0206261
4. Hasse, H., Schmidt, F.K.: Noch eine Begründung der Theorie der höheren Differrentialquotienten in einem algebraischen Funktionenkörper einer Unbestimmten. J. Reine U. Angew. Math. **177**, 223–239 (1937)
5. Hoffmann, D., Kowalski, P.: Integrating Hasse–Schmidt derivations. J. Pure Appl. Algebra **219**, 875–896 (2015)
6. Hoffmann, D., Kowalski, P.: Existentially closed fields with G-derivations. J. Lond. Math. Soc. (2) **93**(3), 590–618 (2016)
7. Matsumura, H.: Integrable derivations. Nagoya Math. J. **87**, 227–245 (1982)
8. Matsumura, H.: Commutative Ring Theory. Cambridge Studies in Advanced Mathematics, vol. 8. Cambridge University Press, Cambridge (1986)
9. Narváez Macarro, L.: Hasse–Schmidt derivations, divided powers and differential smoothness. Ann. Inst. Fourier (Grenoble) **59**(7), 2979–3014 (2009). arXiv:0903.0246
10. Narváez Macarro, L.: On the modules of m-integrable derivations in non-zero characteristic. Adv. Math. **229**(5), 2712–2740 (2012). arXiv:1106.1391
11. Narváez Macarro, L.: Differential Structures in Commutative Algebra. Mini-course at the XXIII Brazilian Algebra Meeting, 27 July–1 August 2014, Maringá, Brazil
12. Narváez Macarro, L.: Rings of differential operators as enveloping algebras of Hasse–Schmidt derivations. arXiv:1807.10193
13. Vojta, P.: Jets via Hasse–Schmidt derivations. In: Diophantine geometry. CRM Series, vol. 4, pp. 335–361. Edizioni della Normale, Pisa (2007). arXiv:math/1201.3594

An Introduction to Resolution of Singularities via the Multiplicity

Diego Sulca and Orlando Villamayor U.

Abstract In these notes we study properties of the multiplicity at points of a variety X over a perfect field. We focus on properties that can be studied using ramification method, such as discriminants and some generalized discriminants that we shall introduce. We also show how these methods lead to an alternative proof of resolution of singularities for varieties over fields of characteristic zero.

1 Introduction

Fix a variety X over a perfect field. The multiplicity function of X is the map

$$\mathrm{mult}_X : X \to \mathbb{N}$$

that assigns to each point $x \in X$ the multiplicity of the local ring $\mathcal{O}_{X,x}$. This function is well known to be upper semi-continuous when $\mathbb{N}$ is given the usual order, so it stratifies X as a finite union of locally closed subsets: *the level sets*

$$F_n(X) := \{x \in X : \mathrm{mult}_X(x) = n\}, \quad n \in \mathbb{N}.$$

It is also well known that the variety X is regular if and only if $\mathrm{mult}_X(x) = 1$ for all $x \in X$. When X is not regular, we denote by $n(X)$ the maximum value of the function mult_X. Since mult_X is upper semi-continuous, for $n = n(X)$ the level set $F_n(X)$, namely the highest multiplicity locus of X, is closed. In these notes we aim to obtain a description of this closed set in a way that will ultimately lead us to its simplification via blow-ups as we shall indicate below.

D. Sulca
CIEM-FAMAF, Universidad Nacional de Córdoba, Córdoba, Argentina
e-mail: sulca@famaf.unc.edu.ar

O. Villamayor U. (✉)
ICMAT-Universidad Autónoma de Madrid, Madrid, Spain
e-mail: villamayor@uam.es

G.-M. Greuel et al. (eds.), *Singularities, Algebraic Geometry, Commutative Algebra, and Related Topics*, https://doi.org/10.1007/978-3-319-96827-8_11

Firstly, a remarkable fact about the multiplicity is that it does not increase when blowing up at equimultiple centers, as stated in the following theorem due to Dade, and later simplified by Orbanz in [16].

Theorem 1.1 (Dade, [6]) *If $X \stackrel{\pi}{\leftarrow} X_1$ is the blow up of X at an irreducible regular center included in a level set $F_n(X)$, then*

$$\operatorname{mult}_{X_1}(x_1) \leq \operatorname{mult}_X(\pi(x_1)), \quad \textit{for all } x_1 \in X_1.$$

In particular, $n(X_1) \leq n(X)$. If $n(X_1) < n(X)$, we say that $X \stackrel{\pi}{\leftarrow} X_1$ is *a reduction of the multiplicity* of X. More generally,

Definition 1.2 A reduction of the multiplicity of X is a sequence of morphisms

$$X = X_0 \stackrel{\pi_1}{\longleftarrow} X_1 \stackrel{\pi_2}{\longleftarrow} \dots \longleftarrow X_{r-1} \stackrel{\pi_r}{\longleftarrow} X_r\ ,$$

where each $X_{i-1} \stackrel{\pi_i}{\leftarrow} X_i$ is the blow-up at an irreducible smooth center included in the maximal multiplicity locus of X_i, and $n(X) = n(X_1) = \cdots = n(X_{r-1}) > n(X_r)$.

Note that if any singular variety over a field k admits a reduction of the multiplicity, then any variety over k admits a resolution of its singularities. An algorithmic reduction of the multiplicity was obtained in [20] for varieties over a field of characteristic zero. Here we shall address the key points of the proof of this result, namely

Theorem 1.3 *If X is a singular variety over a field of characteristic zero, then X admits a reduction of the multiplicity. In particular, X admits a resolution of singularities by blowing up successively at regular equimultiple centers.*

Hironaka proved that there exists a resolution of singularities for any variety over a field of characteristic zero [12]. His proof is existential and makes use of the Hilbert-Samuel function and the notion of normal flatness. In contrast, as was mentioned before, there is an algorithm, or say a constructive procedure, to produce a reduction of the multiplicity of any singular variety over a field of characterisitc zero. This algorithm is based on two fundamental steps: (a) an explicit form of *local presentation* for the maximal multiplicity locus, and (b) a constructive *resolution of pairs*. These steps will be addressed here in these notes following, with some improvements, the presentation in [20] (see also [4]). We remark that while (a) is achieved in any characteristic by using only classical tools from commutative algebra, such as finite extensions and integral closure of ideals, step (b) requires the characteristic to be zero.

Let us explain briefly the general philosophy behind the steps listed above. Firstly there are various ways in which a singular variety X can be presented, and one of them is to view it as a ramified cover of a regular one. In this approach, one defines a finite and surjective morphism, say $X \to V$, where V is a smooth variety. Very classical notions, such as the discriminant, are defined in this context. It is natural to ask if one can extract information from the discriminant so as to simplify the singularities of X. This is the perspective, or say the expectation, with which the reader should go through these notes. For example, assume that V is affine, say with (regular) ring of functions S, and suppose that the ring of functions of X is given by $S[Z]/\langle Z^2 + a_1 Z + a_2\rangle$, where Z is a variable over S. This is a particular case in which X is viewed as a two-fold cover of V, and in this case the discriminant, namely $a_1^2 - 4a_2$, will allow us to describe the singular locus of X, and ultimately to resolve the singularities at least if the characteristic is zero.

The results that will lead us to the proof of Theorem 1.3 will make use of a notion of *pairs*, and their associated closed subsets. The concept of pairs was introduced by Hironaka as an auxiliary tool used in his proof of resolution of singularities over fields of characteristic zero [12] (see also [13]). They will also show up in the alternative proof discussed here, which is based on ramification methods and certain pairs called *generalized discriminants*.

2 An Overview of the Main Results

2.1 Pairs, Transformations of Pairs and Closed Subsets

Fix a smooth variety V over a field k. Given a coherent ideal J on V ($\mathcal{O}_V$-ideal for short) and a positive integer b, we call (J, b) a *pair*. There is a closed subset of V associated with this pair, namely

$$\mathrm{Sing}(J, b) := \{x \in V : \nu_x(J) \geq b\} \subset V.$$

Here $\nu_x(J)$ denotes the order of J_x at the regular local ring $\mathcal{O}_{V,x}$. It is well known that $\mathrm{Sing}(J, b)$ is closed and we refer to [7] for a simple proof of this fact.

If $\mathrm{Sing}(J, b) \neq \emptyset$, then a smooth subvariety of V included in $\mathrm{Sing}(J, b)$ is said to be *a permissible center for* (J, b). Let $V \overset{\pi}{\leftarrow} V_1$ be the blow-up of V at a permissible center Y for (J, b). The fact that Y is included in $\mathrm{Sing}(J, b)$ ensures the existence of a factorization, say

$$J\mathcal{O}_{V_1} = I(H_1)^b J_1,$$

where J_1 is an $\mathcal{O}_{V_1}$-ideal and $H_1 \subset V_1$ is the exceptional hypersurface. The new pair (J_1, b) is called the *transform* of (J, b). Again, there is a closed subset $\mathrm{Sing}(J_1, b) \subset V_1$ associated with (J_1, b), and if this set is not empty, we can repeat the above construction again for a new permissible center. This leads us to the formulation of the following definition.

Definition 2.1 Fix a pair (J, b) on V such that $\mathrm{Sing}(J, b) \neq \emptyset$.

1. A *permissible sequence of blow-ups* for (J, b) is a sequence of the form

$$\begin{array}{ccccccccc} V = V_0 & \xleftarrow{\pi_1} & V_1 & \xleftarrow{\pi_2} & \dots & \xleftarrow{\pi_{r-1}} & V_{r-1} & \xleftarrow{\pi_r} & V_r \\ (J,b) = (J_0,b) & & (J_1,b) & & \dots & & (J_{r-1},b) & & (J_r,b) \end{array},$$

   where $V_{i-1} \leftarrow V_i$ is the blow up of V_{i-1} at a permissible centre for (J_{i-1}, b), and (J_i, b) is the transform of (J_{i-1}, b).
2. A sequence as above is called a resolution of (J, b) if $\mathrm{Sing}(J_r, b) = \emptyset$.

An important auxiliary result used to prove resolution of singularities is that over fields of characteristic zero every pair (J, b) admits a resolution. Moreover, this auxiliary result is rather easy to achieve, and there is an algorithm which produces such a resolution in the sense that given V and (J, b), it provides the first center to blow up, so it produces $V \leftarrow V_1$ and a transform (J_1, b) on V_1. Either $\mathrm{Sing}(J_1, b)$ is empty (in which case we have achieved a resolution for (J, b)), or is not empty and the algorithm produces a new center to blow up, and the blow-up $V_1 \leftarrow V_2$ produces a transform (J_2, b), and so on. The point is that for some index r, $\mathrm{Sing}(J_r, b) = \emptyset$. In addition, the algorithm is such that the exceptional locus of the composition, say $V \leftarrow V_r$, is a union of regular hypersurfaces having normal crossings. Here we will discuss resolution of pairs in a simplified manner, without mentioning the conditions of normal crossings in order to simplify the presentation of the paper. We will only list some properties of this algorithmic approach, detailed proofs can be found in the literature (see for example, [8]). Let us indicate that algorithms of resolutions of pairs have been implemented, and we refer here to

http://www.risc.uni-linz.ac.at/projects/basic/adjoints/blowup,

and also to

https://www.singular.uni-kl.de/Manual/4-0-3/sing_1454.htm

for software that have been developed in this way.

A crucial property of pairs is their role in the *assignment of closed subsets*. To clarify this assertion let us fix a smooth variety V over a field k, and a pair (J, b) on V. We already attached to this pair a closed subset $\mathrm{Sing}(J, b) \subset V$. Recall that if a blow-up $V \leftarrow V_1$ is permissible for (J, b) (Definition 2.1), then a transform (J_1, b) of (J, b) is defined on V_1, so we may say that the pair (J, b) assigns a closed subset $\mathrm{Sing}(J_1, b)$ in the variety V_1 for any blow-up $V \leftarrow V_1$ at a permissible center for (J, b).

We want to identify pairs on V, or say that two pairs on V are equivalent when they define the same closed sets on V and if they also define the same closed sets after permissible blow-ups. To make this precise, we will first introduce two other types of transformations of pairs, apart from blow-ups.

- Restrictions to open subsets: If $U \subset V$ is an open subvariety, the pair (J, b) restricts to a pair $(J|_U, b)$ on U, and the closed subset $\mathrm{Sing}(J|_U, b) \subset U$ is nothing else but $\mathrm{Sing}(J, b) \cap U \subset U$.

- Multiplication by an affine line: If $V[t] := V \times \mathbb{A}^1$ and $\pi : V[t] \to V$ is the projection onto the first component, the pull-back of J, denoted by $J[t]$, defines an $\mathscr{O}_{V[t]}$-ideal on $V[t]$, and it is easy to check that $\mathrm{Sing}(J[t], b) = \pi^{-1}(\mathrm{Sing}(J, b))$.

Summarizing, a pair (J, b) on V defines not only the closed subset $\mathrm{Sing}(J, b) \subset V$ but also closed subsets in other varieties which are obtained from V as result of one of the three transformations we mentioned before. If one iterates these transformations, new varieties and closed subsets will be obtained and we want to underline the fact that these closed subsets are defined by (J, b) via suitable transformations.

Definition 2.2 Let X be an algebraic variety. A local sequence over X is a sequence of morphisms

$$X = X_0 \xleftarrow{\pi_1} X_1 \xleftarrow{\pi_2} \dots \xleftarrow{\pi_{r-1}} X_{r-1} \xleftarrow{\pi_r} X_r \, ,$$

where each $X_{i-1} \overset{\pi_i}{\leftarrow} X_i$ is a morphism of one of the following three types:

(a) an open immersion,
(b) $X_i \cong X_{i-1} \times \mathbb{A}^1$ and π_i is the projection onto the first coordinate,
(c) a blow-up of X_{i-1} at a specified smooth subvariety $Y_{i-1} \subset X_{i-1}$.

Definition 2.3 Fix a smooth variety V over a field k and a pair (J, b) on V. A local sequence over V,

$$V = V_0 \xleftarrow{\pi_1} V_1 \xleftarrow{\pi_2} \dots \xleftarrow{\pi_{r-1}} V_{r-1} \xleftarrow{\pi_r} V_r \, ,$$

is said to be *permissible* for (J, b) if one can define inductively pairs

$$\begin{array}{ccccccccc} V =: V_0 & \xleftarrow{\pi_1} & V_1 & \xleftarrow{\pi_2} & \dots & \xleftarrow{\pi_{r-1}} & V_{r-1} & \xleftarrow{\pi_r} & V_r \\ (J,b) =: (J_0,b) & & (J_1,b) & & \dots & & (J_{r-1},b) & & (J_r,b) \end{array} ,$$

such that for any positive index $i \leq r$, the pair (J_i, b) on V_i is defined from the pair (J_{i-1}, b) on V_{i-1} as follows:

(i) if π_i is of type (a) or type (b) in the notation of Definition 2.2, J_i is the pull-back of J_{i-1};
(ii) if π_i is of type (c), namely, if π_i is a blow-up at a smooth center $Y_{i-1} \subset V_{i-1}$, then Y_{i-1} is actually included in $\mathrm{Sing}(J_{i-1}, b)$ and (J_i, b) is the transform of (J_{i-1}, b).

It follows easily that the pairs (J_i, b) are uniquely determined by (J, b) and the given permissible local sequence. We shall call the sequence of sets

$$\mathrm{Sing}(J_0, b) \subset V_0, \qquad \mathrm{Sing}(J_1, b) \subset V_1, \qquad \dots \qquad \mathrm{Sing}(J_r, b) \subset V_r,$$

the closed subsets defined by (J, b) and the permissible local sequence over V.

This definition enable us to think of a pair on V as an assignment of closed subsets on regular varieties via local sequences over V that are permissible for that pair.

We now introduce two important relations between pairs that arise when considering inclusions between the closed subsets defined by them.

Definition 2.4 Let (J, b) and (J', b') be two pairs on a smooth k-variety V.

1. We say that (J, b) is included in (J', b'), and write
$$(J, b) \subset (J', b'),$$
if any local sequence over V which is permissible for (J, b) (Definition 2.3) is also permissible for (J', b').
2. We say that (J, b) and (J', b') are equivalent and write
$$(J, b) \sim (J', b')$$
if $(J, b) \subset (J', b')$ and $(J', b') \subset (J, b)$.

Remark 2.5 If $(J, b) \subset (J', b')$ and $V = V_0 \leftarrow V_1 \leftarrow \ldots \leftarrow V_r$ is a local sequence over V which is permissible for these two pairs, we claim that

$$\mathrm{Sing}(J_i, b) \subset \mathrm{Sing}(J_i', b'), \quad i = 0, 1, \ldots, r,$$

where (J_i, b) (resp., (J_I', b)) is the pair on V_i defined by (J, b) (resp., (J', b)) and the given local sequence (see Definition 2.3). In fact, let's assume the opposite and let i denote the first index for which $\mathrm{Sing}(J_i, b) \not\subseteq \mathrm{Sing}(J_i', b')$. If $V_i \leftarrow V_i^{(1)}$ is the blow-up of V_i at a closed point in $\mathrm{Sing}(J_i, b) \setminus \mathrm{Sing}(J_i', b')$, we obtain a local sequence $V = V_0 \leftarrow \ldots \leftarrow V_i \leftarrow V_i^{(1)}$ that is permissible for (J, b) but not for (J', b'). This contradicts the relation $(J, b) \subset (J', b')$.

In particular, if $(J, b) \sim (J', b')$, then

$$\mathrm{Sing}(J_i, b) = \mathrm{Sing}(J_i', b'), \quad i = 0, 1, \ldots, r.$$

Remark 2.6 The transformation of pairs in Definition 2.3, (i), are, in particular, pull-backs by smooth morphisms. A different notion of local sequence, and hence of equivalence, arises if we consider arbitrary smooth morphisms. It is proved in section 8 of [3] that the notion of equivalence in Definition 2.4, (2), is not affected by this change.

Example 1 Fix a pair (J, b) over V and a positive integer s. We claim that the pairs (J, b) and (J^s, bs) over V are equivalent. In fact, note that

$$\mathrm{Sing}(J, b) = \{x \in V : \nu_x(J) \geq b\} = \{x \in V : \nu_x(J^s) \geq bs\} = \mathrm{Sing}(J^s, bs).$$

In addition, if $Y \subset V$ is a smooth center included in $\mathrm{Sing}(J, b) = \mathrm{Sing}(J^s, bs)$, after blowing up V at Y we get two transforms:

$$\begin{array}{ccc} V & \longleftarrow & V_1 \\ (J,b) & & (J_1,b) \\ (J^s,bs) & & ((J^s)_1,bs) \end{array}$$

and one readily checks that $(J^s)_1 = (J_1)^s$. In other words, the pairs (J_1, b) and $((J^s)_1, bs)$ are linked by the same relation as before. The same will happen if we restrict V to an open subscheme or if we multiply V by the affine line $\mathbb{A}^1$. This implies that any local sequence over V which is permissible for (J, b) is also permissible for (J^s, sb), and vice versa.

We finally define the intersection of pairs.

Definition 2.7 An intersection of two pairs (J, b) and (K, c) on V is a pair (L, d) such that $(L, d) \subset (J, b)$ and $(L, d) \subset (K, c)$, and it is maximal with this condition. Namely, any other pair included in both (J, b) and (K, c) is actually included in (L, d). An intersection (if it exists) is uniquely determined up to equivalence, and any of them will be denoted by

$$(J, b) \cap (K, c).$$

Proposition 2.8 *The intersection $(J, b) \cap (K, c)$ of the pairs (J, b) and (K, c) on V exists and is given (up to equivalence) by*

$$(L, d) := (J^c + K^b, bc).$$

Moreover, if $V = V_0 \leftarrow V_1 \leftarrow \ldots \leftarrow V_r$ is a local sequence over V that is permissible for (L, d) (so it is also permissible for (J, b) and (K, c)), then

$$\mathrm{Sing}(L_i, d) = \mathrm{Sing}(J_i, b) \cap \mathrm{Sing}(K_i, c), \quad i = 0, 1, \ldots, r,$$

where (L_i, d) (resp., (J_i, b), (K_i, c)) is the pair on V_i defined by (L, d) (resp., (J, b), (K, c)) and the given local sequence (see Definition 2.3*).*

Proof Note that

$$\mathrm{Sing}(J^c + K^b, bc) = \mathrm{Sing}(J^c, bc) \cap \mathrm{Sing}(K^b, bc) = \mathrm{Sing}(J, b) \cap \mathrm{Sing}(K, c),$$

so if $Y \subset V$ is a permissible center for $(J^c + K^b, bc)$, it is so for (J, b) and for (K, c). One readily checks that the transform of $(J^c + K^b, bc)$ by the blow-up of V at Y is $(J_1^c + K_1^b, bc)$, where (J_1, b) and (K_1, c) are respectively the transforms of (J, b) and (K, c). Similarly, if $U \subset V$ is an open subvariety, $((J^c + K^b)|_U, bc) = ((J|_U)^c + (K|_U)^b, bc)$, and if t is a variable over V, $((J^c + K^b)[t], bc) = ((J[t])^c + (K[t])^b, bc)$, as one can easily checks. This already shows that this pattern holds for any local sequence over V which is permissible for $(J^c + K^b, bc)$.

The above argument shows that the pair $(L, d) := (J^c + K^b, bc)$ is included in both (J, b) and (K, c) and that if $V = V_0 \leftarrow V_1 \leftarrow \ldots \leftarrow V_r$ is a local sequence which is permissible for (L, d), then $(L_i, d) = (J_i^c + K_i^d, bc)$ and $\mathrm{Sing}(L_i, d) = \mathrm{Sing}(J_i, b) \cap \mathrm{Sing}(K_i, c)$ for $i = 0, 1, \ldots, r$. Let (L', d') be another pair on V which is included in both (J, b) and (K, c). If $V = V_0 \leftarrow V_1 \leftarrow \ldots \leftarrow V_r$ is a local sequence over V which is permissible for (L', d'), then $\mathrm{Sing}(L_i', d') \subset \mathrm{Sing}(J, b) \cap \mathrm{Sing}(K, c)$ (Proposition 2.5) and this relation already ensures that the given sequence is also permissible for (L, d). □

The intersection of more than two pairs on V, say (J_1, b_1), (J_2, b_2), $\ldots$, (J_n, b_n) is defined similarly, and up to equivalence it is given by

$$(J_1^{b_1'} + J_2^{b_2'} + \ldots + J_n^{b_n'}, b_1 b_2 \ldots b_n),$$

where $b_j' = b_1 b_2 \ldots b_{j-1} b_{j+1} \ldots b_n$.

2.2 *The Role of Pairs in the Simplification of Singularities: Hironaka's Approach*

Pairs were introduced by Hironaka, and used to prove resolution of singularities in characteristic zero [12]. He showed that if we know how to resolve pairs, in the sense of Definition 2.1, one can improve the highest Hilbert Samuel function. Although our goal is to discuss an alternative method, it is instructive to review briefly this fundamental step in Hironaka's theorem, to be formulated in Theorem 2.13.

We mentioned before that by using the multiplicity function we can stratify a variety X into locally closed sets of points with the same multiplicity. Moreover, Theorem 1.1 says that if $X \leftarrow X_1$ is a blow-up at a smooth center included in $F_n(X)$, then $n(X) \geq n(X_1)$, that is, the highest multiplicity of X_1 is at most the highest multiplicity of X.

Hironaka proved that a similar result holds for the stratification given by the Hilbert-Samuel function. Recall that the Hilbert-Samuel function $HS_X : X \to \mathbb{N}^{\mathbb{N}}$ is constructed by setting $HS_X(x)$ the Hilbert-Samuel function of the local ring $\mathcal{O}_{X,x}$ when $x \in X$ is closed, and with a prescribed modification on non-closed points. Bennett proved in [1] that HS_X is upper semi-continuous if $\mathbb{N}^{\mathbb{N}}$ is given the lexicographic order. Let $S = S(X) \in \mathbb{N}^{\mathbb{N}}$ denote the highest value achieved by HS_X, that is, the highest Hilbert function, and let $F_S(X) \subset X$ be the stratum of highest value. A fundamental result introduced by Hironaka and later generalized by Bennett is the following:

Theorem 2.9 (Bennett [1]) *Let $X \leftarrow X_1$ be the blow up at a smooth center $Y \subset F_S(X)$. Then $S(X) \geq S(X_1)$.*

It is therefore natural to formulate a concept of *reduction* in the following terms.

Definition 2.10 A sequence of morphisms $X = X_0 \leftarrow X_1 \leftarrow \dots X_{r-1} \leftarrow X_r$ is said to be a *reduction of the Hilbert-Samuel function* if $S := S(X) = S(X_1) = \dots = S(X_{r-1}) > S(X_r)$ and each $X_{i-1} \leftarrow X_i$ is a blow-up at a smooth center included in $F_S(X_{i-1})$.

2.11 Assume now that X is a closed subvariety of a smooth one (over the fixed base field), say $X \subset V$. If $Y \subset X$ is a smooth closed subvariety, we can blow up both X and V at Y and obtain $X \leftarrow X_1$ and $V \leftarrow V_1$. A property of blow-ups states that in this situation there is a unique closed immersion $X_1 \subset V_1$ which is compatible with $X \subset V$ and the given blow-ups. Usually, it is said that $X_1 (\subset V_1)$ is the *strict transform* of $X \subset V$ by the blow-up $V \leftarrow V_1$. In other words, closed immersion are preserved by blow-ups at centers included in the immersed variety.

2.12 Let X be a closed subvariety of a smooth variety V (over a fixed field). A local sequence over V (Definition 2.2), say

$$V = V_0 \xleftarrow{\pi_1} V_1 \xleftarrow{\pi_2} \dots \xleftarrow{\pi_r} V_r ,$$

is said to induce a local sequence over X if one can define inductively closed immersions $X_i \to V_i$ such that

1. $X_0 = X$ and $X_0 \to V_0$ is the given inclusion.
2. For $i > 0$,
   a. if π_i is an open immersion or if $V_i \cong V_{i-1} \times \mathbb{A}^1$ and π_i is the projection onto the first component, X_i is the inverse image of X_{i-1},
   b. if π_i is a blow-up at a smooth center $Y_{i-1} \subset V_{i-1}$, this center is already included in X_{i-1}, and $X_i \subset V_i$ is the strict transform of X_{i-1}.

Note that, by restriction, we obtain a local sequence over X,

$$X = X_0 \xleftarrow{\pi_1} X_1 \xleftarrow{\pi_2} \dots \xleftarrow{\pi_r} X_r ,$$

and a commutative diagram

$$\begin{array}{ccccccc} V & \xleftarrow{\pi_1} & V_1 & \xleftarrow{\pi_2} & \dots & \xleftarrow{\pi_r} & V_r \\ \uparrow & & \uparrow & & & & \uparrow \\ X & \xleftarrow{\pi_1} & X_1 & \xleftarrow{\pi_2} & \dots & \xleftarrow{\pi_r} & X_r \end{array}$$

where the vertical lines are closed immersions.

We are finally prepared to present the role of pairs within Hironaka's approach.

Theorem 2.13 *Fix a variety X over a perfect field, and let $S = S(X)$ be the highest Hilbert-Samuel function. Then, after restricting X to an étale neighbourhood of a given closed point in $F_S(X)$, there exists a closed immersion $X \subset V$ into a smooth* *variety V and a pair (J, b) on V such that any local sequence over V which is permissible for (J, b) (Definition* 2.3*), say*

$$\begin{array}{ccccccc} V & \xleftarrow{\pi_1} & V_1 & \xleftarrow{\pi_2} & \dots & \xleftarrow{\pi_r} & V_r \\ (J,b) & & (J_1,b) & & \dots & & (J_r,b) \end{array}$$

induces a local sequence over X in the sense of 2.12*, say*

$$X \xleftarrow{\pi_1} X_1 \xleftarrow{\pi_2} \dots \xleftarrow{\pi_r} X_r ,$$

and

$$\mathrm{Sing}(J_i, b) = F_S(X_i), \quad i = 0, \dots, r.$$

Moreover, the pair (J, b) on V with the above property is uniquely determined up to equivalence.

Corollary 2.14 Within the framework of Theorem 2.13, a resolution of the pair (J, b) (Definition 2.1) induces a reduction of the Hilbert-Samuel function of X (Definition 2.10), and vice-versa.

Remark 2.15 In the formulation of Theorem 2.13, we used implicitly the fact that if $\alpha : X' \to X$ is an étale morphism, then for any $x' \in X'$, the local rings $\mathcal{O}_{X',\pi(x')}$ and $\mathcal{O}_{X,x}$ have the same Hilbert-Samuel function. The definition and some properties of étale morphisms will be reviewed in Sect. 5.

The existence of a pair with the properties stated in the previous theorem was proved by Hironaka in [12] at the completion at a given point, whereas the existence in étale topology is due to J.M. Aroca. This was a key step towards a constructive proof of Hironaka's theorem (which is existential).

The construction of such a pair with these properties relies on Hironaka's notion of normal flatness. This was a big step since it is very difficult to extract the equations of the strict transform of an embedded scheme after blowing up at a normally flat center, whereas the law of transformation of a pair is very elementary. However, the construction and existence of this pair require some knowledge on normal flatness and relies also on division theorems. We now address a similar result where normally flat centers are replaced by equimultiple centers, and where pairs arise in a different and quite simple manner.

2.3 The Role of Pairs in the Simplification of the Multiplicity

Given a variety X over a perfect field k, we mentioned at the previous lines that locally in étale topology there is an inclusion $X \subset V$ into a smooth variety for which Theorem 2.13 can be applied. In the approach we follow here, rather than looking into local immersions of X in smooth varieties, we look at X as a ramified covering of a smooth variety. More precisely, we look at finite and surjective morphisms $\delta : X \to V$ such that V is smooth. The generic rank of such a morphism δ is defined as follows: In the case that both varieties are affine and S is the coordinate ring of V and B is that of X, the morphism δ induces a finite extension of rings $S \subset B$ such that S is a smooth k-algebra. If K denotes the quotient field of S, we call

$$n := [B \otimes_S K : K].$$

the generic rank of δ. This definition extends clearly to the case where V and X are not necessarily affine.

Given an arbitrary variety X over a field of characteristic zero, the goal is to resolve its singularities by looking at the multiplicity as main invariant. The following result (see Corollary 3.2) already indicates the role of finite surjective morphisms in the study of the multiplicity at points of X.

Theorem 2.16 (Zariski, see Theorem 3.2) *Let $\delta : X \to V$ be a finite and surjective morphism of varieties, where V is regular. Then the highest multiplicity at points of X is at most the generic rank of δ.*

2.17 We now introduce the setting for the formulation of the first representation theorem for the multiplicity (Theorem 2.18). We fix a finite and surjective morphism $\delta : X \to V$ of affine varieties over a field k such that V is smooth. If S is the coordinate ring of V and B is that of X, by choosing generators for B as an S-algebra, say $B = S[\theta_1, \dots, \theta_N]$ (so that each θ_i integral over S), one obtains an exact sequence

$$0 \to L \to S[Z_1, \dots, Z_N] \to B = S[\theta_1, \dots, \theta_N] \to 0,$$

for some ideal L in $S[Z_1, \dots, Z_N]$. Such presentation provides a closed immersion

$$X \subset W := V \times \mathbb{A}^N.$$

Let $f_i(Z_i) \in S[Z_i]$ be the monic polynomial of lowest degree such that $f_i(\theta_i) = 0$ and let d_i denote its degree. Then $f_i(Z_i) \in S[Z_1, \dots, Z_N]$ and it generates a principal $\mathcal{O}_W$-ideal

$$J_i \subset \mathcal{O}_W.$$

Define the following pair on W:

$$(J, b) := (J_1, d_1) \cap \cdots \cap (J_N, d_N).$$

The following representation theorem will be addressed at the end of Sect. 3.

Theorem 2.18 *The setting being as in* 2.17, *we assume in addition that the highest multiplicity of* X, *say* n, *equals the generic rank of* δ. *Then any local sequence over* W *which is permissible for* (J, b) *(Definition* 2.3*), say*

$$\begin{array}{ccccccc} W = W_0 & \xleftarrow{\pi_1} & W_1 & \xleftarrow{\pi_2} & \dots & \xleftarrow{\pi_r} & W_r \quad , \\ (J,b) & & (J_1,b) & & \dots & & (J_r,b) \end{array}$$

induces a local sequence over X *in the sense of* 2.12, *say*

$$X \xleftarrow{\pi_1} X_1 \xleftarrow{\pi_2} \dots \xleftarrow{\pi_r} X_r \ ,$$

and

$$\operatorname{Sing}(J_i, b) = F_n(X_i).$$

Moreover, any other pair with the previous property is equivalent to (J, b).

Corollary 2.19 Within the setting of Theorem 2.18, any resolution of the pair (J, b) (Definition 2.1) induces a reduction of the multiplicity of X (Definition 1.2), and conversely, any reduction of the multiplicity induces a resolution of the pair (J, b).

Remark 2.20 If X is a variety of highest multiplicity n, and $X \subset W$ is a closed immersion in a smooth variety, then a pair (J, b) on W with the property stated in the theorem will be said *to represent the points of multiplicity* n. Such a pair is necessarily unique up to equivalence.

Remark 2.21 Lipman studies the multiplicity for complex analytic varieties [14]. If Z is a complex analytic variety and $x \in Z$ is a point where the variety has multiplicity n and local dimension d, then at a suitable neighbourhood of x one can construct a finite morphism $Z \to (\mathbb{C}^d, 0)$ with n points in the general fiber. In the formulation of Theorem 2.18 above and of Theorem 2.25 later, the variety X is provided with a finite projection $\delta : X \to V$ onto a smooth variety such that the generic rank of δ equals the maximal multiplicity of X. However, the existence of such projection δ cannot be always ensured, even if we restrict to an affine neighbourhood of a given point $x \in F_n(X)$. In order to fulfill the requirements of these theorems, it will be necessary to consider étale topology. We shall address these topics and prove the following result in Sect. 5.

Theorem 2.22 *Let* X *be a variety over a perfect field, say with highest multiplicity* $n = n(X)$, *and let* $x \in X$ *be a closed point of multiplicity* n. *Then there exists an étale morphism* $\alpha : X' \to X$ *whose image contains* x, *and a finite surjective morphism of varieties* $\delta : X' \to V'$, *where* V' *is smooth and the generic rank of* δ *is* n.

2.4 The Role of Pairs in the Simplification of the Multiplicity in Characteristic Zero

Fix a variety X over a field k. As mention in 2.11, once we fix a closed immersion $X \subset V$, where V is a smooth variety, there is a natural notion of blow-up of such inclusion when blowing up at a smooth center Y included in X. In fact, we get a closed immersion $X_1 \subset V_1$ of the respective blow-ups. We may ask what happens if instead of an immersion of $X \subset V$ we have a finite ramified cover on a smooth variety $X \to V$. As the goal is to try to improve the singularities of X (regardless of immersions or coverings!), it is reasonable to ask if one can preserve a finite covering by blow-ups. The following theorem (to be proved in Theorem 3.21) answers this question.

Theorem 2.23 (Blow-up of a finite morphism) *Let $\delta : X \to V$ be a finite and surjective morphism of varieties, where V is regular. Let n denote the generic rank of δ (so that the highest multiplicity of X is at most n, by Theorem* 2.16*).*

1. *If $F_n(X) \neq \emptyset$, then δ induces a homeomorphism $F_n(X) \equiv \delta(F_n(X))$.*
2. *An irreducible subvariety $Y \subset X$ included in $F_n(X)$ is regular if and only if $\delta(Y)$ is regular. In that case, $\delta|_Y : Y \to \delta(Y)$ is an isomorphism, and if $X \leftarrow X_1$ is the blow-up of X at Y and $V \leftarrow V_1$ is the blow-up of V at $\delta(Y)$, then there is a finite and surjective morphism $\delta_1 : X_1 \to V_1$ such that the resulting diagram*

$$\begin{array}{ccc} X & \longleftarrow & X_1 \\ {\scriptstyle\delta}\downarrow & & \downarrow{\scriptstyle\delta_1} \\ V & \longleftarrow & V_1 \end{array}$$

is commutative. The generic rank of $\delta_1 : X_1 \to V_1$ is again n.

2.24 Let $\delta : X \to V$ be a finite and surjective morphism of varieties, where V is smooth. Assume that the generic rank of δ equals the maximal multiplicity of X, say n. A local sequence over V (Definition 2.2), say $V = V_0 \overset{\pi_1}{\longleftarrow} V_1 \overset{\pi_2}{\longleftarrow} \dots \overset{\pi_r}{\longleftarrow} V_r$, is said to induce a local sequence over X if one can define (inductively) finite and surjective morphisms $\delta_i : X_i \to V_i$ such that

1. $\delta_0 : X_0 \to V_0$ is the given morphism $\delta : X \to V$;
2. and for $i > 0$,

   a. if V_i is an open subscheme of V_{i-1}, then $X_i = \delta_{i-1}^{-1}(V_i)$ and δ_i is the restriction of δ_{i-1}.
   b. if $V_i = V_{i-1} \times \mathbb{A}^1$, then $X_i = X_{i-1} \times \mathbb{A}^1$ and δ_i is the product morphism $\delta_{i-1} \times \mathrm{id}_{\mathbb{A}^1}$.
   c. if V_i is a blow-up of V_{i-1} at a smooth center Y_{i-1}, then Y_{i-1} is already included in $\delta_{i-1}(F_n(X_{i-1}))$ and the variety X_i and the morphism $\delta_i : X_i \to V_i$ are obtained from δ_{i-1} as in Theorem 2.23.

Given the way in which the varieties X_i were defined, there are natural morphisms $\pi'_i : X_i \to X_{i-1}, i = 1, \ldots, r$, such that

$$X \xleftarrow{\pi'_1} X_1 \xleftarrow{\pi'_2} \cdots \xleftarrow{\pi'_{r-1}} X_{r-1} \xleftarrow{\pi'_r} X_r$$

is a local sequence over X (Definition 2.2) and the resulting diagram

$$\begin{array}{ccccccc} X & \longleftarrow & X_1 & \longleftarrow \cdots \longleftarrow & X_{r-1} & \longleftarrow & X_r \\ \downarrow{\scriptstyle \delta} & & \downarrow{\scriptstyle \delta_1} & & \downarrow{\scriptstyle \delta_{r-1}} & & \downarrow{\scriptstyle \delta_r} \\ V & \longleftarrow & V_1 & \longleftarrow \cdots \longleftarrow & V_{r-1} & \longleftarrow & V_r, \end{array}$$

is commutative. Note also that all the morphisms δ_i are finite surjective and have generic rank n. Finally, if all the horizontal lines at the bottom are blow-ups (at smooth centers), so are the horizontal lines at the top row, and the centers of these blow-ups are smooth and included in the maximal multiplicity locus of the respective varieties (Theorem 2.23).

Theorem 2.25 *Let $\delta : X \to V$ be a finite and surjective morphism of varieties, where V is smooth. Assume that the generic rank of δ equals the maximal multiplicity of X, say n. Assume also that the characteristic of the base field does not divide n. Then one can construct a pair (K, d) on V such that any local sequence over V which is permissible for (K, d) (Definition 2.3), say*

$$\begin{array}{ccccccc} V = V_0 & \xleftarrow{\pi_1} & V_1 & \xleftarrow{\pi_2} & \cdots & \xleftarrow{\pi_r} & V_r \\ (K,d) & & (K_1,d) & & \cdots & & (K_n,d) \end{array}$$

induces a local sequence over X, in the sense of 2.24, so that we obtain a commutative diagram

$$\begin{array}{ccccccc} X & \longleftarrow & X_1 & \longleftarrow \cdots \longleftarrow & X_{r-1} & \longleftarrow & X_r\ , \\ \downarrow{\scriptstyle \delta} & & \downarrow{\scriptstyle \delta_1} & & \downarrow{\scriptstyle \delta_{r-1}} & & \downarrow{\scriptstyle \delta_r} \\ V & \longleftarrow & V_1 & \longleftarrow \cdots \longleftarrow & V_{r-1} & \longleftarrow & V_r, \end{array}$$

and

$$\delta_i(F_n(X_i)) = \mathrm{Sing}(K_i, b).$$

Any other pair with the same properties is equivalent to (K, d).

Corollary 2.26 The setting being as in Theorem 2.25, any resolution of the pair (K, d) (Definition 2.3) induces a reduction of the multiplicity of X (Definition 1.2), and vice-versa.

Remark 2.27

(1) Given a finite and surjective morphism $\delta : X \to V$, where V is smooth, if the generic rank of δ, say n, is the maximal multiplicity of X, any pair (K, d) on V satisfying the property stated in Theorem 2.25 is said to represent the shadow of the points of multiplicity n of X.
(2) In contrast with Theorem 2.18, where the dimension of W can be very large, here in Theorem 2.25 the pair (K, d) is defined over a smooth variety of the same dimension as X. This has many advantages from the computational point of view.
(3) *On generalized and weighted discriminants.* The construction of the pair (K, d) is quite explicit. The setting being as in 2.17, let us begin with the case in which $N = 1$, that is, $B = S[\theta] \cong S[Z]/\langle f(Z)\rangle$, where $f(Z) = X^n + a_1X^{n-1} + \cdots + a_0 \in S[Z]$ is the minimal polynomial of θ. We shall prove in Sect. 4 that there are *universal* equations on the coefficients of $f(Z)$ that lead to the pair (K, d). In other words, we will present (K, d) as a *weighted discriminant* (see Proposition 4.7). We end this introduction by giving two examples.

Example 2 For $n = 2$,

$$f(Z) = X^2 + a_1X + a_2 \in S[Z],$$

and we set $H(a_1, a_2) = a_1^2 - 4a_2$. If K is the ideal on V defined by $H(a_1, a_2)$, then the pair $(K, 2)$ fulfills the theorem. So for $n = 2$ we consider the discriminant with "weight 2".

Example 3 For $n = 3$, say

$$f(Z) = X^3 + a_1X^2 + a_2X + a_3 \in S[Z],$$

we set $H(a_1, a_2, a_3) = 3a_2 - a_1^2$ and $G(a_1, a_2, a_3) = -9a_1a_2 + 2a_1^3 + 27a_3$. In this case the pair

$$(K, d) = (\langle H(a_1, a_2, a_3)\rangle, 2) \cap (\langle G(a_1, a_2, a_3)\rangle, 3)$$

fulfills the theorem.

2.28 On the organization of the paper. In Sect. 3 we will address the proof of Theorem 2.18. In Sect. 4 we prove Theorem 2.25. Both Theorems 2.18 and 2.25 are formulated under the assumption that the singular variety can be expressed as a suitable ramified covering of a regular variety. More precisely, under the assumption of Theorem 2.22 which is proved in Sect. 5. The final Sect. 6 is devoted to the proof of Theorem 1.3.

3 Multiplicity, Projection on Smooth Schemes and Blow-Ups

The goal of this section is to prove Theorem 2.18 and also to present the background for the proof of Theorem 2.25, to be addressed in the next section. Given a finite extension $S \subset B$, where S is a smooth domain over a field k, we aim to obtain a description of the maximal multiplicity locus of $\text{Spec}(B)$ by using certain equations arising from the presentation of the S-algebra B, and to show that this description is preserved by blow-ups at regular centers included in the maximal multiplicity locus. All the results of this section are collected in Theorem 3.21, stated so that the proof of the Theorem 2.18 will follow easily.

We start by reviewing some concepts and results from commutative algebra, in particular on the notions of multiplicity and of Hilbert-Samuel functions (see also [11] and [5]).

3.1 Multiplicity and Integral Closure of Ideals

Given a local ring (R, M) and a primary ideal J for M, we will denote by $e_R(J)$ the multiplicity of J. When $J = M$, $e_R(M)$ will be also called the multiplicity of the local ring R. When R is regular, $e_R(M) = 1$. The converse holds if $\hat{R}$, the completion of R at M, is equidimensional and has no embedded components [15, Theorem 40.6], which is the case if R is an excellent local domain (EGA IV, (7.8.3)).

We mentioned in the introduction that the multiplicity along points on a variety defines an upper semi-continuous function. Actually, this result holds for the more general class of equidimensional and pure dimensional excellent schemes. The following result of Nagata is crucial for its proof.

Theorem 3.1 ([15, Theorem 40.1]) *If (R, M) is an equidimensional excellent local ring and $Q \subset R$ is a prime ideal, then $e_{R_Q}(QR_Q) \leq e_R(M)$.*

The following result of Zariski describes the behaviour of the multiplicity with respect to finite extension of rings.

Theorem 3.2 ([21, Theorem 24, p. 297.]) *Let (A, M) be a local domain, say with fraction field K. Given a finite extension $A \subset B$ we set $L := B \otimes_A K$ and denote by $Q_1, \ldots, Q_r$ the maximal ideals of the semi-local ring B. If $\dim B_{Q_i} = \dim A$ for $i = 1, \ldots, r$, then*

$$e_A(M)[L : K] = \sum_{1 \leq i \leq r} e_{B_{Q_i}}(MB_{Q_i})[k_i : k], \tag{1}$$

where k_i is the residue field of B_{Q_i} and k is that of (A, M).

We now review some properties about integral closure of ideals.

Definition 3.3 Fix a ring R and a proper ideal $I \subset R$. An element $r \in R$ is said to be *integral over* I if there are $c_1 \in I,\ c_2 \in I^2,\ \ldots, c_n \in I^n$ (for some n) such that

$$r^n + c_1 r^{n-1} + \ldots + c_n = 0.$$

The set $\overline{I} := \{r \in R : r \text{ is integral over } I\}$ is called the *integral closure* of I. An ideal $J \subset R$ such that $I \subset J$ is said to be integral over I or that I is a *reduction* of J if $J \subset \overline{I}$.

Notice that $I \subseteq \overline{I} \subseteq \sqrt{I}$, so in particular prime ideals coincide with their integral closure, in other words, prime ideals are *integrally closed.*

We now fix an ideal $I \subset R$ and a variable T over R. It is easy to check that $r \in \overline{I}$ if and only if $rT \in R[T]$ is integral over the Rees ring $R[IT] := R \oplus IT \oplus I^2T^2 \oplus \ldots \subset R[T]$ in the usual sense of ring extensions. Hence, $\overline{I}$ coincides with the degree-one part of the integral closure of the ring $R[IT]$ inside $R[T]$ (which is well known to be graded), and so $\overline{I}$ is an ideal. Note also that $R[\overline{I}T]$ is integral over $R[IT]$, so if $r \in R$ is integral over $\overline{I}$, it is already integral over I. In other words, $\overline{\overline{I}} = \overline{I}$.

Lemma 3.4 *Given ideals $I \subset J$ in a noetherian ring R, the following conditions are equivalent.*

(1) *I is a reduction of J.*
(2) *$IJ^n = J^{n+1}$ for some integer $n \geq 0$.*
(3) *There exists an integer n_0 such that $IJ^n = J^{n+1}$ for all integer $n \geq n_0$.*

Proof Notice that (2) and (3) are trivially equivalent so we will only prove the equivalence between the first two conditions. Assume that (1) holds, so I is a reduction of J and hence the extension $R[IT] \subset R[JT]$ is integral. In fact, this extension is finite since R is noetherian and so J is a finitely generated R-module. If $f_1T^{n_1}, \ldots, f_rT^{n_r}$ are homogeneous generators for $R[JT]$ as an $R[IT]$-module, then for $n := \max_{1 \leq i \leq r} n_i$ one readily checks that $J^{n+1} = \sum_{i=1}^{r} I^{n+1-n_i} f_i = IJ^n$, and this gives (2). Conversely, if (2) holds, then the relation $J^{n+1} = IJ^n$ implies that $R[JT]$ is generated as an $R[IT]$-module by $R + JT + \ldots + J^nT^n$. As each J^iT^i is an R-module of finite type, we conclude that $R[JT]$ is a finite module over $R[IT]$, so $R[IT] \subset R[JT]$ is an integral extension, or equivalently, I is a reduction of J. This completes the proof. □

Lemma 3.5 *Given an integral ring extension $R \subset S$ and an ideal $I \subset R$, $\overline{I} = \overline{IS} \cap R$.*

Proof The inclusion $\overline{I} \subseteq \overline{IS} \cap R$ being obvious, we prove the reverse inclusion. Given $r \in \overline{IS} \cap R$, it will be enough to prove that rT is integral over $R[IT]$. Note that $rT \in S[T]$ is integral over $S[(IS)T]$ and this ring is integral over $R[IT]$ since S is integral over R and $S[(IS)T]$ is generated as an $R[IT]$-algebra by S. We conclude that rT is integral over $R[IT]$, namely $r \in \overline{I}$. □

The following result draws a first connection between the theory of multiplicities and that of integral closure of ideals.

Theorem 3.6 (Rees, [18]) *If $I \subset J$ are primary ideals for the maximal ideal in a formally equidimensional local ring (R, M), then J is integral over I if and only if $e_R(I) = e_R(J)$.*

Remark 3.7 Given a local ring (R, M), there is a simple criterion to decide whether an ideal I is a reduction of M. Namely, let $I \subset R$ be a proper ideal and set $L := \frac{I+M^2}{M^2}$, which is a subset of the graded ring $\mathrm{gr}_M(A)$. We claim that I is a reduction of M if and only if

$$\mathrm{gr}_M(R)/\langle L\rangle$$

is zero dimensional. In fact, note that the conditions

1. $IM^n = M^{n+1}$,
2. $M^{n+1}/IM^n = 0$, and
3. $(M^{n+1}/IM^n) \otimes_R R/M = 0$,

are all equivalent. Our claim follows since the component of degree $n+1$ of the graded ring $\mathrm{gr}_M(R)/\langle L\rangle$ is $M^{n+1}/(IM^n + M^{n+2}) = (\frac{M^{n+1}}{IM^n}) \otimes_R R/M$.

3.2 On the Class $\mathscr{T}$

We define a class of finite extensions where Theorems 3.2 and 3.6 can be applied.

Definition 3.8 We denote by $\mathscr{T}$ the class of finite extensions of rings $S \subset B$ satisfying the following conditions:

(1) S is a regular excellent domain and all saturated chains of prime ideals have the same length.
(2) B is equidimensional and has not zero divisors in S.

Note that condition (2) can be replaced by the condition

(2') B is equidimensional and does not have embedded primes, namely, $\mathrm{Min}(B) = \mathrm{Ass}(B)$.

Any finite extension $S \subset B$, where S is a regular domain of finite type over a field and B is equidimensional and has not embedded components, belongs to the class $\mathscr{T}$.

3.9 Fix $S \subset B$ in the class $\mathscr{T}$. The following notions and properties follow easily from the definition.

(a) If K denotes the fraction field of S, the total quotient ring of B is $B \otimes_S K$. We shall call $[B \otimes_S K : K]$ the *generic rank* of the extension.
(b) If S' is a localization of S, the extension $S' \subset B \otimes_S S'$ also belongs to the class $\mathscr{T}$.
(c) If t is a variable over S, the extension $S[t] \subset B[t]$ belongs to the class $\mathscr{T}$.

(d) B is excellent and all saturated chains of prime ideals have the same length $\dim B = \dim S$.
(e) Any subextension $S \subset B_1 (\subset B)$ belongs to the class $\mathscr{T}$.

Corollary 3.10 Let $S \subset B$ be an extension of rings in the class $\mathscr{T}$, let K denote the fraction field of S and set $n := [B \otimes_S K : K]$. Then for any $P \in \mathrm{Spec}(B)$,

$$e_{B_P}(PB_P) \leq n.$$

The equality holds if and only if setting $\mathfrak{p} := P \cap S \in \mathrm{Spec}(S)$ the following three conditions are satisfied.

(1) P is the only prime of B lying above $\mathfrak{p}$, namely $B_P = B_{\mathfrak{p}}$.
(2) The induced inclusion $S_{\mathfrak{p}}/\mathfrak{p}S_{\mathfrak{p}} \subset B_P/PB_P$ is an equality.
(3) $\mathfrak{p}B_P$ is a reduction of PB_P.

Proof The fact that $S_{\mathfrak{p}} \subset B_{\mathfrak{p}}$ belongs to the class $\mathscr{T}$ ensures that the hypothesis of Theorem 3.2 are fulfilled for this extension. As $S_{\mathfrak{p}}$ is regular, so it has multiplicity one, we deduce from the formula in Theorem 3.2 that

$$n \geq e_{B_P}(\mathfrak{p}B_P)[B_P/QB_P : S_{\mathfrak{p}}/\mathfrak{p}S_{\mathfrak{p}}] \geq e_{B_P}(PB_P)[B_P/QB_P : S_{\mathfrak{p}}/\mathfrak{p}S_{\mathfrak{p}}] \geq e_{B_P}(PB_P),$$

and that the equalities hold if and only if conditions (1) and (2) of the corollary hold and $e_{B_P}(\mathfrak{p}B_P) = e_{B_P}(PB_P)$. Finally, as B_P is excellent and equidimensional, it is also formally equidimensional, so we can invoke Theorem 3.6, which asserts that the equality $e_{B_P}(\mathfrak{p}B_P) = e_{B_P}(PB_P)$ is equivalent to the condition (3) of the corollary. □

Given an extension $S \subset B$ in the class $\mathscr{T}$, we denote by

$$F_S(B) \subset \mathrm{Spec}(B)$$

the set of those P such that the multiplicity of B_P coincides with the generic rank of the extension. The previous corollary says that $F_S(B)$ is either empty or it is the maximal multiplicity locus of the scheme $\mathrm{Spec}(B)$. The induced morphism $\delta : \mathrm{Spec}(B) \to \mathrm{Spec}(S)$ is said to be *transversal* if $F_S(B) \neq \emptyset$. As a consequence of the previous corollary we obtain:

Corollary 3.11 If the induced morphism $\delta : \mathrm{Spec}(B) \to \mathrm{Spec}(S)$ is transversal, then $F_S(B) = \delta^{-1}(\delta(F_S(B)))$ and δ induces a homeomorphism

$$F_S(B) \equiv \delta(F_S(B)).$$

Remark 3.12 We point out that if $P \in F_S(B)$, then P contains all the minimal primes of B. In fact, let Q_0 be a minimal prime of B and set $\bar{B} := B/Q_0$. The composition $S \subset B \xrightarrow{\pi} \bar{B}$ is clearly injective and it is an extension in the class $\mathscr{T}$. Let $\bar{P}$ be a prime ideal of $\bar{B}$ dominating $\mathfrak{p} := P \cap S$. Then $\pi^{-1}(\bar{P})$ is a prime ideal of B dominating $\mathfrak{p}$, so $\pi^{-1}(\bar{P}) = P$, by Corollary 3.11. This implies that $P = \pi^{-1}(\bar{P}) \supset \pi^{-1}(0) = Q_0$.

Proposition 3.13 *Let $S \subset B$ be an extension in the class $\mathscr{T}$. Consider intermediate subrings $S \subset B_i \subset B$, $i = 1, \ldots, N$, such that $B = B_1 B_2 \ldots B_N$, and let δ : $\operatorname{Spec}(B) \to \operatorname{Spec}(S)$ and $\delta_i : \operatorname{Spec}(B_i) \to \operatorname{Spec}(S)$ denote the respective induced morphisms. Then*

$$\delta(F_S(B)) = \bigcap_{i=1}^{N} \delta_i(F_S(B_i)).$$

In particular, if δ is transversal, then each δ_i is also transversal.

Proof We first prove the inclusion ($\subseteq$). Given $\mathfrak{p} \in \delta(F_S(B))$, say $\mathfrak{p} = P \cap S$ with $P \in F_S(B)$, we shall show that $P_i := P \cap B_i$ belongs to $F_S(B_i)$ for each index i by proving that $S \subset B_i$ and P_i fulfills the requirement (1), (2) and (3) stated in Corollary 3.10. This will show that $\mathfrak{p} = P_i \cap S \in \delta_i(F_S(B_i))$ for each index i.

When applying Corollary 3.10 to the extension $S \subset B$ and $P \in F_S(B)$ we obtain: (a) $B_\mathfrak{p} = B_P$, so the composition $S_\mathfrak{p} \subset B_{i\,\mathfrak{p}} \subset B_P$ is a finite extension and hence $B_{i\,\mathfrak{p}}$ is local, or equivalently, $B_{i\,\mathfrak{p}} = B_{i\,P_i}$; (b) $S_\mathfrak{p}/\mathfrak{p}S_\mathfrak{p} = B_{i\,P_i}/P_i B_{i\,P_i} = B_P/PB_P$; and (c) $\overline{\mathfrak{p}B_P} = PB_P$. We now apply Proposition 3.5 to the finite extension $B_{i\,P_i} \subset B_P$ and the ideal $\mathfrak{p}B_{i\,P_i}$, and obtain that $\overline{\mathfrak{p}B_{i\,P_i}} = \overline{\mathfrak{p}B_P} \cap B_{i\,P_i} = PB_P \cap B_{i\,P_i} = P_i B_{i\,P_i}$, where the second equality follows by (c).

We now prove the reverse inclusion ($\supseteq$). Fix $\mathfrak{p} \in \operatorname{Spec}(S)$ such that $\mathfrak{p} \in \delta_i(F_S(B_i))$ for all $i = 1, \ldots, N$, and let $P \in \operatorname{Spec}(B)$ be such that $P \cap S = \mathfrak{p}$. We will show that $P \in F_S(B)$ by showing that P fulfills conditions (1), (2) and (3) in Corollary 3.10.

After localizing S, B_i and B at $\mathfrak{p}$ we may assume that S is local with maximal ideal $\mathfrak{p}$. Set $P_i := P \cap B_i$. Note that, by Corollary 3.10, $P_i \in F_S(B_i)$ since $\delta_i(P_i) = \mathfrak{p} \in \delta(F_S(B_i))$. By the same corollary,

(a) B_i is local with maximal ideal P_i,
(b) $S/\mathfrak{p} = B_i/P_i$ and
(c) $\overline{\mathfrak{p}B_i} = P_i$, for each i.

Note that $P_1 B + \ldots + P_N B \subseteq P$. We will show that this is an equality by showing that the extension $S/\mathfrak{p} \subset B/(P_1 B + \ldots + P_N B)$ (induced by $S \subset B$) is an equality. From (b) we obtain that $B_i = S[P_i]$ and hence $B = B_1 B_2 \ldots B_N = S[P_1]S[P_2]\ldots S[P_N] = S[P_1 B + \ldots + P_N B]$; therefore, $S/\mathfrak{p} = B/(P_1 B + \ldots + P_N B)$, as claimed.

As $S/\mathfrak{p}$ is a field it follows that $P_1 B + \ldots + P_N B$ is a maximal ideal of B, and therefore it must be equal to P. Since $P_1, \ldots, P_N$ are uniquely determined by $\mathfrak{p}$, we conclude that P is the unique maximal ideal of B (dominating $\mathfrak{p}$) and we already obtained that $S/\mathfrak{p} = B/P$. It is left to prove that $\overline{\mathfrak{p}B} = P_1 B + \ldots + P_N B$. By (c), each element of $P_i B$ is integral over $\mathfrak{p}B$ and hence $P_1 B + \ldots + P_N B = \overline{\mathfrak{p}B}$. The proof is now complete. □

Proposition 3.14 *Fix $S \subset B$ in the class $\mathcal{T}$, and let K denote the fraction field of S. Given $\theta \in B$ ($\subset B \otimes_S K$), let $f(Z) \in K[Z]$ denote its minimal polynomial, that is, the monic polynomial of lowest degree in $K[Z]$ such that $f(\theta) = 0$ in $B \otimes_S K$. Then $f(Z) \in S[Z]$ and the evaluation map $S[Z] \to S[\theta]$ induces an S-isomorphism $S[\theta] \cong S[Z]/\langle f(Z)\rangle$.*

Proof To prove that $f(Z) \in S[Z]$, it is enough to show that $f(Z) \in S_{\mathfrak{p}}[Z]$ for all prime ideal $\mathfrak{p}$. Therefore, after localizing, we can assume that S is a regular local ring. In particular, S is a unique factorization domain and hence so is $S[Z]$.

Let $g(Z)$ be a monic polynomial in $S[Z]$ of lowest degree such that $g(\theta) = 0$, so in particular $m := \deg(g(Z)) \geq \deg(f(Z)) =: n$. If $m = n$, necessarily $f(Z) = g(Z)$ for otherwise $f(Z) - g(Z)$ would be a polynomial of degree $< n$ that annihilates θ and this would contradict the definition of $f(Z)$. Assume that $m > n$. Choose $s \in S$ such that $sf(Z) \in S[Z]$. We can write $sf(Z) = ch(Z)$, where $c \in S$ and $h(Z) \in S[Z]$ is primitive (the maximum common divisor of its coefficients is 1) of degree n. As $0 = sf(\theta) = ch(\theta) \in B \subset B \otimes_S K$, we obtain that $h(\theta) = 0$. If t is the leading coefficient of $h(Z)$, according to the division algorithm, there exist $q(Z), r(Z) \in S[Z]$ such that $t^{m-n}g(Z) = q(Z)h(Z) + r(Z)$ and either $r(Z) = 0$ or $\deg(r(Z)) < \deg(h(Z)) = \deg(f(Z))$. As clearly $r(\theta) = 0$, necessarily $r(Z) = 0$ and thus $t^{m-n}g(Z) = q(Z)h(Z)$. Given that $h(Z)$ is primitive, we obtain that $h(Z)|g(Z)$ and hence t must be a unit in S since $g(Z)$ is monic. Thus, $t^{-1}h(Z)$ is a monic polynomial of degree $n < m$ which annihilates θ and this contradicts the definition of $g(Z)$. Therefore, $f(Z) = g(Z)$, as we claimed.

For the final part of the proposition, we consider the morphism $S[Z] \to S[\theta]$ given by evaluation at θ. Its kernel includes $\langle f(Z)\rangle$. If $g(Z)$ is any other element in the kernel, using the division algorithm in $S[Z]$, a similar argument as before shows that $f(Z)$ divides $g(Z)$ in $S[Z]$, namely $g(Z) \in \langle f(Z)\rangle$. Thus, $S[\theta] = S[Z]/\langle f(Z)\rangle$. □

Remark 3.15 The setting being as in the previous proposition, if t is a variable over S, one readily checks that the minimal polynomial $f(Z) \in S[Z]$ of an element $\theta \in B$ and the minimal polynomial $\tilde{f}(Z) \in S[t][Z]$ of θ viewed as an element of $B[t]$ coincide.

3.16 We present here the setting and notation to be used in the formulation of forthcoming results. We fix an extension $S \subset B$ in the class $\mathcal{T}$ and an epimorphism of S-algebras

$$\rho : S[Z_1, \ldots, Z_N] \to B,$$

where N is an integer and $Z_1, \ldots, Z_N$ are variables over S. Set $\theta_i := \rho(Z_i)$ and $B_i := S[\theta_i] \subset B$. Let

$$f_i(Z_i) = Z_i^{d_i} + a_{i,1}Z_i^{d_i-1} + \ldots + a_{i,d_i} \in S[Z_i]$$

be the minimal polynomial of θ_i, so that we get an S-isomorphism $B_i \cong S[Z_i]/\langle f_i(Z_i)\rangle$ (Proposition 3.14). Note that there is an inclusion of ideals

$$\langle f_1(Z_1), \dots, f_N(Z_N)\rangle \subseteq \mathrm{Ker}(\rho) \subset S[Z_1, \dots, Z_N].$$

We now set

$$\tilde{B} := \frac{S[Z_1, \dots, Z_N]}{\langle f_1(Z_1), \dots, f_N(Z_N)\rangle} \sim \frac{S[Z_1]}{\langle f_1(Z_1)\rangle} \otimes_S \cdots \otimes_S \frac{S[Z_N]}{\langle f_N(Z_N)\rangle},$$

and identify $\tilde{B}_i := S[Z_i]/\langle f_i(Z_i)\rangle$ with its image in $\tilde{B}$. One readily checks that the composition $S \subset B \to \tilde{B}$ is also an extension in the class $\mathscr{T}$. It follows that for each index $i = 1, \dots, N$ there is a commutative diagram

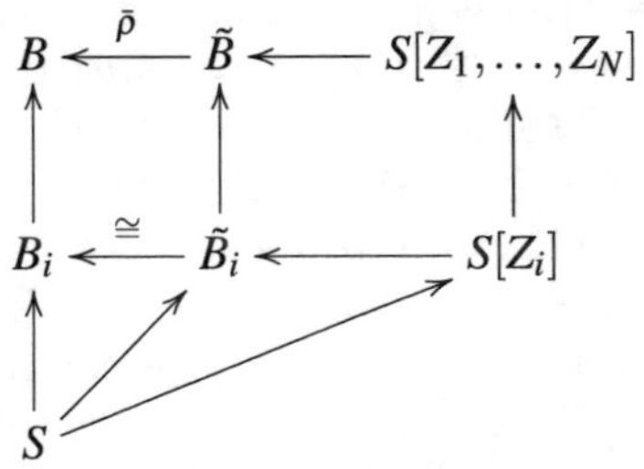

where the horizontal lines are quotient maps and the other ones are inclusions. Finally, let $\delta : \mathrm{Spec}(B) \to \mathrm{Spec}(S)$, $\delta_i : \mathrm{Spec}(B_i) \to \mathrm{Spec}(S)$, $\tilde{\delta} : \mathrm{Spec}(\tilde{B}) \to \mathrm{Spec}(S)$ and $\tilde{\delta}_i : \mathrm{Spec}(\tilde{B}_i) \to \mathrm{Spec}(S)$ denote the respective induced morphisms of schemes.

Proposition 3.17 *In the previous setting, the closed immersion* $\mathrm{Spec}(B) \to \mathrm{Spec}(\tilde{B})$ *induced by* $\bar{\rho} : \tilde{B} \to B$ *restricts to a homeomorphism*

$$F_S(B) \equiv F_S(\tilde{B}).$$

Proof Combining Corollary 3.11 and Proposition 3.13 one obtains homeomorphisms

$$F_S(B) \cong \delta(F_S(B)) = \bigcap \delta_i(F_S(B_i)) = \bigcap \tilde{\delta}_i(F_S(\tilde{B}_i)) = \tilde{\delta}(F_S(\tilde{B})) \cong F_S(\tilde{B}),$$

where the whole composition is induced by the closed immersion $\mathrm{Spec}(B) \to \mathrm{Spec}(\tilde{B})$. □

Corollary 3.18 The setting being as in 3.16, if $\iota : \mathrm{Spec}(B) \to \mathrm{Spec}(S[Z_1, \dots, Z_N])$ is the closed immersion induced by ρ, then $f_i(Z_i)$ has order at most d_i at every point of $\mathrm{Spec}(S[Z_1, \dots, Z_N])$, and

$$\iota(F_S(B)) = \bigcap_{i=1}^{N} \mathrm{Sing}(\langle f_i(Z_i)\rangle, d_i).$$

Proof The first assertion is clear since $f_i(Z_i)$ is a polynomial of degree d_i. We now prove the above equality. We already obtained $\iota(F_S(B)) = F_S(\tilde{B})$. Given $\tilde{P} \in \mathrm{Spec}(\tilde{B}) = \bigcap_{i=1}^N \mathrm{Sing}(\langle f_i(Z_i)\rangle, 1)$, Proposition 3.13 implies that $\tilde{P} \in F_S(\tilde{B})$ if and only if $\tilde{P_i} := \tilde{P} \cap \tilde{B_i} \in F_S(\tilde{B_i})$. As the generic rank of $S \subset B_i$ is d_i, the last assertion holds if and only if the polynomial $f_i(Z_i)$ has order d_i at $\tilde{P_i}$ (viewed as a prime of $S[Z_i]$) for all $i = 1, \dots, N$. Finally, as each $S[Z_i] \to S[Z_1, \dots, Z_N]$ is smooth, the last assertion holds if and if $f_i(Z_i)$ has order d_i at $\tilde{P}$ (viewed as a prime of $S[Z_1, \dots, Z_N]$), namely, $\tilde{P} \in \bigcap_{i=1}^N \mathrm{Sing}(\langle f_i(Z_i)\rangle, d_i)$. □

We now turn to the study of permissible centers.

Proposition 3.19 *Given $S \subset B$ in the class $\mathscr{T}$, if Q is a prime ideal of B such that $Q \in F_S(B)$, then B/Q is regular if and only if $S/(Q \cap S)$ is regular. In such case, $B/Q = S/(Q \cap S)$ and hence $B = S[Q]$ (the S-algebra spanned by Q).*

Proof Fix $Q \in F_S(B)$ and set $\mathfrak{q} := Q \cap S$, so in particular B/Q and $S/\mathfrak{q}$ have the same fraction field (Corollary 3.10). If $S/\mathfrak{q}$ is regular, then $S/\mathfrak{q} \subset B/Q$ is a finite extension with the same fraction field, and hence $S/\mathfrak{q} = B/Q$ since $S/\mathfrak{q}$ is in particular normal.

Conversely, assume that B/Q is regular and let us prove that $S/\mathfrak{q}$ is also regular. After localizing at a prime ideal $\mathfrak{p}$ such that $\mathfrak{q} \subset \mathfrak{p}$ we may assume that S is local with maximal ideal $\mathfrak{p}$ and our task is to prove that $(S/\mathfrak{q}, \mathfrak{p}/\mathfrak{q})$ is a regular local ring. Let P be a maximal ideal of B. As Q is the only minimal prime of $\mathfrak{q}B$ (since $Q \in F_S(B)$) it follows that $Q \subset P$ and hence Theorem 3.1 ensures that $P \in F_S(B)$. Thus, B is local with maximal ideal P and $\mathfrak{p}B$ is a reduction of P, by Corollary 3.10. This shows that P/Q is integral over $\mathfrak{q}B/Q$ and hence $e_{B/Q}(\mathfrak{q}B/Q) = e_{B/Q}(P/Q) = 1$, by Theorem 3.6. In the last equality we used that B/Q is regular. Combining this and Theorem 3.2, applied to the extension $S/\mathfrak{q} \subset B/Q$ (which belongs to the class $\mathscr{T}$), we obtain

$$e_{S/\mathfrak{q}}(\mathfrak{p}/\mathfrak{q})[B_Q/QB_Q : S_\mathfrak{q}/\mathfrak{q}S_\mathfrak{q}] = e_{B/Q}(\mathfrak{p}B/Q)[B_P/P : S/\mathfrak{p}] = 1.$$

Since $S/\mathfrak{q}$ is an excellent local domain, it is unmixed and therefore the last equality and Nagata's theorem [15, Theorem 40.6] ensure that $S/\mathfrak{q}$ is regular. This completes the proof. □

Proposition 3.20 *The setting being as in 3.16, let Q be a prime ideal of B such that $Q \in F_S(B)$ and B/Q is regular, and set $\mathfrak{q} = Q \cap S$. Then there are elements $\lambda_1, \dots, \lambda_N \in S$ such that the following conditions hold.*

(1) $Q = \langle \mathfrak{q}B, \theta_1 - \lambda_1, \dots, \theta_N - \lambda_N\rangle$ *and* $\rho^{-1}(Q) = \langle \mathfrak{q}S[Z_1, \dots, Z_N], Z_1 - \lambda_1, \dots, Z_N - \lambda_N\rangle$.
(2) *If we set $T_i = Z_i - \lambda_i$, so that $S[Z_i] = S[T_i]$, then the expression of $f_i(Z_i)$ as a polynomial in T_i takes the form*

$$f_i(Z_i) = h_i(T_i) = T_i^{d_i} + b_{i,1}T_i^{d_i-1} + \dots + b_{i,d_i-1}T_i + b_{i,d_i},$$

where $b_{i,j} \in \mathfrak{q}^j$. The polynomial $h_i(T)$ is the minimal polynomial of $\theta_i - \lambda_i$.

In particular, these two conditions imply that Q is integral over $\mathfrak{q}B$.

Proof

(1) With the notation as in 3.16, and in accordance with Proposition 3.13, we may assume that $B = \tilde{B} = S[Z_1, \dots, Z_N]/\langle f_1(Z_1), \dots, f_N(Z_N)\rangle$. Given $Q \in F_S(B)$ as above, we set $Q_i := Q \cap B_i$. By Proposition 3.19, $S/\mathfrak{q} = B_i/Q_i = B/Q$ and hence we can choose $\lambda_i \in S$ such that $\theta_i - \lambda_i \in Q_i$. Therefore, when we write $f_i(Z_i)$ as a polynomial in $T_i = Z_i - \lambda_i$, its constant coefficient $f_i(\lambda_i)$ satisfies $f_i(\lambda_i) \equiv f_i(\theta_i) = 0 \mod Q_i$, and hence $f_i(\lambda_i) \in Q_i \cap S = \mathfrak{q}$. One concludes that $\langle f_1(Z_1), \dots, f_N(Z_N)\rangle \subseteq Q' := \langle \mathfrak{q}S[Z_1, \dots, Z_N], Z_1 - \lambda_1, \dots, Z_N - \lambda_N\rangle \subset S[Z_1, \dots, Z_N]$. It is clear that Q' is a prime ideal, and we have shown that Q' includes $\langle f_1(Z_1), \dots, f_N(Z_N)\rangle$; therefore, $\rho(Q')$ is a prime ideal of B. Note also that $Q' \cap S = \mathfrak{q}$, and this implies that $\rho(Q') \cap S = \mathfrak{q}$. It follows that $\rho(Q') = Q$ since Q is the only prime of B dominating $\mathfrak{q}$.

(2) By Corollary 3.18, each $f_i(Z_i)$ has order d_i at Q', and given that $S[Z_i] \to S[Z_1, \dots, Z_N]$ is smooth, $f_i(Z_i)$ has order d_i at $Q' \cap S[Z_i] = \langle \mathfrak{q}S[Z_i], Z_i - \lambda_i\rangle$. Since the latter is a regular prime, we obtain that

$$f_i(Z_i) \in \langle \mathfrak{q}S[Z_i], Z_i - \lambda_i\rangle^{d_i} = \langle \mathfrak{q}S[Z_i - \lambda_i], Z_i - \lambda_i\rangle^{d_i} = \sum_{j=0}^{d_i} \mathfrak{q}^j S[T_i] T^{d_i - j},$$

and this proves that $h_i(T_i)$ has the desired form. It is clear that $h_i(T_i)$ is the minimal polynomial of $\theta_i - \lambda_i$.

The final statement follows now from (1) and (2): in fact $Q = \langle \mathfrak{q}B, \theta_1 - \lambda_1, \dots, \theta_N - \lambda_N\rangle$, and each $\theta_i - \lambda_i$ satisfies the equation $h_i(\theta_i - \lambda_i) = 0$, which is an equation of integral dependence over $\mathfrak{q}B$. Thus, Q is integral over $\mathfrak{q}B$. □

The following theorem is basically a collection of the results we proved previously.

Theorem 3.21 *The setting and notation being as in* 3.16 *and Proposition* 3.20, *we consider the blow-ups*

$$\mathrm{Spec}(S) \xleftarrow{\pi'} V_1\,, \quad \mathrm{Spec}(B) \xleftarrow{\pi} X_1\,, \quad \textit{and} \quad \mathrm{Spec}(S[Z_1, \dots, Z_N]) \xleftarrow{\Pi} W_1,$$

at the (regular) centers defined by $\mathfrak{q} \subset S$, $Q \subset B$ *and* $\tilde{Q} = \langle \mathfrak{q}S[Z_1, \dots, Z_N], Z_1 - \lambda_1, \dots, Z_N - \lambda_N\rangle \subset S[Z_1, \dots, Z_N]$, *respectively. Fix generators* $x_1, \dots, x_r$ *for* $\mathfrak{q}$, *and set*

$$S_t := S\left[\frac{x_1}{x_t}, \dots, \frac{x_r}{x_t}\right] \subset K,$$

where K is the fraction field of S, so that V_1 is covered by the affine charts $\mathrm{Spec}(S_t)$, $t = 1, \dots, r$.

(1) *Fix $t \in \{1, \dots, r\}$. The subring $S_t[\frac{Z_1-\lambda_1}{x_t}, \dots, \frac{Z_N-\lambda_N}{x_t}] \subset K[Z_1, \dots, Z_N]$ is a polynomial ring over S_t; the ring extension*

$$S_t \subset S_t\left[\frac{\theta_1-\lambda_1}{x_t}, \dots, \frac{\theta_N-\lambda_N}{x_t}\right] \quad (\subset B \otimes_S K)$$

belongs to the class $\mathscr{T}$; and the minimal polynomial of $\frac{\theta_i-\lambda_i}{x_t}$, as a polynomial in $S_t[\frac{Z_i-\lambda_i}{x_t}]$, is

$$\begin{aligned} h_i^{(t)}\left(\frac{Z_i-\lambda_i}{x_t}\right) &:= \left(\frac{Z_i-\lambda_i}{x_t}\right)^{d_i} + \frac{b_{i,1}}{x_t}\left(\frac{Z_i-\lambda_i}{x_t}\right)^{d_i-1} + \dots + \frac{b_{i,d_i}}{x_t^{d_i}} \\ &= \frac{h_i(Z_i-\lambda_i)}{x_t^{d_i}} = \frac{f_i(Z_i)}{x_t^{d_i}}. \end{aligned}$$

(2) *The scheme X_1 is obtained by glueing in the natural way the spectrum of the rings $S_t[\frac{\theta_1-\lambda_1}{x_t}, \dots, \frac{\theta_N-\lambda_N}{x_t}] \subset B \otimes_S K$, the morphisms of schemes induced by the extensions $S_t \subset S_t\left[\frac{\theta_1-\lambda_1}{x_t}, \dots, \frac{\theta_N-\lambda_N}{x_t}\right]$ glue together into a finite and dominant morphism $\delta_1 : X_1 \to V_1$, and the resulting diagram*

$$\begin{array}{ccc} \mathrm{Spec}(B) & \xleftarrow{\pi} & X_1 \\ \downarrow{\scriptstyle \delta} & & \downarrow{\scriptstyle \delta_1} \\ \mathrm{Spec}(S) & \xleftarrow{\pi'} & V_1 \end{array}$$

is commutative. Moreover, δ_1 is the unique morphism from X_1 to V_1 making the above diagram commutative.

(3) *The schemes* Spec $\left(S_t\left[\frac{Z_1-\lambda_1}{x_t}, \dots, \frac{Z_N-\lambda_N}{x_t}\right]\right)$ *glue together into an open subscheme U_1 of W_1. The morphisms of schemes induced by the evaluations*

$$\rho_t : S_t\left[\frac{Z_1-\lambda_1}{x_t}, \dots, \frac{Z_N-\lambda_N}{x_t}\right] \to S_t\left[\frac{\theta_1-\lambda_1}{x_t}, \dots, \frac{\theta_N-\lambda_N}{x_t}\right]$$

$$\frac{Z_i-\lambda_i}{x_t} \mapsto \frac{\theta_i-\lambda_i}{x_t},$$

glue together into a closed immersion $X_1 \to U_1$. The composition $\iota : X_1 \to U_1 \subset W$ is also a closed immersion and the induced diagram

$$\begin{array}{ccc} \mathrm{Spec}(S[Z_1,\dots,Z_M]) & \xleftarrow{\Pi} & W_1 \\ \uparrow{\scriptstyle \iota} & & \uparrow{\scriptstyle \iota} \\ \mathrm{Spec}(B) & \xleftarrow{\pi} & X_1 \end{array}$$

is commutative.

(4) *The morphisms of schemes induced by the inclusions* $S_t \subset S_t[\frac{Z_1-\lambda_1}{x_t}, \ldots, \frac{Z_N-\lambda_N}{x_t}]$ *glue together into a smooth morphism* $\beta_1 : U_1 \to V_1$*, and the resulting diagrams*

$$\begin{array}{ccc} X_1 & \xrightarrow{\iota} & U_1 \\ {\scriptstyle \delta_1}\downarrow & \swarrow {\scriptstyle \beta_1} & \\ V_1 & & \end{array} \quad \text{and} \quad \begin{array}{ccc} \mathrm{Spec}(S[Z_1,\ldots,Z_N]) & \xleftarrow{\Pi|_{U_1}} & U_1 \\ \downarrow & & \downarrow {\scriptstyle \beta_1} \\ \mathrm{Spec}(S) & \xleftarrow{\pi'} & V_1 \end{array},$$

are commutative. Moreover, β_1 *is the only morphism from* U_1 *to* V_1 *making the second diagram commutative.*

(5) *Let* $\mathscr{H}_i \subset \mathrm{Spec}(S[Z_1, \ldots, Z_N])$ *be the hypersurface defined by* $f_i(Z_i)$ *and let* $\mathscr{H}_i^{(1)} \subset W_1$ *be the strict transform of* $\mathscr{H}_i$*. Then*

$$\bigcap_{i=1}^{N} \mathscr{H}_i^{(1)} \subset U_1;$$

the ideal $I(\mathscr{H}_i^{(1)})$ *has order at most* d_i *at points of* W_1*; and the restriction of* $\mathscr{H}_i^{(1)}$ *to* $\mathrm{Spec}\left(S_t\left[\frac{Z_1-\lambda_1}{x_t}, \ldots, \frac{Z_M-\lambda_M}{x_t}\right]\right)$ *is defined by* $h_i^{(t)}(\frac{Z_i-\lambda_i}{x_t^{d_i}}) \in S_t[\frac{Z_i-\lambda_i}{x_t}]$*, namely the minimal polynomial of* $\frac{\theta_i-\lambda_i}{x_t}$ *(see (1)). Moreover,* $(I(\mathscr{H}_i^{(1)}), d_i)$ *is the transform of* $(I(\mathscr{H}_i), d_i)$*, and*

$$\iota_1(F_n(X_1)) = \bigcap_{i=1}^{N} \mathrm{Sing}(I(\mathscr{H}_i^{(1)}), d_i),$$

where n is the generic rank of δ*.*

Proof

(1) The first assertion is clear. For the second and the last assertions we first observe that $\theta_i - \lambda_i$ and $\frac{\theta_i-\lambda_i}{x_t}$, when viewed as elements in $B \otimes_S K$, must have minimal polynomials of the same degree. As $h_i^{(t)}(\frac{Z_i-\lambda_i}{x_t}) \in S_t[\frac{Z_i-\lambda_i}{x_t}]$ is monic, has degree d_i and $h_i^{(t)}(\frac{\theta_i-\lambda_i}{x_t}) = \frac{h_i(\theta_i-\lambda_i)}{x_t^{d_t}} = 0$, we conclude that $h_i^{(t)}(\frac{Z_i-\lambda_i}{x_t})$ is in fact the minimal polynomial of $\frac{\theta_i-\lambda_i}{x_t}$. We deduce that $S_t \subset S_t\left[\frac{\theta_1-\lambda_1}{x_t}, \ldots, \frac{\theta_N-\lambda_N}{x_t}\right]$ is a finite extension, and since the latter ring is included in $B \otimes_S K$, condition (3) of Definition 3.8 is fulfilled. According to Remark 3.12, the prime ideal Q contains the minimal primes of B and this implies that $S_t\left[\frac{\theta_1-\lambda_1}{x_t}, \ldots, \frac{\theta_N-\lambda_N}{x_t}\right]$ is equidimensional as it is clearly an affine chart of the blow-up of $\mathrm{Spec}(B)$ at $V(Q)$ (see (2)). Finally, as $\mathfrak{q}$ is a regular

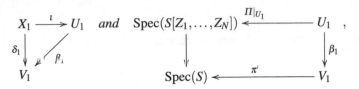

prime of S (Proposition 3.19), S_t satisfies condition (1) of Definition 3.8. We conclude that $S_t \subset S_t\left[\frac{\theta_1-\lambda_1}{x_t},\ldots,\frac{\theta_N-\lambda_N}{x_t}\right]$ belongs to the class $\mathcal{T}$.

(2) By definition, $X_1 = \mathrm{Proj}(B[QT])$, where T is a variable and $B[QT]$ is the Rees ring $B + QT + Q^2T^2 + \ldots$. We claim that X_1 can be covered by using only the affine charts $\mathrm{Spec}(B[QT]_{(x_tT)})$, $t = 1,\ldots,r$, where $B[QT]_{(x_tT)}$ denotes the degree zero part of the localization of the graded ring $B[QT]$ at the homogeneous element x_tT. In fact, any homogeneous prime ideal $\mathcal{P}$ of $B[QT]$ which is not in any of these charts must include $\mathfrak{q}BT$. As $\theta_i - \lambda_i$ satisfies the equation $h_i(\theta_i - \lambda_i) = 0$, one obtains that

$$((\theta_i-\lambda_i)T)^{d_i}+a_{i,1}T((\theta_i-\lambda_i)T)^{d_i-1}+\ldots+a_{i,d_i-1}T^{d_i-1}((\theta_i-\lambda_i)T)+a_{i,d_i}T^{d_i} = 0T^{d_i} \in \mathcal{P}.$$

If follows that $((\theta_i - \lambda_i)T)^{d_i} \in \mathcal{P}$, so $(\theta_i - \lambda_i)T \in \mathcal{P}$. As $Q = \langle \mathfrak{q}B, \theta_1 - \lambda_1, \ldots, \theta_N - \lambda_N\rangle$, we conclude that $\mathcal{P}$ must include the irrelevant ideal, hence it is not an element of $\mathrm{Proj}(B[QT])$.

As each x_t is not a zero divisor of B, there are natural inclusions

$$(B[QT])_{(x_tT)} \subset (B_{x_t}[Q_{x_t}T])_{(x_tT)} \cong B_{x_t} \subset B\otimes_S K, \quad t = 1,\ldots,r,$$

giving the identifications

$$\begin{aligned}(B[QT])_{(x_tT)} &= B\left[\frac{x_1}{x_t},\ldots,\frac{x_r}{x_t},\frac{\theta_1-\lambda_1}{x_t},\ldots,\frac{\theta_N-\lambda_N}{x_t}\right]\\ &= S_t\left[\frac{\theta_1-\lambda_1}{x_t},\ldots,\frac{\theta_N-\lambda_N}{x_t}\right] \subset B\otimes_S K.\end{aligned}$$

It is clear that the morphisms of affine schemes induced by the extensions given in the statement glue together into a morphism $\delta_1 : X_1 \to V_1$ compatible with δ and the blow-ups. In fact, all these extensions are induced by the extension $K \subset B\otimes_S K$.

Note from the construction of X_1 given above that the pull-back by $\pi\delta$ of the sheaf of ideals defined by $\mathfrak{q}$ is the exceptional divisor of the blow-up $\mathrm{Spec}(B) \leftarrow X_1$. Hence, the uniqueness of δ_1 making the given diagram commutative follows from the universal property of the blow-up $\mathrm{Spec}(S) \leftarrow V_1$.

(3) The first assertion is clear. The fact that the morphisms of schemes induced by the morphisms ρ_t glue together follows since the morphisms ρ_t are all induced by the same K-epimorphism $\rho_0 : K[Z_1,\ldots,Z_N] \to B\otimes_S K$, which is given by the evaluation $Z_i \mapsto \theta_i$. The commutativity of the diagram of the statement follows since ρ is also induced by ρ_0. The uniqueness of the morphism $\iota : X_1 \to W_1$ making the given diagram commutative follows from the universal property of the blow-ups. In particular, if $\iota_1' : X_1 \to W_1$ denotes the strict transform (see 2.11), then $\iota_1' = \iota_1$.

(4) Again the existence of β_1 and the commutativity of the diagrams are clear. The assertion about the uniqueness of β_1 follows from the universal property of the blow-up π'.

(5) We first have to prove that there are no points outside U_1 at which each $I(\mathscr{H}_i^{(1)})$ has order ≥ 1. Recall that

$$W_1 = \mathrm{Proj}(S[Z_1, \ldots, Z_N][\tilde{Q}T]) = \mathrm{Proj}(S[T_1, \ldots, T_N][\langle x_1, \ldots, x_r, T_1, \ldots, T_r\rangle T]),$$

where $T_i = Z_i - \lambda_i$, and the complement of U_1 in W_1 consists of the homogeneous prime ideals $\mathscr{P}$ of the Rees ring $S[T_1, \ldots, T_N][\langle x_1, \ldots, x_r, T_1, \ldots, T_r\rangle T]$ that include $\langle x_1 T_1, \ldots, x_r T_r\rangle$ and do not include all the elements $T_1 T, \ldots, T_N T$. Given such a prime ideal $\mathscr{P}$, assume for example that $T_1 T \notin \mathscr{P}$, so we may think of $\mathscr{P}$ as a prime ideal of $S[\frac{T_2}{T_1}, \ldots, \frac{T_N}{T_1}, \frac{x_1}{T_1}, \ldots, \frac{x_r}{T_1}]$ and the condition is that $\mathscr{P}$ contains $\frac{x_1}{T_1}, \ldots, \frac{x_r}{T_1}$. The restriction of $\mathscr{H}_1^{(1)}$ to the affine chart $\mathrm{Spec}(S[\frac{T_2}{T_1}, \ldots, \frac{T_N}{T_1}, \frac{x_1}{T_1}, \ldots, \frac{x_r}{T_1}])$ is defined by $\frac{h_1(T_1)}{T_1^{d_1}} = 1 + \frac{b_{1,1}}{T_1} + \frac{b_{1,2}}{T_1^2} + \ldots + \frac{b_{1,d_1}}{T_1^{d_1}}$. Note that $\frac{b_{1,j}}{T_1^j} \in \mathscr{P}$ and hence $I(\mathscr{H}_1^{(1)})$ has order zero at $\mathscr{P}$.

We proved in Corollary 3.18 that $f_i(Z_i)$ has order d_i at points of $V(\tilde{Q}) \subset \mathrm{Spec}(S[Z_1, \ldots, Z_N])$. Hence $(I(\mathscr{H}_i^{(1)}), d_i)$ is the transform of $(\langle f_i(Z_i)\rangle, d_i)$ and the ideal $I(\mathscr{H}_i^{(1)})|_{\mathrm{Spec}\left(S_t\left[\frac{Z_1-\lambda_1}{x_t}, \ldots, \frac{Z_N-\lambda_N}{x_t}\right]\right)}$ is generated by $\frac{f_i(Z_i)}{x_t^{d_i}} = h_i^{(t)}(\frac{Z_i-\lambda_i}{x_t^{d_i}}) \in S_t[\frac{Z_i-\lambda_i}{x_t}]$, which is the minimal polynomial of $\frac{\theta_i-\lambda_i}{x_t^{d_i}}$. By the results obtained in the previous items and Corollary 3.18 we can ensure that

$$\iota_1(F_n(X_1)) = \left(\bigcap_{i=1}^{N} \mathrm{Sing}(I(\mathscr{H}_i^{(1)}), d_i)\right) \bigcap U_1 = \left(\bigcap_{i=1}^{N} \mathrm{Sing}(I(\mathscr{H}_i^{(1)}), d_i)\right).$$

□

Remark 3.22 In 3.16 we defined $\tilde{B} := S[Z_1, \ldots, Z_N]/\langle f_1(Z_1), \ldots, f_N(Z_N)\rangle$ and proved that $\mathrm{Spec}(\tilde{B}) \subset \mathrm{Spec}(S[Z_1, \ldots, Z_N])$ is a complete intersection subscheme (the intersection of the hypersurfaces defined by $f_1(Z_1), \ldots, f_N(Z_N)$) whose maximal multiplicity locus coincides with that of $\mathrm{Spec}(B)$. The last part of the proof of (5) above also shows that if $\tilde{X}_1$ is the blow-up of $\mathrm{Spec}(\tilde{B})$ at the same center as that of the blow-up of $\mathrm{Spec}(B)$, then $\tilde{X}_1$ is also the intersection of the hypersurfaces $\mathscr{H}_1^{(1)}, \ldots, \mathscr{H}_N^{(1)}$.

Proof of Theorem 2.18 Theorem 3.21 already contains all the ingredients required to prove Theorem 2.18. We only need to reformulate results into global terms. □

Definition 3.23 Let $\mathscr{T}^*$ be the class of finite and surjective morphisms of schemes $\delta : X \to V$, where V is connected and for each affine chart $\mathrm{Spec}(S) \subset V$, if B is the affine ring of $\pi^{-1}(\mathrm{Spec}(S))$, the corresponding ring extension $S \subset B$ belongs to the class $\mathscr{T}$ (see Definition 3.8). The generic rank of such extension $S \subset B$ will also be called the generic rank of δ, which is clearly well defined since V is connected.

Given δ in the class $\mathscr{T}^*$, let $F_\delta(X)$ denote the set of points of multiplicity equal to the generic rank of δ. By Corollary 3.10, either $F_\delta(X)$ is empty or it is the set of points of maximal multiplicity of X. In the latter case, we shall call δ a *transversal morphism*.

Definition 3.24 Fix a morphism $\delta : X \to V$ in the class $\mathscr{T}^*$. A *presentation* for δ consists of the following:

(i) a closed immersion $\iota : X \to U$ into a connected regular scheme U,
(ii) hypersurfaces $\mathscr{H}_1, \ldots, \mathscr{H}_N \subset U$ and positive integers $d_1, \ldots, d_N$,
(iii) a smooth morphism $\beta : U \to V$

such that the following holds.

(1) $I(\mathscr{H}_i)$ has order at most d_i at points of U, the intersection $\bigcap_{i=1}^N \mathscr{H}_i$ is a complete intersection subscheme of U, and

$$\iota(F_\delta(X)) = \bigcap_{i=1}^{N} \mathrm{Sing}(I(\mathscr{H}_i), d_i).$$

(2) The diagram

$$\begin{array}{ccc} X & \xrightarrow{\iota} & U \\ {\scriptstyle\delta}\downarrow & \swarrow{\scriptstyle\beta} & \\ V & & \end{array}$$

is commutative.

(3) At each point $y \in V$, one can find an affine neighbourhood, say $\mathrm{Spec}(S) \subset V$, such that

a. $\beta^{-1}(\mathrm{Spec}(S))$ is of the form $\mathrm{Spec}(S[Z_1, \ldots, Z_N])$,
b. the restriction of β is induced by the inclusion $S \subset S[Z_1, \ldots, Z_N]$, and
c. the restriction of $\mathscr{H}_i$ to $\mathrm{Spec}(S[Z_1, \ldots, Z_N])$ is defined by a monic polynomial in Z_i, say $f_i(Z_i) \in S[Z_i]$, such that if we set $\delta^{-1}(\mathrm{Spec}(S)) =: \mathrm{Spec}(B)$ and $\rho : S[Z_1, \ldots, Z_N] \to B$ denotes the epimorphism induced by the restriction

$$\mathrm{Spec}(B) \stackrel{\iota}{\to} \mathrm{Spec}(S[Z_1, \ldots, Z_N]),$$

then $f_i(Z_i)$ is the minimal polynomial of $\theta_i := \rho(Z_i)$.

A presentation for δ as in the definition will be denoted simply by

$$\begin{array}{ccc} X & \xrightarrow{\iota} & U \supset (\mathscr{H}_1, d_1; \mathscr{H}_2, d_2; \ldots; \mathscr{H}_N, d_N) \\ {\scriptstyle\delta}\downarrow & \swarrow{\scriptstyle\beta} & \\ V & & \end{array}$$

The pair

$$(J,b) := \bigcap_{j=1}^{N} (I(\mathscr{H}_i), d_i),$$

will be called the pair associated with the presentation.

Note that the affine open sets Spec(S) for which condition (3) of the definition is satisfied form a basis for the topology of V.

Remark 3.25 In the case that δ is transversal, we can deduce the following.

(a) The degree of the polynomial $f_i(Z_i)$ given in condition (3) is d_i, by condition (1) and Corollary 3.18.
(b) In the setting of (3) above, assume that $T_1, \dots, T_N$ are elements of $S[Z_1, \dots, Z_N]$ such that $S[Z_1, \dots, Z_N] = S[T_1, \dots, T_N]$ and the restriction of $\mathscr{H}_i$ is defined by a polynomial in T_i, say $g_i(T_i)$, which is also the minimal polynomial of $\rho(T_i)$. We claim that $T_i = u_i Z_i - \lambda_i$ for some unit $u_i \in S^*$ and $\lambda_i \in S$. In fact, by assumption, the ideals $\langle f_i(Z_i)\rangle$ and $\langle g_i(T_i)\rangle$ define the same hypersurface $\mathscr{H}_i$ and hence $g_i(T_i) = \alpha_i f_i(Z_i)$ for some $\alpha_i \in (S[Z_1, \dots, Z_N])^* = S^*$. As $g_i(T_i)$ has degree d_i as a polynomial in T_i, by 3.24 (3) (c), and $f_i(Z_i)$ has degree d_i and is pure in Z_i, the equality $g_i(T_i) = \alpha_i f_i(Z_i)$ ensures that the polynomial T_i is necessarily pure in Z_i of degree one, namely $T_i = u_i Z_i - \lambda_i$ for some $u_i, \lambda_i \in S$. With the same reasoning, Z_i is pure in T_i of degree one. We conclude that u_i is a unit and this proves our claim.

As a consequence, the polynomial $f_i(Z_i)$ is uniquely determined by the presentation up to a change of variable in $S[Z_i]$, namely, a change of the form $T_i = u_i Z_i - \lambda_i$ with $u_i \in S^*$ and $\lambda_i \in S$.

3.26 We now discuss the stability of the presentation by certain kind of transformations. Fix a morphism $\delta : X \to V$ in the class $\mathscr{T}^*$ with a presentation as in Definition 3.24.

(a) *Restriction to open subsets:* If V' is an open subscheme of V, we set $U' := \delta^{-1}(V')$ and the given presentation of $\delta : X \to V$ induces, by restriction, a presentation for $\delta' : X' \to V'$. Namely, we set $U' := \beta^{-1}(V')$ and define β' and $\mathscr{H}'$ respectively as the restriction of β to U' and the restriction of $\mathscr{H}_i$ to U'. The pair associated with this new presentation is just the restriction of the pair of the given presentation for δ.
(b) *Multiplication by an affine line:* By multiplying every term by $\mathbb{A}^1$ we obtain a presentation for $\delta \times 1 : X \times \mathbb{A}^1 \to V \times \mathbb{A}^1$ (see Remark 3.15). The pair associated with this presentation of $\delta \times 1$ is just the pull-back of the pair associated with the given presentation of δ.
(c) *Permissible blow-ups:* Assume now that $\delta : X \to V$ is transversal. Proposition 3.19 implies that there is a one-to-one correspondence between regular centers for X included in $F_\delta(X)$ and regular centers for V included in $\delta(F_\delta(X))$.

Under this correspondence, a regular center $Y \subset V$ included in $\delta(F_\delta(X))$ corresponds to $\delta^{-1}(Y)$, and the restriction $\delta : \delta^{-1}(Y) \to Y$ is an isomorphism of schemes. So we fix a connected regular subscheme $Y \subset V$ included in $\delta(F_\delta(X))$ and consider the blow-ups $V \overset{\pi'}{\leftarrow} V_1$, $X \overset{\pi}{\leftarrow} X_1$, and $U \overset{\Pi}{\leftarrow} W_1$ at Y, $\delta^{-1}(Y)$ and $\iota(\delta^{-1}(Y))$, respectively. There are unique morphisms $\delta_1 : X_1 \to V_1$ and $\iota : X_1 \to W_1$ such that the diagram

$$\begin{array}{ccc} U & \xleftarrow{\ \Pi\ } & W_1 \\ {\scriptstyle\iota}\uparrow & & \uparrow{\scriptstyle\iota_1} \\ X & \xleftarrow{\ \pi\ } & X_1 \\ {\scriptstyle\delta}\downarrow & & \downarrow{\scriptstyle\delta_1} \\ V & \xleftarrow{\ \pi'\ } & V_1 \end{array}$$

is commutative. In fact, ι_1 is a closed immersion and $\iota_1(X_1)$ is the strict transform of $\iota(X)$ by the blow-up Π (see 2.11). The existence and uniqueness of δ_1 is deduced from Theorem 3.21(2). In fact, δ_1 is a morphism in the class $\mathscr{T}^*$. From part (4) of the same theorem, we deduce that there exists an open subscheme $U_1 \subset \Pi^{-1}(U)$ including $\iota_1(X_1)$ and there exists a unique smooth morphism $\beta_1 : U_1 \to V_1$ such that the diagram

$$\begin{array}{ccc} U & \xleftarrow{\ \Pi|_{U_1}\ } & U_1 \\ {\scriptstyle\beta}\downarrow & & \downarrow{\scriptstyle\beta_1} \\ V & \xleftarrow[\ \pi'\ ]{} & V_1 \end{array} \tag{2}$$

is commutative.

Let $\mathscr{H}_i^{(1)} \subset W_1$ denote the strict transform of $\mathscr{H}_i$ by Π. By Definition 3.24 and Theorem 3.21 we deduce that $(I(\mathscr{H}_i^{(1)}), d_i)$ is the transform of the pair $(I(\mathscr{H}_i), d_i)$ by the blow-up Π, $\bigcap_{i=1}^N \mathscr{H}_i^{(1)}$ is a complete intersection subscheme of U_1, and that

$$\begin{array}{ccl} X_1 & \xrightarrow{\ \iota_1\ } & U_1 \supset (\mathscr{H}_1^{(1)}, d_1; \mathscr{H}_2^{(1)}, d_2; \ldots; \mathscr{H}_N^{(1)}, d_N) \\ {\scriptstyle\delta_1}\downarrow & \swarrow{\scriptstyle\beta_1} & \\ V & & \end{array}$$

is in fact a presentation for δ_1. The morphism β_1 is uniquely determined with the requirement that the diagram (2) above is commutative. Note that the pair (J_1, b_1) associated with the presentation of δ_1 is just the transform of the pair associated with the given presentation of δ.

3.27 We finally end the proof of Theorem 2.18. We start with a transversal morphism $\delta : X \to V$, say of generic rank n, which admits a presentation as in Definition 3.24. The claim is that

$$(J, b) := \bigcap_{i=1}^{N} (I(\mathscr{H}_i), d_i)$$

represents the points of multiplicity n (see Remark 2.20).

So we fix a local sequence over U, say $U = W_0 \leftarrow W_1 \leftarrow \ldots \leftarrow W_s$, which is permissible for (J, b). We set $X_0 := X$, $\delta_0 := \delta$, $\tilde{U}_0 = U_0 = W_0 := U$ and $\mathscr{H}_i^{(0)} = \mathscr{H}_i$. Fix an index $1 \leq j \leq s$ and assume that we have defined:

(i) a morphism $\delta_{j-1} : X_{j-1} \to V_{j-1}$ in the class $\mathscr{T}^*$,
(ii) a closed immersion $X_{j-1} \subset \tilde{U}_{j-1}$, open immersions $\tilde{U}_{j-1} \subset U_{j-1}$ and $\tilde{U}_{j-1} \subset W_{j-1}$, so that X_{j-1} is closed in both U_{j-1} and W_{j-1},
(iii) hypersurfaces $\mathscr{H}_1^{(j-1)}, \ldots, \mathscr{H}_N^{(j-1)}$ in $\tilde{U}_{j-1}$ which are restrictions of hypersurfaces in U_{j-1} and in W_{j-1} (we shall use the same symbol $\mathscr{H}_i^{(j-1)}$ to refer to these hypersurfaces when viewed either inside U_{j-1} or inside W_{j-1}), and
(iv) a smooth morphism $\beta_{j-1} : U_{j-1} \to V_{j-1}$,

which are subject to the following conditions.

(a) $I(\mathscr{H}_i^{(j-1)})$ has order at most d_i at points of W_{j-1} and at points of U_{j-1}, and the subscheme $\bigcap_{i=1}^{N} \mathscr{H}_i^{(j-1)}$ is a complete intersection subscheme included in $\tilde{U}_{j-1}$,
(b)

$$\begin{array}{ccl} X_{j-1} & \longrightarrow & U_{j-1} \supset (\mathscr{H}_1^{(j-1)}|_{U_{j-1}}, d_1; \mathscr{H}_2^{(j-1)}|_{U_{j-1}}, d_2; \ldots; \mathscr{H}_N^{(j-1)}|_{U_{j-1}}, d_N) \\ {\scriptstyle \delta_{j-1}} \downarrow & \swarrow {\scriptstyle \beta_{j-1}} & \\ V_{j-1} & & \end{array}$$

is a presentation for δ_{j-1}.

We now define the same collection of objects and arrows with $j-1$ replaced by j as follows:

1. If W_j is an open subscheme of W_{j-1}, we set

$$\tilde{U}_j := W_j \cap \tilde{U}_{j-1}, \qquad V_j := \beta_{j-1}(\tilde{U}_j), \qquad U_j := \beta_{j-1}^{-1}(V_j),$$
$$X_j := \delta_{j-1}^{-1}(V_j), \qquad \delta_j := \delta_{j-1}|_{X_j}, \qquad \beta_j := \beta_{j-1}|_{U_j}$$

and consider the induced presentation of δ_j as defined in (see 3.26, (a)).
2. If $W_j = W_{j-1} \times \mathbb{A}^1$, we multiply everything by $\mathbb{A}^1$.

3. If W_j is a blow-up of W_{j-1} at a regular center $Y_{j-1} \subset \bigcap_{i=1}^{N} \mathrm{Sing}(I(\mathscr{H}_i^{(j-1)}), d_i)$, this center is necessarily included in $F_n(X_{j-1}) \subset X_{j-1}$, by definition of presentation. Thus, both W_j and the blow-up U'_j of U_{j-1} at Y_{j-1} contain the blow-up $\tilde{U}'_j$ of $\tilde{U}_{j-1}$ at Y_{j-1} as an open subscheme. The strict transform $\mathscr{H}_i^{(j)}$ of $\mathscr{H}_i^{(j-1)}$ is defined in both W_j and U'_j. We consider the transform of the given presentation of δ_{j-1} as obtained in 3.26 (c). In particular, we obtain a transversal morphism $\delta_j : X_j \to V_j$ and a smooth morphism $\beta : U_j \to V_j$, where U_j is an open subscheme of U'_j. We finally set $\tilde{U}_j := U_j \cap \tilde{U}'_j$. Note also that X_j is the strict transform of X_{j-1} by the blow-up and it is included in $\tilde{U}_j$.

One easily checks that in any of these cases, we are in the same situation as in the case $j-1$.

It follows by construction that the given local sequence $U = W_0 \leftarrow W_1 \leftarrow \ldots \leftarrow W_s$ induces a local sequence over X, say $X = X_0 \leftarrow X_1 \leftarrow \ldots \leftarrow X_s$, in the sense of 2.12, and that $F_n(X_j) = \bigcap_{i=1}^{N} \mathrm{Sing}(I(\mathscr{H}_i)^{(j)}, d_i)$. This concludes the proof of Theorem 2.18.

3.28 We end this section with a similar analysis but with the perspective that will be used in the next section. Fix a transversal morphism $\delta : X \to V$ in the class $\mathscr{T}^*$, and assume that it admits a presentation as in Definition 3.24. Fix any local sequence over V, say $V = V_0 \xleftarrow{\pi'_1} V_1 \xleftarrow{\pi'_2} \ldots \xleftarrow{\pi'_r} X_r$, which induces a local sequence over X, in the sense of 2.24, so that we obtain a commutative diagram

$$\begin{array}{ccccccccc}
X = X_0 & \longleftarrow & X_1 & \longleftarrow & \cdots & \longleftarrow & X_{r-1} & \longleftarrow & X_r \\
\downarrow{\scriptstyle \delta=\delta_0} & & \downarrow{\scriptstyle \delta_1} & & & & \downarrow{\scriptstyle \delta_{r-1}} & & \downarrow{\scriptstyle \delta_r} \\
V = V_0 & \longleftarrow & V_1 & \longleftarrow & \cdots & \longleftarrow & V_{r-1} & \longleftarrow & V_r,
\end{array}$$

Then for each $j = 0, \ldots, r$ one can define uniquely a presentation

$$\begin{array}{ccc}
X_i & \xrightarrow{\iota_i} & U_i \\
\downarrow{\scriptstyle \delta_i} & \swarrow{\scriptstyle \beta_i} & \\
V_i & &
\end{array} \qquad \supset (\mathscr{H}_1^{(j)}, d_1; \mathscr{H}_2^{(i)}, d_2; \ldots; \mathscr{H}_N^{(j)}, d_N)$$

so that for $j = 0$ this is the given presentation, and for $j > 0$ the presentation of δ_j is obtained from that of δ_{j-1} as in 3.26.

4 On Elimination in Characteristic Zero and the Proof of Theorem 2.25

4.1 We now fix a field k and a positive integer d, and consider extensions of the form $S \subset S[Z]/\langle f(Z)\rangle$, where S is a regular k-algebra and $f(Z) = X^d + a_1 X^{d-1} + \cdots + a_d \in S[Z]$ is a monic polynomial of degree d. Among all these extensions, we shall construct one which is universal in a sense to be specified.

Let $k[X_1, \ldots, X_d, Z]$ be a polynomial ring in $d+1$ variables, and consider the polynomial

$$\begin{aligned} F_d(Z) : &= (Z - X_1)(Z - X_2) \cdots (Z - X_d) \\ &= Z^d + (-1)s_1 Z^{d-1} + \cdots + (-1)^d s_d \in k[s_1, s_2, \ldots, s_d][Z], \end{aligned}$$

where $s_i = s_i(X_1, \ldots, X_d)$ denotes the usual homogeneous symmetric polynomial of degree i. Here $k[s_1, s_2, \ldots, s_d]$ is a subring of $k[X_1, X_2, \ldots, X_d]$ and a theorem of Noether states that

$$k[X_1, X_2, \ldots, X_d]^{\mathbb{S}_d} = k[s_1, s_2, \ldots, s_d],$$

where $\mathbb{S}_d$, the permutation group of d elements, acts on $k[X_1, X_2, \ldots, X_d]$ by permuting the variables. The above equality implies that the extension $k[s_1, s_2, \ldots, s_d] \subset k[X_1, X_2, \ldots, X_d]$ is finite, so $k[s_1, s_2, \ldots, s_n]$ is also a polynomial ring in d variables.

We also get a finite extension by setting:

$$k[s_1, s_2, \ldots, s_d] \subset k[s_1, s_2, \ldots, s_d][Z]/\langle F_d(Z)\rangle, \tag{3}$$

which we claim to be universal: Given any regular k-algebra S and a monic polynomial of degree d, say $f(Z) = Z^d + a_1 Z^{d-1} + \cdots + a_d \in S[Z]$, the finite extension $S \subset S[Z]/\langle f(Z)\rangle$ arises from that in (3) by the change of base ring, or say the morphism of k-algebras

$$\phi : k[s_1, s_2, \ldots, s_d] \to S, \quad s_i \mapsto (-1)^i a_i. \tag{4}$$

Hence in a natural way we can say that (3) is the universal extension of degree d.

4.2 Within the previous setting, we assume now that S is a domain, say with fraction field K. Fix an algebraic closure $K \subset \bar{K}$ and let S' denote the integral closure of S in $\bar{K}$. Then there is a factorization

$$f(Z) = Z^d + a_1 Z^{d-1} + \cdots + a_d = (Z - \Theta_1)(Z - \Theta_2) \cdots (Z - \Theta_d)$$

at $S'[Z]$. Note that the restriction of the morphism of k-algebras, say Φ : $k[X_1, X_2, \ldots, X_d] \to S'$ which maps X_i to Θ_i to the subring $k[s_1, s_2, \ldots, s_d]$ is the morphism $\phi : k[s_1, s_2, \ldots, s_d] \to S$ defined in (4); namely, there is a commutative diagram

$$\begin{array}{ccc} k[X_1,\ldots,X_d] & \xrightarrow{\Phi} & S' \\ \uparrow & & \uparrow \\ k[s_1,\ldots,s_d] & \xrightarrow{\phi} & S \end{array} \tag{5}$$

where the vertical arrows are inclusions.

4.3 We now look for polynomial expressions $G(a_1, \ldots, a_d)$ on the coefficients of $f(Z)$ that are not affected by a change of variable of the form $Z_1 = Z - \lambda$, $\lambda \in S$. Namely, given $\lambda \in S$, if we set $Z_1 = Z - \lambda$, then $S[Z] = S[Z_1]$ and

$$f(Z) = Z^d + a_1 Z^{d-1} + \cdots + a_d = g(Z_1) = Z_1^d + b_1 Z_1^{n-1} + \cdots + b_d = g(Z_1) \in S[Z_1]. \tag{6}$$

The requirement on G is that $G(a_1, \ldots, a_d) = G(b_1, \ldots, b_d)$. An example of an expression with this property is the discriminant. In fact, consider the extension $S \subset S'$ defined in 4.2, and note that

$$f(Z) = (Z - \Theta_1)(Z - \Theta_2) \cdots (Z - \Theta_d) = (Z_1 - (\Theta_1 - \lambda)) \cdots (Z_1 - (\Theta_d - \lambda)) = g(Z_1). \tag{7}$$

This shows that $a_i = (-1)^i s_i(\Theta_1, \ldots, \Theta_d)$, and $b_i = (-1)^i s_i(\Theta_1 - \lambda, \ldots, \Theta_d - \lambda)$ for $i = 1, \ldots, d$. It also shows that the discriminant satisfies

$$\Delta(a_1, \ldots, a_d) = \prod_{i \neq j} (\Theta_i - \Theta_j) = \prod_{i \neq j} ((\Theta_i - \lambda) - (\Theta_j - \lambda)) = \Delta(b_1, \ldots, b_d).$$

4.4 In order to obtain universal expressions on the coefficients of a monic polynomial of degree d that are invariant under a change of variables, we need to come back to the universal setting. We first note that the permutation group $\mathbb{S}_d$ acts on the subring $k[X_i - X_j]_{1 \leq i, j \leq d}$. In particular, we obtain an inclusion of rings

$$k[X_i - X_j]_{1 \leq i, j \leq d}^{\mathbb{S}_d} \subset k[X_1, X_2, \ldots, X_d]^{\mathbb{S}_d} = k[s_1, s_2, \ldots, s_d].$$

As $k[X_i - X_j]_{1 \leq i, j \leq d}$ is also graded, the same holds for the ring of invariants, namely

$$k[X_i - X_j]_{1 \leq i, j \leq d}^{\mathbb{S}_d} = k[G_1(s), \ldots, G_m(s)]$$

where m is a positive integer and each $G_i(s) = G_i(s_1, \dots, s_d)$ is homogeneous, say of degree N_i, when viewed as a polynomial in $k[X_1, X_2, \dots, X_d]$. So $G_i(s)$ is weighted homogeneous in $k[s_1, s_2, \dots, s_d]$, also of degree N_i if each s_i is given weight i.

Lemma 4.5 *In the setting of* 4.4, *any* $G(s_1, \dots, s_n) \in k[G_1(s), \dots, G_m(s)]$ *satisfies*

$$G(a_1, \dots, a_d) = G(b_1, \dots, b_d).$$

Proof Any $G(s) \in k[G_1(s), \dots, G_m(s)]$ is in particular an element of $k[X_i - X_i]_{1 \leq i,j \leq d}$, say $G_i(s) = H_i(X_1 - X_2, X_1 - X_3, \dots, X_1 - X_d)$. We use the commutative diagram (5) with the specialization $X_i \mapsto \Theta_i$, so that s_i is mapped to $(-1)^i a_i$, and obtain that

$$G_i((-1)^1 a_1, \dots, (-1)^d a_d) = H_i(\Theta_1 - \Theta_2, \Theta_1 - \Theta_3, \dots, \Theta_1 - \Theta_d).$$

We now use the diagram (5) again but this time with the specialization $X_i \mapsto \Theta_i - \lambda$, so that s_i is mapped to $(-1)^i b_i$, and obtain that

$$\begin{aligned} G_i((-1)^1 b_1, \dots, (-1)^d b_d) &= H_i((\Theta_1 - \lambda) - (\Theta_2 - \lambda), \dots, (\Theta_1 - \lambda) - (\Theta_d - \lambda)) \\ &= H_i(\Theta_1 - \Theta_2, \dots, \Theta_1 - \Theta_d) = G_i((-1)^1 a_1, \dots, (-1)^d a_d). \end{aligned}$$

As $G_i(s_1, \dots, s_d)$ is weighted homogeneous in $s_1, \dots, s_d$, one gets

$$G_i((-1)^1 a_1, \dots, (-1)^d a_d) = (-1)^{N_i} G_i(a_1, \dots, a_d)$$
$$G_i((-1)^1 b_1, \dots, (-1)^d b_d) = (-1)^{N_i} G_i(b_1, \dots, b_d).$$

The proof of the lemma is now complete. □

4.6 So far we have a ring $k[s_1, \dots, s_d]$ which can be thought of as the ring of polynomial functions on the coefficients of a monic polynomial $f(Z) \in S[Z]$ of degree d. Inside this ring we have $k[G_1(s), \dots, G_m(s)]$, which is the subring of those polynomial expressions that are invariant under any linear change of variables of the form $Z_1 = Z - \lambda$. We shall consider $k[s_1, \dots, s_n]$ as a graded ring, where s_i has weight i. Each $G_i(s)$ is weighted homogeneous, say of degree N_i.

Assume now that the characteristic of the base field k does not divide d (e. g. when k has characteristic zero). We claim that

$$k[s_1, s_2, \dots, s_d] = k[G_1(s), \dots, G_m(s)][s_1]. \tag{8}$$

To check this we express each form $X_j - X_d$ and also s_1 as a linear combination of the basis $\{X_1, \dots, X_d\}$. These expressions define a square matrix which has determinant d. Therefore, $k[X_1, \dots, X_d] = k[X_1 - X_2, X_1 - X_3, \dots, X_1 - X_d][s_1]$

if the characteristic of k does not divide d, and hence

$$k[s_1,\ldots,s_d] = (k[X_1 - X_2,\ldots,X_1 - X_d][s_1])^{\mathbb{S}_d} = k[X_1 - X_2,\ldots,X_1 - X_d]^{\mathbb{S}_d}[s_1]$$
$$= k[G_1(s),\ldots,G_m(s)][s_1].$$

Proposition 4.7 *Let S be a regular domain which is essentially of finite type over a field k, and fix a monic polynomial $f(Z) = Z^d + a_1Z^{d-1} + \ldots + a_d \in S[Z]$, so that we obtain an extension $S \subset B := S[Z]/\langle f(Z)\rangle$ in the class $\mathcal{T}$ (Definition 3.8). Assume that the characteristic of k does not divide d. If $\delta : \mathrm{Spec}(B) \to \mathrm{Spec}(S)$ denotes the induced morphism of schemes, then*

$$\delta(F_S(B)) = \bigcap_{i=1}^{m} \mathrm{Sing}(\langle G_i(a_1,\ldots,a_d)\rangle, N_i),$$

where $G_1,\ldots,G_m$ are defined as in 4.4.

Proof Fix $\mathfrak{p} \in \mathrm{Spec}(S)$. We first assume that $\mathfrak{p} \in \delta(F_S(B))$ and prove that for each $i = 1,\ldots,m$, $G_i(a_1,\ldots,a_d)$ has order $\geq N_i$ at $\mathfrak{p}$. For this purpose, after localizing S at $\mathfrak{p}$, we may assume that S is local with maximal ideal $\mathfrak{p}$. By Corollary 3.10, B has to be local, say with maximal ideal P, and $B/P = S/\mathfrak{p}$. According to Proposition 3.20, there exists $\lambda \in S$ such that if we set $Z_1 = Z - \lambda$, then

$$f(Z) = Z^d + a_1Z^{d-1} + \ldots + a_d = Z_1^d + b_1Z_1^{d-1} + \ldots + b_d \in S[Z] = S[Z_1]$$

with $b_j \in \mathfrak{p}^j$, $j = 1,\ldots,d$. As $G_i(b_1,\ldots,b_d)$ is weighted homogeneous of degree N_i in $b_1,\ldots,b_d$, the fact that $b_j \in \mathfrak{p}^j$ ensures that $G_i(b_1,\ldots,b_d) \in \mathfrak{p}^{N_i}$; therefore $G_i(a_1,\ldots,a_d) \in \mathfrak{p}^{N_i}$ for all $i = 1,\ldots,m$, by Lemma 4.5.

Conversely, we assume now that each $G_i(a_1,\ldots,a_d)$ has order $\geq N_i$ at $\mathfrak{p}$ and prove that $\mathfrak{p} \in \delta(F_S(B))$. Set $\lambda = \frac{a_1}{d}$ and $Z_1 = Z - \lambda$. We obtain

$$f(Z) = Z^d + a_1Z^{d-1} + \ldots + a_d = Z_1^d + b_1Z_1^{d-1} + \ldots + b_d \in S[Z] = S[Z_1]$$

with $b_1 = 0$. By Lemma 4.5, $G_i(b_1,\ldots,b_d) = G(a_1,\ldots,a_d)$, which has order $\geq N_i$ at $\mathfrak{p}$. As $b_1 = 0$, the relation (8) in 4.6 implies that b_j can be written as a weighted homogeneous polynomial of degree j on $G_1(b_1,\ldots,b_d),\ldots,G_m(b_1,\ldots,b_d)$, and hence b_j has order $\geq j$ at $\mathfrak{p}$ for all $j = 1,\ldots,d$. Note that $\tilde{P} := \langle \mathfrak{p}S[Z], Z - \lambda\rangle$ is a prime ideal of $S[Z]$ and we have proved that $f(Z)$ has order $\geq d$ at this prime. If P denotes the image of $\tilde{P}$ in B, we conclude that P is a point of multiplicity d of the scheme $\mathrm{Spec}(B)$, namely $P \in F_S(B)$. It is clear that $P \cap S = \tilde{P} \cap S = \mathfrak{p}$, hence $\mathfrak{p} \in \delta(F_S(B))$. This completes the proof of the proposition. □

4.8 So far we have only considered changes of variable that are of the form $Z_1 = Z - \lambda$. The more general type of a change of variable in $S[Z]$ is of the form $Z_1 = uZ - \lambda$, where $\lambda \in S$ and u is a unit of S. So we fix a unit $u \in S^*$ and consider the change of variable $Z_1 = uZ$. Given a monic polynomial of degree d, say $f(Z) = Z^d + a_1 Z^{d-1} + \ldots + a_d$, one has

$$f(Z) = (\frac{Z_1}{u})^d + a_1(\frac{Z_1}{u})^{d-1} + \ldots + a_{d-1}\frac{Z_1}{u} + a_d,$$

which is not monic, but the associated polynomial

$$u^d f(Z) = Z_1^d + a_1 u Z_1^{d-1} + \ldots + a_{d-1} u^{d-1} + a_d u^d$$

is monic. As $G_i(s_1, \ldots, s_d)$ is weighted homogeneous of degree N_i, we obtain that

$$G_i(a_1 u, a_2 u^2, \ldots, a_d u^d) = u^{N_i} G_i(a_1, \ldots, a_d).$$

With this and Lemma 4.5 we have proved:

Corollary 4.9 Given a monic polynomial $f(Z) = Z^d + a_1 Z^{n-1} + \ldots + a_n \in S[Z]$, the ideal of S spanned by $G_i(a_1, \ldots, a_d)$ is intrinsic to the polynomial $f(Z)$ and independent of any change of variables in $S[Z]$.

4.10 We now turn to the proof of Theorem 2.25. Since there will be fixed positive integers $d_1, \ldots, d_N$ along the proof, it will be convenient to fix some notation. For a positive integer $1 \leq j \leq d$, we shall denote by $s_j^{(d)}$ the usual symmetric polynomial of degree j in the variables $X_1, \ldots, X_d$. For each $i = 1, \ldots, N$, we shall fix weighted homogeneous polynomials

$$G_1^{(d_i)}(s_1^{(d_i)}, \ldots, s_{d_i}^{(d_i)}), \ldots, G_{m_{d_i}}^{(d_i)}(s_1^{(d_i)}, \ldots, s_{d_i}^{(d_i)}) \in k[s_1^{(d_i)}, \ldots, s_{d_i}^{(d_i)}],$$

say of degrees $N_{i1}, \ldots, N_{id_i}$ respectively, which are generators for the ring of invariants $k[X_r - X_s]_{1 \leq r,s \leq d_i}^{\mathbb{S}_{d_i}}$, namely

$$k[X_r - X_s]_{1 \leq r,s \leq d_i}^{\mathbb{S}_{d_i}} = k[G_1^{(d_i)}(s^{(d_i)}), \ldots, G_{m_{d_i}}^{(d_i)}(s^{(d_i)})] \subset k[s_1^{(d_i)}, \ldots, s_{d_i}^{(d_i)}].$$

Here $s^{(d_i)}$ stands for $(s_1^{(d_i)}, \ldots, s_{d_i}^{(d_i)})$.

Proposition 4.11 *Let $\delta : X \to V$ be a transversal morphism in the class $\mathcal{T}^*$, and assume that it admits a presentation as in Definition* 3.24. *Then, for each $i = 1, \ldots, N$, there are $\mathcal{O}_V$-ideals*

$$K_{ij}, \quad j = 1, \ldots, m_{d_i}$$

such that the following hold: Fix an affine open subset $\mathrm{Spec}(S) \subset V$ *for which (3) of Definition* 3.24 *holds, and keep the notation given in there, so that $\beta^{-1}(\mathrm{Spec}(S)) = \mathrm{Spec}(S[Z_1, \ldots, Z_N])$ and the restriction of $\mathcal{H}_i$ to this affine chart is defined by a* *monic polynomial of the form*

$$f_i(Z_i) = Z_i^{d_i} + a_{i1}Z_i^{d_i-1} + \ldots + a_{id_i} \in S[Z_i].$$

The condition is that

$$K_{ij}|_{\mathrm{Spec}(S)} = \langle G_j^{(d_i)}(a_{i1}, \ldots, a_{id_i})\rangle.$$

Moreover, if the characteristic of the base field does not divide $d_1 d_2 \ldots d_N$, then

$$\delta(F_\delta(X)) = \bigcap_{i=1}^{N}\bigcap_{j=1}^{m_{d_i}} \mathrm{Sing}(K_{ij}, N_{ij})$$

Proof By Remark 3.25(b), we know that the polynomials $f_i(Z_i)$ are uniquely determined up to a change of variables of the form $Z_i' = u_i Z_i - \lambda_i$, $u_i \in S^*$ and $\lambda_i \in S$. Thus, the first part of the proposition follows Corollary 4.9. The second part follows by combining Proposition 4.7 and Proposition 3.13 with $B_i = S[Z_i]/\langle f_i(Z_i)\rangle$ (see Proposition 3.14). □

Proposition 4.12 *Let $\delta : X \to V$ be a transversal morphism in the class $\mathscr{T}^*$, and assume it admits a presentation as in Definition 3.24. Fix a regular center Y included in $\delta(F_\delta(X))$ and consider the blow-up $\delta_1 : X_1 \to V_1$ of δ and its presentation defined as in 3.26(c). If $K_{ij}^{(1)}$ is the ideal obtained from this new presentation, then $(K_{ij}^{(1)}, N_{ij})$ is the transform of (K_{ij}, N_{ij}).*

Proof It is enough to consider the affine case, say $X = \mathrm{Spec}(B)$, $V = \mathrm{Spec}(S)$, and $U = \mathrm{Spec}(S[Z_1, \ldots, Z_N])$. According to the previous proposition and Theorem 3.21, in the notation given in the latter, it is enough to prove the following equality of ideals in $S_t[\frac{x_1}{x_t}, \ldots, \frac{x_r}{x_t}]$:

$$x_i^{N_{ij}}\left\langle G_j^{d_i}\left(\frac{b_{i,1}}{x_t}, \ldots, \frac{b_{i,d_i}}{x_t^{d_i}}\right)\right\rangle = \left\langle G_j^{d_i}(b_{i,1}, \ldots, b_{i,d_i})\right\rangle$$

This follows easily as $G_j^{d_i}$ is weighted homogeneous of degree N_{ij}. □

4.13 Let $\delta : X \to V$ be a transversal morphism in the class $\mathscr{T}^*$ and assume that it has a presentation as in Definition 3.24, and consider the ideals (K_{ij}, N_{ij}) obtained from the presentation as in Proposition 4.11. If V' is an open subset of V and if we consider the presentation obtained by restriction as in 3.26(a), it is easy to see that the associated pairs (K_{ij}', N_{ij}') are just the restrictions of the formers. Similarly, if we consider the presentation of $\delta \times 1 : X \times \mathbb{A}^1 \to V \times \mathbb{A}^1$ obtained in 3.26(b), the associated pairs are just the pull-backs of the pairs (K_{ij}, N_{ij}).

4.14 We can now end the proof of Theorem 2.25. The claim is that the pair

$$(K,d) := \bigcap_{i=1}^{N}\bigcap_{j=1}^{m_{d_i}} \mathrm{Sing}(K_{ij}, N_{ij})$$

represents the shadow of points of multiplicity n. This follows from the previous two propositions and the observations in 3.28.

5 Proof of Theorem 2.22

We shall begin this section by recalling the notion of étale morphisms of varieties (and schemes). The main result here is the proof of Theorem 2.22 to be reformulated in Theorem 5.13. This is the first step in the proof of Theorem 1.3, to be addressed in the next and final section.

Fix a variety X. In the formulation of Theorem 2.18 and also of Theorem 2.25, we assumed the existence of a finite and surjective morphism $\delta : X \to V$, where V is smooth and the generic rank of δ is the maximal value of the multiplicity of X. In the case of characteristic zero, where we know of a nice algorithm of resolution of pairs, these theorems ensure that the reduction of the multiplicity is possible. However, as said before, these theorems assume the existence of $\delta : X \to V$ with some prescribed properties. In order to prove the existence of such morphisms δ for a variety X we have to work *locally* and moreover *locally* in étale topology. We will explain the meaning of this assertion, and shall finally justify in the next section why these *local* presentations and resolutions enable us to proof Theorem 1.3, which is stated globally, with no reference to étale topology.

We start by reviewing some concepts and properties related with étale morphisms. Most of the statements are presented without proofs, and we refer to [17] and [10] for proofs and precise statements.

All rings and schemes in this section are supposed to be noetherian.

5.1 *Étale morphisms.* A local ring map $(R, \mathfrak{m}) \to (R', \mathfrak{m}')$ is said to be *étale local* if it is essentially of finite type (namely, it is the composition of a morphism of finite type and a localization) and the following three conditions are satisfied:

(i) $R \to R'$ is flat (in particular injective).
(ii) $\mathfrak{m}R' = \mathfrak{m}'$.
(iii) the field extension $R/\mathfrak{m} \subset R'/\mathfrak{m}'$ is finite and separable.

A ring homomorphism $\varphi : S \to S'$ of finite type is said to be *étale at a prime* $P' \subset S'$ if $S_{P' \cap S'} \to S'_{P'}$ is étale local. The set of points $P' \in \mathrm{Spec}(S')$ at which φ is étale is open (possibly empty). The ring map φ is said to be *étale* if it is étale at every prime ideal of $\mathrm{Spec}(S')$. By the previous observation, we only need to check this condition at maximal ideals.

A morphism of schemes $f : X \to Y$ is said to be étale if it is of finite type and for each $x \in X$ the ring map $\mathcal{O}_{Y,f(x)} \to \mathcal{O}_{X,x}$ is étale local. This condition only needs to be verified at closed points of X.

Étale morphisms are a particular case of smooth morphisms. More precisely, they are the smooth morphisms of relative dimension zero.

5.2 Here are some properties of étale morphisms:

(a) Étale morphisms are open.
(b) Open immersions are étale.
(c) Let $f : X \to Y$ and $g : Y \to Z$ be morphisms and assume that g is étale. Then $g \circ f$ is étale if and only if f is étale.
(d) If $f : X \to Y$ is étale and $g : Y' \to Y$ is any morphism, then $f \times 1_{Y'} : X \times_Y Y' \to Y'$ is étale.
(e) A field extension $k \subset K$ is étale if and only if it is finite and separable.

In particular, if X is a scheme of finite type over a field k and $k \subset k'$ is a finite and separable extension, then the projection $X \times_k k' \to X$ is étale.

5.3 Given a scheme X, étale morphisms $X' \to X$ has to be thought of as open immersions when X is considered with its *étale topology*. Given $x \in X$, an étale neighbourhood of x is an étale morphism $X' \to X$ and a point $x' \in X'$ mapping to x. We shall usually write $(X', x') \to (X, x)$.

An étale covering of X is a collection of étale morphisms $\{X_i \to X\}$ whose images form an open covering of X.

5.4 Fix a local homomorphism $(R, \mathfrak{m}) \to (R', \mathfrak{m}')$ which is flat (so we may assume that $R \subset R'$) and such $\mathfrak{m}R' = \mathfrak{m}'$. Under these assumptions one easily checks that

$$\mathrm{gr}_{\mathfrak{m}'}(R') = \mathrm{gr}_{\mathfrak{m}}(R) \otimes_{R/\mathfrak{m}} (R'/\mathfrak{m}').$$

Here are some consequences of this equality:

(1) R and R' have the same Hilbert-Samuel function, and hence the same dimension and the same multiplicity. In particular, R is regular if and only if R' is regular.
(2) A proper ideal $J \subset R$ is a reduction of $\mathfrak{m}$ if and only if JR' is a reduction of $\mathfrak{m}'$. In fact, if we set $\overline{J} := \frac{J+\mathfrak{m}^2}{\mathfrak{m}^2}$ and $\overline{JR'} := \frac{JR'+\mathfrak{m}'^2}{\mathfrak{m}'^2}$, we obtain that $\overline{JR'} = \overline{J} \otimes_{R/\mathfrak{m}} (R'/\mathfrak{m}')$ and hence

$$\begin{aligned} \mathrm{gr}_{\mathfrak{m}'}(R')/\langle \overline{JR'} \rangle &= \left(\mathrm{gr}_{\mathfrak{m}}(R) \otimes_{R/\mathfrak{m}} (R'/\mathfrak{m}')\right) / \langle \overline{J} \otimes_{R/\mathfrak{m}} (R'/\mathfrak{m}') \rangle \\ &= \left(\mathrm{gr}_{\mathfrak{m}}(R)/\langle \overline{J} \rangle\right) \otimes_{R/\mathfrak{m}} (R'/\mathfrak{m}'). \end{aligned}$$

The conclusion follows now from Remark 3.7.
(3) If $\mathfrak{p}' \in \mathrm{Ass}(R')$, then $\mathfrak{p} := \mathfrak{p}' \cap R \subset \cup_{\mathfrak{q} \in \mathrm{Ass}(R)} \mathfrak{q}$. For if $\mathfrak{p}$ contains a non-zero divisor $x \in R$, the map $R \xrightarrow{x} R$ would be injective and hence, by flatness, $R' \xrightarrow{x} R'$ would be also injective. This is a contradiction since $x \in \mathfrak{p}' \in \mathrm{Ass}(R')$.

(4) Let $R \subset B$ be a finite extension and consider the finite extension $R' \subset B' := B \otimes_R R'$. Assume that there exists a maximal ideal M of B for which $R/\mathfrak{m} = B/M$. Then,

(a) $M' := M \otimes_R R'$ is a maximal ideal of B' and $B'/M' = R'/\mathfrak{m}'$,
(b) If B is local, then so is B',
(c) If $\mathfrak{m}B_M$ is a reduction of MB_M, then $\mathfrak{m}'B'_{M'}$ is a reduction of $M'B_{M'}$.

In fact, given that R' is flat over R, $M \otimes_R R'$ is included in B' and $B'/M' = (B/M) \otimes_R R' = (R/\mathfrak{m}) \otimes_R R' = R'/\mathfrak{m}R' = R'/\mathfrak{m}'$. This shows (a) and implies also that $M' \otimes_R R'$ is the only prime of B' lying above M; thus, $B_M \otimes_R R'$ is local and coincides with $B'_{M'}$. In particular, if B is local then so is B', and this proves (b). We now prove (c). Notice that morphism $B_M \to B'_{M'} = B_M \otimes_R R'$ is flat since it is a base change of a flat morphism, and observe that the maximal ideal of $B'_{M'}$ is the extension of the maximal ideal of B_M. As $\mathfrak{m}B_M$ is a reduction of MB_M, we can apply (2) to the extension $B_M \to B'_{M'}$ and conclude that $\mathfrak{m}B'_{M'}$ is a reduction of $M'B_{M'}$. The proof of (c) follows since $\mathfrak{m}B'_{M'} = (\mathfrak{m}R')B'_{M'} = \mathfrak{m}'B'_{M'}$.

5.5 We now deduce some properties for an étale local ring map $(R, \mathfrak{m}) \to (R', \mathfrak{m}')$.

(1) A prime $\mathfrak{p}'$ of R' is minimal if and only if $\mathfrak{p} := \mathfrak{p}' \cap R$ is minimal. In fact, as the condition of being étale is open, the extension of rings $R_\mathfrak{p} \subset R'_{\mathfrak{p}'}$ is étale and hence they have the same dimension, by 5.4 (1).
(2) If $\mathrm{Ass}(R) = \mathrm{Min}(R)$, then $\mathrm{Ass}(R') = \mathrm{Min}(R')$. This follows from (1) and 5.4(3).
(3) If R is equidimensional and essentially of finite type over a field k, then R' is also equidimensional. In fact, let $\mathfrak{p}'$ be a minimal prime of R' and set $\mathfrak{p} := R \cap \mathfrak{p}'$. Then $R_\mathfrak{p} \subset R'_{\mathfrak{p}'}$ is étale and hence $\kappa(\mathfrak{p}) \subset \kappa(\mathfrak{p}')$ is a finite and separable extension of fields, so they have the same transcendence degree over k. Thus, $R/\mathfrak{p}$ and $R'/\mathfrak{p}'$ have the same dimension.

We now review the concept of henselian rings and that of henselization and strict henselization.

5.6 *Henselian rings.* A local ring $(R, \mathfrak{m}, \kappa)$ is said to be *henselian* if the following two equivalent conditions are satisfied:

(1) Every finite extension of R is a product of local rings.
(2) For any monic polynomial $f(Z) \in R[Z]$, if its reduction modulo $\mathfrak{m}$ decomposes as $\bar{f}(Z) = \bar{g}(Z)\bar{h}(Z)$, where $\bar{g}(Z), \bar{h}(Z) \in \kappa[Z]$ are monic and relatively prime, then $f(Z)$ admits a decomposition in $R[Z]$, say $f(Z) = g(Z)h(Z)$, where $g(Z)$ and $h(Z)$ are monic and their reductions modulo $\mathfrak{m}$ are $\bar{g}(Z)$ and $\bar{h}(Z)$ respectively.

A henselian local ring $(R, \mathfrak{m}, \kappa)$ is called *strictly henselian* if κ is separably closed.
Hensel's lemma states that complete local rings are henselian.

Let $(R, \mathfrak{m}, \kappa) \subset (R', \mathfrak{m}', \kappa')$ be a finite extension of local rings. If R is henselian, condition (1) in the definition implies that R' is also henselian. If, in addition, R is strictly henselian and $\kappa = \kappa'$, then R' is also strictly henselian.

Given a local ring $(R, \mathfrak{m}, \kappa)$ which is not henselian, we may wonder whether R can be embedded as a subring of a henselian or a strictly henselian local ring with some universal properties. This question gives rise to the concept of *henselization* and that of *strict henselization*. For our purpose, we shall only require the latter concept. Here, we refer the reader to [19, Tag 0BSK].

5.7 *Strict henselization.* Let $(R, \mathfrak{m}, \kappa)$ be a local ring. Let $\kappa \subset \kappa^{sep}$ be a separable closure. There exists a commutative diagram

$$\begin{array}{ccc} \kappa & \longrightarrow & \kappa^{sep} \\ \uparrow & & \uparrow \\ R & \longrightarrow & R^{sh} \end{array}$$

with the following properties

1. the map $R \mapsto R^{sh}$ is a local ring map (of noetherian local rings),
2. R^{sh} is strictly henselian and flat over R,
3. R^{sh} is a filtered colimit of étale morphisms $R \to R'$,
4. $\mathfrak{m}R^{sh}$ is the maximal ideal of R^{sh}, and
5. $\kappa^{sep} = R^{sh}/\mathfrak{m}R^{sh}$.

The morphism $R \to R^{sh}$ is uniquely determined by R and $\kappa \subset \kappa^{sep}$ up to isomorphism, and satisfies the following universal property: any commutative diagram as that in the left hand side of the diagram below, where $(R, \mathfrak{m}, \kappa) \to (R', \mathfrak{m}', \kappa')$ is a local homomorphism and R' is strictly henselian, can be completed in a unique way to a commutative diagram as that in the right.

$$\begin{array}{ccccc} \kappa & \longrightarrow & \kappa^{sep} & \longrightarrow & \kappa' \\ \uparrow & & & & \uparrow \\ R & & \longrightarrow & & R' \end{array} \quad \Longrightarrow \quad \begin{array}{ccccc} \kappa & \longrightarrow & \kappa^{sep} & \longrightarrow & \kappa' \\ \uparrow & & \uparrow & & \uparrow \\ R & \longrightarrow & R^{sh} & \longrightarrow & R' \end{array}$$

The ring R^{sh} is uniquely determined up to (non-unique) isomorphism by R. It is called the *strict henselization of* R. More precisely, the strict henselization of R with respect to $\kappa \subset \kappa^{sep}$ is the map $R \to R^{sh}$.

Proposition 5.8 *Let $(R, \mathfrak{m}, \kappa)$ be a local ring, let $R \subset B$ be a finite extension, and assume that there exists a maximal ideal M of B lying over $\mathfrak{m}$ such that $B/M = R/\mathfrak{m} = \kappa$. Fix a separable closure $\kappa \subset \kappa^{sep}$ and consider the strict henselization $R \to R^{sh}$. Then the natural map $B_M \to B_M \otimes_R R^{sh}$ is the strict henselization of B_M (with respect to $\kappa \subset \kappa^{sep}$).*

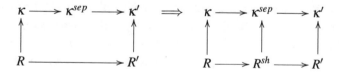

Proof Since R^{sh} is flat over R and $\mathfrak{m}R^{sh}$ is its maximal ideal, we can apply 5.4(4) and assert that $B_M \otimes_R R^{sh}$ is local and rational over R^{sh}. Moreover, it is the localization of $B \otimes_R R^{sh}$ at the maximal ideal $M \otimes_R R^{sh}$. Notice that $B \otimes_R R^{sh}$ is a finite extension of the henselian ring R^{sh}, so $B_M \otimes_R R^{sh}$ is a direct factor of $B \otimes_R R^{sh}$. It follows that $R^{sh} \subset B_M \otimes_R R^{sh}$ is a finite extension of local rings with the same residue field, so $B_M \otimes_R R^{sh}$ is strictly henselian (see 5.6).

The morphism $B_M \to B_M \otimes_R R^{sh}$ is flat, and one easily checks that induced morphism at the level of residue fields is the given extension $\kappa \subset \kappa^{sep}$. Moreover, this morphism is a filtered colimit of étale morphisms of the form $B_M \to B_M \otimes_{S_\mathfrak{m}} S'_{\mathfrak{m}'}$, where $S_\mathfrak{m} \to S'_{\mathfrak{m}'}$ is an étale local morphism. Finally, the maximal ideal of $B_M \otimes_R R^{sh}$ (namely, $MB_M \otimes_R R^{sh}$) is the extension of the maximal ideal of B_M. From 5.7 we conclude that $B_M \to B_M \otimes_R R^{sh}$ is the strict henselization of B_M. □

5.9 Let $(R, \mathfrak{m}, \kappa)$ be a local ring, and let $I \subset \mathfrak{m}$ be an ideal. We consider the strict henselization $R \to R^{sh}$ with respect to a given separable closure $\kappa \subset \kappa^{sep}$. Then the induced morphism $R/I \to R^{sh}/IR^{sh}$ is the strict henselization of R/I. In fact, the ring R^{sh}/IR^{sh} is henselian since it clearly satisfies condition (2) in the definition given in 5.6. Moreover, it is strictly henselian as it has residue field κ^{sep}. Finally, the conditions listed in 5.7 are verified easily.

5.10 Let R be a ring which is not necessarily local, and let $\mathfrak{p} \subset R$ be a prime ideal. Let $\kappa(\mathfrak{p}) \subset \kappa(\mathfrak{p})^{sep}$ be a separable closure. Consider the category of triples $(S, \mathfrak{q}, \phi)$ where $R \to S$ is étale, $\mathfrak{q}$ is a prime lying over $\mathfrak{p}$, and $\phi : \kappa(\mathfrak{q}) \to \kappa(\mathfrak{p})^{sep}$ is a $\kappa(\mathfrak{p})$-algebra map. This category is filtered and

$$(R_\mathfrak{p})^{sh} = \mathrm{colim}_{(S,\mathfrak{q},\phi)} S = \mathrm{colim}_{(S,\mathfrak{q},\phi)} S_\mathfrak{q}, \tag{9}$$

canonically.

5.11 We put this in terms of schemes. Let X be a scheme and fix $x \in X$ and separable closure $\kappa(x) \subset \kappa(x)^{sep}$. The *strict henselization of X at x with respect to $\kappa(x) \subset \kappa(x)^{sep}$* is the composition of the morphism $\mathrm{Spec}(\mathcal{O}_{X,x}^{sh}) \to \mathrm{Spec}(\mathcal{O}_{X,x})$ induced by the strict henselization $\mathcal{O}_{X,x} \to \mathcal{O}_{X,x}^{sh}$ with respect to $\kappa(x) \subset \kappa(x)^{sep}$, followed by the natural morphism $\mathrm{Spec}(\mathcal{O}_{X,x}) \subset X$. It will be denoted simply by $X_x^{sh} \to X$. If we consider the category of triples $(X' \xrightarrow{\alpha} X, x', \phi)$, where $(X', x') \xrightarrow{\alpha} (X, x)$ is an étale neighbourhood of x and $\phi : \kappa(x') \to \kappa(x)^{sep}$ is a $\kappa(x)$-algebra map, then, by (9) above,

$$X_x^{sh} = \lim_{(X' \xrightarrow{\alpha} X, x', \phi)} X'.$$

Proposition 5.12 *Let $S \subset B$ be an extension in the class $\mathscr{T}$ (Definition 3.8) such that S is of finite type over a field k. Fix $P \in F_S(B)$ (if $F_S(B) \neq \emptyset$) and put $\mathfrak{p} := S \cap P$.*

(1) *If $S \to S'$ is an étale morphism and* $\mathrm{Spec}(S')$ *is connected, then the extension $S' \subset B' := B \otimes_S S'$ is in the class $\mathscr{T}$.*

(2) *If* $S_{\mathfrak{p}} \to S_{\mathfrak{p}}^{sh}$ *is a strict henselization, then* $B \otimes_S S_{\mathfrak{p}}^{sh} = B_P \otimes_S S_{\mathfrak{p}}^{sh}$. *This is a local ring and the natural morphism* $B_P \to B \otimes_S S_{\mathfrak{p}}^{sh}$ *is a strict henselization of* B_P.

In addition, if $B = S[\theta_1, \dots, \theta_N]$ *(namely,* $\theta_1, \dots, \theta_N$ *are generators of* B *as an* S*-algebra) and* $f_i(Z) \in S[Z]$ *is the minimal polynomial of* θ_i*, then* $f_i(Z)$ *is also the minimal polynomial over* S' *of* $\theta_i \otimes 1 \in B \otimes_S S'$ *and the minimal polynomial over* $S_{\mathfrak{p}}^{sh}$ *of* $\theta_i \otimes 1 \in B \otimes_S S_{\mathfrak{p}}^{sh}$.

Proof

(1) Note that S' is of finite type over k and it is regular by 5.4 (1). By 5.5, (2) and (3), B' is equidimensional and $\mathrm{Ass}(B') = \mathrm{Min}(B')$. It follows that $S' \subset B'$ is in the class $\mathscr{T}$.

(2) By [9, Theorem 5.3], $S_{\mathfrak{p}}^{sh}$ is an excellent local ring, and it is regular by 5.4 (1). Note that $B \otimes_S S_{\mathfrak{p}}^{sh} = B_{\mathfrak{p}} \otimes_{S_{\mathfrak{p}}} S_{\mathfrak{p}}^{sh} = B_P \otimes_{S_{\mathfrak{p}}} S_{\mathfrak{p}}^{sh}$, where the last equality follows since P is the only prime of B lying over $\mathfrak{p}$ (Corollary 3.10). In fact, $S_{\mathfrak{p}} \subset B_P$ is a finite extension inducing an equality at the residue fields. If follows from Proposition 5.8 that $B_P \to B_P \otimes_{S_{\mathfrak{p}}} S_{\mathfrak{p}}^{sh}$ is the strict henselization of B_P and hence we can write $B_P^{sh} := B \otimes_S S_{\mathfrak{p}}^{sh}$.

In order to complete the proof of (2), it only remains to show that B_P^{sh} is equidimensional and has no zero divisor in $S_{\mathfrak{p}}^{sh}$. Fix non-zero elements $b \in B_P^{sh}$ and $s \in S_{\mathfrak{p}}^{sh}$. Following the notation in 5.10, we can write

$$S_{\mathfrak{p}}^{sh} = \mathrm{colim}_{(S',\mathfrak{p}',\phi')}\, S'_{\mathfrak{p}'},$$

and hence

$$B_P^{sh} = B_P \otimes_{S_{\mathfrak{p}}} S_P^{sh} = \mathrm{colim}_{(S',\mathfrak{p}',\phi')}\, B_P \otimes_{S_{\mathfrak{p}}} S'_{\mathfrak{p}'}.$$

Therefore, there is a suitable étale local ring map $S_{\mathfrak{p}} \to S'_{\mathfrak{p}'}$ such that $s \in S'_{\mathfrak{p}'}$ and $b \in B_{\mathfrak{p}} \otimes_{S_{\mathfrak{p}}} S'_{\mathfrak{p}'}$. Since by (1) the extension $S' \to B \otimes_S S'$ is in the class $\mathscr{T}$, we obtain that $bs \neq 0$ in $B_P \otimes_{S_{\mathfrak{p}}} S'_{\mathfrak{p}'}$. Thus $bs \neq 0$ in B_P^{sh} since $B_P \otimes_{S_{\mathfrak{p}}} S'_{\mathfrak{p}'} \to B_P \otimes_{S_{\mathfrak{p}}} S_{\mathfrak{p}}^{sh}$ is flat and hence injective.

We now prove that B_P^{sh} is equidimensional. If $\tilde{\mathfrak{q}}$ is a minimal prime ideal of B_P^{sh}, it is finitely generated since B_P^{sh} is noetherian, and hence, by the arguments in the previous paragraph, there exists an étale local ring map $S_{\mathfrak{p}} \to S'_{\mathfrak{p}'}$ such that the generators of $\tilde{\mathfrak{q}}$ belong to $B_{\mathfrak{p}} \otimes_{S_{\mathfrak{p}}} S'_{\mathfrak{p}'}$. Thus, if we set $\mathfrak{q}' := \tilde{\mathfrak{q}} \cap (B_{\mathfrak{p}} \otimes_{S_{\mathfrak{p}}} S'_{\mathfrak{p}'})$, it is clear that $\mathfrak{q}' B_P^{sh} = \tilde{\mathfrak{q}}$. Hence $B_P^{sh}/\tilde{\mathfrak{q}}$ is the strict henselization of $(B_{\mathfrak{p}} \otimes_{S_{\mathfrak{p}}} S'_{\mathfrak{p}'})/\mathfrak{q}'$, by 5.9, and so these two rings must have the same dimension, by 5.4 (1). Finally, $\mathfrak{q}'$ is a minimal prime of $B_{\mathfrak{p}} \otimes_{S_{\mathfrak{p}}} S'_{\mathfrak{p}'}$ since $\mathfrak{q}' \cap S'_{\mathfrak{p}'} = (\tilde{\mathfrak{q}} \cap S_{\mathfrak{p}}^{sh}) \cap S'_{\mathfrak{p}'}$ and $\tilde{\mathfrak{q}} \cap S_{\mathfrak{p}}^{sh} = 0$, where the last equality follows since we proved that $S_{\mathfrak{p}}^{sh}$ has no zero divisors of B_P^{sh}. We conclude that $S_{\mathfrak{p}}^{sh} \subset B_P^{sh}$ belongs to the class $\mathscr{T}$.

We now prove the final part of the proposition. After replacing S by $S_{\mathfrak{p}}$, we may assume that S is local. Its maximal ideal will be denoted by $\mathfrak{m}$. Then B is also local and its maximal ideal will be denoted by M. By Proposition 3.20, after replacing θ_i by $\theta_i - \lambda_i$ for some $\lambda_i \in S$, we may assume that $M = \langle \mathfrak{m}B, \theta_1, \ldots, \theta_N \rangle$, where the minimal polynomial of the θ_i undergoes the change of variables $Z_i' = Z_i - \lambda_i$. Notice that, as B is local, the subring $S[\theta_i]$ is also local and has multiplicity the generic rank of $S \subset S[\theta_i]$. The latter number coincides with the degree of $f_i(Z)$ since by Proposition 3.14 there is an S-isomorphism $S[\theta_i] \cong S[Z]/\langle f_i(Z) \rangle$. We now let $S \to S'$ be either an étale local morphism or a strict henselization. Notice that $S[\theta_i] \otimes_S S'$ is identified with the subring $S'[\theta_i \otimes 1]$ of $B \otimes_S S'$ since $S \subset S'$ is flat. Notice also that $S[\theta_i] \subset S[\theta_i] \otimes_S S'$ is étale if $S \to S'$ is étale or is a strict henselization if $S \to S'$ is a strict henselization, by part (1) and (2) applied to $B = S[\theta_i]$. In particular, the multiplicity of $S'[\theta_i \otimes 1]$ is equal to the multiplicity of $S[\theta_i]$ (see 5.4(1)). We repeat the previous argument and conclude that if $\tilde{f}_i(Z) \in S'[Z]$ is the minimal polynomial of $\theta_i \otimes 1$, its degree equals the multiplicity of $S'[\theta_i \otimes 1]$, so $f_i(Z)$ and $\tilde{f}_i(Z)$ are polynomials of the same degree. As $f_i(Z)$ clearly annihilates $\theta_i \otimes 1$, we conclude that $f_i(Z) = \tilde{f}_i(Z)$. □

We are now ready to prove Theorem 2.22, which is restated as:

Theorem 5.13 *Let X be a variety, say of dimension d over a perfect field k. Then for any closed point $x \in X$ of highest multiplicity, there exists an étale neighbourhood $\alpha : (X', x') \to (X, x)$ of x and a transversal morphism $\delta : X' \to V'$ in the class $\mathscr{T}^*$.*

Proof

(1) The field extension $k \subset \kappa(x)$ is finite and separable since k is perfect and x is closed; thus, the projection $X \times_k \kappa(x) \to X$ is étale. Note that $X \times_k \kappa(x)$ is equidimensional and has no embedded components since X is integral (see 5.5, (2) and (3)). Therefore, we may replace X by $X \times_k \kappa(x)$, k by $\kappa(x)$ and x by any point in the fiber, and assume that x is rational and that X is now equidimensional and has not embedded components.

(2) Next, after restricting to an affine neighbourhood of x, we may assume that X is affine, say $X = \mathrm{Spec}(B)$, with B of the form $B = \frac{k[X_1,\ldots,X_D]}{I}$ for some ideal I, and since x is rational, after a change of variables, we may assume that $I \subseteq \langle X_1, \ldots, X_D \rangle$ and that the point x is in correspondence with the maximal ideal $M := \langle X_1, \ldots, X_D \rangle / I$.

(3) By Noether's normalization lemma (or from the proof of it), we can define a non-empty open k-subvariety $U \subset GL_D$ such that for any $\Phi \in U(\bar{k})$, if we set $(T_1, \ldots, T_D) := (X_1, \ldots, X_D) \cdot \Phi$, then the composed map $\bar{k}[T_1, \ldots, T_d] \subset \bar{k}[T_1, \ldots, T_D] = \bar{k}[X_1, \ldots, X_D] \to B \otimes_k \bar{k}$ is injective and finite.

We now consider the localization of $k[X_1, \ldots, X_D]$ at $\langle X_1, \ldots, X_D \rangle$ and consider the associated graded ring $k[x_1, \ldots, x_D]$, where $x_i := \mathrm{Inn}(X_i)$. Note that $\mathrm{gr}_{MB_M}(B_M)$ is a d-dimensional quotient of $k[x_1, \ldots, x_D]$. Note also that $\mathrm{gr}_{M'B'_{M'}}(B'_{M'}) = \mathrm{gr}_{MB_M}(B_M) \otimes_k k'$ for any algebraic extension $k \subset k'$, where

$B' := B \otimes_k k'$ and $M' := M' \otimes_k k'$. Again, by Noether's normalization lemma, there is a non-empty open k-subvariety $U' \subset GL_D$ such that for any $\Psi \in U'(\bar{k})$, if we set $(t_1, \ldots, t_D) := (x_1, \ldots, x_D) \cdot \Psi$, then the composed morphism $\bar{k}[y_1, \ldots, y_d] \subset \bar{k}[y_1, \ldots, y_D] = \bar{k}[x_1, \ldots, x_D] \to \mathrm{gr}_{MB_M} \otimes_k \bar{k}$ is finite and injective.

Therefore, after replacing k by a finite extension, we may assume that $(U \cap U')(k) \neq \emptyset$. Choose $\Phi \in (U \cap U')(k)$, so if we set $(T_1, \ldots, T_D) := (X_1, \ldots, X_D) \cdot \Phi$, then

(i) The composition $S := k[T_1, \ldots, T_d] \subset k[Y_1, \ldots, Y_D] \to B$ is injective and finite. We identify S with its image in B. Set $N := D - d$ and let θ_i denote the image of T_{d+i} at B, for $i = 1, \ldots, N$, so that we have $B = S[\theta_1, \ldots, \theta_N]$ and each θ_i is integral over S. We also set $\mathfrak{m} := \langle T_1, \ldots, T_d \rangle \subset S$, and with this notation we have $M = \langle \mathfrak{m}B, \theta_1, \ldots, \theta_N \rangle$.

(ii) If t_i denotes the class of T_i at the associated graded ring of $k[T_1, \ldots, T_D]_{\langle T_1, \ldots, T_D \rangle}$, then $k[t_1, \ldots, t_d] \subset k[t_1, \ldots, t_D] = k[x_1, \ldots, x_D] \to \mathrm{gr}_{MB_M}(B_M)$ is finite and injective. This implies that $\mathrm{gr}_{MB_M}(B_M)/\langle t_1, \ldots, t_d \rangle$ is zero-dimensional and hence, by Remark 3.7, $\mathfrak{m}B_M$ is a reduction of MB_M.

In the previous discussion, we may have changed the base field by a finite extension. According to 5.5, we can still assume that B is equidimensional and that $\mathrm{Min}(B) = \mathrm{Ass}(B)$.

Notice that the resulting extension $S \subset B$ is finite, $B/M = S/\mathfrak{m}$ and $\mathfrak{m}B_M$ is a reduction of MB_M. We would be able to apply Corollary 3.10 in order to conclude that $S \subset B$ is transversal if we knew that M is the only prime of B lying over $\mathfrak{m}$, but we can not ensure that the latter is true. Nevertheless, we are not so far from the conclusion. It will be enough to apply a suitable étale base change at S in order to fulfil all the conditions in Corollary 3.10.

(4) The notation being as before, let $f_i(Z) \in S[Z]$ denote the minimal polynomial of θ_i. There is an S-isomorphism $S[\theta_i] \cong S[Z]/\langle f_i(Z) \rangle$, by Proposition 3.14. The (rational) maximal ideal M of B dominates $S[\theta_i]$ at the (rational) maximal ideal $M_i := \langle \mathfrak{m}S[\theta]_i, \theta_i \rangle = \langle \mathfrak{m}S[Z], Z \rangle / \langle f_i(Z) \rangle$, which in turn dominates $\mathfrak{m}$. As mentioned before, M_i might not be the only maximal ideal of $S[\theta_i]$ dominating $\mathfrak{m}$ and hence the reduction of $f_i(Z)$ modulo $\mathfrak{m}$, has a factorization of the form

$$\bar{f}_i(Z) = \bar{g}_i(Z)\bar{h}_i(Z) \in (S/\mathfrak{m})[Z],$$

where $\bar{g}_i(Z) = \bar{Z}^r$ for some r and $\bar{h}_i(Z)$ is monic and relatively prime with $\bar{g}_i(Z)$.

If we consider $f(Z)$ as a polynomial over the strict henselization $(S_{\mathfrak{m}})^{sh}$ of $S_{\mathfrak{m}}$, then we obtain a factorization

$$f_i(Z) = g_i(Z)h_i(Z) \in (S_{\mathfrak{m}})^{sh}[Z],$$

where $g_i(Z)$ and $h_i(Z)$ are monic and they induce $\bar{g}_i(Z)$ and $\bar{h}_i(Z)$, for all $i = 1, \ldots, N$. By 5.10, there exists an étale morphism $S \to S'$ and a maximal ideal $\mathfrak{m}'$ of S' lying over $\mathfrak{m}$ such that the previous factorization can be performed at $S'[Z]$. It follows that

$$S'_{\mathfrak{m}'}[Z]/\langle f_i(Z)\rangle = S'_{\mathfrak{m}'}[Z]/\langle g_i(Z)\rangle \times S'_{\mathfrak{m}'}[Z]/\langle h_i(Z)\rangle$$

and after replacing S' by a suitable localization of the form S'_h, we actually have factorization

$$S[\theta_i] \otimes_S S' = S'[Z]/\langle f_i(Z)\rangle = S'[Z]/\langle g_i(Z)\rangle \times S'[Z]/\langle h_i(Z)\rangle.$$

Set $B'_i := S'[Z]/\langle g_i(Z)\rangle$, which also coincides with the localization $(S'[Z]/\langle f_i(Z)\rangle)_{h_i(Z)}$, so in particular $S[\theta_i] \subset B'_i$ is étale. The extension $S' \subset B'_i$ is clearly finite, and by construction, the closed fiber of $S_{\mathfrak{m}'} \subset (B'_i)_{\mathfrak{m}} = S'_{\mathfrak{m}'}[Z]/\langle g_i(Z)\rangle$ has only one point, which is rational. If we consider $S[\theta_i] \otimes_S S'$ as a subring of $B \otimes_S S'$, and denote by $\tilde{h}_i \in S[\theta_i] \otimes_S S' = S'[Z]/\langle f_i(Z)\rangle$ the class of $h_i(Z)$, then clearly

$$\begin{aligned} B' := (B \otimes_S S')_{\tilde{h}_1,\ldots,\tilde{h}_N} &= (S[\theta_1] \otimes_S S')_{\tilde{h}_1} (S[\theta_2] \otimes_S S')_{\tilde{h}_2} \cdots (S[\theta_N] \otimes_S S')_{\tilde{h}_N} \\ &= B'_1 B'_2 \cdots B'_N. \end{aligned}$$

In particular, B' is finite over S' and the composition $B \to B \otimes_S S' \to B'$ is étale. Note that $(B')_{\mathfrak{m}'} = (B'_1)_{\mathfrak{m}'}(B'_2)_{\mathfrak{m}'} \cdots (B'_N)_{\mathfrak{m}'}$ and hence $(B')_{\mathfrak{m}'}$ is local and rational over $\mathfrak{m}'$ as each $(B'_i)_{\mathfrak{m}}$ is local and rational over $\mathfrak{m}'$. One easily checks that $(B')_{\mathfrak{m}'} = B_M \otimes_{S_{\mathfrak{m}}} S'_{\mathfrak{m}'}$. By 5.4 (2), $\mathfrak{m}' B'_{\mathfrak{m}'}$ is a reduction of the maximal ideal of $B_{\mathfrak{m}'}$. We can now apply Corollary 3.10 to conclude that the generic rank of $S' \subset B'$ is equal to the multiplicity of $B'_{\mathfrak{m}'}$. Finally, Proposition 5.12 ensures that $S' \subset B'$ is in the class $\mathscr{T}$. □

6 On Some Properties of Constructive Resolution and the Proof of Theorem 1.3

In this final section we sketch briefly how we can make use of Theorems 2.18 and 2.25 to prove Theorem 1.3. This final step relies on properties of the algorithm of resolution of pairs (of basic objects) that we discuss below, which hold only in characteristic zero.

6.1 Recall that given a pair (J, b) over a smooth variety V, a resolution of (J, b) consists of a sequence of blow-ups $V = V_0 \leftarrow V_1 \leftarrow \ldots \leftarrow V_r$ at smooth centers $Y_i \subset V_i$, and pairs (J_i, b) on V_i with $J_0 = J$ such that

1. $Y_i \subset \mathrm{Sing}(J_i, b)$ for $0 \leq i < r$,
2. (J_i, b) is the transform of (J_{i-1}, b) for $0 < i \leq r$.
3. $\mathrm{Sing}(J_r, b) = \emptyset$.

Over fields of characteristic zero it is known that there are algorithms to produce a resolution of a given pair. More precisely, for each positive integer D there is a totally ordered set $(\Omega_D, \leq)$, so that given any pair (J, b) over a smooth variety V of dimension D, there is an upper-semi-continuous function $\Gamma_0 : \mathrm{Sing}(J, b) \to \Omega_D$, with the property that $\underline{Max}(\Gamma_0)$ (which is closed) is a regular center in $V = V_0$. Here $\underline{Max}(\Gamma_0)$ denotes the set of points at which Γ_0 attains its maximum value. This enables us to define a blow up at this prescribed center, together with a transform of (J, b), say

$$\begin{array}{ccc} V & \longleftarrow & V_1 \\ (J,b) & & (J_1,b) \end{array}$$

and again, as V_1 is D-dimensional, either $\mathrm{Sing}(J_1, b) = \emptyset$ or the algorithm provides us of a function, say $\Gamma_1 : \mathrm{Sing}(J_1, b) \to \Omega_D$, and $\underline{Max}(\Gamma_1)$ is closed and regular in V_1. In this way, given any pair (J, b) over a D-dimensional variety V of characteristic zero, these functions on Ω_D (or say the algorithm) provides us of a sequence of transformations (a very precise one!) as in 2.1, and the point is that for some integer r we get that $\mathrm{Sing}(J_r, b) = \emptyset$. This is called the *constructive resolution of pairs* (also called constructive resolution of *basic objects*). We refer to [8] or [2] for more details and properties of this algorithm (check in particular 4.12 in [2] for the notion of equivalence of pairs used in these notes).

6.2 We list now some properties of constructive resolution to be used in our discussion:

(A) There is a first property of compatibility of constructive resolution with equivalence that we state in the following terms: Let (J, b) and (J', b') be two pairs over the same smooth variety V of dimension D, and assume that the pairs are equivalent in the sense of Definition 2.4. In such a case $\mathrm{Sing}(J, b) = \mathrm{Sing}(J', b')$. The algorithm provides us of two functions defined on the same set, say $\Gamma_0 : \mathrm{Sing}(J, b) \to \Omega_D$ and $\Gamma'_0 : \mathrm{Sing}(J', b') \to \Omega_D$ and the property is that both functions coincide.

In particular $\underline{Max}(\Gamma_0) = \underline{Max}(\Gamma'_0)$, so both pairs are transformed into the same variety, say V_1. On the other hand the two transforms, say (J_1, b) and (J'_1, b') are equivalent on V_1. So again: $\mathrm{Sing}(J_1, b) = \mathrm{Sing}(J'_1, b')$ and the functions $\Gamma_1 : \mathrm{Sing}(J_1, b) \to \Omega_D$ and $\Gamma'_1 : \mathrm{Sing}(J'_1, b') \to \Omega_D$ coincide.

In other words, the constructive resolution does not distinguish between two pairs (J, b) and (J', b') over the same V if they are equivalent.

(B) Fix a smooth D-dimensional variety V and a pair (J, b), let $\beta : V' \to V$ be étale. Then V' is also smooth and D-dimensional. There is a natural pullback, say, (J', b) at V', and clearly $\mathrm{Sing}(J', b) = \beta^{-1}(\mathrm{Sing}(J, b))$.

So we have two pairs on two different D-dimensional schemes, and two functions $\Gamma_0 : \mathrm{Sing}(J, b) \to \Omega_D$ and $\Gamma_0' : \mathrm{Sing}(J', b') \to \Omega_D$. A property of the algorithmic resolution is that

$$\Gamma_0(\beta(x)) = \Gamma_0'(x).$$

for any $x \in \mathrm{Sing}(J', b)$. In particular $\beta^{-1}(\underline{Max}(\Gamma_0)) = \underline{Max}(\Gamma_0')$ so we get a commutative diagram of blow ups with étale vertical morphisms, say:

$$\begin{array}{ccc} V' & \longleftarrow & V_1' \\ {\scriptstyle \beta}\downarrow & & \downarrow{\scriptstyle \beta_1} \\ V & \longleftarrow & V_1 \end{array} \tag{10}$$

and, in particular, (J_1', b) (the transform of (J', b)) is the pullback of (J_1, b) via β_1. So again $\Gamma_1(\beta(x)) = \Gamma_1'(x)$ for any $x \in \mathrm{Sing}(J_1', b)$, and the same arguments as before apply, and therefore one can iterate this discussion. This property is the crucial point in proving compatibility of constructive resolution with étale topology.

(C) Assume now that V and V' are D-dimensional smooth varieties and that an equidimensional scheme X can be embedded as a closed subscheme in both V and V'. If (J, b) is a pair on V and (J', b') is a pair on V' and if they both represent the points of highest multiplicity of X, say $F_n(X)$ (for example in the setting of Theorem 2.18), then they undergo the same constructive resolution: Namely, if $\Gamma_0 : \mathrm{Sing}(J, b) = F_n(X) \to \Omega_D$ and $\Gamma_0' : \mathrm{Sing}(J', b') = F_n(X) \to \Omega_D$ are the functions provided by the algorithm when applied to the pairs (J, b) and (J', b') respectively, then

$$\Gamma_0 = \Gamma_0'.$$

In particular, $\underline{Max}(\Gamma_0) = \underline{Max}(\Gamma_0')$ and this is a regular center of X, V and V'. If X_1 is the blow-up of X at this center, V_1 is that of V and V_1' is that of V', then X_1 is naturally embedded in both V_1 and in V_1'. The transform (J_1, b) of (J, b) and the transform (J_1', b) of (J', b) represent the points of multiplicity n of X_1 (so that $\mathrm{Sing}(J, b) = \mathrm{Sing}(J', b') = F_n(X_1)$) and the two functions $\Gamma_1 : \mathrm{Sing}(J_1, b) \to \Omega_D$ and $\Gamma_1' : \mathrm{Sing}(J_1', b') \to \Omega_D$ coincide again. And the same holds after iteration.

Remark 6.3

(1) A first consequence of (B) is that the constructive resolution of (J', b) over V', is the fiber product of $V' \to V$ with the constructive resolution of (J, b)

over V. Note here that if V' is an open set in V, the inclusion, or say an open immersion, is an étale morphism and the latter property states that the constructive resolution of a pair is compatible with open restriction on V.

(2) Fix an embedding $X \subset W$ and a pair (J, b) over the smooth k-variety W as in Theorem 2.18. In that theorem we denoted by n the highest multiplicity at points of X, and the pair (J, b) was constructed so as to describe the points of multiplicity n. Let D be the dimension of W and assume that k is a field of characteristic zero. Let

$$\begin{array}{ccccccccc} W = W_0 & \stackrel{\pi_1}{\longleftarrow} & W_1 & \stackrel{\pi_2}{\longleftarrow} & \dots & \stackrel{\pi_r}{\longleftarrow} & W_r & \stackrel{\pi_r}{\longleftarrow} & W_{r+1} \ , \\ (J,b) & & (J_1,b) & & \dots & & (J_r,b) & & (J_{r+1},b) \end{array} \qquad (11)$$

be now the constructive resolution defined by the functions on $(\Omega_D, \leq)$ (where $W_i \leftarrow W_{i+1}$ is the blow up at $\underline{Max}(\Gamma_i)$); so $\mathrm{Sing}(J_{r+1}, b) = \emptyset$.

Recall that

$$\mathrm{Sing}(J_i, b) = F_n(X_i) \subset W_i \text{ for } i = 1, \dots, r + 1, \qquad (12)$$

in particular we get a function $\Gamma_i : F_n(X_i) \to \Omega_D$, and the induced sequence

$$\begin{array}{ccccccccc} X & \stackrel{\pi_1}{\longleftarrow} & X_1 & \stackrel{\pi_2}{\longleftarrow} & \dots & \stackrel{\pi_r}{\longleftarrow} & X_r & \stackrel{\pi_{r+1}}{\longleftarrow} & X_{r+1} \\ F_n(X) & & F_n(X_1) & & \dots & & F_n(X_r) & & F_n(X_{r+1}) \end{array} \qquad (13)$$

is a simplification of the multiplicity, obtained by blowing up successively on $\underline{Max}(\Gamma_i)$ for $i = 1, \dots, r$. In other words, the constructive resolution of basic objects leads to the simplification of the multiplicity of X by blowing up at regular equimultiple centers.

(3) If (J', b') and (J, b) are two pairs which fulfill the conditions in Theorem 2.18, then both are clearly equivalent. In particular, the property in 6.2(A) ensures that they induce the same simplification of the multiplicity (13); moreover, they induce the same functions $\Gamma_i : F_n(X_i) \to \Omega_D$ for $i = 1, \dots, r$.

(4) In the formulation of Theorem 2.18 we view X as a variety included in a *smooth* variety over a field k, and we have indicated, in the previous lines, that the upper semi continuous functions on $(\Omega_D, \leq)$, or say the constructive resolution of basic objects, leads to a reduction of the multiplicity. However both Theorem 2.18 and the constructive resolution hold in more generality, say for schemes X included in suitable regular schemes, which are not necessarily smooth over a field k. For example, if we fix a closed point $x \in X \subset W$ with W smooth over k, and if X and W are replaced by the completions at such point, then Theorem 2.18 still holds at this setting, and constructive resolution of (J, b) can be stated over the new regular ambient scheme with the same properties discussed above. In other words functions on $(\Omega_D, \leq)$ are defined in this context and produce a reduction of the multiplicity.

Remark 6.4 Along these notes we have dealt with pairs on smooth k-varieties V. However, as indicated above pairs and constructive resolution of pairs can be stated at suitable but more general regular schemes. Moreover, the notion of pair attached to the Hilbert Samuel function in the sense of Theorem 2.13, appears for the first time in [12] where the pair is defined at a *complete* local regular ring containing a field. It was proved later that given a variety X included in a smooth variety V and a closed point $x \in X \subset V$, a pair can be attached to the Hilbert Samuel function (always in the sense of Theorem 2.13) at the regular scheme $\mathrm{Spec}(\tilde{S})$, where $\tilde{S}$ is an inductive limit of étale morphisms $S' \to S''$ and S is the algebra of functions on W The scheme $\mathrm{Spec}(\tilde{S})$ is called the strict henselization of W at x. $\mathrm{Spec}(\tilde{S})$ is a regular scheme and all the notions related with pairs, like equivalence, permissible local sequences, constructive resolutions, etc, can be extended to this context. Moreover, in characteristic zero the algorithm of resolution of pairs, together with all the properties listed in 6.2, can be extended to pairs in this context.

There is a natural morphism $\mathrm{Spec}(\tilde{S}) \to V$ which factors though (V', x') for any étale morphism $\beta : V' \to V$ mapping x' to x. So for example the property discussed in 6.2(B), stated there for a pair (J, b) on a smooth D-dimensional variety V and for $\beta : V' \to V$ étale, also holds for $\mathrm{Spec}(\tilde{S}) \to V$.

6.5 Proof of Theorem 1.3. We now turn to the proof of Theorem 1.3, which will be done in several steps. We fix a reduced equidimensional algebraic scheme X over a field of characteristic zero k, say of dimension d and without embedded components. We denote by n the highest multiplicity at points of X. Assume that $n > 1$.

Since X is noetherian, there exists an integer D and a finite open covering $\{U_i\}$ of X such that U_i can be embedded as a closed subscheme of $\mathbb{A}^D_k$. We aim to define a function

$$\Gamma_0^X : F_n(X) \to \Omega_D \tag{14}$$

so that $\underline{Max}(\Gamma_0^X) \subset X$ is a regular center and the blow-up of X at this center is the first step in the reduction of the multiplicity of X.

Following the proof of Theorem 5.13, one can find étale morphisms $\alpha_i : X_i \to X$, $i = 1, \dots, s$, so that $\{\mathrm{Im}(\alpha_i) \cap F_n(X)\}_{i=1}^s$ cover $F_n(X)$, and such that for each index i:

1. There exists a transversal morphism $\delta_i : X_i \to V_i$ in the class $\mathscr{T}^*$.
2. There is a presentation of δ_i of the form

$$\begin{array}{ccc} X_i & \longrightarrow & V_i \times \mathbb{A}^N \\ {\scriptstyle \delta_i}\downarrow & \swarrow & \\ V_i & & \end{array} \qquad \supset (\mathscr{H}_{i1}, d_{i1}; \mathscr{H}_{i2}, d_{i2}; \dots; \mathscr{H}_{iN}, d_{iN}) \tag{15}$$

where $N = D - d$.

Recall that there is a pair associated to this presentation, namely the pair $(J_i, b_i) := \bigcap_{j=1}^{N} (I(\mathscr{H}_{ij}), d_{ij})$ on $V_i \times \mathbb{A}^N$, which represents the points of multiplicity n of X_i. Hence for each index i, the algorithm of resolution of pairs provides us of a function

$$\Gamma_0^{X_i} : F_n(X_i) \to \Omega_D. \tag{16}$$

which is used to define the first blow-up in the reduction of multiplicity of X_i. The function $\Gamma_0^X : F_n(X) \to \Omega_D$ to be defined in the next lines will be constructed so that it is compatible with each $\Gamma_0^{X_i}$. We discuss now this property and show why it leads to the proof of our Theorem.

Fix any index i such that $x \in \alpha_i(X_i)$ and fix $x_i \in X_i$ mapping to x. Notice that $x_i \in F_n(X_i)$ and recall that $F_n(X_i)$ can be identified with $\delta_i(F_n(X_i))$ and that $\kappa(x_i) = \kappa(\delta_i(x_i))$.

Consider a strict henselization, say $(V_i)^{sh}_{\delta_i(x_i)} \to V_i$ of V_i at $\delta_i(x_i)$. In the commutative diagram

$$\begin{array}{ccc} X_i & \longleftarrow & X_i \times_{V_i} (V_i)^{sh}_{\delta_i(x_i)} \\ \big\downarrow & & \big\downarrow \\ V_i & \longleftarrow & (V_i)^{sh}_{\delta_i(x_i)} \end{array}$$

the vertical line on the right is a transversal morphism in the class $\mathscr{T}^*$ and the top row is a strict henselization of X_i at x_i, which also coincides with a strict henselization of X at x (Proposition 5.12). Hence $X_x^{sh} = X_i \times_{V_i} (V_i)^{sh}_{\delta_i(x_i)}$. The isomorphism class of X_x^{sh} is completely determined by $\mathscr{O}_{X,x}$.

By taking the natural pull-back of the diagram and presentation (15) with $V_i \leftarrow (V_i)^{sh}_{\delta_i(x_i)}$, we obtain

$$\begin{array}{ccc} X_x^{sh} = (X_i)^{sh}_{x_i} & \longrightarrow & (V_i)^{sh}_{\delta_i(x_i)} \times \mathbb{A}^N \qquad \supset (\tilde{\mathscr{H}}_{i1}, d_{i1}; \tilde{\mathscr{H}}_{i2}, d_{i2}; \ldots; \tilde{\mathscr{H}}_{iN}, d_{iN}) \\ {\scriptstyle \delta_i^{sh}} \big\downarrow & \swarrow & \\ (V_i)^{sh}_{\delta_i(x_i)} & & \end{array} \tag{17}$$

which, in turn, is a presentation of $\delta_i{}^{sh}$ as a consequence of Proposition 5.12. We denote by $(\tilde{J}_i, \tilde{b}_i)$ the pair on $(V_i)^{sh}_{\delta_i(x_i)} \times \mathbb{A}^N$ associated with this presentation, namely, $(\tilde{J}_i, \tilde{b}_i) := \bigcap_{j=1}^{N} (I(\tilde{\mathscr{H}}_{ij}), d_{ij})$.

Since we know that $(\tilde{J}_i, \tilde{b}_i)$ represents the maximal multiplicity of the scheme X_x^{sh}, the algorithm of resolution of pairs provides a function

$$\Gamma_0^{sh} : F_n(X_x^{sh}) \to \Omega_D.$$

From 6.2 (C) and the discussion in 6.4, we conclude that this function depends only on the integer D and on X_x^{sh}, but not on the transversal morphism $\delta_i : X_i \to V_i$ nor on the given presentation for δ_i. Therefore, we can define

$$\Gamma_0^X(x) := \Gamma_0^{X^{sh}}(\bar{x}) \in \Omega_D$$

where $\bar{x}$ is the closed point of X_x^{sh}, as this value is independent of the choice of the index i such that $x \in \alpha_i(X_i)$ and it is also independent of the point x_i that we chose in the fiber of x.

In this way we have defined a function $\Gamma_0^{X_0} : X \to \Omega_D$ whose value at a point $x \in X$ depends only on the isomorphism class of $\mathcal{O}_{X,x}$.

In summary, we have defined a function $\Gamma_0^X : F_n(X) \to \Omega_D$ such that for each index i there is a commutative diagram

$$\begin{array}{ccc} F_n(X_i) & \xrightarrow{\Gamma_0^{X_i}} & \Omega_D \, . \\ \downarrow & \nearrow_{\Gamma_0^X} & \\ F_n(X) & & \end{array} \qquad (18)$$

As the images of $F_n(X_i) \to F_n(X)$ form a covering of $F_n(X)$, we obtain that Γ_0^X is upper-semicontinuous, and for each index i

$$\underline{Max}(\Gamma_0^{X_i}) = \alpha_i^{-1}(\underline{Max}(\Gamma_0^X)).$$

This implies that $\underline{Max}(\Gamma_0^X)$ is a regular center of X. If $X^{(1)}$ is the blow-up of X at this center and $X_i^{(1)}$ is the blow-up of X_i at $\underline{Max}(\Gamma_0^{X_i}) = \alpha_i^{-1}(\underline{Max}(\Gamma_0^X))$, we obtain a Cartesian diagram

$$\begin{array}{ccc} X_i & \longleftarrow & X_i^{(1)} \\ {\scriptstyle \alpha_i}\downarrow & & \downarrow{\scriptstyle \alpha_i^{(1)}} \\ X & \longleftarrow & X^{(1)}. \end{array}$$

We can repeat the previous discussion starting from $X^{(1)}$ and the étale morphisms $\{\alpha_i^{(1)} : X_i^{(1)} \to X^{(1)} \mid i = 1, \dots, s\}$ and arrive to the same conclusion. Therefore, the algorithmic reduction of the multiplicity of each X_i induces a reduction of the multiplicity of X. This completes the proof of the theorem.

References

1. Bennett, B.M.: On the characteristic functions of a local ring. Ann. Math. **91**, 25–87 (1970)
2. Benito, A., Encinas, S., Villamayor, O.: Some natural properties of constructive resolution of singularities. Asian J. Math. **15**, 141–192 (2011)
3. Bravo, A., García Escamilla, M.L., Villamayor U, O.E.: On Rees algebras and invariants for singularities over perfect fields. Indiana Univ. Math. J. **61**(3), 1201–1251 (2012)
4. Bravo, A., Villamayor, O.E.: On the behavior of the multiplicity on schemes: stratification and blow ups. In: Ellwood, D., Hauser, H., Mori, S., Schicho, J. (eds.) The Resolution of Singular Algebraic Varieties. Clay Institute Mathematics Proceedings, vol. 20, pp. 81–207 (340pp.). AMS/CMI, Providence (2014)
5. Cossart, V., Jannsen, U., Saito, S.: Canonical embedded and non-embedded resolution of singularities for excellent two-dimensional schemes. Prepint, arXiv:0905.2191
6. Dade, E.C.: Multiplicity and Monoidal transformations. Thesis, Princeton University (1960)
7. Dietel, B.: A refinement of Hironaka's group schemes for an extended invariant. Thesis, Universität Regensburg (2014)
8. Encinas, S., Villamayor, O.E.: A course on constructive desingularization and equivariance. In: Hauser, H., Lipman, J., Oort, F., Quiros, A. (eds.) Resolution of Singularities. A Research Textbook in Tribute to Oscar Zariski. Progress in Mathematics, vol. 181, pp. 47–227. Birkhauser, Basel (2000)
9. Greco, S.: Two theorems on excellent rings. Nagoya Math. J. **60**, 139–149 (1976)
10. Grothendieck, A., Dieudonné, J.: Éléments de Géométrie Algébrique, IV Étude locale des schémas et des morphismes de schémas. Publ. Math. Inst. Hautes Etudes Sci. **24** (1965)
11. Herrmann, M., Ikeda, S., Orbanz, U.: Equimultiplicity and Blowing up. Springer, Berlin/Heidelberg (1988)
12. Hironaka, H.: Resolution of singularities of an algebraic variety over a field of characteristic zero I–II. Ann. Math. **79**, 109–326 (1964)
13. Hironaka, H.: Introduction to the Theory of Infinitely Near Singular Points. Memorias de Matematica del Instituto Jorge Juan, vol. 28. Concejo Superior de Investigaciones Científicas, Madrid (1974)
14. Lipman, J.: Equimultiplicity, reduction, and blowing up. In: Draper, R.N. (eds.) Commutative Algebra. Lecture Notes in Pure and Applied Mathematics, vol. 68, pp. 111–147. Marcel Dekker, New York (1982)
15. Nagata, M.: Local Rings. Interscience Tracts in Pure and Applied Mathematics, vol. 13. J. Wiley, New York (1962)
16. Orbanz, U.: Multiplicities and Hilbert functions under blowing up. Manuscripta Math. **36**(2), 179–186 (1981)
17. Raynaud, M.: Anneaux Lacaux Henséliens. Lecture Notes in Mathematics, vol. 169. Springer, Berlin/Heidelberg/New York (1970)
18. Rees, D.: $\mathfrak{a}$-transforms of local rings and a theorem of multiplicity. Math. Proc. Camb. Philos. Soc. **57**, 8–17 (1961)
19. The Stacks Project Authors: Stacks Project. http://stacks.math.columbia.edu (2017)
20. Villamayor U, O.: Equimultiplicity, algebraic elimination and blowing-up. Adv. Math. **262**, 313–369 (2014)
21. Zariski, O., Samuel, P.: Commutative Algebra, vol. 2. Van Nostrand, Princeton (1960)

Platonic Surfaces

Brenda Leticia De La Rosa-Navarro, Gioia Failla, Juan Bosco Frías-Medina, Mustapha Lahyane, and Rosanna Utano

Dedicated to Professor Antonio Campillo-López (http://www.singacom.uva.es/campillo/index.html) on the occasion of his 65th birthday

Abstract We define the notion of Platonic surfaces. These are anticanonical smooth projective rational surfaces defined over any fixed algebraically closed field of arbitrary characteristic and having the projective plane as a minimal model with very nice geometric properties. We prove that their Cox rings are finitely generated. In particular, they are extremal and their effective monoids are finitely generated. Thus, these Platonic surfaces are built from points of the projective plane which are in good position. It is worth noting that not only their Picard number may be

B. L. De La Rosa-Navarro
Facultad de Ciencias, Universidad Autónoma de Baja California, Ensenada, Baja California, México
e-mail: brenda.delarosa@uabc.edu.mx

G. Failla
Università Mediterranea di Reggio Calabria, Reggio Calabria, Italy
e-mail: gioia.failla@unirc.it

J. B. Frías-Medina
Instituto de Matemáticas, Universidad Nacional Autónoma de México. Área de la Investigación Científica. Circuito Exterior, Ciudad Universitaria, Coyoacán, Ciudad de México, Mexico

Unidad Académica de Matemáticas, Universidad Autónoma de Zacatecas, Zacatecas, Zac, México

M. Lahyane (✉)
Instituto de Física y Matemáticas (IFM), Universidad Michoacana de San Nicolás de Hidalgo (UMSNH), Morelia, Michoacán, México
e-mail: lahyane@ifm.umich.mx

R. Utano
Dipartimento di Scienze Matematiche e Informatiche. Scienze Fisiche e Scienze della Terra, Università di Messina, Messina, Italy
e-mail: utano@unime.it

G.-M. Greuel et al. (eds.), *Singularities, Algebraic Geometry, Commutative Algebra, and Related Topics*, https://doi.org/10.1007/978-3-319-96827-8_12

big but also an anticanonical divisor may have a very large number of irreducible components.

1 Introduction

Given a constellation $\mathscr{C}$ with three origins in the projective plane $\mathbb{P}^2$ (see Sect. 2.2 for the definition of a constellation), the smooth projective surface X obtained as the blow up of the projective plane $\mathbb{P}^2$ at the points of $\mathscr{C}$ may have an infinite number of smooth rational curves of self-intersection -1, see [19, Proposition (3.2), page 142], [40, Proposition, page 538], [29, Lemma I.4, page 411], [33, Theorem 1.1, page 214], [34, Corollary 1.2, page 1594] and more explicitly in the examples in [35, Section 5, page 110]. Hence, the effective monoid $M(X)$ of X is not finitely generated. Here, the effective monoid of a surface S means the monoid of effective divisor classes on S modulo numerical equivalence. In this paper, we give a numerical condition on the points of $\mathscr{C}$ such that $M(X)$ is finitely generated assuming only that these points belong to a triangle and that the base field is algebraically closed of characteristic p, where p may not be equal to zero. Here, a triangle in $\mathbb{P}^2$ is the union of three non-concurring lines.

Since the singular locus of a projective plane cubic consisting of three lines may be three nodes (recall that a node is an ordinary multiple point of multiplicity two), we will denote, hereafter, X_2 the surface obtained as the blow up of the projective plane $\mathbb{P}^2$ at the support of a zero dimensional scheme Z when this latter (i.e., the support of Z) is contained in the degenerate cubic with three singular points. Here, the support of Z may contain some of the elements of the above singular locus and also whose elements may be infinitely near to each other. By convention henceforward, any point of the exceptional divisor of the blow up of a given surface S at a closed point $P \in S$ will be called a point of the first neighborhood of P. Our main results are Theorems 1, 2, Corollary 2, Proposition 2 and Theorem 3 below.

2 Preliminaries

2.1 General Notions

Let S be a smooth projective rational surface defined over an algebraically closed field of arbitrary characteristic. A canonical divisor on S, and the Néron-Severi group of S will be denoted by K_S and $NS(S)$, respectively. There is an intersection form on $NS(S)$ induced by the intersection of divisors on S, it will be denoted by a dot, that is, for x and y in $NS(S)$, $x \cdot y$ is the intersection number of x and y (see [30] and [2]).

The following result known as the Riemann-Roch theorem for smooth projective rational surfaces is stated using the Serre duality.

Lemma 1 *Let D be a divisor on a smooth projective rational surface S having an algebraically closed field of arbitrary characteristic as a ground field. Then the following equality holds:*

$$h^0(S, \mathcal{O}_S(D)) - h^1(S, \mathcal{O}_S(D)) + h^0(S, \mathcal{O}_S(K_S - D)) = 1 + \frac{1}{2}(D^2 - D \cdot K_S).$$

$\mathcal{O}_S(D)$ is an invertible sheaf associated canonically to the divisor D.

Here, we recall some standard results. see [20, 21, 28] and [30] and the references therin. A divisor class x modulo numerical equivalence on a smooth projective rational surface S is effective respectively numerically effective, nef in short, if an element of x is an effective, respectively numerically effective, divisor on S. Here a divisor D on S is nef if the integer $D \cdot C$ is nonnegative for every integral curve C on S. Now, we start with some properties which follow from a successive iterations of blowing up closed points of a smooth projective rational surface.

Lemma 2 *Let $\pi^\star : NS(X) \to NS(Y)$ be the natural group homomorphism on Néron-Severi groups induced by a given birational morphism $\pi : Y \to X$ of smooth projective rational surfaces. Then $\pi^\star$ is an injective intersection-form preserving map of free abelian groups of finite rank. Furthermore, it preserves the dimensions of cohomology groups, the effective divisor classes and the numerically effective divisor classes.*

Proof See [27, Lemma II.1, page 1193]. □

Lemma 3 *Let x be an element of the Néron-Severi group $NS(X)$ of a smooth projective rational surface X. The effectiveness or the nefness of x implies the noneffectiveness of $k_X - x$, where k_X denotes the element of $NS(X)$ which contains a canonical divisor on X. Moreover, The nefness of x implies that the self-intersection of x is greater than or equal to zero.*

Proof See [27, Lemma II.2, page 1193]. □

The following result is also needed. We recall that a (-1)-curve, respectively a (-2)-curve, is a smooth rational curve of self-intersection -1, respectively -2.

Lemma 4 *The monoid of effective divisor classes modulo algebraic equivalence on a smooth projective rational surface X having an effective anticanonical divisor is finitely generated if and only if X has only a finite number of (-1)-curves and only a finite number of (-2)-curves.*

Proof See [35, Corollary 4.2, page 109]. □

2.2 Constellations

Let S be a smooth projective surface defined over an algebraically closed field of arbitrary characteristic and let P be a closed point of S. By a *constellation* $\mathcal{C}$ with origin the point P, we mean the following: $\mathscr{C} = \cup_{i=0}^{i=n} \mathscr{C}_i$, where $\mathscr{C}_0$ is the set consisting of only the point P, $\mathscr{C}_1$ is a nonempty finite subset of the exceptional divisor of the blow up the surface S at the point P, and by induction $\mathscr{C}_{i+1}$ is a nonempty finite subset of the exceptional locus of the blow up the surface containing the points of $\mathscr{C}_i$ at these ones for every $i = 0, \ldots, n-1$. Here, n is a positive integer. Moreover, the union of two (respectively, three) constellations with different origins will be called a constellation with two (respectively, three) origins.

2.3 *Platonic Surfaces*

Definition 1 Let S be a smooth projective rational surface defined over an algebraically closed field of arbitrary characteristic. S is Platonic if its monoid of effective divisor classes is finitely generated and if it has the projective plane as a minimal model such that the projection of its effective anticanonical divisor is a triangle.

Example 1 The projective plane and toric surfaces having the projective plane as a minimal model are Platonic.

Remark 1 Some Platonic surfaces can be constructed using [7, Theorem 4.1, p. 6].

Remark 2 Knowing the geometry of Platonic surfaces extends the knowledge of geometries of the rational surfaces with an anticanonical cycle studied by Looijenga in [38], and also the surfaces studied in [4, 5, 11, 12, 14, 15, 17, 22, 24, 41, 42] and [18].

The following lemma gives the possibility to avoid the special cases where the values of γ, δ, ρ, ϱ, μ, and ν (see Sect. 3 for the notation) and the values of x, y and z may vanish. Recall that a surface is always assumed to be smooth and projective (not necessarily rational).

Lemma 5 *Assume there exists a proper birational morphism from the surface V to a normal surface W. Then, the finite generation of the monoid $M(V)$ of effective divisor classes of V implies the finite generation of the monoid $M(W)$ of effective divisor classes of W.*

It is a direct consequence of the following stronger result:

Lemma 6 *The finite generation of the monoid $M(V)$ of effective divisor classes of a surface V implies that there are only a finite number of proper birational morphism from V to any given normal surface modulo isomorphisms.*

Proof See [37, Lemma 26, p.1223]. □

3 Geometry of Platonic Rational Surfaces

This section is devoted to deal with the study of the geometry of the surface X_2 that we have mentioned in the introduction. For its explicit construction, we need the following notation: let P, Q and R be the three nodes of the divisor $L_1 + L_2 + L_3$, here $L_1 \bigcap L_2 = \{P\}$, $L_1 \bigcap L_3 = \{Q\}$ and $L_2 \bigcap L_3 = \{R\}$, where L_i is a fixed line in the projective plane for $i = 1, 2$ and 3. Denote by $\mathscr{C}_P, \mathscr{C}_Q, \mathscr{C}_R, \mathscr{C}_{P_1^{L_1}}, \ldots, \mathscr{C}_{P_x^{L_1}}$, $\mathscr{C}_{P_1^{L_2}}, \ldots, \mathscr{C}_{P_y^{L_2}}, \mathscr{C}_{P_1^{L_3}}, \ldots,$ and $\mathscr{C}_{P_z^{L_3}}$ be the mutually disjoint constellations with only one origin and whose points gives X_2 after blowing up the projective plane $\mathbb{P}^2$ at them (see below).

- $\mathscr{C}_P = \{P, P^1, P^1_{\star 1}, \ldots, P^1_{\star\gamma}, P^2, P^2_{\star 1}, \ldots, P^2_{\star\delta}\}$, here P^1 (respectively P^2) is the intersection of the first neighborhood of the point P and the strict transform of line L_1 (respectively the strict transform of the line L_2); and for $k = 1, \ldots, \gamma$ the point $P^1_{\star k}$ is the point of the first neighborhood of $P^1_{\star k-1}$ that belongs to the strict transform of L_1 with the convention $P^1_{\star 0} = P^1$. Similarly for $k = 1, \ldots, \delta$ the point $P^2_{\star k}$ is the point of the first neighborhood of $P^2_{\star k-1}$ that belongs to the strict transform of L_2 with the convention $P^2_{\star 0} = P^2$.
- $\mathscr{C}_Q = \{Q, Q^1, Q^1_{\star 1}, \ldots, Q^1_{\star\rho}, Q^3, Q^3_{\star 1}, \ldots, Q^3_{\star\varrho}\}$, here Q^1 (respectively Q^3) is the intersection of the first neighborhood of the point Q and the strict transform of line L_1 (respectively the strict transform of the line L_3); and for $k = 1, \ldots, \rho$ the point $Q^1_{\star k}$ is the point of the first neighborhood of $Q^1_{\star k-1}$ that belongs to the strict transform of L_1 with the convention $Q^1_{\star 0} = Q^1$. Similarly for $k = 1, \ldots, \varrho$ the point $Q^3_{\star k}$ is the point of the first neighborhood of $Q^3_{\star k-1}$ that belongs to the strict transform of L_3 with the convention $Q^3_{\star 0} = Q^3$.
- $\mathscr{C}_R = \{R, R^2, R^2_{\star 1}, \ldots, R^2_{\star\mu}, R^3, R^3_{\star 1}, \ldots, R^3_{\star\nu}\}$, here R^2 (respectively R^3) is the intersection of the first neighborhood of the point R and the strict transform of line L_2 (respectively the strict transform of the line L_3); and for $k = 1, \ldots, \mu$ the point $R^2_{\star k}$ is the point of the first neighborhood of $R^2_{\star k-1}$ that belongs to the strict transform of L_2 with the convention $R^2_{\star 0} = R^2$. Similarly for $k = 1, \ldots, \nu$ the point $R^3_{\star k}$ is the point of the first neighborhood of $R^3_{\star k-1}$ that belongs to the strict transform of L_3 with the convention $R^3_{\star 0} = R^3$.
- For $i = 1, \ldots, x$, $\mathscr{C}_{P_i^{L_1}} = \{P_i^{L_1}, P_{i\star 1}^{L_1}, \ldots, P_{i\star\rho_i^1}^{L_1}\}$, here $P_{i\star j}^{L_1}$ is the unique point of the intersection of the first neighborhood of the point $P_{i\star j-1}^{L_1}$ and the strict transform of the line L_1 with the convention $P_{i\star 0}^{L_1} = P_i^{L_1} \in L_1$ for any $j = 1, \ldots, \rho_i^1$, and $P_i^{L_1} \neq P_s^{L_1}$ for any $s = 1, \ldots, x$.
- For $i = 1, \ldots, y$, $\mathscr{C}_{P_i^{L_2}} = \{P_i^{L_2}, P_{i\star 1}^{L_2}, \ldots, P_{i\star\rho_i^2}^{L_2}\}$, here $P_{i\star j}^{L_2}$ is the unique point of the intersection of the first neighborhood of the point $P_{i\star j-1}^{L_2}$ and the strict transform of the line L_2 with the convention $P_{i\star 0}^{L_2} = P_i^{L_2} \in L_2$ for any $j = 1, \ldots, \rho_i^2$, and $P_i^{L_2} \neq P_s^{L_2}$ for any $s = 1, \ldots, y$.

- For $i = 1, \ldots, z$, $\mathscr{C}_{P_i^{L_3}} = \{P_i^{L_3}, P_{i\star 1}^{L_3}, \ldots, P_{i\star\rho_i^3}^{L_3}\}$, here $P_{i\star j}^{L_3}$ is the unique point of the intersection of the first neighborhood of the point $P_{i\star j-1}^{L_3}$ and the strict transform of the line L_3 with the convention $P_{i\star 0}^{L_3} = P_i^{L_3} \in L_3$ for any $j = 1, \ldots, \rho_i^3$, and $P_i^{L_3} \neq P_s^{L_3}$ for any $s = 1, \ldots, z$.

Consequently, the Néron-Severi group $NS(X_2)$ of X_2 is a lattice which is equipped with the following integral basis:

$$(\mathscr{E}_0; -\mathscr{E}_P, -\mathscr{E}_{P^1}, -\mathscr{E}_{P^1_{\star 1}}, \ldots, -\mathscr{E}_{P^1_{\star\gamma}}, -\mathscr{E}_{P^2}, -\mathscr{E}_{P^2_{\star 1}}, \ldots, -\mathscr{E}_{P^2_{\star\delta}},$$

$$-\mathscr{E}_Q, -\mathscr{E}_{Q^1}, -\mathscr{E}_{Q^1_{\star 1}}, \ldots, -\mathscr{E}_{Q^1_{\star\rho}}, -\mathscr{E}_{Q^2}, -\mathscr{E}_{Q^2_{\star 1}}, \ldots, -\mathscr{E}_{Q^2_{\star\varrho}},$$

$$-\mathscr{E}_R, -\mathscr{E}_{R^2}, -\mathscr{E}_{R^2_{\star 1}}, \ldots, -\mathscr{E}_{R^2_{\star\mu}}, -\mathscr{E}_{R^3}, -\mathscr{E}_{R^3_{\star 1}}, \ldots, -\mathscr{E}_{R^3_{\star\nu}},$$

$$-\mathscr{E}_{P_1^{L_1}}, \ldots, -\mathscr{E}_{P^{L_1}_{x\star\rho_x^1}}, -\mathscr{E}_{P_1^{L_2}}, \ldots, -\mathscr{E}_{P^{L_2}_{y\star\rho_y^2}}, -\mathscr{E}_{P_1^{L_3}}, \ldots, -\mathscr{E}_{P^{L_3}_{z\star\rho_z^3}}),$$

which is defined by:

- $\mathscr{E}_0$ is the class of a general line in the projective plane,
- $\mathscr{E}_P$ (respectively $\mathscr{E}_{P^1}, \mathscr{E}_{P^2}$) is the class of the exceptional divisor corresponding to the point blown-up P (respectively P^1, P^2), and for $k = 1, \ldots, \gamma$, $\mathscr{E}_{P^1_{\star k}}$ is the class of the exceptional divisor corresponding to the point blown-up $P^1_{\star k}$.

  Similarly, for $k = 1, \ldots, \delta$, $\mathscr{E}_{P^2_{\star k}}$ is the class of the exceptional divisor corresponding to the point blown-up $P^2_{\star k}$,
- $\mathscr{E}_Q$ (respectively $\mathscr{E}_{Q^1}, \mathscr{E}_{Q^3}$) is the class of the exceptional divisor corresponding to the point blown-up Q (respectively Q^1, Q^3), and for $k = 1, \ldots, \rho$, $\mathscr{E}_{Q^1_{\star k}}$ is the class of the exceptional divisor corresponding to the point blown-up $Q^1_{\star k}$.

  Similarly, for $k = 1, \ldots, \varrho$, $\mathscr{E}_{Q^3_{\star k}}$ is the class of the exceptional divisor corresponding to the point blown-up $Q^3_{\star k}$,
- $\mathscr{E}_R$ (respectively $\mathscr{E}_{R^2}, \mathscr{E}_{R^3}$) is the class of the exceptional divisor corresponding to the point blown-up R (respectively R^2, R^3), and for $k = 1, \ldots, \mu$, $\mathscr{E}_{R^2_{\star k}}$ is the class of the exceptional divisor corresponding to the point blown-up $R^2_{\star k}$.

  Similarly, for $k = 1, \ldots, \nu$, $\mathscr{E}_{R^3_{\star k}}$ is the class of the exceptional divisor corresponding to the point blown-up $R^3_{\star k}$,
- For $i = 1, \ldots, x$, $\mathscr{E}_{P_i^{L_1}}$ is the class of the exceptional divisor corresponding to the point blown-up $P_i^{L_1}$, and for $k = 1, \ldots, \rho_i^1$, $\mathscr{E}_{P^{L_1}_{i\star k}}$ is the class of the exceptional divisor corresponding to the point blown-up $P^{L_1}_{i\star k}$,

- For $i = 1, \ldots, y$, $\mathscr{E}_{P_i^{L_2}}$ is the class of the exceptional divisor corresponding to the point blown-up $P_i^{L_2}$, and for $k = 1, \ldots, \rho_i^2$, $\mathscr{E}_{P_{i\star k}^{L_2}}$ is the class of the exceptional divisor corresponding to the point blown-up $P_{i\star k}^{L_2}$,
- For $i = 1, \ldots, z$, $\mathscr{E}_{P_i^{L_3}}$ is the class of the exceptional divisor corresponding to the point blown-up $P_i^{L_3}$, and for $k = 1, \ldots, \rho_i^3$, $\mathscr{E}_{P_{i\star k}^{L_3}}$ is the class of the exceptional divisor corresponding to the point blown-up $P_{i\star k}^{L_3}$.

Thus, any class of a divisor on X_2 in $NS(X_2)$ can be represented by a tuple with integer entries such as:

$$(d; a_P, a_{P^1}, a_{P^1_{\star 1}}, \ldots, a_{P^1_{\star\gamma}}, a_{P^2}, a_{P^2_{\star 1}}, \ldots, a_{P^2_{\star\delta}},$$

$$a_Q, a_{Q^1}, a_{Q^1_{\star 1}}, \ldots, a_{Q^1_{\star\rho}}, a_{Q^3}, a_{Q^3_{\star 1}}, \ldots, a_{Q^3_{\star\varrho}},$$

$$a_R, a_{R^2}, a_{R^2_{\star 1}}, \ldots, a_{R^2_{\star\mu}}, a_{R^3}, a_{R^3_{\star 1}}, \ldots, a_{R^3_{\star\nu}},$$

$$a_{P_1^{L_1}}, \ldots, a_{P^{L_1}_{x\star\rho_x^1}}, a_{P_1^{L_2}}, \ldots, a_{P^{L_2}_{y\star\rho_y^2}}, a_{P_1^{L_3}}, \ldots, a_{P^{L_3}_{z\star\rho_z^3}}).$$

The following lemma gives the irreducible components of an anticanonical divisor $-K_{X_2}$ on X_2. In particular, the surface X_2 is anticanonical, see [27].

Lemma 7 *With notation as above, the class of $-K_{X_2}$ in $NS(X_2)$ is:*

$$-\mathscr{K}_{X_2} =$$

$$\left(\mathscr{E}_0 - \mathscr{E}_P - \mathscr{E}_{P^1} - \mathscr{E}_{P^1_{\star 1}} - \ldots - \mathscr{E}_{P^1_{\star\gamma}} - \mathscr{E}_Q - \mathscr{E}_{Q^1} - \mathscr{E}_{Q^1_{\star 1}} - \ldots - \mathscr{E}_{Q^1_{\star\rho}} - \sum_{i=1}^{i=x}(\sum_{k=0}^{k=\rho_i^1} \mathscr{E}_{P^{L_1}_{i\star k}})\right) +$$

$$\left(\mathscr{E}_0 - \mathscr{E}_P - \mathscr{E}_{P^2} - \mathscr{E}_{P^2_{\star 1}} - \ldots - \mathscr{E}_{P^2_{\star\delta}} - \mathscr{E}_R - \mathscr{E}_{R^2} - \mathscr{E}_{R^2_{\star 1}} - \ldots - \mathscr{E}_{R^2_{\star\mu}} - \sum_{i=1}^{i=y}(\sum_{k=0}^{k=\rho_i^2} \mathscr{E}_{P^{L_2}_{i\star k}})\right) +$$

$$\left(\mathscr{E}_0 - \mathscr{E}_Q - \mathscr{E}_{Q^3} - \mathscr{E}_{P^2_{\star 1}} - \ldots - \mathscr{E}_{Q^3_{\star\varrho}} - \mathscr{E}_R - \mathscr{E}_{R^3} - \mathscr{E}_{R^3_{\star 1}} - \ldots - \mathscr{E}_{R^3_{\star\nu}} - \sum_{i=1}^{i=z}(\sum_{k=0}^{k=\rho_i^3} \mathscr{E}_{P^{L_3}_{i\star k}})\right) +$$

$$(\mathscr{E}_P - \mathscr{E}_{P^1} - \mathscr{E}_{P^2}) + (\mathscr{E}_Q - \mathscr{E}_{Q^1} - \mathscr{E}_{Q^3}) + (\mathscr{E}_R - \mathscr{E}_{R^2} - \mathscr{E}_{R^3}) +$$

$$\left(\mathscr{E}_{P^1} - \mathscr{E}_{P^1_{\star 1}}\right) + \left(\mathscr{E}_{P^1_{\star 1}} - \mathscr{E}_{P^1_{\star 2}}\right) + \ldots + \left(\mathscr{E}_{P^1_{\star\gamma-1}} - \mathscr{E}_{P^1_{\star\gamma}}\right) + \mathscr{E}_{P^1_{\star\gamma}} +$$

$$\left(\mathscr{E}_{P^2} - \mathscr{E}_{P^2_{\star 1}}\right) + \left(\mathscr{E}_{P^2_{\star 1}} - \mathscr{E}_{P^2_{\star 2}}\right) + \ldots + \left(\mathscr{E}_{P^2_{\star\delta-1}} - \mathscr{E}_{P^2_{\star\delta}}\right) + \mathscr{E}_{P^2_{\star\delta}} +$$

$$\left(\mathscr{E}_{Q^1} - \mathscr{E}_{Q^1_{\star 1}}\right) + \left(\mathscr{E}_{Q^1_{\star 1}} - \mathscr{E}_{Q^1_{\star 2}}\right) + \ldots + \left(\mathscr{E}_{Q^1\star\rho-1} - \mathscr{E}_{Q^1_{\star\rho}}\right) + \mathscr{E}_{Q^1_{\star\rho}} +$$

$$\left(\mathscr{E}_{Q^3} - \mathscr{E}_{Q^3_{\star 1}}\right) + \left(\mathscr{E}_{Q^3_{\star 1}} - \mathscr{E}_{Q^3_{\star 2}}\right) + \ldots + \left(\mathscr{E}_{Q^3\star\varrho-1} - \mathscr{E}_{Q^3_{\star\varrho}}\right) + \mathscr{E}_{Q^3_{\star\varrho}} +$$

$$\left(\mathscr{E}_{R^2} - \mathscr{E}_{R^2_{\star 1}}\right) + \left(\mathscr{E}_{R^2_{\star 1}} - \mathscr{E}_{R^2_{\star 2}}\right) + \ldots + \left(\mathscr{E}_{R^2\star\mu-1} - \mathscr{E}_{R^2_{\star\mu}}\right) + \mathscr{E}_{R^2_{\star\mu}} +$$

$$\left(\mathscr{E}_{R^3} - \mathscr{E}_{R^3_{\star 1}}\right) + \left(\mathscr{E}_{R^3_{\star 1}} - \mathscr{E}_{R^3_{\star 2}}\right) + \ldots + \left(\mathscr{E}_{R^3\star\nu-1} - \mathscr{E}_{R^3_{\star\nu}}\right) + \mathscr{E}_{R^3_{\star\nu}}.$$

Proof Straightforward consequence from [30, Proposition 3.3, page 387]. □

Here we determine the set of trivial (−2)-curves on X_2. By definition, a (-2)-curve N on X_2 is said to be trivial if its class $\mathscr{N}$ in $NS(X_2)$ satisfies the following equality: $\mathscr{N} \cdot \mathscr{E}_0 = 0$.

Lemma 8 *With notation as above. For any given positive integers $x, y, z, \gamma, \delta, \rho, \varrho, \mu, \nu$, and $\rho^1_i, \rho^2_j, \rho^3_k$, where $i = 1, \ldots, x$, $j = 1, \ldots, y$, and $k = 1, \ldots, z$, the classes of trivial (−2)-curves on X_2 in $NS(X_2)$ are:*

1. $(\mathscr{E}_{P^1_{\star i}} - \mathscr{E}_{P^1_{\star(i+1)}})$ *for any* $i = 0, \ldots, (\gamma - 1)$,
2. $(\mathscr{E}_{P^2_{\star i}} - \mathscr{E}_{P^2_{\star(i+1)}})$ *for any* $i = 0, \ldots, (\delta - 1)$,
3. $(\mathscr{E}_{Q^1_{\star i}} - \mathscr{E}_{Q^1_{\star(i+1)}})$ *for any* $i = 0, \ldots, (\rho - 1)$,
4. $(\mathscr{E}_{Q^3_{\star i}} - \mathscr{E}_{Q^3_{\star(i+1)}})$ *for any* $i = 0, \ldots, (\varrho - 1)$,
5. $(\mathscr{E}_{R^2_{\star i}} - \mathscr{E}_{R^2_{\star(i+1)}})$ *for any* $i = 0, \ldots, (\mu - 1)$,
6. $(\mathscr{E}_{R^3_{\star i}} - \mathscr{E}_{R^3_{\star(i+1)}})$ *for any* $i = 0, \ldots, (\nu - 1)$,
7. $(\mathscr{E}_{P^{L_1}_{i\star j}} - \mathscr{E}_{P^{L_1}_{i\star(1+j)}})$ *for any* $j = 0, \ldots, (\rho^1_i - 1)$ *and for any* $i = 1, \ldots, x$,
8. $(\mathscr{E}_{P^{L_2}_{i\star j}} - \mathscr{E}_{P^{L_2}_{i\star(1+j)}})$ *for any* $j = 0, \ldots, (\rho^2_i - 1)$ *and for any* $i = 1, \ldots, y$, *and*
9. $(\mathscr{E}_{P^{L_3}_{i\star j}} - \mathscr{E}_{P^{L_3}_{i\star(1+j)}})$ *for any* $j = 0, \ldots, (\rho^3_i - 1)$ *and for any* $i = 1, \ldots, z$.

Proof It is obvious that any element of the list is the class of a trivial (−2)-curve on X_2. Conversely, any trivial (−2)-curve on X_2 which is different from the members of the list would imply that it is the zero divisor, hence its self-intersection would be zero. So, the only trivial (−2)-curves on X_2 are the ones of the list. □

Similarly, here, we determine the set of trivial (−1)-curves on X_2. By definition, a (−1)-curve E on X_2 is said to be trivial if its class $\mathscr{E}$ in $NS(X_2)$ satisfies the following equality: $\mathscr{E} \cdot \mathscr{E}_0 = 0$.

Lemma 9 *With notation as above. For any given positive integers $n, m, \gamma, \delta, \rho, \varrho, \mu, \nu$, and $\rho^1_i, \rho^2_j, \rho^3_k$, where $i = 1, \ldots, x$, $j = 1, \ldots, y$, and $k = 1, \ldots, z$, the classes of trivial (−1)-curves on X_2 in $NS(X_2)$ are:*

1. $\mathscr{E}_{P^1_{\star\gamma}}, \mathscr{E}_{P^2_{\star\delta}}, \mathscr{E}_{Q^1_{\star\rho}}, \mathscr{E}_{Q^3_{\star\varrho}}, \mathscr{E}_{R^2_{\star\mu}}, \mathscr{E}_{R^3_{\star\nu}}$,
2. $\mathscr{E}_{P^{L_1}_{i\star\rho^1_i}}$ *for any* $i = 1, \ldots, x$,

3. $\mathscr{E}_{P^{L_2}_{i\star\rho_i^2}}$ *for any* $i = 1, \ldots, y$, *and*
4. $\mathscr{E}_{P^{L_3}_{i\star\rho_i^3}}$ *for any* $i = 1, \ldots, z$.

Proof It is obvious that any element of the list is a the class of a trivial (-1)-curve on X_2. Conversely, any trivial (-1)-curve on X_2 which is different from the members of the list would imply that it is the zero divisor, hence its self-intersection would be zero. So, the only trivial (-1)-curves on X_2 are the ones of the list. □

Remark 3 One may also determine the trivial curves on X_2 of self-intersection less than -2, i.e., those integral curves C with $C^2 < -2$ and whose classes $\mathscr{C}$ in $NS(X_2)$ satisfy $\mathscr{C} \cdot \mathscr{E}_0 = 0$.

Our main result in this section is:

Theorem 1 *With the same notation as above,* X_2 *is Platonic if the following inequality holds:*

$$\left(1 - \frac{1}{x + \sum_{k=1}^{k=x} \rho_k^1} - \frac{1}{y + \sum_{k=1}^{k=y} \rho_k^2} - \frac{1}{z + \sum_{k=1}^{k=z} \rho_k^3}\right) < 0.$$

Proof According to Lemma 4, we need only to prove that the set of (-1)-curves on X_2 and the set of (-2)-curves on X_2 are both finite:

3.1 X_2 Has Only a Finite Number of (-1)-Curves

We first show that X_2 holds only a finite number of (-1)-curves. To do so, let E be a (-1)-curve on X_2 and assume that its class in $NS(X_2)$ is determined by the following tuple of integers:

$$\begin{gathered}(d; a_P, a_{P^1}, a_{P^1_{\star 1}}, \ldots, a_{P^1_{\star\gamma}}, a_{P^2}, a_{P^2_{\star 1}}, \ldots, a_{P^2_{\star\delta}},\\ a_Q, a_{Q^1}, a_{Q^1_{\star 1}}, \ldots, a_{Q^1_{\star\rho}}, a_{Q^3}, a_{Q^3_{\star 1}}, \ldots, a_{Q^3_{\star\varrho}},\\ a_R, a_{R^2}, a_{R^2_{\star 1}}, \ldots, a_{R^2_{\star\mu}}, a_{R^3}, a_{R^3_{\star 1}}, \ldots, a_{R^3_{\star\nu}},\\ a_{P^{L_1}_1}, \ldots, a_{P^{L_1}_{x\star\rho_x^1}}, a_{P^{L_2}_1}, \ldots, a_{P^{L_2}_{y\star\rho_y^2}}, a_{P^{L_3}_1}, \ldots, a_{P^{L_3}_{z\star\rho_z^3}}).\end{gathered}$$

Using Lemma 5, we may assume the positivity of all the integers $\gamma, \delta, \rho, \varrho, \mu, \nu, x$, y and z. From $E \cdot K_{X_2} = -1$, one may obtain the following equality:

$$3d - a_P - a_{P^1} - a_{P^1_{\star 1}} - \ldots - a_{P^1_{\star\gamma}} - a_{P^2} - a_{P^2_{\star 1}} - \ldots - a_{P^2_{\star\delta}} -$$

$$a_Q - a_{Q^1} - a_{Q^1_{\star 1}} - \ldots - a_{Q^1_{\star\rho}} - a_{Q^3} - a_{Q^3_{\star 1}} - \ldots - a_{Q^3_{\star\varrho}} -$$

$$a_R - a_{R^2} - a_{R^2_{\star 1}} - \ldots - a_{R^2_{\star\mu}} - a_{R^3} - a_{R^3_{\star 1}} - \ldots - a_{R^3_{\star\nu}} -$$

$$a_{P_1^{L_1}} - \ldots - a_{P^{L_1}_{x\star\rho^1_x}} - a_{P_1^{L_2}} - \ldots - a_{P^{L_2}_{y\star\rho^2_y}} - a_{P_1^{L_3}} - \ldots - a_{P^{L_3}_{z\star\rho^3_z}} = 1.$$

Since the trivial (-1)-curves on X_2 are finite in number, we may moreover assume that E is not trivial (see Lemma 9), then from the following equality which is equivalent to the last one:

$$\left(d - a_P - a_{P^1} - a_{P^1_{\star 1}} - \ldots - a_{P^1_{\star\gamma}} - a_Q - a_{Q^1} - a_{Q^1_{\star 1}} - \ldots - a_{Q^1_{\star\rho}} - \sum_{i=1}^{i=x}\left(\sum_{k=0}^{k=\rho^1_i} a_{P^{L_1}_{i\star k}}\right)\right) +$$

$$\left(d - a_P - a_{P^2} - a_{P^2_{\star 1}} - \ldots - a_{P^2_{\star\delta}} - a_R - a_{R^2} - a_{R^2_{\star 1}} - \ldots - a_{R^2_{\star\mu}} - \sum_{i=1}^{i=y}\left(\sum_{k=0}^{k=\rho^2_i} a_{P^{L_2}_{i\star k}}\right)\right) +$$

$$\left(d - a_Q - a_{Q^3} - a_{P^2_{\star 1}} - \ldots - a_{Q^3_{\star\varrho}} - a_R - a_{R^3} - a_{R^3_{\star 1}} - \ldots - a_{R^3_{\star\nu}} - \sum_{i=1}^{i=z}\left(\sum_{k=0}^{k=\rho^3_i} a_{P^{L_3}_{i\star k}}\right)\right) +$$

$$\left(a_P - a_{P^1} - a_{P^2}\right) + \left(a_Q - a_{Q^1} - a_{Q^3}\right) + \left(a_R - a_{R^2} - a_{R^3}\right) +$$

$$\left(a_{P^1} - a_{P^1_{\star 1}}\right) + \left(a_{P^1_{\star 1}} - a_{P^1_{\star 2}}\right) + \ldots + \left(a_{P^1_{\star\gamma-1}} - a_{P^1_{\star\gamma}}\right) + a_{P^1_{\star\gamma}} +$$

$$\left(a_{P^2} - a_{P^2_{\star 1}}\right) + \left(a_{P^2_{\star 1}} - a_{P^2_{\star 2}}\right) + \ldots + \left(a_{P^2_{\star\delta-1}} - a_{P^2_{\star\delta}}\right) + a_{P^2_{\star\delta}} +$$

$$\left(a_{Q^1} - a_{Q^1_{\star 1}}\right) + \left(a_{Q^1_{\star 1}} - a_{Q^1_{\star 2}}\right) + \ldots + \left(a_{Q^1\star\rho-1} - a_{Q^1_{\star\rho}}\right) + a_{Q^1_{\star\rho}} +$$

$$\left(a_{Q^3} - a_{Q^3_{\star 1}}\right) + \left(a_{Q^3_{\star 1}} - a_{Q^3_{\star 2}}\right) + \ldots + \left(a_{Q^3\star\varrho-1} - a_{Q^3_{\star\varrho}}\right) + a_{Q^3_{\star\varrho}} +$$

$$\left(a_{R^2} - a_{R^2_{\star 1}}\right) + \left(a_{R^2_{\star 1}} - a_{R^2_{\star 2}}\right) + \ldots + \left(a_{R^2\star\mu-1} - a_{R^2_{\star\mu}}\right) + a_{R^2_{\star\mu}} +$$

$$\left(a_{R^3} - a_{R^3_{\star 1}}\right) + \left(a_{R^3_{\star 1}} - a_{R^3_{\star 2}}\right) + \ldots + \left(a_{R^3\star\nu-1} - a_{R^3_{\star\nu}}\right) + a_{R^3_{\star\nu}} = 1,$$

one may obtain only one of the four cases below:

Case 1:

$$d-\sum_{i=1}^{i=x}(\sum_{k=0}^{k=\rho_i^1} a_{P^{L_1}_{i\star k}})=1,$$

$$d-\sum_{i=1}^{i=y}(\sum_{k=0}^{k=\rho_i^2} a_{P^{L_2}_{i\star k}})=0,$$

$$d-\sum_{i=1}^{i=z}(\sum_{k=0}^{k=\rho_i^3} a_{P^{L_3}_{i\star k}})=0,$$

$$a_P=a_{P^1}=a_{P^2}=a_Q=a_{Q^1}=a_{Q^3}=a_R=a_{R^2}=a_{R^3}=0,$$

$$a_{P^1}=a_{P^1_{\star 1}}=a_{P^1_{\star 2}}=\ldots=a_{P^1_{\star\gamma-1}}=a_{P^1_{\star\gamma}}=a_{P^2}=a_{P^2_{\star 1}}=a_{P^2_{\star 2}}=\ldots=a_{P^2_{\star\delta-1}}=a_{P^2_{\star\delta}}=0,$$

$$a_{Q^1}=a_{Q^1_{\star 1}}=a_{Q^1_{\star 2}}=\ldots=a_{Q^1{\star\rho-1}}=a_{Q^1_{\star\rho}}=a_{Q^3}=a_{Q^3_{\star 1}}=a_{Q^3_{\star 2}}$$
$$=\ldots=a_{Q^3{\star\varrho-1}}=a_{Q^3_{\star\varrho}}=0, \text{ and}$$

$$a_{R^2}=a_{R^2_{\star 1}}=a_{R^2_{\star 2}}=\ldots=a_{R^2{\star\mu-1}}=a_{R^2_{\star\mu}}=a_{R^3}=a_{R^3_{\star 1}}=a_{R^3_{\star 2}}=\ldots=a_{R^3{\star\nu-1}}=a_{R^3_{\star\nu}}=0.$$

Case 2:

$$d-\sum_{i=1}^{i=x}(\sum_{k=0}^{k=\rho_i^1} a_{P^{L_1}_{i\star k}})=0,$$

$$d-\sum_{i=1}^{i=y}(\sum_{k=0}^{k=\rho_i^2} a_{P^{L_2}_{i\star k}})=1,$$

$$d-\sum_{i=1}^{i=z}(\sum_{k=0}^{k=\rho_i^3} a_{P^{L_3}_{i\star k}})=0,$$

$$a_P=a_{P^1}=a_{P^2}=a_Q=a_{Q^1}=a_{Q^3}=a_R=a_{R^2}=a_{R^3}=0,$$

$$a_{P^1} = a_{P^1_{\star 1}} = a_{P^1_{\star 2}} = \ldots = a_{P^1_{\star\gamma-1}} = a_{P^1_{\star\gamma}} = a_{P^2} = a_{P^2_{\star 1}} = a_{P^2_{\star 2}} = \ldots = a_{P^2_{\star\delta-1}} = a_{P^2_{\star\delta}} = 0,$$

$$a_{Q^1} = a_{Q^1_{\star 1}} = a_{Q^1_{\star 2}} = \ldots = a_{Q^1 \star\rho-1} = a_{Q^1_{\star\rho}} = a_{Q^3} = a_{Q^3_{\star 1}} = a_{Q^3_{\star 2}}$$
$$= \ldots = a_{Q^3 \star\varrho-1} = a_{Q^3_{\star\varrho}} = 0, \text{ and}$$

$$a_{R^2} = a_{R^2_{\star 1}} = a_{R^2_{\star 2}} = \ldots = a_{R^2 \star\mu-1} = a_{R^2_{\star\mu}} = a_{R^3} = a_{R^3_{\star 1}} = a_{R^3_{\star 2}} = \ldots = a_{R^3 \star\nu-1} = a_{R^3_{\star\nu}} = 0.$$

Case 3:

$$d - \sum_{i=1}^{i=x}\Big(\sum_{k=0}^{k=\rho_i^1} a_{P^{L_1}_{i\star k}}\Big) = 0,$$

$$d - \sum_{i=1}^{i=y}\Big(\sum_{k=0}^{k=\rho_i^2} a_{P^{L_2}_{i\star k}}\Big) = 0,$$

$$d - \sum_{i=1}^{i=z}\Big(\sum_{k=0}^{k=\rho_i^3} a_{P^{L_3}_{i\star k}}\Big) = 1,$$

$$a_P = a_{P^1} = a_{P^2} = a_Q = a_{Q^1} = a_{Q^3} = a_R = a_{R^2} = a_{R^3} = 0,$$

$$a_{P^1} = a_{P^1_{\star 1}} = a_{P^1_{\star 2}} = \ldots = a_{P^1_{\star\gamma-1}} = a_{P^1_{\star\gamma}} = a_{P^2} = a_{P^2_{\star 1}} = a_{P^2_{\star 2}} = \ldots = a_{P^2_{\star\delta-1}} = a_{P^2_{\star\delta}} = 0,$$

$$a_{Q^1} = a_{Q^1_{\star 1}} = a_{Q^1_{\star 2}} = \ldots = a_{Q^1 \star\rho-1} = a_{Q^1_{\star\rho}} = a_{Q^3} = a_{Q^3_{\star 1}} = a_{Q^3_{\star 2}}$$
$$= \ldots = a_{Q^3 \star\varrho-1} = a_{Q^3_{\star\varrho}} = 0, \text{ and}$$

$$a_{R^2} = a_{R^2_{\star 1}} = a_{R^2_{\star 2}} = \ldots = a_{R^2 \star\mu-1} = a_{R^2_{\star\mu}} = a_{R^3} = a_{R^3_{\star 1}} = a_{R^3_{\star 2}} = \ldots = a_{R^3 \star\nu-1} = a_{R^3_{\star\nu}} = 0.$$

Case 4:

There exist three integers u, v, and w such that:
$0 \leq u \leq \max(2+\gamma, 2+\rho)$, $0 \leq v \leq \max(2+\delta, 2+\mu)$ and $0 \leq w \leq \max(2+\varrho, 2+\nu)$, and

$$d - u - \sum_{i=1}^{i=x}(\sum_{k=0}^{k=\rho_i^1} a_{P_{i\star k}^{L_1}}) = 0,$$

$$d - v - \sum_{i=1}^{i=y}(\sum_{k=0}^{k=\rho_i^2} a_{P_{i\star k}^{L_2}}) = 0, \text{ and}$$

$$d - w - \sum_{i=1}^{i=z}(\sum_{k=0}^{k=\rho_i^3} a_{P_{i\star k}^{L_3}}) = 0.$$

Assume that we are in the Case 1. Then consider the following parameters:

$$\alpha_{P_{i\star k}^{L_1}} = a_{P_{i\star k}^{L_1}} - \frac{(d-1)}{x + \sum_{j=1}^{j=x} \rho_k^1} \text{ for any } i = 1, \ldots, x \text{ and for any } k = 0, \ldots, \rho_i^1,$$

$$\beta_{P_{i\star k}^{L_2}} = a_{P_{i\star k}^{L_2}} - \frac{d}{y + \sum_{j=1}^{j=y} \rho_k^2} \text{ for any } i = 1, \ldots, y \text{ and for any } k = 0, \ldots, \rho_i^2,$$

$$\xi_{P_{i\star k}^{L_3}} = a_{P_{i\star k}^{L_3}} - \frac{d}{z + \sum_{j=1}^{j=z} \rho_k^3} \text{ for any } i = 1, \ldots, z \text{ and for any } k = 0, \ldots, \rho_i^3.$$

Using the fact that E is a (-1)-curve on X_2, we obtain the following equality:

$$\sum_{i=1}^{i=x}\sum_{k=0}^{k=\rho_i^1}(\alpha_{P_{i\star k}^{L_1}})^2 + \sum_{i=1}^{i=y}\sum_{k=0}^{k=\rho_i^2}(\beta_{P_{i\star k}^{L_2}})^2 + \sum_{i=1}^{i=z}\sum_{k=0}^{k=\rho_i^3}(\xi_{P_{i\star k}^{L_3}})^2$$
$$= 1 + d^2 - \frac{(d-1)^2}{x+\sum_{k=1}^{k=x}\rho_k^1} - \frac{d^2}{y+\sum_{k=1}^{k=y}\rho_k^2} - \frac{d^2}{z+\sum_{k=1}^{k=z}\rho_k^3}.$$

Hence the following inequality holds:

$$1 + d^2 - \frac{(d-1)^2}{x+\sum_{k=1}^{k=x}\rho_k^1} - \frac{d^2}{y+\sum_{k=1}^{k=y}\rho_k^2} - \frac{d^2}{z+\sum_{k=1}^{k=z}\rho_k^3} \geq 0.$$

Consequently, by assumption, the degree d is bounded. So there are only a finite number of (-1)-curves on X_2 that satisfy the case one. Similarly, we prove that there are only a finite number of (-1)-curves that satisfy either the Case 2, or the Case 3. So we are left with the Case 4.

To handle this case, Then consider the following parameters:

$$\alpha_{P^{L_1}_{i\star k}} = a_{P^{L_1}_{i\star k}} - \frac{(d-u)}{x+\sum_{j=1}^{j=x}\rho^1_k} \text{ for any } i=1,\ldots,x \text{ and for any } k=0,\ldots,\rho^1_i,$$

$$\beta_{P^{L_2}_{i\star k}} = a_{P^{L_2}_{i\star k}} - \frac{(d-v)}{y+\sum_{j=1}^{j=y}\rho^2_k} \text{ for any } i=1,\ldots,y \text{ and for any } k=0,\ldots,\rho^2_i,$$

$$\xi_{P^{L_3}_{i\star k}} = a_{P^{L_3}_{i\star k}} - \frac{(d-w)}{z+\sum_{j=1}^{j=z}\rho^3_k} \text{ for any } i=1,\ldots,z \text{ and for any } k=0,\ldots,\rho^3_i.$$

Using the fact that $E^2=-1$, we obtain the following equality:

$$\sum_{i=1}^{i=x}\sum_{k=0}^{k=\rho^1_i}(\alpha_{P^{L_1}_{i\star k}})^2 + \sum_{i=1}^{i=y}\sum_{k=0}^{k=\rho^2_i}(\beta_{P^{L_2}_{i\star k}})^2 + \sum_{i=1}^{i=z}\sum_{k=0}^{k=\rho^3_i}(\xi_{P^{L_3}_{i\star k}})^2 =$$

$$1+d^2-u-v-w-\frac{(d-u)^2}{x+\sum_{k=1}^{k=x}\rho^1_k}-\frac{(d-v)^2}{y+\sum_{k=1}^{k=y}\rho^2_k}-\frac{(d-w)^2}{z+\sum_{k=1}^{k=z}\rho^3_k}.$$

Hence, the following inequality holds:

$$1+d^2-u-v-w-\frac{(d-u)^2}{x+\sum_{k=1}^{k=x}\rho^1_k}-\frac{(d-v)^2}{y+\sum_{k=1}^{k=y}\rho^2_k}-\frac{(d-w)^2}{z+\sum_{k=1}^{k=z}\rho^3_k} \geq 0.$$

Consequently, by assumption, the degree d is bounded. So there are only a finite number of (-1)-curves on X_2 that satisfy the case four.

In conclusion, under our numerical condition, X_2 has only a finite number of (-1)-curves.

3.2 X_2 Has Only a Finite Number of (-2)-Curves

Without loss of generality, we assume that every integer γ, δ, ρ, ϱ, μ, ν, x, y and z is positive (see Lemma 5). Now we proceed to prove that there are only a finite number of (-2)-curves on X_2. Let $\mathscr{N}$ be the class in $NS(X_2)$ of a (-2)-curve N, and its corresponding tuple of integers is as follows:

$$(d; a_P, a_{P^1}, a_{P^1_{\star 1}}, \ldots, a_{P^1_{\star\gamma}}, a_{P^2}, a_{P^2_{\star 1}}, \ldots, a_{P^2_{\star\delta}},$$

$$a_Q, a_{Q^1}, a_{Q^1_{\star 1}}, \ldots, a_{Q^1_{\star \rho}}, a_{Q^3}, a_{Q^3_{\star 1}}, \ldots, a_{Q^3_{\star \varrho}},$$

$$a_R, a_{R^2}, a_{R^2_{\star 1}}, \ldots, a_{R^2_{\star \mu}}, a_{R^3}, a_{R^3_{\star 1}}, \ldots, a_{R^3_{\star \nu}},$$

$$a_{P_1^{L_1}}, \ldots, a_{P^{L_1}_{x\star\rho^1_x}}, a_{P_1^{L_2}}, \ldots, a_{P^{L_2}_{y\star\rho^2_y}}, a_{P_1^{L_3}}, \ldots, a_{P^{L_3}_{z\star\rho^3_z}}).$$

Since $N \cdot K_{X_2} = 0$, we have the following the equality:

$$3d - a_P - a_{P^1} - a_{P^1_{\star 1}} - \ldots - a_{P^1_{\star\gamma}} - a_{P^2} - a_{P^2_{\star 1}} - \ldots, - a_{P^2_{\star\delta}} -$$

$$a_Q - a_{Q^1} - a_{Q^1_{\star 1}} - \ldots - a_{Q^1_{\star\rho}} - a_{Q^3} - a_{Q^3_{\star 1}} - \ldots - a_{Q^3_{\star\varrho}} -$$

$$a_R - a_{R^2} - a_{R^2_{\star 1}} - \ldots - a_{R^2_{\star\mu}} - a_{R^3} - a_{R^3_{\star 1}} - \ldots - a_{R^3_{\star\nu}} -$$

$$a_{P_1^{L_1}} - \ldots - a_{P^{L_1}_{x\star\rho^1_x}} - a_{P_1^{L_2}} - \ldots - a_{P^{L_2}_{y\star\rho^2_y}} - a_{P_1^{L_3}} - \ldots - a_{P^{L_3}_{z\star\rho^3_z}} = 0.$$

Which in turn is equivalent to:

$$\left(d - a_P - a_{P^1} - a_{P^1_{\star 1}} - \ldots - a_{P^1_{\star\gamma}} - a_Q - a_{Q^1} - a_{Q^1_{\star 1}} - \ldots - a_{Q^1_{\star\rho}} - \sum_{i=1}^{i=x}(\sum_{k=0}^{k=\rho^1_i} a_{P^{L_1}_{i\star k}})\right) +$$

$$\left(d - a_P - a_{P^2} - a_{P^2_{\star 1}} - \ldots - a_{P^2_{\star\delta}} - a_R - a_{R^2} - a_{R^2_{\star 1}} - \ldots - a_{R^2_{\star\mu}} - \sum_{i=1}^{i=y}(\sum_{k=0}^{k=\rho^2_i} a_{P^{L_2}_{i\star k}})\right) +$$

$$\left(d - a_Q - a_{Q^3} - a_{P^2_{\star 1}} - \ldots - a_{Q^3_{\star\varrho}} - a_R - a_{R^3} - a_{R^3_{\star 1}} - \ldots - a_{R^3_{\star\nu}} - \sum_{i=1}^{i=z}(\sum_{k=0}^{k=\rho^3_i} a_{P^{L_3}_{i\star k}})\right) +$$

$$\left(a_P - a_{P^1} - a_{P^2}\right) + \left(a_Q - a_{Q^1} - a_{Q^3}\right) + \left(a_R - a_{R^2} - a_{R^3}\right) +$$

$$\left(a_{P^1} - a_{P^1_{\star 1}}\right) + \left(a_{P^1_{\star 1}} - a_{P^1_{\star 2}}\right) + \ldots + \left(a_{P^1_{\star\gamma-1}} - a_{P^1_{\star\gamma}}\right) + a_{P^1_{\star\gamma}} +$$

$$\left(a_{P^2} - a_{P^2_{\star 1}}\right) + \left(a_{P^2_{\star 1}} - a_{P^2_{\star 2}}\right) + \ldots + \left(a_{P^2_{\star\delta-1}} - a_{P^2_{\star\delta}}\right) + a_{P^2_{\star\delta}} +$$

$$\left(a_{Q^1} - a_{Q^1_{\star 1}}\right) + \left(a_{Q^1_{\star 1}} - a_{Q^1_{\star 2}}\right) + \ldots + \left(a_{Q^1\star\rho-1} - a_{Q^1_{\star\rho}}\right) + a_{Q^1_{\star\rho}} +$$

$$\left(a_{Q^3} - a_{Q^3_{\star 1}}\right) + \left(a_{Q^3_{\star 1}} - a_{Q^3_{\star 2}}\right) + \ldots + \left(a_{Q^3\star\varrho-1} - a_{Q^3_{\star\varrho}}\right) + a_{Q^3_{\star\varrho}} +$$

$$\left(a_{R^2} - a_{R^2_{\star 1}}\right) + \left(a_{R^2_{\star 1}} - a_{R^2_{\star 2}}\right) + \ldots + \left(a_{R^2{\star\mu-1}} - a_{R^2_{\star\mu}}\right) + a_{R^2_{\star\mu}} +$$

$$\left(a_{R^3} - a_{R^3_{\star 1}}\right) + \left(a_{R^3_{\star 1}} - a_{R^3_{\star 2}}\right) + \ldots + \left(a_{R^3{\star\nu-1}} - a_{R^3_{\star\nu}}\right) + a_{R^3_{\star\nu}} = 0,$$

Assuming furthermore that N is not trivial (see Lemma 8), the last equality implies the following equalities:

$$d - \sum_{i=1}^{i=x}(\sum_{k=0}^{k=\rho_i^1} a_{P^{L_1}_{i\star k}}) = 0,$$

$$d - \sum_{i=1}^{i=y}(\sum_{k=0}^{k=\rho_i^2} a_{P^{L_2}_{i\star k}}) = 0,$$

$$d - \sum_{i=1}^{i=z}(\sum_{k=0}^{k=\rho_i^3} a_{P^{L_3}_{i\star k}}) = 0,$$

$$a_P = a_{P^1} = a_{P^2} = a_Q = a_{Q^1} = a_{Q^3} = a_R = a_{R^2} = a_{R^3} = 0,$$

$$a_{P^1} = a_{P^1_{\star 1}} = a_{P^1_{\star 2}} = \ldots = a_{P^1_{\star\gamma-1}} = a_{P^1_{\star\gamma}} = a_{P^2} = a_{P^2_{\star 1}} = a_{P^2_{\star 2}} = \ldots = a_{P^2_{\star\delta-1}} = a_{P^2_{\star\delta}} = 0,$$

$$a_{Q^1} = a_{Q^1_{\star 1}} = a_{Q^1_{\star 2}} = \ldots = a_{Q^1{\star\rho-1}} = a_{Q^1_{\star\rho}} = a_{Q^3} = a_{Q^3_{\star 1}} = a_{Q^3_{\star 2}}$$
$$= \ldots = a_{Q^3{\star\varrho-1}} = a_{Q^3_{\star\varrho}} = 0, \text{ and}$$

$$a_{R^2} = a_{R^2_{\star 1}} = a_{R^2_{\star 2}} = \ldots = a_{R^2{\star\mu-1}} = a_{R^2_{\star\mu}} = a_{R^3} = a_{R^3_{\star 1}} = a_{R^3_{\star 2}} = \ldots = a_{R^3{\star\nu-1}} = a_{R^3_{\star\nu}} = 0.$$

Hence, we get the following three equations:

$$d - \sum_{i=1}^{i=x}(\sum_{k=0}^{k=\rho_i^1} a_{P^{L_1}_{i\star k}}) = 0,$$

$$d - \sum_{i=1}^{i=y}(\sum_{k=0}^{k=\rho_i^2} a_{P^{L_2}_{i\star k}}) = 0, \text{ and}$$

$$d - \sum_{i=1}^{i=z}(\sum_{k=0}^{k=\rho_i^3} a_{P^{L_3}_{i\star k}}) = 0.$$

Consider now the following parameters:

$$\alpha_{P^{L_1}_{i\star k}} = a_{P^{L_1}_{i\star k}} - \frac{d}{x+\sum_{j=1}^{j=x}\rho^1_k} \text{ for any } i=1,\ldots,x \text{ and for any } k=0,\ldots,\rho^1_i,$$

$$\beta_{P^{L_2}_{i\star k}} = a_{P^{L_2}_{i\star k}} - \frac{d}{y+\sum_{j=1}^{j=y}\rho^2_k} \text{ for any } i=1,\ldots,y \text{ and for any } k=0,\ldots,\rho^2_i,$$

$$\xi_{P^{L_3}_{i\star k}} = a_{P^{L_3}_{i\star k}} - \frac{d}{z+\sum_{j=1}^{j=z}\rho^3_k} \text{ for any } i=1,\ldots,z \text{ and for any } k=0,\ldots,\rho^3_i.$$

Using the fact that the self-intersection of N is equal to -2, one may obtain the following equality:

$$\sum_{i=1}^{i=x}\sum_{k=0}^{k=\rho^1_i}(\alpha_{P^{L_1}_{i\star k}})^2 + \sum_{i=1}^{i=y}\sum_{k=0}^{k=\rho^2_i}(\beta_{P^{L_2}_{i\star k}})^2 + \sum_{i=1}^{i=z}\sum_{k=0}^{k=\rho^3_i}(\xi_{P^{L_3}_{i\star k}})^2 = 2+$$

$$d^2\left(1-\frac{1}{x+\sum_{k=1}^{k=x}\rho^1_k}-\frac{1}{y+\sum_{k=1}^{k=y}\rho^2_k}-\frac{1}{z+\sum_{k=1}^{k=z}\rho^3_k}\right).$$

Henceforth, the last equation implies the following inequality:

$$2+d^2\left(1-\frac{1}{x+\sum_{k=1}^{k=x}\rho^1_k}-\frac{1}{y+\sum_{k=1}^{k=y}\rho^2_k}-\frac{1}{z+\sum_{k=1}^{k=z}\rho^3_k}\right) \geq 0.$$

The latter inequality, combined with our assumption, confirms that the degree d of N is bounded. More precisely, we have the following inequality for d:

$$d \leq \frac{\sqrt{2}}{\sqrt{\left(-1+\frac{1}{x+\sum_{k=1}^{k=x}\rho^1_k}+\frac{1}{y+\sum_{k=1}^{k=y}\rho^2_k}+\frac{1}{z+\sum_{k=1}^{k=z}\rho^3_k}\right)}}. \tag{1}$$

From the above and as a general conclusion, X_2 holds only a finite number of (-1)-curves as well as a finite number of (-2)-curves. So we are done. □

Remark 4 It appears clearly from our proof of Theorem 1 that there are only a few (-2)-curves on X_2 that go through the two singular points of our degenerate plane cubic

Here we recover the main result of [13], see [13, Theorem 1., page 2]:

Corollary 1 *With the same notation as above. For* $\gamma = \delta = \rho = \varrho = \mu = \nu = 0$ *and for* $\rho_1^1 = \ldots = \rho_x^1 = \rho_1^2 = \ldots = \rho_y^2 = \rho_1^3 = \ldots = \rho_z^3 = 0$, $M(X_2)$ *is finitely generated if the following inequality holds:*

$$1 - \frac{1}{x} - \frac{1}{y} - \frac{1}{z} < 0.$$

3.3 Regularity of nef Divisors on X_2

Recall that a divisor D on a surface S is said to be numerically effective if the intersection number of D and of C is nonnegative for every integral curve C on S. The aim of this subsection is to prove the following result:

Theorem 2 *With notation as above and with the assumption*

$$\left(1 - \frac{1}{x + \sum_{k=1}^{k=x} \rho_k^1} - \frac{1}{y + \sum_{k=1}^{k=y} \rho_k^2} - \frac{1}{z + \sum_{k=1}^{k=z} \rho_k^3}\right) < 0.$$

Then the first cohomology group of any numerically effective divisor on X_2 vanishes.

Proof It is sufficient to prove only that every numerically effective divisor on X_2 which is orthogonal to K_{X_2} is the zero divisor ([25, Theorem III.1, page 1197]). To do so, let D be a numerically effective divisor on X_2 and let

$$(d; a_P, a_{P^1}, a_{P^1_{\star 1}}, \ldots, a_{P^1_{\star\gamma}}, a_{P^2}, a_{P^2_{\star 1}}, \ldots, a_{P^2_{\star\delta}},$$

$$a_Q, a_{Q^1}, a_{Q^1_{\star 1}}, \ldots, a_{Q^1_{\star\rho}}, a_{Q^3}, a_{Q^3_{\star 1}}, \ldots, a_{Q^3_{\star\varrho}},$$

$$a_R, a_{R^2}, a_{R^2_{\star 1}}, \ldots, a_{R^2_{\star\mu}}, a_{R^3}, a_{R^3_{\star 1}}, \ldots, a_{R^3_{\star\nu}},$$

$$a_{P_1^{L_1}}, \ldots, a_{P^{L_1}_{x\star\rho^1_x}}, a_{P_1^{L_2}}, \ldots, a_{P^{L_2}_{y\star\rho^2_y}}, a_{P_1^{L_3}}, \ldots, a_{P^{L_3}_{z\star\rho^3_z}}).$$

be its corresponding tuple of integers in $NS(X_2)$. From the equality $D \cdot K_{X_2} = 0$ and Lemma 7, one may obtain the following equalities:

$$d - \sum_{i=1}^{i=x}(\sum_{k=0}^{k=\rho_i^1} a_{P^{L_1}_{i\star k}}) = 0,$$

$$d - \sum_{i=1}^{i=y}(\sum_{k=0}^{k=\rho_i^2} a_{P^{L_2}_{i\star k}}) = 0,$$

$$d-\sum_{i=1}^{i=z}(\sum_{k=0}^{k=\rho_i^3} a_{P_{i\star k}^{L_3}})=0,$$

$$a_P=a_{P^1}=a_{P^2}=a_Q=a_{Q^1}=a_{Q^3}=a_R=a_{R^2}=a_{R^3}=0,$$

$$a_{P^1}=a_{P^1_{\star 1}}=a_{P^1_{\star 2}}=\ldots=a_{P^1_{\star\gamma-1}}=a_{P^1_{\star\gamma}}=a_{P^2}=a_{P^2_{\star 1}}=a_{P^2_{\star 2}}=\ldots=a_{P^2_{\star\delta-1}}=a_{P^2_{\star\delta}}=0,$$

$$a_{Q^1}=a_{Q^1_{\star 1}}=a_{Q^1_{\star 2}}=\ldots=a_{Q^1\star\rho-1}=a_{Q^1_{\star\rho}}=a_{Q^3}=a_{Q^3_{\star 1}}=a_{Q^3_{\star 2}}$$
$$=\ldots=a_{Q^3\star\varrho-1}=a_{Q^3_{\star\varrho}}=0,\ \text{and}$$

$$a_{R^2}=a_{R^2_{\star 1}}=a_{R^2_{\star 2}}=\ldots=a_{R^2\star\mu-1}=a_{R^2_{\star\mu}}=a_{R^3}=a_{R^3_{\star 1}}=a_{R^3_{\star 2}}=\ldots=a_{R^3\star\nu-1}=a_{R^3_{\star\nu}}=0.$$

Hence, we get the following three equations:

$$d-\sum_{i=1}^{i=x}(\sum_{k=0}^{k=\rho_i^1} a_{P_{i\star k}^{L_1}})=0,$$

$$d-\sum_{i=1}^{i=y}(\sum_{k=0}^{k=\rho_i^2} a_{P_{i\star k}^{L_2}})=0,\ \text{and}$$

$$d-\sum_{i=1}^{i=z}(\sum_{k=0}^{k=\rho_i^3} a_{P_{i\star k}^{L_3}})=0.$$

Consider now the following parameters:

$$\alpha_{P_{i\star k}^{L_1}}=a_{P_{i\star k}^{L_1}}-\frac{d}{x+\sum_{j=1}^{j=x}\rho_k^1}\ \text{for any } i=1,\ldots,x \text{ and for any } k=0,\ldots,\rho_i^1,$$

$$\beta_{P_{i\star k}^{L_2}}=a_{P_{i\star k}^{L_2}}-\frac{d}{y+\sum_{j=1}^{j=y}\rho_k^2}\ \text{for any } i=1,\ldots,y \text{ and for any } k=0,\ldots,\rho_i^2,$$

$$\xi_{P_{i\star k}^{L_3}}=a_{P_{i\star k}^{L_3}}-\frac{d}{z+\sum_{j=1}^{j=z}\rho_k^3}\ \text{for any } i=1,\ldots,z \text{ and for any } k=0,\ldots,\rho_i^3.$$

On the other hand, since D is numerically effective and using Lemma 3, we have the nonnegativity of the integer D^2. Hence, we obtain that:

$$\sum_{i=1}^{i=x}\sum_{k=0}^{k=\rho_i^1}(\alpha_{P_{i\star k}^{L_1}})^2+\sum_{i=1}^{i=y}\sum_{k=0}^{k=\rho_i^2}(\beta_{P_{i\star k}^{L_2}})^2+\sum_{i=1}^{i=z}\sum_{k=0}^{k=\rho_i^3}(\xi_{P_{i\star k}^{L_3}})^2\leq$$

$$d^2\left(1-\frac{1}{x+\sum_{k=1}^{k=x}\rho_k^1}-\frac{1}{y+\sum_{k=1}^{k=y}\rho_k^2}-\frac{1}{z+\sum_{k=1}^{k=z}\rho_k^3}\right)$$

which in turn, by assumption, proves that d vanishes. So we are done. □

As a consequence, we are able to compute the dimension of any complete linear system associated to any given numerically effective divisor:

Corollary 2 *With notation as above and with the assumption*

$$\left(1-\frac{1}{x+\sum_{k=1}^{k=x}\rho_k^1}-\frac{1}{y+\sum_{k=1}^{k=y}\rho_k^2}-\frac{1}{z+\sum_{k=1}^{k=z}\rho_k^3}\right)<0.$$

Let D be a nef divisor on X_2, then the dimension of the complete linear system $|D|$ associated to D is equal to $\frac{1}{2}(D^2-D\cdot K_{X_2})$.

Proof Apply Lemmas 1, 3 and Theorem 2.

Remark 5 The proof of Theorem 2 enables us to say that the points of the constellation that leads to our Platonic surfaces are in good position. See [23] for more details about this concept, and for example [12, 21, 25, 26] and [36, 37] for more configurations of points that are in good position.

4 Finite Generation of the Cox Rings of Platonic Surfaces

The aim of this section is to prove that the Cox rings of our Platonic surfaces are finitely generated.

Here we recall some concepts and facts, see [10]. Given a toric variety defined over the field of complex numbers $\mathbb{C}$, Cox associated to such variety a $\mathbb{C}$-algebra called the homogeneous coordinate ring, see [8]. In the case of the projective space, this algebra is nothing but the classical homogeneous coordinate ring, which is a polynomial ring. Moreover, he proved that this algebra is always a polynomial ring, and henceforth is a finitely generated $\mathbb{C}$-algebra for any toric variety. Later on, in [31] Hu and Keel offered a generalization of such algebra to any projective variety Y, defined over a fixed algebraically closed field k, and they called it the Cox ring of such variety (nowadays, known also as the total coordinate ring). Following [31], the definition of Cox ring is considered over an algebraically closed field for any projective variety Y and it is

$$Cox(Y) = \bigoplus_{(n_1,\ldots,n_r)\in\mathbb{Z}^r} H^0(Y, \mathcal{O}_Y(L_1^{n_1} \otimes \ldots \otimes L_r^{n_r})),$$

where $(L_1, \ldots, L_r)$ is a basis of the $\mathbb{Z}$-module $Pic(Y)$ of classes of invertible sheaves on Y modulo isomorphisms under the tensor product, and we have assumed that the linear and numerical equivalences on the group of Cartier divisors on Y are the same (for example, such assumption is satisfied for the smooth projective rational surfaces). For a survey on Cox rings, a good reference is [32], and for recent and related works on the finite generation of Cox rings of surfaces, one may look at [1, 3, 6, 16, 43] and [17]. For example in [43], Testa, Várilly-Alvarado and Velasco proved the finite generation of the Cox ring for any smooth projective rational surface whose anticanonical Iitaka dimension is equal to two. However, this k-algebra may fail to be finitely generated for other kinds of varieties. For example, the Cox ring of the blow up of the projective plane $\mathbb{P}^2$ at more than nine points in general position is not finitely generated (since such rational surface has an infinite number of integral curves with negative self-intersection), see [39, Theorem 4a, page 283] and [16, Theorem 1, page 94]. One of the main interests of having the finite generation of the Cox ring of a $\mathbb{Q}$-factorial higher dimensional projective variety X is that the Minimal Model Program can be carried out from any divisor on X, for more details see [31, Proposition 2.9, page 342].

Our approach to deal with the Cox rings of our Platonic surfaces is to use the geometric criterion given in [9]. It implies that for an extremal smooth projective surface, the finite generation of its Cox ring is equivalent to the finite generation of its effective monoid (see [9, Theorem 21, page 1137]). For the precise meaning of extremal and the effective monoid, see below. We should mention that not every smooth projective rational surface is extremal, see [10, Example 2].

Now, we present two special monoids and introduce the notion of extremal surfaces. For more details about these surfaces, one may take a look at [9] and [10].

Definition 2 Let X be a smooth projective surface.

1. The *characteristic monoid* $Char(X)$ of X is the set of elements x in $NS(X)$ such that there exists an effective divisor on X whose associated complete linear system is base point free and whose class in $NS(X)$ is equal to x.
2. The *monoid of fractional base point free effective classes* $[Char(X) : Nef(X)]$ of X is the set of elements y in $NS(X)$ such that there exists a positive integer n with $ny \in Char(X)$.

Proposition 1 *Let X be a smooth projective surface. The following hold:*

1. *The sets $Char(X)$ and $[Char(X) : Nef(X)]$ are monoids.*
2. $Char(X) \subseteq [Char(X) : Nef(X)] \subseteq Nef(X)$.

Proof See [9, Lemma 12, p. 5]. □

Definition 3 A smooth projective surface X is *extremal* if the following equality $[Char(X) : Nef(X)] \cap M(X) = Nef(X) \cap M(X)$ holds. Here, $Nef(X)$ stands for the nef divisor classes in $NS(X)$.

Our Platonic surfaces are extremal:

Proposition 2 *With notation as above and with the assumption*

$$\left(1 - \frac{1}{x + \sum_{k=1}^{k=x} \rho_k^1} - \frac{1}{y + \sum_{k=1}^{k=y} \rho_k^2} - \frac{1}{z + \sum_{k=1}^{k=z} \rho_k^3}\right) < 0.$$

Then X_2 is extremal.

Proof It is sufficient to show that $|2D|$ is base point free for every nef divisor on X_2. For let D be a nonzero nef divisor on X_2. Since $2D$ is a nonzero nef divisor such that $-K_{X_2} \cdot (2D)$ is greater than one, we obtain, using [27, Theorem III.1, p. 1197], that the associated complete linear system of $2D$ is base point free. Thus, the class of D in $NS(X_2)$ is an element in $[Char(X_2) : Nef(X_2)]$, so we are done. □

Finally, we are able to prove the finite generation of the Cox rings of our Platonic surfaces.

Theorem 3 *With notation as above and with the assumption*

$$\left(1 - \frac{1}{x + \sum_{k=1}^{k=x} \rho_k^1} - \frac{1}{y + \sum_{k=1}^{k=y} \rho_k^2} - \frac{1}{z + \sum_{k=1}^{k=z} \rho_k^3}\right) < 0.$$

Then the Cox ring of X_2 is finitely generated.

Proof Apply Theorem 1, Proposition 2 and [9, Theorem 14, page 5]. □

Acknowledgements We are very grateful to the anonymous Referees for their comments and suggestions regarding this work. This research paper was partially supported by a grant from the group GNSAGA of INdAM, and another one from the Coordinación de Investigación Científica de la Universidad Michoacana de San Nicolás de Hidalgo during 2017 and 2018. De La Rosa-Navarro was supported by "Programa para el Desarrollo Profesional Docente, para el Tipo Superior" under the Grant Number UABC-PTC-558. Frías-Medina acknowledges the financial support of "Fondo Institucional de Fomento Regional para el Desarrollo Científico, Tecnológico y de Innovación", FORDECYT 265667, during 2017.

References

1. Artebani, M., Laface, A.: Cox rings of surfaces and the anticanonical Iitaka dimension. Adv. Math. **226**(6), 5252–5267 (2011)
2. Barth, W., Peters, C., Van de Ven, A.: Compact Complex Surfaces. Springer. Berlin (1984)
3. Batyrev, V., Popov, O.: The Cox ring of a Del Pezzo surface. In: Poonen, B., Tschinkel, Y. (eds.) Arithmetic of Higher-Dimensional Algebraic Varieties, pp. 85–103. Birkhäuser, Boston (2004)

4. Campillo, A., Piltant O., Reguera-López, A.J.: Cones of curves and of line bundles on surfaces associated with curves having one place at infinity. Proc. Lond. Math. Soc. (3) **84**(3), 559–580 (2002)
5. Campillo, A., Piltant, O., Reguera, A.J.: Cones of curves and of line bundles at "infinity". J. Algebra. **293**, 503–542 (2005)
6. Castravet, A.M., Tevelev, J.: Hilbert's 14th problem and Cox rings. Compos. Math. **142**(6), 1479–1498 (2006)
7. Cerda-Rodríguez, J.A., Failla, G., Lahyane, M., Osuna-Castro, O.: Fixed loci of the anticanonical complete linear systems of anticanonical rational surfaces. Balkan J. Geom. Appl. **17**(1), 1–8 (2012)
8. Cox, D.: The homogeneous coordinate ring of a toric variety. J. Algebraic Geom. **4**(1), 17–50 (1995)
9. De La Rosa Navarro, B.L., Frías Medina, J.B., Lahyane, M., Moreno Mejía, I., Osuna Castro, O.: A geometric criterion for the finite generation of the Cox ring of projective surfaces. Rev. Mat. Iberoam. **31**(4), 1131–1140 (2015)
10. De La Rosa-Navarro, B.L., Frías-Medina, J.B., Lahyane, M.: Rational surfaces with finitely generated Cox rings and very high Picard numbers. Rev. R. Acad. Cienc. Exactas Fís. Nat. Ser. A Math. RACSAM. **111**(2), 297–306 (2017)
11. Failla, G., Lahyane, M., Molica Bisci, G.: On the finite generation of the monoid of effective divisor classes on rational surfaces of type (n, m). Atti Accademia Peloritana Pericolanti Cl. Sci. Fis., Mat. Natur. **LXXXIV** (2006). https://doi.org/10.1478/C1A0601001
12. Failla, G., Lahyane, M., Molica Bisci, G.: The finite generation of the monoid of effective divisor classes on Platonic rational surfaces. In: Singularity Theory, pp. 565–576. World Scientific Publication, Hackensack (2007)
13. Failla, G., Lahyane, M., Molica Bisci, G.: Rational surfaces of Kodaira type IV. Boll. Unione Mat. Ital., Sezione B. (8) **10**(3), 741–750 (2007)
14. Galindo, C., Monserrat, F.: On the cone of curves and of line bundles of a rational surface. Int. J. Math. **15**(4), 393–407 (2004)
15. Galindo, C., Monserrat, F.: The cone of curves associated to a plane configuration. Comment. Math. Helv. **80**(1), 75–93 (2005)
16. Galindo, C., Monserrat, F.: The total coordinate ring of a smooth projective surface. J. Algebra. **284**(1), 91–101 (2005)
17. Galindo, C., Monserrat, F.: The cone of curves and the Cox ring of rational surfaces given by divisorial valuations. Adv. Math. **290**, 1040–1061 (2016)
18. Galindo, C., Hernando, F., Monserrat, F.: The log-canonical threshold of a plane curve. Math. Proc. Camb. Philos. Soc. **160**(3), 513–535 (2016)
19. Harbourne, B.: Blowings-up of $\mathbb{P}^2$ and their blowings-down. Duke Math. J. **52**(1), 129–148 (1985)
20. Harbourne, B.: Complete linear systems on rational surfaces. Trans. Am. Math. Soc. **289**(1), 213–226 (1985)
21. Harbourne, B.: The geometry of rational surfaces and Hilbert functions of points in the plane. In: Proceedings of the 1984 Vancouver Conference in Algebraic Geometry, CMS Conference Proceedings, vol. 6, pp. 95–111. American Mathematical Society, Providence (1986)
22. Harbourne, B., Lang, W.E.: Multiple fibers on rational surfaces. Trans. Am. Math. Soc. **307**(1), 205–223 (1988)
23. Harbourne, B.: Hilbert functions of points in good position in $\mathbb{P}^2$. The Curves Seminar at Queen's, vol. VI (Kingston, ON, 1989), Exp. No. G, 8 pp., Queen's Papers in Pure and Applied Mathematics, vol. 83. Queen's University, Kingston (1989)
24. Harbourne, B.: Free resolutions of fat point ideals on $\mathbf{P}^2$. J. Pure Appl. Algebra. **125**(1–3), 213–234 (1998)
25. Harbourne, B.: Points in good position in $\mathbf{P}^2$. Zero-dimensional schemes (Ravello, 1992), pp. 213–229. de Gruyter, Berlin (1994)
26. Harbourne, B.: Rational surfaces with $K^2 > 0$. Proc. Am. Math. Soc. **124**(3), 727–733 (1996)

27. Harbourne, B.: Anticanonical rational surfaces. Trans. Am. Math. Soc. **349** (3), 1191–1208 (1997)
28. Harbourne, B.: Global aspects of the geometry of surfaces. Ann. Univ. Paedagog. Crac. Stud. Math. **9**, 5–41 (2010)
29. Harbourne, B., Miranda, R.: Exceptional curves on rational numerically elliptic surfaces. J. Algebra. **128**, 405–433 (1990)
30. Hartshorne, R.: Algebraic Geometry. Graduate Texts in Mathematics. Springer, New York/Heidelberg (1977)
31. Hu, Y., Keel, S.: Mori dream spaces and GIT. Michigan Math. J. **48**, 331–348 (2000)
32. Laface, A., Velasco, M.: A survey on Cox rings. Geom. Dedicata. **139**, 269–287 (2009)
33. Lahyane, M.: Rational surfaces having only a finite number of exceptional curves. Math. Z. **247** (1), 213–221 (2004)
34. Lahyane, M.: Exceptional curves on smooth rational surfaces with $-K$ not nef and of self-intersection zero. Proc. Am. Math. Soc. **133**, 1593–1599 (2005)
35. Lahyane, M., Harbourne, B.: Irreducibility of -1-classes on anticanonical rational surfaces and finite generation of the effective monoid. Pac. J. Math. **218**(1), 101–114 (2005)
36. Lahyane, M., Failla G., Moreno Mejía, I.: On the vanishing of cohomology of divisors on nonsingular rational surfaces. Int. J. Contemporary Mathematical Sciences **3**(21–24), 1031–1040 (2008)
37. Lahyane, M.: On the finite generation of the effective monoid of rational surfaces. J. Pure Appl. Algebra. **214**, 1217–1240 (2010)
38. Looijenga, E.: Rational surfaces with an anticanonical cycle. Ann. Math. (2). **114**(2), 267–322 (1981)
39. Nagata, M.: On rational surfaces. II. Mem. Coll. Sci. Univ. Tokyo Ser. Math. **33** (2), 271–293 (1960)
40. Miranda, R., Persson, U.: On extremal rational elliptic surfaces. Math. Z. **193**, 537–558 (1986)
41. Monserrat, F.: Lins Neto's examples of foliations and the Mori cone of blow-ups of $\mathbf{P}^2$. Bull. Lond. Math. Soc. **43**, 335–346 (2011)
42. Monserrat, F.: Fibers of pencils of curves on smooth surfaces. Int. J. Math. **22**(10), 1433–1437 (2011)
43. Testa, D., Várilly-Alvarado, A., Velasco, M.: Big rational surfaces. Math. Ann. **351**(1), 95–107 (2011)

Coverings of Rational Ruled Normal Surfaces

Enrique Artal Bartolo, José Ignacio Cogolludo-Agustín, and Jorge Martín-Morales

Abstract In this work we use arithmetic, geometric, and combinatorial techniques to compute the cohomology of Weil divisors of a special class of normal surfaces, the so-called rational ruled toric surfaces. These computations are used to study the topology of cyclic coverings of such surfaces ramified along $\mathbb{Q}$-normal crossing divisors.

1 Introduction

The main purpose of this paper is the study of a special type of rational normal surfaces generalizing Hirzebruch surfaces, that carry a rational fibration. These are called rational ruled toric surfaces (see Sect. 3 for a precise definition).

The first goal is to study the Picard group of these rational surfaces. Note that, while the Picard group of smooth rational surfaces is finitely generated and torsion free, this last property may be lost in some cases for toric ruled surfaces.

The second goal is to study the cohomology of the sheaves $\mathcal{O}_S(D)$ for any Weil divisor D of S.

The final purpose is to apply this in order to calculate the first Betti number of cyclic covers of S with a prescribed ramification divisor D.

This will be done in several steps:

- Compute the Euler characteristic of $\mathcal{O}_S(D)$ (Lemma 3.9). Using the Riemann-Roch formula for normal surfaces (see Blache [2]) one can obtain the Euler characteristic of $\mathcal{O}_S(D)$ in terms of intersection numbers and a correction term coming from the singular locus $\operatorname{Sing}(S)$.

E. Artal Bartolo (✉) · J. I. Cogolludo-Agustín
Departamento de Matemáticas, IUMA, Universidad de Zaragoza, Zaragoza, Spain
e-mail: artal@unizar.es; jicogo@unizar.es

J. Martín-Morales
Centro Universitario de la Defensa-IUMA, Academia General Militar, Zaragoza, Spain
e-mail: jorge@unizar.es; http://cud.unizar.es/martin

G.-M. Greuel et al. (eds.), *Singularities, Algebraic Geometry, Commutative Algebra, and Related Topics*, https://doi.org/10.1007/978-3-319-96827-8_13

- Compute $H^0(S, \mathcal{O}_S(D))$ (Sects. 4 and 5). Using the contributions of Sakai [9] and the theory of weighted blow-ups, we can express these groups as subspaces of 0-cohomology groups for line bundles in $\mathbb{P}^2$, defined by imposing weighted-multiplicity conditions at some points. These computations can be translated into the counting of integer points inside rational polygons, that is, one whose vertices are rational points (as in Sect. 4 (14)).
- Calculate $H^2(S, \mathcal{O}_S(D))$ (Sects. 4 and 5). Using Serre duality, $H^2(S, \mathcal{O}_S(D))$ will be obtained as the dual of $H^0(S, \mathcal{O}_S(K_S - D))$, where K_S is a canonical divisor.
- Calculate $b_1(\tilde{S})$ of a cyclic cover of S (Sect. 6). This is done using Esnault-Viehweg type of formulas adapted to the quotient singularity case.

With these data, complete computations will be carried out in Sect. 4 for biruled surfaces (admitting sections with vanishing self-intersection) and in Sect. 5 for uniruled surfaces.

The main result of our paper in this direction is Theorem 4.2 for biruled surfaces, which states that given any divisor, at most one cohomology group does not vanish. This is still true for uniruled surfaces as shown in Theorems 5.1 and 5.2 when the divisor is located in some special *regions*, as it is well known to be the case for smooth ruled surfaces different from $\mathbb{P}^1 \times \mathbb{P}^1$.

As mentioned above, the main application of this computation is the study of Betti numbers of the n-cyclic coverings of such a surface S, which are ramified at $\mathbb{Q}$-normal crossing divisors of S. This is done by extending classical results of Esnault-Viehweg to the quotient singularity case, where this Betti number computation of the cyclic cover is reduced to the computation of 1-cohomology groups of some divisors associated with the ramification divisor D and to some divisor H such that D and dH are linearly equivalent. This information is finer than the one required in the smooth case. The main subtlety in the normal case comes from the fact that the codimension 1 ramification locus does not determine the codimension 2 ramification.

This is the first step of a more ambitious project, including the extension of Esnault-Viehweg results and their application to more general ruled surfaces having quotient singular points at two disjoint sections. The reason behind these computations is the complete study of the monodromy of Lê-Yomdine singularities where these computations are the key missing points.

The paper is organized as follows. In Sect. 2, we recall basic tools in this theory such as weighted blow-ups, Blache-Sakai theory of normal surfaces, the behavior of Picard groups by birational morphisms, and the Riemann-Roch formula with its correction terms for surfaces with quotient singular points. Section 3 is devoted to analyzing the properties of toric ruled surfaces, including the description of their Picard groups. In Sects. 4 and 5, the cohomology of Weil divisors of toric biruled and uniruled surfaces is studied by using combinatorial and arithmetical tools. The extension of Esnault-Viehweg results is presented in Sect. 6, which ends with some illustrative examples.

During the preparation of this work, we have combined geometric and algebraic techniques in a spirit close to one we believe is characteristic of Antonio Campillo's to whom we dedicate this work.

2 Preliminaries

2.1 *Quotient Singularities and Weighted Blow-Ups*

We will denote by μ_d the group of d-roots of unity in $\mathbb{C}^*$.

Definition 2.1 A complex algebraic surface S has a *cyclic singular point of type* $\frac{1}{d}(a,b)$ at $0 \in S$ if $(S,0)$ is locally is isomorphic to the germ $(\mathbb{C}^2/\mu_d, 0)$ obtained by the action $\zeta \cdot (x,y) := (\zeta^a x, \zeta^b y)$ on $\mathbb{C}^2$.

From now on, $\frac{1}{d}(a,b)$ will denote the germ of the surface at such a singular point.

Remark 2.2 Even though we do not require a priori any condition on the integers d, a, b other than $d > 0$, there is no actual restriction in assuming that $u := \gcd(d,a,b)$ is 1 since $\frac{1}{d}(a,b) = \frac{u}{d}(\frac{a}{u}, \frac{b}{u})$. If $u = 1$, we can also assume that $v := \gcd(a,b)$ is also 1 since $\frac{1}{d}(a,b) = \frac{1}{d}(\frac{a}{v}, \frac{b}{v})$. Finally, note that if $w := \gcd(d,a)$, then the map $(x,y) \mapsto (x, y^w)$ induces an isomorphism $\frac{1}{d}(a,b) \to \frac{w}{d}(\frac{a}{w}, b)$. Hence, we may also assume $\gcd(d,a) = \gcd(d,b) = 1$.

Notation 2.3 *Sometimes it is useful to keep track of the divisors appearing in the definition of* $\frac{1}{d}(a,b)$, *and write* $\frac{1}{d}(a_A, b_B)$, *where* A, B *are Weil divisors whose* local equation in $\mathbb{C}^2$ *are* $x = 0$ *and* $y = 0$, *respectively as in Definition* 2.1.

Example 2.4 The first examples are weighted projective spaces. Let us start with *weighted* projective lines. Let $\omega := (p,q)$ be coprime positive integers. We consider in $\mathbb{C}^2 \setminus \{0\}$ the equivalence relation given by $(x,y) \sim (t^p x, t^q y)$, for $t \in \mathbb{C}^*$. Its quotient is the ω-weighted projective line $\mathbb{P}^1_\omega$ and its elements are denoted by $[x:y]_\omega$. This space can be understood using *charts*. Let

$$\begin{array}{ccc} \mathbb{C}_x \xrightarrow{\sigma_x} \mathbb{P}^1_\omega & \qquad & \mathbb{C}_y \xrightarrow{\sigma_y} \mathbb{P}^1_\omega \\ x \longmapsto [x:1]_\omega & & y \longmapsto [1:y]_\omega \end{array}$$

The images of these maps cover $\mathbb{P}^1_\omega$ but they are not actually charts, and they define charts if we consider the quotients $\mathbb{C}_x/\mu_q$ and $\mathbb{C}_y/\mu_p$. Note that $\mathbb{C}_x/\mu_q \to \mathbb{C}$, $[x] \mapsto x^q$, and $\mathbb{C}_y/\mu_p \to \mathbb{C}$, $[y] \mapsto y^p$, are isomorphisms, and $\mathbb{P}^1_\omega$ is actually $\mathbb{P}^1$.

2.1.1 Weighted Blow-Ups of Smooth Points

Let $P \in S$ be a smooth point, and let A, B two divisors which are normal crossing at P; let us suppose that we have local coordinates such that $A : x = 0$ and $B : y = 0$. Let $\omega := (p, q)$ be coprime positive integers. The (p, q)-weighted blowing-up of P (relative to A, B) is a map $\rho : S_{P,\omega} \to S$, which is an isomorphism outside P and the model near P is as follows:

$$\begin{array}{rcl} \hat{\mathbb{C}}^2_\omega := \{((x, y), [u : v]) \in \mathbb{C}^2 \times \mathbb{P}^1_\omega \mid x^q v^p = y^p u^q\} & \xrightarrow{\rho} & \mathbb{C}^2 \\ ((x, y), [u : v]) & \longmapsto & (x, y) \end{array} \tag{1}$$

Let $E := \rho^{-1}(0)$; it is abstractly isomorphic to $\mathbb{P}^1_\omega \cong \mathbb{P}^1$. Note that $S_{P,\omega}$ may have singular points if either p or q are greater than 1. Let us study for that $\hat{\mathbb{C}}^2_\omega$. It will be covered by two maps

$$\begin{array}{ccc} \mathbb{C}^2_x & \xrightarrow{\tau_x} & \hat{\mathbb{C}}^2_\omega \\ (x, y) & \longmapsto & ((xy^p, y^q), [x : 1]_\omega) \end{array} \qquad \begin{array}{ccc} \mathbb{C}^2_y & \xrightarrow{\tau_y} & \hat{\mathbb{C}}^2_\omega \\ (x, y) & \longmapsto & ((x^p, x^q y), [1 : y]_\omega) \end{array} \tag{2}$$

If we want these maps to be isomorphisms onto the image, we need to replace $\mathbb{C}^2_x$ by $\frac{1}{q}(p_A, -1_E)$ and $\mathbb{C}^2_y$ by $\frac{1}{p}(-1_E, q_B)$. Using the properties shown in [1], we have that $E^2 = -\frac{1}{pq}$. Let us keep the same notation for the strict transforms. Then,

$$A \cdot E = \frac{1}{q}, \quad B \cdot E = \frac{1}{p}, \quad (A \cdot A)_{S_{P,\omega}} = (A \cdot A)_S - \frac{p}{q}, \quad (B \cdot B)_{S_{P,\omega}} = (B \cdot B)_S - \frac{q}{p}.$$

2.1.2 Weighted Blow-Ups of Singular Points: Special Case

Quotient singular points do also admit weighted blow-ups. We present the two examples which will be used later.

Let $P \in S$ be a singular point of type $\frac{1}{d}(p_A, q_B)$, and let A, B two divisors which are $\mathbb{Q}$-normal crossing at P, with *local coordinates* such that $A : x = 0$ and $B : y = 0$, with $\gcd(p, q) = \gcd(d, p) = \gcd(d, q) = 1$. Denote $\omega := (p, q)$; the (p, q)-weighted blowing-up of P (relative to A, B) is a map $\rho : S_{P,\omega} \to S$, which is an isomorphism outside P and the model near P is a quotient of (1). We cover this space with two maps as in (2). Let us consider the origin of $\mathbb{C}^2_x$; besides the action of μ_q, the following commutative diagram holds:

$$
\begin{array}{ccccc}
(x,y) & \longmapsto & & & (xy^p, y^q) \\
\Big\downarrow & & \begin{array}{ccc} \mathbb{C}^2_x & \xrightarrow{\tau_x} & \mathbb{C}^2 \\ \downarrow & & \downarrow \\ \mathbb{C}^2_x & \xrightarrow{\tau_x} & \mathbb{C}^2 \end{array} & & \Big\downarrow \quad \Longrightarrow \left(\begin{array}{c|cc} q & p & -1 \\ d & 0 & 1 \end{array}\right) \cong \frac{1}{q}(p,-d) \\
(x,\zeta_d y) & \longmapsto & & & (x(\zeta_d y)^p, (\zeta_d y)^q)
\end{array}
$$

where $\left(\begin{array}{c|cc} q & p & -1 \\ d & 0 & 1 \end{array}\right)$ represents the quotient of $\mathbb{C}^2$ by the group $\mu_q \times \mu_d$ indicated by the right hand side of the matrix as in Definition 2.1. Hence, in $E := \rho^{-1}(0) \cong \mathbb{P}^1$ we have two points of type $\frac{1}{q}(p_A, -d_E)$ and $\frac{1}{p}(-d_E, q_B)$. The following holds:

$$
\begin{aligned}
E^2 = -\frac{d}{pq} \quad A \cdot E = \frac{1}{q}, \quad B \cdot E = \frac{1}{p}, \quad (A \cdot A)_{S_{P,\omega}} \\
= (A \cdot A)_S - \frac{p}{dq}, \quad (B \cdot B)_{S_{P,\omega}} = (B \cdot B)_S - \frac{q}{dp}.
\end{aligned} \tag{3}
$$

Example 2.5 Let $\omega := (p, q, r)$ be a triple of pairwise coprime positive integer numbers. Let $\mathbb{P}^2_\omega$ be the projective plane associated to ω, i.e. the quotient of $\mathbb{C}^3 \setminus \{0\}$ by the action $t \cdot (x, y, z) := (t^p x, t^q y, t^r z)$, whose elements will be denoted by $[x : y : z]_\omega$. Let $X \subset \mathbb{P}^2_\omega$ be the curve defined by $x = 0$, and define in the same way the curves Y, Z.

Consider the map $\rho : \mathbb{P}^2 \to \mathbb{P}^2_\omega$ given by $\rho([x : y : z]) := [x^p : y^q : z^r]_\omega$ (of degree pqr). Note that X, Y, Z are Weil divisors and

$$
X \cdot Y = \frac{1}{r}, \quad Y \cdot Z = \frac{1}{p}, \quad X \cdot Z = \frac{1}{q}, \quad X^2 = \frac{p}{qr}, \quad Y^2 = \frac{q}{pr}, \quad X^2 = \frac{r}{pq}.
$$

Let us give another way to construct this weighted projective plane. Let us fix $\alpha, \beta \in \mathbb{Z}_{>0}$ such that $pq + r = p\beta + q\alpha$. Such integers exist because of their coprimality and the properties of the semigroup generated by p, q. Let us consider the weighted blow-ups of $\mathbb{P}^2$ with the weights of the left part of Fig. 1. Let S be the surface in the middle which, has four singular points; the self-intersection of the strict transform of Z is given by the choice of α, β. Note that this surface can be obtained from $\mathbb{P}^2_\omega$ using the weighted blow-ups with weights indicated in the right part of Fig. 1. Note that $\frac{1}{p}(1, \alpha) = \frac{1}{p}(q, r)$ and $\frac{1}{q}(1, \beta) = \frac{1}{q}(p, r)$.

2.2 *The Projection Formula*

This section will mainly follow the results presented in [9] for normal surfaces, where proofs can be found. Recall that $\operatorname{div}(S)$ is the free abelian group with basis the

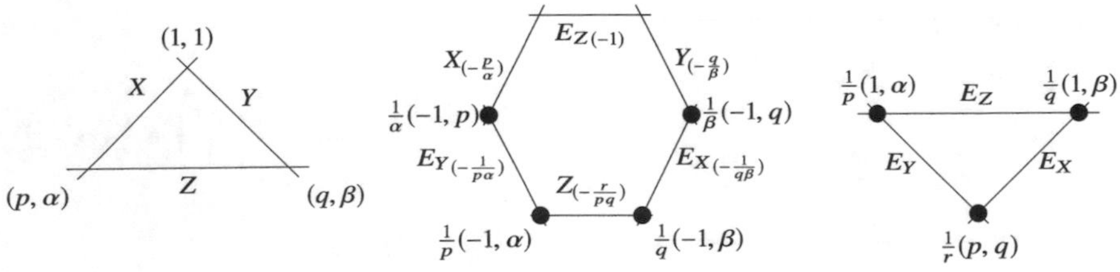

Fig. 1 Birational map from $\mathbb{P}^2$ to $\mathbb{P}^2_\omega$

irreducible projective curves in S, i.e., the irreducible Weil divisors (in the smooth case the concepts of Weil and Cartier divisors coincide). Given a divisor $D \in \operatorname{div}(S)$ on a projective normal surface S and $S_0 := S \setminus \operatorname{Sing}(S)$, where $i : S_0 \hookrightarrow S$ is the inclusion, the divisor D defines a coherent reflexive sheaf of rank one $\mathcal{O}_S(D) = i_*(\mathcal{O}_{S_0}(D|_{S_0}))$ which can be given as $\pi_*(\mathcal{O}_{\bar{S}}(\bar{D}))$, where $\bar{D}$ is the strict transform of D by a resolution π; let $\mathcal{E}$ be the *exceptional locus*, that is, the collection of all the irreducible exceptional divisors of π. By definition, for any $\mathbb{Q}$-divisor $D = \sum_{i=1}^n a_i D_i \in \operatorname{div}_{\mathbb{Q}}(S) = \mathbb{Q} \otimes_{\mathbb{Z}} \operatorname{div}(S)$ we consider $\mathcal{O}_S(D) := \mathcal{O}_S(\lfloor D \rfloor)$, where $\lfloor D \rfloor = \sum \lfloor a_i \rfloor D_i$. The definition of $\pi^* D$ is given as

$$\pi^* D = \bar{D} + \sum_{E \in \mathcal{E}} m_E E$$

where $\bar{D}$ is the strict transform of D and $m_E \in \mathbb{Q}$ are rational numbers satisfying $\bar{D} \cdot E' + \sum_{E' \in \mathcal{E}} m_E E \cdot E' = 0$. Since the intersection matrix $M_{\mathcal{E}} = (E \cdot E')_{E, E' \in \mathcal{E}}$ is negative definite, the m_E's are unique. One has the following generalization of the projection formula.

Proposition 2.6 (Projection formula [9, Thm. 2.1]) *If* $D \in \operatorname{div}_{\mathbb{Q}}(S)$ *and* $\pi : \bar{S} \to S$ *a resolution of singularities of* S*, then*

$$\pi_*(\mathcal{O}_{\bar{S}}(\pi^* D)) \cong \mathcal{O}_S(D).$$

Proposition 2.7 *Let* S *be a normal surface and* $D \in \operatorname{div}_{\mathbb{Q}}(S)$*. Then, the cohomology group* $H^0(S, \mathcal{O}_S(D))$ *can be identified with* $\{h \in \mathbb{C}(S) \mid (h) + D \geq 0\}$.

Proof The statement is true for an integral divisor on a smooth variety. Let $\pi : \bar{S} \to S$ be a resolution of S. Then, by Proposition 2.6 (the projection formula) one has $\mathcal{O}_S(D) = \pi_* \mathcal{O}_{\bar{S}}(\pi^* D) = \pi_* \mathcal{O}_{\bar{S}}(\lfloor \pi^* D \rfloor)$ and thus $H^0(S, \mathcal{O}_S(D)) = H^0(\bar{S}, \mathcal{O}_{\bar{S}}(\lfloor \pi^* D \rfloor))$. Since the latter is an integral divisor in a smooth variety,

$$H^0(\bar{S}, \mathcal{O}_{\bar{S}}(\lfloor \pi^* D \rfloor)) = \{h \in \mathbb{C}(\bar{S}) \mid (h) + \lfloor \pi^* D \rfloor \geq 0\}.$$

The condition $(h) + \lfloor \pi^* D \rfloor \geq 0$ is equivalent to $(h) + \pi^* D \geq 0$ because (h) is an integral divisor. Writing $h = \pi^* h_1$ where $h_1 \in \mathbb{C}(S)$, one can split the previous condition $(\pi^*h_1) + \pi^*D \geq 0$ into two different ones,

$$(h_1) + D \geq 0 \quad \text{and} \quad \operatorname{ord}_E\left((\pi^*h_1) + \pi^*D\right) \geq 0, \tag{4}$$

for all $E \in \mathcal{E}$.

Note that if $A = (a_{ij})_{i,j}$ is a negative definite real matrix such that $a_{ij} \geq 0$, $i \neq j$, then $-A^{-1}$ has all non-negative entries. This implies that the pull-back of an effective divisor is also an effective divisor. Hence one can easily observe that second condition in (4) is implied by the first one and the proof is complete. □

In more generality, given $\varphi : \Sigma \to S$ a birational morphism between normal surfaces and $\mathcal{E}$ its exceptional locus, one can define $\varphi_*(C)$ for any given irreducible divisor $C \in \operatorname{div}(\Sigma)$ as 0 if $C \in \mathcal{E}$ or $\varphi(C)$ otherwise. This can be extended by linearity to $\operatorname{div}_{\mathbb{Q}}(\Sigma)$. Also, if $D \in \operatorname{div}_{\mathbb{Q}}(S)$, then $\varphi^*(D)$ is defined as in the case of a resolution π since for φ the intersection matrix of exceptional irreducible divisors is still negative definite. One has the following generalization of the projection formula.

Proposition 2.8 [9, Thm. 6.2]) *If $D \subset \operatorname{div}_{\mathbb{Q}}(S)$ and $\varphi : \Sigma \to S$ is a birational morphism between normal surfaces and Z an effective divisor supported on the exceptional locus $\mathcal{E}$ of φ, then*

$$\varphi_*(\mathcal{O}_\Sigma(\varphi^*D + Z)) \cong \mathcal{O}_S(D).$$

2.3 *Picard Group for Normal Surfaces*

The Picard group $\operatorname{Pic}(S)$ of a smooth projective surface S is the quotient of $\operatorname{div}(S)$ by the subgroup of linear divisors, i.e. the divisors of the non-zero meromorphic functions on S. The following exact sequence holds:

$$0 \to \mathbb{C}^* \to \mathbb{C}(S)^* \to \operatorname{div}(S) \to \operatorname{Pic}(S) \to 0.$$

Two divisors $D_1, D_2 \in \operatorname{div}(S)$ that define the same class in $\operatorname{Pic}(S)$ are called *linearly equivalent* and denoted $D_1 \sim D_2$. Linear equivalence can be extended to $\operatorname{div}_{\mathbb{Q}}(S)$ as follows: $D_1, D_2 \in \operatorname{div}_{\mathbb{Q}}(S)$ are linearly equivalent if $D_1 - D_2 \in \operatorname{div}(S)$ and $D_1 - D_2 \sim 0$. Analogously, two divisors $D_1, D_2 \in \operatorname{div}(S)$ are called *numerically equivalent* and denoted $D_1 \overset{\text{num}}{\sim} D_2$ if $D_1 \cdot C = D_2 \cdot C$ for any $C \in \operatorname{div}(S)$. Numerical equivalence can also be extended to $\mathbb{Q}$-divisors.

Let $\mathcal{D}$ be a subset of Weil divisors of S (a normal projective surface) and let $\operatorname{div}(S, \mathcal{D})$ be the free abelian subgroup of $\operatorname{div}(S)$ with basis $\mathcal{D}$. Let $\mathbb{C}_{\mathcal{D}}(S)^*$ be the multiplicative subgroup of functions h such that $(h) \in \operatorname{div}(S, \mathcal{D})$. It is clear that the cokernel of the divisor map $\mathbb{C}_{\mathcal{D}}(S)^* \to \operatorname{div}(S, \mathcal{D})$ is a subgroup of $\operatorname{Pic}(S)$. In particular, if the images of $\mathcal{D}$ in $\operatorname{Pic}(S)$ form a generating system, then the divisors of a generating system of $\mathbb{C}_{\mathcal{D}}(S)^*$ are a complete system of relations for

$\operatorname{Pic}(S)$; conversely, given a generating system $\mathcal{D}$ and a complete system of relations for $\operatorname{Pic}(S)$, the functions coming from these relations are a generating system of $\mathbb{C}_{\mathcal{D}}(S)^*$.

Example 2.9 Let $S = \mathbb{P}^2$; the irreducible divisors are the irreducible plane curves. Let H be any line; for any curve C_d of degree d, it is well known that $dH - C_d$ is a linear divisor, i.e., H generates $\operatorname{Pic}(\mathbb{P}^2)$. Since $K_H^*(\mathbb{P}^2) = \mathbb{C}^*$, this implies $\operatorname{Pic}(\mathbb{P}^2) = \mathbb{Z}H$ as it is well known. With the same ideas $\operatorname{Pic}(\mathbb{P}^1 \times \mathbb{P}^1)$ is isomorphic to $\mathbb{Z}^2$ generated by the factors.

Proposition 2.10 *Let $\varphi : \Sigma \to S$ be a birational morphism; let $\mathcal{D}$ be a set of divisors of S generating* $\operatorname{Pic}(S)$ *and let $\mathcal{R}$ be a set of relations such that* $\operatorname{Pic}(S) \cong \operatorname{div}(S, \mathcal{D})/\langle\mathcal{R}\rangle$. *For each $R \in \mathcal{R}$, let $h_R \in K_{\mathcal{D}}^*(S) \subset K^*(S)$ be a function such that $R = (h_R)$. Let $\tilde{h}_R = \varphi^* h_R \in K^*(\Sigma)$. Then, if $\mathcal{E}$ is the set of exceptional locus of φ and $\tilde{\mathcal{D}}$ is the set of strict transforms of the divisors in $\mathcal{D}$, then*

$$\operatorname{Pic}(\Sigma) \cong \operatorname{div}(\Sigma, \mathcal{E} \cup \tilde{\mathcal{D}})/\langle(\tilde{h}_R) \mid R \in \mathcal{R}\rangle$$

Proof Let us consider the natural map $\operatorname{div}(\Sigma, \mathcal{E} \cup \tilde{\mathcal{D}}) \to \operatorname{Pic}(\Sigma)$. Let us consider $D \in \operatorname{div}(\Sigma)$. Note that $D = \varphi^*\varphi_* D + E_D$ where $E_D \in \operatorname{div}(\Sigma, \mathcal{E})$. There is a divisor $D_1 \in \operatorname{div}(S, \mathcal{D})$ such that $\varphi_* D \sim D_1$ and as a consequence $D \sim \varphi^* D_1 + \tilde{E}$ where $\tilde{E} \in \operatorname{div}(\Sigma, \mathcal{E})$ and $\varphi^* D_1 \in \operatorname{div}(\Sigma, \tilde{E})$. Hence the above natural map is surjective.

The relations are given by functions in $\mathbb{C}(\Sigma)^*$ whose divisors are in $\operatorname{div}(\Sigma, \mathcal{E} \cup \tilde{\mathcal{D}})$ and those ones are generated by $h_R = \varphi^* h_R$, $R \in \mathcal{R}$. □

Remark 2.11 Let $R = \sum_{D \in \mathcal{D}} m_D D$; let us denote $\tilde{D}$ its strict transform. Then

$$(\tilde{h}_R) = \sum_{D \in \mathcal{D}} m_D \varphi^*(D) = \sum_{D \in \mathcal{D}} m_D \tilde{D} + \sum_{E \in \mathcal{E}} m_E E;$$

the coefficients m_E are integers, while it is possible that the coefficient of E in each particular $\varphi^*(D)$ is non integer.

The proposition above has a simple converse.

Proposition 2.12 *Let $\varphi : \Sigma \to S$ be a composition of weighted blow-ups; let $\mathcal{E}$ be the exceptional locus of φ and let $\mathcal{D}$ be a set of divisors of Σ generating* $\operatorname{Pic}(S)$ *and containing $\mathcal{E}$; let $\mathcal{R}$ be a set of relations such that* $\operatorname{Pic}(\Sigma) \cong \operatorname{div}(\Sigma, \mathcal{D})/\langle\mathcal{R}\rangle$. *Let $\check{\mathcal{D}}$ be the set of φ_*-images of the elements of $\mathcal{D} \setminus \mathcal{E}$. Let us express $R \in \mathcal{R}$ as $R = \sum_{D \in \mathcal{D}} m_D(R) D$. Then*

$$\operatorname{Pic}(S) \cong \operatorname{div}(S, \check{\mathcal{D}}) \Big/ \left\langle \sum_{D \in \mathcal{D} \setminus \mathcal{E}} m_D(R) \pi_*(D) \,\middle|\, R \in \mathcal{R} \right\rangle$$

Proof It is easily seen that $D_1, D_2 \in \operatorname{div}(\Sigma)$ such that $D_1 \sim D_2$ then $\varphi_* D_1 \sim \varphi_* D_2$. Hence, it is clear that $\operatorname{div}(S, \check{\mathcal{D}})$ generates $\operatorname{Pic}(S)$. Note also that $\mathbb{C}_{\mathcal{D}}(\Sigma)^*$ equals $\mathbb{C}_{\check{\mathcal{D}}}(S)^*$, for we need $\mathcal{E} \subset \mathcal{D}$. □

Remark 2.13 The condition $\mathcal{E} \subset \mathcal{D}$ is essential in the hypotheses of the proposition; without it, the result does not hold.

Example 2.14 With the notations of Example 2.5, we can express

$$\operatorname{Pic}(\mathbb{P}^2) = \langle X, Y, Z \mid X = Y = Z \rangle = \langle X, Y, Z \mid X = Y, qX = qZ, pY = pZ \rangle$$

From Proposition 2.10

$$\begin{gathered}\operatorname{Pic}(S) = \langle X, Y, Z, E_X, E_Y, E_Z \mid X + E_Z + pE_Y = Y + E_Z + qE_X, \\ qX + qE_Z + pqE_Y = qZ + q\alpha E_Y + q\beta E_X, pY + pE_Z + pqE_X = pZ + p\alpha E_Y + p\beta E_X \rangle = \\ \langle X, Y, Z, E_X, E_Y, E_Z \mid X + pE_Y = Y + qE_X, qX + qE_Z + p\beta E_Y = qZ + rE_Y + q\beta E_X, \\ pY + pE_Z + q\alpha E_X = pZ + p\alpha E_Y + rE_X \rangle.\end{gathered}$$

With Proposition 2.12 we obtain

$$\operatorname{Pic}(\mathbb{P}^2_\omega) = \langle E_X, E_Y, E_Z \mid pE_Y = qE_X, qE_Z = rE_Y, pE_Z = rE_X \rangle$$

which is the standard presentation of $\operatorname{Pic}(\mathbb{P}^2_\omega)$.

2.4 The Canonical Cycle for Q-Resolutions and the Riemann-Roch Formula

A canonical cycle K_S on a normal projective surface S can be obtained as the direct image of a canonical cycle in a resolution of S; in particular, $K_S \in \operatorname{div}(S)$: Analogously to the smooth case, a canonical cycle on a surface S with quotient singularities obtained after blowing-up satisfies the adjunction formula (c.f. [4, p. 716])

$$-K_S \cdot C = C^2 + \chi^{\mathrm{orb}}(C), \tag{5}$$

where $\chi^{\mathrm{orb}}(C)$ is the orbifold Euler characteristic of a *smooth* irreducible divisor C, which in our case is simply,

$$\chi^{\mathrm{orb}}(C) = 2 - 2g(C) + \sum_{\substack{P \in C \\ S_P = \frac{1}{d_P}(1,\cdot)}} \left(\frac{1}{d_P} - 1\right),$$

the point P runs on all singular points of the surface on C and d_P denotes its order as a quotient singularity. Sometimes we need only the numerical properties of a canonical divisor. This is way a *canonical cycle* can be defined as any cycle $Z_K \in \operatorname{div}_{\mathbb{Q}}(S)$ satisfying the numerical conditions (5), that is, $Z_K \cdot C = C^2 + \chi^{\text{orb}}(C)$ for all *smooth* irreducible divisors C. Note that, even if a canonical cycle had integer coefficients, it may not be an anti-canonical divisor. The anti-canonical divisor is numerically equivalent to the canonical cycle $-K_S \overset{\text{num}}{\sim} Z_K$.

A useful tool to calculate $h^i(\mathcal{O}_S(D)) := h^i(S, \mathcal{O}_S(D))$ is the Riemann-Roch formula for normal surfaces

$$\chi(S, \mathcal{O}_S(D)) - \chi(S, \mathcal{O}_S) = \frac{1}{2} D(D + Z_K) - \Delta_S(-D), \tag{6}$$

where $\Delta_S(-D)$ is a correction term that deserves some special attention. This tool is combined with Serre Duality [2, Theorem 3.1], which implies that $h^0(S, \mathcal{O}_S(D)) = h^2(S, \mathcal{O}_S(K_S - D))$.

Example 2.15 Consider $S = \mathbb{P}^2_w$, the weighted projective plane of weight $w = (w_0, w_1, w_2)$. The canonical divisor $K_S = -X_0 - X_1 - X_2$ has degree $-|w|$ for $|w| := w_0 + w_1 + w_2$ – see [6]. If $-|w| < k < 0$, then $\chi_S(k) := \chi(S, \mathcal{O}_S(k)) = 0$ and (5) becomes

$$\sum_{\{i_0,i_1,i_2\}=\{0,1,2\}} \Delta_{\frac{1}{w_{i_0}}(w_{i_1}, w_{i_2})}(k + |w|) = 1 + \frac{k(k + |w|)}{2\bar{w}} =: g_{w,k}, \tag{7}$$

where $\bar{w} = w_0 w_1 w_2$.

2.5 The Correction Term Δ_S

The correction term $\Delta_S(D)$ of the Riemann-Roch formula for normal surfaces (6) has been considered in different contexts (cf. [2, 3, 8]). Here we will consider the quotient surface singularity case, where $\Delta_S(D) = \sum_{P \in \operatorname{Sing} S} \Delta_P(D)$ is the sum of locally defined invariants at the singular points of S of local type $\mathbb{C}^2/G$ for a finite group G. More explicitly, $\Delta_P(D)$ is a rational map

$$\Delta_P(D) : \operatorname{Weil}_P(S)/\operatorname{Cart}_P(S) \longrightarrow \mathbb{Q}$$

defined on the local Weil divisors (formal finite combinations of local irreducible curves) vanishing on the local Cartier divisors (associated with principal ideals). This map is explicitly described in [5] for cyclic surface singularities. The quotient $\operatorname{Weil}_P(S)/\operatorname{Cart}_P(S)$ is isomorphic to the (multiplicative) group $\operatorname{Hom}(G, \mathbb{C}^*)$ of characters on G, which in the cyclic case is the group of d-roots of unity, for $d := |G|$. Once a choice of this root of unity ζ_d is given, then we can use the notation $\Delta_P(k)$ to describe $\Delta_P(\zeta^k)$ and the standard notation $S_P = \frac{1}{d}(1, p)$ for a local cyclic surface singularity $\mathbb{C}^2/\mu_d$ given by the action $\mu_d(x, y) = (\zeta_d x, \zeta_d^p y)$.

Proposition 2.16 *Let $k \in \mathbb{Z}$ and $p, q \in \mathbb{Z}_{\geq 0}$ with $d = p + q > 0$ and $r, s \in \{0, 1\}$. Then*

$$\Delta_{\frac{1}{d}(1,p)}(k - (r-1)p) + \Delta_{\frac{1}{d}(1,q)}(k - (s-1)q) = \frac{(1-d)(r+s-2)}{2d} + \left\{\frac{-k}{d}\right\}(r+s-1) \tag{8}$$

In particular,

(1) $\Delta_{\frac{1}{d}(1,p)}(k) + \Delta_{\frac{1}{d}(1,q)}(k+q) = \frac{d-1}{2d}$,
(2) $\Delta_{\frac{1}{d}(1,p)}(k) + \Delta_{\frac{1}{d}(1,q)}(k) = \{\frac{-k}{d}\}$.

Proof Note that k can be considered modulo d and hence it is enough to show the result for $k = 1, \ldots, d$. We will use formula (7) repeatedly for the projective planes $S_i = \mathbb{P}^2_{w_i}$ for $w_1 = (d, 1, p)$, $w_2 = (d, 1, q)$, and $w_3 = (1, p, q)$. Note that

$$0 = \chi_{S_1}(k - 1 - d - rp) = g_{w_1,k-1-d-rp} - \Delta_{\frac{1}{d}(1,p)}(k - (r-1)p) - \Delta_{\frac{1}{p}(1,q)}(k)$$

$$0 = \chi_{S_2}(k - 1 - d - sq) = g_{w_2,k-1-d-sq} - \Delta_{\frac{1}{d}(1,q)}(k - (s-1)q) - \Delta_{\frac{1}{q}(1,p)}(k)$$

$$0 = \chi_{S_3}(k - 1 - d) = g_{w_3,k-1-d} - \Delta_{\frac{1}{p}(1,q)}(k) - \Delta_{\frac{1}{q}(1,p)}(k).$$

Subtracting the third equation from the sum of the first two, one obtains

$$g_{w_1,k-1-d-rp} + g_{w_2,k-1-d-sq} - g_{w_3,k-1-d} = \Delta_{\frac{1}{d}(1,p)}(k-(r-1)p) + \Delta_{\frac{1}{d}(1,q)}(k-(s-1)q).$$

Also note that

$$\begin{aligned} &g_{w_1,k-1-d-rp} + g_{w_2,k-1-d-sq} - g_{w_3,k-1-d} \\ &= \frac{pr^2 + qs^2 + (q - 2k + 1)r + (p - 2k + 1)s + 2k - 2}{2d} \\ &= \frac{(d - 2k + 1)(r + s) + 2k - 2}{2d} \end{aligned}$$

since $r^2 = r$ and $s^2 = s$.

Note that this equals the right-hand side of (8) in case $k \in \{1, \ldots, d\}$. The result follows for general $k \in \mathbb{Z}$ using that $-\frac{\bar{k}}{d} = \left\{-\frac{k}{d}\right\} - 1$, where $\bar{k}$ denotes the remainder class of k in $\{1, \ldots, d\}$. □

Example 2.17 In the particular case $k = 0$, notice that $\Delta_{\frac{1}{d}(1,p)}(p) = \frac{d-1}{2d}$, since $\Delta_{\frac{1}{d}(1,q)}(0) = 0$. Multiplying by $\bar{p}$ (which amounts to choosing a different primitive d-root of unity) one obtains $\Delta_{\frac{1}{d}(\bar{p},1)}(1) = \Delta_{\frac{1}{d}(1,\bar{p})}(1) = \frac{d-1}{2d}$. Since this is true for any p prime with d, one obtains $\Delta_{\frac{1}{d}(1,p)}(1) = \frac{d-1}{2d}$.

In particular, if $d = 2$, then $\Delta_{\frac{1}{2}(1,1)}(0) = 0$ and $\Delta_{\frac{1}{2}(1,1)}(1) = \frac{1}{4}$.

3 Rational Ruled Toric Surfaces

Following [10] a *rational ruled fibration* over a normal projective surface S is a surjective morphism $\pi : S \to \mathbb{P}^1$ whose generic fiber is isomorphic to $\mathbb{P}^1$. Moreover, the fibration is called *minimal* if its fibers are irreducible. In this spirit one can define

Definition 3.1 A *rational ruled toric surface* is a projective normal surface S with quotient singularities with and a minimal rational ruled fibration $\pi : S \to \mathbb{P}^1$ with two marked fibers F_1, F_2 and two marked disjoint sections S_1, S_2. The toric structure of S comes from the two-dimensional torus (projecting onto $\mathbb{P}^1 \setminus \{\pi(F_1), \pi(F_2)\} = \mathbb{C}^*$ with $\mathbb{C}^*$-fiber given by $\pi^{-1}(P) \setminus \{S_1 \cap \pi^{-1}(P), S_2 \cap \pi^{-1}(P)\}$), four 1-dimensional tori (the two fibers and the two sections outside their intersection) and the 4 vertices $F_i \cap S_j$ (containing the singularities of S).

The following statement is an immediate consequence of Theorem 1.2 in [10].

Lemma 3.2 *Let $C \subset S$ be a a singular fiber of a rational ruled toric surface. Then $\#(C \cap \operatorname{Sing} S) = 2$ and the two points are of type $\frac{1}{d}(1, m)$ and $\frac{1}{d}(1, -m)$.*

If the surface S is smooth, we are considering a Hirzebruch surface Σ_n, two fibers, the negative self-intersection section (if $n > 0$) and a section with self-intersection n (two null self-intersections if $n = 0$).

Example 3.3 Let us fix three lines A, B, C in general position in $\mathbb{P}^2$, and let $p := A \cap B$. Let $\pi : \Sigma_1 \to \mathbb{P}^2$ be the blowing-up of p. Then, the projection from p induces a rational ruled toric surface with the fibers A, B and the sections C, E (where E is the exceptional divisor of π).

Example 3.4 The above example can be easily generalized. Keep the notation of Example 2.5.

If $p_z := X \cap Y$, then $(\mathbb{P}^2_\omega, p_z)$ is a singular point of type $\frac{1}{r}(p_X, q_Y)$. The (p, q)-weighted blow-up of p_z is a map $\rho : \Sigma \to \mathbb{P}^2_\omega$ which is a rational ruled toric surface for the fibers X, Y and the sections Z, E; there are two singular points on E of type

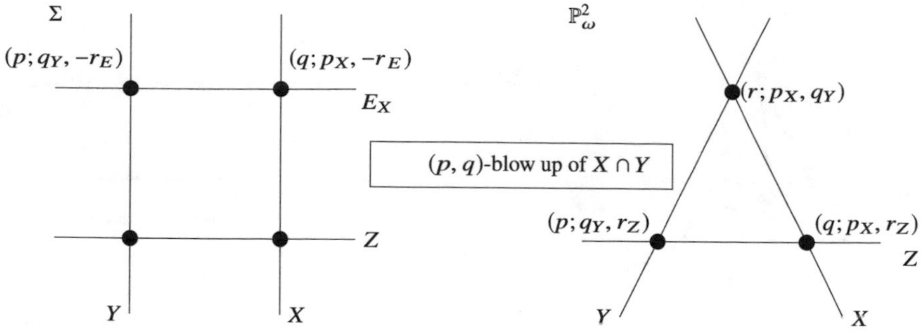

Fig. 2 Weighted blow-up in $\mathbb{P}^2_\omega$

$\frac{1}{q}(p_X, -r_E)$ and $\frac{1}{p}(q_Y, -r_E)$ (Fig. 2). The following holds:

$$E^2 = -\frac{r}{pq} = -Z^2 \quad X^2 = Y^2 = 0.$$

Note the symmetries in the singulars point lying on a fiber.

Proposition 3.5 *Let S be a rational ruled toric surface. Let S_+, S_- be the sections and F_1, F_2 be the fibers.*

(1) *The types of the singular points $F_i \cap S_\pm$ are of the form $\frac{1}{d_i}(1, \pm n_i)$ corresponding to the equations of $S_\pm$ and F_i respectively.*
(2) $S_+^2 = -S_-^2 =: r$.
(3) $\frac{n_1}{d_1} + \frac{n_2}{d_2} - r \in \mathbb{Z}$.

Proof The first statement is a direct consequence of Lemma 3.2. □

Theorem 3.6 *For any $r \in \mathbb{Q}_{\geq 0}$, $0 \leq n_i < d_i$ ($d_i \in \mathbb{N}$, $n_i \in \mathbb{Z}_{\geq 0}$, $\gcd(n_i, d_i) = 1$) there is a unique rational ruled toric surface with these invariants.*

For the proof, we are going to introduce the concept of weighted Nagata transformations.

3.1 Weighted Nagata Transformations

Let us consider Σ a smooth ruled surface (with base $\mathbb{P}^1$ for simplicity) and let $P \in \Sigma$. The Nagata transformation of Σ based on P is a birational map constructed as follows. First, we blow up the point P to obtain a surface $\hat{\Sigma}$; let $F \subset \Sigma$ the fiber containing P; note that $(F \cdot F)_\Sigma = 0$ and hence $(F \cdot F)_{\hat{\Sigma}} = -1$. By Castelnuovo criterion, we can contract F to obtain a new surface $\tilde{\Sigma}$ which is also ruled. Let us assume that $\Sigma \cong \Sigma_n$, $n \geq 0$; if P belongs to a section with non-positive self-intersection then $\tilde{\Sigma} \cong \Sigma_{n+1}$; such a section is unique if $n > 0$ and the condition

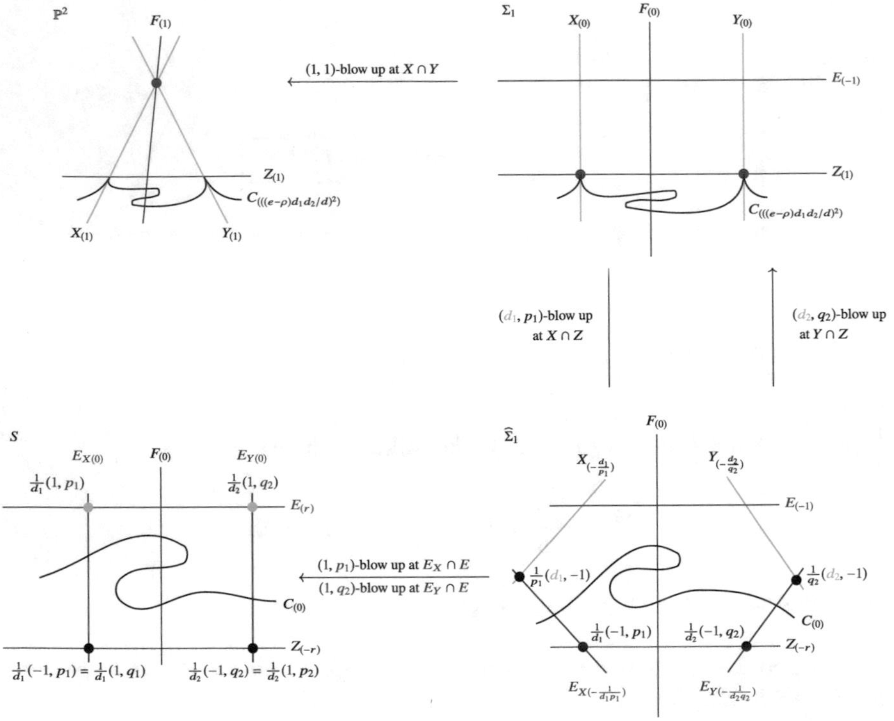

Fig. 3 Construction of $S_{r,(d_1,p_1),(d_2,q_2)}$

is always satisfied if $n = 0$. If the condition is not satisfied, then $\tilde{\Sigma} \cong \Sigma_{n-1}$. We are going to generalize this notion; the generalization will work for singular ruled surfaces and it will allow for smooth ones changes $n \to n \pm k$, with $k > 1$.

3.2 Construction of the Surfaces $S_{(d_1,d_2,p_1,q_2,r)}$

One can produce these surfaces starting from the Hirzebruch surface Σ_1 and after performing two weighted Nagata transformations as follows (see Fig. 3).

As a brief explanation of Fig. 3, we have considered two positive integers $p_1, q_2 \in \mathbb{Z}_{>0}$ such that

$$\frac{p_1}{d_1} + \frac{q_2}{d_2} - r = 1. \tag{9}$$

Hence,

$$d_2 p_1 + d_1 q_2 = d_1 d_2 + r d_1 d_2 \geq d_1 d_2. \tag{10}$$

In order to find p_1 and q_2 let us start with the equality $\frac{n_1}{d_1} + \frac{n_2}{d_2} - r = 1 - k \in \mathbb{Z}$ from Proposition 3.5 where $0 < n_i < d_i$ and $r \geq 0$. Under these conditions $k \geq 0$. Then, for instance, $p_1 := n_1$ and $q_2 := n_2 + kd_2 > 0$ satisfy (9).

The purpose is to construct the rational ruled toric surface S associated with d_i, n_i, r in Proposition 3.5. Self-intersection of divisors is shown in parenthesis. Also, in order to simplify notation, each blow-up and each singular point are color coded in order to specify the direction. For instance, $\frac{1}{d}(a, b)$ represents a cyclic singular point of order d whose action in local coordinates is given by $(x, y) \mapsto (\zeta^a x, \zeta^b)$ where $\{x = 0\}$ (resp. $\{y = 0\}$) is the local equation the divisor colored in red (resp. blue) – see Notation 2.3. Analogously, a weighted blow up of type (a, b) means *relative* to A, B where A (resp. B) is the divisor colored in red (resp. blue). The resulting self-intersections after blow-ups and the types of the singular points on exceptional divisors are a consequence of the discussion in Sect. 2.1.2 and the intersections formulas summarized in (3).

Remark 3.7 If $r = \frac{\rho}{e}$ and $\frac{d_1}{d_2} = \frac{d_1'}{d_2'}$ irreducible fractions, where $d_i' = \frac{d_i}{d}$ and $d := \gcd(d_1, d_2)$, then $d_1'd_2'|e$. If $r \in \mathbb{Z}$, then $d = d_1 = d_2$ and $n_1 + n_2 = d$ and S can be obtained as the quotient by μ_d of a smooth ruled surface which is $\mathbb{P}^1 \times \mathbb{P}^1$ if $r = 0$; the action of μ_d in an affine chart isomorphic to $\mathbb{C}^2$ is given by $\zeta \cdot (x, y) := (\zeta x, \zeta^{n_1} y)$. When $r = 0$ the surface will be called *biruled* (since two rulings exist); otherwise the surface is called *uniruled*.

Even though it is not necessary for the construction of S, note that the strict transform of the rational curve $C = \{x^{ep_1d_2'}y^{eq_2d_1'} - z^{(e-\rho)d_1d_2/d}\} \subset \mathbb{P}^2$ – see (10) and Remark 3.7 – is a multi-section of the ruling with zero self-intersection and intersecting the smooth fiber F at $(e - \rho)d_1d_2/d$ points.

Remark 3.8 Conceptually, the surface $S_{(d,p)}$ can also be obtained as the quotient of $\mathbb{P}^1 \times \mathbb{P}^1$ by the cyclic action of the multiplicative group μ_d as $\zeta \cdot ([x : y], [s : t]) = ([x : \zeta^p y], [s : \zeta^q t])$. However, the construction presented here will be more convenient for calculation purposes.

Proof of Theorem 3.6 Let us start with $\mathbb{P}^1 \times \mathbb{P}^1$. A sequence of weighted Nagata transformations yield the desired ruled toric surface. Moreover, from any ruled toric surface, the *inverse* sequence of Nagata transformations produces $\mathbb{P}^1 \times \mathbb{P}^1$. □

3.3 The Picard Group of S

Recall that the Picard group $\operatorname{Pic}(S)$ of a surface S is defined as the divisor class group factored out by linear equivalence, that is, $D_1 \sim D_2$ if $D_1 - D_2 = \operatorname{supp}(f)$ for a rational global function $f \in \mathbb{C}(S)$. Given a list of invariants $I = (d_1, d_2, p_1, q_2, r)$ we will follow Sect. 3.2 to construct the rational ruled toric S_I and use Propositions 2.10 and 2.12 to describe a presentation of the Picard groups of S_I and all the intermediate surfaces obtained after each blow-up.

$$
\begin{aligned}
\operatorname{Pic}(\mathbb{P}^2) &= \langle X, Y, Z, F : X = Y = Z = F \rangle \cong \mathbb{Z}Z.\\
\operatorname{Pic}(\Sigma_1) &= \langle X, Y, Z, F, E : X + E = Y + E = Z = F + E \rangle \cong \mathbb{Z}Z \times \mathbb{Z}E.\\
\operatorname{Pic}(\widehat{\Sigma}_1) &= \left\langle X, E_X, Y, E_Y, Z, F, E : \begin{cases} X + d_1 E_X = Y + d_2 E_Y = F, \\ Z + p_1 E_X + q_2 E_Y = F + E \end{cases} \right\rangle \\
&\cong \mathbb{Z}Z \times \mathbb{Z}E \times \mathbb{Z}E_X \times \mathbb{Z}E_Y.\\
\operatorname{Pic}(S) &= \left\langle E_X, E_Y, Z, F, E : \begin{cases} d_1 E_X = d_2 E_Y = F, \\ Z + p_1 E_X + q_2 E_Y = F + E \end{cases} \right\rangle \\
&\cong \langle Z, F, E_X, E_Y \mid F = d_1 E_X = d_2 E_Y \rangle \cong \mathbb{Z}\, Z \times \mathbb{Z}\, E_X \times \tfrac{\mathbb{Z}}{d\mathbb{Z}}\, T,
\end{aligned}
\tag{11}
$$

where $T = \frac{d_1}{d} E_X - \frac{d_2}{d} E_Y$ and $d := \gcd(d_1, d_2)$. The intersection matrix with respect to the generating system $\{Z, F, E_X, E_Y\}$ is given as

$$
M_S = \begin{bmatrix} r & 1 & \frac{1}{d_1} & \frac{1}{d_2} \\ 1 & 0 & 0 & 0 \\ \frac{1}{d_1} & 0 & 0 & 0 \\ \frac{1}{d_2} & 0 & 0 & 0 \end{bmatrix}
$$

and the kernel of this bilinear form is generated by $F - d_1 E_X$, $F - d_2 E_Y$, and the torsion divisor T which is also numerically equivalent to zero. In particular $\frac{d_1}{d} E_X$ and $\frac{d_2}{d} E_Y$ are numerically equivalent, however they are not linearly equivalent unless $d = 1$.

Moreover, let Γ be the free subgroup generated by Z, F; then, the quotient of the Picard group by Γ is isomorphic to the direct product $\mathbb{Z}/d_1 \times \mathbb{Z}/d_2$ generated by the classes of E_X, E_Y. Hence any element D of the Picard group of S can be uniquely represented as

$$
D \sim aZ + bF + \alpha E_X + \beta E_Y, \quad a, b, \alpha, \beta \in \mathbb{Z}, \quad 0 \le \alpha < d_1, \quad 0 \le \beta < d_2.
$$

3.4 The Canonical Cycle of S

Following the discussion in Sect. 2.4 the canonical cycle is numerically equivalent to

$$
Z_K \overset{\text{num}}{\sim} 2Z + rF + E_X + E_Y. \tag{12}
$$

Using Riemann-Roch one can calculate $\chi(\mathcal{O}_S(D)) := \chi(S, \mathcal{O}_S(D))$ as follows

Lemma 3.9 *Let $D \sim aZ + bF + \alpha E_X + \beta E_Y \in \mathrm{Pic}(S)$, where $0 \le \alpha < d_1$ and $0 \le \beta < d_2$, then*

$$\begin{aligned}\chi(\mathcal{O}_S(D)) = 1 + \tfrac{1}{2}a(b+r) + \left(b + \tfrac{\alpha}{d_1} + \tfrac{\beta}{d_2} - ar\right)\tfrac{(a+2)}{2} + \tfrac{a(\alpha+1)}{2d_1} + \tfrac{a(\beta+1)}{2d_2}\\ -\Delta_{\frac{1}{d_1}(1,q_1)}(-\alpha - aq_1) - \Delta_{\frac{1}{d_2}(1,p_2)}(-\beta - ap_2)\\ -\Delta_{\frac{1}{d_1}(1,p_1)}(-\alpha) - \Delta_{\frac{1}{d_2}(1,q_2)}(-\beta).\end{aligned}$$

Proof By the discussion in the preceding section

$$D(D+Z_K) = (a,b,\alpha,\beta)M_S\begin{pmatrix} a+2\\ b+r\\ \alpha+1\\ \beta+1\end{pmatrix} = a(b+r)$$

$$+\left(b + \frac{\alpha}{d_1} + \frac{\beta}{d_2} - ar\right)(a+2) + \frac{a(\alpha+1)}{d_1} + \frac{a(\beta+1)}{d_2}$$

The rest is an immediate consequence of the Riemann-Roch formula (6). □

4 Rational Biruled Toric Surfaces

The purpose of this section is to consider the special case $r = 0$, since this case has a special behavior in terms of cohomology of line bundles. The main result of this section is Theorem 4.2. It states that if $r = 0$, then all cohomology of line bundles is concentrated in only one degree, all degrees are possible, and there are precise formulas for these dimensions as they coincide with the Euler characteristic of the line bundle $\mathcal{O}(D)$, calculated in Lemma 3.9. Now we want to study the global sections $H^0(S, \mathcal{O}_S(D))$.

Recall from Remark 3.7 that any rational biruled toric surface S_I can be characterized by two numbers $I = (d, p)$, so that $d = d_1 = d_2$, $p = p_1 = p_2$, $q = q_1 = q_2 = d - p$, $r = 0$. Let us consider $D = aZ + bF + \alpha E_X + \beta E_Y \in \mathrm{Pic}(S)$, where $0 \le \alpha, \beta < d$. Note that $\hat{\rho}^*(D) = aZ + bF + \alpha E_X + \beta E_Y + \frac{\alpha}{d}X + \frac{\beta}{d}Y$ and hence

$$\lfloor \hat{\rho}^*(D) \rfloor = aZ + bF + \alpha E_X + \beta E_Y = (a+b)Z - bE + (\alpha + bp)E_X + (\beta + bq)E_Y \in \mathrm{Pic}(\hat{\Sigma}_1).$$

On the other hand

$$\rho^*((a+b)Z) = (a+b)\rho^*(Z) = (a+b)(Z + pE_X + qE_Y) \in \mathrm{Pic}(\hat{\Sigma}_1).$$

Therefore

$$\lfloor \hat{\rho}^*(D) \rfloor = (a+b)\rho^*(Z) - bE - (ap-\alpha)E_X - (aq-\beta)E_Y.$$

Using the projection formula

$$H^0(S, \mathcal{O}_S(D)) = H^0(\hat{\Sigma}_1, \mathcal{O}(\lfloor \hat{\rho}^*(D) \rfloor))$$

$$= \rho_* H^0(\hat{\Sigma}_1, \mathcal{O}_{\hat{\Sigma}_1}(\rho^*((a+b)Z) - bE - (ap-\alpha)E_X - (aq-\beta)E_Y))$$

$$= \left\{ h \in H^0(\mathbb{P}^2, \mathcal{O}_{\mathbb{P}^2}((a+b)Z)) \;\middle|\; \begin{array}{l} \operatorname{mult}_{(1,1)} h(X,Y,1) \geq b, \\ \operatorname{mult}_{(d,q)} h(1,Y,Z) \geq aq-\beta, \\ \operatorname{mult}_{(d,p)} h(X,1,Z) \geq ap-\alpha \end{array} \right\}$$

$$= \langle X^u Y^v Z^{a+b-u-v} \in \mathbb{C}[X,Y,Z]_{a+b} \mid u+v \geq b, \quad -pb-\alpha \leq qu-pv \leq qb+\beta \rangle \tag{13}$$

A straightforward calculation – see Fig. 4 – shows that

$$h^0(S, \mathcal{O}_S(D)) = \begin{cases} \binom{a+b+2}{2} - \binom{b+1}{2} - \#T_\beta - \#T_\alpha & \text{if } b \geq 0, a \geq 0 \\ \binom{a+1}{2} - \#T_\beta - \#T_\alpha & \text{if } b = -1, a \geq 0, \text{ and } \alpha+\beta \geq d \\ 0 & \text{otherwise.} \end{cases} \tag{14}$$

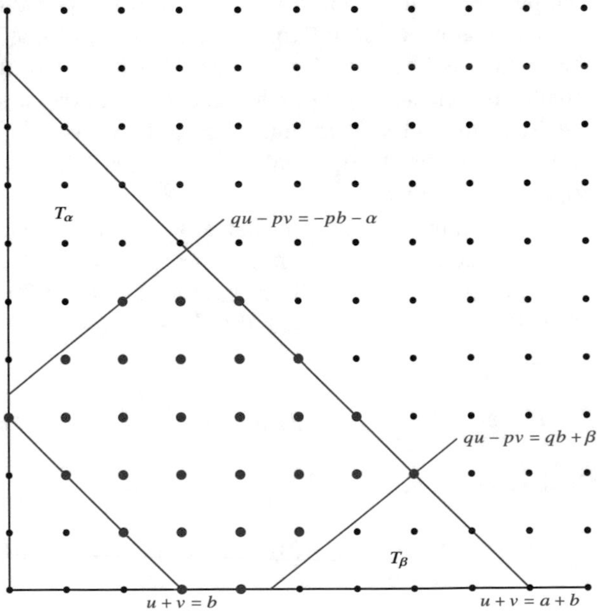

Fig. 4 Monomials in $H^0(\mathcal{O}_S(D))$, $D = 6Z + 3F + 2E_X + 6E_Y$, $(d, p, q) = (9, 5, 4)$

Let us calculate the number of integer points in the triangles $T_\beta = \{(u, v) \in \mathbb{Z}^2_{\geq 0} \mid qu - pv > qb + \beta, u + v \leq a + b\}$ and $T_\alpha = \{(u, v) \in \mathbb{Z}^2_{\geq 0} \mid qu - pv < -pb - \alpha, u + v \leq a + b\}$. For T_β consider the following change of coordinates:

$$i = qu - pv - qb - \beta - 1 \geq 0, \quad j = v \geq 0, \quad k = a + b - u - v \geq 0.$$

Note that $\tilde{T}_\beta = \{(i, j, k) \in \mathbb{Z}^3_{\geq 0} \mid i + dj + qk = qa - \beta - 1\}$ is such that $(i, j, k) \in \tilde{T}_\beta$ if and only if $(a + b - j - k, j) \in T_\beta$ and $(u, v) \in T_\beta$ if and only if $(qu - pv - qb - \beta - 1, v, a + b - u - v) \in \tilde{T}_\beta$.

Then $\#T_\beta = \#\tilde{T}_\beta = h^0(\mathbb{P}^2_w, \mathcal{O}_{\mathbb{P}^2_w}(qa - \beta - 1))$, for $w = (d, 1, q)$. Using the adjunction formula

$$\begin{aligned}\chi(\mathcal{O}_{\mathbb{P}^2_w}(qa - \beta - 1)) &= g_{w,qa-\beta-1} - \textstyle\sum_P \Delta_P(|w| + qa - \beta - 1)\\ &= \tfrac{(qa-\beta-1)(qa-\beta-1+|w|)}{2\bar{w}} + 1 - \Delta_{\frac{1}{d}(1,q)}(|w| + qa - \beta - 1) - \Delta_{\frac{1}{q}(1,d)}(|w| + qa - \beta - 1)\\ &= \tfrac{(qa-\beta-1)(qa-\beta+d+q)}{2dq} + 1 - \Delta_{\frac{1}{d}(1,q)}(-\beta + (a+1)q) - \Delta_{\frac{1}{q}(1,p)}(-\beta + p),\end{aligned} \tag{15}$$

where $|w| = 1 + d + q$. Since

$$qa - \beta - 1 - (-|w|) = q(a + 1) + (d - \beta) > 0$$

have $h^1(\mathbb{P}^2_w, \mathcal{O}_{\mathbb{P}^2_w}(qa-\beta-1)) = h^2(\mathbb{P}^2_w, \mathcal{O}_{\mathbb{P}^2_w}(qa-\beta-1)) = 0$. The vanishing of h^1 is a general property of weighted-projective spaces – see for instance [6, p. 39]. The vanishing of h^2 follows from Serre's duality since $-|w| = -1-q-d < qa-\beta-1$. Therefore, $\chi(\mathcal{O}_{\mathbb{P}^2_w}(qa - \beta - 1)) = h^0(\mathcal{O}_{\mathbb{P}^2_w}(qa - \beta - 1))$.

Analogously for T_α

$$\#T_\alpha = \frac{(pa-\alpha-1)(pa-\alpha+d+p)}{2dp} + 1 - \Delta_{\frac{1}{d}(1,p)}(-\alpha + (a+1)p) - \Delta_{\frac{1}{p}(1,q)}(-\alpha + q).$$

and hence (14) becomes

$$\begin{aligned}h^0(S, \mathcal{O}_S(D)) =& \binom{a+b+2}{2} - \binom{b+1}{2} - 2 - \frac{(qa-\beta-1)(qa-\beta+d+q)}{2dq} - \frac{(pa-\alpha-1)(pa-\alpha+d+p)}{2dp}\\ &+ \Delta_{\frac{1}{d}(1,q)}(-\beta + (a+1)q) + \Delta_{\frac{1}{q}(1,p)}(-\beta + p)\\ &+ \Delta_{\frac{1}{d}(1,p)}(-\alpha + (a+1)p) + \Delta_{\frac{1}{p}(1,q)}(-\alpha + q).\end{aligned} \tag{16}$$

if $b \geq -1$ and $h^0(S, \mathcal{O}_S(D)) = 0$ otherwise.

Lemma 4.1 *Consider* $D = aZ + bF + \alpha E_X + \beta E_Y \in \mathrm{Pic}(S)$, *where* $0 \leq \alpha, \beta < d$, *then*

(1) *if either* $a \leq -1$ *or* $b \leq -2$, *then* $h^0(\mathcal{O}_S(D)) = 0$,
(2) *if* $b = -1$, $a \geq 0$, *and* $\alpha + \beta < d$, *then* $h^0(\mathcal{O}_S(D)) = 0$,
(3) *if either* $a \geq -1$, $b \geq 1$, *or* $a + b \geq -2$, *then* $h^2(\mathcal{O}_S(D)) = 0$,

Proof Part (1) is a direct consequence of (14). For part (3) note that $K - D \sim a'Z + b'F + \alpha' E_X + \beta' E_Y$, where $a' = -(a+2)$, $b' = -(b+1) - d\left(\left[\frac{\alpha'}{d}\right] + \left[\frac{\beta'}{d}\right]\right)$, $\alpha' = d - \alpha - p - 1$, and $\beta' = d - \beta - q - 1$. In particular $-(b+3) \leq b' \leq -(b+1)$. Therefore, if either $a \geq -1$, $b \geq 1$ or $a + b \geq -2$ one has $a' \leq -1$, $b' \leq -2$, or $a' + b' \leq -1$. Hence, by (1) and Serre's Duality this implies $h^2(\mathcal{O}_S(D)) = h^0(\mathcal{O}_S(K - D)) = 0$. □

Theorem 4.2 *Consider* $D \sim aZ + bF + \alpha E_X + \beta E_Y \in \mathrm{Pic}(S)$, *where* $0 \leq \alpha, \beta < d$, *then* $H^*(S, \mathcal{O}_S(D))$ *is either trivial or concentrated in only one degree* $k = k(D)$, *for which*

$$h^k(S, \mathcal{O}_S(D)) = (-1)^k \chi(S, \mathcal{O}_S(D)).$$

Moreover, if $a = -1$, *then* $\mathcal{O}_S(D)$ *is acyclic. For the remaining cases, k can be obtained according to the following:*

	$a > -1$	$a < -1$
$\begin{cases} b > -1 \\ b = -1 \text{ and } \alpha + \beta \geq d \end{cases}$	$k = 0$	$k = 1$
$\begin{cases} b < -1 \\ b = -1 \text{ and } \alpha + \beta < d \end{cases}$	$k = 1$	$k = 2$

Proof We shall prove the moreover part which implies the first part of the statement. It will be done on a case by case basis. The most involved case being $a, b > -1$.

(1) Case $a \geq -1$, $b > -1$: Note that $h^2(\mathcal{O}_S(D)) = 0$ is a consequence of Lemma 4.1. Hence it is enough to show $h^0(\mathcal{O}_S(D)) = \chi(\mathcal{O}_S(D))$ using Riemann-Roch's formula. From Lemma 3.9 for $d_1 = d_2 = d$, $p_1 = p_2 = p$, $q_1 = q_2 = q$, and $r = 0$ one has

$$\begin{aligned} \chi(\mathcal{O}_S(D)) = 1 + & \tfrac{((a+1)(db+\alpha+\beta+1)-1)}{d} \\ & -\Delta_{\frac{1}{d}(1,p)}(-\alpha) - \Delta_{\frac{1}{d}(1,q)}(-\beta) \\ & -\Delta_{\frac{1}{d}(1,p)}(-\beta - pa) - \Delta_{\frac{1}{d}(1,q)}(-\alpha - qa) \end{aligned} \tag{17}$$

First we will combine the values of Δ_P, $P \in \operatorname{Sing}(S)$ appearing in (16) and (17). The following are a consequence of Proposition 2.16.

$$\Delta_{\frac{1}{d}(1,q)}(qa-\beta+q)+\Delta_{\frac{1}{d}(1,p)}(-\beta-pa)=\tfrac{d-1}{2d}$$

$$\Delta_{\frac{1}{d}(1,p)}(pa-\alpha+p)+\Delta_{\frac{1}{d}(1,q)}(-\alpha-qa)=\tfrac{d-1}{2d}.$$

Also, using the adjunction formula

$$\chi(\mathcal{O}_{\mathbb{P}^2_{(1,d,p)}}(-\alpha-1-p))=g_{(1,d,p),-\alpha-1-p}-\Delta_{\frac{1}{p}(1,q)}(-\alpha+q)-\Delta_{\frac{1}{d}(1,p)}(-\alpha)$$

$$\chi(\mathcal{O}_{\mathbb{P}^2_{(1,d,q)}}(-\beta-1-q))=g_{(1,d,q),-\beta-1-q}-\Delta_{\frac{1}{q}(1,p)}(-\beta+p)-\Delta_{\frac{1}{d}(1,q)}(-\beta)$$

On the other hand $h^i(\mathbb{P}^2_{(1,d,p)}, \mathcal{O}_{\mathbb{P}^2_w}(-\alpha-1-p))=0$, for $i=0$ (since $-\alpha-1-p<0$) for $i=1$ (general property of quasi-projective planes), and for $i=2$ (using Serre's duality plus the fact that $-|w|=-1-d-p<-\alpha-1-p$), then $\chi(\mathcal{O}_{\mathbb{P}^2_{(1,d,p)}}(-\alpha-1-p))=0$. Analogously $\chi(\mathcal{O}_{\mathbb{P}^2_{(1,d,q)}}(-\beta-1-q))=0$. Using the formula for $h^0(\mathcal{O}_S(D))$ for the case $b\geq 0$, $a\geq 0$ one obtains

$$\begin{aligned}h^0(\mathcal{O}_S(D))-\chi(\mathcal{O}_S(D))=&\binom{a+b+2}{2}-\binom{b+1}{2}-\tfrac{(qa-\beta-1)(qa-\beta+d+q)}{2dq}\\&-\tfrac{(pa-\alpha-1)(pa-\alpha+d+p)}{2dp}-1-\tfrac{((a+1)(db+\alpha+\beta+1)-1)}{d}\\&+\tfrac{d-1}{d}+\tfrac{(\alpha+p+1)(\alpha-d)}{2dp}+\tfrac{(\beta+q+1)(\beta-d)}{2dq}=0.\end{aligned}$$

The last equality is a straightforward calculation.

(2) Case $b<-1\leq a$: note that $h^2(\mathcal{O}_S(D))=h^0(\mathcal{O}_S(D))=0$ by Lemma 4.1.

(3) Case $a\leq -1$, $b<-1$: first note that $b<-1$ implies $h^0(\mathcal{O}_S(D))=0$ by Lemma 4.1. Also $a\leq -1$ implies $-(a+2)\geq -1$, and $b<-1$ implies $-(b+1)>-1$. Hence by Serre's duality $h^1(\mathcal{O}_S(D))=h^1(\mathcal{O}_S(K-D))=0$ and case (1).

(4) Case $a<-1<b$: again $a\leq -1$ implies $h^0(\mathcal{O}_S(D))=0$. Also $b\geq 0$ implies $h^2(\mathcal{O}_S(D))=0$ according to case (1). then $h^1(\mathcal{O}_S(D))=-\chi(\mathcal{O}_S(D))$.

(5) Case $b=-1$, $\alpha+\beta\geq d$, and $a\geq -1$: the proof of case (1) is also valid in this situation.

(6) Case $b=-1$, $\alpha+\beta<d$, and $a\geq -1$: according to Lemma 4.1(3) $h^2(\mathcal{O}_S(D))=0$ (since $a\geq -1$). To prove $h^0(\mathcal{O}_S(D))=0$ we distinguish two cases: if $a\geq 0$ one uses Lemma 4.1(2) and if $a=-1$ one can use Lemma 4.1(1).

(7) Case $b=-1$, $\alpha+\beta\geq d$, and $a\leq -1$: according to Lemma 4.1(1) $h^0(\mathcal{O}_S(D))=0$ (since $a\leq -1$). To prove $h^2(\mathcal{O}_S(D))=0$ we will use Serre's duality over $K-D\sim a'Z+b'F+\alpha'E_X+\beta'E_Y$ as described in the proof of Lemma 4.1 to show that $h^0(\mathcal{O}_S(K-D))=0$. Note that $a\leq -1$ implies $a'\geq -1$. If $a'=1$, then one can use Lemma 4.1(1) to show that

$h^0(\mathcal{O}_S(K-D)) = 0$. Hence we will assume $a' \geq 0$. We will distinguish several cases. If both $\alpha \geq q$, and $\beta \geq p$, then $b' = -2$ and thus $h^0(\mathcal{O}_S(K-D)) = 0$ by Lemma 4.1(1). Otherwise $b' = -1$, which is obtained only if either $\alpha \geq q$ and $\beta < p$ or $\alpha < q$ and $\beta \geq p$. In either case $\alpha' + \beta' = 2d - \alpha - \beta - 2 < d$. The result then follows from Lemma 4.1(2).

(8) Case $b = -1$, $\alpha + \beta < d$, and $a \leq -1$: according to Lemma 4.1(1) $h^0(\mathcal{O}_S(D)) = 0$ (since $a \leq -1$). To prove $h^1(\mathcal{O}_S(D)) = 0$ we will use Serre's duality over $K - D \sim a'Z + b'F + \alpha' E_X + \beta' E_Y$ as described in the proof of Lemma 4.1 to show that $h^1(\mathcal{O}_S(K-D)) = 0$. Note that $b = -1$ and $\alpha + \beta < d$ implies $b' = 0$, and hence case (1) implies $h^1(\mathcal{O}_S(K-D)) = 0$.

The case $a = -1$, $b \neq -1$ is a consequence of cases (1)–(2). □

Remark 4.3 The case $b = -1$ is special and the nullity of $\chi(\mathcal{O}_S(D))$ (which implies acyclicity by Theorem 4.2) depends on the values of α and β. For instance, if S is the rational ruled toric surface given by $(d, p, q, r) = (5, 3, 2, 0)$, $D_1 \sim Z - F + 3E_X + 2E_Y$, $D_2 \sim Z - F + 3E_X + E_Y$, and $D_3 \sim Z - F + 2E_X + E_Y$, then

$$
\begin{array}{ccc}
h^0(\mathcal{O}_S(D_1)) = \chi(\mathcal{O}_S(D_1)) = 1 & h^1(\mathcal{O}_S(D_1)) = 0 & h^2(\mathcal{O}_S(D_1)) = 0 \\
h^0(\mathcal{O}_S(D_2)) = 0 & h^1(\mathcal{O}_S(D_2)) = 0 & h^2(\mathcal{O}_S(D_2)) = 0 \\
h^0(_S(D_3)) = 0 & h^1(\mathcal{O}_S(D_3)) = -\chi(\mathcal{O}_S(D_3)) = 1 & h^2(\mathcal{O}_S(D_3)) = 0
\end{array}
$$

Example 4.4 Consider the rational ruled toric surface $S_{(d,p)}$ with sections Z and E and $E^2 = Z^2 = 0$ as in Fig. 3. Note that despite $dE_Y = dE_Y = F$, the divisors E_X and E_Y are not equivalent. This can be deduced from the homology calculations, since $D = E_X - E_Y = -F + E_X + (d-1)E_Y$ satisfies $\chi(\mathcal{O}_S(D)) = 0$ however, $\chi(\mathcal{O}_S) = 1$. Note that in particular the Euler characteristic of $\mathcal{O}_S(D)$ is sensitive to the torsion of Pic(S). In the Riemann-Roch formula (6) the first summand $\frac{1}{2}D(D + Z_K)$ is only numerical, however Δ_S is not. In particular $\Delta_S(-D) = 1$ but $\Delta_S(dD) = \Delta_S(0) = 0$.

5 Rational Uniruled Toric Surfaces

Let S be a toric ruled surface, where a section has two quotient singular points $\frac{1}{d_i}(1, n_i)$, $i = 1, 2$, with $\gcd(d_i, n_i) = 1$, $0 < n_i \leq d_i$, and self-intersection $r > 0$ such that $\frac{n_1}{d_1} + \frac{n_2}{d_2} - r = 1 - k \in \mathbb{Z}$ $(k \geq 0)$. Since the value of n_i is only well defined modulo d_i, one can consider $q_2 := n_2 \geq n_1$, $p_1 = n_1 + kd_1$, $p_2 = d_2 - q_2$, and $q_1 = d_1 - n_1$, in which case

$$\frac{p_1}{d_1} + \frac{q_2}{d_2} - r = 1.$$

Hence,

$$d_2 p_1 + d_1 q_2 = d_1 d_2 + r d_1 d_2 > d_1 d_2. \tag{18}$$

Similarly to the discussion in Sect. 4 note that $H^0(S, \mathcal{O}_S(D))$ is isomorphic to the linear subspace V_0 with basis

$$\{X^u Y^v \mid u, v \geq 0,\ b \leq u+v \leq a+b,\ q_2 u - p_2 v \leq p_2 b + \beta,\ q_1 u - p_1 v \geq -p_1 b - \alpha\}.$$

Let us consider necessary conditions for V_0 to be different from $\{0\}$. It is easily seen that both $a + b \geq -1$ and $a \geq -1$ are required. The last two conditions are determined by the lines $\ell_1 = \{p_1 v - q_1 u = p_1 b + \alpha\}$ and $\ell_2 = \{q_2 u - p_2 v = q_2 b + \beta\}$; their intersection point is in the line

$$u + v = b + \frac{b + \frac{\alpha}{d_1} + \frac{\beta}{d_2}}{r}.$$

In particular, we can also assume $b \geq -1$. This point is

$$\left(\frac{\frac{p_1}{d_1} b + \frac{\alpha}{d_1}}{r} + \frac{p_1 \beta - q_2 \alpha}{d_1 d_2 r}, \frac{\frac{q_2}{d_2} b + \frac{\beta}{d_2}}{r} - \frac{p_1 \beta - q_2 \alpha}{d_1 d_2 r} \right).$$

Note that in general, ℓ_1 and ℓ_2 are not parallel anymore, since their slopes are $m(\ell_i) = \frac{q_i}{p_i}$ and $q_2 p_1 - q_1 p_2 = q_2 p_1 - (d_1 - p_1)(d_2 - q_2) = p_1 d_2 + q_2 d_1 - d_1 d_2 > 0$ by (18).

Two cases will be distinguished:

5.1 *Case k* = 0

In this case one can check that both slopes $m(\ell_i)$ are positive. Hence for values $a < \frac{b+\frac{\alpha}{d_1}+\frac{\beta}{d_2}}{r}$ – see Fig. 5a – note that $h^0(\mathcal{O}_S(D))$ depends on a, however if $a \geq \frac{b+\frac{\alpha}{d_1}+\frac{\beta}{d_2}}{r}$ – see Fig. 5b –, then $h^0(\mathcal{O}_S(D))$ is independent of a.

Theorem 5.1 *Assume $D \sim aZ + bF + \alpha E_X + \beta E_Y$, where $a \geq 0$ and $b \geq -1$, then*

- *if* $\frac{b+\frac{\alpha}{d_1}+\frac{\beta}{d_2}}{r} > a \geq 0$, $b \geq -1$, *then*

$$h^0(\mathcal{O}_S(D)) = \chi(\mathcal{O}_S(D)),\ h^1(\mathcal{O}_S(D)) = h^2(\mathcal{O}_S(D)) = 0$$

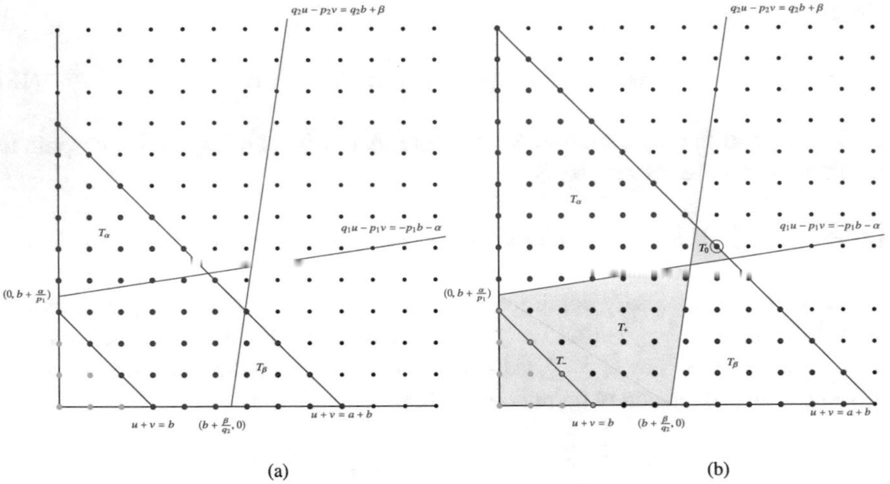

Fig. 5 Monomials in $H^0(\mathcal{O}_S(D))$ case $k = 0$. (**a**) Case $a < \frac{b+\frac{\alpha}{d_1}+\frac{\beta}{d_2}}{r}$. (**b**) Case $a > \frac{b+\frac{\alpha}{d_1}+\frac{\beta}{d_2}}{r}$

- *otherwise* $\chi(\mathcal{O}_S(D))$ *is given as in Lemma 3.9,* $h^2(\mathcal{O}_S(D)) = 0$ *and*

$$h^1(\mathcal{O}_S(D)) = h^0(\mathcal{O}_S(D)) - \chi(\mathcal{O}_S(D)) = h^0(\mathbb{P}^2_w; \mathcal{O}(\rho)),$$

where $w = (d_1, d_2, rd_1d_2)$ *and* $\rho = rd_1d_2a - d_1d_2b - d_1\beta - d_2\alpha \geq 0$ *is quadratic in* a.

Proof For the first part one can apply the following formula

$$h^0(\mathcal{O}_S(D)) = \binom{a+b+2}{2} - \binom{b+1}{2} - \#T_\alpha - \#T_\beta \tag{19}$$

– in Fig. 5a the dimension is given by the solid blue dots. An analogous proof to that of Theorem 4.2 case (1) using (19) and Lemma 3.9 shows that

$$h^0(\mathcal{O}_S(D)) - \chi(\mathcal{O}_S(D)) = \frac{(d_1d_2r + d_1d_2 - d_2p_1 - d_1q_2)(a+1)a}{2d_1d_2},$$

which is zero by (18). The result follows direct calculation of $h^0(\mathcal{O}_S(K - D)) = h^2(\mathcal{O}_S(D)) = 0$ (by Serre's duality).

For the second part one has to apply a different formula for $h^0(\mathcal{O}_S(D))$ which is independent of a, namely

$$h^0(\mathcal{O}_S(D)) = T_+ + T_- - \binom{b+1}{2}$$

– in Fig. 5b the dimension is given by the solid blue dots. However, note that the formula for $\chi(\mathcal{O}_S(D))$ given as in Lemma 3.9 is quadratic in a. An alternative way to see this is using the triangles T_α and T_β from the previous paragraph. The formula $\chi(\mathcal{O}_S(D)) = \binom{a+b+2}{2} - \binom{b+1}{2} - \#T_\alpha - \#T_\beta$ is still true, however the right-hand side now equals $h^0(\mathcal{O}_S(D)) - \#T_0$. Since $h^2(\mathcal{O}_S(D)) = 0$ as above, one obtains $h^1(\mathcal{O}_S(D)) = \#T_0$. Using the techniques introduced in Sect. 4 one can rewrite the number of integer points in a triangle as the dimension of the space of global sections of some weighted projective space. In this case one can check that

$$\begin{aligned}\#T_0 &= h^0(\mathbb{P}^2_w, \mathcal{O}(\rho)) \\ &= g_{w,\rho} - \Delta_{\frac{1}{d_1}(d_2, rd_1d_2)}(d_2\alpha - rd_1d_2a) - \Delta_{\frac{1}{d_2}(d_1, rd_1d_2)}(d_1\beta - rd_1d_2a) \\ &\quad -\Delta_{\frac{1}{rd_1d_2}(d_1,d_2)}(d_1\beta + d_2\alpha - d_1d_2b - rd_1d_2a),\end{aligned}$$

where $w = (d_1, d_2, rd_1d_2)$ and $\rho = rd_1d_2a - d_1d_2b - d_1\beta - d_2\alpha \geq 0$ by hypothesis. □

5.2 *Case k > 0*

In this case one can check that $-1 < m(\ell_1) < 0$ and $m(\ell_2) > 0$. Hence again, for values $a < \frac{b+\frac{\alpha}{d_1}+\frac{\beta}{d_2}}{r}$ – see Fig. 5a – note that $h^0(\mathcal{O}_S(D))$ depends on a, however if $a \geq \frac{b+\frac{\alpha}{d_1}+\frac{\beta}{d_2}}{r}$ – see Fig. 5b –, then $h^0(\mathcal{O}_S(D))$ is independent of a.

The case $a \geq \frac{b+\frac{\alpha}{d_1}+\frac{\beta}{d_2}}{r}$ however has a new situation that we cover in this result.

Theorem 5.2 *Assume $D \sim aZ + bF + \alpha E_X + \beta E_Y$, where $b \geq -1$, $a \geq \frac{b+\frac{\alpha}{d_1}+\frac{\beta}{d_2}}{r} \geq 0$, and $-q_1a \geq d_1b + \alpha$, then*

$$h^0(\mathcal{O}_S(D)) = h^0(\mathbb{P}^2_{w_1}, \mathcal{O}(\rho_1)) - h^0(\mathbb{P}^2_{w_2}, \mathcal{O}(\rho_2)) - \binom{b+1}{2},$$

where $w_1 = (1, p_1, -q_1)$, $\rho_1 = bp_1 + \alpha$, $w_2 = (-q_1, q_2, rd_1d_2)$, and $\rho_2 = q_2d_1b + \alpha q_2 + \beta q_1$.

6 Cyclic Branched Coverings on Rational Ruled Toric Surfaces

Let S be a rational ruled toric surface as in Definition 3.1 and consider $\pi : \widetilde{S} \to S$ a cyclic branched covering of n sheets. Assume the ramification set is given by a $\mathbb{Q}$-normal crossing Weil divisor $D = \sum_{i=1}^r m_iD_i$ linearly equivalent to nH for some Weil divisor H. One is interested in studying the monodromy of the covering acting on $H^1(\widetilde{S},\mathbb{C}) = H^0(\widetilde{S},\Omega^1_{\widetilde{S}}) \oplus H^1(\widetilde{S},\mathcal{O}_{\widetilde{S}})$. Since both summands are complex conjugated the decomposition becomes

$$H^1(\widetilde{S},\mathbb{C}) = H^1(\widetilde{S},\mathcal{O}_{\widetilde{S}}) \oplus \overline{H^1(\widetilde{S},\mathcal{O}_{\widetilde{S}})}.$$

In the smooth case, Esnault and Viehweg [7] proved that $H^1(\widetilde{S},\mathcal{O}_{\widetilde{S}}) = H^1(S,\pi_*\mathcal{O}_{\widetilde{S}})$ and provided a precise description of the sheaf $\pi_*\mathcal{O}_{\widetilde{S}}$ giving its Hodge structure compatible with the monodromy of the covering. In a forthcoming paper we will prove that the same is true for surfaces with abelian quotient singularities. In particular, the following result holds.

Theorem 6.1 *Using the previous notation,*

$$H^1(\widetilde{S},\mathcal{O}_{\widetilde{S}}) = \bigoplus_{k=0}^{n-1} H^1(S,\mathcal{O}_S(L^{(k)})), \qquad L^{(k)} = -kH + \sum_{i=1}^{r}\left\lfloor \frac{km_i}{n}\right\rfloor D_i,$$

where the monodromy of the cyclic covering acts on $H^q(S,\mathcal{O}_S(L^{(k)}))$ by multiplication by $e^{\frac{2\pi i k}{n}}$.

Remark 6.2 Note that any covering can be encoded by the data (S, D, H, n) such that $D, H \in \operatorname{Pic}(S)$ and $D \sim nH$. If $\operatorname{Pic}(S)$ is torsion free, e.g. if S is smooth and rational, then the divisor H is uniquely determined by (S, D, n). Otherwise, this is no longer true, both if S is rational – see Example 7.2 – or even if S is smooth – for instance an Enriques surface S where $D = 0$, $n = 2$, and H can be chosen to be either $H = 0$ or $H = K_S$ the canonical divisor.

6.1 The Divisor $L^{(k)}$

Let us white D in terms of its irreducible components $D = \sum_{i=1}^r m_i D_i$ and suppose $D_i \sim a_i Z + \alpha_i E_X + \beta_i E_Y$ and $H \sim zZ + e_x E_X + e_y E_Y$ for some $a_i, \alpha_i, \beta_i, z, e_x, e_y \in \mathbb{Z}$. Recall $F \cdot E_X = F \cdot E_Y = 0$ and $F \cdot Z = 1$. Assume without loss of generality that D is an effective divisor. Since D is linearly equivalent to nH, $D \cdot F = nH \cdot F$ and hence $\sum_{i=1}^r m_i a_i = nz$. Analogously, using $D \cdot Z = nH \cdot Z$ and the fact that $Z^2 = 0$, $Z \cdot E_X = Z \cdot E_Y = \frac{1}{d}$, one obtains $\sum_{i=1}^r m_i(\alpha_i + \beta_i) = n(e_x + e_y)$.

The divisor $L^{(k)} = -kH + \sum_{i=1}^r \left\lfloor \frac{km_i}{n}\right\rfloor D_i \in \operatorname{Pic}(S)$ can be rewritten as $u_k Z + v_k E_X + w_k E_Y$, where the coefficients are given by

$$u_k = -kz + \sum_{i=1}^{r}\left\lfloor \frac{km_i}{n}\right\rfloor a_i, \quad v_k = -ke_x + \sum_{i=1}^{r}\left\lfloor \frac{km_i}{n}\right\rfloor \alpha_i, \quad w_k = -ke_y + \sum_{i=1}^{r}\left\lfloor \frac{km_i}{n}\right\rfloor \beta_i.$$

Note that

$$u_k = \sum_{i=1}^{r}\left(-\frac{km_i}{n}+\left\lfloor\frac{km_i}{n}\right\rfloor\right)a_i, \qquad v_k+w_k = \sum_{i=1}^{r}\left(-\frac{km_i}{n}+\left\lfloor\frac{km_i}{n}\right\rfloor\right)(\alpha_i+\beta_i). \tag{20}$$

Since all D_i's are effective divisors, $a_i = D_i \cdot F \geq 0$ and $\alpha_i + \beta_i = D_i \cdot (dZ) \geq 0$. In particular, $u_k \leq 0$ and $v_k + w_k \leq 0$ for all $k = 0, \ldots, n-1$.

6.2 The First Cohomology Group of $\mathcal{O}_S(L^{(k)})$

According to Theorem 6.1, one is interested in computing $h^1(S, \mathcal{O}_S(L^{(k)}))$. To do so we will apply Theorem 4.2 to the divisor $L^{(k)}$. First one needs to divide v_k and w_k by d, that is, $v_k = c_{k,1}d + r_{k,1}$ and $w_k = c_{k,2}d + r_{k,2}$, so that $L^{(k)} \sim u_k Z + (c_{k,1}+c_{k,2})F + r_{k,1}E_X + r_{k,2}E_Y$. Four different cases are discussed.

1. $u_k < 0$ and $v_k + w_k < 0$. Then $c_{k,1} + c_{k,2} = \frac{v_k+w_k}{d} - \frac{r_{k,1}+r_{k,2}}{d} < 0$. If either $u_k = -1$; $u_k < -1$ and $c_{k,1} + c_{k,2} < -1$; or $u_k < -1$, $c_{k,1} + c_{k,2} = -1$, and $r_{k,1} + r_{k,2} < d$, then $h^1(S, \mathcal{O}_S(L^{(k)})) = 0$. Note that the case $u_k < -1$, $c_{k,1} + c_{k,2} = -1$, and $r_{k,1} + r_{k,2} \geq d$ cannot occur.
2. $u_k = 0$ and $v_k + w_k = 0$. Then $c_{k,1} + c_{k,2} = -\frac{r_{k,1}+r_{k,2}}{d}$ can be either zero or minus one. In both cases, the first cohomology group vanishes.
3. $u_k = 0$ and $v_k + w_k < 0$. Then as above $c_{k,1} + c_{k,2} < 0$. The case $c_{k,1} + c_{k,2} = -1$ and $r_{k,1} + r_{k,2} \geq d$ is not compatible with $v_k + w_k < 0$. Thus $h^1(S, \mathcal{O}_S(L^{(k)})) = -\chi(S, \mathcal{O}_S(L^{(k)})) = -1 - \left\lfloor \frac{v_k}{d} \right\rfloor - \left\lfloor \frac{w_k}{d} \right\rfloor$. The latter formula holds by Lemma 6.3, see below. In particular, if $v_k + w_k = -1$, then the dimension is zero.
4. $u_k < 0$ and $v_k + w_k = 0$. The sheaf $L^{(k)}$ is acyclic for $u_k = -1$. If $u_k < -1$, then $h^1(S, \mathcal{O}_S(L^{(k)})) = -\chi(S, \mathcal{O}_S(L^{(k)})) = -1 - \left\lfloor \frac{u_k - v_k p'}{d} \right\rfloor - \left\lfloor \frac{v_k p'}{d} \right\rfloor$, where $p'p \equiv 1 \mod d$, see Lemma 6.3. Note that the previous formula still holds for $u_k = -1$, since in such a case it vanishes.

Lemma 6.3 *$\chi(S, \mathcal{O}_S(v_k E_X + w_k E_Y)) = 1 + \left\lfloor \frac{v_k}{d} \right\rfloor + \left\lfloor \frac{w_k}{d} \right\rfloor$ and $\chi(S, \mathcal{O}_S(u_k Z + v_k(E_X - E_Y))) = 1 + \left\lfloor \frac{u_k - v_k p'}{d} \right\rfloor + \left\lfloor \frac{v_k p'}{d} \right\rfloor$.*

Proof As above, using the divisions $v_k = c_{k,1}d + r_{k,1}$ and $w_k = c_{k,2}d + r_{k,2}$, one writes the divisor $v_k E_X + w_k E_Y$ as $(c_{k,1}+c_{k,2})F + r_{k,1}E_X + r_{k,2}E_Y$. By Lemma 3.9,

$$\chi(S, \mathcal{O}_S(v_k E_X + w_k E_Y)) = 1 + c_{k,1} + c_{k,2} + \frac{r_{k,1}}{d} + \frac{r_{k,2}}{d} - \Delta_{\frac{1}{d}(1,q)}(-r_{k,1})$$
$$- \Delta_{\frac{1}{d}(1,p)}(-r_{k,2}) - \Delta_{\frac{1}{d}(1,p)}(-r_{k,1}) - \Delta_{\frac{1}{d}(1,q)}(-r_{k,2}).$$

Note that by definition $c_{k,1} = \lfloor \frac{v_k}{d} \rfloor$ and $c_{k,2} = \lfloor \frac{w_k}{d} \rfloor$. The first part of the statement follows since, by Proposition 2.16, $\Delta_{\frac{1}{d}(1,p)}(-r_{k,i}) + \Delta_{\frac{1}{d}(1,q)}(-r_{k,i}) = \frac{r_{k,i}}{d}$, $i = 1, 2$. For the second part, one uses the relation calculated in (11) to rewrite the divisor $u_k Z + v_k(E_X - E_Y) \in \operatorname{Pic}(S)$ as $(u_k - v_k p')Z + v_k p' E$. The symmetry of the surface S completes the proof. □

The previous discussion can be summarized in the following result.

Proposition 6.4 *The first cohomology group of the sheaf $\mathcal{O}_S(L^{(k)})$, being $L^{(k)} = u_k Z + v_k E_X + w_k E_Y$, is*

$$h^1(S, \mathcal{O}_S(L^{(k)})) = \begin{cases} -1 - \lfloor \frac{v_k}{d} \rfloor - \lfloor \frac{w_k}{d} \rfloor & u_k = 0,\ v_k + w_k \le -2, \\ -1 - \left\lfloor \frac{u_k - v_k p'}{d} \right\rfloor - \left\lfloor \frac{v_k p'}{d} \right\rfloor & v_k + w_k = 0,\ u_k \le -2, \\ 0 & \textit{otherwise}. \end{cases}$$

6.3 Decomposition of $H^1(\widetilde{S}, \mathcal{O}_{\widetilde{S}})$

Note that $u_k = 0$ happens only when $\frac{k m_i}{n} \in \mathbb{Z}$ for all $i \in I_1 := \{i \mid a_i \neq 0\}$, see (20), or equivalently, when $\frac{n}{\gcd(n, \{m_i\}_{i \in I})}$ divides k. Let us denote by $n_1 = \gcd(n, \{m_i\}_{i \in I_1})$ and $k = \frac{n}{n_1} k_1$, where k_1 runs from 0 to $n_1 - 1$. Analogously, consider $I_2 := \{i \mid \alpha_i + \beta_i \neq 0\}$, then $v_k + w_k \neq 0$ holds if and only if $\frac{n}{\gcd(n, \{m_i\}_{i \in I_2})}$ divides k – denote by $n_2 = \gcd(n, \{m_i\}_{i \in I_2})$, $k = \frac{n}{n_2} k_2$, $k_2 = 0, \ldots, n_2 - 1$. The direct sum in Theorem 6.1 splits into two parts as follows,

$$H^1(\widetilde{S}, \mathcal{O}_{\widetilde{S}}) = \bigoplus_{k_1=0}^{n_1-1} H^1(S, \mathcal{O}_S(L^{(\frac{nk_1}{n_1})})) \oplus \bigoplus_{k_2=0}^{n_2-1} H^1(S, \mathcal{O}_S(L^{(\frac{nk_2}{n_2})})). \tag{21}$$

Let $\pi_j : \widetilde{S}_j \to S$, $j = 1, 2$, be the branched covering of n_j sheets whose ramification set is given by $D = \sum_{i=1} m_i D_i$ with associated divisor $H_j = \frac{n}{n_j} H$. A simple observation shows that the first (resp. second) direct sum in (21) coincides with $H^1(\widetilde{S}_1, \mathcal{O}_{\widetilde{S}_1})$ (resp. $H^1(\widetilde{S}_2, \mathcal{O}_{\widetilde{S}_2})$) and the decomposition is compatible with the monodromy of π_1 (resp. π_2), since $e^{\frac{2\pi i k}{n}} = e^{\frac{2\pi i k_1}{n_1}}$ (resp. $e^{\frac{2\pi i k}{n}} = e^{\frac{2\pi i k_2}{n_2}}$).

We have just proven the next result.

Theorem 6.5 *The characteristic polynomial of the monodromy of $\pi : \widetilde{S} \to S$, $\pi_1 : \widetilde{S}_1 \to S$, and $\pi_2 : \widetilde{S}_2 \to S$ satisfy*

$$p_{H^1(\widetilde{S}, \mathcal{O}_{\widetilde{S}})}(t) = p_{H^1(\widetilde{S}_1, \mathcal{O}_{\widetilde{S}_1})}(t) \cdot p_{H^1(\widetilde{S}_2, \mathcal{O}_{\widetilde{S}_2})}(t),$$

and moreover the equality can be understood at the level of its Hodge structure. Using Proposition (6.4), *explicit formulas for the characteristic polynomials* $p_{H^1(\tilde{S}_j,\mathcal{O}_{\tilde{S}_j})}(t)$ *can be calculated.*

Some examples will appear in the following section.

7 Examples

Example 7.1 Denote by $\tilde{S}_i$ the covering of S_I with $I = (d_1, d_2, p_1, q_2, r)$, where $d_1 = d_2 = 12$, $p_1 = 1, q_2 = 11, r = 0$, given by the data $n = 12$ and $D = F \sim 12H_i$ for $H_i = E_X + i(E_X - E_Y)$, $i = 0, \ldots, 11$. Due to the symmetry of the surface S_I, the covers $\tilde{S}_i$ and $\tilde{S}_{11-i}$ are isomorphic. In these cases $L_i^{(k)} = -k(E_X + i(E_X - E_Y))$ and $h^1(L_i^{(k)}) = -1 - \lfloor\frac{-k-ki}{12}\rfloor - \lfloor\frac{ki}{12}\rfloor$. A straightforward calculation shows

$h^1(L_i^{(k)})$	$k=1$	2	3	4	5	6	7	8	9	10	11
$i=0$	0	0	0	0	0	0	0	0	0	0	0
$i=1$	0	0	0	0	0	0	0	1	1	1	1
$i=2$	0	0	0	0	1	0	0	0	1	1	1
$i=3$	0	0	0	0	0	0	1	0	0	1	1
$i=4$	0	0	0	0	1	0	0	1	0	1	1
$i=5$	0	0	0	0	0	0	1	0	1	0	1

Using formula (21) this shows that $h^1(\tilde{S}_0) = 0$, $h^1(\tilde{S}_3) = h^1(\tilde{S}_5) = 6$, and $h^1(\tilde{S}_2) = h^1(\tilde{S}_4) = 8$. This already distinguishes the cases $i = 0$, $\{3, 5\}$ and $\{2, 4\}$. In the remaining cases their characteristic polynomial of the monodromy are different and hence the covers $\tilde{S}_3 \to S_3$ and $\tilde{S}_5 \to S_5$ (resp. $\tilde{S}_2 \to S_2$ and $\tilde{S}_4 \to S_4$) can be distinguished.

Example 7.2 Analogously to Example 7.1 for the case $d_1 = d_2 = 5$, $p_1 = 1, q_2 = 4, r = 0$. A straightforward calculation shows

$h^1(L_i^{(k)})$	$k=0$	$k=1$	$k=2$	$k=3$	$k=4$
$i=0$	0	0	0	0	0
$i=1$	0	0	0	1	1
$i=2$	0	0	1	0	1

Again, using formula (21), $h^1(\tilde{S}_0) = 0$, $h^1(\tilde{S}_1) = h^1(\tilde{S}_2) = 4$. This already distinguishes the cases $i = 0$ and $i \neq 0$. However, note that the different eigenspaces of the monodromy action as described in Theorem 6.1 distinguishes the covers $\tilde{S}_i \to S_i$ for the cases $i \in \{0, 1, 2\}$.

Example 7.3 We consider a surface S obtained as follows. Start with $\mathbb{P}^1 \times \mathbb{P}^1$ with two 0-sections S_0, S_∞ and four fibers F_j, $1 \leq j \leq 4$. We perform four weighted Nagata transformations; we start with $(1, 2)$-blow-ups over $S_0 \cap F_j$, $j = 1, 2$, and $S_\infty \cap F_j$, $j = 3, 4$. Blowing-down the strict transforms of the fibers, we obtain a new surface with 8 points of type $\frac{1}{2}(1, 1)$ in the intersection points of the strict transforms of S_0, S_∞ (we keep their notation) with the four new fibers $\tilde{F}_j$, $1 \leq j \leq 4$. Note that $S_0^2 = S_\infty^2 = 0$. Using Propositions 2.10 and 2.12 we obtain

$$\operatorname{Pic}(S) = \langle S_0, F, \tilde{F}_1, \dots, \tilde{F}_4 \mid 2\tilde{F}_1 = F, \dots, 2\tilde{F}_4 = F \rangle \cong \mathbb{Z}^2 \times (\mathbb{Z}/2)^3$$

where F is a generic fiber. We want to study the connected double coverings where $D = 0$; as for H we can choose any 2-torsion divisor. We can consider two types of such divisors, namely $H_1 = \tilde{F}_1 - \tilde{F}_2$ or $H_2 = \tilde{F}_1 - \tilde{F}_2 + \tilde{F}_3 - \tilde{F}_4$. For H_j the divisor $L_j^{(1)} = H_j$, i.e., if we want to compute the first Betti number of those coverings, then we need $H^1(S, \mathcal{O}_S(L_j^{(1)}))$. Note that these two divisors are numerically trivial and the values of Δ-invariant are 0 for the singular points not in the support of H_j and $\frac{1}{4}$ for the other ones. Hence

$$\chi(S, \mathcal{O}_S(H_1)) = 1 - 4\frac{1}{4} = 0, \qquad \chi(S, \mathcal{O}_S(H_2)) = 1 - 8\frac{1}{4} = -1.$$

Using Serre duality and the ideas of the previous sections, we note that $H^0(S, \mathcal{O}_S(H_j)) = H^2(S, \mathcal{O}_S(H_j)) = 0$. Hence, $h^1(S, \mathcal{O}_S(H_1)) = 0$ and $h^1(S, \mathcal{O}_S(H_2)) = 1$. Hence the double covering associated with the divisor H_1 has first Betti number 0, as expected since it is again a ruled surface with 8 singular double points as S. The first Betti number of the double covering associated with the divisor H_2 is 2; in fact it is the product of an elliptic curve and $\mathbb{P}^1$.

Acknowledgements The authors are partially supported by MTM2016-76868-C2-2-P.

References

1. Artal, E., Martín-Morales, J., Ortigas-Galindo, J.: Intersection theory on abelian-quotient V-surfaces and $\mathbf{Q}$-resolutions. J. Singul. **8**, 11–30 (2014)
2. Blache, R.: Riemann-Roch theorem for normal surfaces and applications. Abh. Math. Sem. Univ. Hamburg **65**, 307–340 (1995)
3. Brenton, L.: On the Riemann-Roch equation for singular complex surfaces. Pac. J. Math. **71**(2), 299–312 (1977)
4. Chen, W.: Orbifold adjunction formula and symplectic cobordisms between lens spaces. Geom. Topol. **8**, 701–734 (2004)
5. Cogolludo-Agustín, J.I., Martín-Morales, J.: The space of curvettes of quotient singularities and associated invariants. Preprint available at arXiv:1503.02487v1 [math.AG] (2015)
6. Dolgachev, I.: Weighted projective varieties. In: Group Actions and Vector Fields (Vancouver, B.C., 1981). Lecture Notes in Mathematics, vol. 956, pp. 34–71. Springer, Berlin (1982)

7. Esnault, H., Viehweg, E.: Revêtements cycliques. In: Algebraic Threefolds (Varenna, 1981). Lecture Notes in Mathematics, vol. 947, pp. 241–250. Springer, Berlin/New York (1982)
8. Laufer, H.B.: Normal Two-Dimensional Singularities. Annals of Mathematics Studies, vol. 71. Princeton University Press/University of Tokyo Press, Princeton/Tokyo (1971)
9. Sakai, F.: Weil divisors on normal surfaces. Duke Math. J. **51**(4), 877–887 (1984)
10. Sakai, F.: Ruled fibrations on normal surfaces. J. Math. Soc. Jpn. **40**(2), 249–269 (1988)

Ulrich Bundles on Veronese Surfaces

Laura Costa and Rosa Maria Miró-Roig

Abstract It is a longstanding problem to determine whether the d-uple Veronese embedding of $\mathbb{P}^k$ supports a rank r Ulrich bundle. In this short note, we explicitly determine the integers d and r such that rank r Ulrich bundles on $\mathbb{P}^2$ for the Veronese embedding $\mathcal{O}(d)$ exist and, in particular, we solve Conjecture A.1 in Coskun and Genc (Proc Am Math Soc 145:4687–4701, 2017).

1 Introduction

Let Y be a smooth projective variety of dimension n and H a very ample divisor on Y. A vector bundle $\mathcal{E}$ on Y is an *Ulrich bundle* if it satisfies $H^i(Y, \mathcal{E}(-iH)) = 0$ for all $i > 0$ and $H^j(Y, \mathcal{E}(-(j+1)H)) = 0$ for all $j < n$. Equivalently, $\mathcal{E}$ is an Ulrich bundle if it satisfies $H^i(Y, \mathcal{E}(-tH)) = 0$ for all $i \geq 0$ and t in the range $1 \leq t \leq n$. Its pushforward to $\mathbb{P}^m$ via the inclusion of Y given by the ample line bundle $\mathcal{O}_Y(H)$ has a linear resolution or, equivalently, for any linear projection $Y \to \mathbb{P}^n$, its pushforward is trivial (see [6], Proposition 2.1).

Ulrich bundles have made a first appearance in commutative algebra, being associated to maximal Cohen-Macaulay graded modules with maximal number of generators, [13]. In algebraic geometry, their importance was underlined in [1] and [6], where a relation between their existence and the representations of Cayley–Chow forms was found. They have also important applications in liaison theory, singularity theory, moduli spaces and Boij-Söderberg theory. For instance, the existence of rank one (resp. rank two) Ulrich bundles on a hypersurface is related with its representation as a determinantal (resp. Pfaffian) variety. In spite of the increasing interest on Ulrich bundles due to their applications and in spite that in the literature there are few examples (see [4–7, 10, 11] and [12] and references quoted there), multitude of theoretical questions related to these bundles remain

L. Costa (✉) · R. M. Miró-Roig
Departament de Matemàtiques i Informàtica, Facultat de Matemàtiques i Informàtica, Barcelona, Spain
e-mail: costa@ub.edu; miro@ub.edu

G.-M. Greuel et al. (eds.), *Singularities, Algebraic Geometry, Commutative Algebra, and Related Topics*, https://doi.org/10.1007/978-3-319-96827-8_14

open. As was mentioned in [6] the following natural challenging problem is of great importance:

Problem Is every smooth variety $Y \subset \mathbb{P}^n$ the support of an Ulrich bundle? If so, what is the smallest possible rank for such a bundle?

In this short note, we study the existence problem for Ulrich bundles on Veronese surfaces. More precisely, we consider $\mathbb{P}^2$ together with the Veronese d-uple embedding $\mathbb{P}^2 \to \mathbb{P}^{\binom{d+2}{d}-1}$ defined by means of the very ample line bundle $\mathcal{O}_{\mathbb{P}^2}(d)$ and we want to determine the couples $(r, d) \in \mathbb{Z}^2$ such that $(\mathbb{P}^2, \mathcal{O}(d))$ carries a rank r Ulrich bundle. In fact, after some preliminaries stated in Sect. 2, in Sect. 3 we prove the following result

Theorem 1 *Assume $d, r \geq 2$. A rank r Ulrich bundle $\mathcal{E}$ on $(\mathbb{P}^2, \mathcal{O}(d))$ exists if and only if $r(d-1) \equiv 0$ (mod. 2). In addition, for $d \geq 3$ and $r(d-1) \equiv 0$ (mod. 2), there is a stable (hence undecomposable) rank r Ulrich bundle $\mathcal{E}$ on $(\mathbb{P}^2, \mathcal{O}(d))$, it has natural cohomology and a minimal free resolution of the following type:*

$$0 \longrightarrow \mathcal{O}^{\frac{r}{2}(d-1)}(-2) \longrightarrow \mathcal{O}^{\frac{r}{2}(d+1)}(-1) \longrightarrow \mathcal{E}(-d) \longrightarrow 0.$$

In particular, we completely solve the following conjecture stated in [3].

Conjecture 1 There exists a rank 3 Ulrich bundle on $(\mathbb{P}^2, \mathcal{O}(d))$ for all odd degrees $d \geq 3$.

2 Preliminaries

In this section we briefly review the definition and basic properties of Ulrich bundles.

Definition 1 Let $X \subset \mathbb{P}^N$ be an n dimensional smooth projective variety embedded by a very ample line bundle $\mathcal{O}_X(1)$. A vector bundle $\mathcal{E}$ on X is an **initialized Ulrich** bundle if $H^q(\mathcal{E}(j)) = 0$ for $0 < q < n$ and $j \in \mathbb{Z}$, $H^0(\mathcal{E}(-1)) = 0$ and $h^0(\mathcal{E}) = deg(X) rank(\mathcal{E})$.

We recall the following proposition due to Eisenbud and Schreyer and we refer to [6] and [2] for more details.

Theorem 2 *Let $X \subseteq \mathbb{P}^N$ be an n-dimensional smooth projective variety and let $\mathcal{E}$ be an initialized (i.e. $H^0(\mathcal{E}(-1)) = 0$ and $H^0(\mathcal{E}) \neq 0$) vector bundle on X. The following conditions are equivalent:*

(i) *$\mathcal{E}$ is Ulrich.*
(ii) *$\mathcal{E}$ admits a linear $\mathcal{O}_{\mathbb{P}^N}$-resolution of the form:*

$$0 \longrightarrow \mathcal{O}_{\mathbb{P}^N}(-N+n)^{a_{N-n}} \longrightarrow \ldots \longrightarrow \mathcal{O}_{\mathbb{P}^N}(-1)^{a_1} \longrightarrow \mathcal{O}_{\mathbb{P}^N}^{a_0} \longrightarrow \mathcal{E} \longrightarrow 0$$

with $a_0 = \deg(X) \times rank(\mathcal{E})$ and $a_i = \binom{N-n}{i} a_0$.

(iii) $H^q(\mathcal{E}(-q)) = 0$ *for* $q > 0$ *and* $H^q(\mathcal{E}(-q-1)) = 0$ *for* $q < n$.
(iv) $H^q(\mathcal{E}(-i)) = 0$ *for all* q *and* $1 \leq i \leq n$.
(v) *For some (resp. all) finite linear projections* $\pi : X \longrightarrow \mathbb{P}^n$*, the sheaf* $\pi_*\mathcal{E}$ *is* $\mathcal{O}^t_{\mathbb{P}^n}$ *for some* t.

In particular, initialized Ulrich bundles are 0*-regular and therefore they are globally generated.*

Proof See [6], Proposition 2.1 and [2], Theorem 1. □

The definition of Ulrich bundle depends on the polarization $\mathcal{O}_X(1)$ of the underlying variety X and the change of the polarization can affect the existence of Ulrich bundles. In this work, we study the existence of Ulrich bundles of arbitrary rank on $\mathbb{P}^n$ with respect to the polarization $\mathcal{O}_{\mathbb{P}^n}(d)$ for any integer $d \geq 1$. From now on, for any positive integer d, we denote by $(\mathbb{P}^n, \mathcal{O}(d))$ the n-dimensional *Veronese variety* defined as the image of the Veronese embedding $\mathbb{P}^n \longrightarrow \mathbb{P}^{\binom{n+d}{d}-1}$ given by the very ample line bundle $\mathcal{O}_{\mathbb{P}^n}(d)$ and we address the following problem:

Problem 1 To determine the triples $(n, d, r) \in \mathbb{Z}^3$ such that $(\mathbb{P}^n, \mathcal{O}(d))$ carries a rank r Ulrich bundle.

Definition 2 We say that a vector bundle $\mathcal{E}$ on $\mathbb{P}^n$ has **natural cohomology** if for any integer i at most one cohomology group $H^j(\mathbb{P}^n, \mathcal{E}(i)) \neq 0$ for $0 \leq j \leq n$.

As an example of vector bundle with natural cohomology we have the cotangent bundle $\Omega^1_{\mathbb{P}^n}$. In this paper vector bundles with natural cohomology will play an important role since in [6], Eisenbud and Schreyer proved that any Ulrich bundle $\mathcal{E}$ on $(\mathbb{P}^n, \mathcal{O}(d))$ has *natural cohomology* and they obtained a criteria to find triples (n, d, r) such that there is no rank r Ulrich bundles $\mathcal{E}$ on $(\mathbb{P}^n, \mathcal{O}(d))$ (see[6], Corollary 5.3). Other restrictions on the Chern classes of rank r Ulrich bundles $\mathcal{E}$ on $(\mathbb{P}^n, \mathcal{O}(d))$ come from the following result:

Proposition 1 *Let* $\mathcal{E}$ *be a rank* r *Ulrich bundle on* $(\mathbb{P}^n, \mathcal{O}(d))$*. Then,* $\mathcal{E}$ *has a resolution of the following type:*

$$0 \longrightarrow \mathcal{O}^{a_n}(-n) \longrightarrow \cdots \longrightarrow \mathcal{O}^{a_2}(-2) \longrightarrow \mathcal{O}^{a_1}(-1) \longrightarrow \mathcal{E}(-d) \longrightarrow 0,$$

where

$$a_j = -rd^n \sum_{k=0}^{j} (-1)^{j-k} \binom{n+1}{k} \binom{n-1+\frac{j-k}{d}}{n}.$$

Proof See [3], Corollary 4.3 and [8], Theorem 4.3.

In spite of this result, the problem 1 is rather open and few results are known. We summarize them here.

Since the cohomology of line bundles on $\mathbb{P}^n$ is well known (Bott's formula), it is easy to check that $\mathcal{E}$ is an Ulrich line bundle on $(\mathbb{P}^n, \mathcal{O}_{\mathbb{P}^n}(d))$ if and only if $d = 1$

and $\mathscr{E} \cong \mathscr{O}_{\mathbb{P}^n}$ or $n = 1$ and $\mathscr{E} \cong \mathscr{O}_{\mathbb{P}^1}(d-1)$ So, from now on we will assume $r \geq 2$. Also according to Horrocks (see also [8], Corollary 4.4), a rank r vector bundle on $(\mathbb{P}^n, \mathscr{O}_{\mathbb{P}^n}(1))$ is an initialized Ulrich bundle if and only if $\mathscr{E} \cong \mathscr{O}^r_{\mathbb{P}^n}$. Hence, we will also assume that $d \geq 2$. In [2], Proposition 4, Beauville showed that for any integers $n, d \geq 1$ there exists an Ulrich bundle $\mathscr{E}$ of rank $n!$ on the d-uple Veronese embedding of $\mathbb{P}^n$ using the quotient map $p : (\mathbb{P}^1)^n \longrightarrow Sym^n\mathbb{P}^1 = \mathbb{P}^n$ of degree $n!$ and in [6], Corollary 5.7 Eisenbud and Schereyer proved that there is a unique indecomposable homogeneous bundle on $\mathbb{P}^n$ that is an Ulrich bundle for the d-uple, namely $S_\lambda Q$ where Q is the universal rank n quotient bundle on $\mathbb{P}^n$ and λ the partition with $\lambda = ((d-1)(n-1), (d-1)(n-2), \cdots, (d-1), 0)$. It has rank $d^{\binom{n}{2}}$.

So, the existence of Ulrich bundles on the d-uple Veronese embedding of $\mathbb{P}^n$ is well known. Unfortunately as n or d increase $n!$ and $d^{\binom{n}{2}}$ grow very fast and we do not yet know the smallest possible rank of an Ulrich bundle $\mathscr{E}$ on the d-uple Veronese embedding of $\mathbb{P}^n$ for $n \geq 4$ and $d \geq 2$. For $(n, d) = (1, d)$ or $(n, 1)$ the smallest rank is 1, for $(n, d) = (2, d)$ the smallest rank is 2 (see Theorem 3), and for $(n, d) = (3, d)$ the smallest rank is 2 if and only if d is not equivalent to 0 (mod 3) (see [6], Proposition 5.10). So, we are led to pose the following problem

Problem 2 What is the smallest possible rank of an Ulrich bundle $\mathscr{E}$ on the d-uple Veronese embedding of $\mathbb{P}^n$?

3 Ulrich Bundles on Veronese Surfaces

In this section we consider the Veronese embedding $\mathbb{P}^2 \longrightarrow \mathbb{P}^{\binom{d+2}{2}-1}$ given by the very ample line bundle $\mathscr{O}_{\mathbb{P}^2}(d)$. Its image is the Veronese surface of degree d^2 that we denote by $(\mathbb{P}^2, \mathscr{O}(d))$. The goal of this section is to discuss the existence of undecomposable rank r Ulrich bundles on $(\mathbb{P}^2, \mathscr{O}(d))$. More precisely, we completely determine the pairs $(d, r) \in \mathbb{Z}^2$ such that $(\mathbb{P}^2, \mathscr{O}(d))$ carries a rank r Ulrich bundle and, in particular, we solve Conjecture A.1 in [3].

Ulrich line bundles on $(\mathbb{P}^2, \mathscr{O}(d))$ exist only for $d = 1$ and, for $d = 1$, any rank r Ulrich bundle is a direct sum of r copies of $\mathscr{O}_{\mathbb{P}^2}$. So, from now on, we assume $d \geq 2$. The following characterization of Ulrich bundles on $(\mathbb{P}^2, \mathscr{O}(d))$ will be essential for proving the main result of this note.

Proposition 2 *Let $\mathscr{E}$ be a rank r vector bundle on $(\mathbb{P}^2, \mathscr{O}(d))$. It holds:*

(i) *If $\mathscr{E}$ is an Ulrich bundle, then it has a resolution of the following type:*

$$0 \longrightarrow \mathscr{O}^{\frac{r}{2}(d-1)}(-2) \longrightarrow \mathscr{O}^{\frac{r}{2}(d+1)}(-1) \longrightarrow \mathscr{E}(-d) \longrightarrow 0.$$

(ii) *If $\mathscr{E}$ has natural cohomology and a resolution of the following type:*

$$0 \longrightarrow \mathscr{O}^{\frac{r}{2}(d-1)}(-2) \longrightarrow \mathscr{O}^{\frac{r}{2}(d+1)}(-1) \longrightarrow \mathscr{E}(-d) \longrightarrow 0 \quad (1)$$

then, $\mathscr{E}$ is an Ulrich bundle.

Proof

(i) See[3], Corollary 4.3 or [8], Corollary 4.6.
(ii) It follows from [8], Corollary 4.6, taking into account that for vector bundles with resolution (1) and natural cohomology, the induced map

$$H^2(\mathcal{O}^{\frac{r}{2}(d-1)}(-d-2)) \longrightarrow H^2(\mathcal{O}^{\frac{r}{2}(d+1)}(-d-1))$$

has maximal rank, i.e. it is injective or surjective.

Notice that it follows from Proposition 2 that an Ulrich vector bundle on $(\mathbb{P}^2, \mathcal{O}(d))$ for even d, must have even rank. For $d = 2$, by [3], Theorem 5.2 and Theorem 5.4, there exists a unique rank 2 Ulrich bundle and higher rank Ulrich bundles for $d = 2$ are direct sum of this unique rank 2 Ulrich bundle. Also in [3], Coskun and Genc proved that for $d > 2$ even there exist Ulrich bundles for any even rank r and for $d > 2$ odd they conjectured that Ulrich bundles exist for any rank r. More precisely, they stated the following conjecture [3], Conjecture A.1):

Conjecture 2 There exists a rank 3 Ulrich bundle on $(\mathbb{P}^2, \mathcal{O}(d))$ for all odd degrees $d \geq 3$.

Using Macaulay2, Coskun and Genc demonstrated that rank 3 Ulrich bundles exist on $(\mathbb{P}^2, \mathcal{O}(d))$ for odd degrees up to $d = 43$.

Our next result completely solves the existence of rank r Ulrich bundles on Veronese surfaces $(\mathbb{P}^2, \mathcal{O}(d))$ and, in particular, proves the above conjecture.

Theorem 3 *Assume $d, r \geq 2$. A rank r Ulrich bundle $\mathcal{E}$ on $(\mathbb{P}^2, \mathcal{O}(d))$ exists if and only if $r(d-1) \equiv 0$ (mod. 2). In addition, for $d \geq 3$ and $r(d-1) \equiv 0$ (mod. 2), there is a stable (hence undecomposable) rank r Ulrich bundle $\mathcal{E}$ on $(\mathbb{P}^2, \mathcal{O}(d))$, it has natural cohomology and a minimal free resolution of the following type:*

$$0 \longrightarrow \mathcal{O}^{\frac{r}{2}(d-1)}(-2) \longrightarrow \mathcal{O}^{\frac{r}{2}(d+1)}(-1) \longrightarrow \mathcal{E}(-d) \longrightarrow 0.$$

Proof We analyze separately the different values of d.

For $d = 2$ then r is necessarily even and, by [3], Theorems 5.2 and 5.4, the tangent bundle $T_{\mathbb{P}^2}$ is the unique rank 2 Ulrich bundle on $(\mathbb{P}^2, \mathcal{O}(2))$ and $T^s_{\mathbb{P}^2}$ is the unique rank $2s$ Ulrich bundle on $(\mathbb{P}^2, \mathcal{O}(2))$.

For even $d > 2$, again r is necessarily even and by [3], Theorem 6.1 there exist rank $2s$ stable (hence undecomposable) Ulrich bundles on $(\mathbb{P}^2, \mathcal{O}(d))$ for all $s \geq 1$.

Let us assume d is odd. By [3], Theorem 6.2, for $d \geq 3$ odd there exist stable Ulrich bundles of any rank $r \geq 2$ on $(\mathbb{P}^2, \mathcal{O}(d))$ if and only if there exists a rank 3 stable Ulrich bundle on $(\mathbb{P}^2, \mathcal{O}(d))$. So, our goal will be to prove the existence of rank 3 Ulrich bundles on $(\mathbb{P}^2, \mathcal{O}(d))$ for all odd integers $d \geq 3$. To this end, we write $d = 2k + 1$ with $k \geq 1$ and we denote by $\mathcal{M} := \mathcal{M}_{\mathbb{P}^2}(3; 0, \frac{3}{2}k(k+1))$ the moduli space of rank 3 stable (in the sense of Mumford-Takemoto) vector bundles $\mathcal{F}$ on $\mathbb{P}^2$ with Chern classes $c_1(\mathcal{F}) = 0$ and $c_2(\mathcal{F}) = \frac{3}{2}k(k+1)$. By [9], Corollary 2.4.2, a generic element $\mathcal{F}$ of the moduli space $\mathcal{M}$ has natural cohomology and a resolution of the following type

$$0 \longrightarrow \mathcal{O}^{t-\alpha_-}(-s-1) \oplus \mathcal{O}^{\alpha_-}(-s) \longrightarrow \mathcal{O}^{t+3-\alpha_+}(-s+1) \oplus \mathcal{O}^{\alpha_+}(-s) \longrightarrow \mathcal{F} \longrightarrow 0$$

where

$$\begin{aligned} s &= \max\{\rho \in \mathbb{Z} \mid 3\rho^2 + 2c_1(\mathcal{F})\rho - 3\rho \leq 2c_2(\mathcal{F}) - c_1(\mathcal{F})^2 + c_1(\mathcal{F}) - 1\} \\ \alpha_+ &= \max(0, 2c_2(\mathcal{F}) - c_1(\mathcal{F})^2 + 3 - 3s^2 - 2c_1(\mathcal{F})s) \\ \alpha_- &= \max(0, -2c_2(\mathcal{F}) + c_1(\mathcal{F})^2 - 3 + 3s^2 + 2c_1(\mathcal{F})s) \\ t &= (3s + c_1(\mathcal{F}) - 3 + 2c_2(\mathcal{F}) - c_1(\mathcal{F})^2 + 3 - 3s^2 - 2c_1(\mathcal{F})s)/2. \end{aligned}$$

We easily check that

$$s = k, \quad \alpha_+ = 3(k+1), \quad \alpha_- = 0, \text{ and } t = 3k.$$

Therefore, the resolution of $\mathcal{F}$ has the following shape:

$$0 \longrightarrow \mathcal{O}^{3k}(-k-1) \longrightarrow \mathcal{O}^{3k+3}(-k) \longrightarrow \mathcal{F} \longrightarrow 0$$

and, applying Proposition 2, we conclude that $\mathcal{E} := \mathcal{F}(k-1+d)$ is a rank 3 Ulrich bundle on $(\mathbb{P}^2, \mathcal{O}(d))$ for any odd $d = 2k+1$.

Acknowledgements The authors dedicate this note to Antonio Campillo for his 65th birthday in recognition to his contributions to Algebraic Geometry in Spain. The authors are partially supported by MTM2016-78623-P.

References

1. Beauville, A.: Determinantal hypersurfaces. Michigan Math. J. **48**, 39–64 (2000)
2. Beauville, A.: An Introduction to ulrich bundles. Eur. J. Math. (2017). https://doi.org/10.1007/s40879-017-0154-4
3. Coskun, E., Genc, O.: Ulrich bundles on veronese surfaces. Proc. Am. Math. Soc. **145**, 4687–4701 (2017)
4. Costa, L., Miró-Roig, R.M.: GL(V)-invariant Ulrich bundles on Grassmannians. Math. Ann. **361**(1–2), 443–457 (2015)
5. Costa, L., Miró-Roig, R.M., Pons-Llopis, J.: The representation type of Segre varieties. Adv. Math. **230**, 1995–2013 (2012)
6. Eisenbud, D., Schreyer, F.-O. (with an appendix by J. Weyman): Resultants and Chow forms via exterior syzygies. J. Am. Math. Soc. **16**, 537–579 (2003)
7. Kleppe, J.O., Miró-Roig, R.M.: On the normal sheaf of determinantal varieties. J. Reine Angew. Math. Ahead of Print Journal. https://doi.org/10.1515/crelle-2014-0041
8. Lin, Z.: Ulrich bundles on projective spaces. Preprint arXiv 1703.06424
9. Maggesi, M.: Some results on holomorphic vector bundles over projective spaces. Tesi di Dottorato, Università di Firenze (1999)
10. Miró-Roig, R.M.: The representation type of rational normal scrolls. Rendiconti del Circolo Matematico di Palermo **62**, 153–164 (2013)

11. Miró-Roig, R.M., Pons-Llopis, J.: Representation type of rational ACM surfaces $X \subset \mathbb{P}^4$. Algebr. Represent. Theory **16**, 1135–1157 (2013)
12. Miró-Roig, R.M., Pons-Llopis, J.: N-dimensional Fano varieties of wild representation type. J. Pure Appl. Alg. **218**, 1867–1884 (2014)
13. Ulrich, B.: Gorenstein rings and modules with high numbers of generators. Math. Z. **188**, 23–32 (1984)

Multiple Structures on Smooth on Singular Varieties

M. R. Gonzalez-Dorrego

Dedicated to Professor Antonio Campillo

MSC: 14B05, 14E15, 32S25, 14J17, 14J30, 14J35, 14J40, 14J70

Let k an algebraically closed field, $char\ k = 0$. Let C be an irreducible nonsingular curve such that $rC = S \cap F, r \in \mathbb{N}$, where S and F are two surfaces in $\mathbb{P}^3$ and all the singularities of F are of the form $z^p = x^{ps} - y^{ps}$, p prime, $s \in \mathbb{N}$. We prove that C can never pass through such kind of singularities of a surface, unless $r = pa, a \in \mathbb{N}$. These singularities are Kodaira singularities. We also study multiplicity-r structures on higher-dimensional varieties, where $r \in \mathbb{N}$. Let F be a $(N-1)$-fold in $\mathbb{P}^N$. Let Z be a reduced irreducible nonsingular $(n-1)$-dimensional variety such that $rZ = X \cap F$, where X is a normal n-fold in $\mathbb{P}^N$, with certain type of singularities, like lincar compound Vp singularity or (d,l) complete intersection compound Vp singularity, and such that $Z \cap \mathrm{Sing}(X) \neq \emptyset$. We prove that Z can never pass through such kind of singularities of a n-fold, unless $r = pa$, $a \in \mathbb{N}$. The case with $p = 3$ was studied in [2].

We would like to thank the Department of Mathematics at the University of Toronto for their hospitality during the preparation of this manuscript and Mark Spivakovsky for useful discussions.

Definition 1 Let X be an n-dimensional normal variety and P a point of X. Let P be an n-fold isolated singularity (that is, the spectrum of an equicharacteristic local noetherian ring of Krull dimension n, without zero divisors, whose closed point P is singular). Let $\pi : \tilde{X} \to X$ be the desingularization of X at P. The **genus** of a normal singularity P is defined to be $\dim_k (R^{n-1}\pi_* \mathcal{O}_{\tilde{X}})_P$. If the genus is 0, the singularity is said to be **rational**. If the genus is 1, it is **elliptic**.

M. R. Gonzalez-Dorrego (✉)
Departamento de Matemáticas, Universidad Autónoma de Madrid, Madrid, Spain
e-mail: mrosario.gonzalez@uam.es

G.-M. Greuel et al. (eds.), *Singularities, Algebraic Geometry, Commutative Algebra, and Related Topics*, https://doi.org/10.1007/978-3-319-96827-8_15

Notation 2 *Let F be a reduced surface and P a point of F. Let (F, P) be a surface singularity (that is, the spectrum of an equicharacteristic local ring of Krull dimension 2 whose closed point P is singular). Let $\pi : \tilde{F} \to F$ be the minimal desingularization of F at P.*

Definition 3 Let X be a non-singular complex surface. Let $\Delta \in \mathbb{C}$ a small open disc around the origin.

- Let $\Phi : S \to \Delta$ be surjective proper holomorphic map. If the fibers $S_t = \Phi^{-1}(t)$, $t \neq 0$, are non-singular connected curves of genus g, we call it a **pencil of curves** of genus g. S_0 is called the **singular fiber.**
- Let $\Phi : S \to \Delta$ a pencil of curves of genus g which has reduced components. Let $P_1, \ldots, P_r \in Supp(S_0)$ be non-singular points of S_0. Let $\sigma : S' \to S$ be a finite number of blowing-ups at $P_1, \ldots, P_r$. Let $\psi : M \to X$ be a holomorphic map from an open neighbourhood M of the strict transform of $Supp(S_0)$ in S' which defines a resolution of X. A normal surface singularity isomorphic to a singularity obtained in this way is called **a Kodaira singularity** of genus g (or associated to Φ) [6, 2.1, 2.2].

Definition 4 Let F be a reduced surface with a singular point at O. We shall denote by (V_{a_0,a_1,a_2}, O) a singularity analytically isomorphic to

$$\{(x_0, x_1, x_2) \in U \subset k^3 \text{ such that } {x_0}^{a_0} + {x_1}^{a_1} = {x_2}^{a_2}\}$$

with $2 < a_0 \leq a_1 \leq a_2$.

Notation 5 *Let* $l = gcd(a_0, a_1, a_2)$, $l_i = \frac{gcd(a_j,a_k)}{l}$, $\alpha_i = \frac{a_i}{l_j l_k l}$, $i, j, k \in \{0, 1, 2\}$, $i \neq j, i \neq k, j \neq k$.

Proposition 6 *Let* $k = \mathbb{C}$. (V_{a_0,a_1,a_2}, O), $2 \leq a_0 \leq a_1 \leq a_2$, *is a Kodaira singularity if and only if* $\alpha_0\alpha_1 l_2 \leq \alpha_2$.

Proof [4, Prop.4.4].

Lemma 7 *Let* $k = \mathbb{C}$. $(V_{p,ps,ps}, O)$, *p prime,* $s \in \mathbb{N}$ *is a Kodaira singularity.*

Proof According to (0.5), $l = p$, $l_0 = s$, $l_1 = l_2 = 1$, $\alpha_0 = \alpha_1 = \alpha_2 = 1$. By Proposition 0.6, $(V_{p,ps,ps}, O)$, $s \in \mathbb{N}$ is a Kodaira singularity since $\alpha_0\alpha_1 l_2 = 1 \leq \alpha_2 = 1$.

Definition 8

- We call **maximal cycle** $Z_{\tilde{F}}$ the cycle $Z_{\tilde{F}} = \sum m_i E_i$, defined by the divisorial part of $\mathcal{M}\mathcal{O}_{\tilde{F}}$, where $\mathcal{M}$ is the maximal ideal $Max\mathcal{O}_{F,P}$ of $\mathcal{O}_{F,P}$; the E_i are the irreducible components of dimension 1 of the exceptional fiber $\pi^{-1}(P)$ and the m_i are nonnegative integers. A component E_j such that $m_j = 1$ is called a reduced component of the cycle.

- Consider positive cycles $Z = \sum r_i E_i$, $r_i \geq 0$ such that $(Z.E_i) \leq 0$, for all i. The unique smallest cycle Z satisfying $(Z.E_i) \leq 0$, for all i, is called **the fundamental cycle** of $\tilde{F}$.

The fundamental cycle for $(\mathbf{V_{p,ps,ps}}, \mathbf{O})$, p prime, $\mathbf{s} \in \mathbb{N}$.

Let us denote $x_0 = z$, $x_1 = y$, $x_2 = x$. We consider the surface singularity (F, O) given by $z^p = x^{ps} - y^{ps}$. We want to desingularize F at O.

Applying to $z^p = x^{ps} - y^{ps}$ the change $x_{(1)} = \frac{x}{y}$, $z_{(1)} = \frac{z}{y}$, $y_{(1)} = y$, we obtain in $F_{(1)}$ that $z_{(1)}{}^p = y_{(1)}{}^{ps-p}(x_{(1)}{}^{ps} - 1)$; $\pi_1 : F_{(1)} --- > F$, $E = \pi_1{}^{-1}(O)$ exceptional divisor given by $y_{(1)} = 0$.

Let us apply to $z_{(1)}{}^p - y_{(1)}{}^{ps-p}(x_{(1)}{}^{ps} - 1)$ the change $x_{(2)} = x_{(1)}$, $z_{(2)} = \frac{z_{(1)}}{y_{(1)}}$, $y_{(2)} = y_{(1)}$.

We obtain $z_{(2)}{}^p = y_{(2)}{}^{ps-2p}(x_{(2)}{}^{ps} - 1)$ in $F_{(2)}$; $\pi_2 : F_{(2)} --- > F_{(1)}$, $G = \pi_2{}^{-1}(E)$ exceptional divisor given by $y_{(2)} = 0$. We repeat the process $s - 1$ times until we arrive at $z_{(s)}{}^p = x_{(s)}{}^{ps} - 1$ in $F_{(s)}$; $\pi_s : F_{(s)} --- > F_{(s-1)}$, $D = \pi_s{}^{-1}(E^{(s-1)})$ exceptional divisor.

The fundamental cycle is $Z = D$. Its self-intersection is $D.D = -p$. To know the genus of D we apply Riemann-Hurwitz Formula. We obtain $2g - 2 = p(0 - 2) + (p - 1)(ps)$, thus $g = (p - 1)(\frac{1}{2}(ps) - 1)$. We shall write the weighted dual graph of the fundamental cycle as

$$\overset{1}{\bullet}$$
$$[(p-1)(\tfrac{1}{2}(ps)-1)]$$

1 denotes that D is a reduced component of the fundamental cycle;
$[(p-1)(\frac{1}{2}(ps)-1)]$ denotes that the genus of D is $(p-1)(\frac{1}{2}(ps)-1)$; $D^2 = -p$.

Proposition 9 *The families of smooth curves on a normal surface singularity are in one-to-one correspondence with the reduced components of the maximal cycle of its minimal desingularization π.*

Proof [3, 1.14].

Proposition 10 *For a surface singularity of the type (V_{a_0,a_1,a_2}, O), the maximal cycle of π and the fundamental cycle of its weighted dual graph coincide if and only if $\alpha_2 \geq l_2$.*

Proof [4, Th. 3.6].

Corollary 11 *For a surface singularity of the type $(V_{p,ps,ps}, O)$, p prime, $s \in \mathbb{N}$, the maximal cycle of π coincides with the fundamental cycle of its weighted dual graph.*

Proof It follows from Proposition 10 since $\alpha_2 = 1 \geq l_2 = 1$.

Theorem 12 *Let C be an irreducible nonsingular curve such that $rC = S \cap F$, where S and F are surfaces in $\mathbb{P}^3$ and $r \in \mathbb{N}$. Then C cannot pass through any singular point $(V_{p,ps,ps}, O)$, p prime, $s \in \mathbb{N}$ of F, unless $r = pa$, $a \in \mathbb{N}$.*

Proof By Proposition 9 and Corollary 11, if an irreducible nonsingular curve C passes through a singularity of F, then its strict transform must intersect transversally only one exceptional divisor of multiplicity one in the fundamental cycle. Since our curve C is not a Cartier divisor, we consider rC, which is a Cartier divisor; thus, its preimage in the minimal resolution of F must have intersection 0 with each exceptional divisor. Let $\tilde{F}$ be the minimal resolution of F. Consider the fundamental cycle $Z = D$ of a surface singularity of the type $(V_{p,ps,ps}, O)$, p prime, $s \in \mathbb{N}$, (see (1.1)). If C would pass through such a singularity, the total transform of rC in $\tilde{F}$, for r, $a \in \mathbb{N}$, is the cycle

$$\underset{[(p-1)(\frac{1}{2}(ps)-1)]}{a \bullet} - - - - \text{or}$$

This cycle has intersection 0 with D only if $r = pa$, $a \in \mathbb{N}$. Thus, there are no C passing through a surface singularity of the type $(V_{p,ps,ps}, O)$, p prime, $s \in \mathbb{N}$, unless $r = pa$, $a \in \mathbb{N}$.

Definition 13

- Let $(\mathscr{O}_{X,P}, \mathscr{M}_P)$ be the local ring of a point $P \in X$ of a k-scheme. Let $V \subset \mathscr{M}_P$ be a finite dimensional k-vector space which generates $\mathscr{M}_P$ as an ideal of $\mathscr{O}_{X,P}$. By a **general hyperplane through P** we mean the subscheme H defined in a suitable open neighbourhood U of P by the ideal $(v)\mathscr{O}_X$, where $v \in V$ is a k-point of a certain dense Zariski open set in V. [5, (2.5)]. By a **general linear variety of codimension r through P** we mean the subscheme $L \subset U$ defined in a suitable open neighbourhood U of P by the ideal $(v_1, \ldots, v_r)\mathscr{O}_X$, where $v_1, \ldots, v_r \in V$ are k-points of a certain dense Zariski open set in V.
- Let X be a singular n-fold. We say that a point $Q \in \mathrm{Sing}(X)$ is **a general point of** Sing(**X**) if, for a general hyperplane H such that $Q \in H$ and for some divisorial resolution $f : V \to X$, the preimage $f^{-1}(Q)$ of Q and the strict transform $f_*^{-1}(X \cap H)$ satisfy $f^{-1}(Q) \subset f_*^{-1}(X \cap H)$.

  [1, 2.4].

Example 14 If Q is a nonsingular point of $\mathrm{Sing}(X)$ and both X and V are topological products with $\mathrm{Sing}(X)$ locally near Q, then Q is a general point of $\mathrm{Sing}(X)$.

Definition 15 Let X be a singular n-fold embedded in $\mathbb{P}^N$. $\dim \mathrm{Sing}(X) > 0$

- Let H be a hyperplane in $\mathbb{P}^N$ such that $H \cap \mathrm{Sing}(X) \neq \emptyset$. Denote $X \cap H$ by X_0. We say that X_0 is **a general hyperplane section meeting** Sing(**X**) if it is irreducible and, for some divisorial resolution $f : V \to X$, the total transform $f^*(X_0)$ is equal to the strict transform $f_*^{-1}(X_0)$.

- Let L_{r+1} be a linear variety of codimension $r+1$, $0 \leq r \leq n-3$, in $\mathbb{P}^N$ such that $L_{r+1} \cap \text{Sing}(X) \neq \emptyset$. Denote $X \cap L_{r+1}$ by W_r, $0 \leq r \leq n-3$. We say that W_r is **a general linear section meeting** Sing(**X**) if it is irreducible and, for some divisorial resolution $f: V \to X$, the total transform $f^*(W_r)$ is equal to the strict transform $f_*^{-1}(W_r)$.

Definition 16 Let X be a n-fold in $\mathbb{P}^N$. A point $P \in X$ is called a **linear compound Vp singularity or a lcVp point**, p prime, if, for a general linear variety W of codimension $n-2$ through P, $P \in W$, $X \cap W$ is a singularity analytically isomorphic to $V_{p,ps,ps}$, p prime, $s \in \mathbb{N}$.

Notation 17 *Fix a point $P \in Sing(X)$. Let L_{r+1} be a general linear variety of codimension $r+1$ in $\mathbb{P}^N$, $0 \leq r \leq n-3$ such that $Sing(X) \cap L_{r+1} \neq \emptyset$. Let $W_r = X \cap L_{r+1}$.*

Proposition 18 *Let X be a quasiprojective scheme over any field. Let L_{l+1} be a general linear variety of codimension $l+1$ in $\mathbb{P}^N$, $0 \leq l < N-1$. Let $W_l = X \cap L_{l+1}$. Let $r \in \mathbb{N}$. If X satisfies Serre's condition S_r, then so does W_l. Thus, if X is a normal variety, then so is W_l.*

Proof [1, 3.4].

Definition 19 Let us consider the d-uple embedding $\rho_d : \mathbb{P}^N \to \mathbb{P}^M$, $M = \binom{N+d}{d} - 1$. Let Y be a complete intersection nonsingular variety in $\mathbb{P}^N$ defined by l equations $\sum_{i=0}^{M} a_{ij} y_i$, $1 \leq j \leq l$, $1 \leq l \leq N-1$, where all the y_i, $0 \leq i \leq M$, are monomials of degree d in $x_0, \ldots, x_N$, for some $l, d \in \mathbb{N}$. We call Y a **(d, l) complete intersection nonsingular variety**.

Proposition 20 *Let X be a normal singular variety of dimension n in $\mathbb{P}^N$. Let Y be a (d,l) complete intersection nonsingular variety such that $Y \cap Sing(X) \neq \emptyset$. Let us consider the d-uple embedding $\rho_d : \mathbb{P}^N \to \mathbb{P}^M$, $M = \binom{N+d}{d} - 1$. We have that, as abstract varieties, $X \cap Y = \rho_d(X) \cap \tilde{Y}$, where $\tilde{Y}$ is a linear variety in $\mathbb{P}^M$ obtained from Y using ρ_d.*

Proof [2, §8].

Definition 21 Let X be a normal singular variety of dimension n in $\mathbb{P}^N$. Let Y be a (d,l) complete intersection nonsingular variety in $\mathbb{P}^N$ such that $Y \cap \text{Sing}(X) \neq \emptyset$. Let us consider the d-uple embedding $\rho_d : \mathbb{P}^N \to \mathbb{P}^M$, $M = \binom{N+d}{d} - 1$. We have that, as abstract varieties, $X \cap Y = \rho_d(X) \cap \tilde{Y}$, where $\tilde{Y}$ is a linear variety in $mathbb{P}^M$ obtained from Y using ρ_d. We say that Y is **a (d, l) general complete intersection nonsingular variety in $\mathbb{P}^N$ meeting** Sing **X** if $\tilde{Y}$ is a general linear variety in $\mathbb{P}^M$ meeting Sing $\rho_d(X)$, according to Definition 15.

Definition 22 Let X be a n-fold. A point $P \in X$ is called a **(d, l) complete intersection compound Vp singularity or a dlcicVp point** if, for a general (d,l) complete intersection nonsingular variety Y of codimension $n-2$ through P, $P \in Y$, $X \cap Y$ is a singularity analytically isomorphic to $V_{p,ps,ps}$, p prime, $s \in \mathbb{N}$.

Proposition 23 *Let Z be a reduced irreducible nonsingular $(n-1)$-dimensional variety such that $rZ = X \cap F$, $r \in \mathbb{N}$, where X is an n-fold and F is a $(N-1)$-fold in $\mathbb{P}^N$, X normal with a lcVp singularity P and such that $Z \cap Sing(X) \neq \emptyset$. Let W_{n-3} be a general linear variety as in (17). Z has empty intersection with a lcVp singularity of X such that $W_{n-3} \cap X = Q$, where Q is a surface singularity analytically isomorphic to $V_{p,ps,ps}$, p prime, $s \in \mathbb{N}$, unless $r = pa$, $a \in \mathbb{N}$.*

Proof Let us define recursively W_t. Let H_{t+1}, $0 \leq t \leq n-4$, be a general hyperplane meeting $\text{Sing}(W_t) \cap F$. Given $rZ = X \cap F$, $r \in \mathbb{N}$, $r \geq 2$, we intersect it with H_t, $0 \leq t \leq n-3$, as follows: $rZ \cap H_0 \cap \cdots \cap H_{n-3} = F \cap X \cap H_0 \cap \cdots \cap H_{n-3}$. We obtain a nonsingular curve C such that $rC = F \cap X \cap H_0 \cap \cdots \cap H_{n-3}$ and that $C \cap \text{Sing}(W_{n-3}) \neq \emptyset$. We apply Theorem 12 to obtain the result.

Corollary 24 *Let Z be a reduced irreducible nonsingular $(n-1)$-dimensional variety such that $rZ = X \cap F$, where X is an n-fold and F is a $(N-1)$-fold in $\mathbb{P}^N$, X normal with a dlcicVp singularity P and such that $Z \cap Sing(X) \neq \emptyset$. Let Y be a general (d, l) complete intersection nonsingular variety of codimension $n-2$ as in (19). Then Z has empty intersection with a dlcicVp singularity of X such that $Y \cap X = Q$, where Q is a surface singularity analytically isomorphic to $V_{p,ps,ps}$, $s \in \mathbb{N}$, p prime, unless $r = pa$, $a \in \mathbb{N}$.*

Proof Let us consider the d-uple embedding $\rho_d : \mathbb{P}^N \to \mathbb{P}^M$, $M = \binom{N+d}{d} - 1$. Let Y defined as in (19). We have that, as abstract varieties, $X \cap Y = \rho_d(X) \cap \tilde{Y}$, where $\tilde{Y}$ is a linear variety in $\mathbb{P}^M$. To obtain the result we follow the proof of Proposition 23 substituting W_{n-3} by $\tilde{Y}$.

Definition 25 Let $\pi : (\tilde{X}, E) \to (X, O)$ minimal resolution of the normal surface singularity (X, O). A divisor $\tilde{K}$ with support in $E = \cup_{i=1}^{n} E_i$ is called the **canonical cycle** if

$$-\tilde{K}.E_i = K.E_i,\ 1 \leq i \leq n,$$

where K is the canonical divisor of $\tilde{X}$. If the canonical cycle has integer coefficients, then the singularity (X, O) is called **numerically Gorenstein**. The canonical line bundle is topologically trivial.

Proposition 26 *Let $(X, O) = (V_{p,ps,ps}, O)$, p prime, $s \in \mathbb{N}$. The canonical cycle is $\tilde{K} = (s(p-1)-1)D$, $s(p-1)-1 \in \mathbb{N}$.*

Proof Let the canonical cycle be $\tilde{K} = aD$. By the adjunction formula $D(D + K) = 2g - 2$, where K is the canonical divisor and g is the genus of D which is $\frac{(p-1)ps}{2} - (p-1)$. Moreover $D^2 = -p$. Thus, $\tilde{K} = (s(p-1)-1)D$, $s(p-1) - 1 \in \mathbb{N}$.

References

1. Gonzalez-Dorrego, M.R.: Smooth double subvarieties on singular varieties. RACSAM **108**, 183–192 (2014)
2. Gonzalez-Dorrego, M.R.: Smooth double subvarieties on singular varieties, III. Banach Center Publ. **108**, 85–93 (2016)
3. Gonzalez-Sprinberg, G., Lejeune-Jalabert, M.: Families of smooth curves on singularities and wedges. Ann. Pol. Math. **LXVII**(2), 179–190 (1997)
4. Konno, K., Nagashima, D.: Maximal ideal cycles over normal surface singularities of Brieskorn type. Osaka J. Math. **49**, 225–245 (2012)
5. Reid, M.: Canonical threefolds. In: Beauville, A. (ed.) Journées de Géometrie Algébrique et Singularités, Angers 1979. Sijthoff and Noordhoff (1980), pp. 273–310
6. Tomaru, T.: Complex surface singularities and degenerations of compact complex curves. Demonstratio Math. **XLIII**(2), 339–359 (2010)

Smoothness in Some Varieties with Dihedral Symmetry and the DFT Matrix

Santiago López de Medrano

Para Antonio Campillo, en sus 65 años.

Abstract We study the smoothness question for some families of real and complex varieties with cyclic or dihedral symmetry. This question is related to deep properties of the Vandermonde matrix on the roots of unity, also known as the Discrete Fourier Transform matrix. We present some partial results on these questions.

1 Introduction

For many years I have studied with different collaborators the topology of transverse intersections of quadrics in $\mathbb{R}^n$. I will try to give a synthesis of the main results.

Starting from the most general ones, we have a characterization of the compact smooth manifolds which are transverse intersections of quadrics or transverse intersections of ellipsoids [9]. More specific intersections that are sometimes easier to describe completely are the transverse intersections of centered ellipsoids

$$Q_i(x) = 0, \quad i = 1, \ldots, m$$
$$\Sigma_{i=1}^n x_i^2 = 1$$

(where the Q_i are quadratic forms) or of coaxial ellipsoids, meaning that the quadratic forms are simultaneously diagonalizable:

$$\Sigma_{i=1}^n A_i x_i^2 = 0$$
$$\Sigma_{i=1}^n x_i^2 = 1$$

(where $A_i \in \mathbb{R}^m$).

S. López de Medrano (✉)
Instituto de Matemáticas, Universidad Nacional Autónoma de México, México City, México
e-mail: santiago@matem.unam.mx

G.-M. Greuel et al. (eds.), *Singularities, Algebraic Geometry, Commutative Algebra, and Related Topics*, https://doi.org/10.1007/978-3-319-96827-8_16

These are relatively easier to study because they have more symmetry: if Z is the previous intersection, changing the sign of x_i leaves invariant Z so there is an action of $\mathbb{Z}_2^n$ on them with quotient $Z \cap \mathbb{R}_+^n$ which is homeomorphic to the convex polyope P:

$$\Sigma_{i=1}^n A_i r_i = 0$$

$$\Sigma_{i=1}^n r_i = 1$$

$$r_i \geq 0, \quad i = 1, \ldots, n$$

One can recover Z topologically from P by reflecting it in all coordinate hyperplanes $x_i = 0$. So the convex polytope P and the configuration of coefficients are two combinatorial objects (called *Gale dual*) that contain all the information about Z. Furthermore, any polytope can be realized as the polytope P of such an intersection. When the intersection is transverse the polytope is of a special type called *simple*. This already suggests that a complete description of the topological types is close to the complete classification of the simple polytopes, a goal that looks out of reach.

Nevertheless, we have been able along the years to describe topologically large families of them:

(A) For $m = 2$ we have a complete description of all the transversal intersections, including the non-diagonal ones [11].

For the diagonal ones the configurations are collections of points in the plane, where the basic ones are the vertices of a regular polygon of n sides. So we can write their equations in terms of roots of unity:

$$\Sigma_{i=1}^n \rho^i x_i^2 = 0$$

$$\Sigma_{i=1}^n x_i^2 = 1$$

where $\rho = e^{2\pi i/n}$.

The rest of the objects are constructed by taking each of the coefficients with a multiplicity.

The first step for the topological description of these basic objects is to decide their smoothness. This follows very easily from the general criterion (see Sect. 5): they are smooth if, and only if, n is odd.

The second step is to describe the topology for all smooth cases and the proof (which includes all the generic non-diagonal cases) is quite long [11, 15, 16]. For the basic configurations it is a connected sum of n copies of $S^{(n-3)/2} \times S^{(n-3)/2}$; for the configurations with multiplicity it can be a triple product of spheres or a connected sum of spheres of different dimensions.

(B) For $m = n - 3$ we get surfaces by realizing the n-gon as the polytope of an intersection. These were first constructed by Hirzebruch as an unpublished side result to [13]. The connected ones have genus $g = 2^{n-3}(n-4)+1$. Other genera can be realized as transverse intersections of centered, non-coaxial ellipsoids [10].

(C) For $m > 2$, the generalization of the basic examples in (A) are all those intersections which are connected up to the middle dimension and which are constructed from the dual of a special type of polytope (called *neighbourly*). When taken with multiplicities they form a large subset of the possible cases and when the dimension is as least 5, it can be proved that they all are connected sums of products of spheres of different dimensions [8].

(D) Simple operations with the polytopes, such as products or truncation of a vertex or another face give rise to many new examples with known topology [8].

More recently, explicit equations were constructed in [6] of the equations for the surfaces in (B) such that the associated polygon is regular. One form of these equations is as follows:

$$\Sigma_{i=1}^{n} \rho^{ki} x_i^2 = 0, \quad k = 2, \ldots, [n/2]$$

$$\Sigma_{i=1}^{n} x_i^2 = 1$$

The homogeneous equations are of the same type as those in the basic example in (A), but different roots of unity are used in each equation, including powers of ρ from 2 to $[n/2]$. So the equations of this example and those of the basic example in (A) are complementary within the natural range $1, \ldots, [n/2]$ (observe that higher powers up to $n-1$ would give the conjugates of equations already considered and the equation with the n-th power would contradict that of the sphere). Observe that in both cases the configuration of coefficients, the variety and the polytope all have dihedral symmetry.[1]

Moreover, there is a strange *duality* between the examples: surprisingly, the configuration of the coefficients of the first one (the n-gon) coincides with the polytope of the second *and viceversa*: it turns out that the polytope of the first one coincides combinatorially with the configuration of the coefficients of the second one! However, there is an asymmetry: the second variety is always smooth, while the first one is smooth for n odd only.

A natural generalization in immediate: take $K \subset [1, \ldots, [n/2]]$ and $Z(K)$ be the variety.

$$\Sigma_{i=1}^{n} \rho^{ki} x_i^2 = 0, \quad k \in K$$

$$\Sigma_{i=1}^{n} x_i^2 = 1$$

[1] An abstract version of this construction was studied in [3], giving results about the topology of the quotient surfaces under the cyclic group.

The matrix of coefficients consists of rows of the Vandermonde matrix $V_n = (\rho^{ij})$, the $n \times n$ Vandermonde matrix evaluated on the n-th roots of unity, also known as the *Discrete Fourier Transform* (DFT) matrix.

In many particular examples one can check that the same type of duality holds between the combinatorial objects associated to $Z(K)$ and those of the complementary $Z(K^c)$, while the smoothness of one of them seems to be related to a different regularity property of the other.

This is an interesting phenomenon that is worth understanding. Furthermore, together with the examples obtained by including multiplicities on the basic ones this is an interesting new family of intersections of ellipsoids that contains and connects the families (A) and (B) and whose topological description could increase our knowledge about this subject.

But this family turns out to be much more difficult to study than the previous ones: just the first step (deciding which are the smooth cases, which in the previous families was easily solved) happens to be an extremely difficult problem in number theory involving the minors of V_n. We have up to now only partial information about this question. The subsequent steps (understanding the precise nature of the observed duality and describing the topology of the smooth examples) look also like two extremely difficult tasks.

The analogous family of projective algebraic varieties with cyclic or dihedral symmetry

$$\Sigma_{i=1}^{n} \rho^{ki} z_i^d = 0, \quad k \in K, \quad d > 1,$$

where K is a subset of $[1, \ldots, n]$ of cardinality m, is an interesting first step for the understanding the above questions. In this family, the relation between the smoothness of the family and the minors of V_n is simpler: smoothness is equivalent to the property that for the $m \times m$ matrix of coefficients of the equations, all $m \times m$ minors are non zero. This property is relatively simpler from both the theoretical and the computational point of view than the corresponding property in the real case. Furthermore, the duality in this case between complementary examples holds neatly: an example is smooth if, and only if, the complementary one is smooth.

This family is not only a toy example for the understanding of the real case, but it has life of its own: the simpler results about their smoothness already give examples of smooth intersections of hypersurfaces with cyclic or dihedral symmetry that extend to other smooth intersections of hypersurfaces diffeomorphic to them, and in some cases one obtains towers of such varieties all with the same symmetries. Also, the question of their smoothness is connected to some problems of interest in some engineering areas, such as signal recognition or self-correcting codes.

In this article we present the known results about the minors V_n and show how the simplest ones give interesting examples, including *towers* of them. We also include a conjecture about the topology of the intersections of real quadrics in one of these towers. We leave for another article the exploration of other more elaborate results.

Sections 2 and 3 describe the known results about minors of V_n including some proofs. Section 4 contains some implications of the above to the intersections of complex hypersurfaces while Sect. 5 describes those implications for the intersections of real ellipsoids, including a conjecture about the topology of some towers of them.

Recently we discovered, looking through the literature citing Tao [21] and its sequels, some papers about signal recognition dealing with the same type of problems but under a different name: matrices we call hyperbolic are known in this field as *full spark matrices*.

In particular, in [1] there are some results about the case $g = 1$ of our Theorem 4. We also learned there about [2], where a version of our Theorem 2 is proved. In a subsequent article we will describe the relations between those results and ours.

2 Jacobi Formula for the Co-rank

In this section A is a non singular $n \times n$ matrix with entries in a field and $B = A^{-1}$.

Let I, J be subsets of $[1, \ldots, n]$ and I^c, J^c their complements.

We denote by $M[I, J]$ the submatrix of a matrix M consisting of the entries M_{ij} with $i \in I$ and $j \in J$. When M is divided into blocks, we will denote them by

$$M = \begin{pmatrix} M_{11} & M_{12} \\ M_{21} & M_{22} \end{pmatrix}$$

Theorem 1 *Assume I and J have the same cardinality. Then*

(i) *Determinant Jacobi formula:*

$$|A|\,|B[J^c, I^c]| = \pm |A[I, J]|$$

(ii) *Co-rank formula: The co-rank of the square matrix $A[I, J]$ is equal to the co-rank of the complementary submatrix $B[J^c, I^c]$ of the inverse matrix B. In particular, if $A[I, J]$ is invertible then so is $B[J^c, I^c]$.*

Proof For both parts, we follow the last proof of (ii) in [18]. A geometric proof was produced for us by Mike Shub.

Without loss of generality we can assume (using permutations of rows and of columns) that $I = J = [1, \ldots, k]$. We decompose our matrices into blocks as above, where M_{11} is a $k \times k$ submatrix.

We have to show that the co-rank of A_{11} is the same as the co-rank of B_{22}.

From the equality $AB = I$ we obtain that $A_{11}B_{12} + A_{12}B_{22} = 0$ and $A_{21}B_{12} + A_{22}B_{22} = I_{22}$. Then it follows that:

$$A \begin{pmatrix} I_{11} & B_{12} \\ 0 & B_{22} \end{pmatrix} = \begin{pmatrix} A_{11} & 0 \\ A_{21} & I_{22} \end{pmatrix}$$

Since A is non-singular the other two matrices have the same co-rank which coincides with the co-ranks of A_{11} and B_{22} and part (ii) is proved.

From the last equality we obtain the equality of determinants:

$$|A||B_{22}| = |A_{11}|$$

that yields part (i) of the theorem, where the sign $\pm$ comes from the signs of the permutations taking I and J into $[1, \ldots, k]$.

3 Minors of the Vandermonde V_n on the n-th Roots of Unity

Our main interest is in the Vandermonde matrix $V = V_n = (\rho^{ij})$ on the powers of ρ, a primitive n-th root of unity. This matrix is also known as the Discrete Fourier Transform matrix.

More precisely, we are interested in properties of the square submatrices of V_n. In the present paper we deal only with the question of which of those submatrices are non-singular and we will prove some partial results on this question. But a second, more subtle question about their convexity properties, for which we only know the implications of the first one, is actually more relevant for the results in Sect. 5.

A weaker question is, given I, when does it happen that $V[I, J]$ is non-singular for every J of the same size as I? When this happens, we will say that I is *hyperbolic*.

We do not have a complete answer to any of these questions, but we will present many partial results, examples and situations for which this condition is valid. This will be the main question for the geometric consequences we will discuss in later sections.

3.1 First Results

Proposition 1 *For every k, the interval* $[1, \ldots, k]$ *is hyperbolic and so is any interval obtained from it by translation* (mod p).

Proof For the initial interval, since V is symmetric, $V[[1, \ldots, k], J] = V[J, [1, \ldots, k]]$ and the latter is the Vandermonde matrix on the roots ρ^j for $j \in J$. Since these roots are all different, its determinant is non-zero.

This result extends to all collections I which are equivalent to $[1, \ldots, k]$ through simple operations: multiplication of columns by powers of ρ and substitution of ρ by another primitive n-th root of unity. In particular, for all the translates of the initial interval and the proposition is proved.

We formalize this equivalence relation by introducing the *Affine Galois group:* Consider the group of n-th roots of unity $\{\rho^i, i = 0, \ldots, n-1\}$ which we identify with the group of indices $i \mod n$ in $\mathbb{Z}_n$.

Let $\mathbb{Z}_n^*$ denote the group of units of the ring $\mathbb{Z}_n$ and G_n the Affine Galois group of transformations of $\mathbb{Z}_n$ of the form

$$\phi_{u,v}(i) = ui + v$$

where $u \in \mathbb{Z}_n^*$ and $v \in \mathbb{Z}_n$.

To any subset $I \subset Z_n$ with m elements, $\phi_{u,v}$ associates a new subset $\phi_{u,v}(I)$. The interest of this action is that if we consider I as a family of rows of V_n, then the submatrix with rows in I will be equivalent to the corresponding submatrix with rows in $\phi_{u,v}(I)$.

A superficial understanding of this action already gives some theoretical results and simplifies many computations. But a deeper understanding is needed for the same purposes.

3.2 *The Complementarity Theorem*

Since V is symmetric and a scalar multiple of a unitary matrix, the results of Sect. 2 give:

Theorem 2

(1) *A square submatrix $V[I, J]$ is non-singular if, and only if, the complementary submatrix is non-singular.*
(2) *A set $I \subset [1, \dots, n]$ is hyperbolic if, and only if, the complementary subset I^c is hyperbolic.*
(3) *More generaly, the co-rank of a square matrix $V[I, J]$ is equal to the co-rank of the complementary submatrix $V[I^c, J^c]$.*

The proof of this theorem is due to Matthias Franz. Together with the results of the previous sections this result gives various examples of hyperbolicity.

3.3 *The Case $n = p$ Prime: Chebotaryov's Theorem*

Theorem 3 (Chebotaryov 1926)[2] *If n is prime, all minors of the Vandermonde matrix V_n are non-zero.*

We describe the idea of the proof in the next section, which extends easily to the case where n is a prime power.

The condition in the theorem is also necessary: if $n = ab$ is composite then there are always null 2×2 minors. In fact, if $n = ab$ then the $a \times b$ submatrix with row indices $b, 2b, \dots, ab$ and column indices $a, 2a, \dots, ba$ has rank 1.

[2]For the history of this theorem see [20]. Recent proofs appear in [7, 21], among many others.

This leads to a simple characterization of which 2×2 minors are zero. We still do not have a criterion for when a 3×3 minor is zero.

3.4 Some Cases Where n Is a Prime Power, $n = p^k$, p Odd

For any I of cardinality m we define the Dieudonné-Tao number:

$$DT(I) = \frac{\Pi_{i,i' \in I, i<i'}(i' - i)}{\Pi_{i,i',1 \leq i < i' \leq m}(i' - i)}$$

Proposition 2

(i) *$iDT(I)$ is always a positive integer.*
(ii) *If $n = p^k$ and $V[I, J]$ is singular, then $DT(I)$ is a multiple of p.*

Proof The proof follows the steps of those by Dieudonné and Tao for $k = 1$:

(1) If the minor $V[I, J]$ is singular, this means that the matrix V_x with entries $x_i^j, i \in I, j \in J$ has determinant zero when we substitute x_i by ρ^i.

On the other hand, since V_x is skew-symmetric with respect to the variables $x_i, i \in I$ we can factor

$$|V_x| = \Pi_{i,i' \in I. i>i'}(x_i - x_{i'}) S_I(x_i, i \in I)$$

where S_I is a polynomial with integer coefficients in the variables $x_i, i \in I$. But when we substitute the $x_i, i \in I$ by the different roots ρ^i the product in this expression is non-zero, so

$$S_I(\rho^i, i \in I) = 0$$

The polynomial $S_I(y^i, i \in I)$ in the single variable y vanishes at $y = \rho$, so it must be the product of the minimal polynomial $M(y)$ of ρ and a polynomial with integer coefficients.

This means, when $n = p^k$, that $S(1, \ldots, 1)$ is a multiple of $M(1) = p$.

(2) On the other hand, Dieudonné and Tao give proofs (using different computations) of the following equality:

$$DT(I) = S(1, \ldots, 1)$$

This shows that $DT(I)$ is always an integer and, when $n = p^k$, it is a multiple of $M(1) = p$ and the proposition is proved.

Observe that if n has at least two different prime factors, even if $\left|V[I, J]\right| = 0$ we cannot conclude by this argument anything about $DT(I)$ (but see next section).

Many interesting hyperbolic examples I for $n = p^k$ appear when no difference between the elements of I is a multiple of p:

(1) Any subset of $[1, \ldots, p]$ is hyperbolic.
(2) Any subset of $[1, \ldots, 2p]$ with p odd, consisting of indices of the same parity, is hyperbolic.
(3) For any two disjoint subsets I_1 and I_2 of $[1, \ldots, p]$, the set $I_1 \cup (I_2 + rp)$, for any $r \in \mathbb{N}$, is hyperbolic.

A more elaborate example, where p divides some differences of the elements of I, is the following:

(4) If ℓ_1, ℓ_2 are positive integers and g is a non-negative integer, let I_g consist of the integers $1, \ldots, \ell_1, , \ell_1+g+1, \ldots, \ell_1+\ell_2+g$, i.e., two intervals of lengths ℓ_1 and ℓ_2 separated by a gap of length g.

One can compute the order of p in $DT(I_g)$ under certain circumstances. Here we will only consider the simplest one:

First observe that if a set I is bounded by p^2, then no difference $i' - i$ is divisible by p^2, so the order of p in $\Pi_{i,i' \in I, i' > i}(i' - i)$ is equal to the number of pairs $i' > i$ in I which are congruent mod p. So we will assume that

$$\ell_1 + \ell_2 + g \le p^2$$

Finally, let $\ell_i = s_i p + r_i$ for $i = 1, 2$ and $g := sp + g_1$ with integers s_i, r_i, r satisfying $s_i, s \ge 0$ and $0 \le r_i, g_1 < p$. One can express $DT(I_g)$ in terms of these parameters, but in order to avoid long formulas and computations we will only state now the cases where $DT(I_g)$ is not divisible by p:

Theorem 4 *If $\ell_1 + \ell_2 + g \le p^2$ and*

(i) $r_1 + r_2 \le p$ *and* $0 \le g_1 \le p - r_1 - r_2$, *or*
(ii) $r_1 + r_2 > p$ *and* $2p - r_1 - r_2 \le g_1 \le p$,

then $DT(I_g)$ is not divisible by p, so I_g is hyperbolic.

Proof Separate the elements of I_g according to their classes mod p. Then all pairs within a class $c \in \mathbb{Z}_n$ have $i - i'$ divisible by p, so the contribution of this class to the order of p in the numerator of $DT(I_g)$ is $\binom{s(c)}{2}$ where $s(c)$ denotes the cardinality of the class c. Since no pair in different classes contributes to it, the order of p in the numerator of $DT(I_g)$ is the sum of those combinatorial numbers.

Therefore, *for two values of g whose sums are equal, the order of p in $DT(I_g)$ is the same.* Also, if this happens for $g = g_1$ it happens also for any g within the assumed range, since it is obtained from the case g_1 by adding a fixed number s for each congruence class. So we assume $g = g_1$ in the rest of the proof.

Now,in case (i), $s(c)$ is as follows: to begin with, in each class c there are s_i elements coming from $\ell_i = s_i p + r_i$, so we have always at least $s_1 + s_2$ in each class. Then, if c is between 1 and r_1, or between $r_1 + g + 1$ and $r_1 + g + r_2$ there is an extra element coming from the residues r_1 or r_2. So $s(c) = s_1 + s_2 + 1$ for these classes and $s(c) = s_1 + s_2$ for the rest of them, *and this is independent of* g *in its range* $0 \leq g \leq p - r_1 - r_2$, so the order of p in $DT(I_g)$ is the same within that range. Since it includes $g = 0$ for which that order is the same as that of an interval (which is 0), then that order is 0 for all of them, $DT(I_g)$ is not divisible by p and I_g is hyperbolic.

In case (ii), one has classes of size $s_1 + s_2 + 1$ and $s_1 + s_2 + 2$ and again one can check the number of classes of each of the two sizes is independent of g within the stated range, which includes $g = p$ for which the order of p in $DT(I_g)$ is 0. So it is also 0 for all g in that range and again I_g is hyperbolic. So Theorem 4 is proved.

To have an image of the proof one can draw the graph of $s(c)$ as a function of c in the different cases.

Remark 1

1. For the values of g outside the interval considered in cases (i) or (ii), a more precise computation shows that the order of p in $DP(I_g)$ starts growing with g up to a point and then starts descending until it reaches 0 and stays so for the corresponding interval.
2. The result is true for other arbitrarily large gaps, for example, those where the second interval is far from any multiple of p^2. Actually, the graph of the order of p in $DT(I_g)$ as a function of g shows zones with higher and higher peaks that eventually reduce to 0 and rest there for an interval before growing again. We show here the graph for $p = 23$ and g up to 10,000.

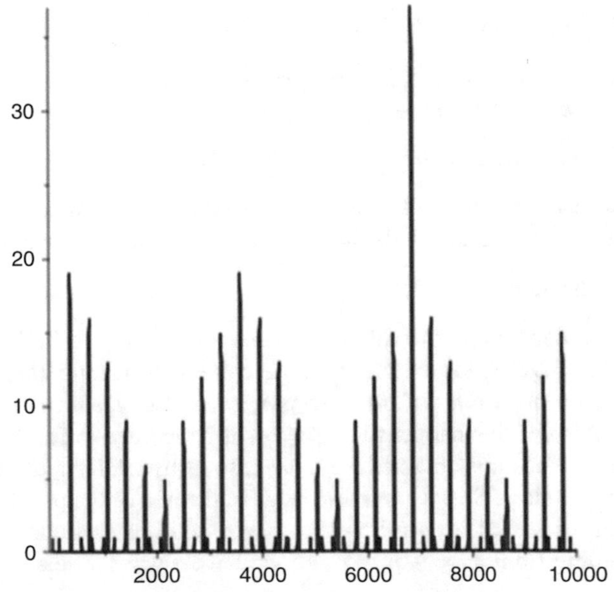

3. Also we can consider three or more intervals of different lengths with different gaps between them. The condition on those gaps can be easily guessed looking at the graph of the sizes of the different classes mod p, as above, the simplest case being when the sum of the residues mod p of the lengths of the intervals and gaps is less that p.
4. There are several promising variations of the above. And it must be stressed that the condition $p|DT(I)$ is necessary for I to be non-hyperbolic but is a long way from being sufficient, so many of the above examples can be hyperbolic even when the hypotheses fail. This gives many more examples of hyperbolic sets.
5. In [1] the authors study some cases where $g = 1$.

3.5 The Murty-Whang Criterion

In [19] the above $DT(I)$ criterion is generalized to all n, using Representation Theory:

If I is non-hyperbolic then $DT(I)$ is a linear combination of the prime factors of n with non-negative integer coefficients.

This allows a generalization of the examples in the previous section to situations where p odd is the smallest prime number dividing n (with restrictive extra hypotheses):

(1') Any subset of $[1, \ldots, p]$ is hyperbolic.
(2') Any subset of $[1, \ldots, 2p]$ consisting of indices of the same parity is hyperbolic, assuming that every other prime factor of n is greater that $2p - 1$.
(3') For any two disjoint subsets of I_1 and I_2 of $[1, \ldots, p]$, the set $I_1 \cup (I_2 + kp)$ for $k \in \mathbb{N}$ is hyperbolic, provided every other prime factor of n is an upper bound for it.
(4') I_g in the theorem in the last section is hyperbolic with the same hypotheses if, further, every other prime factor of n is greater that $p^2 - 1$.

4 Some Complex Varieties with Cyclic and Dihedral Symmetry

4.1 Some Complex Varieties with Cyclic Symmetry

Let ξ be an n-th root of unity, $d > 1$, an integer and consider the hypersurface in $\mathbb{C}P^{n-1}$ given by the equation:

$$\Sigma_{j=1}^{n} \xi^j z_j^d = 0$$

This variety is clearly smooth and invariant under a cyclic permutation of the coordinates z_j. Our varieties will be intersections of such hypersurfaces for different n-th roots of unity but same[3] degree $d > 1$:

Let ρ be a *primitive* root of unity, $K \subset [1, \ldots, n]$ with m elements, $d > 1$ and $\mathcal{Z}(K)$ the projective variety with equations

$$\Sigma_{j=1}^{n} \rho^{ij} z_j^d = 0, \ i \in K$$

This is again invariant under the cyclic group and also invariant under the obvious action of $\mathbb{Z}_d^{n-1}$. Combining both actions we get a group of order nd^{n-1}. We will not consider this extended action for the moment, although it will be present in all the examples in this section.

For equations like these there is a simple criterion for smoothness:

Lemma 1 *An intersection of hypersurfaces $V \subset \mathbb{C}P^{n-1}$ given by equations of the form*

$$\Sigma_{j=1}^{n} A_j z_j^d = 0$$

with $A_j \in \mathbb{C}^m$ and $d > 1$ is smooth if, and only if, the $m \times n$ matrix A with columns A_j satisfies the property

(H) *All*[4] *the $m \times m$ minors of A are different from zero.*

Proof Assume one $m \times m$ minor is zero with columns in $J \subset [1, \ldots, n]$. This means that there are constants α_j for $j \in J$, not all zero, such that

$$\Sigma_{j \in J} \alpha_j A_j = 0.$$

A point $z \in \mathbb{C}^n$ such that z_j is a d-th root of α_j for $j \in J$ and is 0 otherwise satisfies

$$\Sigma_{j=1}^{n} A_j z_j^d = \Sigma_{j \in J} \alpha_j A_j = 0$$

so it defines a point in $V \subset \mathbb{C}P^{n-1}$ where the jacobian matrix of the equations has columns $dz_j^{d-1} A_j$ which are zero for $j \notin J$ and a scalar multiple of A_j for $j \in J$. Since the latter are linearly dependent, the jacobian matrix at z has rank less than m and z is a singular point of V.

Reciprocally, if z is a singular point of V, let J be the set of indices such that $z_j \neq 0$.

[3]In the case with equations of different degrees, it is not clear how smoothness could be expressed in terms of the matrix V_n.

[4]Compare with the linear case $d = 1$ in which it is sufficient for smoothness that *one* of the minors is non-zero.

If J has at least m elements, take any collection $J' \subset J$ of m of them. The columns $dz_j^{d-1}A_j$ for $j \in J'$ of the jacobian matrix of the equations are linearly dependent and so are the columns A_j of A for $j \in J'$ which are scalar multiples of them since all the corresponding factors dz_j^{d-1} are non-zero and (H) fails.

If J has less than m elements then the equations themselves

$$\Sigma_{j\in J} z_j^d A_j = 0$$

show that the A_j with $j \in J$ are linearly dependent and so is any collection of m columns of A that contain them. Again (H) fails.

Applying this to our symmetric variety $\mathcal{Z}(K)$ we see that it is smooth if, and only if, the minors $V[K, J]$ of the matrix V_n are non-zero for all $J \subset [1, \dots, n]$ with m elements. From the results of the last section we obtain:

Theorem 5

(1) *If n is prime, all the 2^n intersections $\mathcal{Z}(K)$ are smooth.*
(2) *For all n and m, all the intersections $\mathcal{Z}(K)$, where $K = [1, \dots, m]$ are smooth.*
(3) *For all n, $\mathcal{Z}(K)$ is smooth if, and only if, the complementary $\mathcal{Z}(K^c)$ is smooth.*
(4) *If K is any of the hyperbolic examples described in the previous section, then $\mathcal{Z}(K)$ is smooth.*

To these we should add all the K equivalent to the ones described in the theorem under the action of the affine Galois group, in particular any interval $mod\ p$. This gives many smooth examples as well as some non-smooth ones, but still does not give a simple criterion to decide if a given one is smooth or not.

We can get some general results if we change the question: given a smooth algebraic variety X, is it diffeomorphic to one that has cyclic symmetry?

If X is a smooth complete intersection of m hypersurfaces in $\mathbb{C}P^n$, all of degree d, we have a way to attack it: construct one of our examples with cyclic symmetry having the same degrees. If this one is also smooth, then it is diffeomorphic to X and we are finished! This is a consequence of

Thom's principle *Any two smooth complete intersections of m hypersurfaces in $\mathbb{C}P^n$ with the same degrees are diffeomorphic.*

This is because in the complex space of all collections of m hypersurfaces of a given degree, the set of those that are not transverse is given by algebraic equations and therefore has *real* codimension at least 2. Then any two smooth ones can be joined by a path consisting only of smooth ones and are therefore diffeomorphic.

So we are free to choose which equations to use and there is a good choice: take the first m rows of the Vandermonde matrix V_n:

Theorem 6 *Any smooth complete intersection of hypersurfaces of degree d in $\mathbb{C}P^{n-1}$ is diffeomorphic to a smooth complete intersection of the same type that is invariant under the natural cyclic group of order n.*

There are various results about the topology of smooth complete intersections from which one can determine in many cases the topology of these smooth varieties ([4, 14], among others.)

In the particular case $m = n - 2$ we can describe completely the topology of a complete intersection:

Theorem 7 *For the surface S of Euler characteristic χ the following are equivalent (cf. [10]):*

(i) χ is of the form

$$\chi = d^{n-2}(n - d(n-2))$$

for some positive integers $n, d > 1$.

(ii) S is diffeomorphic to a smooth complete intersection of hypersurfaces of degree d in $\mathbb{C}P^{n-1}$.

(iii) S is diffeomorphic to such a smooth complete intersection in $\mathbb{C}P^{n-1}$ invariant under the natural cyclic group of order n.

The equivalence of (i) and (ii) follows from the formula for the Euler characteristic of an algebraic curve which is a smooth complete intersection of hypersurfaces of degrees $d_1, d_2, \ldots, d_{n-2}$ in $\mathbb{C}P^{n-1}$:

$$\chi = d_1 \ldots d_{n-2}(n - (d_1 + \cdots + d_{n-2}))$$

This wonderful formula is not as well-known as it deserves. One of its consequences is that the Riemann surfaces of genus 2 and 7, for example, cannot be realized as smooth complete intersections. We learned it from a paper by Grujic [12], then found it again in one by Libgober and Wood [14]. See also [10]. A geometric proof was produced for us by Enrique Artal (private communication).

4.2 Some Complex Varieties with Dihedral Symmetry

The natural way to upgrade the above cyclic actions to a dihedral action is by asking that the variety is also invariant under conjugation.

To obtain that, we assume that for every equation we have also the one with conjugate coefficients. In other words: if $i \in K$ then also $n - i \in K$. We will say in this case that K is *balanced*. So we have a much smaller set of examples of varieties with dihedral symmetry. Nevertheless, all the general smoothness results are valid for balanced varieties and we get many smooth types of intersections of hypersufaces with dihedral symmetry:

Theorem 8

(1) *If n is prime, all the $2^{\lfloor n/2 \rfloor + 1}$ balanced intersections $\mathcal{Z}(K)$ are smooth.*

(2) *For all n and m, all the intersections $\mathcal{Z}(K)$, where K is a balanced interval, are smooth.*

(3) *For all n, the complementary $\mathcal{Z}(K^c)$ of a balanced intersection $\mathcal{Z}(K)$ is also balanced, so one of them is smooth if, and only if, the other one is.*
(4) *Many of the examples of Sect. 3.4 can be made to be balanced and give smooth examples. In particular, the examples I_g with two intervals of the same size, can be translated to be balanced for appropriate values of the gap g.*

To apply this to the existence of smooth actions on smooth complete intersections of hypersurfaces of degree d, we need to see if there are enough balanced ones of a given degree and dimension. We obtain about three fourths of them:

Theorem 9 *If n and m are not both even, then every smooth complete intersection of m hypersurfaces of degrees d in $\mathbb{C}P^{n-1}$ admits a smooth dihedral action.*

This is because, if $n = 2k+1$, there are balanced submatrices of any even number of rows around the center, with K an interval from $[k, k+1]$ up to $[1, \ldots, 2k]$. The complementary intervals I^c are also balanced so we get smooth varieties of all codimensions m.

When $n = 2k$ we get again balanced submatrices of any odd number of rows around the center, with K an interval from $[k]$ up to $[1, \ldots, 2k-1]$. But in this case the complementary K^c is again odd unless K is empty.

For example, if n is even and $m = 2$, there are no balanced examples that are smooth, so the same goes for $m = n-2$ and we get in this case no smooth algebraic curves with dihedral symmetry. For other even values of n and m we do not know any smooth balanced examples. Nevertheless, for n odd we can take K to be the complementary of the interval $[(n-1)/2, (n+1)/2]$ to get a smooth balanced system. This gives about half of the surfaces with cyclic symmetry:

Theorem 10 *The surface of Euler characteristic*

$$\chi = d^{n-2}(n - d(n-2))$$

for $n > 1$ odd is diffeomorphic to a smooth complete intersection of hypersurfaces of degree d in $\mathbb{C}P^{n-1}$ invariant under the natural dihedral group.

There are many known examples of Riemann surfaces with cyclic or dihedral symmetry (see [5]), so our examples could be just new constructions of known cases. Maybe the interest of our examples lies in the fact that we realize them embedded in standard linear actions of specific dimensions and forming part of towers of varieties with the same groups acting.

5 Intersections of Real Affine Ellipsoids with Dihedral Symmetry

For the real analogs of the previous complex varieties

$$\Sigma_{j=1}^{n} A_j x_j^d = 0$$

where now $A_j \in \mathbb{R}^m$, to describe their topology we can restrict to the case $d = 2$, even in the non-smooth case:

For d odd, the homeomorphism of $\mathbb{R}^n$ to itself sending x_i to x_i^d restricts to a homeomorphism between the variety and the linear variety corresponding to $d = 1$, whose topology is evident.

For $d = 2k$, it is the homeomorphism of $\mathbb{R}^n$ to itself sending x_i to $sign(x_i)x_i^k$ that establishes a homeomorphism between the variety and the variety corresponding to $d = 2$. So for the topological description it is enough to consider this case.

The smoothness condition for a system with d even is now weaker than hyperbolicity: if we have a real linear dependence between m or less columns of the matrix of coefficients of a system

$$\Sigma_{j \in J} \alpha_j A_j = 0,$$

we can not obtain a critical point of the variety as in Sect. 4 by taking the d-th roots of the α_j, *unless they are all non-negative*. So now what we have to avoid is that there are m or less columns in the matrix such that a *positive* linear combination of them is zero. Or, equivalently, to ask that no convex combination of them is zero. This condition is called Weak Hyperbolicity:

(WH) *An $m \times n$ real matrix is called weakly hyperbolic if the origin in $\mathbb{R}^m$ is not a convex combination of m of its columns.*

A detailed proof of the equivalence between (WH) and smoothness for $d = 2$ can be found in [17]. It works for any even d, and equally well for the projective version and for the affine one outside the origin or intersected with the unit sphere. It is in the last case that progress on the topological description has been obtained in large families of them (described in the introduction), so it is there that we can hope to apply the results of Sect. 3.

Now we can look at the examples with dihedral symmetry mentioned in the introduction:

Take $K \subset [1, \ldots, [n/2]]$ of cardinality m and $Z(K)$ be the variety

$$\Sigma_{i=1}^n \rho^{ki} x_i^2 = 0, \quad k \in K$$

$$\Sigma_{i=1}^n x_i^2 = 1$$

Each of the homogeneous equations is actually a pair of real equations (taking real and imaginary parts of the coefficients), with the exception, when n is even and $n/2 \in K$, of the last one which has real coefficients 1 and -1. So, if we denote by $\hat{m}$ the number of real *homogeneous* equations, $\hat{m} = 2m$ in general case and $\hat{m} = 2m - 1$ in the exceptional case.

The $\hat{m} \times n$ matrix of coefficients of this system is equivalent by real row operations to the balanced submatrix of V_n formed by adding to the rows their complex conjugate rows. So the question of the weak hyperbolicity of these two matrices are equivalent.

The theory of weak hyperbolicity of submatrices of V is still to be developed. But we can use the fact that this property is implied by hyperbolicity and use all the positive results of Sect. 3 to obtain results about the smoothness of $Z(K)$. It must be remembered that for a smooth $Z(K)$ its matrix may not be hyperbolic. This would mean that its complex version would have only imaginary singularities.

The complementary matrix should be taken inside the range $[1, \ldots, [n/2]]$ of rows so even after taking the balanced equations it corresponds to complementarity in the range $[1 \ldots, n-1]$. So it does not correspond to complementarity inside the range $[1 \ldots, n]$ for which we understand the duality of smoothness. The complementary of hyperbolicity in this case is a property that can be called *affine hyperbolicity*, meaning that no collection of $\hat{m}+1$ columns lie in an affine subspace. The property that is complementary (in this new sense) to weak hyperbolicity has not been characterized so far. Still, the duality between the combinatorial objects associated to $Z(K)$ and $Z(K^c)$ describe in the introduction is a solid fact that needs to be understood.

We summarize the main positive results about the smoothness of $Z(K)$:

Theorem 11

(1) *For all n, if I is an interval containing $[n/2]$, then $Z(K)$ is smooth.*
(2) *If n is prime, all the $2^{[n/2]}$ intersections $Z(K)$ are smooth.*
(3) *If $n = p^k$ and K is an interval, then $Z(K)$ is smooth, provided K and its conjugate interval satisfy the conditions of Theorem 4 with $\ell_1 = \ell_2$ and g the gap between them.*

Case 1 looks interesting because for each n we have actually a tower of varieties with dihedral symmetry, namely, all those corresponding to the intervals $[k, \ldots, [n/2]]$ for k from 1 to $[n/2]$. The varieties in this tower have even dimensions going from 0 to $n-2$. They give also towers of polytopes with the same symmetry and dimensions. These are for the moment the only examples for which we have explored their topology:

For the cases $k = 1$, $k = 2$ and $k = [n/2]$ their topology is known from the results (A) and (B) mentioned in the introduction: they are all connected sums of sphere products, connected up to the middle dimension.

Long computations of faces of the corresponding polytopes give always that their duals are neighbourly, which means that the varieties are again connected sums of sphere products, connected up to the middle dimension. The number of summands for small k grows very fast with n (recall the genus of the surfaces in part B) of the introduction.

One can conjecture that this is true for all n. If this is so, we would have interesting towers of highly connected varieties and of neighborly polytopes with dihedral symmetry. The proof of this conjecture seems to imply an even deeper understanding of the properties of the minors of the DFT matrix.

Acknowledgements Conversations with Enrique Artal, Pablo Barrera, Shirley Bromberg, Peter Bürgisser, Sylvain Cappell, Marc Chaperon, Antonio Costa, Genaro de la Vega, Javier Elizondo,

Matthias Franz, Ignacio Luengo, Mike Shub, Denis Sullivan, Yuri Tschinkel, Luis Verde, Alberto Verjovsky, Felipe Zaldívar and Adrián Zepeda have been very helpful.

Special mention is deserved by Matthias Franz who has dedicated a great effort to the Vandermonde minors question. Long discussions with him have clarified many aspects of the theory (and of the present paper) and the proof of Theorem 2 is due to him. Much joint work is still in progress and will certainly be the object of a joint future paper more closely related to his interests and point of view.

This work was partially supported by a Papiit-UNAM grant IN111415.

References

1. Achanta, H.K., Biswas, S., Dasgupta, B.N., Dasgupta, S., Jacob, M., Mudumbai, R.: The spark of Fourier matrices: Connections to vanishing sums and coprimeness. Digital Signal Process. **61**, 76–85 (2017)
2. Alexeev, B., Cahill, J., Mixon, D.G.: Full spark frames. J. Fourier Anal. Appl. **18**(6), 1167–1194 (2012)
3. Al-Raisi, A.: Equivariance, Module Structure, Branched Covers, Strickland Maps and Cohomology Related to the Polyhedral Product Functor. Ph.D. Thesis, University of Rochester (2014)
4. Browder, W.: Complete intersections and the Kervaire invariant. Lect. Notes Math. **763**, 88–108 (1979)
5. Bujalance, E., Cirre, F.J., Gamboa, J.M., Gromadzki, G.: Symmetries of Compact Riemann Surfaces. Lecture Notes in Mathematics, vol. 2007, pp. xx+158. Springer, Berlin/Heidelberg (2010)
6. de la Vega, G., López de Medrano, S.: Generalizing the May-Leonard system to any number of species. In: Proceedings of the International Conference Dynamical Systems: 100 Years After Poincaré, Gijón, Spain, Springer Proceedings in Mathematics and Statistics, vol. 54, pp. 395–407 (2013)
7. Dieudonné, J.: Une propriété des racines de l'unité, Collection of articles dedicated to Alberto González Domínguez on his sixty-fifth birthday. Rev. Un. Mat. Argentina **25**, 1–3 (1970/1971)
8. Gitler, S., López de Medrano, S.: Intersections of quadrics, moment-angle manifolds and connected sums. Geom. Topol. **17**(3), 1497–1534 (2013)
9. Gómez Gutiérrez, V., López de Medrano, S.: Stably parallelizable compact manifolds are complete intersections of quadrics. Publicaciones Preliminares del Instituto de Matemáticas, UNAM (2004)
10. Gómez Gutiérrez, V., López de Medrano, S.: Surfaces as complete intersections. In: Riemann and Klein Surfaces, Automorphisms, Symmetries and Moduli Spaces. Contemporary Mathematics, vol. 629, pp. 171–180. AMS, Providence (2014)
11. Gómez Gutiérrez, V., López de Medrano, S.: Topology of the intersections of quadrics II. Bol. Soc. Mat. Mex. **20**(2), 237–255 (2014)
12. Grujić, V.N.: ξ_y-characteristics of projective complete intersections. Publicatons de l'Institut Mathématique, nouvelle série **74**(88), 19–23 (2003)
13. Hirzebruch, F.: Arrangements of lines and algebraic surfaces. In: Arithmetic and Geometry. Progress in Mathematics 36, vol. II, pp. 113–140. Boston: Birkhauser (1983)
14. Libgober, S., Wood, J.W.: Differentiable structures I. Topology **21**(4), 469–482 (1982)
15. López de Medrano, S.: The space of Siegel leaves of a holomorphic vector field. In: Holomorphic Dynamics (Mexico, 1986). Lecture Notes in Mathematics, vol. 1345, pp. 233–245 Springer, Berlin (1988)
16. López de Medrano, S.: The topology of the intersection of quadrics in $\mathbb{R}^n$. In: Algebraic Topology (Arcata, 1986). Lecture Notes in Mathematics, vol. 1370, pp. 280–292. Springer, Berlin (1989)

17. López de Medrano, S.: Singular intersections of quadrics I. In: Singularities in Geometry, Topology, Foliations and Dynamics- A Celebration of the 60th Birthday of José Seade. Trends in Mathematics, pp. 155–170. Birhäuser, Basel (2016)
18. https://mathoverflow.net/questions/87877/jacobis-equality-between-complementary-minors-of-inverse-matrices
19. Murty, M.R., Whang, J.P.: The uncertainty principle and a generalization of a theorem of Tao. Linear Algebra Appl. (2012). https://doi.org/doi:10.1016/j.laa.2012.02.009
20. Stevenhagen, P., Lenstra, H.W.: Chebotarëv and his density theorem. Math. Intelligencer **18**(2), 26–36 (1996). Springer
21. Tao, T.: An uncertainty principle for cyclic groups of prime order. Math. Res. Lett. **12**(1), 121–127 (2005)

The Greedy Algorithm and the Cohen-Macaulay Property of Rings, Graphs and Toric Projective Curves

Argimiro Arratia

For Antonio Campillo and Miguel Angel Revilla

Abstract It is shown in this paper how a solution for a combinatorial problem obtained from applying the greedy algorithm is guaranteed to be optimal for those instances of the problem that, under an appropriate algebraic representation, satisfy the Cohen-Macaulay property known for rings and modules in Commutative Algebra. The choice of representation for the instances of a given combinatorial problem is fundamental for recognizing the Cohen-Macaulay property. Departing from an exposition of the general framework of simplicial complexes and their associated Stanley-Reisner ideals, wherein the Cohen-Macaulay property is formally defined, a review of other equivalent frameworks more suitable for graphs or arithmetical problems will follow. In the case of graph problems a better framework to use is the edge ideal of Rafael Villarreal. For arithmetic problems it is appropriate to work within the semigroup viewpoint of toric geometry developed by Antonio Campillo and collaborators.

1 Introduction

A greedy algorithm is one of the simplest strategies to solve an optimization problem. It is based on a step-by-step selection of a candidate solution that seems best at the moment, that is a local optimal solution, in the hope that in the end this

A. Arratia (✉)
Department of Computer Science and Barcelona Graduate School of Mathematics, Universitat Politècnica de Catalunya, Barcelona, Spain
e-mail: argimiro@cs.upc.edu

G.-M. Greuel et al. (eds.), *Singularities, Algebraic Geometry, Commutative Algebra, and Related Topics*, https://doi.org/10.1007/978-3-319-96827-8_17

process leads to a global optimal solution. Greedy algorithms do not always output (global) optimal solutions, but for many optimization problems they do.

A seminal result of Jack Edmonds [7], states that for a greedy algorithm to output optimal solutions it is necessary and sufficient that the input is a *matroid*, regardless of the weight function associated to the optimization problem. A matroid is a simplicial complex with the additional *exchange property*. Korte and Lovazs have extended Edmonds result to cover other type of weight functions and weaker combinatorial objects as input. Specifically they have slimmed the matroid by removing the subclusiveness property and named the new object a *greedoid* [13].

The intuition behind Edmonds result is to view the problem of determining the correctness of the greedy algorithm as a localization problem: in order to know if greedy works for some set H, it suffices to look at some discrete partition of H and check if greedy works for each of the parts then it should work for H. This is a classical working paradigm for the algebraic geometer (see, e.g., [15]), namely to solve problems locally and translate solutions globally, and vice versa. Matroids fit in very well this local–global pattern, which is best viewed in the realm of Commutative Algebra as follows:

> As a simplicial complex Δ over some ground set S, a matroid is such that for every subset W of S, the induced subcomplex $\Delta_W := \{F \in \Delta : F \subseteq W\}$ has an associated ideal in some ring of polynomials which is Cohen-Macaulay. By extension we say then that each Δ_W is Cohen-Macaulay [18].

This suggests that in order to guarantee optimal solutions from the greedy algorithm we should start with those instances of the input that are Cohen-Macaulay or piece–wise Cohen-Macaulay (as matroids).

Now, a related question is how to recognize those Cohen-Macaulay instances for a given combinatorial problem. This is a question about the choice of representation, since there is more than one way to associate to a simplicial complex some module that encodes its algebraic properties, like being Cohen-Macaulay. It also has to do with the chosen ground set of the simplicial complexes. The selection of the type of simplicial complex and associated module should be determined by the type of combinatorial problem to which we apply some form of greedy algorithm.

For a general ground set S identified with an initial segment of the natural numbers, the standard module to associate is the Stanley-Reisner ring over some ring of polynomials [18]. However, for problems on graphs there is a natural ideal to associate, which is the *edge ideal* [20], equivalent to a Stanley-Reisner ideal over a particular simplicial complex whose faces correspond to the independent sets of the graph. For arithmetic problems, an alternative framework is given by the numeric semigroups viewpoint of Toric Geometry [3, 4], which serves as a bridge between affine and projective toric varieties to and from polytopes and simplicial complexes. We will illustrate in the following sections the usefulness of these algebraic ideas for ascertaining the correctness of greedy algorithms.

2 Greedy Algorithms and Simplicial Complexes

Let S be a finite set and Δ a collection of subsets of S. We say that (S, Δ) is a *simplicial complex* if it verifies the following two conditions:

(S1) For all $s \in S$, $\{s\} \in \Delta$.
(S2) $F \subseteq G \in \Delta$ implies $F \in \Delta$.

S is called the *ground* set and is usually identified with $[n] := \{1, \ldots, n\}$, in which case one refers to the simplicial complex as Δ. The elements of Δ are called *faces*, and the maximal elements of Δ, with respect to $\subseteq$, are called *facets* (or Δ-maximal sets). The dimension of the simplicial complex Δ, denoted $\dim(\Delta)$, is the maximum dimension of its faces, where the dimension of face F is $\dim(F) = |F| - 1$, where $|F|$ denotes the cardinality of F.

Given a simplicial complex Δ over the ground set S, and given a *linear* weight function $f : S \to \mathbb{R}_{\geq 0}$, we can extend f to 2^S (the set of subsets of S) by defining for each $A \subseteq S$, $f(A) = \sum_{a \in A} f(a)$. We state the general form of an optimization problem as a maximization problem.

Definition 1 The Optimization Problem for (S, Δ) and weight function $f : S \to \mathbb{R}_{\geq 0}$, denoted f-OPT(Δ), is the following:

To find a set $A \in \Delta$ with maximum f-weight.

Observe that if f is linear, or at least *monotone* (i.e. $A \subseteq B$ implies $f(A) \leq f(B)$), then the optimization problem f-OPT(Δ) reads:

To find a facet of Δ with maximum f-weight.

From now on we assume that f is linear.

Definition 2 The greedy algorithm associated to Optimization Problem for (S, Δ) and weight f, denoted GREEDY, is presented in Fig. 1.

Remark 1 Monotonicity (or positive linearity) of f is needed for inducing a partial order in S and sorting makes sense. The algorithm always terminates because S is

```
GREEDY
Input: (S,Δ) and f : S → ℝ≥0
A ← ∅
   sort S in nonincreasing order by weight f
   while S ≠ ∅ do
       choose a ∈ S in the nonincreasing order by f
       S ← S − {a}
       if A ∪ {a} ∈ Δ then A ← A ∪ {a}
   end while
end
Output: A.
```

Fig. 1 Algorithm GREEDY

finite, and it will always output a non empty set A which is contained in a facet. The complexity of GREEDY will mostly depend on the membership test in line 6. Note that f can be a positive constant function. In this case we can select the elements in S in any order, and the correctness of GREEDY have to be determined for any of the possible orders of selecting the equally weighted elements. Apart from this difficulty, the consideration of constant f is useful to treat under the GREEDY scheme optimization problems where instances are not explicitly weighted and we turn them into weighted problems by assigning equal constant weight to every element. This we will do in Sect. 4.

Definition 3 We say that the algorithm GREEDY for (S, Δ) and f *correctly solves* the associated optimization problem if it gives as output a set A such that: (i) A is Δ-maximal (a facet) and (ii) for all $B \in \Delta$ $(f(A) \geq f(B))$.

We begin by showing that we need no extra assumptions about Δ (other than to be a simplicial complex) to guarantee that the output of GREEDY is a Δ-maximal set.

Proposition 1 *The output of* GREEDY *is a* Δ*-maximal set.*

Proof Let A be the output and suppose A is not maximal. Then there is a $C \in \Delta$ such that $A \subset C$. Let $x \in C - A$, then $A \cup \{x\} \subseteq C \in \Delta$, and hence $A \cup \{x\} \in \Delta$. But this x must have been considered at some step in the algorithm and should have been placed in A, so $x \in A$, a contradiction. □

Since the previous result holds regardless of the weight function f, what then we really need to guarantee is that

GREEDY outputs an A of maximum f-weight

For some weight functions (e.g., constant functions), one way to achieve this is to impose on Δ the stronger condition of being *pure*, that is

all Δ-maximal elements have same dimension

For a more ample spectrum of weight functions (e.g. linear), GREEDY achieves optimal solutions for inputs where the pureness condition can be localized. This is the point of matroids, and by extension of Cohen-Macaulay complexes.

3 Matroids and Cohen-Macaulay Complexes

A *matroid* is a simplicial complex (S, Δ) which, in addition, verifies the following principle:

Principle of Exchange (PE): If $A, B \in \Delta$ and $|A| < |B|$ then there exists $x \in B - A$ such that $A \cup \{x\} \in \Delta$.

The Principle of Exchange is equivalent to a *localization* of the property of being pure[1]:

$$\boxed{\text{PE} \iff \forall W \subseteq S,\ \Delta_W := \{F \in \Delta : F \subseteq W\} \text{ is pure}}$$

Moreover, the Principle of Exchange is equivalent to the correctness of GREEDY. This is Edmond's result on matroids and the greedy algorithm [7], but see [17, Theorem 12.5, p. 285] for a textbook exposition of this important result in optimization. We collect all these facts in the following theorem (and using an updated notation and terminology from Commutative Algebra).

Theorem 1 ([7]) *Given a simplicial complex Δ over a ground set S, the following statements are equivalent:*

(i) GREEDY *correctly solves the optimization problem for (S, Δ) and any (linear) weight function.*
(ii) *The Principle of Exchange (i.e. (S, Δ) is a matroid).*
(iii) $\forall W \subseteq S$, *the induced subcomplex Δ_W is pure.* □

We shall see next that the localization of pureness is equivalent to a localization of the Cohen-Macaulay property. Thus, matroids are locally Cohen-Macaulay complexes, and we can view the correctness of GREEDY in the world of Cohen-Macaulay rings, and extensions, which we argue here to be an appropriate algebraic framework (if not the correct one) to understand the workings of GREEDY.

3.1 *Greedy on Locally Cohen-Macaulay Complexes*

Given a simplicial complex Δ over the ground set $[n] := \{1, \ldots, n\}$, and given $\mathcal{K} := k[x_1, \ldots, x_n]$, the polynomial ring in n variables over some field k, the Stanley–Reisner ideal of Δ is the square free monomial ideal $I_\Delta = \langle \mathbf{x}^A : A \notin \Delta \rangle \subseteq \mathcal{K}$, where $\mathbf{x}^A := x_{i_1} x_{i_2} \cdots x_{i_r}$ with $A = \{i_1, \ldots, i_r\} \subseteq [n]$. The Stanley-Reisner ring (or *face ring*) is the quotient $R_\Delta = \mathcal{K}/I_\Delta$. From the correspondence between Δ and I_Δ, the latter can be characterized as

$$I_\Delta = \bigcap_{\substack{F \in \Delta \\ F \text{ a facet}}} M^{F^c} \tag{1}$$

where M^{F^c} is the monomial prime ideal corresponding to the non-face $F^c = [n] \setminus F$; in other words, $M^{F^c} = \langle x_i : i \notin F \rangle$. Equation (1) gives an *irreducible*

[1]The reader is encouraged to prove this equivalence.

decomposition of I_Δ; as M^{F^c}, for F a facet (a maximal face), is an *irreducible component* provided it is not redundant (i.e. cannot be deleted).

The dimension of the face ring, $\dim(R_\Delta)$, can be defined as $\dim(R_\Delta) = \dim(\Delta) + 1$, and therefore[2]

$$\dim(R_\Delta) = \max[|F| : F \in \Delta] \tag{2}$$

The codimension of R_Δ, $\operatorname{codim}(R_\Delta)$, can be defined as the *smallest number of generators of any irreducible component of* I_Δ (see [16, §5.5]). Other two useful measures of dimension are: (1) the *projective dimension* of R_Δ, $\operatorname{pd}(R_\Delta)$, as the length of a minimal resolution of R_Δ; and (2) the *depth* of R_Δ, $\operatorname{depth}(R_\Delta)$, as the maximal length of a regular sequence on R_Δ. All these forms of dimension are related as follows:

$$\operatorname{pd}(R_\Delta) \geq \operatorname{codim}(R_\Delta) \text{ and } \dim(R_\Delta) \geq \operatorname{depth}(R_\Delta).$$

Now, the ideal I_Δ (or equivalently the ring R_Δ) is *Cohen-Macaulay* if and only if $\operatorname{pd}(R_\Delta) = \operatorname{codim}(R_\Delta)$ if and only if $\operatorname{depth}(R_\Delta) = \dim(R_\Delta)$.

The simplicial complex Δ is Cohen-Macaulay if its face ring R_Δ is Cohen-Macaulay.

Remark 2 Technically the Cohen-Macaulay property depends on the choice of the field k, because computing regular sequences involves finding non zero divisors of certain quotient modules, and being a divisor or not depends on the characteristic of the field. Hence, we will always assume that our field k is of characteristic 0. For further simplicity the reader can assume that k is the field of real numbers.

A consequence of the above definitions and facts about the Stanley-Reisner ring R_Δ is the following result (cf. [2, Corollary 5.1.5] or [16, p. 114]):

Proposition 2 *If the face ring $\mathcal{K}/I_\Delta$ is Cohen-Macaulay then all irreducible components of I_Δ have equal cardinality.* □

It follows from the above result that a Cohen-Macaulay simplicial complex is pure. The converse is true locally, that is, a locally pure complex (i.e. a matroid) is locally Cohen-Macaulay [18, Proposition 3.1].

We then have the following characterization of the correctness of the greedy algorithm in terms of the Cohen-Macaulay property:

Theorem 2 *Given a simplicial complex Δ over a ground set S, the following statements are equivalent:*

(i) GREEDY *correctly solves the optimization problem for (S, Δ) and any (linear) weight function.*

[2]The dimension of a finitely generated ring is the maximum cardinality of an algebraically independent set. This is equivalent in R_Δ to $\dim(\Delta) + 1$ and Eq. (2).

(ii) $\forall W \subseteq S$, Δ_W *is pure (i.e.* (S, Δ) *is a matroid).*
(iii) $\forall W \subseteq S$, Δ_W *is Cohen-Macaulay.*

Proof The equivalence of (i) and (ii) is Theorem 1, and the equivalence of (ii) and (iii) is Proposition 3.1 of [18]. □

4 Maximum Independent Set and Cohen-Macaulay Graphs

The Maximum Independent Set problem (or MIS) is the optimization problem that asks for a largest subset of vertices in a graph which are pairwise non-adjacent. A dual problem to MIS is the Minimum Vertex Cover (or MVC), which asks for the smallest subset of vertices in a graph where all the edges have at least one endpoint. Both problems are related to each other by the following equivalence: *a set of vertices is a vertex cover if, and only if, its complement is an independent set.* The MIS (and the MVC) problem is **NP**-complete [10], and in view of its general intractability various greedy algorithms have been proposed for obtaining approximate solutions. A common feature of many of these greedy strategies for finding solutions to the MIS is to select vertices in some order with respect to their degrees (i.e. number of incident edges) and remove them and their adjacent vertices at each step. Although, in general, these strategies based on vertex selection have a poor approximation ratio under a worst case analysis (cf. [17, §17.1] or [6]), some of these work for some classes of graphs, meaning that they do provide us with the optimal solution (an independent set of maximum possible cardinality). We shall see that these vertex selection greedy strategies (dependable on the order of selecting the vertices) work in general for Cohen-Macaulay graphs.

First we shall fix some notation and terminology on graphs.

Definition 4 We denote graphs as $G = \langle V(G), E(G)\rangle$ or $\langle [n], E(G)\rangle$, where the vertex set $V(G)$ of cardinality n is identified with the labelling set $[n] := \{1, 2, \ldots, n\}$, and $E(G) \subseteq \{\{i, j\} : i, j \in V(G)\}$ is the set of edges. The complement of G, denoted G^c, is a graph with same set of vertices as G and edge set $E(G^c) := \{\{i, j\} : \{i, j\} \notin E(G)\}$. Given a vertex $a \in V(G)$, the neighbourhood of a is the set $N_G(a) = \{b : \{a, b\} \in E(G)\}$, and its degree is denoted $\deg(a)$. For a subset of vertices $W \subseteq V(G)$, the graph induced by W has as vertices the set W and as edges all those in $E(G)$ among pairs of vertices in W. The graph induced by W is formally denoted G_W.

All throughout this paper a graph is always simple and undirected. Given a graph G, a subset C of $V(G)$ is a *clique* if for all distinct pairs $i, j \in C$, $\{i, j\} \in E(G)$. An independent set of G is a subset $M \subseteq V(G)$ such that for all $i, j \in M$, $\{i, j\} \notin E(G)$. The independent set M is *maximal* if no extension of M is an independent set. A *maximum* independent set (MIS) is a (maximal) independent set of greatest possible cardinality. A vertex cover of G is a subset $C \subseteq V(G)$ such that for all $\{i, j\} \in E(G)$, $C \cap \{i, j\} \neq \emptyset$. C is *minimal* if no proper subset of C is a vertex cover of G, and is a *minimum* vertex cover (MVC) if it has smallest possible cardinality.

It is usually said that G is *unmixed* if all of its minimal vertex covers have the same size.

Next, we shall denote *the vertex selection strategy following the rule* ω *for selecting vertices* as VERTEXSELECT[ω], and which proceeds as follows:

VERTEXSELECT[ω] : select a vertex in the order established by the rule ω, remove its neighbours and repeat the selection procedure in the reduced graph. The output is the set of selected vertices.

By construction the output of VERTEXSELECT[ω] is a maximal independent set. We shall first deal with two simple rules for selecting vertices:

ω_L: choose the vertex with largest degree first;
ω_S: choose the vertex with smallest degree first.

Note that for both rules, selection of vertices is always possible, so both variants of the VERTEXSELECT[ω] algorithm terminate. Also, under these rules, vertices are sorted in the order given by their degrees, breaking ties at each step of the algorithm by using the implicit order given initially by the numeric labelling of the vertices.

An example where the algorithm VERTEXSELECT[ω] outputs an optimal solution, regardless of ω, is the graph in Fig. 2. For this graph an optimal output obtained using the rule ω_S of the smallest degree first is the pair {5, 1} but it could also be: {2, 5}, or {3, 4}, where the last two pairs can also be obtained with the rule ω_L of the largest degree first.

On the other hand, Fig. 3 shows a graph where VERTEXSELECT[ω_S] gives the optimal solution $\{1, 4, 6\}$, whilst VERTEXSELECT[ω_L] gives $\{3, 4\}$, an independent set which is not maximum.

We will see next how to associate a suitable simplicial complex to a graph in order to analyze the correctness of VERTEXSELECT[ω] through the lens of Commutative Algebra.

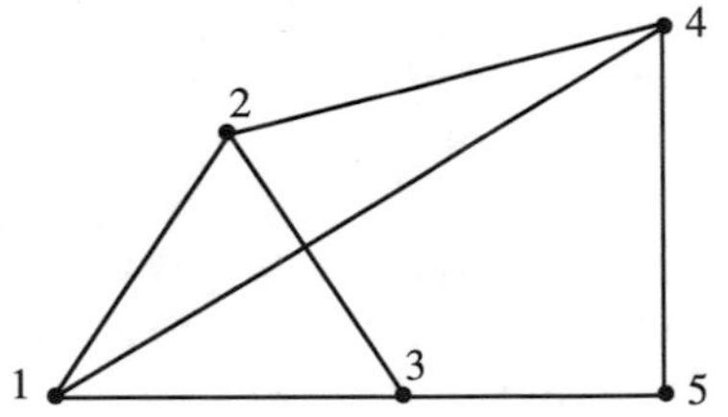

Fig. 2 A graph for which VERTEXSELECT[ω] gives optimal solution for $\omega \in \{\omega_S, \omega_L\}$

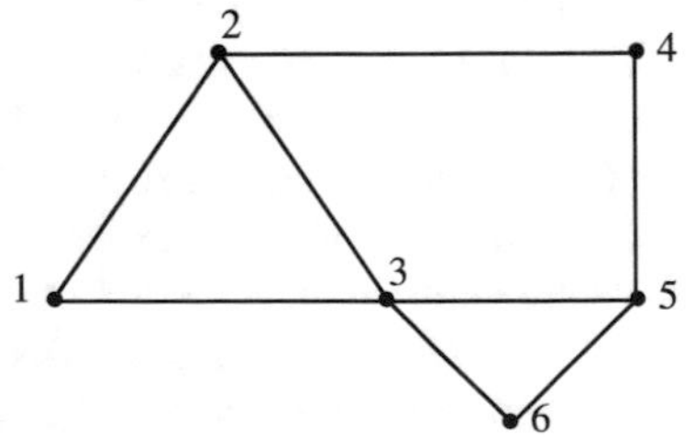

Fig. 3 A graph for which VERTEXSELECT[ω_S] gives optimal solution, but VERTEXSELECT[ω_L] does not

4.1 The Edge Ideal and Correctness of VERTEXSELECT

Let G be a graph on $[n]$ and edge set $E(G)$; let $\mathscr{K} = k[x_1, \ldots, x_n]$, with k a field. To G one can associate the *edge ideal* $I(G) \subseteq \mathscr{K}$, which is generated by all square–free monomials $x_i x_j$ with $(i, j) \in E(G)$. This edge ideal seems to have been first defined by Rafael Villarreal in [20].

To the complementary graph of G, G^c, one can associate the *clique complex* $\kappa(G^c)$ where a face of dimension d is a clique of G^c of size $d+1$; that is, any subset $A \subseteq [n]$ is in $\kappa(G^c)$ iff $\forall i \forall j (i, j \in A \rightarrow (i, j) \in E(G^c))$.

Hence, the Stanley–Reisner monomial ideal associated to $\kappa(G^c)$, namely $I_{\kappa(G^c)}$, is exactly the edge ideal $I(G)$, and the following definition is sound:

Definition 5 A graph $G = \langle [n], E(G) \rangle$ is Cohen-Macaulay over a field k, if the edge ideal $I(G) = \langle x_i x_j : (i, j) \in E(G) \rangle$ (and equivalently, the ideal $I_{\kappa(G^c)}$) is Cohen-Macaulay over $\mathscr{K} = k[x_1, \ldots, x_n]$, that is, $\mathscr{K}/I(G)$ (or $\mathscr{K}/I_{\kappa(G^c)}$) is Cohen-Macaulay ring.

Cohen-Macaulay graphs are extensively studied in [20]. For example, the only cycles that are Cohen-Macaulay are those of three or five vertices.

Now, if $F \subset [n]$ is a maximal clique in G^c then F is a maximal *independent set* in G and $F^c := [n] \setminus F$ is a minimal *vertex cover* in G of cardinality $n - |F|$, and vice versa. From this combinatorial equivalence we can derive the following facts:

Fact 1: Given a graph G, another simplicial complex often associated to G is $\Delta(G)$ consisting of all independent sets of G. By the previous equivalence one sees that $\Delta(G) = \kappa(G^c)$. Thus, the edge ideal $I(G) = I_{\Delta(G)}$.

Fact 2: By Eq. (1) the irreducible components of the edge ideal $I(G)$ are obtained from the intersection of monomial prime ideals of the form $\langle x_i : i \in C \rangle$ such that C is a minimal vertex cover of G.

Fact 3: Using Eq. (2), we get that the dimension of the face ring of $\Delta(G)$, $\dim(R_{\Delta(G)}) = \dim(\mathscr{K}/I(G))$, is equal to the maximum cardinality of a clique in the complementary graph G^c, or equivalently, the maximum cardinality of an independent set in G. On the other hand, $\operatorname{codim}(R_{\Delta(G)})$ is equal to the minimum cardinality of a vertex cover in G. This is by definition of codimension and simply noting that (max. cardinality of independent set in G) + (min. cardinality of vertex cover in G) = n.

Theorem 3 *If the graph G is Cohen-Macaulay then, for any of the degree-based rules $\omega \in \{\omega_S, \omega_L\}$, the algorithm* VERTEXSELECT$[\omega]$ *outputs an optimal solution on input G.*

Proof By Proposition 2 all irreducible components of $I(G) = I_{\Delta(G)}$ have same cardinality, which by Fact 2 means that G is unmixed, and since VERTEXSELECT$[\omega]$ always output a maximal solution, this solution is optimal. □

The converse of Theorem 3 is not true. We have seen that for the graph G in Fig. 2 the algorithm VERTEXSELECT$[\omega]$ outputs an optimal solution, for either rule

ω_S or ω_L, because G is unmixed. However, this graph G is not Cohen-Macaulay: the dimension of its face ring, $\dim(k[x_1, \ldots, x_5]/I(G)) = 2$, but its depth is $\text{depth}(k[x_1, \ldots, x_5]/I(G)) = 1$.

We can use Theorem 3 to produce a polynomial-time deterministic test for the failure of the Cohen-Macaulay property in graphs.

An algorithmic criterion for showing that a graph is not Cohen-Macaulay. Given a graph G as input, run VERTEXSELECT[ω] for both rules ω_S and ω_L, and check that the outputs are of different cardinalities.

By this criterion the graph shown in Fig. 3 is not Cohen-Macaulay.

Moreover, it is instructive to see how VERTEXSELECT[ω] fits in the general greedy scheme (Definition 2). Take as weights on the set of vertices a counting function, i.e. $f : V(G) \to \mathbb{R}$ given by $f(v) = 1$. Then VERTEXSELECT[ω] is the GREEDY algorithm with this counting function, and ω_S and ω_L give particular orderings for selecting the vertices (cf. Remark 1). In fact, for the graph in Fig. 2 it does not matter in which order the vertices are taken, GREEDY will always give the optimal independent set (of size 2) *for this particular weight function*. However, GREEDY will fail to give the maximum weighted, and maximum independent, set of vertices if the weights are given by $f(v) = \deg(v)$ and now we select vertices ordered from highest to lowest degree (breaking ties by using the order given by their labelling). For this weight f, in the graph of Fig. 2, we get as solution $\{1, 5\}$ whose deg-weight is $3 + 2 = 5$, whilst the MIS of maximum weight of 6 is $\{3, 4\}$. For the graph in Fig. 3 and $f(v) = \deg(v)$, the greedy solution is $\{3, 4\}$, which has maximum weight of 6, but is not a MIS. The MIS of maximum weight in this case is $\{1, 4, 6\}$. By Theorem 2, it follows that neither of these graphs is a matroid. This can also be seen by observing that the subgraph induced by $W = \{3, 4, 5\}$ has as set of facets $\{\{3, 4\}, \{5\}\}$, so it is not pure.

5 Chordal Graphs and Shellable Simplicial Complexes

A special subclass of Cohen-Macaulay complexes are the shellable simplicial complexes [18]. The associated edge ideal allows to refine our analysis of correctness of greedy algorithms for the MIS problem for more interesting classes of graphs, like the chordal graphs.

Definition 6 (Shellable simplicial complex) A simplicial complex Δ is called *shellable* if the facets of Δ can be arranged in a linear order $F_1, \ldots, F_m$ such that for each pair i, j, $1 \leq i < j \leq m$, there exists some $v \in F_j \setminus F_i$ and some $k < j$ such that $F_j \setminus F_k = \{v\}$. Such an ordering of the facets is called a *shelling order*.

A pure shellable simplicial complex is Cohen-Macaulay [2, Theorem 5.1.13]. The notion of shellability (as in the previous definition) can also be applied to non pure complexes, as it is done in [1].

A graph G is *shellable* if the simplicial complex $\Delta(G)$ is shellable (in the non pure sense). The notion of shellable graph was introduced and studied in [19], and the remarkable main result contained in that paper is that *chordal graphs are shellable*.

Theorem 4 ([19]) *If G is a chordal graph then G is shellable.* □

Recall that a graph is *chordal* if every cycle in it of length >3 has a *chord*, where a chord is an edge between two non-consecutive vertices. As an example, drawing an edge among vertices 3 and 4 in the graph of Fig. 3 makes that graph chordal. Observe that with that extra edge, applying VERTEXSELECT[ω_S] to the extended graph still gives the optimal solution $\{1, 4, 6\}$, whilst VERTEXSELECT[ω_L] gives $\{3\}$, so by our Theorem 3 this chordal graph is not Cohen-Macaulay.

We seek to answer the following two questions for chordal graphs:

1. What version of our VERTEXSELECT[ω] algorithm may work correctly for chordal graphs?
2. How to algebraically characterize the correctness of VERTEXSELECT[ω]?

The answer to question 1 is found implicitly in a combinatorial characterization of chordality due to Fulkerson and Gross [9]:

> **Fulkerson-Gross criterion:** Given a graph G, to know if it is chordal, it is necessary and sufficient that the following procedure eliminates all the vertices of G: search for a simplicial vertex in G, and if one is found, suppress it and repeat the procedure in the reduced graph.

Recall that a vertex v is simplicial if the subgraph induced by v and its neighbours, $\{v\} \cup N(v)$, form a clique. Observe that the Fulkerson-Gross simplicial vertices selection procedure establishes an ordering of the vertices, which has been named a *perfect elimination ordering*. A perfect elimination ordering of a graph is an ordering of its vertices such that each vertex in the ordering and its neighbours following it in the order form a clique. The Fulkerson-Gross criterion translates then to the following statement:

> A graph G is chordal iff G has a perfect elimination ordering

Now, let us denote by ω_{pe} a perfect elimination ordering $v_1, \ldots, v_n$ of the vertices, and let VERTEXSELECT[ω_{pe}] be VERTEXSELECT with the Fulkerson-Gross selection strategy consisting in selecting a simplicial vertex (if it exists), remove its neighbours and repeat the procedure in the reduced graph, until no simplicial vertex is found or there are no more vertices to select. The vertices collected this way form an independent set, but observe that a priori there is no guarantee that it will be a maximal independent set because the algorithm may stop at some step where it does not finds any more simplicial vertices. Of course it will be guaranteed to end with a solution to MIS if the input graph is chordal. Thus we have an answer to question 1 in the following theorem.

Theorem 5 VERTEXSELECT[ω_{pe}] *correctly solves the MIS problem for input graph G if and only if G is chordal.* □

Back to our graph in Fig. 3 with the extra edge $\{3, 4\}$ to turn it chordal, VERTEXSELECT[ω_{pe}] on input this graph will output the maximum independent set $\{1, 4, 6\}$.

There is more to say about chordal graphs in Commutative Algebra. A celebrated theorem by Fröberg [8] states that:

> The edge ideal $I(G)$ of a graph G has a linear resolution if and only if its complementary graph G^c is chordal.

On the other hand, the following is known (cf. [16]):

> If an edge ideal $I(G)$ of a graph G has linear quotients, then $I(G)$ has a linear resolution.

The missing ingredient to answer question 2 was found in the recent work [11, 12], where the new notion of strongly shellable is introduced.

Definition 7 (Strongly shellable [12]) A simplicial complex Δ is called *strongly shellable* if its facets can be arranged in a linear order $F_1, \ldots, F_m$ in such a way that for each pair $i < j$, there exists $k < j$, such that $|F_j \setminus F_k| = 1$ and $F_i \cap F_j \subseteq F_k \subseteq F_i \cup F_j$. Such an ordering of facets is called a *strong shelling order.*

By definition strongly shellable simplicial complexes are shellable. Less obvious is the fact that matroids are (pure) strongly shellable [11].

Now, for a finite simple graph G, with vertex set $V(G)$ and edge set $E(G)$, a strong shelling order $\succ$ on the edge set $E(G)$ (corresponding to a strong shelling order of the edge ideal $I(G)$), means that whenever we have two disjoint edges $E_i \succ E_j$, then we can find some $E_k \succ E_j$ that intersects both E_i and E_j non-trivially. If such strong shelling order on $E(G)$ exists, one says that G is *edgewise strongly shellable.*

Putting together the known facts on chordal graphs, linear quotients and linear resolutions of edge ideals, with this new combinatorial property of strongly shellable, Guo, Shen and Wu have shown the following beautiful result:

Theorem 6 ([12]) *Let G be a finite simple graph. Then the following conditions are equivalent:*

(i) *G is edgewise strongly shellable.*
(ii) *The edge ideal $I(G)$ has linear quotients.*
(iii) *The edge ideal $I(G)$ has a linear resolution.*
(iv) *The complement graph G^c is chordal.*
(v) *The complement graph G^c has a perfect elimination ordering.* □

Turning this around our greedy (algorithmic) interests, we obtain the following answer to question 2:

Theorem 7 *Let G be a finite simple graph. Then the following conditions are equivalent:*

(i) VERTEXSELECT[ω_{pe}] *is correct on* G.
(ii) G *is chordal.*
(iii) G *has a perfect elimination ordering.*
(iv) *The complement graph* G^c *is edgewise strongly shellable.*
(v) *The edge ideal of the complement graph* $I(G^c)$ *has linear quotients.*
(vi) *The edge ideal of the complement graph* $I(G^c)$ *has a linear resolution.* □

There is certainly much more to research on the correctness of greedy algorithms for the MIS problem on other classes of graphs. Take for example the *Petersen* graph. This graph is not chordal; VERTEXSELECT[ω] with ordering by degrees gives different sizes of independent sets; hence it is not Cohen-Macaulay.

6 Coin-Exchange Problems and Cohen-Macaulay Toric Projective Curves

Given a set of coin values $M = \{e_1, \ldots, e_h\}$ and a target value $B > 0$, the Coin-Exchange problem asks for the minimum number of coins whose values sum up to B. We will always assume that $e_1 = 1$ so that all values can be attained, and that $e_1 < e_2 < \ldots < e_h$. Thus, formally, the Coin-Exchange problem (a.k.a. *optimal representation problem*) reads:

Input: Integers $1 = e_1 < e_2 < \ldots < e_h$, and $B > 0$

Output: h-tuple of non negative integers $(a_1, a_2, \ldots, a_h)$ such that $B = \sum_{i=1}^{h} a_i e_i$ and $\sum_{i=1}^{h} a_i$ is minimized.

The h-tuple $(a_1, a_2, \ldots, a_h)$ is called a representation of B, and the one that minimizes $\sum_{i=1}^{h} a_i$ is called an optimal representation, and we write $\mathrm{opt}(B; e_1, e_2, \ldots, e_h)$ for the minimum value $\sum_{i=1}^{h} a_i$. If the coin system is clear from the context we might just write $\mathrm{opt}(B)$.

The Coin-Exchange problem is a form of the Knapsack problem, and so it is in general NP-complete (cf. Integer Knapsack in [10]). This justifies the design of heuristics for obtaining approximate solutions, that for some cases may find the optimal. One such heuristics is the following greedy algorithm.

Greedy strategy for optimal Coin-Exchange. A greedy strategy to solve the optimal representation problem (or Coin-Exchange) works as follows: Initially consider an empty h-tuple α. Then, repeat the following steps until $B = 0$: find the largest index i such that coin $e_i \leq B$, add 1 in the ith entry of α, and replace B by $B - e_i$.

The output $\alpha := (a_1, a_2, \ldots, a_h)$ obtained with this greedy strategy will be called the *greedy representation* of B, and write greed$(B; e_1, e_2, \ldots, e_h)$, or simply greed$(B)$, for the value $\sum_{i=1}^{h} a_i$ obtained by this algorithm.

Throughout the rest of this section the "greedy algorithm" refers to the above greedy strategy for the optimal Coin-Exchange problem.

Here is a list of some obvious facts that will be useful later:

(i) For a set of two coins $\{1, e_2\}$, greed$(B) =$ opt(B) for all B.

(ii) For any coin value $e_j \in \{e_1, e_2, \ldots, e_h\}$ and for every $B \geq e_j$,

$$\text{opt}(B) \leq \text{opt}(B - e_j) + 1, \tag{3}$$

with equality if coin e_j is used in a minimal representation of B. (Since $B = B - e_j + e_j$, an optimal representation of $B - e_j$ gives a representation of B adding 1 to the jth term.)

(iii) The value of the greedy representation bounds the optimal value for the Coin-Exchange:

$$\text{greed}(B; e_1, e_2, \ldots, e_h) \geq \text{opt}(B; e_1, e_2, \ldots, e_h) \tag{4}$$

We can assume we have the following implementation of the greedy algorithm: for each $i = h, h-1, \ldots, 2, 1$, let $a_i = \lfloor B/e_i \rfloor$, and set B to be $B - e_i a_i$. This produces the greedy representation in time $O(h \log e_h)$. This is polynomial in h. (Observe that we could have $e_h = 2^h$.)

An example for which the greedy algorithm fails to output an optimal representation of the Coin-Exchange problem is given by the system $\{1, 3, 4\}$ and the target $B = 6$. The greedy representation for this value is (2, 0, 1) of size greed(6) $= 3$, while the optimal representation is (0, 2, 0) of size opt(6) $= 2$. On the other hand, for the US coin system (of cents) $\{1, 5, 10, 25, 50, 100\}$ or the Eurozone coin system $\{1, 2, 5, 10, 20, 50, 100, 200\}$, the greedy algorithm will always produce an optimal representation for any given value. A formal proof of these facts will be possible after we endow this numerical problem with the algebraic structure appropriate for applying the theory of Cohen-Macaulay rings to the analysis of correctness of the greedy algorithm.

6.1 *The Algebraic Framework*

We will now go through the work in [5] for the necessary background on toric projective curves and their Cohen-Macaulay characterization.

Given a finite set of integers $M_h = \{1 = e_1 < e_2 < \ldots < e_h\}$, consider the subsemigroup S of $\mathbb{N}^2$ generated by $(0, e_h), (e_1, e_h - e_1), \ldots, (e_h, e_h - e_h)$, that is,

$$S = \langle (0, e_h), (e_j, e_h - e_j) : j = 1, 2, \ldots, h \rangle$$

and consider its projection on the second coordinate of $\mathbb{N}^2$:

$$S' = \langle e_h, e_h - e_1, e_h - e_2, \ldots, e_h - e_{h-1}, 0 \rangle$$

For each $i = 1, 2, \ldots, e_h$, let c_i be the smallest integer in S' such that $c_i = e_h - i($ mod $e_h)$, i.e. $c_i = (e_h - i) + t \cdot e_h$ for some $t \geq 0$. For completeness, set $c_0 = 0$, and observe that $c_{e_i} = e_h - e_i$. Next, fix a field k and consider the semigroup algebra $k[S] := \bigoplus_{\alpha \in S} k\chi^\alpha$, with product given by $\chi^\alpha \cdot \chi^\beta = \chi^{\alpha+\beta}$ for $\alpha, \beta \in S$. The $\mathbb{N}$-grading on $k[S]$ given by $\deg(\chi^\alpha) = |\alpha|$, where $|\alpha| = b + c$ if $\alpha = (b, c)$, makes $k[S]$ the homogeneous coordinate algebra of a toric projective curve C_h. The curve C_h is *arithmetically Cohen-Macaulay* if the following property holds:

$$\boxed{\text{If } \alpha \in \mathbb{Z}^2, \alpha + (e_h, 0) \in S, \alpha + (0, e_h) \in S, \text{ then } \alpha \in S.} \qquad (5)$$

Within this algebraic framework, Campillo and Revilla [5] have shown the following results that lead to a nice characterization of the correctness of the greedy strategy for the Coin-Exchange problem.

Proposition 3 *The toric projective curve C_h is arithmetically Cohen-Macaulay if and only if for all $i : 0 \leq i < e_h$, $(i, c_i) \in S$, where c_i is as defined above.* □

Proposition 4 *C_h is arithmetically Cohen-Macaulay if and only if for all $B \geq e_h$, $opt(B) = opt(B - e_h) + 1$.* □

Theorem 8 *For each $j \leq h$, set $M_j = \{1 = e_1, \ldots, e_j\}$ and let C_j be the corresponding toric projective curve.*

(i) *If the greedy algorithm correctly solves the Coin-Exchange problem for the coin system M_j, then C_j is arithmetically Cohen-Macaulay.*
(ii) *If C_1, C_2, ..., C_j are arithmetically Cohen-Macaulay, then the greedy algorithm correctly solves the Coin-Exchange problem for M_j.* □

Additionally the following characterization of opt(B) in terms of the subsemigroup $S = \langle (0, e_h), (e_j, e_h - e_j) : j = 1, 2, \ldots, h \rangle$ will be useful (this was originally established in [3]).

Proposition 5 *For all $B \geq 0$, if c is the least integer such that $(B, c) \in S$ then $opt(B) = (B + c)/e_h$.* □

6.2 *Finding an Optimal Solution for Coin-Exchange Effectively Within the Algebraic Framework*

Theorem 8 give us a way of showing the correctness of the greedy algorithm for a given coin system M_h: one has to check that all the local toric projective curves C_3, C_4, ..., C_h are Cohen-Macaulay (where each curve C_j is associated to the system

$M_j = \{e_1, \ldots, e_j\}$). However, checking that each curve C_j is Cohen-Macaulay goes through checking membership in the associated semigroup S (Proposition 2), and this latter check depends on the universal property (5) to hold. So this strategy seems hard, in principle.

To find an effective way for checking that the greedy algorithm outputs an optimal solution for any given coin system, the idea is to actually check for those values where the greedy algorithm does not give optimal solutions, for it happens that *the set of witness where greedy fails for a given coin system is finite*. This remarkable and useful result was obtained by Kozen and Zaks in [14] through combinatorial arguments, but we shall derive it using the algebraic tools of Campillo and Revilla laid out in the previous section.

Theorem 9 ([14]) *Given a system of coins $M_h = \{1 = e_1 < e_2 < \ldots < e_h\}$. If for some B, greed$(B) >$ opt(B), then the smallest such B satisfies*

$$e_3 + 1 < B < e_h + e_{h-1}.$$

Proof If $B < e_3$, we only need the subset $\{1, e_2\}$ of the coin system and for this $\text{greed}(B) = \text{opt}(B)$. For $B = e_3, e_3 + 1$, $\text{greed}(B) = \text{opt}(B) = 1, 2$.

For the other bound, let $B \geq e_h + e_{h-1}$, and assume that for all $x < B$, $\text{greed}(x) = \text{opt}(x)$. Let $i < e_h$ such that $B = te_h + i$, for some integer t (i.e. $B \equiv i \mod e_h$). Let c_i be the least integer in S' such that $c_i \equiv e_h - i \mod e_h$. Then $(i, c_i) \in S$ and $\text{opt}(i) = (i + c_i)/e_h$. Also $(B, c_i) \in S$ and

$$\text{opt}(B) \cdot e_h = B + c_i = te_h + i + c_i,$$

hence

$$\text{opt}(B) = t + \frac{i + c_i}{e_h} = t + \text{opt}(i)$$

Now, by assumption and the fact that $i < e_h$ we conclude that

$$\text{opt}(B) = t + \text{opt}(i) = t + \text{greed}(i) = \text{greed}(te_h + i) = \text{greed}(B). \quad \square$$

Next, in order to avoid computing $\text{opt}(B)$, Kozen and Zaks use the previous theorem to characterize greedy optimally solely in terms of $\text{greed}(x)$. Again we give a proof of this fact using the algebraic setting for the Coin-Exchange problem:

Corollary 1 ([14]) *Given a system of coins $M_h = \{1 = e_1 < e_2 < \ldots < e_h\}$, the greedy algorithm is correct for M_h if and only if $\forall B \in (e_3 + 1, e_h + e_{h-1})$, $\forall c \in \{e_3, \ldots, e_h\}$ $(c < B \longrightarrow greed(B) \leq greed(B - c) + 1)$.*

Proof ($\Rightarrow$) Consider $B \in (e_3 + 1, e_h + e_{h-1})$ and e_j, for $j \in \{3, \ldots, h\}$, such that $e_j < B$. By hypothesis, Theorem 8 (i), Proposition 4 and Eq. (4),

$$\text{greed}(B) = \text{opt}(B) = \text{opt}(B - e_j) + 1 \leq \text{greed}(B - e_j) + 1$$

($\Leftarrow$) If there is a B such that $\mathrm{greed}(B) > \mathrm{opt}(B)$, by Theorem 9 the smallest such B lies in $(e_3 + 1, e_h + e_{h-1})$. Let e_j be a coin used in the minimal representation of B. Then

$$\mathrm{opt}(B - e_j) = \mathrm{opt}(B) - 1 < \mathrm{greed}(B) - 1 \leq \mathrm{greed}(B - e_j)$$

contradicting that B is smallest witness of the failure of the greedy algorithm. □

Using Corollary 1 one can effectively check that for the following coin systems the greedy algorithm always output optimal solutions:

1. US coin system.
2. Eurozone coin system.
3. Fibonacci coin system $\{1, 2, 3, 5, 8, 13, 21, 44\}$.

On the other hand, the toric projective curves associated to each of these systems are examples of Cohen-Macaulay curves.

Acknowledgements The author acknowledges the support of the Ministerio de Economía, Industria y Competitividad, Spain, project MACDA [TIN2017-89244-R] and of the Generalitat de Catalunya, project MACDA [SGR2014-890]

References

1. Björner, A., Wachs, M.: Shellable nonpure complexes and posets I. Trans. Am. Math. Soc. **348**(4), 1299–1327 (1996)
2. Bruns, W., Herzog, J.: Cohen-Macaulay Rings. Cambridge University Press, Cambridge (1993)
3. Campillo, A., Gimenez, P.: Syzygies of affine toric varieties. J. Algebra **225**, 142–161 (2000)
4. Campillo, A., Pisón, P.: Toric mathematics from semigroup viewpoint. In: Granja, A., et al. (eds.) Ring Theory and Algebraic Geometry. Lecture Notes in Pure and Applied Mathematics, vol. 221, pp. 95–112. CRC Press, Boca Raton (2001)
5. Campilllo, A., Revilla, M.A.: Coin exchange algorithms and toric projective curves. Commun. Algebra **29**(7), 2985–2989 (2001)
6. Dinur, I., Safra, S.: On the hardness of approximating minimum vertex-cover. Ann. Math. **162**(1), 439–485 (2005)
7. Edmonds, J.: Matroids and the greedy algorithm. Math. Program. **1**, 127–136 (1971)
8. Fröberg, R.: On Stanley-Reisner rings. Topics Algebra **26**(2), 57–70 (1990). Banach Center Publications
9. Fulkerson, D.R., Gross, O.A.: Incidence matrices and interval graphs. Pac. J. Math. **15**, 835–855 (1965)
10. Garey, M.R., Johnson, D.S.: Computers and Intractability. Freeman, San Francisco (1979)
11. Guo, J., Shen, Y.H., Wu, T.: Strong shellability of simplicial complexes. arXiv preprint arXiv:1604.05412 (2016)
12. Guo, J., Shen, Y.H., Wu, T.: Edgewise strongly shellable clutters. J. Algebra Appl. **17**(1) (2018)
13. Korte, B., Lovász, L.: Greedoids and linear objective functions. SIAM J. Algebraic Disc. Methods **5**(2), 229–238 (1984)
14. Kozen, D., Zaks, S.: Optimal bounds for the change-making problem. Theor. Comput. Sci. **123**, 377–388 (1994)

15. Kunz, E.: Introduction to Commutative Algebra and Algebraic Geometry. Birkhäuser, Basel (1985)
16. Miller, E., Sturmfels, B.: Combinatorial Commutative Algebra. Springer, New York (2005)
17. Papadimitriou, C., Steiglitz, K.: Combinatorial Optimization, Algorithms and Complexity. Dover, Mineola (1998)
18. Stanley, R.: Combinatorics and Commutative Algebra, 2nd edn. Birkhäuser, Boston (1996)
19. van Tuyl, A., Villarreal, R.: Shellable graphs and sequentially Cohen–Macaulay bipartite graphs. J. Combin. Theory Ser. A **115**(5), 799–814 (2008)
20. Villarreal, R.: Cohen-Macaulay graphs. Manuscripta Math. **66**(3), 277–293 (1990)

Binomial Ideals and Congruences on $\mathbb{N}^n$

Laura Felicia Matusevich and Ignacio Ojeda

Dedicated to Professor Antonio Campillo on the occasion of his 65th birthday.

Abstract A *congruence* on $\mathbb{N}^n$ is an equivalence relation on $\mathbb{N}^n$ that is compatible with the additive structure. If $\Bbbk$ is a field, and I is a *binomial ideal* in $\Bbbk[X_1, \ldots, X_n]$ (that is, an ideal generated by polynomials with at most two terms), then I induces a congruence on $\mathbb{N}^n$ by declaring $\mathbf{u}$ and $\mathbf{v}$ to be equivalent if there is a linear combination with nonzero coefficients of $\mathbf{X}^{\mathbf{u}}$ and $\mathbf{X}^{\mathbf{v}}$ that belongs to I. While every congruence on $\mathbb{N}^n$ arises this way, this is not a one-to-one correspondence, as many binomial ideals may induce the same congruence. Nevertheless, the link between a binomial ideal and its corresponding congruence is strong, and one may think of congruences as the underlying combinatorial structures of binomial ideals. In the current literature, the theories of binomial ideals and congruences on $\mathbb{N}^n$ are developed separately. The aim of this survey paper is to provide a detailed parallel exposition, that provides algebraic intuition for the combinatorial analysis of congruences. For the elaboration of this survey paper, we followed mainly (Kahle and Miller Algebra Number Theory 8(6):1297–1364, 2014) with an eye on Eisenbud and Sturmfels (Duke Math J 84(1):1–45, 1996) and Ojeda and Piedra Sánchez (J Symbolic Comput 30(4):383–400, 2000).

The first author was partially supported by NSF grant DMS-1500832.

The second author was partially supported by the project MTM2015-65764-C3-1, National Plan I+D+I, and by Junta de Extremadura (FEDER funds) – FQM-024.

L. F. Matusevich
Mathematics Department, Texas A&M University, College Station, TX, USA
e-mail: laura@math.tamu.edu

I. Ojeda (✉)
Departamento de Matemáticas, Universidad de Extremadura, Badajoz, Spain
e-mail: ojedamc@unex.es

G.-M. Greuel et al. (eds.), *Singularities, Algebraic Geometry, Commutative Algebra, and Related Topics*, https://doi.org/10.1007/978-3-319-96827-8_18

1 Preliminaries

In this section we introduce our main objects of study: binomial ideals and monoid congruences, and recall some basic results.

Throughout this article, $\Bbbk[\mathbf{X}] := \Bbbk[X_1, \ldots, X_n]$ is the commutative polynomial ring in n variables over a field $\Bbbk$. In what follows we write $\mathbf{X}^{\mathbf{u}}$ for $X_1^{u_1} X_2^{u_2} \cdots X_n^{u_n}$, where $\mathbf{u} = (u_1, u_2, \ldots, u_n) \in \mathbb{N}^n$, where here and henceforth, $\mathbb{N}$ denotes the set of nonnegative integers.

1.1 Binomial Ideals

In this section we begin our study of binomial ideals. First of all, we recall that a **binomial** in $\Bbbk[\mathbf{X}]$ is a polynomial with at most two terms, say $\lambda\mathbf{X}^{\mathbf{u}} + \mu\mathbf{X}^{\mathbf{v}}$, where $\lambda, \mu \in \Bbbk$ and $\mathbf{u}, \mathbf{v} \in \mathbb{N}^n$. We emphasize that, according to this definition, monomials are binomials.

Definition 1 A **binomial ideal** of $\Bbbk[\mathbf{X}]$ is an ideal of $\Bbbk[\mathbf{X}]$ generated by binomials.

Throughout this article, we assume that the base field $\Bbbk$ is algebraically closed. The reason for this is that some desirable results are not valid over an arbitrary field. These include the characterization of binomial prime ideals (Theorem 5), and the fact that associated primes of binomial ideals are binomial (see, e.g. Proposition 10). This failure can be seen even in one variable: the ideal $\langle X^2 + 1\rangle \subset \mathbb{R}[X]$ is prime, but does not conform to the description in Theorem 5; the ideal $\langle X^3 - 1\rangle \subset \mathbb{R}[X]$ has the associated prime $\langle X^2+X+1\rangle$, which is not binomial. It is also worth noting that the characteristic of $\Bbbk$ plays a role when studying binomial ideals, as can be seen by the different behaviors presented by $\langle X^p - 1\rangle \subset \Bbbk[X]$ depending on whether the characteristic of $\Bbbk$ is p.

The following result is an invaluable tool when studying binomial ideals.

Proposition 1 *Let $I \subset \Bbbk[\mathbf{X}]$ be an ideal. The following are equivalent:*

(a) *I is a binomial ideal.*
(b) *The reduced Gröbner basis of I with respect to any monomial order on $\Bbbk[\mathbf{X}]$ consists of binomials.*
(c) *A universal Gröbner basis of I consists of binomials.*

Proof If I has a binomial generating set, the S-polynomials produced by a step in the Buchberger algorithm are necessarily binomials.

Since the Buchberger algorithm for computing Gröbner bases respects the binomial condition, Gröbner techniques are particularly effective when working with these objects. In particular, it can be shown that some important ideal theoretic operations preserve binomiality. For instance, it is easy to show that eliminating variables from binomial ideals results in binomial ideals.

Corollary 1 *Let I be a binomial ideal of $\Bbbk[\mathbf{X}]$. The elimination ideal $I \cap \Bbbk[X_i \mid i \in \sigma]$ is a binomial ideal for every nonempty subset $\sigma \subset \{1, \ldots, n\}$.*

Proof The intersection is generated by a subset of the reduced Gröbner basis of I with respect to a suitable lexicographic order.

Example 1 Let $\varphi : \Bbbk[X, Y, Z] \to \Bbbk[T]$ be the $\Bbbk-$algebra morphism such that

$$X \to T^3, Y \to T^4 \text{ and } Z \to T^5.$$

It is known that $\ker(\varphi) = \langle X - T^3, Y - T^4, Z - T^5 \rangle \cap \Bbbk[X, Y, Z]$. As a consequence of Corollary 1, $\ker(\varphi)$ is a binomial ideal. In fact, $\ker(\varphi)$ is the ideal generated by $\{Y^2 - XZ, X^2Y - Z^2, X^3 - YZ\}$, as can be checked by executing the following code in Macaulay2 [8]:

```
R = QQ[X,Y,Z,T]
I = ideal(X-T^3,Y-T^4,Z-T^5)
eliminate(T,I)
```

Taking ideal quotients is a fundamental operation in commutative algebra. We can now show that some ideal quotients of binomial ideals are binomial.

Corollary 2 *If I is a binomial ideal of $\Bbbk[\mathbf{X}]$, and $\mathbf{X}^{\mathbf{u}}$ is a monomial, then $(I : \mathbf{X}^{\mathbf{u}})$ is a binomial ideal.*

Proof Recall that if $\{f_1, \ldots, f_\ell\}$ is a system of generators for $I \cap \langle \mathbf{X}^{\mathbf{u}} \rangle$, then $\{f_1/\mathbf{X}^{\mathbf{u}}, \ldots, f_\ell/\mathbf{X}^{\mathbf{u}}\}$ is a system of generators for $(I : \mathbf{X}^{\mathbf{u}})$. Thus, the binomiality of $(I : \mathbf{X}^{\mathbf{u}})$ follows if we show that $I \cap \langle \mathbf{X}^{\mathbf{u}} \rangle$ is binomial.

Introducing an auxiliary variable T, we have that

$$I \cap \langle \mathbf{X}^{\mathbf{u}} \rangle = (TI + (1 - T)\langle \mathbf{X}^{\mathbf{u}} \rangle) \cap \Bbbk[\mathbf{X}].$$

Since $TI + (1 - T)\langle \mathbf{X}^{\mathbf{u}} \rangle$ is a binomial ideal, Corollary 1 implies that $I \cap \langle \mathbf{X}^{\mathbf{u}} \rangle$ is also binomial, as we wanted.

We remark that the ideal quotient of a binomial ideal by a binomial is not necessarily binomial, and neither is the ideal quotient of a binomial ideal by a monomial ideal. When taking colon with a single binomial, the above proof breaks because the product of two binomials is not a binomial in general; indeed,

$$\big(\langle X^3 - 1 \rangle : \langle X - 1 \rangle\big) = \langle X^2 + X + 1 \rangle \subset \Bbbk[X].$$

In the case of taking ideal quotient by a monomial ideal, say $J = \langle \mathbf{X}^{\mathbf{u}_1}, \ldots \mathbf{X}^{\mathbf{u}_r} \rangle$, instead of a single monomial, what makes the argument invalid is that the ideal $(I : J)$ is equal to $\cap_{i=1}^{r}(I : \langle \mathbf{X}^{\mathbf{u}_i} \rangle)$, and the intersection of binomial ideals is not necessarily binomial, as the following shows: $\langle X - 1 \rangle \cap \langle X - 2 \rangle = \langle X^2 - 3X + 2 \rangle \subset \Bbbk[X]$.

1.2 Graded Algebras

Gradings play a big role when studying binomial ideals. The main result of this section is that a ring is a quotient of a polynomial ring by a binomial ideal if and only if it has a special kind of grading (Theorem 1).

Recall that a $\Bbbk-$algebra of finite type R is *graded* by a finitely generated commutative monoid S if R is a direct sum

$$R = \bigoplus_{\mathbf{a} \in S} R_{\mathbf{a}}$$

of $\Bbbk-$vector spaces and the multiplication of R satisfies the rule $R_{\mathbf{a}} R_{\mathbf{a}'} = R_{\mathbf{a}+\mathbf{a}'}$.

Example 2 Observe that $\Bbbk[\mathbf{X}] = \bigoplus_{\mathbf{u} \in \mathbb{N}^n} \mathrm{Span}_{\Bbbk}\{\mathbf{X}^{\mathbf{u}}\}$.

Remark 1 Let I be *any* ideal of $\Bbbk[\mathbf{X}]$ and let π be the canonical projection of $\Bbbk[\mathbf{X}]$ onto $R := \Bbbk[\mathbf{X}]/I$. Let S be the set of all one-dimensional subspaces $\mathrm{Span}_{\Bbbk}\{\pi(\mathbf{X}^{\mathbf{u}})\}$ of R; if the kernel of π contains monomials, we adjoin to S the symbol ∞ associated to the monomials in $\ker(\pi)$. The set S is a commutative monoid with the operation

$$\mathrm{Span}_{\Bbbk}\{\pi(\mathbf{X}^{\mathbf{u}})\} + \mathrm{Span}_{\Bbbk}\{\pi(\mathbf{X}^{\mathbf{v}})\} = \mathrm{Span}_{\Bbbk}\{\pi(\mathbf{X}^{\mathbf{u}}\mathbf{X}^{\mathbf{v}})\} = \mathrm{Span}_{\Bbbk}\{\pi(\mathbf{X}^{\mathbf{u}+\mathbf{v}})\}$$

and identity element $\mathrm{Span}_{\Bbbk}\{1\} = \mathrm{Span}_{\Bbbk}\{\pi(\mathbf{X}^{\mathbf{0}})\}$. Note that if $\mathbf{X}^{\mathbf{v}} \in \ker(\pi)$, then for any other monomial $\mathbf{X}^{\mathbf{u}}$, $\mathbf{X}^{\mathbf{u}}\mathbf{X}^{\mathbf{v}} \in \ker(\pi)$. In other words,

$$\mathrm{Span}_{\Bbbk}\{\pi(\mathbf{X}^{\mathbf{u}})\} + \infty = \infty.$$

We point out that the set $\{\mathrm{Span}_{\Bbbk}\{\pi(X_1)\}, \ldots, \mathrm{Span}_{\Bbbk}\{\pi(X_n)\}\}$ generates S as a monoid. There is a natural $\Bbbk-$vector space surjection

$$\bigoplus_{\substack{\mathrm{Span}_{\Bbbk}\{\pi(\mathbf{X}^{\mathbf{u}})\} \in S \\ \mathbf{X}^{\mathbf{u}} \notin \ker \pi}} \mathrm{Span}_{\Bbbk}\{\pi(\mathbf{X}^{\mathbf{u}})\} \to R. \tag{1}$$

We observe that if (1) is an isomorphism of $\Bbbk-$vector spaces, then R is **finely graded** by S, meaning that R is $S-$graded and every graded piece has dimension at most 1.

The following result provides the first link between binomial ideals and monoids.

Theorem 1 *A $\Bbbk-$algebra R of finite type admits a presentation of the form $\Bbbk[\mathbf{X}]/I$, where I is a binomial ideal, if and only if R can be finely graded by a finitely generated commutative monoid.*

Proof First assume that R admits a grading of the given type by a finitely generated commutative monoid S. Let $f_1, \ldots, f_n$ be $\Bbbk-$algebra generators of R. Without loss of generality, we may assume that $f_1, \ldots, f_n$ are homogeneous. Denote $\mathbf{a}_i$ the degree of f_i, for $i = 1, \ldots, n$. Since R is (finely) graded by the monoid generated by $\{\mathbf{a}_1, \ldots, \mathbf{a}_n\}$, we may assume that S is generated by $\{\mathbf{a}_1, \ldots, \mathbf{a}_n\}$.

Give $\Bbbk[\mathbf{X}]$ an S−grading by setting the degree of X_i to be $\mathbf{a}_i$, and consider the surjection $\Bbbk[\mathbf{X}] \to R$ given by $X_i \mapsto f_i$, which is a graded ring homomorphism. The kernel of this map is a homogeneous ideal of $\Bbbk[\mathbf{X}]$, and is therefore generated by homogeneous elements. On the other hand, by the fine grading condition, for any two monomials $\mathbf{X}^{\mathbf{u}}, \mathbf{X}^{\mathbf{v}} \in \Bbbk[\mathbf{X}]$ with the same S−degree, neither of which maps to zero in R, there is a scalar $\lambda \in \Bbbk^*$ such that the binomial $\mathbf{X}^{\mathbf{u}} - \lambda \mathbf{X}^{\mathbf{v}} \in \Bbbk[\mathbf{X}]$ maps to zero in R. Thus, the kernel of the above surjection is generated by binomials.

Conversely, by Remark 1, it suffices to show that the map (1) is injective. We need to prove that if Σ is a nonempty subset of $\{\mathrm{Span}_{\Bbbk}\{\pi(\mathbf{X}^{\mathbf{u}})\} \in S \mid \mathbf{X}^{\mathbf{u}} \notin \ker \pi\}$, then the image of Σ in R is linearly independent. This follows if we show that if $f = \sum_{i=1}^{r} \lambda_i \mathbf{X}^{\mathbf{u}_i} \in I$ with $\lambda_1, \ldots, \lambda_r \in \Bbbk^*$ and $\mathbf{X}^{\mathbf{u}_i} \notin I$ for all $1 \leq i \leq r$, then there exist $1 \leq j \leq r$ and $\lambda \in \Bbbk^*$ such that $\mathbf{X}^{\mathbf{u}_1} - \lambda \mathbf{X}^{\mathbf{u}_j} \in I$ (in other words, $\pi(\mathbf{X}^{\mathbf{u}_1}) = \pi(\mathbf{X}^{\mathbf{u}_j})$). To see this, note that since I is a binomial ideal, it has a $\Bbbk$−vector space basis consisting of binomials, and therefore we can write $f = \sum_{i=1}^{\ell} \mu_i B_i$, where $\mu_1, \ldots, \mu_\ell \in \Bbbk^*$ and each B_i is a binomial in I with two terms, neither of which is in I (the latter by the assumption on f). The monomial $\mathbf{X}^{\mathbf{u}_1}$ must appear in at least one of the binomials $B_1, \ldots, B_\ell$, say B_{i_1}. Of course, the second monomial appearing in B_{i_1} has the same image under π as $\mathbf{X}^{\mathbf{u}_1}$. If this second monomial in B_{i_1} is one of the $\mathbf{X}^{\mathbf{u}_2}, \ldots, \mathbf{X}^{\mathbf{u}_r}$, we are done. Otherwise, the second term of B_{i_1} must appear in another of the binomials B_i, say B_{i_2}. Note that both monomials in B_{i_2} have the same image under π as $\mathbf{X}^{\mathbf{u}_1}$. If the second monomial of B_{i_2} is one of the $\mathbf{X}^{\mathbf{u}_2}, \ldots, \mathbf{X}^{\mathbf{u}_r}$, again, we are done. Otherwise, continue in the same manner. Since we only have finitely many binomials to consider, this process must stop, and produce a monomial $\mathbf{X}^{\mathbf{u}_j}$ such that $\pi(\mathbf{X}^{\mathbf{u}_1}) = \pi(\mathbf{X}^{\mathbf{u}_j})$.

2 Congruences on Monoids and Binomial Ideals

We now start our study of monoid congruences, and their relationship to binomial ideals. We show how binomial ideals induce congruences, and how any congruence can arise this way. We also address the question of when two different binomial ideals give rise to the same congruence.

Definition 2 Let S be a commutative monoid. A **congruence** $\sim$ on S is an equivalence relation on S which is additively closed: $\mathbf{a} \sim \mathbf{b} \Rightarrow \mathbf{a} + \mathbf{c} \sim \mathbf{b} + \mathbf{c}$ for $\mathbf{a}, \mathbf{b}$ and $\mathbf{c} \in S$.

The following result, which follows directly from the definition, gives a first indication that congruences on commutative monoids are analogous to ideals in commutative rings.

Proposition 2 *If $\sim$ is a congruence on a commutative monoid S, then $S/\sim$ is a commutative monoid.* □

Let $\phi : S \to S'$ be a monoid morphism.The **kernel of ϕ** is defined as

$$\ker \phi := \{(\mathbf{a}, \mathbf{b}) \in S \times S \mid \phi(\mathbf{a}) = \phi(\mathbf{b})\}.$$

Note that if ϕ is a monoid morphism, the relation on S determined by $\ker \phi \subset S \times S$ is actually a congruence. Moreover, every congruence on S arises in this way: if $\sim$ is a congruence on S, then $\sim$ can be recovered as the congruence induced by the kernel of the natural surjection $S \to S/\!\sim$.

We write $\mathrm{cong}(S) \subset \mathscr{P}(S \times S)$ for the set of congruences on S ordered by inclusion. (Here $\mathscr{P}$ indicates the power set.) We say that S is **Noetherian** if every nonempty subset of $\mathrm{cong}(S)$ has a maximal element (equivalently, $\mathrm{cong}(S)$ satisfies the ascending chain condition). The following is an important result in monoid theory.

Theorem 2 *A commutative monoid S is Noetherian if and only if S is finitely generated.*

The fact that a Noetherian monoid is finitely generated is the hard part of the proof. It is due to Budach [4], and is the main result in Chapter 5 in Gilmer's book [7], where it appears as Theorem 5.10. Brookfield has given a short and self contained proof in [2]. We will just provide a proof of the converse, namely, that finitely generated monoids are Noetherian (see [7, Theorem 7.4]), after Theorem 3.

Set S be a commutative monoid finitely generated by $\mathscr{A} = \{\mathbf{a}_1, \ldots, \mathbf{a}_n\}$. The monoid morphism

$$\pi : \mathbb{N}^n \longrightarrow S;\ \mathbf{e}_i \longmapsto \mathbf{a}_i,\ i = 1, \ldots, n, \tag{2}$$

where $\mathbf{e}_i$ denotes the element in $\mathbb{N}^n$ whose i−th coordinate is 1 with all other coordinates 0, is surjective and gives a **presentation**

$$S = \mathbb{N}^n/\!\sim$$

by simply taking $\sim = \ker \pi$. Unless stated otherwise, we write $[\mathbf{u}]$ for the class of $\mathbf{u} \in \mathbb{N}^n$ modulo $\sim$.

Remark 2 In what follows, all monoids considered are commutative and finitely generated.

Given a monoid S, the **semigroup algebra** $\Bbbk[S] := \bigoplus_{\mathbf{a} \in S} \mathrm{Span}_{\Bbbk}\{\chi^{\mathbf{a}}\}$ is the direct sum with multiplication $\chi^{\mathbf{a}}\chi^{\mathbf{b}} = \chi^{\mathbf{a}+\mathbf{b}}$. (This terminology is in wide use, even though the algebra $\Bbbk[S]$ would be more precisely named a "monoid algebra".)

Theorem 3 *Let $\mathscr{A} = \{\mathbf{a}_1, \ldots, \mathbf{a}_n\}$ be a generating set of a monoid S, and consider the presentation map $\pi : \mathbb{N}^n \to S$ induced by $\mathscr{A}$. We define a map of semigroup algebras*

$$\hat{\pi} : \Bbbk[\mathbb{N}^n] = \Bbbk[\mathbf{X}] \to \Bbbk[S]\,; \quad \mathbf{X}^{\mathbf{u}} \mapsto \chi^{\pi(\mathbf{u})}. \tag{3}$$

Let

$$I_{\mathscr{A}} := \langle \mathbf{X}^{\mathbf{u}} - \mathbf{X}^{\mathbf{v}} \mid \pi(\mathbf{u}) = \pi(\mathbf{v}) \rangle \subseteq \Bbbk[\mathbf{X}]. \tag{4}$$

Then $\ker \hat{\pi} = I_{\mathscr{A}}$*, so that* $\Bbbk[S] \cong \Bbbk[\mathbf{X}]/I_{\mathscr{A}}$*. Moreover,* $I_{\mathscr{A}}$ *is spanned as a* $\Bbbk-$*vector space by* $\{\mathbf{X}^{\mathbf{u}} - \mathbf{X}^{\mathbf{v}} \mid \pi(\mathbf{u}) = \pi(\mathbf{v})\}$.

Proof By construction, $I_{\mathscr{A}} \subseteq \ker \hat{\pi}$. To prove the other inclusion, give $\Bbbk[\mathbf{X}]$ an $S-$grading by setting $\deg(X_i) = \pi(\mathbf{e}_i) = \mathbf{a}_i$. Then the map $\hat{\pi}$ is graded (considering $\Bbbk[S]$ with its natural $S-$grading), and therefore its kernel is a homogeneous ideal of $\Bbbk[\mathbf{X}]$. Note that $\mathbf{X}^{\mathbf{u}}$ and $\mathbf{X}^{\mathbf{v}}$ have the same $S-$degree if and only if $\pi(\mathbf{u}) = \pi(\mathbf{v})$.

We observe that $\ker \hat{\pi}$ contains no monomials, so any polynomial in $\ker \hat{\pi}$ has at least two terms. Let f be a homogeneous element of $\ker \hat{\pi}$. Then there are $\lambda, \mu \in \Bbbk^*$ and $\mathbf{u}, \mathbf{v} \in \mathbb{N}^n$ such that $f = \lambda \mathbf{X}^{\mathbf{u}} + \mu \mathbf{X}^{\mathbf{v}} + g$, with g a homogeneous polynomial with two fewer terms than f. Since f is homogeneous, we have that $\pi(\mathbf{u}) = \pi(\mathbf{v})$, and therefore $\mathbf{X}^{\mathbf{u}} - \mathbf{X}^{\mathbf{v}} \in I_{\mathscr{A}} \subset \ker \hat{\pi}$. Then $f - \lambda(\mathbf{X}^{\mathbf{u}} - \mathbf{X}^{\mathbf{v}})$ is a homogeneous element of $\ker \hat{\pi}$, and has fewer terms than f. Continuing in this manner, we conclude that $f \in I_{\mathscr{A}}$. Since $\ker \hat{\pi}$ is a homogeneous ideal, we see that $I_{\mathscr{A}} \supseteq \ker \hat{\pi}$, and therefore $I_{\mathscr{A}} = \ker \hat{\pi}$.

For the final statement, we note that any binomial ideal in $\Bbbk[\mathbf{X}]$ is spanned as a $\Bbbk-$vector space by the set of all of its binomials. Since $I_{\mathscr{A}}$ contains no monomials and is $S-$graded, any binomial in $I_{\mathscr{A}}$ is of the form $\mathbf{X}^{\mathbf{u}} - \lambda \mathbf{X}^{\mathbf{v}}$, where $\lambda \in \Bbbk^*$ and $\pi(\mathbf{u}) = \pi(\mathbf{v})$. But then $\mathbf{X}^{\mathbf{u}} - \mathbf{X}^{\mathbf{v}} \in I_{\mathscr{A}}$, and again using that $I_{\mathscr{A}}$ contains no monomials, we see that $\lambda = 1$. This implies that $\{\mathbf{X}^{\mathbf{u}} - \mathbf{X}^{\mathbf{v}} \mid \pi(\mathbf{u}) = \pi(\mathbf{v})\}$ is the set of all binomials of $I_{\mathscr{A}}$, which implies that it is a $\Bbbk-$spanning set for this ideal.

We are now ready to prove that finitely generated monoids are Noetherian.

Proof (Proof of Theorem 2, reverse implication) Let S be a finitely generated monoid, and consider a presentation $S = \mathbb{N}^n/\sim$, where $\sim$ is a congruence on $\mathbb{N}^n$. In this proof, for $\mathbf{u} \in \mathbb{N}^n$, we denote by $[\mathbf{u}]$ the equivalence class of $\mathbf{u}$ with respect to $\sim$.

Let $\approx$ be a congruence on S, and let $\simeq$ be the congruence on $\mathbb{N}^n$ given by setting the equivalence class of $\mathbf{u} \in \mathbb{N}^n$ with respect to $\simeq$ to be the set $\bigcup_{\{\mathbf{v} \in \mathbb{N}^n | [\mathbf{u}] \approx [\mathbf{v}]\}} [\mathbf{v}]$. Then the congruence $\simeq$ is such that $S/\approx \, = \mathbb{N}^n/\simeq$.

Now let $\approx_1$ and $\approx_2$ be two congruences on S and consider the natural surjections $\pi_i : \mathbb{N}^n \to \mathbb{N}^n/\simeq_i$ for $i = 1, 2$. Then if $\approx_1 \subseteq \approx_2$ (as subsets of $S \times S$), we have that $I_{\mathscr{A}_1} \subseteq I_{\mathscr{A}_2}$, where these ideals are defined as in (4) by considering the generating sets $\mathscr{A}_i = \{\pi_i(\mathbf{e}_j) \mid j = 1, \ldots, n\}$, $i = 1, 2$, respectively. We conclude that Noetherianity of the monoid S follows from the fact that $\Bbbk[\mathbf{X}]$ is a Noetherian ring.

In order to continue to explore the correspondence between congruences and binomial ideals, we introduce some terminology.

Definition 3 A binomial ideal is said to be **unital** if it is generated by binomials of the form $\mathbf{X}^{\mathbf{u}} - \lambda \mathbf{X}^{\mathbf{v}}$ with λ equal to either 0 or 1. A binomial ideal is said to be **pure** if does not contain any monomial.

Corollary 3 *A relation* $\sim$ *on* $\mathbb{N}^n$ *is a congruence if and only if there exists a pure unital ideal* $I \subset \Bbbk[\mathbf{X}]$ *such that* $\mathbf{u} \sim \mathbf{v} \Longleftrightarrow \mathbf{X}^{\mathbf{u}} - \mathbf{X}^{\mathbf{v}} \in I$.

Proof If $\sim$ is a congruence on $\mathbb{N}^n$, then $\mathbb{N}^n/\sim$ is a (finitely generated) monoid. Consider the natural surjection $\pi : \mathbb{N}^n \to \mathbb{N}^n/\sim$, and let $\mathscr{A} = \{\pi(\mathbf{e}_1), \ldots, \pi(\mathbf{e}_n)\}$. Use this information to construct $I_{\mathscr{A}}$ as in (4). By Theorem 3 and its proof, the ideal $I_{\mathscr{A}}$ satisfies the required conditions.

For the converse, let I a pure unital ideal of $\Bbbk[\mathbf{X}]$ such that $\mathbf{u} \sim \mathbf{v} \Longleftrightarrow \mathbf{X}^{\mathbf{u}} - \mathbf{X}^{\mathbf{v}} \in I$. Clearly, $\sim$ is reflexive and symmetric. For transitivity, it suffices to observe that $\mathbf{X}^{\mathbf{u}} - \mathbf{X}^{\mathbf{w}} = (\mathbf{X}^{\mathbf{u}} - \mathbf{X}^{\mathbf{v}}) + (\mathbf{X}^{\mathbf{v}} - \mathbf{X}^{\mathbf{w}}) \in I$, for every $\mathbf{u}$, $\mathbf{v}$ and $\mathbf{w}$ such that $\mathbf{u} \sim \mathbf{v}$ and $\mathbf{v} \sim \mathbf{w}$. Finally, as I is an ideal, it follows that $\mathbf{X}^{\mathbf{w}}(\mathbf{X}^{\mathbf{u}} - \mathbf{X}^{\mathbf{v}}) = \mathbf{X}^{\mathbf{u}+\mathbf{w}} - \mathbf{X}^{\mathbf{v}+\mathbf{w}} \in I$, for every $\mathbf{X}^{\mathbf{u}} - \mathbf{X}^{\mathbf{v}} \in I$ and $\mathbf{X}^{\mathbf{w}} \in \Bbbk[\mathbf{X}]$. We conclude that $\sim$ is a congruence.

We review some examples of pure unital binomial ideals and their associated congruences. We remark in particular that different binomial ideals may give rise to the same congruence.

Example 3

(i) The ideal $I = \langle X - Y \rangle \subset \Bbbk[X, Y]$ defines a congruence $\sim$ on $\mathbb{N}^2$ with $\mathbb{N}^2/\!\sim = \mathbb{N}$.
(ii) The ideal $I = \langle X - Y, Y^2 - 1 \rangle \subset \Bbbk[X, Y]$ defines a congruence $\sim$ on $\mathbb{N}^2$ such that $\mathbb{N}^2/\!\sim = \mathbb{Z}/2\mathbb{Z}$.
(iii) The ideal $I = \langle X^2 - Y^2 \rangle \subset \Bbbk[X, Y]$ defines a congruence $\sim$ on $\mathbb{N}^2$ such that $\mathbb{N}^2/\!\sim$ is isomorphic to the submonoid S of $\mathbb{Z} \oplus \mathbb{Z}/2\mathbb{Z}$ generated by $(1, 0)$ and $(1, 1)$.
(iv) Consider the monoid $S = \{0, a, b\}$ where the sum is defined as follows:

+	0	a	b
0	0	a	b
a	a	b	b
b	b	b	b

The ideal $I = \langle X - Y, Y^3 - Y^2 \rangle \subset \Bbbk[X, Y]$ determines a congruence $\sim$ on $\mathbb{N}^2$ such that $S \cong \mathbb{N}^2/\!\sim$.

An arbitrary binomial ideal J of $\Bbbk[\mathbf{X}]$ induces a congruence $\sim_J$ on $\mathbb{N}^n$ defined as

$$\mathbf{u} \sim_J \mathbf{v} \Longleftrightarrow \text{there exists } \lambda \in \Bbbk^* \text{ such that } \mathbf{X}^{\mathbf{u}} - \lambda \mathbf{X}^{\mathbf{v}} \in J. \tag{5}$$

Note that this ideal defines the same congruence as the pure unital binomial ideal

$$I = \langle \mathbf{X}^{\mathbf{u}} - \mathbf{X}^{\mathbf{v}} \mid \text{there exists } \lambda \in \Bbbk^* \text{ such that } \mathbf{X}^{\mathbf{u}} - \lambda \mathbf{X}^{\mathbf{v}} \in J \rangle.$$

Example 4

(i) Let $J = \langle X - Y, Y^2\rangle \subset \mathbb{k}[X, Y]$. The congruence $\sim_J$ induced by J on $\mathbb{N}^2$ is exactly the same that one in Example 3(iv).
(ii) The congruence $\sim_{\langle X,Y\rangle}$ on $\mathbb{N}^2$ is the same as the induced by $I = \langle X - Y, X - X^2\rangle$ on $\mathbb{N}^2$. Note that $\langle X, Y\rangle$ is a monomial ideal, while I contains no monomials.

If a binomial ideal I contains monomials, then the exponents of all monomials in I form a single equivalence class in the congruence $\sim_I$. This equivalence class satisfies an absorption property, as in the definition below.

Definition 4 A non-identity element ∞ in a monoid S is **nil** if $\mathbf{a} + \infty = \infty$, for all $\mathbf{a} \in S$.

For example, the "formal" element ∞ introduced in Remark 1 is nil, since it corresponds to the monomial class. Note that a monoid S can have at most one nil element: if $\infty, \infty' \in S$ are both nil, then $\infty + \infty' = \infty'$ because ∞' is nil, and $\infty' + \infty = \infty$ because ∞ is nil. Since S is commutative, $\infty = \infty'$.

As we have noted above, if I is a binomial ideal that contains monomials, then the class of monomial exponents is a nil element for the congruence $\sim_I$. The converse of this assertion is false: if J is a binomial ideal containing monomials, then the ideal I produced by Corollary 3 for the congruence $\sim_J$ has no monomials and has a nil element (since J contains monomials, and therefore $\sim_J$ does). On the other hand, if $\sim$ is a congruence on $\mathbb{N}^n$ with a nil element ∞, then there exists a binomial ideal J in $\mathbb{k}[\mathbf{X}]$ that contains monomials, and such that $\sim = \sim_J$. To see this, let I be the ideal produced by Corollary 3 for $\sim$, and consider $J = I + \langle \mathbf{X}^{\mathbf{e}} \mid [\mathbf{e}] = \infty\rangle$, noting that adding this particular monomial ideal does not change the underlying congruence. We make this more precise in Proposition 3.

Proposition 3 *Let $I \subset \mathbb{k}[\mathbf{X}]$ be a binomial ideal. If J is a binomial ideal of $\mathbb{k}[\mathbf{X}]$ such that $I \subset J$ and $\sim_J = \sim_I$, then $\mathbb{N}^n/\sim_I$ has a nil ∞ and $J = I + \langle \mathbf{X}^{\mathbf{e}} \mid [\mathbf{e}] = \infty\rangle$.*

Proof As $I \subset J$, there is a binomial $\mathbf{X}^{\mathbf{u}} - \lambda\mathbf{X}^{\mathbf{v}} \in J \setminus I$. Since $\sim_I = \sim_J$, necessarily $\mathbf{X}^{\mathbf{u}}, \mathbf{X}^{\mathbf{v}} \in J$; in particular $\mathbb{N}^n/\sim_J = \mathbb{N}^n/\sim_I$ has a nil ∞. We claim that the ideal J is equal to $I + \langle \mathbf{X}^{\mathbf{e}} \mid [\mathbf{e}] = \infty\rangle$. To see that J contains $I + \langle \mathbf{X}^{\mathbf{e}} \mid [\mathbf{e}] = \infty\rangle$, we note that $I \subset J$. Also, we know that J contains a monomial $\mathbf{X}^{\mathbf{u}}$, and so $[\mathbf{u}] = \infty$. If $\mathbf{e} \in \mathbb{N}^n$ is such that $[\mathbf{e}] = \infty = [\mathbf{u}]$, then $\mathbf{X}^{\mathbf{u}} - \mu\mathbf{X}^{\mathbf{e}} \in J$ for some $\mu \in \mathbb{k}^*$, and since $\mathbf{X}^{\mathbf{u}} \in J$, we see that $\mathbf{X}^{\mathbf{e}} \in J$. For the reverse inclusion, it is enough to see that any binomial in J belongs to $I + \langle \mathbf{X}^{\mathbf{e}} \mid [\mathbf{e}] = \infty\rangle$. But as before, if $\mathbf{X}^{\mathbf{u}} - \lambda\mathbf{X}^{\mathbf{v}} \in J \setminus I$, then $\mathbf{X}^{\mathbf{u}}, \mathbf{X}^{\mathbf{v}} \in J$, and therefore $[\mathbf{u}] = [\mathbf{v}] = \infty$, because a monoid can have at most one nil element.

A **monoid ideal** E of $\mathbb{N}^n$ is a proper subset such that $E + \mathbb{N}^n \subseteq E$; Fig. 1 shows a typical example.

Let $E \subseteq \mathbb{N}^n$ be a monoid ideal of $\mathbb{N}^n$. The **Rees congruence** on $\mathbb{N}^n$ modulo E is the correspondence $\sim$ on $\mathbb{N}^n$ defined by $\mathbf{u} \sim \mathbf{v} \iff \mathbf{u} = \mathbf{v}$ or both $\mathbf{u}$ and $\mathbf{v} \in E$.

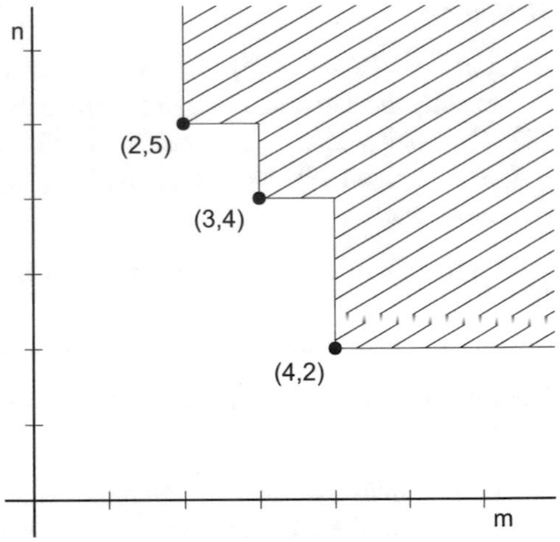

Fig. 1 The integer points in shaded area form a monoid ideal of $\mathbb{N}^2$

Notice that the Rees congruence on $\mathbb{N}^n$ modulo E is the same as the induced by $\sim_{M_E}$ with $M_E = \langle \mathbf{X}^{\mathbf{e}} \mid \mathbf{e} \in E\rangle$.

Monoid ideals and nil elements are related as follows.

Lemma 1 *Let S be a monoid. Then S has a nil element if and only if for any presentation $\mathbb{N}^n/\sim$ of S there exists a monoid ideal E of $\mathbb{N}^n$ such that $\sim$ contains the Rees congruence on $\mathbb{N}^n$ modulo E. In this case, $[\mathbf{e}] = \infty$, for any $\mathbf{e} \in E$.*

Proof Let $\mathbb{N}^n/\sim$ be a presentation of S given by a monoid surjection $\pi : \mathbb{N}^n \to S$.

For the direct implication, assume that $\infty \in S$ is a nil. Then $E := \pi^{-1}(\infty)$ is a monoid ideal of $\mathbb{N}^n$. Indeed, given $\mathbf{u} \in \mathbb{N}^n$ and $\mathbf{e} \in E$ we have that

$$\pi(\mathbf{e}+\mathbf{u}) = \pi(\mathbf{e}) + \pi(\mathbf{u}) = \infty + \pi(\mathbf{u}) = \infty,$$

so that $\mathbf{e}+\mathbf{u} \in E$. Note that $E \neq \mathbb{N}^n$ since nil elements are nonzero. Moreover, by construction, if $\mathbf{e}, \mathbf{e}' \in E$, then $\mathbf{e} \sim \mathbf{e}'$, which means that $\sim$ contains the Rees congruence on $\mathbb{N}^n$ modulo E.

Conversely, let E be a monoid ideal of $\mathbb{N}^n$ such that $\sim$ contains the Rees congruence on $\mathbb{N}^n$ modulo E. We claim that the class $[\mathbf{e}]$ (for any $\mathbf{e} \in E$) is a nil in $S = \mathbb{N}^n/\sim$. To see this, let $\mathbf{e} \in E$, $\mathbf{u} \in \mathbb{N}^n$. Then $\pi(\mathbf{e}) + \pi(\mathbf{u}) = \pi(\mathbf{e}+\mathbf{u})$. Since E is a monoid ideal, $\mathbf{e}+\mathbf{u} \in E$. This implies that $\mathbf{e} \sim \mathbf{e}+\mathbf{u}$ (or equivalently, $\pi(\mathbf{e}) = \pi(\mathbf{e}+\mathbf{u})$) because $\sim$ contains the Rees congruence modulo E. To complete the proof of our claim, we need to show that $[\mathbf{e}]$ ($\mathbf{e} \in E$) is not the zero class. This follows from the fact that $E \neq \mathbb{N}^n$.

Our next goal is to prove Theorem 4, which is a more precise version of Proposition 3. With that result in hand, we will be able to introduce the binomial ideal associated to a congruence in Definition 6.

Definition 5 An **augmentation ideal** for a given binomial ideal $I \subset \Bbbk[\mathbf{X}]$ is a maximal ideal of the form

$$I_{\text{aug}} := \langle X_i - \lambda_i \mid \lambda_i \in \Bbbk^*,\ i = 1, \ldots, n\rangle$$

such that $I \cap I_{\text{aug}}$ is a binomial ideal.

We point out that, given a binomial ideal I, an augmentation ideal for I may or may not exist (see [10, Example 9.13] for a binomial ideal without an augmentation ideal). The following result is the $\mathbb{N}^n$-version of [10, Theorem 9.12].

Theorem 4 *If $I_\ell \supset \ldots \supset I_0$ is a chain of distinct binomial ideals of $\Bbbk[\mathbf{X}]$ inducing the same congruence on $\mathbb{N}^n$, then $\ell \leq 1$. Moreover, if $\ell = 1$ then I_0 is pure and I_1 is not: $I_0 = I_1 \cap I_{\text{aug}}$ for an augmentation ideal for I_1.*

Proof By Proposition 3, all we need to show is that if $\ell = 1$, then $I_0 = I_1 \cap I_{\text{aug}}$, where I_{aug} is an augmentation ideal for I_1. Denote by $\sim$ the congruence induced by I_0 (and I_1).

Assume $\ell = 1$, so that I_0 does not have monomials, and I_1 does. In particular, we may select a monomial $\mathbf{X}^{\mathbf{e}} \in I_1$, and its equivalence class $[\mathbf{e}]$ with respect to $\sim$ is a nil element, that we denote ∞. For each $1 \leq i \leq n$, consider the monomial $X_i = \mathbf{X}^{\mathbf{e}_i}$. Since $[\mathbf{e}_i] + \infty = \infty$, there exists $\lambda_i \in \Bbbk^*$ such that $X_i\mathbf{X}^{\mathbf{e}} - \lambda_i\mathbf{X}^{\mathbf{e}} \in I_0$ (because I_0 and I_1 induce the same congruence). We now define $I_{\text{aug}} = \langle X_i - \lambda_i \mid i = 1, \ldots, n\rangle$, and claim that $I_1 \cap I_{\text{aug}} = I_0$, which in particular shows that I_{aug} is an augmentation ideal for I_1.

By construction, $(I_0 : \mathbf{X}^{\mathbf{e}}) \supseteq I_{\text{aug}}$. Note that $(I_0 : \mathbf{X}^{\mathbf{e}}) \neq \langle 1\rangle$, as I_0 contains no monomials. Thus, since I_{aug} is maximal, $(I_0 : \mathbf{X}^{\mathbf{e}}) = I_{\text{aug}}$, and we conclude that I_{aug} contains I_0. This, and $I_1 \supset I_0$, imply that $I_1 \cap I_{\text{aug}} \supseteq I_0$. Moreover, $I_{\text{aug}} \not\supseteq I_1$ because I_1 has monomials, while I_{aug} does not. Consequently $I_1 \supsetneq I_1 \cap I_{\text{aug}} \supseteq I_0$. Now the equality $I_1 \cap I_{\text{aug}} = I_0$ will follow from Proposition 3 if we show that $I_1 \cap I_{\text{aug}}$ is binomial (since the fact that I_1 and I_0 induce the same congruence $\sim$ implies that the congruence induced by $I_1 \cap I_{\text{aug}}$ is also $\sim$). To see that $I_1 \cap I_{\text{aug}}$ is binomial, we use the argument from [5, Corollary 1.5]. Introduce an auxiliary variable t, and consider the binomial ideal $J = I_0 + tI_{\text{aug}} + (1 - t)\langle \mathbf{X}^{\mathbf{u}} \mid [\mathbf{u}] = [\mathbf{e}]\rangle \subset \Bbbk[\mathbf{X}, t]$. Since, by Proposition 3, $I_1 = I_0 + \langle \mathbf{X}^{\mathbf{u}} \mid [\mathbf{u}] = \infty = [\mathbf{e}]\rangle$, we have that $J \cap \Bbbk[\mathbf{X}] = I_1 \cap I_{\text{aug}}$. Now apply Corollary 1.

Example 5 If $I_1 = \langle X - Y, Y^2\rangle \subset \Bbbk[X, Y]$, then $I_0 = I_1 \cap \langle X - 1, Y - 1\rangle = \langle X - Y, Y^3 - Y^2\rangle$. This can be verified as follows.

```
R = QQ[X,Y];
I1 = ideal(X-Y, Y^2);
Iaug = ideal(X-1,Y-1);
I0 = intersect(I1,Iaug);
mingens I0;
```

Note that these ideals already appeared in Examples 4(i) and 3(iv).

Remark 3 The previous results highlight one way in which two different binomial ideals in $\mathbb{k}[\mathbf{X}]$ induce the same congruence on $\mathbb{N}^n$, namely if one contains the other, the congruence has a nil element, and the larger ideal contains monomials corresponding to the nil class, while the smaller ideal has no monomials.

There is another way to produce binomial ideals inducing the same congruence. Let I be a binomial ideal in $\mathbb{k}[\mathbf{X}]$, and let $\mu_1, \ldots, \mu_n \in \mathbb{k}^*$. Consider the ring isomorphism $\mathbb{k}[\mathbf{X}] \to \mathbb{k}[\mathbf{X}]$ given by $X_i \mapsto \mu_i X_i$ for $i = 1, \ldots, n$. (This kind of isomorphism is known as **rescaling the variables**.) Then the image of I is a binomial ideal, which induces the same congruence as I. Indeed, the effect on I of rescaling the variables is to change the coefficients of the binomials in I by a nonzero multiple, which does not alter the exponents of those monomials.

In Theorem 4, the ideal I_0 can be made unital by rescaling the variables, by using that $\mathbb{k}$ is algebraically closed if necessary. The ideal obtained this way equals the ideal introduced in (4).

We are now ready to introduce the binomial ideal associated to a congruence in $\mathbb{N}^n$.

Definition 6 Given a congruence $\sim$ on $\mathbb{N}^n$, denote by $I_\sim$ the unital binomial ideal of $\mathbb{k}[\mathbf{X}]$ which is maximal among all proper binomial ideals inducing $\sim$. We say that $I_\sim$ is the **binomial ideal associated** to $\sim$.

To close this section, we introduce one final notion.

Definition 7 Let $\sim_1$ and $\sim_2$ be congruences on $\mathbb{N}^n$. The **intersection** $\sim$ of $\sim_1$ and $\sim_2$, denoted $\sim = \sim_1 \cap \sim_2$, is the congruence on $\mathbb{N}^n$ defined by $\mathbf{u} \sim \mathbf{v}$ if and only if $\mathbf{u} \sim_1 \mathbf{v}$ and $\mathbf{u} \sim_2 \mathbf{v}$.

From the point of view of equivalence relations, the equivalence classes of $\sim_1 \cap \sim_2$ form a partition of $\mathbb{N}^n$ which is the common refinement of the partitions induced by $\sim_1$ and $\sim_2$. The following result motivates the use of the intersection notation and terminology: the intersection of congruences corresponds to the ideal generated by the binomials in the intersection of their associated binomial ideals.

Proposition 4 *Let $\sim$, $\sim_1$ and $\sim_2$ be congruences on $\mathbb{N}^n$ whose associated ideals in* $\mathbb{k}[\mathbf{X}]$ (Definition 6) *are $I_\sim$, $I_{\sim_1}$ and $I_{\sim_2}$, respectively. Then $\sim = \sim_1 \cap \sim_2$ if and only if $I_\sim \subseteq I_{\sim_1} \cap I_{\sim_2}$, and the equality holds if and only if $I_{\sim_1} \cap I_{\sim_2}$ is a binomial ideal.*

Proof The statement $\mathbf{u} \sim \mathbf{v}$ if and only if $\mathbf{u} \sim_1 \mathbf{v}$ and $\mathbf{u} \sim_2 \mathbf{v}$ is exactly the same as $\mathbf{X}^\mathbf{u} - \mathbf{X}^\mathbf{v} \in I_\sim$ if and only if $\mathbf{X}^\mathbf{u} - \mathbf{X}^\mathbf{v} \in I_{\sim_1}$ and $\mathbf{X}^\mathbf{u} - \mathbf{X}^\mathbf{v} \in I_{\sim_2}$. The direct implication of the last statement follows from Theorem 4 and its converse is trivially true because $I_\sim$ is a binomial ideal.

The following example illustrates the last statement above.

Example 6 Let $\sim_1$ and $\sim_2$ be the congruences on $\mathbb{N}^2$ such that $\mathbf{u} \sim_1 \mathbf{v}$ if $\mathbf{u} - \mathbf{v} \in \mathbb{Z}(2, -2)$ and $\mathbf{u} \sim_2 \mathbf{v}$ if $\mathbf{u} - \mathbf{v} \in \mathbb{Z}(3, -3)$, respectively. The binomial ideals of $\mathbb{Q}[X, Y]$ associated to $\sim_1$ and $\sim_2$ are $I_{\sim_1} = \langle X^2 - Y^2 \rangle$ and $I_{\sim_2} = \langle X^3 - Y^3 \rangle$, respectively. Clearly, the binomial ideal associated to $\sim = \sim_1 \cap \sim_2$ is $I_\sim = \langle X^6 - Y^6 \rangle$. Whereas, $I_{\sim_1} \cap I_{\sim_2} = \langle X^4 + X^3Y - XY^3 - Y^4 \rangle$:

```
R = QQ[X,Y];
I1 = ideal(X^2-Y^2);
I2 = ideal(X^3-Y^3);
intersect(I1,I2);
```

3 Toric, Lattice and Mesoprime Ideals

This section is devoted to the (finitely generated abelian) monoids contained in a group.

Let $(G, +)$ be a finitely generated abelian group and let $\mathscr{A} = \{\mathbf{a}_1, \ldots, \mathbf{a}_n\}$ be a given subset of G, we consider the subsemigroup S of G generated by $\mathscr{A}$, that is to say,

$$S = \mathbb{N}\mathbf{a}_1 + \ldots + \mathbb{N}\mathbf{a}_n.$$

Since $0 \in \mathbb{N}$, the semigroup S is actually a monoid. We may define a surjective monoid map as follows

$$\deg_{\mathscr{A}} : \mathbb{N}^n \longrightarrow S; \quad \mathbf{u} = (u_1, \ldots, u_n) \longmapsto \deg_{\mathscr{A}}(\mathbf{u}) = \sum_{i=1}^{n} u_i \mathbf{a}_i. \tag{6}$$

In the literature, this map is called the factorization map of S and accordingly, the fiber $\deg_{\mathscr{A}}^{-1}(\mathbf{a})$ is called the set of factorizations of $\mathbf{a} \in S$.

Clearly $\deg_{\mathscr{A}}(-)$ determines a congruence on $\mathbb{N}^n$; in fact, it is the congruence on $\mathbb{N}^n$ whose presentation map is precisely $\deg_{\mathscr{A}}(-)$ (cf. (2)). Therefore, if $\widehat{\deg_{\mathscr{A}}}$ is the map defined in (3), namely,

$$\widehat{\deg_{\mathscr{A}}} : \Bbbk[\mathbb{N}^n] = \Bbbk[\mathbf{X}] \to \Bbbk[S]\,; \quad \mathbf{X}^{\mathbf{u}} \mapsto \chi^{\deg_{\mathscr{A}}(\mathbf{u})},$$

by Theorem 3, we have that $I_{\mathscr{A}} = \ker(\widehat{\deg_{\mathscr{A}}})$ is spanned as a $\Bbbk-$vector space by the set of binomials

$$\{\mathbf{X}^{\mathbf{u}} - \mathbf{X}^{\mathbf{v}} \mid \mathbf{u}, \mathbf{v} \in \mathbb{N}^n \text{ with } \deg_{\mathscr{A}}(\mathbf{u}) = \deg_{\mathscr{A}}(\mathbf{v})\}. \tag{7}$$

Observe that $\Bbbk[\mathbf{X}]$ is $S-$graded via $\deg(X_i) = \mathbf{a}_i$, $i = 1, \ldots, n$. This grading is known as the $\mathscr{A}-$**grading** on $\Bbbk[\mathbf{X}]$. The semigroup algebra $\Bbbk[S] = \oplus_{\mathbf{a} \in S} \mathrm{Span}_{\Bbbk}\{\chi^{\mathbf{a}}\}$ also has a natural $S-$grading. Under these gradings, the map of semigroup algebras $\widehat{\deg_{\mathscr{A}}}$ is a graded map. Hence, the ideal $I_{\mathscr{A}} = \ker(\widehat{\deg_{\mathscr{A}}})$ is $S-$homogeneous.

Proposition 5 *Use the notation introduced above, and assume that* $\mathbf{a}_1, \ldots, \mathbf{a}_n$ *are nonzero. The following are equivalent:*

(a) *The fibers of map* $\deg_{\mathscr{A}}(-)$ *are finite.*
(b) $\deg_{\mathscr{A}}^{-1}(\mathbf{0}) = \{(0, \ldots, 0)\}$.
(c) $S \cap (-S) = \{0\}$, *that is to say,* $\mathbf{a} \in S$ *and* $-\mathbf{a} \in S \Rightarrow \mathbf{a} = \mathbf{0}$.
(d) *The relation* $\mathbf{a}' \preceq \mathbf{a} \Longleftrightarrow \mathbf{a}' - \mathbf{a} \in S$ *is a partial order on* S.

Proof Before we proceed with the proof, we note that if one of the $\mathbf{a}_i$ is zero, then this result is false. For example, let $G = \mathbb{Z}$, $\mathscr{A} = \{a_1 = 0, a_2 = 1\}$. Then $S = \mathbb{N}$, for which (c) and (d) hold, but $\deg_{\mathscr{A}}(-)$ does not satisfy either (a) or (b).

(a)⇒(b) If $\mathbf{u} \in \deg_{\mathscr{A}}^{-1}(\mathbf{0})$, then for every $\ell \in \mathbb{N}$, $\ell\mathbf{u} \in \deg_{\mathscr{A}}^{-1}(\mathbf{0})$. If $\mathbf{u} \neq (0, \ldots, 0)$, then $\deg_{\mathscr{A}}^{-1}(\mathbf{0})$ is infinite.

(a)⇐(b) Dickson's Lemma states that any nonempty subset of $\mathbb{N}^n$ has finitely many minimal elements with respect to the partial order given by coordinatewise $\leq$. Suppose that $\deg_{\mathscr{A}}^{-1}(\mathbf{a})$ is infinite. Then by Dickson's Lemma there exists $\mathbf{u} \in \deg_{\mathscr{A}}^{-1}(\mathbf{a})$ which is not minimal, and therefore there is also $\mathbf{v} \in \deg_{\mathscr{A}}^{-1}(\mathbf{a})$ such that $\mathbf{v} \leq \mathbf{u}$ coordinatewise. We conclude that $\mathbf{u} - \mathbf{v} \in \mathbb{N}^n$ is a nonzero element of $\deg_{\mathscr{A}}^{-1}(\mathbf{0})$.

(b)⇒(c) Let $\mathbf{u}, \mathbf{v} \in \mathbb{N}^n$ be such that $\deg_{\mathscr{A}}(\mathbf{u}) = \mathbf{a}$ and $\deg_{\mathscr{A}}(\mathbf{v}) = -\mathbf{a}$. Then $\deg_{\mathscr{A}}(\mathbf{u}+\mathbf{v}) = \mathbf{0}$, so that $\mathbf{u}+\mathbf{v} = (0, \ldots, 0)$, and therefore $\mathbf{u} = \mathbf{v} = (0, \ldots, 0)$, which implies that $\mathbf{a} = \mathbf{0}$.

(b)⇐(c) Let $\mathbf{u} \in \deg_{\mathscr{A}}^{-1}(\mathbf{0})$. If $\mathbf{u}_1, \mathbf{u}_2 \in \mathbb{N}^n$ are such that $\mathbf{u} = \mathbf{u}_1 + \mathbf{u}_2$, then by (c) $\deg_{\mathscr{A}}(\mathbf{u}_1) = \deg_{\mathscr{A}}(\mathbf{u}_2) = \mathbf{0}$. Repeatedly applying this argument, we conclude that if $\mathbf{u} \neq 0$, so in particular it has a nonzero coordinate, then there exists $1 \leq i \leq n$ such that $\mathbf{a}_i = \deg_{\mathscr{A}}(\mathbf{e}_i) = \mathbf{0}$, a contradiction.

(c)⇔(d) The relation $\preceq$ is always reflexive and transitive. The fact that $\preceq$ is antisymmetric is equivalent to (c).

Remark 4 If the conditions of Proposition 5 hold, the monoid S generated by $\mathscr{A}$ is said to be **positive**. When S is positive, $\mathfrak{m} = \langle X_1, \ldots, X_n \rangle$ is the only $S-$homogeneous maximal ideal in $\Bbbk[\mathbf{X}]$. Recall that a graded ideal $\mathfrak{m}$ in a graded ring R is a **graded maximal ideal** or ***maximal ideal** if the only graded ideal properly containing $\mathfrak{m}$ is R itself. Graded rings with a unique graded maximal ideal are known as **graded local rings** or ***local rings**. Many results valid for local rings are also valid for graded local rings, starting with Nakayama's Lemma. In particular, the *minimal free resolution* of any finitely generated $\mathscr{A}-$graded $\Bbbk[\mathbf{X}]-$module is well-defined (see [3, Section 1.5] and [1]).

All the monoids in this section are contained in a group. The next result characterize the condition for a monoid to be contained in a group. To state it we need to introduce the following concepts.

Definition 8 Let $\sim$ be a congruence on $\mathbb{N}^n$. We will say that $\mathbf{a} \in \mathbb{N}^n/\sim$ is **cancellable** if $\mathbf{b} + \mathbf{a} = \mathbf{c} + \mathbf{a} \Rightarrow \mathbf{b} = \mathbf{c}$, for all $\mathbf{b}, \mathbf{c} \in \mathbb{N}^n/\sim$. A monoid is said to be **cancellative** if all its elements are cancellable. A congruence $\sim$ on $\mathbb{N}^n$ is cancellative if the monoid $\mathbb{N}^n/\sim$ is cancellative.

In the part (iv) of Example 3, an example of non-cancellative monoid is exhibited.

Proposition 6 *A (finitely generated commutative) monoid is contained in a group if and only if it is cancellative.*

Proof The direct implication is clear. Conversely, if $S = \mathbb{N}^n/\sim$ is a cancellative finitely generated commutative monoid, then $\sim$ can be extended on $\mathbb{Z}^n$ as follows: $\mathbf{u} \sim \mathbf{v}$ if $\mathbf{u}+\mathbf{e} \sim \mathbf{v}+\mathbf{e}$ for some (any) $\mathbf{e} \in \mathbb{N}^n$ such that $\mathbf{u}+\mathbf{e}$ and $\mathbf{v}+\mathbf{e} \in \mathbb{N}^n$. Since $G = \mathbb{Z}^n/\sim$ has a natural group structure and $S \subseteq G$, we are done.

The above result shows that our definition of cancellative congruence is equivalent to the usual one (see [7, p. 44]).

3.1 Toric Ideals and Toric Congruences

Suppose now that G is torsion-free and let $G(\mathscr{A})$ denote the subgroup of G generated by $\mathscr{A}$. Since G is torsion-free, then $G \cong \mathbb{Z}^m$, for some m. Thus, the semigroup S is isomorphic to a subsemigroup of $\mathbb{Z}^m$. In this case, S is said to be an **affine semigroup** and the ideal $I_{\mathscr{A}}$ is called the **toric ideal** associated to $\mathscr{A}$.

Without loss of generality, we may assume that $\mathbf{a}_i \in \mathbb{Z}^d$, for every $i = 1, \ldots, n$, with $d = \operatorname{rank}(G(\mathscr{A})) \leq m$. Moreover, one can prove that, if $\mathscr{A}$ generates a positive monoid (see Remark 4), there exists a monoid isomorphism under which $\mathbf{a}_i$ is mapped to an element of $\mathbb{N}^n$, $i = 1, \ldots, n$ (see, e.g. [3, Proposition 6.1.5]), which justifies the use of the term "positive".

Lemma 2 *If $\mathscr{A} = \{\mathbf{a}_1, \ldots, \mathbf{a}_n\} \subset \mathbb{Z}^d$, then $I_{\mathscr{A}}$ is prime.*

Proof By hypothesis, we have that $\Bbbk[S]$ is isomorphic to the subring $\Bbbk[\mathbf{t}^{\mathbf{a}_1}, \ldots, \mathbf{t}^{\mathbf{a}_n}]$ of the Laurent polynomial ring $\Bbbk[\mathbb{Z}^d] = \Bbbk[t_1^{\pm}, \ldots, t_d^{\pm}]$; in particular, $\Bbbk[S] \cong \Bbbk[\mathbf{X}]/I_{\mathscr{A}}$ is a domain. Therefore $I_{\mathscr{A}}$ is prime.

Theorem 5 *Let I be a binomial ideal of $\Bbbk[\mathbf{X}]$. The ideal I is prime if and only if there exists $\mathscr{A} = \{\mathbf{a}_1, \ldots, \mathbf{a}_r\} \subset \mathbb{Z}^d$ such that*

$$I = I_{\mathscr{A}}\,\Bbbk[\mathbf{X}] + \langle X_{r+1}, \ldots, X_n\rangle,$$

up to permutation and rescaling of variables.

Proof Suppose that I is prime. If I contains monomials, there exists a set of variables, say $X_{r+1}, \ldots, X_n$ (by permuting variables if necessary), such that I is equal to $I'\,\Bbbk[\mathbf{X}] + \langle X_{r+1}, \ldots, X_n\rangle$ where I' is a pure prime binomial ideal of $\Bbbk[X_1, \ldots, X_r]$. Therefore, without loss of generality, we may suppose $I = I'$ and $r = n$. Now, by Theorem 1, $\Bbbk[\mathbf{X}]/I \cong \Bbbk[S] = \bigoplus_{\mathbf{a}\in S} \operatorname{Span}_{\Bbbk}\{\chi^{\mathbf{a}}\}$, for some commutative monoid S generated by $\mathscr{A} = \{\mathbf{a}_1, \ldots, \mathbf{a}_n\}$. Recall that the above isomorphism maps X_i to $\lambda_i \chi^{\mathbf{a}_i}$ for some $\lambda_i \in \Bbbk^*$, $i = 1, \ldots, n$. So, by rescaling variables if necessary, we may assume $\lambda_i = 1$ for every i. Now, if S is not contained in a group, by Proposition 6, there exist $\mathbf{a}, \mathbf{a}'$ and $\mathbf{b} \in S$ such that $\mathbf{a} + \mathbf{b} = \mathbf{a}' + \mathbf{b}$ and $\mathbf{a} \neq \mathbf{a}'$. Thus, $\mathbf{X}^{\mathbf{v}}(\mathbf{X}^{\mathbf{u}} - \mathbf{X}^{\mathbf{u}'}) \in I$, but $\mathbf{X}^{\mathbf{u}} - \mathbf{X}^{\mathbf{u}'} \notin I$, where $\mathbf{u} \in \deg_{\mathscr{A}}^{-1}(\mathbf{a})$, $\mathbf{u}' \in \deg_{\mathscr{A}}^{-1}(\mathbf{a}')$ and $\mathbf{v} \in \deg_{\mathscr{A}}^{-1}(\mathbf{b})$. So, since I is prime, we have that $\mathbf{X}^{\mathbf{v}} \in I$ which is a contradiction. On other hand, if $G(S)$ has torsion, there exist two different elements $\mathbf{a}$ and $\mathbf{a}' \in S$ such that $n\mathbf{a} = n\mathbf{a}'$ for some $n \in \mathbb{N}$. Therefore $\mathbf{X}^{n\mathbf{u}} - \mathbf{X}^{n\mathbf{u}'} \in I$, where $\mathbf{u} \in \deg_{\mathscr{A}}^{-1}(\mathbf{a})$ and $\mathbf{u}' \in \deg_{\mathscr{A}}^{-1}(\mathbf{a}')$. Since $\Bbbk$ is algebraically closed and I is prime, $\mathbf{X}^{\mathbf{u}} - \zeta_n \mathbf{X}^{\mathbf{u}'} \in I$, where ζ_n is a $n-$th root of unity; in particular, $\mathbf{a} = \mathbf{a}'$ which is a contradiction. Putting all this together, we conclude that S is an affine semigroup.

The opposite implication is a direct consequence of Lemma 2.

Definition 9 A congruence $\sim$ on $\mathbb{N}^n$ is said to be **toric** if the ideal $I_\sim$ is prime.

The following result proves that our definition agrees with the one given in [10].

Corollary 4 *A congruence $\sim$ on $\mathbb{N}^n$ is toric if and only if the non-nil elements of $\mathbb{N}^n/\sim$ form an affine semigroup.*

Proof The direct implication follows from Theorem 5. Conversely, we assume that the non-nil elements of $\mathbb{N}^n/\sim$ form an affine semigroup S. In this case, we have that $[\mathbf{u}] + [\mathbf{v}] = \infty$ implies $[\mathbf{u}] = \infty$ or $[\mathbf{v}] = \infty$, for every $\mathbf{u}$ and $\mathbf{v} \in \mathbb{N}^n$, because S is contained in a group and groups have no nil element. Therefore, since $\mathbb{N}^n/\sim$ is generated by the classes $[\mathbf{e}_i]$ modulo $\sim$, $i = 1, \ldots, n$, we obtain that S is generated by $\mathscr{A} = \{[\mathbf{e}_i] \neq \infty \mid i = 1, \ldots, n\}$ Now, applying Theorem 5 again, we conclude that $I_\sim$ is a prime ideal.

3.2 *Lattice Ideals and Cancellative Congruences*

Consider now a subgroup $\mathscr{L}$ of $\mathbb{Z}^n$ and define the following congruence $\sim$ on $\mathbb{N}^n$:

$$\mathbf{u} \sim \mathbf{v} \Longleftrightarrow \mathbf{u} - \mathbf{v} \in \mathscr{L}.$$

Clearly, $\mathbb{N}^n/\sim$ is contained in the group $\mathbb{Z}^n/\mathscr{L}$ and the associated ideal $I_\sim$ is equal to

$$I_{\mathscr{L}} := \{\mathbf{X}^{\mathbf{u}} - \mathbf{X}^{\mathbf{v}} \mid \mathbf{u} - \mathbf{v} \in \mathscr{L}\}.$$

The subgroups of $\mathbb{Z}^n$ are also called lattices. This justifies the term "lattice" in the following definition.

Definition 10 Let $\mathscr{L}$ be a subgroup of $\mathbb{Z}^n$ and $\rho : \mathscr{L} \to \Bbbk^*$ be a group homomorphism. The lattice ideal corresponding to $\mathscr{L}$ and ρ is

$$I_{\mathscr{L}}(\rho) := \langle \mathbf{X}^{\mathbf{u}} - \rho(\mathbf{u} - \mathbf{v})\mathbf{X}^{\mathbf{v}} \mid \mathbf{u} - \mathbf{v} \in \mathscr{L} \rangle.$$

An ideal I of $\Bbbk[\mathbf{X}]$ is called a **lattice ideal** if there is subgroup $\mathscr{L} \subset \mathbb{Z}^n$ and a group homomorphism $\rho : \mathscr{L} \to \Bbbk^*$ such that $I = I_{\mathscr{L}}(\rho)$.

Observe that the ideal $I_{\mathscr{L}}$ above is a lattice ideal for the group homomorphism $\rho : \mathscr{L} \to \Bbbk^*$ such that $\rho(\mathbf{u}) = 1$, for every $\mathbf{u} \in \mathscr{L}$. Moreover, given a subgroup $\mathscr{L}$ of $\mathbb{Z}^n$, we have that the congruence on $\mathbb{N}^n$ defined by a lattice ideal $I_{\mathscr{L}}(\rho)$ is the same as the congruence on $\mathbb{N}^n$ defined by $I_{\mathscr{L}}$, for every group homomorphism $\rho : \mathscr{L} \to \Bbbk^*$.

Let us characterize the cancellative congruences on $\mathbb{N}^n$ in terms of their associated binomial ideals. In order to do this, we first recall the following result from [5].

Proposition 7 ([5, Corollary 2.5]) *If I is a pure binomial ideal of $\Bbbk[\mathbf{X}]$, then there is a unique group morphism $\rho : \mathscr{L} \subseteq \mathbb{Z}^n \to \Bbbk^*$ such that $I : (\prod_{i=1}^n X_i)^\infty = I_{\mathscr{L}}(\rho)$.*

Observe that from Proposition 7, it follows that no monomial is a zero divisor modulo a lattice ideal.

Corollary 5 *A congruence $\sim$ on $\mathbb{N}^n$ is cancellative if and only if $I_\sim$ is a lattice ideal.*

Proof By Proposition 6, $\sim$ is cancellative if and only if $\mathbb{N}^n/\sim$ is contained in a group G. Thus, the natural projection $\pi : \mathbb{N}^n \to \mathbb{N}^n/\sim$ can be extended to a group homorphism $\bar{\pi} : \mathbb{Z}^n \to G$ whose restriction to $\mathbb{N}^n$ is π. Since the kernel, $\mathscr{L}$, of $\bar{\pi}$ is a subgroup of $\mathbb{Z}^n$ that defines the same congruence as $\sim$, we conclude that both ideals $I_\sim$ and $I_{\mathscr{L}}$ are equal. For the converse, we first note that the congruence on $\mathbb{N}^n$ defined by a lattice ideal $I_{\mathscr{L}}(\rho)$ is the same as the congruence on $\mathbb{N}^n$ defined by $I_{\mathscr{L}}$, for every group homomorphism $\rho : \mathscr{L} \subset \mathbb{Z}^n \to \Bbbk^*$ (see the comment after Eq. (5)). Now, it suffices to note that if $I_\sim = I_{\mathscr{L}}$ for some subgroup $\mathscr{L}$ of $\mathbb{Z}^n$, then $\mathbb{N}^n/\sim$ is contained in $\mathbb{Z}^n/\mathscr{L}$.

Observe that a lattice ideal $I_{\mathscr{L}}$ is not prime in general. Indeed, $I = \langle X^2 - Y^2 \rangle$ is a lattice ideal corresponding to the subgroup of $\mathbb{Z}^2$ generated by $(2, -2)$ which is clearly not prime. Let us give a necessary and sufficient condition for a lattice ideal to be prime.

Definition 11 Let $\mathscr{L}$ be subgroup of $\mathbb{Z}^n$ and set

$$\mathrm{Sat}(\mathscr{L}) := (\mathbb{Q} \otimes_{\mathbb{Z}} \mathscr{L}) \cap \mathbb{Z}^n = \{\mathbf{u} \in \mathbb{Z}^n \mid d\,\mathbf{u} \in \mathscr{L} \text{ for some } d \in \mathbb{Z}\}.$$

Clearly, $\mathrm{Sat}(\mathscr{L})$ is subgroup of $\mathbb{Z}^n$ and it is called the **saturation** of $\mathscr{L}$. We say that $\mathscr{L}$ is **saturated** if $\mathscr{L} = \mathrm{Sat}(\mathscr{L})$.

Proposition 8 *A lattice ideal $I_{\mathscr{L}}(\rho)$ is prime if and only if $\mathscr{L}$ is saturated.*

Proof By using the same argument as in the proof of Corollary 5, we obtain that $\mathbb{Z}^n/\sim = \mathbb{Z}^n/\mathscr{L}$, where $\sim$ the congruence defined by $I_{\mathscr{L}}(\rho)$ on $\mathbb{Z}^n$. Now, since

$\mathbb{Z}^n/\mathscr{L}$ is the group generated by $\mathbb{N}^n/\sim$, and $\mathbb{Z}^n/\mathscr{L}$ is torsion-free if and only if $\mathscr{L}$ is saturated, we obtain the desired equivalence.

Notice that the congruence defined by $\mathscr{L}$ is contained in the congruence defined by $\mathrm{Sat}(\mathscr{L})$. In fact, $\mathrm{Sat}(\mathscr{L})$ defines the smallest toric congruence on $\mathbb{N}^n$ containing the congruence defined by $\mathscr{L}$ on $\mathbb{N}^n$. Therefore, we may say each cancellative congruence *has exactly one toric congruence associated.*

The primary decomposition of a lattice ideal $I_{\mathscr{L}}(\rho)$ can be completely described in terms of $\mathscr{L}$ and ρ. Let us reproduce this result. For this purpose, we need additional notation.

Definition 12 If p is a prime number, we define $\mathrm{Sat}_p(\mathscr{L})$ and $\mathrm{Sat}'_p(\mathscr{L})$ to be the largest sublattices of $\mathrm{Sat}(\mathscr{L})$ containing L such that $\mathrm{Sat}_p(\mathscr{L})/\mathscr{L}$ has order a power of p and $\mathrm{Sat}'_p(\mathscr{L})/\mathscr{L}$ has order relatively prime to p. If $p = 0$, we adopt the convention that $\mathrm{Sat}_p(\mathscr{L}) = \mathscr{L}$ and $\mathrm{Sat}'_p(\mathscr{L}) = \mathrm{Sat}(\mathscr{L})$.

Theorem 6 ([5, Corollaries 2.2 and 2.5]) *Let* $\mathrm{char}(\Bbbk) = p \geq 0$ *and consider a group morphsim* $\rho : \mathscr{L} \subseteq \mathbb{Z}^n \to \Bbbk^*$. *If the order of* $\mathrm{Sat}'_p(\mathscr{L})/\mathscr{L}$ *is* g, *there are* g *distinct group morphisms* $\rho_1, \ldots, \rho_g$ *extending* ρ *to* $\mathrm{Sat}'_p(\mathscr{L})$ *and for each* $j \in \{1, \ldots, g\}$ *a unique group morphism* ρ'_j *extending* ρ *to* $\mathrm{Sat}(\mathscr{L})$. *Moreover, there is a unique group morphism* ρ' *extending* ρ *to* $\mathrm{Sat}_p(\mathscr{L})$. *The radical, associated primes and minimal primary decomposition of* $I_{\mathscr{L}}(\rho) \subset \Bbbk[\mathbf{X}]$ *are:*

$$\sqrt{I_{\mathscr{L}}(\rho)} = I_{\mathrm{Sat}_p(\mathscr{L})}(\rho'),$$

$$\mathrm{Ass}(\Bbbk[\mathbf{X}]/I_{\mathscr{L}}(\rho)) = \{I_{\mathrm{Sat}(\mathscr{L})}(\rho'_j) \mid j = 1, \ldots, g\}$$

and

$$I_{\mathscr{L}}(\rho) = \bigcap_{j=1}^{g} I_{\mathrm{Sat}'_p(\mathscr{L})}(\rho_j)$$

where $I_{\mathrm{Sat}'_p(\mathscr{L})}(\rho_j)$ *is* $I_{\mathrm{Sat}(\mathscr{L})}(\rho'_j)$*-primary. In particular, if* $p = 0$, *then* $I_{\mathscr{L}}(\rho)$ *is a radical ideal. The associated primes* $I_{\mathrm{Sat}(\mathscr{L})}(\rho'_j)$ *of* $I_{\mathscr{L}}(\rho)$ *are all minimal and have the same codimension* $\mathrm{rank}(\mathscr{L})$.

3.3 Mesoprime Ideals and Prime Congruences

Given $\delta \subseteq \{1, \ldots, n\}$, set $\mathbb{N}^\delta := \{(u_1, \ldots, u_n) \in \mathbb{N}^n \mid u_i = 0, \text{ for all } i \notin \delta\}$ and define $\mathbb{Z}^\delta$ as the subgroup of $\mathbb{Z}^n$ generated by $\mathbb{N}^\delta$. Morover, if $\delta = \varnothing$, by convention, then $\mathbb{Z}^\delta = \{\mathbf{0}\} \subset \mathbb{Z}^n$.

Definition 13 Given $\delta \subseteq \{1, \ldots, n\}$ and a group homomorphism $\rho : \mathcal{L} \subseteq \mathbb{Z}^\delta \to \Bbbk^*$, a $\delta-$**mesoprime ideal** is an ideal of the form

$$I_{\mathcal{L}}(\rho) + \mathfrak{p}_{\delta^c}$$

with $\mathfrak{p}_{\delta^c} := \langle X_j \mid j \notin \delta \rangle$. By convention, $\mathfrak{p}_{\varnothing^c} = \langle X_1, \ldots, X_n \rangle$ and $\mathfrak{p}_{\varnothing} = \langle 0 \rangle$.

Example 7

(i) The ideal $\langle X_1^{17} - 1, X_2 \rangle \subset \Bbbk[X_1, X_2]$ is mesoprime for $\delta = \{1\}$
(ii) By Theorem 5, every binomial prime ideal is mesoprime, for a suitable δ.
(iii) Lattice ideals are mesoprime for $\delta = \{1, \ldots, n\}$.

Due to Theorem 6, a mesoprime ideal can be understood as a condensed expression that includes all the information necessary to produce the primary decomposition of the ideal simply by using arithmetic arguments.

Observe that the congruence on $\mathbb{N}^n$ defined by $I_{\mathcal{L}} + \mathfrak{p}_{\delta^c}$ is the same as the congruence defined by $I_{\mathcal{L}}(\rho) + \mathfrak{p}_{\delta^c}$, for every $\delta \subseteq \{1, \ldots, n\}$ and every group homomorphism $\rho : \mathcal{L} \subseteq \mathbb{Z}^\delta \to \Bbbk^*$.

Lemma 3 *Let $\delta \subseteq \{1, \ldots, n\}$. If I is a $\delta-$mesoprime ideal, then $I : X_i = I$, for all $i \in \delta$. Equivalently, $I : (\prod_{i \in \delta} X_i)^\infty = I$.*

Proof If I is a $\delta-$mesoprime ideal, there exists $\rho : \mathcal{L} \subseteq \mathbb{Z}^\delta \to \Bbbk^*$ such that $I = I_{\mathcal{L}}(\rho) + \mathfrak{p}_{\delta^c}$. Let $X_i f \in I,\ i \in \delta$. We want to show that $f \in I$. So, without loss of generality, we may assume that no term of f lies in $\mathfrak{p}_{\delta^c}$. In this case, $X_i f \in I_{\mathcal{L}}(\rho)$. Now, by Proposition 7, we conclude that $f \in I_{\mathcal{L}}(\rho)$, and hence $f \in I$.

Definition 14 A congruence $\sim$ on $\mathbb{N}^n$ is said to be **prime** if the ideal $I_\sim$ is mesoprime for some $\delta \subseteq \{1, \ldots, n\}$.

Let us prove that this notion of prime congruence is the same as the usual one (see [7, p. 44]).

Proposition 9 *A congruence $\sim$ on $\mathbb{N}^n$ is **prime** if and only if every element of $\mathbb{N}^n/\sim$ is either nil or cancellable.*

Proof If $\sim$ is prime congruence on $\mathbb{N}^n$, then there exist $\delta \subseteq \{1, \ldots, n\}$ and a subgroup $\mathcal{L} \subseteq \mathbb{Z}^\delta$ such that $I_\sim = I_{\mathcal{L}} + \mathfrak{p}_{\delta^c}$. Let $[\mathbf{u}]$ be non-nil and let $[\mathbf{v}]$ and $[\mathbf{w}] \in \mathbb{N}^n/\sim$ be such that $[\mathbf{v}] + [\mathbf{u}] = [\mathbf{v} + \mathbf{u}] = [\mathbf{w} + \mathbf{u}] = [\mathbf{w}] + [\mathbf{u}]$. In particular, $\mathbf{X}^{\mathbf{u}}(\mathbf{X}^{\mathbf{v}} - \mathbf{X}^{\mathbf{w}}) = \mathbf{X}^{\mathbf{v}+\mathbf{u}} - \mathbf{X}^{\mathbf{w}+\mathbf{v}} \in I_\sim$. Since $[\mathbf{u}]$ is non-nil, $\mathbf{X}^{\mathbf{u}}$ does not belong to $I_\sim$. Therefore, $\mathbf{u} \in \{X_i\}_{i \in \delta}$ and, by Lemma 3, $\mathbf{X}^{\mathbf{v}} - \mathbf{X}^{\mathbf{w}} \in I_\sim$, that is, $[\mathbf{v}] = [\mathbf{w}]$. So $[\mathbf{u}]$ is cancellable.

Conversely, suppose that every element of $\mathbb{N}^n/\sim$ is either nil or cancellable, set $\delta = \{i \in \{1, \ldots, n\} : [\mathbf{e}_i] \text{ is cancellable}\}$. Clearly, $j \notin \delta$ if and only if $X_j \in I_\sim$. So, there exist a binomial ideal J in $\Bbbk[\{X_i\}_{i \in \delta}]$ such that $I_\sim = J\,\Bbbk[\mathbf{X}] + \mathfrak{p}_{\delta^c}$ (if $\delta = \varnothing$, take $J = \langle 0 \rangle$). Moreover, since $[\mathbf{e}_i]$ is cancellable for every $i \in \delta$, if $\mathbf{X}^{\mathbf{e}_i} f = X_i f \in J$, for some $i \in \delta$, then $f \in J$. Thus, by Proposition 7, J is lattice ideal of $\Bbbk[\{X_i\}_{i \in \delta}]$ and, consequently, $J\,\Bbbk[\mathbf{X}]$ is a lattice ideal. Therefore, $I_\sim = J\,\Bbbk[\mathbf{X}] + \mathfrak{p}_{\delta^c}$ is a $\delta-$mesoprime ideal and we are done.

4 Cellular Binomial Ideals

In this section we study the so-called cellular binomial ideals defined by D. Eisenbud and B. Sturmfels in [5]. Cellular binomial ideals play a central role in the theory of primary decomposition of binomials ideals (see [5] and also [6, 13, 14]). As in the previous section, we will determine the congruences on $\mathbb{N}^n$ corresponding to those ideals. We will also outline an algorithm to compute a decomposition of a binomial ideal into cellular binomial ideals which will produce (primary) decompositions of the corresponding congruences.

Let us start by defining the notion of cellular ideal.

Definition 15 A proper ideal I of $\Bbbk[\mathbf{X}]$ is **cellular** if, for some $\delta \subseteq \{1, \ldots, n\}$, we have that

(a) $I : (\prod_{i \in \delta} X_i)^\infty = I$; equivalently $I : X_i = I$, for every $i \in \delta$,
(b) there exists $d_i \in \mathbb{N}$ such that $X_i^{d_i} \in I$, for every $i \notin \delta$.

In this case, we say that I is cellular with respect to δ or, simply, $\delta-$cellular. By convention, the $\varnothing-$cellular ideals are the binomial ideals whose radical is $\langle X_1, \ldots, X_n \rangle$.

Observe that an ideal I of $\Bbbk[\mathbf{X}]$ is cellular if, and only if, every variable of $\Bbbk[\mathbf{X}]$ is either a nonzerodivisor or nilpotent modulo I. In particular, prime, lattice, mesoprime and primary ideals are cellular.

The following proposition establishes the relationship between cellular binomial and mesoprime ideals.

Proposition 10 *Let $\delta \subseteq \{1, \ldots, n\}$. If I is a $\delta-$cellular binomial ideal in $\Bbbk[\mathbf{X}]$, there exists a group morphism $\rho : \mathscr{L} \subseteq \mathbb{Z}^\delta \to \Bbbk^*$ such that*

(a) $(I \cap \Bbbk[\{X_i\}_{i \in \delta}])\, \Bbbk[\mathbf{X}] = I_{\mathscr{L}}(\rho)$.
(b) $I + \mathfrak{p}_{\delta^c} = I_{\mathscr{L}}(\rho) + \mathfrak{p}_{\delta^c}$.
(c) $\sqrt{I + \mathfrak{p}_{\delta^c}} = \sqrt{I_{\mathscr{L}}(\rho)} + \mathfrak{p}_{\delta^c}$.
(d) $\sqrt{I} = \sqrt{I_{\mathscr{L}}(\rho)} + \mathfrak{p}_{\delta^c}$.

In particular, the radical of a cellular binomial ideal is a mesoprime ideal, and the minimal associated primes of I are binomial.

Proof If $\delta = \varnothing$, then $I \cap \Bbbk[\{X_i\}_{i \in \delta}] = 0$ and it suffices to take $\rho : \{\mathbf{0}\} \to \Bbbk^*$; $\mathbf{0} \mapsto 1$. So, assume without loss of generality that $\delta \neq \varnothing$.

In order to prove part (a), we first note that $J := I \cap \Bbbk[\{X_i\}_{i \in \delta}]$ is binomial by Corollary 1, and that $J : (\prod_{i \in \delta} X_i)^\infty = J$ by the definition of cellular ideal. Thus, $J\, \Bbbk[\mathbf{X}] : (\prod_{i=1}^n X_i)^\infty = J\, \Bbbk[\mathbf{X}]$ and, by Proposition 7, there is a unique group morphism $\rho : \mathscr{L} \subseteq \mathbb{Z}^\delta \to \Bbbk^*$ such that $J\, \Bbbk[\mathbf{X}] = I_{\mathscr{L}}(\rho)$.

Part (b) is an immediate consequence of (a).

By part (b) and according to the properties of the radical, we have that

$$\sqrt{I + \mathfrak{p}_{\delta^c}} = \sqrt{I_{\mathscr{L}}(\rho) + \mathfrak{p}_{\delta^c}} = \sqrt{\sqrt{I_{\mathscr{L}}(\rho)} + \mathfrak{p}_{\delta^c}} \supseteq \sqrt{I_{\mathscr{L}}(\rho)} + \mathfrak{p}_{\delta^c} \supseteq \sqrt{I_{\mathscr{L}}(\rho)} + \mathfrak{p}_{\delta^c}.$$

On other hand, given $f \in \sqrt{I_{\mathscr{L}}(\rho)} + \mathfrak{p}_{\delta^c}$, we can write $f = h + \sum_{i \notin \delta} g_i X_i$ where $h^e \in I_{\mathscr{L}}(\rho)$ for some $e > 0$. Now, since $I_{\mathscr{L}}(\rho) \subseteq I_{\mathscr{L}}(\rho) + \mathfrak{p}_{\delta^c} = I + \mathfrak{p}_{\delta^c}$, we have that $f^e = \left(h + \sum_{i \notin \delta} g_i X_i\right)^e \in I + \mathfrak{p}_{\delta^c}$, that is to say, $f \in \sqrt{I + \mathfrak{p}_{\delta^c}}$. Thus, we obtain that $\sqrt{I + \mathfrak{p}_{\delta^c}} = \sqrt{I_{\mathscr{L}}(\rho)} + \mathfrak{p}_{\delta^c}$, as claimed in (c).

For part (d), we observe that

$$\sqrt{I_{\mathscr{L}}(\rho) + \mathfrak{p}_{\delta^c}} = \sqrt{I_{\mathscr{L}}(\rho) + \langle X_i^{d_i} \mid i \notin \delta \rangle},$$

and that

$$I_{\mathscr{L}}(\rho) + \langle X_i^{d_i} \mid i \notin \delta \rangle = (I \cap \Bbbk[\{X_i\}_{i \in \delta}])\, \Bbbk[\mathbf{X}] + \langle X_i^{d_i} \mid i \notin \delta \rangle \subseteq I \subseteq I + \mathfrak{p}_{\delta^c},$$

for every $d_i \geq 1$, $i \notin \delta$. Therefore, taking radicals, by part (c) we conclude that $\sqrt{I_{\mathscr{L}}(\rho) + \mathfrak{p}_{\delta^c}} = \sqrt{I} = \sqrt{I + \mathfrak{p}_{\delta^c}}$.

Now, the last statements are direct consequences of the definition of mesoprimary ideal and Theorem 6.

In the following definition we introduce the concept of primary congruence on $\mathbb{N}^n$. We prove that our notion of primary congruence is equivalent to the one given in [7, p. 44].

Definition 16 A congruence $\sim$ on $\mathbb{N}^n$ is said to be **primary** if the ideal $I_\sim$ is cellular.

Definition 17 Let $\sim$ be a congruence on $\mathbb{N}^n$. An element $\mathbf{a} \in \mathbb{N}^n/\sim$ is said to be **nilpotent** if $d\,\mathbf{a}$ is nil, for some $d \in \mathbb{N}$.

Proposition 11 *A congruence $\sim$ on $\mathbb{N}^n$ is primary if and only if every element of $\mathbb{N}^n/\sim$ is nilpotent or cancellable.*

Proof If $\sim$ is a primary congruence on $\mathbb{N}^n$, the binomial associated ideal $I_\sim$ is $\delta-$cellular for some $\delta \subseteq \{1, \ldots, n\}$. Let $[\mathbf{u}]$ be a non-nilpotent element of $\mathbb{N}^n/\sim$. Given $[\mathbf{v}]$ and $[\mathbf{w}] \in \mathbb{N}^n/\sim$ such that $[\mathbf{v}] + [\mathbf{u}] = [\mathbf{v} + \mathbf{u}] = [\mathbf{w} + \mathbf{u}] = [\mathbf{w}] + [\mathbf{u}]$, we have that $\mathbf{X}^{\mathbf{u}}(\mathbf{X}^{\mathbf{v}} - \mathbf{X}^{\mathbf{w}}) \in I_\sim$. Since $[\mathbf{u}]$ is not nilpotent, $(\mathbf{X}^{\mathbf{u}})^d \notin I_\sim$, for every $d \in \mathbb{N}$. Therefore, no variable X_i with $i \notin \delta$ divides $\mathbf{X}^{\mathbf{u}}$ and, by the definition of cellular ideal, we conclude that $I_\sim : \mathbf{X}^{\mathbf{u}} = I_\sim$; in particular, $\mathbf{X}^{\mathbf{v}} - \mathbf{X}^{\mathbf{w}} \in I_\sim$, that is, $[\mathbf{v}] = [\mathbf{w}]$, and hence $[\mathbf{u}]$ is cancellable.

Conversely, suppose that every element of $\mathbb{N}^n/\sim$ is nilpotent or cancellable. Set $\delta = \{i \in \{1, \ldots, n\} : [\mathbf{e}_i] \text{ is cancellable}\}$. Clearly, $j \in \delta$ if and only if X_j is a nonzerodivisor modulo $I_\sim$ and $j \notin \delta$ if and only if $X_j^{d_j} \in I_\sim$, for some $d_j \geq 1$. Therefore, $I_\sim$ is a $\delta-$cellular ideal (see the paragraph just after Definition 15).

As a consequence, if $\sim$ is a primary congruence on $\mathbb{N}^n$, then, by Proposition 10, $J := \sqrt{I_\sim}$ is a mesoprime ideal. Therefore, associated to $\sim$ there is one and only one prime congruence, $\sim_J$, obtained by removing nilpotent elements.

4.1 Cellular Decomposition of Binomial Ideals

Definition 18 A cellular decomposition of an ideal $I \subseteq \Bbbk[\mathbf{X}]$ is an expression of I as an intersection of cellular ideals with respect to different $\delta \subseteq \{1, \ldots, n\}$, say

$$I = \bigcap_{\delta \in \Delta} \mathscr{C}_\delta, \tag{8}$$

for some subset Δ of the power set of $\{1, \ldots, n\}$. Moreover, the cellular decomposition (8) is said to be minimal if $\mathscr{C}'_\delta \not\supseteq \bigcap_{\delta \in \Delta \setminus \{\delta'\}} \mathscr{C}_\delta$ for every $\delta' \in \Delta$; in this case, the cellular component $\mathscr{C}_\delta$ is said to be a $\delta-$cellular component of I.

Example 8 Every minimal primary decomposition of a monomial ideal $I \subseteq \Bbbk[\mathbf{X}]$ into monomial ideals is a minimal cellular decomposition of I. Consequently, there is non-uniqueness for cellular decomposition in general: consider for instance the following cellular (primary) decomposition

$$\langle X^2, XY \rangle = \langle X \rangle \cap \langle X^2, XY, Y^n \rangle,$$

where n can take any positive integral value.

Cellular decompositions of an ideal I of $\Bbbk[\mathbf{X}]$ always exist. A simple algorithm for cellular decomposition of binomial ideals can be found in [13, Algorithm 2], this algorithm forms part of the `binomials` package developed by T. Kahle and it is briefly described below. The interested reader may consult [9] and [13] for further details.

The following result is the key for producing cellular decompositions of binomial ideals into binomial ideals.

Lemma 4 *Let I be a proper binomial ideal in $\Bbbk[\mathbf{X}]$. If I is not cellular then there exists $i \in \{1, \ldots, n\}$ and a positive integer d such that $I = (I : X_i^d) \cap (I + \langle X_i^d \rangle)$, with $I : X_i^d$ and $I + \langle X_i^d \rangle$ binomial ideals strictly containing I.*

Proof If I is not cellular, there exists at least one variable X_i which is zerodivisor and not nilpotent modulo I. Then, by the Noetherian property of $\Bbbk[\mathbf{X}]$, there is a positive integer d such that $I : \langle X_i^d \rangle = I : \langle X_i^e \rangle$ for every $e \geq d$. We claim that I decomposes as $(I : X_i^d) \cap (I + \langle X_i^d \rangle)$. Indeed, let $f \in (I : X_i^d) \cap (I + \langle X_i^d \rangle)$ and let $f = g + hX_i^d$ for some $g \in I$. Then $X_i^d f = X_i^d g + hX_i^{2d}$ and, thus $hX_i^{2d} = X_i f - X_i g \in I$. That is, $h \in I : \langle X_i^{2d} \rangle = I : \langle X_i^d \rangle$. Hence, $hX_i^d \in I$ and, consequently, $f \in I$.

It remains to see that both $I : X_i^d$ and $I + \langle X_i^d \rangle$ are binomial ideals which strictly contain I. On the one hand, the ideal $I + \langle X_i^d \rangle$ is binomial and I is strictly contained in it, as X_i is not nilpotent modulo I. On the other hand, $I : X_i^d$ is binomial by Corollary 2, and I is strictly contained in $I : X_i^d$ because X_i is a zerodivisor modulo I.

Now, by Lemma 4, if I is not a cellular ideal then we can find two new proper ideals strictly containing I. If these ideals are cellular then we are done. Otherwise, we can repeat the same argument with these new ideals, getting strictly increasing chains of binomial ideals. Since $\Bbbk[\mathbf{X}]$ is a Noetherian ring, each one of these chains has to be stationary. So, in the end, we obtain a (redundant) cellular decomposition of I. Observe that this process does not depend on the base field.

Example 9 Consider the binomial ideal $I = \langle X^4Y^2 - Z^6, X^3Y^2 - Z^5, X^2 - YZ\rangle$ of $\mathbb{Q}[X, Y, Z]$. By using [13, Algorithm 2] we obtain the following cellular decomposition, $I = I_1 \cap I_2 \cap I_3$, where

$$\begin{aligned} I_1 &= \langle Y - Z, X - Z\rangle \\ I_2 &= \langle Z^2, XZ, X^2 - YZ\rangle \\ I_3 &= \langle X^2 - YZ, XY^3Z - Z^5, XZ^5 - Z^6, Z^7, Y^7\rangle. \end{aligned}$$

```
loadPackage "Binomials";
R = QQ[X,Y,Z];
I = ideal(X^4*Y^2-Z^6,X^3*Y^2-Z^5,X^2-Y*Z);
binomialCellularDecomposition I
```

As a final conclusion we may notice the following:

Corollary 6 *Let $\sim$ be a congruence on $\mathbb{N}^n$. A primary decomposition of $\sim$ can be obtained by computing a cellular decomposition of $I_\sim$.*

Proof It is a direct consequence of Proposition 4 by the definition of primary congruence.

5 Mesoprimary Ideals

The main objective of this section is to analyze the mesoprimary ideals and their corresponding congruence. Mesoprimary ideals were introduced by Thomas Kahle and Ezra Miller in [10] as an intermediate construction between cellular and primary binomial ideals. Kahle and Miller proved combinatorially that every cellular binomial can be decomposed into finitely many mesoprimary ideals over an arbitrary field. However, not every decomposition of a binomial ideal as an intersection of mesoprimary ideals is a mesoprimary decomposition in the sense of Kahle and Miller. These mesoprimary decompositions feature refined combinatorial requirements, and currently there is no algorithm available to compute them. Despite of this, mesoprimary decompositions have been successfully used to solve open problems (see [11] and [12]).

The following preparatory result will be helpful in understanding what our mesoprimary ideals are.

Proposition 12 *Let I be a δ−cellular binomial ideal in $\Bbbk[\mathbf{X}]$. If $\mathbf{X}^{\mathbf{u}} \in \Bbbk[\{X_i\}_{i \notin \delta}] \setminus I$, then $I : \mathbf{X}^{\mathbf{u}}$ is a δ−cellular binomial ideal.*

Proof First of all, we note that $I : \mathbf{X}^{\mathbf{u}} \neq \langle 1 \rangle$ because $\mathbf{X}^{\mathbf{u}} \notin I$. Moreover, we have that $I : \mathbf{X}^{\mathbf{u}}$ is binomial by Corollary 2. Now, since $I : (\prod_{i \in \delta} X_i)^\infty = I$, then

$$(I : \mathbf{X}^{\mathbf{u}}) : (\prod_{i \in \delta} X_i)^\infty = (I : (\prod_{i \in \delta} X_i)^\infty) : \mathbf{X}^{\mathbf{u}} = I : \mathbf{X}^{\mathbf{u}}.$$

And, clearly, for every $i \notin \delta$, $X_i^{d_i} \in I : \mathbf{X}^{\mathbf{u}}$ for some $d_i \geq 1$ because $I \subseteq I : \mathbf{X}^{\mathbf{u}}$. Putting all this together, we conclude that $I : \mathbf{X}^{\mathbf{u}}$ is a δ−cellular binomial ideal.

If I is a δ−cellular binomial ideal, then the ideal $(I : \mathbf{X}^{\mathbf{u}}) + \mathfrak{p}_{\delta^c}$ is δ−mesoprime by Propositions 12 and 10(b). Moreover, there exists $d_i \geq 1$ such that $X_i^{d_i} \in I$ for each $i \notin \delta$. Thus there are finitely many mesoprime ideals of the form $(I : \mathbf{X}^{\mathbf{u}}) + \mathfrak{p}_{\delta^c}$. These are the so-called mesoprimes associated to I:

Definition 19 Let I be a δ−cellular binomial ideal in $\Bbbk[\mathbf{X}]$. We will say that $I_{\mathscr{L}}(\rho) + \mathfrak{p}_{\delta^c}$ is a mesoprime ideal associated to I if there exist a monomial $\mathbf{X}^{\mathbf{u}} \in \Bbbk[\{X_i\}_{i \notin \delta}]$ such that

$$\big((I : \mathbf{X}^{\mathbf{u}}) \cap \Bbbk[\{X_i\}_{i \in \delta}]\big)\, \Bbbk[\mathbf{X}] = I_{\mathscr{L}}(\rho).$$

Now we may introduce the notion of mesoprimary ideal.

Definition 20 A binomial ideal is said to be **mesoprimary** if it is cellular and it has only one associated mesoprime ideal. A congruence $\sim$ on $\mathbb{N}^n$ is mesoprimary if $I_\sim$ is a mesoprimary ideal of $\Bbbk[\mathbf{X}]$

The following lemma clarifies the notion of mesoprimary ideal.

Lemma 5 *A δ−cellular binomial ideal I in $\Bbbk[\mathbf{X}]$ is mesoprimary if and only if $(I : \mathbf{X}^{\mathbf{u}}) \cap \Bbbk[\{X_i\}_{i \notin \delta}] = I \cap \Bbbk[\{X_i\}_{i \notin \delta}]$, for all $\mathbf{X}^{\mathbf{u}} \in \Bbbk[\{X_i\}_{i \notin \delta}] \setminus I$.*

Proof It suffices to note that I has two different associated mesoprimes if and only if there exists $\mathbf{X}^{\mathbf{u}} \in \Bbbk[\{X_i\}_{i \notin \delta}]$ such that $(I : \mathbf{X}^{\mathbf{u}}) \cap \Bbbk[\{X_i\}_{i \notin \delta}] \neq I \cap \Bbbk[\{X_i\}_{i \notin \delta}]$ because, in this case, by Proposition 10, $(I : \mathbf{X}^{\mathbf{u}}) + \mathfrak{p}_{\delta^c}$ and $I + \mathfrak{p}_{\delta^c}$ are two different associated mesoprimes to I.

Definition 21 Let $\sim$ be a congruence on $\mathbb{N}^n$. An element $\mathbf{a} \in \mathbb{N}^n/\sim$ is said to be **partly cancellable** if $\mathbf{a} + \mathbf{b} = \mathbf{a} + \mathbf{c} \neq \infty \Rightarrow \mathbf{b} = \mathbf{c}$, for all cancellable $\mathbf{b}, \mathbf{c} \in \mathbb{N}^n$

Proposition 13 *A congruence $\sim$ on $\mathbb{N}^n$ is mesoprimary if and only if it is primary and every element in $\mathbb{N}^n/\sim$ is partly cancellable.*

Proof If $\sim$ is a mesoprimary congruence on $\mathbb{N}^n$, then $I = I_\sim$ is δ−cellular for some $\delta \subseteq \{1, \ldots, n\}$. Thus, $\sim$ is primary. Moreover, $(I : \mathbf{X}^{\mathbf{u}}) \cap \Bbbk[\{X_i\}_{i \in \delta}] = I \cap \Bbbk[\{X_i\}_{i \in \delta}]$, for all $\mathbf{X}^{\mathbf{u}} \in \Bbbk[\{X_i\}_{i \notin \delta}] \setminus I$ (equivalently, for all $\mathbf{u} \in \mathbb{N}^n$ such that $[\mathbf{u}]$ is nilpotent and it is not a nil). Therefore, if $[\mathbf{u}] \in \mathbb{N}^n/\sim$ is nilpotent and $[\mathbf{v}]$, $[\mathbf{w}]$ are cancellable elements such that $[\mathbf{u}] + [\mathbf{v}] = [\mathbf{u}] + [\mathbf{w}] \neq \infty$, then

$$\mathbf{X}^{\mathbf{v}} - \mathbf{X}^{\mathbf{w}} \in (I : \mathbf{X}^{\mathbf{u}}) \cap \Bbbk[\{X_i\}_{i\in\delta}] = I \cap \Bbbk[\{X_i\}_{i\in\delta}],$$

that is to say $[\mathbf{v}] = [\mathbf{w}]$. So, $[\mathbf{u}]$ is partly cancellative.

Conversely, suppose that $\sim$ is primary congruence on $\mathbb{N}^n$ such that every element in $\mathbb{N}^n/\sim$ is partly cancellable. Since $\sim$ is primary, we have that $I_\sim$ is $\delta-$cellular, by setting $\delta = \{i \in \{1, \ldots, n\} : [\mathbf{e}_i] \text{ is cancellable}\}$. Now, if $\mathbf{X}^{\mathbf{u}} \in \Bbbk[\{X_i\}_{i\notin\delta}] \setminus I$, we have that $[\mathbf{u}]$ is partly cancellable. Thus, for every $\mathbf{X}^{\mathbf{v}} - \mathbf{X}^{\mathbf{w}} \in \Bbbk[\{X_i\}_{i\in\delta}]$, we have that $\mathbf{X}^{\mathbf{u}}(\mathbf{X}^{\mathbf{v}} - \mathbf{X}^{\mathbf{w}}) \in I \Rightarrow \mathbf{X}^{\mathbf{v}} - \mathbf{X}^{\mathbf{w}} \in I$. Therefore, $(I : \mathbf{X}^{\mathbf{u}}) \cap \Bbbk[\{X_i\}_{i\in\delta}] \subseteq I \cap \Bbbk[\{X_i\}_{i\in\delta}]$. Now, since the opposite inclusion is always fulfilled, by Lemma 5, we are done.

There are other intermediate constructions between cellular and primary ideals, such as the unmixed decomposition (see [5, 13] and, more recently, [6]). The following example shows that unmixed cellular binomial ideals are not mesoprimary. Recall that an unmixed cellular binomial ideal is a cellular binomial ideal with no embedded associated primes (see [13, Proposition 2.4]).

Example 10 Consider the unmixed cellular binomial $I \subset \Bbbk[X, Y]$ generated by $\{X^2 - 1, Y(X - 1), Y^2\}$. The ideal I is not mesoprimary, because

$$(I : Y) \cap \Bbbk[X] = \langle X - 1\rangle \neq \langle X^2 - 1\rangle = I \cap \Bbbk[X].$$

```
loadPackage "Binomials";
R = QQ[X,Y]
I = ideal(X^2-1,Y*(X-1),Y^2)
cellularBinomialAssociatedPrimes I
eliminate(I:Y,Y)
eliminate(I,Y)
```

We end this section by exhibiting the statement of Kahle and Miller which describes the primary decomposition of a mesoprimary ideal, in order to give an idea of how useful would be to have an algorithm for the mesoprimary decomposition of a cellular binomial ideal.

Proposition 14 ([10, Corollary 15.2 and Proposition 15.4]) *Let I be a (δ-cellular) mesoprimary ideal, and denote by $I_{\mathscr{L}}(\rho)$ the lattice ideal $I \cap \Bbbk[\{X_i\}_{i\in\delta}]$. The associated primes of I are exactly the (minimal) primes of its associated mesoprime $I + \mathfrak{p}_{\delta^c}$. Moreover, if $I_{\mathscr{L}}(\rho) = \cap_{j=1}^{g} I_j$ is the primary decomposition of $I_{\mathscr{L}}(\rho)$ from Theorem 7, then*

$$I = \bigcap_{j=1}^{g}(I + I_j)$$

is the primary decomposition of I.

Notice that the hypothesis $\Bbbk$ algebraically closed is only needed when Theorem 7 is applied.

Acknowledgements We thank the anonymous referees for their detailed suggestions and comments, which have greatly improved this article. The present paper is based on a course of lectures delivered by the second author at the EACA's Third International School on Computer Algebra and Applications https://www.imus.us.es/EACASCHOOL16/. He thanks the organizers for giving him that opportunity.

References

1. Briales, E. Campillo, A. Marijuán, C. Pisón, P. Combinatorics of syzygies for semigroup algebra. Collect. Math. **49**, 239–256 (1998)
2. Brookfield, G.: Commutative Noetherian semigroups are finitely generated. Semigroup Forum **66**(2), 323–327 (2003)
3. Bruns, W., Herzog, J.: Cohen-Macaulay Rings. Cambridge Studies in Advanced Mathematics, vol. 39. Cambridge University Press, Cambridge (1993)
4. Budach, L.: Struktur Noetherscher kommutativer Halbgruppen. Monatsb. Deutsch. Akad. Wiss. **6**, 85–88 (1964)
5. Eisenbud, D., Sturmfels, B.: Binomial ideals. Duke Math. J. **84**(1), 1–45 (1996)
6. Eser, Z.S., Matusevich, L.F.: Decompositions of cellular binomial ideals. J. Lond. Math. Soc. **94**, 409–426 (2016)
7. Gilmer, R.: Commutative Semigroup Rings. Chicago Lectures in Mathematics. University of Chicago Press, Chicago (1984)
8. Grayson, D.R., Stillman, M.E.: Macaulay2, a software system for research in algebraic geometry. Available at http://www.math.uiuc.edu/Macaulay2/
9. Kahle, T.: Decompositions of binomial ideals. J. Softw. Algebra Geom. **4**(1), 1–5 (2012)
10. Kahle, T., Miller, E.: Decompositions of commutative monoid congruences and binomial ideals. Algebra Numb. Theory **8**(6), 1297–1364 (2014)
11. Kahle, T., Miller, E., O'Neill, C.: Irreducible decomposition of binomial ideals. Compos. Math. **152**, 1319–1332 (2016)
12. Matusevich, L.F., O'Neill C.: Some algebraic aspects of mesoprimary decomposition. J. Pure Appl. Algebra (2018) to appear
13. Ojeda, I., Piedra Sánchez, R.: Cellular binomial ideals. Primary decomposition of binomial ideals. J. Symbolic Comput. **30**(4), 383–400 (2000)
14. Ojeda, I.: Binomial canonical decompositions of binomial ideals. Commun. Algebra **39**(10), 3722–3735 (2011)

The K-Theory of Toric Schemes Over Regular Rings of Mixed Characteristic

G. Cortiñas, C. Haesemeyer, M. E. Walker, and C. A. Weibel

Abstract We show that if X is a toric scheme over a regular commutative ring k then the direct limit of the K-groups of X taken over any infinite sequence of nontrivial dilations is homotopy invariant. This theorem was previously known for regular commutative rings containing a field. The affine case of our result was conjectured by Gubeladze. We prove analogous results when k is replaced by an appropriate K-regular, not necessarily commutative k-algebra.

Cortiñas' research was supported by Conicet and partially supported by grants UBACyT 20021030100481BA, PIP 112-201101-00800CO, PICT 2013-0454, and MTM2015-65764-C3-1-P (Feder funds).

Haesemeyer's research was partially supported by the University of Melbourne Research Grant Support Scheme and ARC Discovery Project grant DP170102328.

Walker's research was partially supported by a grant from the Simons Foundation (#318705).

Weibel's research was supported by NSA and NSF grants.

G. Cortiñas (✉)
Department of Matemática-Inst. Santaló, FCEyN, Universidad de Buenos Aires, Buenos Aires, Argentina
e-mail: gcorti@dm.uba.ar

C. Haesemeyer
School of Mathematics and Statistics, University of Melbourne, Melbourne, VIC, Australia
e-mail: christian.haesemeyer@unimelb.edu.au

M. E. Walker
Department of Mathematics, University of Nebraska – Lincoln, Lincoln, NE, USA
e-mail: mark.walker@unl.edu

C. A. Weibel
Department of Mathematics, Rutgers University, New Brunswick, NJ, USA
e-mail: weibel@math.rutgers.edu

G.-M. Greuel et al. (eds.), *Singularities, Algebraic Geometry, Commutative Algebra, and Related Topics*, https://doi.org/10.1007/978-3-319-96827-8_19

1 Introduction

Let $A = (A, \cdot)$ be a commutative monoid. For each integer $c \geq 2$, the c-th power map $\theta_c : A \to A$, $\theta_c(a) = a^c$, is an endomorphism of A; it is called a *dilation*. For any ring k, θ_c induces an endomorphism of the monoid ring $k[A]$, its K-theory $K_*(k[A])$, and its homotopy K-theory $KH_*(k[A])$.

The affine *Dilation Theorem* says that the monoid of dilations acts nilpotently on the reduced K-theory of $k[A]$, at least when A is a submonoid of a torsionfree abelian group, A has no nontrivial units and k is an appropriately regular ring.

Case (a) of the following theorem verifies a conjecture of Gubeladze [19, 1.1].

Dilation Theorem 1.1 *Let A be a submonoid of a torsionfree abelian group. Assume that A has no non-trivial units. Let $\mathfrak{c} = (c_1, c_2, \dots)$ be a sequence of integers with $c_i \geq 2$ for all i. If Λ is any of the rings listed below, then the following canonical map is an isomorphism:*

$$K_*(\Lambda) \xrightarrow{\cong} \varinjlim_{\theta_c} K_*(\Lambda[A]). \tag{1}$$

(a) *a regular commutative ring;*
(b) *a commutative C^*-algebra;*
(c) *an associative regular ring that admits the structure of a flat algebra over a regular commutative ring k_0 of finite Krull dimension;*
(d) *an associative ring that is K-regular and that admits the structure of a flat algebra over a regular commutative ring k_0 of finite Krull dimension having infinite residue fields; or*
(e) *an associative ring that is K-regular and that admits the structure of a flat algebra over a regular commutative ring k_0 of finite Krull dimension such that for every prime $\mathfrak{p} \subset k_0$ having finite residue field, the ring $\Lambda \otimes_{k_0} k'$ is K-regular for every étale $(k_0)_{\mathfrak{p}}$-algebra k'.*

Recall that an associative ring Λ is called (right) *regular* if it is (right) Noetherian and every finitely generated (right) Λ module has finite projective dimension. It is called *K-regular* if the canonical map $K_*(\Lambda) \to K_*(\Lambda[x_1, \dots, x_n])$ is an isomorphism for all $n \geq 0$. Also recall that, up to canonical isomorphism, a (unital) commutative C^*-algebra is the ring $C(T)$ of complex-valued continuous functions on a compact Hausdorff space T. All rings considered in this article are unital; in particular, nonunital C^*-algebras are not considered.

Gubeladze proved the Dilation Theorem for regular $\mathbb{Q}$-algebras in [19] and [20]; it was established for regular k of positive characteristic by the authors in [9]. Gubeladze also verified the theorem when A is a "simplicial" monoid for all commutative regular rings k in [18]. Further, his result [17, Theorem 3.2.2] implies that if A is simplicial and Λ is a not necessarily commutative K-regular ring, then the map (1) is an isomorphism whenever $\mathfrak{c}$ is a constant prime sequence $\mathfrak{c} = (p, p, \dots)$.

Remark Nilpotence for the constant prime sequences $\mathfrak{c} = (p, p, \dots)$ implies nilpotence for all constant sequences $\mathfrak{c} = (n, n, \dots)$ because $(c_1)_*(c_2)_* = (c_2)_*(c_1)_*$. However, it doesn't imply nilpotence for sequences containing infinitely many primes.

As we showed in [10], it is useful to pass from abelian monoids to pointed abelian monoids (adding an element '0'), and to generalize even further to *monoid schemes*, associating the affine monoid scheme $\mathrm{MSpec}(A)$ to a pointed monoid A. A monoid scheme X has a *k-realization* X_k over any commutative ring k; if $X = \mathrm{MSpec}(A)$ then X_k is $\mathrm{Spec}(k[A])$. Again there are dilation maps $\theta_c : X \to X$, defined locally as the c-th power map on affine open subschemes, and θ_c induces dilations of both X_k and its K-theory.

Even if Λ is a noncommutative ring, we can still make sense out of $K(X_\Lambda)$, although the scheme X_Λ is not defined. If $X = \mathrm{MSpec}(A)$ then $K(X_\Lambda)$ is $K(\Lambda[A])$; in general, the spectrum $K(X_\Lambda)$ may either be defined using Zariski descent on X, or equivalently as the K-theory of the dg category $\mathrm{perf}(X_k) \otimes_k^{\mathbb{L}} \Lambda$; see Example 5.3 for details.

Theorem 1.2 below is the *Dilation Theorem* for monoid schemes; the hypotheses on X are satisfied whenever it is a toric monoid scheme or, more generally, a partially cancellative torsionfree (pctf) monoid scheme of finite type. (The definitions of all these terms are recalled below, at the end of this introduction.) They are also satisfied by $X = \mathrm{MSpec}(A)$ when A is a monoid of finite type satisfying the assumptions in Theorem 1.1. Observe also that since every abelian monoid is the filtered colimit of its finitely generated submonoids, and since K-theory commutes with such colimits, the finitely generated case of Theorem 1.1 implies the general case. We remark also that for A and Λ as in Theorem 1.1, $K_*(\Lambda) = KH_*(\Lambda) = KH_*(\Lambda[A])$. Thus Theorem 1.1 follows from Theorem 1.2.

Theorem 1.2 *Let X be a separated, partially cancellative, torsionfree monoid scheme of finite type, let $\mathfrak{c} = (c_1, c_2, \dots)$ be a sequence of integers with $c_i \geq 2$ for all i and let Λ be an associative ring satisfying one of (a) through (e) in Theorem* 1.1. *Then the canonical map is an isomorphism:*

$$\varinjlim_{\theta_c} K_*(X_\Lambda) \xrightarrow{\cong} \varinjlim_{\theta_c} KH_*(X_\Lambda). \tag{2}$$

The particular case of this theorem when Λ is a commutative regular ring containing a field was proven by the authors in [8] and [9], and was used to verify Gubeladze's conjecture for these rings.

We remark that for commutative noetherian rings Λ, Theorem 1.2 is equivalent to the affine Dilation Theorem 1.1, by Zariski descent. The affine case, however, seems to be no easier to prove. Our proof of the Dilation Theorem follows the strategy used in [9], using Theorem 1.2 to deduce 1.1. The basic outline is as follows:

Step 1. We show that the singularities of any nice monoid scheme may be resolved by a finite sequence of blow-ups along smooth normally flat centers. This is based upon a theorem of Bierstone-Milman (see [1]). Even if we begin with a monoid, the blow-ups of its monoid scheme take us out of the realm of (affine) monoids. This step is carried out in Sect. 3, and is preceded by a short calculation with monoids in Sect. 2. This step is independent of Λ.

Step 2. In [10], we introduced the notion of *cdh* descent for presheaves on pctf monoid schemes, and showed that the functor $X \mapsto KH(X_\Lambda)$ from pctf monoid schemes to spectra satisfies *cdh* descent when Λ is a commutative regular ring of finite Krull dimension that contains a field. In Sect. 4, we review and use Step 1 to modify this proof, replacing the assumptions on Λ given above by the assumptions that Λ is a flat associative algebra over a commutative regular ring of finite Krull dimension all of whose residue fields are infinite; see Corollaries 4.6 and 5.4. Also in this section, we apply the arguments of [9] to prove that both the fiber of the Jones-Goodwillie Chern character $K(X_\Lambda) \otimes \mathbb{Q} \to HN(X_{\Lambda\otimes\mathbb{Q}})$ and the fiber of the p-local cyclotomic trace with $\mathbb{Z}/p^n$-coefficients to the pro-spectrum $\{TC^\nu(X_\Lambda; p)\}_\nu$ satisfy *cdh* descent whenever Λ is as in Theorem 1.1 (see Propositions 5.7 and 5.10).

Step 3. For a presheaf of spectra E on monoid schemes, write $\mathscr{F}_E$ for the homotopy fiber of the map from E to its *cdh*-fibrant replacement. From Step 2 we conclude that if Λ is a flat K-regular algebra over a commutative regular ring k_0 of finite Krull dimension all of whose residue fields are infinite, then $\mathscr{F}_K$ is equivalent to the fiber of the map $K \to KH$, its rationalization is equivalent to $\mathscr{F}_{HN(-\otimes\mathbb{Q})}$, and for each n the cofiber $\mathscr{F}_K/p^n$ of multiplication by p^n is equivalent to the pro-spectrum $\{\mathscr{F}_{E_\nu}\}$ for $E_\nu = TC^\nu(-;p)/p^n$. Making use of the constructions developed in [9, Section 6], we prove that for an arbitrary ring Λ, monoid scheme X and $\nu \geq 1$, both $\mathscr{F}_{HN(-\otimes\mathbb{Q})}(X)$ and $\mathscr{F}_{TC^\nu(-;p)}(X)$ become contractible upon taking the colimit over any infinite sequence of non-trivial dilations. The case of $TC^\nu(-;p)$ is considered in Sect. 6 (see Theorem 6.3) and that of $HN(-\otimes\mathbb{Q})$ in Sect. 7 (see Corollary 7.3).

Step 4. Theorem 1.2 is finally obtained in Sect. 8: Theorem 8.5 gives parts (c) and (e) and the other parts are consequences. The particular case when Λ is commutative and contains a field is [9, Theorem 8.3]. The case when Λ is flat over a commutative regular ring k_0 of finite Krull dimension all of whose residue fields are infinite follows using Steps 2–3 and the argument of the proof of [9, Theorem 8.3]. We show further that, if k_0 has finite Krull dimension and $\Lambda \supset k_0$ satisfies the hypothesis of the theorem, then the map (2) is an isomorphism if and only if this happens for every local ring of k_0 (Lemma 8.1). Then we show that for k_0 a regular local ring, the case when the residue field is infinite implies the case of finite residue field. This establishes parts (c) and (e) of the theorem. Part (d) is a special case of part (e) and part (b) follows from the fact that commutative C^*-algebras are K-regular [5, Theorem 8.2]. Finally we observe that every commutative regular ring Λ with Spec(Λ) connected is a flat extension of either a finite field or a localization of the ring of integers; this proves (a).

Some notation Given a presheaf E from some full subcategory of the category of monoid schemes to spectra (or spaces, or chain complexes, or equivariant spectra or equivariant spaces) and a sequence $\mathfrak{c} = (c_1, c_2, c_3, \dots)$ of integers larger than

1, we write E^c for the (sequential) homotopy colimit $\varinjlim_{\theta_{c_i}} E$, again a presheaf of spectra (or spaces, or ...) on the same category of monoid schemes. If our presheaf is obtained by sending the monoid scheme X to the spectrum (or space, etc.) $E(X_\Lambda)$ for a ring Λ we will sometimes (when no confusion is possible) abuse notation and write E^c for the presheaf $X \mapsto \varinjlim_{\theta_{c_i}} E(X_\Lambda)$.

Monoid terminology Unless otherwise stated, a *monoid* is a pointed abelian monoid written multiplicatively, i.e., it is an abelian monoid with unit 1 and a basepoint 0 satisfying $a \cdot 0 = 0$ for all $a \in A$. If B is an unpointed monoid, adjoining an element 0 yields a pointed monoid B_+. If A is pointed monoid, we write A' for the subset $A \setminus \{0\}$. We say that A is *cancellative* if $ab = ac$ implies $b = c$ or $a = 0$ for any $a, b, c \in A$. In particular, if A is cancellative, then A' is an unpointed submonoid of A. We say A is *torsionfree* if whenever $a^n = b^n$ for $a, b \in A$ and some $n \geq 1$, we have $a = b$. A monoid is cancellative and torsionfree if and only if it is a submonoid of the pointed monoid T_+ associated to a torsionfree abelian group T.

An *ideal* I of a monoid A is a subset containing 0 and satisfying $AI \subseteq I$. In this case, the quotient monoid A/I is defined by collapsing I to 0; the product in A/I is the unique one making the canonical surjection $A \twoheadrightarrow A/I$ a morphism. A monoid A is said to be *partially cancellative* if it is a quotient C/I of a cancellative monoid C by some ideal I; if C is both cancellative and torsionfree, we say that A is *pctf*.

The prime ideals in A form a space $\mathrm{MSpec}(A)$, with a sheaf of monoids. A *monoid scheme* is a space X with a sheaf of monoids $\mathscr{A}$ that is locally isomorphic to $\mathrm{MSpec}(A)$ for some A. A closed subscheme Z is *equivariant* if its structure sheaf is $\mathscr{A}/\mathscr{I}$ for a sheaf of ideals $\mathscr{I}$. A monoid scheme is of *finite type* if it admits a finite open cover by affine monoid schemes associated to finitely generated monoids.

2 Free A-Sets

We begin by recalling standard notation for pointed sets. If X, Y are pointed sets, we write $X \bigvee Y$ and $X \wedge Y$ for their pointed coproduct and smash product, respectively.

Given a monoid A, an *A-set* is a pointed set X, with basepoint 0, together with a function $A \wedge X \to X$, written $(a, x) \mapsto a \cdot x$, such that $a \cdot (a' \cdot x) = aa' \cdot x$ for all $a, a' \in A$ and $x \in X$. The k-realization $k[X]$ of an A-set is a $k[A]$-module. The monoid A is an A-set in an obvious way, and given any collection of A-sets, their wedge sum is again an A-set. A subset B of an A-set X is a *basis* if $X = \bigvee_{b \in B} Ab$ and if $ab = a'b'$, for $a, a' \in A'$ and $b, b' \in B$, implies that either $b = b'$ and $a = a'$ or $a = a' = 0$. We say that an A-set X is *free* if it has a basis, or equivalently if it is isomorphic to a wedge sum of copies of A, $\bigvee_{i \in I} A$.

The goal of this section is to prove the following result.

Theorem 2.1 *Let A be a cancellative monoid, X a finitely generated A-set and k a commutative ring. If $k[X]$ is a free $k[A]$-module, then X is a free A-set.*

Example 2.2 Suppose that $A = G_+$ for an abelian group G. Then $X \setminus \{0\}$ is a disjoint union of orbits G/H_i, and $k[X]$ is the direct sum of the $k[G/H_i]$. If $k[X]$ is a free $k[A]$-module, then each H_i is trivial, so X is a free A-set.

The proof requires a sequence of preliminary results.

Let $A^\times$ denote the group of units of A and define the quotient monoid $\overline{A} = A/\sim$ where $a \sim b$ if and only if $a = ub$ for some $u \in A^\times$. Similarly, $\overline{X} = X \wedge_A \overline{A}$ is the $\overline{A}$-set $X/\sim$, where $x \sim y$ if and only if $x = uy$ for some $u \in A^\times$.

Lemma 2.3 *For any monoid A and any commutative ring k, $k[\overline{A}] \cong k[A] \otimes_{k[A^\times]} k$. If $k[X]$ is free as a $k[A]$-module, then $k[\overline{X}]$ is free as a $k[\overline{A}]$-module.*

Proof The kernel I of $k[A^\times] \to k$ is generated by $\{u - 1 : u \in A^\times\}$; we have to show that the surjection $k[A]/Ik[A] \to k[\overline{A}]$ is an isomorphism. Choose a set of representatives $\{a_i\}$ for $\sim$, so that A is the disjoint union of 0 and the sets $A^\times a_i$. Then $k[A]$ is the direct sum of the $k[A^\times a_i]$, and $k[A]/Ik[A]$ is the direct sum of the $k[a_i]$, i.e., $k[\overline{A}]$. The result for free modules follows from the equation below (see [10, paragraph below diagram 1.8])

$$k[\overline{X}] = k[X \wedge_A \overline{A}] \cong k[X] \otimes_{k[A]} k[\overline{A}].$$

□

We will say that an A-set X is *cancellative* if $ax = bx$ implies $a = b$ for any nonzero $x \in X$ and $a, b \in A$.

Lemma 2.4 *Assume A is a cancellative monoid and X is a cancellative A-set. Then X is free as an A-set if and only if $\overline{X}$ is free as an $\overline{A}$-set.*

Proof The forward direction holds since $\overline{\bigvee_I A} = \bigvee_I \overline{A}$. Conversely, suppose that $\overline{X}$ is free as an $\overline{A}$-set with basis $\overline{B}$. Let B be a subset of X given by choosing one representative for each element of $\overline{B}$. It is clear that B generates X, and is an A-basis of X because X is cancellative: if $ab = a'b$ in X then $a = a'$. □

Remark 2.5 In Lemma 2.4, the hypothesis that X be cancellative is needed, as the example $A = \{0, 1, -1\}$ and $X = \{0, 1\}$ shows; in this case $\overline{A} = \overline{X}$.

Remark 2.6 If A is cancellative and has no non-trivial units, we define a partial ordering on A by $b \le c$ if $c = ab$ for some a in A. This partial ordering respects the group operation on A: if $a \le b$ then $ac \le bc$ for all $c \in A$.

The group completion G_+ of such a monoid A is also partially ordered: $h \le g$ if $g = ah$ for some a in A. The inclusion $A \subset G_+$ respects the partial ordering.

We say that a ring R is A-*graded* if $R = \bigoplus_{a \in A} R_a$ with $R_a R_b \subseteq R_{ab}$ and $R_0 = \{0\}$. If A has no non-trivial units, then $R_{\neq 1} = \bigoplus_{a \neq 1} R_a$ is an ideal of R, with $R_1 = R/R_{\neq 1}$.

Example 2.7 For any monoid A and ring k, the monoid ring $k[A]$ is A-graded with $k[A]_a := k \cdot a$. If A has no non-trivial units, $k[A]_{\neq 1} = k[\mathfrak{m}]$ where $\mathfrak{m} = \mathfrak{m}_A$ is the unique maximal ideal of A.

Lemma 2.8 *Let A be a cancellative monoid with no non-trivial units. Suppose R is an A-graded ring. If T is a finitely generated nonzero A-graded R-module such that $T_0 = 0$, then $R_{\neq 1} T \neq T$.*

Proof Set $S = \{a \in A' : T_a \neq 0\}$, and suppose that T is generated by homogenous elements $x_1, \ldots, x_l$ of degrees $s_1, \ldots, s_l$ ($s_i \in S$). Let S_0 denote the set of minimal elements in $\{s_1, \ldots, s_l\}$ with respect to the partial ordering in Remark 2.6. Then every element of S_0 is minimal in S. If $s \in S$, there is a non-zero homogeneous element t in T_s. Write $t = \sum_i r_i x_i$. Grouping the sum by common degrees, we may assume the r_i are homogeneous and that $r_i = 0$ unless $s = |r_i x_i|$. Since $t \neq 0$, there is an i so that $s = |r_i x_i| \geq s_i$. This shows that $s \geq s_0$ for some $s_0 \in S_0$.

Fix $s \in S_0$ and let t be a non-zero element of T_s. If $t \in R_{\neq 1} T$ then $t = \sum_i r_i x_i$ with $r_i \in R_{\neq 1}$. As before, we may assume all the r_i are homogenous and nonzero, and that $r_i x_i \in R_s$ for all i. Since $r_i \in R_{\neq 1}$, we have $|r_i| > 1$ and so $|r_i x_i| > 1|x_i| = s_i$. This contradicts the minimality of s, showing that $T \neq R_{\neq 1} T$. □

Corollary 2.9 *Let A be a cancellative monoid with no nontrivial units. Suppose R is an A-graded ring and T is a finitely generated, free graded R-module. If Y is any set of homogeneous elements in T whose image in $T/R_{\neq 1}T$ is a basis of the free R_1-module $T/R_{\neq 1}T$, then Y is a basis of T as an R-module.*

Proof The classical proof applies; here are the details:

First note that Y must be finite, since it maps mod $R_{\neq 1}$ to a basis of a free, finitely generated R_1-module. Let L denote the free graded R-module on the set Y. The canonical map $\pi : L \to T$ is graded and $L_0 = T_0 = 0$; hence its kernel K and cokernel C are both graded modules, and $K_0 = C_0 = 0$. Since applying $- \otimes_R R/R_{\neq 1}$ to this map results in an isomorphism by assumption, $C/R_{\neq 1}C = 0$. By Lemma 2.8, $C = 0$ and hence π is surjective. Since T is free graded, π splits (in the graded category). In particular, K is a direct summand of a finitely generated free module and hence finitely generated. It follows that $K/R_{\neq 1}K = 0$ as well, whence $K = 0$ using the Lemma again. □

Lemma 2.10 *Suppose A is a monoid, X is an A-set and $S \subseteq A$ is a multiplicatively closed subset. Then the natural homomorphism of $k[A]$-modules $S^{-1}k[X] \to k[S^{-1}X]$ is an isomorphism.*

Proof The inverse homomorphism is obtained from the natural map of A-sets $S^{-1}X \to S^{-1}k[X]$ by adjunction. □

Lemma 2.11 *If A is a cancellative monoid and X an A-set so that $k[X]$ is a free $k[A]$-module, then X is cancellative.*

Proof Consider the group completion of A, which is isomorphic to G_+ for some abelian group G. Since $k[A] \subseteq k[G]$ and $k[X]$ is free, we have $k[X] \subseteq k[X] \otimes_{k[A]} k[G]$. By Lemma 2.10 and [10, Lemma 5.5], $k[X] \otimes_{k[A]} k[G] = k[X_G]$, where X_G is the localization of X at A'. It follows that $X \subseteq X_G$, and the latter is a free G_+-set by Example 2.2. Thus, X is a cancellative A-set as asserted. □

Proof of Theorem 2.1 By Lemma 2.11, X is a cancellative A-set. Thus Lemmas 2.3 and 2.4 imply that we may assume that A has no non-trivial units.

Define a partial ordering on X by $x \leq y$ if and only if $y = ax$ for some $a \in A$. The reflexive and transitive properties are clear. Since X is cancellative, it is also symmetric: if $x \leq y$ and $y \leq x$ there are $a, b \in A$ with $y = ax$, $x = by$ and hence $x = abx$. This implies that $ab = 1$; since A has no nontrivial units, we have $a = b = 1$ and hence $x = y$. (When $X = A$, this is the partial ordering of Remark 2.6.) As X is finitely generated, there is a finite set B of minimal elements of X with respect to the partial ordering. In fact, B is the unique smallest collection of generators of X. We will prove that B forms a basis of X.

Let F denote the free graded A-set on B. We need to show that the canonical map $F \xrightarrow{\pi} X$ sending $a[b]$ to ab is a bijection. The map is surjective since B generates X by construction. Passing to k-realization gives a surjection of graded $k[A]$-modules, $k[F] \twoheadrightarrow k[X]$. By Corollary 2.9, with $R = k[A]$ and $Y = B$, it suffices to prove that this map becomes an isomorphism after applying $- \otimes_{k[A]} k$, where $k = k[A]/k[A]_{\neq 1}$; this will show that $k[F] \to k[X]$ and hence $F \to X$, are injective. By construction, there is a canonical isomorphism $k[F] \otimes_{k[A]} k \cong k[B_+]$, and, by Example 2.7,

$$k[X] \otimes_{k[A]} k = k[X] \otimes_{k[A]} k[A/\mathfrak{m}_A] = k[X/\mathfrak{m}_A X].$$

Since the natural map $B_+ \to X/\mathfrak{m}_A X$ induced by π is an isomorphism, the induced map $k[B_+] \to k[X/A_+X]$ is also an isomorphism. Thus B is a basis of X. □

3 Normal Flatness for Monoid Schemes

Recall from [10, 6.4] that a finitely generated pointed monoid A is *smooth* if A is the wedge product of a finitely generated free pointed monoid T by Γ_+, where Γ is a finitely generated free abelian group. That is, for some d and r:

$$A = \langle t_1, \ldots, t_d, s_1, 1/s_1, \ldots, s_r, 1/s_r \rangle.$$

Note that $k[A]$ is the Laurent polynomial ring $k[t_1, \ldots, t_d, s_1, 1/s_1, \ldots, s_r, 1/s_r]$ for every ring k.

More generally, we say that a monoid scheme X is smooth if it is of finite type and each stalk monoid $\mathscr{A}_x$ is smooth.

Recall that a commutative ring R is said to be *normally flat* along an ideal J if each quotient J^n/J^{n+1} is a projective module over the ring R/J. A scheme is said to be normally flat along a closed subscheme if locally the coordinate ring is normally flat along the ideal defining it.

Definition 3.1 A monoid A is said to be *normally flat* along an ideal I if each quotient I^n/I^{n+1} is a free A/I-set. Let X be a monoid scheme of finite type with structure sheaf $\mathscr{A}$; we say that X is *normally flat* along an equivariant closed subscheme Z if for every $x \in X$ the monoid $\mathscr{A}_x$ is normally flat along the stalk of the ideal defining Z.

Proposition 3.2 *Let I be a finitely generated ideal in a monoid A, with A/I smooth. The following are equivalent:*

(1) *A is normally flat along I.*
(2) *$k[A]$ is normally flat along the ideal $k[I]$ for every commutative ring k.*
(3) *$\mathbb{Q}[A]$ is normally flat along the ideal $\mathbb{Q}[I]$.*

Proof The k-realization functor commutes with colimits, since it is left adjoint to the forgetful functor, so $k[I]^n/k[I]^{n+1} \cong k[I^n/I^{n+1}]$. Hence (1) implies (2). It is clear that (2) implies (3). Suppose that (3) holds, and write J for $\mathbb{Q}[I]$. Then each quotient $J^n/J^{n+1} \cong \mathbb{Q}[I^n/I^{n+1}]$ is a projective module over the ring $\mathbb{Q}[A]/J = \mathbb{Q}[A/I]$, and, since all projective $\mathbb{Q}[A/I]$-modules are free [29], Theorem 2.1 implies that each A/I-set I^n/I^{n+1} is free. Hence (1) holds. □

Corollary 3.3 *Let Z be a smooth equivariant closed subscheme of X, a monoid scheme of finite type. The following are equivalent:*

(1) *X is normally flat along Z.*
(2) *X_k is normally flat along Z_k for every commutative ring k.*
(3) *$X_{\mathbb{Q}}$ is normally flat along $Z_{\mathbb{Q}}$.*

Proof Again, it suffices to prove that (3) implies (1). As the conditions are local, it suffices to pick $x \in X$ and assume that $X = \text{MSpec}(A)$ and $Z = \text{MSpec}(A/I)$ for $A = \mathscr{A}_x$, with A/I smooth. This case is covered by Proposition 3.2. □

Theorem 3.4 *Let X be a separated cancellative torsionfree monoid scheme of finite type, embedded as an equivariant closed subscheme in a smooth toric scheme. Then there is a sequence of blow-ups along smooth equivariant centers $Z_i \subset X_i$, $0 \le i \le n-1$,*

$$Y = X_n \to \cdots \to X_0 = X$$

such that Y is smooth and each X_i is normally flat along Z_i. In addition, for any commutative ring k, each $(X_i)_k$ is normally flat along $(Z_i)_k$.

Proof The case $k = \mathbb{Q}$ of Theorem 14.1 in our paper [10] states that there is a sequence of blowups along smooth equivariant centers $Z_i \subset X_i$, such that each

$(X_i)_\mathbb{Q}$ is normally flat along $(Z_i)_\mathbb{Q}$. By Corollary 3.3, each X_i is normally flat along Z_i, and each $(X_i)_k$ is normally flat along $(Z_i)_k$. □

4 A Descent Theorem for Functors via Realizations

In this section, we recall the notion of cdh descent, establish a technical result generalizing [10, Theorem 11.3] to the present context, and use it to promote several results in *op. cit.* to our situation.

In more detail, let $\mathcal{M}_{\text{pctf}}$ denote the category of separated pctf monoid schemes of finite type. Fix a commutative regular ring k_0 of finite Krull dimension all of whose residue fields are infinite, and suppose $k_0 \subseteq k$ is a flat extension of commutative rings. We will show that if F is a presheaf of spectra on the category Sch/k of separated k-schemes essentially of finite presentation that satisfies a weak version of cdh descent (see Definition 4.3), then the presheaf $\mathcal{F}$ on $\mathcal{M}_{\text{pctf}}$ defined by $\mathcal{F}(X) = F(X_k)$ satisfies cdh descent; see Theorem 4.5.

We first recall the necessary definitions from [10] and [34]. By a *cd structure* on a category, we mean a family of distinguished commutative squares

$$\begin{array}{ccc} D & \longrightarrow & Y \\ \downarrow & & \downarrow{\scriptstyle p} \\ C & \xrightarrow{\ e\ } & X. \end{array} \qquad (3)$$

Definition 4.1 A cartesian square (3) in $\mathcal{M}_{\text{pctf}}$ is called

1. an *abstract blow-up square* if p is proper, e is an equivariant closed immersion, and p maps the open complement $Y \setminus D$ isomorphically onto $X \setminus C$;
2. a *Zariski square* if p and e form an open cover of X;
3. a *cdh square* if it is either a Zariski or an abstract blow-up square.

The cdh topology on $\mathcal{M}_{\text{pctf}}$ is the topology generated by the cdh squares. It is a bounded, complete and regular cd structure by [10, Theorems 12.7 and 12.8].

A presheaf of spectra $\mathcal{F}$ on $\mathcal{M}_{\text{pctf}}$ satisfies the *Mayer-Vietoris property* for some family $\mathcal{C}$ of cartesian squares if $\mathcal{F}(\emptyset) = *$ and the application of $\mathcal{F}$ to each member of the family gives a homotopy cartesian square of spectra.

Definition 4.2 Let $\mathcal{F}$ be a presheaf of spectra on $\mathcal{M}_{\text{pctf}}$. We say that $\mathcal{F}$ satisfies *cdh descent* on $\mathcal{M}_{\text{pctf}}$ if the canonical map $\mathcal{F}(X) \to \mathbb{H}_{\text{cdh}}(X, \mathcal{F})$ is a weak equivalence of spectra for all X. Here $\mathbb{H}_{\text{cdh}}(-, \mathcal{F})$ is the fibrant replacement of $\mathcal{F}$ in the model structure of [23] and [24]. By [10, Proposition 12.10], this is equivalent to $\mathcal{F}$ satisfying the Mayer-Vietoris property for the cdh structure; i.e., for both the family of Zariski squares and the family of abstract blow-up squares.

Here is a restatement of [10, Definition 13.8].

Definition 4.3 Let k be a commutative ring and let Sch/k be the category of separated k-schemes, essentially of finite presentation. A presheaf of spectra F on Sch/k is said to satisfy *weak cdh descent* if it satisfies the Mayer-Vietoris property for all open covers, finite abstract blow-ups, and blow-ups along regularly embedded subschemes.

Example 4.3.1 The presheaf $\mathscr{K}H$ satisfies weak cdh descent. This was observed in [10, Example 13.11]; the hypothesis there that the ring k be Noetherian is not needed. Indeed, the Mayer-Vietoris property for finite abstract blow-ups (excision for ideals and invariance under nilpotent extensions) is proved in [35] for general rings, and Mayer-Vietoris for open covers as well as Mayer-Vietoris for blow-ups along regular sequences applies to the category of quasi-separated and quasi-compact schemes.

We will see other examples in Theorems 4.5 and 5.2 below.

Theorem 4.4 *Let k be a commutative regular ring of finite Krull dimension, with infinite residue fields, and F a presheaf of spectra on Sch/k. If F satisfies weak cdh descent on Sch/k, then the presheaf $\mathscr{F}(X) = F(X_k)$ satisfies cdh descent on $\mathscr{M}_{pctf}$.*

Proof The proof is almost identical to the proof of Theorem 14.3 in our earlier paper [10], which assumed that k contains a field.

We merely point out the adjustments necessary for the proof to work for k. Although the proof in *loc. cit.* makes no direct use of the hypothesis that k contains a field, this hypothesis is buried in the references to Lemma 13.9 and Theorem 14.2 of [10]. (Theorem 13.3 and Proposition 13.6 are referred to, but apply as stated.)

Although the hypothesis of Lemma 13.9 is that k is a commutative regular domain containing an infinite field, the proof goes through if we only assume that k is a commutative regular ring of finite Krull dimension such that every residue field of k is infinite. The hypothesis that k is regular is needed so that X_k is a Cohen-Macaulay scheme for every toric monoid scheme X (by [22]). The finite Krull dimension hypothesis is needed in Lemma 13.9 in order that the local-global spectral sequence converges (and so the proof of Lemma 13.9 is flawed as written; see the remark below for the correction).

The last hypothesis, that every residue field of k is infinite, is needed to use Lemma 13.7 of [10]. Without this hypothesis, we would have to pass to finite extensions of each local ring of k to get a minimal reduction generated by a regular sequence.

Finally, the hypothesis in Theorem 14.2 that k contains a field is required to make use of the Bierstone-Milman theorem [10, Theorem 14.1]. Replacing it by the more general Theorem 3.4 in this paper makes the proof go through. □

Remark 4.4.1 As explained in the proof above, the hypothesis that k have finite Krull dimension is missing in Lemma 13.9 and Theorem 14.3 of [10]. This however does not affect the main results of [10], since K-theory commutes with filtering colimits and, by Popescu's theorem [30], every regular ring containing a field is a filtering colimit of regular rings of finite Krull dimension.

Theorem 4.5 *Let k_0 be a commutative regular ring of finite Krull dimension all of whose residue fields are infinite, let $i : k_0 \subset k$ be a flat extension of commutative rings. If F is a presheaf of spectra on Sch/k satisfying weak cdh descent on Sch/k, then the presheaf $\mathscr{F}_k$ on $\mathscr{M}_{pctf}$ defined by $\mathscr{F}_k(X) = F(X_k)$ satisfies cdh descent.*

Proof Let i_*F denote the direct image presheaf on Sch/k_0, defined by $i_*F(S) = F(S_k)$. Since the flat basechange i_* preserves open immersions, closed immersions, surjective morphisms, finite morphisms, regular closed immersions, and blow-ups, it also preserves open covers, finite abstract blow-ups, and blow-ups along regular closed immersions. Because F satisfies weak cdh descent on Sch/k, i_*F satisfies weak cdh descent on Sch/k_0. The result is now immediate from Theorem 4.4. □

Given a commutative ring k, let $\mathscr{K}_k$ and $\mathscr{K}H_k$ denote the presheaves of spectra on $\mathscr{M}_{\text{pctf}}$ sending X to $K(X_k)$ and $KH(X_k)$, respectively. Here is the analogue of [10, Corollary 14.5].

Corollary 4.6 *Let k_0 be a commutative regular ring of finite Krull dimension all of whose residue fields are infinite, and let k be a commutative flat k_0-algebra. Then the presheaf of spectra $\mathscr{K}H_k$ satisfies cdh descent on $\mathscr{M}_{pctf}$. If in addition k is K-regular, then the map*

$$\mathbb{H}_{\text{cdh}}(X, \mathscr{K}_k) \to \mathbb{H}_{\text{cdh}}(X, \mathscr{K}H_k)$$

is a weak equivalence for all X in $\mathscr{M}_{pctf}$.

Proof As observed in Example 4.3.1, the presheaf $\mathscr{K}H$ on Sch/k satisfies weak cdh descent. Thus the first statement follows from Theorem 4.5.

Next assume that k is K-regular. We have to prove that the map of the corollary is a weak equivalence; we mimick the argument we used in [10, Example 13.11]. By [10, Theorem 11.1], every cdh covering in $\mathscr{M}_{\text{pctf}}$ admits a refinement consisting of smooth monoid schemes. If X is smooth, then X_k is locally the spectrum of a Laurent polynomial ring over k. Since k is assumed to be K-regular, we conclude that for smooth X, $\mathscr{K}_k(X) \to \mathscr{K}H_k(X)$ is an equivalence, using Mayer-Vietoris for open covers and the Fundamental Theorem of K-theory. The result for general X in $\mathscr{M}_{\text{pctf}}$ now follows from the Mayer-Vietoris property for $\mathbb{H}_{\text{cdh}}(X, \mathscr{K}_k)$ and resolution of singularities for monoid schemes. □

5 Presheaves of Spectra and dg Categories

Let k_0 be a commutative ring and let E be a functor from small dg k_0-categories to spectra. We may use E in two different ways to obtain a functor on Sch $/k_0$. First, by regarding a k_0-algebra as a dg category with just one object, we may restrict

E to a functor of commutative k_0-algebras. This restriction induces a presheaf $\mathscr{E}$ on Sch/k_0 by mapping $S \mapsto E(\mathscr{O}(S))$ on Sch/k_0, and its fibrant replacement is $\mathbb{H}_{\mathrm{zar}}(-, \mathscr{E})$. On the other hand, we may simply compose E with the functor that sends a scheme S to the dg k_0-category $\mathrm{perf}(S)$ of perfect complexes on S, obtaining the functor $E(\mathrm{perf}(-))$; see [7, Example 2.7] or [28, Section 2.4] for a precise definition.

Following Sections 3 and 5 of [25], we say E is dg *Morita invariant* if it sends dg Morita equivalences to weak equivalences; we say that E *localizing* if it sends short exact sequences of dg categories to fibration sequences.

Lemma 5.1 *If E is a functor from small dg k_0-categories to spectra that is dg Morita invariant and localizing, then the functors $\mathbb{H}_{\mathrm{zar}}(-, \mathscr{E})$ and $E(\mathrm{perf}(-))$ are equivalent.*

Proof For a commutative ring R, the functor $R \to \mathrm{perf}(R)$ is a dg Morita equivalence and thus $E(R) \xrightarrow{\sim} E(\mathrm{perf}(R))$ is an equivalence. It follows that the natural transformation of presheaves on Sch/k_0 from $E(\mathscr{O}(-))$ to $E(\mathrm{perf}(\mathscr{O}(-)))$ is an equivalence Zariski locally. The induced functor on fibrant replacements is thus also an equivalence. Since E is localizing, $S \mapsto E(\mathrm{perf}(S))$ satisfies Zariski descent (and even Nisnevich descent) by [31, Theorem 3.1]. We thus get a pair of natural equivalences

$$\mathbb{H}_{\mathrm{zar}}(S, \mathscr{E}) \xrightarrow{\sim} \mathbb{H}_{\mathrm{zar}}(S, E(\mathrm{perf}(-))) \xleftarrow{\sim} E(\mathrm{perf}(S))$$

for all $S \in \mathrm{Sch}/k_0$. □

From now on, if $S \in \mathrm{Sch}/k_0$ and E is a functor from small dg k_0-categories that is dg Morita invariant and localizing, by $E(S)$ we shall always mean $E(\mathrm{perf}(S))$.

Given two small dg k_0-categories $\mathscr{A}$ and $\mathscr{B}$, we write $\mathscr{A} \otimes_{k_0} \mathscr{B}$ for their dg tensor product (as defined in [25, Sec. 2.3]).

Theorem 5.2 *Let k_0 be a commutative ring, E a functor from small dg k_0-categories to spectra and Λ a flat (not necessarily commutative) k_0-algebra. Assume*

1. *E is dg Morita invariant and*
2. *E is localizing.*

Then the presheaf E_Λ of spectra on Sch/k_0, sending S to

$$E_\Lambda(S) = E(\mathrm{perf}(S) \otimes_{k_0} \Lambda),$$

satisfies the Mayer-Vietoris property with respect to open covers (Zariski descent) and blow-ups of regularly embedded subschemes.

Assume furthermore that

3. *the restriction of E to k_0-algebras satisfies excision for ideals and is invariant under nilpotent ring extensions.*

Then E_Λ *satisfies weak cdh descent on* Sch $/k_0$. *In particular, by Theorem* 4.4*, the functor* $\mathscr{E}_\Lambda(X) = E_\Lambda(X_{k_0})$ *satisfies cdh descent on* $\mathscr{M}_{pctf}$.

Proof Because Λ is flat, tensoring with Λ preserves dg Morita equivalences and short exact sequences of dg categories. Hence E_Λ is both dg Morita invariant and localizing whenever E is. In particular it satisfies Zariski descent, by the discussion above. The Mayer-Vietoris property for blow-ups along regularly embedded subschemes also follows from hypotheses (1) and (2) using [7, Lemma 1.5], as pointed out in [6, Theorem 3.2] and implicitly in [7, Theorem 2.10]. To finish the proof we must show that if E also satisfies (3) then it has the Mayer-Vietoris property for finite abstract blow-ups. As observed in the proof of [7, Theorem 3.12], the latter property follows from excision for ideals, invariance under nilpotent ring extensions and Zariski descent. Tensoring with Λ preserves Milnor squares of k_0-algebras; because E satisfies excision for ideals, so does E_Λ. If I is a nilpotent ideal, so is $I \otimes_{k_0} \Lambda$, so hypothesis (3) implies that E_Λ is invariant under nilpotent ring extensions. □

Remark 5.2.1 If E is a dg Morita invariant, localizing functor from small dg k_0-categories to spectra, and k is a flat commutative k_0-algebra, then $E(S_k)$ is naturally equivalent to the $E_k(S)$ of Theorem 5.2. Indeed, the result holds when $S = \mathrm{Spec}(R)$ is an affine scheme, since $R_k \to \mathrm{perf}(R_k)$ and $\mathrm{perf}(R) \otimes_{k_0} k \to \mathrm{perf}(R_k)$ are dg Morita equivalences; the result for arbitrary schemes follows from the fact that, by Theorem 5.2, both E_k and $E(-_k)$ satisfy Zariski descent. In particular, it follows that the functors $\mathscr{E}(X) = E(X_k)$ and $\mathscr{E}_k(X) = E_k(X_{k_0})$ on $\mathscr{M}_{\mathrm{pctf}}$ are naturally equivalent.

Motivated by all this, we shall often write $E(X_\Lambda)$ for $\mathscr{E}_\Lambda(X)$ when Λ is a flat but not necessarily commutative k_0-algebra, even if no scheme X_Λ is defined. By Lemma 5.1, $E(X_\Lambda)$ is equivalent to the functor defined by Zariski descent from the spectra $E(\Lambda[A])$ on the affine opens $\mathrm{MSpec}(A)$ of X.

Example 5.3 For any dg category $\mathscr{A}$, abusing notation a bit, we also write perf($\mathscr{A}$) for the (unenriched) category whose morphisms are $\mathscr{A}$-module homomorphisms; i.e., the category whose hom sets are the zero cycles of underlying dg category. Then perf($\mathscr{A}$) may be regarded as a Waldhausen category; the weak equivalences are quasi-isomorphisms and the cofibrations are $\mathscr{A}$-module homomorphisms which admit retractions as homomorphisms of graded $\mathscr{A}$-modules; see [25, Sec. 5.2] for example.

We define $K(\mathscr{A})$ to be the (Waldhausen) K-theory of perf($\mathscr{A}$). As pointed out in *loc. cit.,* this functor is localizing and Morita invariant, and its restriction to k_0-algebras is naturally equivalent to the usual K-theory of algebras. Theorem 5.2 applies to KH, which satisfies (3), showing that $KH_\Lambda(S)$ satisfies *cdh* descent on Sch/k_0, and $\mathscr{K}H_\Lambda(X)$ satisfies *cdh* descent on $\mathscr{M}_{\mathrm{pctf}}$.

Corollary 5.4 *Let* k_0 *be a commutative regular ring of finite Krull dimension all of whose residue fields are infinite, and let* Λ *be a flat* K*-regular associative*

k_0-algebra. Then the presheaf of spectra $\mathcal{KH}_\Lambda$ satisfies cdh descent on $\mathcal{M}_{pctf}$, and the maps

$$\mathcal{KH}_\Lambda(X) \to \mathbb{H}_{cdh}(X, \mathcal{KH}_\Lambda) \leftarrow \mathbb{H}_{cdh}(X, \mathcal{K}_\Lambda)$$

are weak equivalences for all X in $\mathcal{M}_{pctf}$.

Proof The left arrow is a weak equivalence by Example 5.3. The proof that the other arrow is an equivalence is exactly like the corresponding proof of Corollary 4.6. □

Example 5.5 The Hochschild homology, cyclic homology and negative cyclic homology of a dg category $\mathcal{A}$ are defined using the mixed complex of perf($\mathcal{A}$); see [25, Sec. 5.3]. Moreover, the restrictions of these functors to flat algebras (resp., schemes) are isomorphic to the usual homology theories of algebras (resp., of schemes).

Notation 5.6 *Given a presheaf of spectra $\mathcal{F}$ on some category we will write $\mathcal{F} \otimes \mathbb{Q}$ for the rationalization of $\mathcal{F}$ and $\mathcal{F}/n$ for the cofiber of multiplication by n on $\mathcal{F}$. We make an exception for the presheaf of spectra $HN(-_\Lambda)$ on $\mathcal{M}_{pctf}$ sending X to negative cyclic homology $HN(X_\Lambda)$. By $HN_{\Lambda\mathbb{Q}}(X)$ we will mean $HN(X_{\Lambda\otimes\mathbb{Q}})$, which is defined as in Remark 5.2.1. The Jones-Goodwillie Chern character is a natural morphism on algebras, from $K(\Lambda) \otimes \mathbb{Q}$ to $HN(\Lambda \otimes \mathbb{Q})$; see [15, p. 351]. It extends to a natural transformation of functors on dg categories, and hence a natural morphism $\mathcal{K}_\Lambda(X) \otimes \mathbb{Q} \to HN_{\Lambda\mathbb{Q}}(X)$ on monoid schemes (see [3, Example 9.10], [26, Section 4.4] for the Chern character for dg categories).*

Proposition 5.7 *Let k_0 be commutative regular ring of finite Krull dimension all of whose residue fields are infinite, and let Λ be a flat, associative k_0-algebra that is K-regular. For any monoid scheme X in $\mathcal{M}_{pctf}$, the following square of spectra is homotopy cartesian:*

$$\begin{array}{ccc} \mathcal{K}_\Lambda(X)\otimes\mathbb{Q} & \longrightarrow & \mathcal{KH}_\Lambda(X)\otimes\mathbb{Q} \\ \downarrow & & \downarrow \\ HN_{\Lambda\mathbb{Q}}(X) & \longrightarrow & \mathbb{H}_{cdh}(X, HN_{\Lambda\mathbb{Q}}). \end{array}$$

Proof By Corollary 5.4 the presheaves in the right column satisfy cdh-descent on $\mathcal{M}_{pctf}$. Thus to prove the assertion, we need to show that the homotopy fiber F of the left vertical map satisfies cdh descent. Since $\mathcal{K}_\Lambda \otimes \mathbb{Q}$ and $HN_{\Lambda\mathbb{Q}}$ satisfy Zariski descent, so does F. Given Examples 5.3 and 5.5, Theorem 5.2 shows that it suffices to show that F (the corresponding homotopy fiber on Sch/k_0) satisfies excision for ideals and invariance under nilpotent extensions. The first of these properties is the main theorem of [4]; the second is the scheme version of Goodwillie's [15]. □

The analogue of the Jones-Goodwillie Chern character in characteristic p is the cyclotomic trace; it is a compatible family of morphisms $tr^\nu : \mathcal{K}(\Lambda) \to TC^\nu(\Lambda; p)$, where Λ is an associative ring and the pro-spectrum $\{TC^\nu(\Lambda; p)\}_\nu$ is

p-local topological cyclic homology. For each $n \geq 1$, the cyclotomic trace induces a map $\mathscr{K}(\Lambda)/p^n \to TC^\nu(\Lambda; p)/p^n$. Letting ν vary, we get a strict map of pro-spectra. Blumberg and Mandell showed in [2, 1.1, 1.2, 7.1, 7.3] that TC^ν extends to a functor on dg categories that is localizing, Morita invariant and satisfies Zariski descent on schemes.

As shown in [2] the cyclotomic trace extends to a natural transformation of functors on dg categories; in particular, if X is a monoid scheme and Λ is an associative, flat algebra over some commutative ring k_0, we obtain a natural map $tr^\nu : \mathscr{K}(X_\Lambda)/p^n \to TC^\nu(X_\Lambda; p)$.

Recall from [13] (or [10, Section 14]) that a strict map $\{E^\nu\} \to \{F^\nu\}$ of pro-spectra is said to be a *weak equivalence* if for every q the induced map $\{\pi_q(E^\nu)\} \to \{\pi_q(F^\nu)\}$ is an isomorphism of pro-abelian groups. Let

$$\begin{array}{ccc} \{E^\nu\} & \longrightarrow & \{F^\nu\} \\ \downarrow & & \downarrow \\ \{G^\nu\} & \longrightarrow & \{H^\nu\} \end{array} \tag{4}$$

be a square diagram of strict maps of pro-spectra. We say that (4) is *homotopy cartesian* if the canonical map from the upper left pro-spectrum to the level-wise homotopy limit of the other terms is a weak equivalence.

The presheaves of pro-spectra we shall consider come from pro-presheaves of spectra, that is, from inverse systems of presheaves of spectra. Let $\{F^\nu\}$ be a pro-presheaf of spectra on Sch/k; for X in $\mathscr{M}_{\text{pctf}}$, write $\mathscr{F}^\nu(X)$ for $F^\nu(X_k)$.

The pro-analogue of $\{F^\nu\}$ (or $\{\mathscr{F}^\nu\}$) having the Mayer-Vietoris property for a family of squares is the obvious one, as are the notions 4.2, 4.3 of $\{F^\nu\}$ having cdh descent or weak cdh descent. Here is the pro-analogue of Theorem 4.4.

Theorem 5.8 *Let $\{F^\nu\}$ be a pro-presheaf of spectra on Sch/k, where k is a commtative regular ring of finite Krull dimension all of whose residue fields are infinite. If $\{F^\nu\}$, regarded as a presheaf of pro-spectra, satisfies weak cdh descent on Sch/k, then the presheaf $\{\mathscr{F}^\nu\}$ defined by $\mathscr{F}^\nu(X) = F^\nu(X_k)$ satisfies cdh descent on $\mathscr{M}_{pctf}$.*

Proof We modify the proof of Theorem 4.4 (which is in turn a modification of the proof of [10, 14.3]). By construction, each $\mathbb{H}_{\text{cdh}}(-, \mathscr{F}^\nu)$ satisfies cdh descent. As noted in the proof of Theorem 4.4, the proof of Theorem 13.9 in [10] goes through to show that the presheaf of pro-spectra $\{\mathscr{F}^\nu\}$ has the Mayer-Vietoris property for "nice" blow-up squares (Definition 13.5 in [10]). Now each $\mathscr{F}^\nu$ has the Mayer-Vietoris property for Zariski squares and smooth blow-up squares, so by [10, 12.13] each $\mathscr{F}^\nu$ satisfies $scdh$ descent (terminology of [10, 12.12]). Hence each $\{\mathscr{F}^\nu(X)\} \to \{\mathbb{H}_{\text{cdh}}(X, \mathscr{F}^\nu)\}$ is a weak equivalence of pro-spectra by Theorem 14.2 of [10]. □

Corollary 5.9 *Assume that k_0 is a commutative regular ring of finite Krull dimension all of whose residue fields are infinite, and that Λ is a flat unital associative*

k_0-algebra. Let $\{E^\nu\}$ be an inverse system of functors from the category of small dg k_0-categories to spectra such that each E^ν is localizing and dg Morita invariant.

If the restriction of $\{E^\nu\}$ to k_0-algebras, regarded as a functor to pro-spectra, satisfies excision for ideals and is invariant under nilpotent ring extensions, then the presheaf of pro-spectra $\{E^\nu_\Lambda\}$, defined by

$$E^\nu_\Lambda(S) = E^\nu(\mathrm{perf}(S) \otimes_{k_0} \Lambda),$$

satisfies weak cdh descent on Sch/k_0.

By Theorem 5.8, the functor $\{\mathscr{E}^\nu_\Lambda\}$ satisfies cdh descent on $\mathscr{M}_{pctf}$.

Proof The proof of Theorem 5.2 applies, using Theorem 5.8 in place of Theorem 4.4. □

Proposition 5.10 *Let k_0 be a commutative regular ring of finite Krull dimension all of whose residue fields are infinite, and let Λ be a flat associative k_0-algebra that is K-regular. For any monoid scheme X in $\mathscr{M}_{pctf}$, any prime p and all $n > 0$, the following square of strict maps of pro-spectra is homotopy cartesian.*

$$\begin{array}{ccc} \mathscr{K}_\Lambda/p^n(X) & \longrightarrow & \mathscr{K}H_\Lambda/p^n(X) \\ \downarrow & & \downarrow \\ \{TC^\nu(X_\Lambda;p)/p^n\}_\nu & \longrightarrow & \{\mathbb{H}_{\mathrm{cdh}}(X_\Lambda, TC^\nu(-;p)/p^n)\}_\nu. \end{array}$$

The vertical maps are induced by the cyclotomic trace.

Proof As in Proposition 5.7, it suffices to prove that the homotopy fiber $\{\mathscr{F}^\nu_\Lambda(X)\}$ of the left vertical map satisfies cdh descent as a functor on $\mathscr{M}_{\mathrm{pctf}}$, in the sense that $\{F^\nu\} \to \{\mathbb{H}_{\mathrm{cdh}}(-;\mathscr{F}^\nu)\}$ is a weak equivalence of pro-spectra. (This uses the fact that pro-abelian groups form an abelian category.)

For each dg category $\mathscr{A}$, let $F^\nu(\mathscr{A})$ denote the homotopy fiber of $K/p^n(\mathscr{A}) \to TC^\nu(\mathscr{A};p)/p^n$. Each F^ν is localizing and dg Morita invariant; these properties are inherited from K/p^n and $TC^\nu(-;p)/p^n$. To see that $\{F^\nu_\Lambda\}$ satisfies weak descent on Sch/k, and hence that $\{\mathscr{F}^\nu_\Lambda\}$ satisfies cdh descent on $\mathscr{M}_{\mathrm{pctf}}$, it suffices by Corollary 5.9 to show that the restriction of the pro-presheaf $\{F^\nu\}$ to algebras, regarded as a functor to pro-spectra, satisfies excision for ideals and invariance under nilpotent ring extensions. Excision for ideals is [12, Theorem 1]; if $I \lhd R$ is an ideal and $f : R \to S$ is a ring homomorphism mapping I bijectively to an ideal of S, then the map $\{F^\nu(R, I)\} \to \{F^\nu(S, f(I))\}$ is a weak equivalence. Invariance under nilpotent ring extensions follows from McCarthy's theorem [27] as strengthened in [11, Theorem 2.1.1]: if I is nilpotent then $\{F^\nu(R, I)\}$ is weakly equivalent to a point. □

6 $\widetilde{\Omega}$ and Dilated Cyclic Homologies

Here we briefly recall the results of Section 6 of our paper [9]. These results do not require the existence of a base field. We use the notation for sequential homotopy colimits over dilations introduced at the end of the Introduction.

The *cyclic bar construction* $N^{\mathrm{cy}}(A)$ of a pointed monoid A is the cyclic set whose underlying set of n-simplices is $A \wedge \cdots \wedge A$ ($n+1$ factors), with t_n being rotation of the entries to the left. It is A-graded: for $a \in A$, a simplex $(\alpha_0, \dots, \alpha_n)$ is in $N^{\mathrm{cy}}(A; a)$ if $\prod \alpha_i = a$. In [9, 3.1], we introduced the *dilated* cyclic bar construction

$$\widetilde{N}^{\mathrm{cy}}(A) = \bigvee_{a \in A} N^{\mathrm{cy}}(A\big[\tfrac{1}{a}\big]; a),$$

where the element a in $N^{\mathrm{cy}}(A\big[\frac{1}{a}\big]; a)$ refers to the element $\frac{a}{1}$ of $A\big[\frac{1}{a}\big]$. Thus an n-simplex of $\widetilde{N}^{\mathrm{cy}}(A)$ is given by $(a; \alpha_0, \dots, \alpha_n)$ where $a \in A$, $\alpha_0, \dots, \alpha_n \in A\big[\frac{1}{a}\big]$, and the equation $\alpha_0 \cdots \alpha_n = \frac{a}{1}$ holds in $A\big[\frac{1}{a}\big]$.

The geometric realizations of both $N^{\mathrm{cy}}(A)$ and $\widetilde{N}^{\mathrm{cy}}(A)$ are $\mathbb{S}^1$*-spaces*, meaning they have a continuous action of the circle group $\mathbb{S}^1$.

We now assume that A is a quotient $\tilde{A}/I$ of a cancellative monoid $\tilde{A}$ (i.e., A is *partially cancellative*), and that A is *reduced* in the sense that $a^n = 0$ implies $a = 0$ (this is equivalent to the assertion that if $a, b \in A$ satisfy $a^n = b^n$ for all $n > 1$ then $a = b$; see [9, 1.6]). Such a monoid is naturally contained in a monoid A_{sn} that is *seminormal*, meaning that whenever $x, y \in A_{\mathrm{sn}}$ satisfy $x^3 = y^2$, then there is a (unique) $z \in A_{\mathrm{sn}}$ such that $x = z^2$ and $y = z^3$. In fact, $A \to A_{\mathrm{sn}}$ is universal with respect to maps from A to seminormal monoids; see [9, Proposition 1.15].

Definition 6.1 ([9, 4.1]) For a pc monoid A, we define $\widetilde{\Omega}_A$ to be the $\mathbb{S}^1$-space

$$\widetilde{\Omega}_A = |\widetilde{N}^{\mathrm{cy}}(A_{\mathrm{sn}})|.$$

Since $A \mapsto A_{\mathrm{sn}}$ is a functor, the assignment $(X, \mathscr{A}) \mapsto \widetilde{\Omega}_{\mathscr{A}} = \widetilde{\Omega}_{\mathscr{A}(X)}$ yields a presheaf on $\mathscr{M}_{\mathrm{pctf}}$. Moreover, there is a natural map $|N^{\mathrm{cy}}(A)| \to \widetilde{\Omega}_A$ of $\mathbb{S}^1$-spaces.

Given a sequence $\mathfrak{c} = (c_1, c_2, \dots)$ of integers with $c_i \ge 2$ for all i, it is shown in [9, 3.6] that $|N^{\mathrm{cy}}(A)|^{\mathfrak{c}} \xrightarrow{\simeq} |\widetilde{\Omega}_A|^{\mathfrak{c}}$ is an $\mathbb{S}^1$-homotopy equivalence.

The $\mathbb{S}^1$-equivariant smash product $\widetilde{\Omega}_{\mathscr{A}} \wedge T$ of $\widetilde{\Omega}_{\mathscr{A}}$ with any $\mathbb{S}^1$-spectrum T is a presheaf of $\mathbb{S}^1$-spectra on $\mathscr{M}_{\mathrm{pctf}}$. For every integer $r \ge 1$, we write $\widetilde{\Omega}^{T,r}$ for the presheaf of fixed-point spectra $X \mapsto (\widetilde{\Omega}_{\mathscr{A}(X)} \wedge T)^{C_r}$ on $\mathscr{M}_{\mathrm{pctf}}$ (where C_r is the cyclic subgroup of $\mathbb{S}^1$ having r elements), and $\mathbb{H}_{\mathrm{zar}}(-, \widetilde{\Omega}^{T,r})$ for its fibrant replacement for the Zariski topology. The following result was proven in [9, Theorem 6.2].

Theorem 6.2 *For any $\mathbb{S}^1$-spectrum T and integer $r \ge 1$, the presheaf of spectra $\mathbb{H}_{\mathrm{zar}}(-, \widetilde{\Omega}^{T,r})$ satisfies cdh descent on $\mathscr{M}_{\mathrm{pctf}}$.*

Hesselholt and Madsen proved in [21, Theorem 7.1] that for any associative ring Λ and monoid A, there is a natural equivalence of cyclotomic spectra,

$$TH(\Lambda) \wedge |N^{cy}(A)| \xrightarrow{\simeq} TH(\Lambda[A]),$$

where TH is topological Hochschild homology. Combining this with Theorem 6.2, we showed in [9, Corollary 6.6] that the dilated topological cyclic homology $X \mapsto TC^{\nu}(X_k; p)^{\mathfrak{c}}$ satisfies cdh-descent on $\mathcal{M}_{\text{pctf}}$. Here is a more general statement.

Theorem 6.3 ([9, 6.6]) *Let $\mathfrak{c} = (c_1, c_2, \dots)$ be a sequence of integers with $c_i \geq 2$ for all i. For any associative ring Λ and integer $\nu \geq 1$, the spectrum-valued functor*

$$X \mapsto TC^{\nu}(X_{\Lambda}; p)^{\mathfrak{c}}$$

satisfies cdh descent on $\mathcal{M}_{pctf}$.

Proof The proofs of Theorem 6.5 and Corollary 6.6 in [9] go through with $TH(k)$ replaced by $TH(\Lambda)$. □

7 Descent for Hochschild and Cyclic Homology

We write $HH(\Lambda)$ for the Eilenberg-MacLane spectrum associated to the absolute Hochschild homology of a ring Λ (i.e., relative to the base ring $\mathbb{Z}$) and similarly for cyclic and negative cyclic homology. Since the Hochschild complex of Λ is a cyclic complex, $HH(\Lambda)$ is an $\mathbb{S}^1$-spectrum.

Remark 7.1 The Hochschild homology we are using in this section is the classical version defined by the bar complex of an algebra. There is also a derived version (see [14, IV]), and this is the one that coincides with the Hochschild homology of dg categories used in the rest of this paper. The results of this section we need going forward (Corollaries 7.3 and 7.4) deal with $\mathbb{Q}$-algebras; for such algebras, the derived and classical definitions agree.

Theorem 7.2 *Let $\mathfrak{c} = (c_1, c_2, \dots)$ be a sequence of integers with each $c_i \geq 2$. For any associative ring Λ, the spectrum-valued functors*

$$X \mapsto HH(X_{\Lambda})^{\mathfrak{c}} \quad \textit{and} \quad X \mapsto HC(X_{\Lambda})^{\mathfrak{c}}$$

satisfy cdh descent on $\mathcal{M}_{pctf}$.

Proof By Theorem 6.2 with $T = HH(\Lambda)$ and $r = 1$, $\mathbb{H}_{\text{zar}}(-, HH(\Lambda) \wedge \widetilde{\Omega})$ satisfies cdh descent on $\mathcal{M}_{\text{pctf}}$. Since $|N^{cy}|^{\mathfrak{c}} \xrightarrow{\simeq} |\widetilde{\Omega}|^{\mathfrak{c}}$, we see that $\mathbb{H}_{\text{zar}}(-, HH(\Lambda) \wedge |N^{cy}|)^{\mathfrak{c}}$ also satisfies cdh descent. This is equivalent to $HH(-_{\Lambda})^{\mathfrak{c}} \simeq \mathbb{H}_{\text{zar}}(-, HH(-_{\Lambda}))^{\mathfrak{c}}$ because there is a natural weak equivalence of spectra

$$HH(\Lambda) \wedge |N^{cy}(A)| \simeq HH(\Lambda[A]).$$

The HH assertion now follows. The HC assertion follows from this, together with the SBI sequence connecting cyclic homology and Hochschild homology (see [36, 9.6.11]). In more detail, let $\mathbb{HH}$ and $\mathbb{H}$ denote the presheaves $\mathbb{HH}(X) = \mathbb{H}_{\mathrm{cdh}}(X, HH(-_\Lambda)^{\mathfrak{c}})$ and $\mathbb{H}(X) = \mathbb{H}_{\mathrm{cdh}}(X, HC(-_\Lambda)^{\mathfrak{c}})$, and abbreviate $HC_n(X_\Lambda)^{\mathfrak{c}}$ as $HC_n^{\mathfrak{c}}$. By induction on n using the diagram with exact rows

$$\begin{array}{ccccccccc}
HC^{\mathfrak{c}}_{n-1} & \longrightarrow & HH_n(X_\Lambda)^{\mathfrak{c}} & \longrightarrow & HC^{\mathfrak{c}}_n & \longrightarrow & HC^{\mathfrak{c}}_{n-2} & \longrightarrow & HH_{n-1}(X_\Lambda)^{\mathfrak{c}} \\
\downarrow{\cong} & & \downarrow{\cong} & & \downarrow & & \downarrow{\cong} & & \downarrow{\cong} \\
\pi_{n-1}\mathbb{H}(X) & \longrightarrow & \pi_n\mathbb{HH}(X) & \longrightarrow & \pi_n\mathbb{H}(X) & \longrightarrow & \pi_{n-2}\mathbb{H}(X) & \longrightarrow & \pi_{n-1}\mathbb{HH}(X),
\end{array}$$

we see that the middle map $\pi_n HC(X_\Lambda)^{\mathfrak{c}} \to \pi_n\mathbb{H}(X)$ is an isomorphism for all n. Hence $HC(X_\Lambda)^{\mathfrak{c}} \to \mathbb{H}(X)$ is a homotopy equivalence for all X. □

Corollary 7.3 *Let $\mathfrak{c} = (c_1, c_2, \dots)$ be a sequence of integers with each $c_i \geq 2$. For any associative $\mathbb{Q}$-algebra Λ, the spectrum-valued functor*

$$X \mapsto HN(X_\Lambda)^{\mathfrak{c}}$$

satisfies cdh descent on $\mathscr{M}_{\mathrm{pctf}}$.

Proof As in the proof of Corollary 3.13 in [7], the un-dilated presheaf HP satisfies weak cdh descent on Sch/k. By Theorem 4.5, $HP(-_\Lambda)$ satisfies cdh descent on $\mathscr{M}_{\mathrm{pctf}}$. Therefore the dilated presheaf $X \mapsto HP(X_\Lambda)^{\mathfrak{c}}$ also satisfies cdh descent on $\mathscr{M}_{\mathrm{pctf}}$. Using the SBI sequence for HN and HP, together with Theorem 7.2, it follows that $X \mapsto HN(X_\Lambda)^{\mathfrak{c}}$ satisfies cdh descent on $\mathscr{M}_{\mathrm{pctf}}$. □

Corollary 7.4 *Let k_0 be commutative regular ring of finite Krull dimension all of whose residue fields are infinite, and let Λ be a flat K-regular k_0-algebra. Then*

$$K(X_\Lambda)^{\mathfrak{c}} \otimes \mathbb{Q} \xrightarrow{\simeq} KH(X_\Lambda)^{\mathfrak{c}} \otimes \mathbb{Q}.$$

Proof Combine Proposition 5.7 and Corollary 7.3 (the latter applied using the $\mathbb{Q}$-algebra $\Lambda \otimes \mathbb{Q}$). □

8 Main Theorem

Lemma 8.1 *Let X be a monoid scheme and $\mathfrak{c} = (c_1, c_2, \dots)$ a sequence of integers with $c_i \geq 2$ for all i. Let k be a commutative Noetherian ring of finite Krull* *dimension, and let* Λ *be any associative* k*-algebra. Suppose that for every prime ideal* $\wp$ *of* k*, the natural map*

$$\phi_{k_\wp} : \mathcal{K}^c(X_{\Lambda_\wp}) \longrightarrow \mathcal{K}H^c(X_{\Lambda_\wp})$$

is a weak equivalence. Then ϕ_k *is a weak equivalence.*

Proof Write $\mathcal{F}$ for the presheaf on $S = \mathrm{Spec}(k)$ sending the open $U \subset S$ to the homotopy fiber of $\phi_{\mathcal{O}(U)}$. The hypothesis implies that the Zariski sheaves $a_{\mathrm{zar}}\pi_q\mathcal{F}$ associated to $\pi_q\mathcal{F}$ vanish on S. Because the source and target, viewed as functors of U, both satisfy Zariski descent [32, 8.1], so does $\mathcal{F}$. Since k is Noetherian and of finite Krull dimension, there is a spectral sequence converging to $\pi_*\mathcal{F}(k)$ with $E_2^{p,q} = H^p_{\mathrm{zar}}(\mathrm{Spec}\, k, a_{\mathrm{zar}}\pi_{-q}\mathcal{F}) = 0$; see [37, Theorem V.10.11]. Hence $\mathcal{F}(k) \simeq *$ and ϕ_k is a weak homotopy equivalence. □

Remark 8.2 An argument similar to that in the proof of Lemma 8.1 shows that if k is Noetherian of finite Krull dimension and $\phi_{k^h_\wp}$ is a weak equivalence for the henselization of every local ring of k, then ϕ_k is a weak equivalence. One simply has to substitute the Nisnevich for the Zariski topology and use [37, Remark V.10.11.1].

Lemma 8.3 *Let* $(R, \mathfrak{m})$ *be a commutative local ring with finite residue field. For each integer* $l > 0$*, there is a finite étale extension* $R \to R'$ *of rank* l *with* $(R', \mathfrak{m}R')$ *local.*

Proof Let $\mathfrak{m}$ be the maximal ideal of R and set $k = R/\mathfrak{m}$. Because k is finite, there is separable field extension k'/k of degree l. Pick a primitive element $\alpha \in k'$ for this field extension, so that $k' = k(\alpha)$, and let $\bar{f} \in k[x]$ be the minimum polynomial of α. Let $f \in R[x]$ be a monic lift of $\bar{f}$ and set $R' = R[x]/f$. Then $R \to R'$ is a finite flat extension of rank l and R' is local because $R'/\mathfrak{m}R' = k'$ is a field. Since the field extension is separable, $\bar{f}'(\alpha) \neq 0$ and hence $f'(x)$ is a unit in R'. This proves $R \to R'$ is an étale extension. □

Recall from the introduction that an associative ring Γ is (right) regular if it is (right) Noetherian and every finitely generated (right) module has finite projective dimension.

Lemma 8.4 *Let* Γ *be an associative algebra over a commutative ring* k *and* $f \in k[x]$ *a monic polynomial such that the derivative* df/dx *is invertible in* $k' = k[x]/\langle f\rangle$*. If* Γ *is regular, then so is* $\Gamma' = \Gamma \otimes_k k'$*.*

Proof First of all, Γ' is Noetherian since it is finite as a module over Γ, which is Noetherian by assumption. We must show every finitely generated right Γ'-module M has finite projective dimension. This is clear for M of the form $N \otimes_k k'$ for some finitely generated Γ-module N, since $k \to k'$ is flat and we are assuming that Γ is regular. If M is any Γ'-module, write $M' = M \otimes_k k'$; the multiplication map $\mu : M' \to M$ is a surjective homomorphism of Γ'-modules. On the other hand, because k' is a separable k-algebra, there exists an idempotent $e = \sum_i x_i \otimes y_i \in k' \otimes_k k'$ such that $\sum_i x_i y_i = 1 \in k'$ and such that $(x \otimes 1 - 1 \otimes x)e = 0$ for all $x \in k'$. One checks that the map $s : M \to M'$, $m \mapsto me$ is Γ'-linear and that $\mu s = 1_M$. Thus M is a direct summand of M', and a direct summand of a module of finite projective dimension is itself of finite projective dimension. □

We are now ready to prove our main theorem.

Theorem 8.5 *Let k_0 be a commutative regular ring of finite Krull dimension, Λ a flat (not necessarily commutative) k_0-algebra, and assume that one of the following three conditions holds:*

1. *Λ is regular,*
2. *Λ is K-regular and all of the residue fields of k_0 are infinite, or*
3. *Λ is K-regular and for every prime $\mathfrak{p} \subset k_0$ whose residue field is finite and every étale $(k_0)_\mathfrak{p}$-algebra k', the ring $\Lambda \otimes_{k_0} k'$ is K-regular.*

Let X be a pctf monoid scheme of finite type, and $\mathfrak{c} = (c_1, c_2, \dots)$ any sequence of integers with $c_i \geq 2$ for all i. Then the natural map

$$K^\mathfrak{c}(X_\Lambda) \to KH^\mathfrak{c}(X_\Lambda)$$

is a weak equivalence.

Proof Both regularity and K-regularity localize under central nonzero divisors [37, Lemma V.8.5]. It follows from this and from Lemma 8.1 that we may assume that k_0 is local.

Suppose first that the residue fields of k_0 are infinite. Let $\mathscr{F}_K$ be the homotopy fiber of the map $K \to KH$. We must show that the homotopy groups of $\mathscr{F}_K^\mathfrak{c}(X_\Lambda)$ vanish. By Corollary 7.4 we have $\pi_*(\mathscr{F}_K^\mathfrak{c}(X_\Lambda)) \otimes \mathbb{Q} = 0$. Hence the groups $\pi_*(\mathscr{F}_K^\mathfrak{c}(X_\Lambda))$ are torsion. We will be done if we show that multiplication by any prime p induces an isomorphism on $\pi_*(\mathscr{F}_K^\mathfrak{c}(X_\Lambda))$, or equivalently, that $\pi_*(\mathscr{F}_K^\mathfrak{c}(X_\Lambda)/p) = 0$. This follows from Proposition 5.10 and Theorem 6.3, using [9, Lemma 8.2], as in the proof of [9, Theorem 8.3]. This proves the result if (2) holds.

Now assume that the residue field $k_0/\mathfrak{m}$ of k_0 is finite and that either (1) or (3) hold. Let $\mathscr{F}^\mathfrak{c}(k_0)$ be the fiber of the map $K^\mathfrak{c}(X_\Lambda) \to KH^\mathfrak{c}(X_\Lambda)$. We must show that the homotopy groups of $\mathscr{F}^\mathfrak{c}(k_0)$ vanish. Let l be a prime number. By Lemma 8.3 there exists a tower of finite, étale extensions $k_0 \subset k_1 \subset \cdots$ of the form $k_{i+1} = k_i[x]/\langle f_i \rangle$ with df_i/dx invertible in k_{i+1}, such that $(k_n, \mathfrak{m}k_n)$ is local and has rank l^n over k_0, for all n.

Set $k' = \bigcup_n k_n$; it is Noetherian by [16, 0_{III} 10.3.1.3]. Then $(k', \mathfrak{m}k')$ is regular local, with infinite residue field, and $k' \subseteq \Lambda \otimes_{k_0} k'$ is flat. If (3) holds, then $\Lambda \otimes_{k_0} k'$ is K-regular by hypothesis. If (1) holds, then by Lemma 8.4, $\Lambda_n = \Lambda \otimes_{k_0} k_n$ is regular and thus K-regular for each n, whence again we conclude that $\Lambda \otimes_{k_0} k'$ is K-regular. Hence

$$0 = \pi_* \mathscr{F}^c(k') = \operatorname*{colim}_n \pi_* \mathscr{F}^c(k_n).$$

Since $\Lambda \to \Lambda \otimes_{k_0} k_n$ is finite and flat, there is a natural transfer map $K(X_{\Lambda \otimes_{k_0} k_n}) \to K(X_\Lambda)$ such that the composition $K(X_\Lambda) \to K(X_{\Lambda \otimes_{k_0} k_n}) \to K(X_\Lambda)$ induces multiplication by l^n on homotopy groups, for all n. By naturality, K^c also admits such a transfer map. Likewise, KH-theory and hence KH^c-theory have such transfer maps and they are compatible with the map $K^c \to KH^c$. We obtain a map $\mathscr{F}^c(k_n) \to \mathscr{F}^c(k_0)$ such that the composition $\mathscr{F}^c(k_0) \to \mathscr{F}^c(k_n) \to \mathscr{F}^c(k_0)$ induces multiplication by l^n on homotopy groups. It follows that the kernel of $\pi_* \mathscr{F}^c(k_0) \to \pi_* \mathscr{F}^c(k')$ is an l-primary torsion group. Since this occurs for every prime l, we must have $\pi_*(\mathscr{F}^c(k_0)) = 0$. □

Theorem 8.5 implies Theorem 1.2:

Proof of Theorem 1.2 Cases (c) and (e) of Theorem 1.2 are cases (1) and (3) of Theorem 8.5, and case (d) is a special case of (e). If k is a regular commutative ring and Spec(k) is connected, then either k is flat over a field k_0 or it is flat over the ring of integers. Hence it satisfies the hypothesis of Theorem 8.5. If k is an arbitrary commutative regular ring, then it is a finite product of regular rings with connected Spec. Thus case (a) of Theorem 1.2 is proved.

If Λ is a commutative C^*-algebra, then it is flat over the field $k_0 = \mathbb{C}$ and satisfies the K-regularity hypothesis by [5, Theorem 8.1]. Thus case (b) of Theorem 1.2 also follows from Theorem 8.5. □

Remark 8.7 If in Theorem 8.5 we assume that Λ is commutative Noetherian and K-regular then every étale extension of Λ is K-regular by van der Kallen's theorem [33, Theorem 3.2]. In particular the assumption that $\Lambda \otimes_{k_0} k'$ is K-regular for every étale k_0-algebra k' is superfluous in this case.

Remark 8.8 Theorems 1.1 and 1.2 for commutative C^*-algebras can alternatively be derived from Gubeladze's main result of [20] using [5, Theorem 7.7].

Acknowledgements This article is part of a collection of papers published in honour of Antonio Campillo's 65th birthday. The first named author is very grateful to him for all his help over many years; the authors dedicate this paper to him.

References

1. Bierstone, E., Milman, P.: Desingularization of toric and binomial varieties. J. Alg. Geom. **15**, 443–486 (2006)
2. Blumberg, A., Mandell, M.: Localization theorems in topological Hochschild homology and topological cyclic homology. Geom. Topol. **16**, 1053–1120 (2012)

3. Cisinski, D.-C., Tabuada, G.: Symmetric monoidal structure on non-commutative motives. J. K-Theory **9**(2), 201–268 (2012)
4. Cortiñas, G.: The obstruction to excision in K-theory and cyclic homology. Invent. Math. **164**, 143–173 (2006)
5. Cortiñas, G., Thom, A.: Algebraic geometry of topological spaces I. Acta Math. **209**, 83–131 (2012)
6. Cortiñas, G., Rodríguez Cirone, E.: Singular coefficients in the K-theoretic Farrell – Jones conjecture. Algebr. Geom. Topol. **16**, 129–147 (2016)
7. Cortiñas, G., Haesemeyer, C., Schlichting, M., Weibel, C.: Cyclic homology, cdh-cohomology and negative K-theory. Ann. Math. **167**, 549–563 (2008)
8. Cortiñas, G., Haesemeyer, C., Walker, M., Weibel, C.: The K–theory of toric varieties. Trans. AMS **361**, 3325–3341 (2009)
9. Cortiñas, G., Haesemeyer, C., Walker, M., Weibel, C.: The K-theory of toric varieties in positive characteristic. J. Topol. **7**, 247–286 (2014)
10. Cortiñas, G., Haesemeyer, C., Walker, M., Weibel, C.: Toric varieties, monoid schemes and cdh descent. J. Reine Angew. Math. **698**, 1–54 (2015)
11. Geisser, T., Hesselholt, L.: On the K-theory and topological cyclic homology of smooth schemes over a discrete valuation ring. Trans. AMS **358**, 131–145 (2006)
12. Geisser, T., Hesselholt, L.: Bi-relative algebraic K-theory and topological cyclic homology. Invent. Math. **166**, 359–395 (2006)
13. Geisser, T., Hesselholt, L.: On the vanishing of negative K-groups. Math. Ann. **348**, 707–736 (2010)
14. Goodwillie, T.: Cyclic homology, derivations, and the free loopspace. Topology **24**, 187–215 (1985)
15. Goodwillie, T.: Relative algebraic K-theory and cyclic homology. Ann. Math. **124**, 347–402 (1986)
16. Grothendieck, A.: Élements de Géométrie Algébrique, III. Étude cohomologique des faisceaux cohérents (première partie). Publications Mathématiques, vol. 11, 167pp. IHES, Paris (1961)
17. Gubeladze, J.: Geometric and algebraic representations of commutative cancellative monoids. Proc. A. Razmadze Math. Inst. **113**, 31–81 (1995)
18. Gubeladze, J.: Higher K-theory of toric varieties. K-Theory **28**, 285–327 (2003)
19. Gubeladze, J.: The nilpotence conjecture in K-theory of toric varieties. Inventiones Math. **160**, 173–216 (2005)
20. Gubeladze, J.: Global coefficient ring in the nilpotence conjecture. Trans. AMS **136**, 499–503 (2008)
21. Hesselholt, L., Madsen, I.: On the K-theory of finite algebras over Witt vectors of perfect fields. Topology **36**, 29–101 (1997)
22. Hochster, M.: Rings of invariants of tori, Cohen-Macaulay rings generated by monomials, and polytopes. Ann. Math. **96**, 318–337 (1972)
23. Jardine, J.F.: Simplicial presheaves. J. Pure Appl. Algebra **47**, 35–87 (1987)
24. Jardine, J.F.: Generalized étale cohomology theories. Birkhäuser Verlag, Basel (1997)
25. Keller, B.: On differential graded categories. In: Proceedings of Barcelona ICM, vol. II, pp. 151–190. European Mathematical Society, Zürich (2006)
26. McCarthy, R.: The cyclic homology of an exact category. J. Pure Appl. Algebra **93**(3), 251–296 (1994)
27. McCarthy, R.: Relative algebraic K-theory and topological cyclic homology. Acta Math. **179**, 197–222 (1997)
28. Rodríguez Cirone, E.: A strictly-functorial and small dg-enhancement of the derived category of perfect complexes. Preprint, arXiv 1502.06573 (2015)
29. Swan, R.G.: Projective modules over Laurent polynomial rings. Trans. AMS **237**, 111–120 (1978)
30. Swan, R.G.: Néron-Popescu desingularization. In: Algebra and Geometry (Taipei, 1995). Lectures on Algebraic Geometry, vol. 2, pp. 135–192. International Press, Cambridge (1998)
31. Tabuada, G.: E_n-regularity implies E_{n-1}-regularity. Documenta Math. **19**, 121–139 (2014)

32. Thomason, R.W., Trobaugh, T: Higher algebraic K-theory of schemes and of derived categories. In: The Grothendieck Festschrift, vol. III. Progress in Mathematics, vol. 88, pp. 247–435. Birkhäuser, Boston (1990)
33. van der Kallen, W.: Descent for the K-theory of polynomial rings. Math. Z. **191**(3), 405–415 (1986)
34. Voevodsky, V.: Homotopy theory of simplicial sheaves in completely decomposable topologies. J. Pure Appl. Algebra **214**, 1384–1398 (2010)
35. Weibel, C.A.: Homotopy algebraic K-theory. In: Algebraic K-Theory and Algebraic Number Theory (Honolulu 1987). Contemporary Mathematics, vol. 83, pp. 461–488. AMS, Providence (1989)
36. Weibel, C.A.: An Introduction to Homological Algebra. Cambridge University Press, Cambridge (1994)
37. Weibel, C.A.: The K-Boook. An Introduction to Algebraic K-Theory. Graduate Studies in Mathematics, vol. 145. AMS, Providence (2013)

On Finite and Nonfinite Generation of Associated Graded Rings of Abhyankar Valuations

Steven Dale Cutkosky

Dedicated to Professor Antonio Campillo on the occasion of his 65th birthday

Abstract We consider the condition that the associated graded ring of a local ring along an Abhyankar valuation is finitely generated. We characterize when this holds for regular local rings of domension two, and give some examples of maximal rank Abhyankar valuations for which the associated graded ring is not finitely generated on two dimensional normal local rings and regular local rings of dimension larger than two. We characterize two dimensional normal local rings for which all divisorial valuations which dominate the ring have a finitely generated associated graded ring.

1 Introduction

Suppose that R is a Noetherian local domain with maximal ideal m_R which is dominated by a valuation ν of the quotient field of R. Let Φ_ν be the value group of ν and V_ν be the valuation ring of ν, with maximal ideal m_ν. The condition that ν dominates R is that $R \subset V_\nu$ and $R \cap m_\nu = m_R$. The semigroup of ν on R is

$$S^R(\nu) = \{\nu(f) \mid 0 \neq f \in R\}.$$

The associated graded ring of R along ν is

$$\mathrm{gr}_\nu(R) = \bigoplus_{\gamma \in \Phi_\nu} \mathcal{P}_\gamma(R)/\mathcal{P}_\gamma^+(R) = \bigoplus_{\gamma \in S^R(\nu)} \mathcal{P}_\gamma(R)/\mathcal{P}_\gamma^+(R) \tag{1}$$

S. D. Cutkosky (✉)
Department of Mathematics, University of Missouri, Columbia, MO, USA
e-mail: cutkoskys@missouri.edu

G.-M. Greuel et al. (eds.), *Singularities, Algebraic Geometry, Commutative Algebra, and Related Topics*, https://doi.org/10.1007/978-3-319-96827-8_20

which is defined by Teissier in [9]. Here

$$\mathscr{P}_\gamma(R) = \{f \in R \mid \nu(f) \geq \gamma\} \text{ and } \mathscr{P}^+_\gamma(R) = \{f \in R \mid \nu(f) > \gamma\}$$

are valuation ideals in R. This ring plays an important role in local uniformization of singularities [9] and [10].

A necessary condition for $\mathrm{gr}_\nu(R)$ to be a finitely generated R/m_R-algebra is that $S^\nu(R)$ be a finitely generated semigroup. Since the group generated by $S^\nu(R)$ is the value group Φ_ν, we have that a necessary condition for $\mathrm{gr}_\nu(R)$ to be a finitely generated R/m_R-algebra is that Φ_ν be a finitely generated group. In the case that the residue field of the valuation ring is equal to the residue field of R, then $\mathrm{gr}_\nu(R)$ is a finitely generated R/m_R-algebra if and only if the semigroup $S^R(\nu)$ is a finitely generated semigroup (Remark 2.1).

We have seen that we can only have finite generation of $\mathrm{gr}_\nu(R)$ as an R/m_R-algebra if Φ_ν is a finitely generated group, so we will restrict to this case, namely that Φ_ν is a finitely generated group.

When ν is a divisorial valuation dominating R, the valuation group $\Phi_\nu \cong \mathbb{Z}$, so both Φ_ν and $S^\nu(R)$ are finitely generated, and V_ν is a finitely generated transcendental extension of R/m_R, there is not a simple obstruction to finite generation of $\mathrm{gr}_\nu(R)$. However, in this situation, there is a very clear picture of when finite generation of such $\mathrm{gr}_\nu(R)$ always holds or always does not hold.

If R is a regular local ring of dimension two and ν is a divisorial valuation of the quotient field that dominates R, then $\mathrm{gr}_\nu(R)$ is a finitely generated R/m_R-algebra (Corollary 3.11 [5], Proposition 3.11 [6] or Theorem 8.6 [8]). For normal two dimensional complete local rings, we have the following necessary and sufficient condition, which we deduce from Theorem 4 [3].

Theorem 1.1 *Suppose that R is a complete normal local domain of dimension two. Then $\mathrm{gr}_\nu(R)$ is a finitely generated R/m_R-algebra for all divisorial valuations ν which dominate R if and only if the divisor class group $\mathrm{Cl}(R)$ of R is a torsion group.*

In [2], an example is given to show that in regular local rings R of dimension ≥ 3, there exists a divisorial valuation ν which dominates R such that $\mathrm{gr}_\nu(R)$ is not a finitely generated R/m_R-algebra.

We show that for regular local rings of dimension two, if the valuation group Φ_ν is finitely generated and the residue field of ν is a finitely generated field extension of R/m_R, we have that $\mathrm{gr}_\nu(R)$ is a finitely generated R/m_R-algebra.

Proposition 1.2 *Suppose that R is a two dimensional regular local ring and ν is a valuation dominating R such that Φ_ν is a finitely generated group and V_ν/m_ν is a finitely generated extension field of R/m_R. Then $\mathrm{gr}_\nu(R)$ is a finitely generated R/m_R-algebra.*

If R is a two dimensional regular local ring, such that the algebraic closure of the residue field R/m_R is not finite over R/m_R, then examples are given in [7], showing that there exist discrete valuations ν dominating R such that V_ν/m_ν is an algebraic but not finite extension of R/m_R and $\mathrm{gr}_\nu(R)$ is not a finitely generated R/m_R-algebra.

However, we have the following general statement when the residue field of R is algebraically closed.

Proposition 1.3 *Suppose that R is a two dimensional regular local ring with algebraically closed residue field and ν is a valuation dominating R. Then $\mathrm{gr}_\nu(R)$ is a finitely generated R/m_R-algebra if and only if Φ_ν is a finitely generated group.*

A fundamental inequality for a valuation ν dominating a Noetherian local ring is Abhyankar's inequality ([1] or Appendix 2, [11]):

$$\dim R \geq \mathrm{trdeg}_{R/m_R} V_\nu/m_\nu + \text{ rat rank } \nu. \tag{2}$$

The rational rank, rat rank ν, of ν is the dimension of $\Phi_\nu \otimes_{\mathbb{Z}} \mathbb{Q}$ as a rational vector space, which is greater than or equal to the rank of ν, which is the Krull dimension of V_ν. If equality holds in (2), then $\Phi_\nu \cong \mathbb{Z}^r$ as an (unordered) group, where $r =$ rat rank ν and V_ν/m_ν is a finitely generated extension field of R/m_R. Such a valuation, where we have equality in Abhyankar's inequality, is called an Abhyankar valuation. A divisorial valuation which dominates R is a particular example of a (rational rank 1) Abhyankar valuation.

If R is a two dimensional regular local ring and ν is an Abhyankar valuation which dominates R, then we always have finite generation.

Proposition 1.4 *Suppose that R is a two dimensional regular local ring and ν is an Abhyankar valuation which dominates R. Then $\mathrm{gr}_\nu(R)$ is a finitely generated R/m_R-algebra.*

Proposition 1.4 is an immediate consequence of Proposition 1.2.

The conclusions of Proposition 1.4 do not extend to most normal local rings of dimension two or to regular local rings of dimension ≥ 3. This already follows from our analysis of divisorial valuations given above, as a divisorial valuation which dominates R is an Abhyankar valuation. Divisorial valuations have minimal rank 1 (their value group is $\mathbb{Z}$).

We give some examples of Abhyankar valuations of maximal rank dominating some normal local rings of dimension two and some regular local rings of arbitrary dimension ≥ 3 such that $\mathrm{gr}_\nu(R)$ is not a finitely generated R/m_R-algebra, in Examples 3.1 and 3.2.

It is however known that if we blow up enough along an Abhyankar valuation, then we always obtain finite generation, as has been shown by Teissier [10].

Theorem 1.5 (Theorem 7.21 [10]) *Suppose that R is a complete equicharacteristic Noetherian local ring with algebraically closed residue field, and ν is a valuation dominating R. Then there exists a birational extension $R \to R'$ of local rings such that R' is dominated by ν and $\mathrm{gr}_\nu(R')$ is a finitely generated $R'/m_{R'}$-algebra if and only if ν is an Abhyankar valuation.*

The conclusions of the theorem require the two assumptions that R is complete and that R/m_R is algebraically closed. Theorem 1.5 is not difficult to prove in the case that embedded resolution of singularities is known to hold, but it is much more difficult to prove in the case where embedded resolution of singularities is not known to exist.

2 Proofs of the Finite Generation Results

Remark 2.1 Suppose that $V_\nu/m_\nu = R/m_R$. Then $\mathrm{gr}_\nu(R)$ is isomorphic as a graded R/m_R-algebra to the group algebra $R/m_R[t^\gamma \mid \gamma \in S^R(\nu)]$. Thus (with the assumption that $V_\nu/m_\nu = R/m_R$) we have that $\mathrm{gr}_\nu(R)$ is a finitely generated R/m_R-algebra if and only if $S^\nu(R)$ is a finitely generated semigroup.

Proof of Theorem 1.1 Let ν be a divisorial valuation which dominates R, and let

$$I_n(\nu) = \{f \in R \mid \nu(f) \geq n\}.$$

Then $\mathrm{gr}_\nu(R) = \bigoplus_{n\geq 0} I_n(\nu)/I_{n+1}(\nu)$. By Theorem 4 [3], we have that $\mathrm{Cl}(R)$ is a torsion group if and only if $\bigoplus_{n\geq 0} I_n(\nu)$ is a finitely generated R-algebra for all divisorial valuations ν dominating R. It remains to show that for all divisorial valuations ν dominating R,

$$\begin{array}{l}\bigoplus_{n\geq 0} I_n(\nu) \text{ is a finitely generated } R\text{-algebra if and only if}\\ \mathrm{gr}_\nu(R) \text{ is a finitely generated } R/m_R\text{-algebra.}\end{array} \tag{3}$$

The "if" direction of (1) is immediate. We will prove the "only if" direction. Suppose that $\mathrm{gr}_\nu(R)$ is a finitely generated R/m_R-algebra. Then there exists a positive integer r and $\sigma_i \in \mathbb{N}$ for $1 \leq i \leq r$ such that $F_i \in I_{\sigma_i} \setminus I_{\sigma_i+1}$ for $1 \leq i \leq r$ and the residues $\overline{F}_i \in I_{\sigma_i}(\nu)/I_{\sigma_i+1}(\nu)$ generate $\bigoplus_{n\geq 0} I_n(\nu)/I_{n+1}(\nu)$ as an R/m_R-algebra. Let J_n be the ideal in R

$$J_n = (F_1^{i_1} \cdots F_r^{i_r} \mid i_1\sigma_1 + \cdots + i_r\sigma_r \geq n).$$

We will show that $J_n = I_n(\nu)$ for all n. Suppose that $n \in \mathbb{Z}_+$ and $h \in I_n(\nu)$. Then there exists $s_n \in \mathbb{N}$ and $i_1^k(n), \ldots, i_r^k(n) \in \mathbb{N}$ and $r^n_{i_1^k(n),\ldots,i_r^k(n)} \in R$ for $1 \leq k \leq s_n$ such that

$$i_1^k(n)\sigma_1 + \cdots + i_r^k(n)\sigma_r = n$$

for all k, and

$$h - \sum_{k=1}^{s_n} r^n_{i_1^k(n),\ldots,i_r^k(n)} F_1^{i_1^k(n)} \cdots F_r^{i_r^k(n)} \in I_{n+1}(\nu).$$

Iterating, we construct a Cauchy sequence in $I_n(\nu)$, which converges to h,

$$h = \sum_{j=n}^{\infty} \sum_{k=1}^{s_j} r^j_{i_1^k(j),\ldots,i_r^k(j)} F_1^{i_1^k(j)} \cdots F_r^{i_r^k(j)}.$$

Thus $I_n(\nu) \subset J_n + m_R I_n(\nu)$ as $F_1^{i_1} \cdots F_r^{i_r} \in m_R I_n(\nu)$ if

$$i_1\sigma_1 + \cdots + i_r\sigma_r > n + \max\{\sigma_1, \ldots, \sigma_r\}.$$

By Nakayama's Lemma, $I_n(\nu) = J_n$, and so $\bigoplus_{n \geq 0} I_n(\nu)$ is generated as an R-algebra by the classes of $F_1, \ldots, F_r$.

Proof of Proposition 1.2 We begin by stating a result on the construction of generating sequences in a two dimensional regular local ring from [4]. This theorem is a generalization of the algorithm to construct generating sequences in regular local rings of dimension two with algebraically closed residue fields in [8]. If $t_1, \ldots, t_r$ are in a group H, then $G(t_1, \ldots, t_r)$ will denote the subgroup generated by $t_1, \ldots, t_r$.

Theorem 2.2 (Theorem 4.2 [4]) *Suppose that R is a regular local ring of dimension two, with maximal ideal m_R and residue field $\mathfrak{k} = R/m_R$. For $f \in R$, let $\overline{f}$ or $[f]$ denote the residue of f in $\mathfrak{k}$. Suppose that CS is a coefficient set of R. Suppose that ν is a valuation of the quotient field of R dominating R. Let $L = V_\nu/m_\nu$ be the residue field of the valuation ring V_ν of ν. For $f \in V_\nu$, let $[f]$ denote the class of f in L. Suppose that x, y are regular parameters in R. Then there exist $\Omega \in \mathbb{Z}_+ \cup \{\infty\}$ and $P_i \in m_R$ for $i \in \mathbb{Z}_+$ with $i < \min\{\Omega + 1, \infty\}$ such that $P_0 = x$, $P_1 = y$ and for $1 \leq i < \Omega$, there is an expression*

$$P_{i+1} = P_i^{n_i} + \sum_{k=1}^{\lambda_i} c_k P_0^{\sigma_{i,0}(k)} P_1^{\sigma_{i,1}(k)} \cdots P_i^{\sigma_{i,i}(k)} \tag{4}$$

with $n_i \geq 1$, $\lambda_i \geq 1$,

$$0 \neq c_k \in CS \tag{5}$$

for $1 \leq k \leq \lambda_i$, $\sigma_{i,s}(k) \in \mathbb{N}$ for all s, k, $0 \leq \sigma_{i,s}(k) < n_s$ for $s \geq 1$. Further,

$$n_i \nu(P_i) = \nu(P_0^{\sigma_{i,0}(k)} P_1^{\sigma_{i,1}(k)} \cdots P_i^{\sigma_{i,i}(k)})$$

for all k.

For all $i \in \mathbb{Z}_+$ with $i < \Omega$, the following are true:

(1) *$\nu(P_{i+1}) > n_i \nu(P_i)$.*
(2) *Suppose that $r \in \mathbb{N}$, $m \in \mathbb{Z}_+$, $j_k(l) \in \mathbb{N}$ for $1 \leq l \leq m$ and $0 \leq j_k(l) < n_k$ for $1 \leq k \leq r$ are such that $(j_0(l), j_1(l), \ldots, j_r(l))$ are distinct for $1 \leq l \leq m$, and*

$$\nu(P_0^{j_0(l)} P_1^{j_1(l)} \cdots P_r^{j_r(l)}) = \nu(P_0^{j_0(1)} \cdots P_r^{j_r(1)})$$

for $1 \le l \le m$. *Then*

$$1, \left[\frac{P_0^{j_0(2)} P_1^{j_1(2)} \cdots P_r^{j_r(2)}}{P_0^{j_0(1)} P_1^{j_1(1)} \cdots P_r^{j_r(1)}}\right], \ldots, \left[\frac{P_0^{j_0(m)} P_1^{j_1(m)} \cdots P_r^{j_r(m)}}{P_0^{j_0(1)} P_1^{j_1(1)} \cdots P_r^{j_r(1)}}\right]$$

are linearly independent over $\mathfrak{k}$.

(3) *Let*

$$\overline{n}_i = [G(\nu(P_0), \ldots, \nu(P_i)) : G(\nu(P_0), \ldots, \nu(P_{i-1}))].$$

Then $\overline{n}_i$ *divides* $\sigma_{i,i}(k)$ *for all* k *in (4). In particular,* $n_i = \overline{n}_i d_i$ *with* $d_i \in \mathbb{Z}_+$

(4) *There exists* $U_i = P_0^{w_0(i)} P_1^{w_1(i)} \cdots P_{i-1}^{w_{i-1}(i)}$ *for* $i \ge 1$ *with* $w_0(i), \ldots, w_{i-1}(i) \in \mathbb{N}$ *and* $0 \le w_j(i) < n_j$ *for* $1 \le j \le i-1$ *such that* $\nu(P_i^{\overline{n}_i}) = \nu(U_i)$ *and if*

$$\alpha_i = \left[\frac{P_i^{\overline{n}_i}}{U_i}\right]$$

then

$$b_{i,t} = \left[\sum_{\sigma_{i,i}(k) = t\overline{n}_i} c_k \frac{P_0^{\sigma_{i,0}(k)} P_1^{\sigma_{i,1}(k)} \cdots P_{i-1}^{\sigma_{i,i-1}(k)}}{U_i^{(d_i - t)}}\right] \in \mathfrak{k}(\alpha_1, \ldots, \alpha_{i-1})$$

for $0 \le t \le d_i - 1$ *and*

$$f_i(u) = u^{d_i} + b_{i,d_i-1} u^{d_i - 1} + \cdots + b_{i,0}$$

is the minimal polynomial of α_i *over* $\mathfrak{k}(\alpha_1, \ldots, \alpha_{i-1})$.

The algorithm terminates with $\Omega < \infty$ *if and only if either*

$$\overline{n}_\Omega = [G(\nu(P_0), \ldots, \nu(P_\Omega)) : G(\nu(P_0), \ldots, \nu(P_{\Omega-1}))] = \infty \tag{6}$$

or

$$\begin{aligned} &\overline{n}_\Omega < \infty \text{ (so that } \alpha_\Omega \text{ is defined as in 4)) and} \\ &d_\Omega = [\mathfrak{k}(\alpha_1, \ldots, \alpha_\Omega) : \mathfrak{k}(\alpha_1, \ldots, \alpha_{\Omega-1})] = \infty. \end{aligned} \tag{7}$$

If $\overline{n}_\Omega = \infty$, *set* $\alpha_\Omega = 1$.

Let x, y be regular parameters in R such that $\nu(x) = \nu(m_R)$. Let $P_0 = x$, $P_1 = y, P_2, \ldots$ be the generating sequence in R constructed in Theorem 2.2. Then either $\{\mathrm{in}_\nu(x)\} \cup \{\mathrm{in}_\nu(P_i) \mid n_i > 0\}$ minimally generate $\mathrm{gr}_\nu(R)$ as an R/m_R-algebra, or there exists $g \in R$ such that $\{\mathrm{in}_\nu(x)\} \cup \{\mathrm{in}_\nu(P_i) \mid n_i > 0\} \cup \{\mathrm{in}_\nu(g)\}$ minimally generate $\mathrm{gr}_\nu(R)$ as an R/m_R-algebra by Theorems 4.11 and 4.12 [4]. So by (3), (6) and (7) of Theorem 2.2, if $\mathrm{gr}_\nu(R)$ is not a finitely generated R/m_R-algebra, then there exist infinitely many i such that $1 < \overline{n}_i d_i = n_i < \infty$. The integer $\overline{n}_i$ can be strictly between 1 and ∞ for only finitely many i since Φ_ν is a finitely generated group. Further, d_i can be strictly between 1 and ∞ for only finitely many i since the algebraic closure of R/m_R in V_ν/m_ν is a finite extension of R/m_R because V_ν is a finitely generated field extension of R/m_R. Thus there are only finitely many P_i such that $n_i > 1$, and so $\mathrm{gr}_\nu(R)$ is a finitely generated R/m_R-algebra.

Proof of Proposition 1.3 if V_ν/m_ν is not algebraic over R/m_R, then equality holds in Abhyankar's inequality (2), and so $\Phi_\nu \cong \mathbb{Z}$ and V_ν/m_ν is a finitely generated extension field of R/m_R. Thus $\mathrm{gr}_\nu(R)$ is a finitely generated R/m_R-algebra by Proposition 1.2.

Now suppose that V_ν/m_ν is algebraic over R/m_R. Then $V_\nu/m_\nu = R/m_R$ as R/m_R is algebraically closed. If Φ_ν is a finitely generated group, we then have that $\mathrm{gr}_\nu(R)$ is a finitely generated R/m_R-algebra by Proposition 1.2.

Conversely, if $\mathrm{gr}_\nu(R)$ is a finitely generated R/m_R-algebra, we have that $S^\nu(R)$ is a finitely generated semigroup by Remark 2.1, and so Φ_ν is a finitely generated group.

3 Examples of Nonfinite Generation

Example 3.1 There exists an example of a two dimensional normal hypersurface singularity whose local ring A is dominated by a rank 2 Abhyankar valuation ν such that $\mathrm{gr}_\nu(A)$ is not a finitely generated A/m_A-algebra.

We now give the construction of Example 3.1. Let k be an algebraically closed field of characteristic $\neq 2$, which we assume has positive transcendence degree over the prime field if $\mathrm{char}\, k > 2$. Let F be a cubic form in the polynomial ring $k[x_1, x_2, x_3]$ such that $Z(F) \subset \mathbb{P}^2$ is nonsingular, so $E = \mathrm{Proj}(k[x_1, x_2, x_3]/(f)$ is an elliptic curve with associated very ample divisor $\mathcal{O}_E(1)$. Let $R = k[x_1, x_2, x_3]/(F)$ and $S = \mathrm{Spec}(R)$. Let $\pi : X \to S$ be the blow up of the maximal ideal $\mathfrak{m} = (x_1, x_2, x_3)$. The morphism π is a resolution of singularities (as can be seen by considering X as the strict transform of S in the blow up of $\mathfrak{m}$ in $\mathrm{Spec}(k[x_1, x_2, x_3]))$. Now $X = \mathrm{Proj}(\bigoplus_{n \geq 0} \mathfrak{m}^n)$, so $\mathcal{O}_X(1) = \mathfrak{m}\mathcal{O}_X$ and the scheme theoretic fiber of $\mathfrak{m}$ is

$$\pi^{-1}(\mathfrak{m}) = X \times_S \mathrm{Spec}(R/\mathfrak{m}) = \mathrm{Proj}(\bigoplus_{n \geq 0} \mathfrak{m}^n/\mathfrak{m}^{n+1}).$$

Now

$$\mathrm{gr}_{\mathfrak{m}}(R) = \bigoplus_{n \geq 0} \mathfrak{m}^n/\mathfrak{m}^{n+1} \cong k[x_1, x_2, x_3]/(F)$$

is graded by $\deg x_1 = \deg x_2 = \deg x_3 = 1$ and so $\mathfrak{m}\mathcal{O}_X = \mathcal{O}_X(-E)$ and $\pi^{-1}(\mathfrak{m}) \cong E$.

We have

$$\mathcal{O}_X(-nE)/\mathcal{O}_X(-(n+1)E) \cong \mathcal{O}_E(n)$$

for all $n \geq 0$ and we have short exact sequences

$$0 \to \mathcal{O}_E(m+n) \to \mathcal{O}_{(n+1)E}(m) \to \mathcal{O}_{nE}(m) \to 0 \tag{8}$$

for all $n \geq 0$ and $m \in \mathbb{Z}$. We have that $H^1(E, \mathcal{O}_E(m)) = 0$ for all $m \geq 1$ since $\deg \mathcal{O}_E(1) = 3$ and E is an elliptic curve. Thus by induction on n in (8),

$$H^1(X, \mathcal{O}_{nE}(m)) = 0$$

for all $n \geq 1$ and $m \geq 1$. By the formal function theorem, where $\hat{R}$ is the $\mathfrak{m}$-adic completion of R, we thus have that $H^1(X, \mathcal{O}_X(m)) \otimes_R \hat{R} = 0$ for all $m \geq 1$. Thus

$$H^1(X, \mathcal{O}_X(m)) = 0 \text{ for all } m \geq 1$$

since R is graded. Thus we have surjections

$$\Gamma(X, \mathcal{O}_X(-mE)) \cong \Gamma(X, \mathcal{O}_X(m)) \to \Gamma(E, \mathcal{O}_E(m)) \tag{9}$$

for all $m \geq 0$.

Let $q \in E$ be a point such that the degree 0 invertible sheaf

$$\mathcal{O}_E(1) \otimes \mathcal{O}_E(-3q) \text{ has infinite order in the Jacobian of } E. \tag{10}$$

Such a q exists because of our assumption that our algebraically closed field k has positive transcendence degree over the prime field if $\operatorname{char} k > 2$.

Let K be the quotient field of R. Let $f \in \mathcal{O}_{X,q}$ be a local equation of E. For nonzero $g \in \mathcal{O}_{X,q}$, define

$$\omega(g) = m \text{ if } g = f^m h \text{ where } f \nmid h \text{ in } \mathcal{O}_{X,q}$$

and define

$$\nu(g) = (m, \operatorname{ord}_q \overline{h}) \in (\mathbb{Z}^2)_{\text{lex}} \tag{11}$$

if $\overline{h}$ is the residue of h in $\mathcal{O}_{E,q} = \mathcal{O}_{X,q}/(f)$. We have that ω extends to a divisorial valuation of K which dominates R and ν extends to a rank 2 Abhyankar valuation of K which dominates R, and is composite with ω.

Since R is normal, the valuation ideal of R

$$I_m(\omega) = \{g \in R \mid \omega(g) \geq m\} = \Gamma(X, \mathcal{O}_X(-mE)) \tag{12}$$

for all $m \geq 0$. By (12), (9) and (11), we have that. given $m \geq 1$ and $r \geq 0$, there exists $g \in R$ such that $\nu(g) = (m, r)$ if and only if the natural inclusion

$$\Gamma(E, \mathcal{O}_E(m) \otimes \mathcal{O}_E(-(r+1)q)) \to \Gamma(E, \mathcal{O}_E(m) \otimes \mathcal{O}_E(-rq)) \text{ is not an equality.} \tag{13}$$

Since $\deg \mathcal{O}_E(m) \otimes \mathcal{O}_E(-rq) = 3m - r$, by the Riemann Roch theorem and (10), we have for $m, r \geq 0$ not both zero,

$$h^0(E, \mathcal{O}_E(m) \otimes \mathcal{O}_E(-rq)) = \begin{cases} 3m - r & \text{if } 3m - r \geq 1 \\ 0 & \text{if } 3m - r \leq 0 \end{cases}$$

Thus for $m \geq 0$ and $r \geq 0$ not both zero, we have that (13) is not an equality if and only if $3m - r \geq 1$. Thus the semigroup

$$S^R(\nu) = \{\nu(g) \mid g \in R\} = \{(a, b) \in (\mathbb{N}^2)_{\text{lex}} \mid 3a - b > 0\}$$

which is not a finitely generated semigroup. Let $A = R_m$. Since k is algebraically closed and $V_\nu/m_\nu = k$, we have that the associated graded ring $\text{gr}_\nu(A) \cong k[t^{S^R(\nu)}]$ which is not a finitely generated k-algebra.

Example 3.2 Let k be an algebraically closed field of characteristic $\neq 2$ which has positive transcendence degree over the prime field if $\text{char}\, k > 2$, and let $B = k[x_1, \ldots, x_n]_{(x_1, \ldots, x_n)}$, the localization of the polynomial ring over k in $n \geq 3$ variables. Then there exists a rank n (Abhyankar) valuation σ which dominates B such that $\text{gr}_\sigma(B)$ is not a finitely generated $k = B/m_B$-algebra.

Let $T = k[x_1, \ldots, x_n]$ and suppose that $0 \neq f \in T$. Write $f = x_n^{a_n} g$ where $g \in T$ and $x_n \nmid g_n$. Let f_{n-1} be the residue of g_n in $k[x_1, \ldots, x_n]/(x_n) \cong k[x_1, \ldots, x_{n-1}]$. Write $f_{n-1} = x_{n-1}^{a_{n-1}} g_{n-1}$ with $g_{n-1} \in k[x_1, \ldots, x_{n-1}]$ and $x_{n-1} \nmid g_{n-1}$. Inductively define g_i by $f_i = x_i^{a_i} g_i$ and f_{i-1} to be the residue of g_i in $k[x_1, \ldots, x_{i-1}] = k[x_1, \ldots, x_n]/(x_i, \ldots, x_n)$ for $i \geq 4$. Let $F \in k[x_1, x_2, x_3]$ be the cubic form, $R = k[x_1, x_2, x_3]/(F)$ be the ring and ν be the Abhyankar valuation of Example 3.1. Write $f_3 = F^{a_3} g_3$ where $F \nmid g_3$. Define f_2 to be the residue of g_3 in R. Define

$$\sigma(f) = (a_n, a_{n-1}, \ldots, a_3, \nu(f_2)) \in (\mathbb{Z}^n)_{\text{lex}}.$$

The function σ extends to a rank n Abhyankar valuation which dominates B whose residue field is k. We have that $S^B(\sigma) = (\mathbb{N}^{2-1} \times S^R(\nu))_{\text{lex}}$ is not a finitely generated semigroup. Thus $\text{gr}_\sigma(B) = k[t^{S^B(\sigma)}]$ is not a finitely generated $k = B/m_B$-algebra.

Acknowledgements This research was partially supported by NSF.

References

1. Abhyankar, S.: On the valuations centered in a local domain. Am. J. Math. **78**, 321–348 (1956)
2. Cossart, V., Galindo, C., Piltant, O.: Un exemple effectif de gradué non noetherien associé à une valuation divisorielle. Ann. Inst. Fourier **50**, 105–112 (2000)
3. Cutkosky, S.D.: On unique and almost unique factorization of complete ideals II. Inv. Math. **98**, 59–74 (1989)
4. Cutkosky, S.D., Vinh, P.A.: Valuation semigroups of two dimensional local rings. Proc. Lond. Math. Soc. **108**, 350–384 (2014)
5. Göhner, H.: Semifactorizality and Muhly's condition (N) in two dimensional local rings. J. Algebra **34**, 403–429 (1975)
6. Lipman, J.: On complete ideals in regular local rings. In: Algebraic Geometry and Commutative Algebra in Honor of M. Nagata, pp. 203–231. Kinokuniya, Tokyo (1988)
7. Sandal, S.D.: Irrational behavior of algebraic discrete valuations. J. Algebra **447**, 530–547 (2016)
8. Spivakovsky, M.: Valuations in function fields of surfaces. Am. J. Math. **112**, 107–156 (1990)
9. Teissier, B.: Valuations, deformations and toric geometry. In: Kuhlmann, F.V., Kuhlmann, S., Marshall, M. (eds.) Valuation Theory and Its Applications II. Fields Institute Communications, vol. 33, pp. 361–459. American Mathematical Society, Providence (2003)
10. Teissier, B.: Overweight deformations of affine toric varieties and local uniformization. In: Campillo, A., Kehlmann, F.-V., Teissier, B. (eds.) Valuation Theory in Interaction. Proceedings of the Second International Conference on Valuation Theory, Segovia-El Escorial, 2011. Congress Reports Series, Sept 2014. European Mathematical Society Publishing House, Zürich, pp. 474–565 (2014)
11. Zariski, O., Samuel, P.: Commutative Algebra, vol. II. Van Nostrand, Princeton (1960)

Symbolic Powers of Monomial Ideals and Cohen-Macaulay Vertex-Weighted Digraphs

Philippe Gimenez, José Martínez-Bernal, Aron Simis, Rafael H. Villarreal, and Carlos E. Vivares

Dedicated to Professor Antonio Campillo on the occasion of his 65th birthday

Abstract In this paper we study irreducible representations and symbolic Rees algebras of monomial ideals. Then we examine edge ideals associated to vertex-weighted oriented graphs. These are digraphs having no oriented cycles of length two with weights on the vertices. For a monomial ideal with no embedded primes we classify the normality of its symbolic Rees algebra in terms of its primary components. If the primary components of a monomial ideal are normal, we present a simple procedure to compute its symbolic Rees algebra using Hilbert bases, and give necessary and sufficient conditions for the equality between its ordinary and symbolic powers. We give an effective characterization of the Cohen–Macaulay vertex-weighted oriented forests. For edge ideals of transitive weighted oriented graphs we show that Alexander duality holds. It is shown that edge ideals of weighted acyclic tournaments are Cohen–Macaulay and satisfy Alexander duality.

P. Gimenez
Facultad de Ciencias, Instituto de Investigación en Matemáticas de la Universidad de Valladolid (IMUVA), Valladolid, Spain
e-mail: pgimenez@agt.uva.es

J. Martínez-Bernal · R. H. Villarreal (✉) · C. E. Vivares
Departamento de Matemáticas, Centro de Investigación y de Estudios Avanzados del IPN, México City, México
e-mail: jmb@math.cinvestav.mx; vila@math.cinvestav.mx; cevivares@math.cinvestav.mx

A. Simis
Departamento de Matemática, Universidade Federal de Pernambuco, Recife, PE, Brazil
e-mail: aron@dmat.ufpe.br

G.-M. Greuel et al. (eds.), *Singularities, Algebraic Geometry, Commutative Algebra, and Related Topics*, https://doi.org/10.1007/978-3-319-96827-8_21

1 Introduction

Let $R = K[x_1, \ldots, x_n]$ be a polynomial ring over a field K and let $I \subset R$ be a monomial ideal. The *Rees algebra* of I is

$$R[It] := R \oplus It \oplus \cdots \oplus I^k t^k \oplus \cdots \subset R[t],$$

where t is a new variable, and the *symbolic Rees algebra* of I is

$$R_s(I) := R \oplus I^{(1)}t \oplus \cdots \oplus I^{(k)}t^k \oplus \cdots \subset R[t],$$

where $I^{(k)}$ is the k-th symbolic power of I (see Definition 2).

One of the early works on symbolic powers of monomial ideal is [35]. Symbolic powers of ideals and edge ideals of graphs where studied in [1]. A method to compute symbolic powers of radical ideals in characteristic zero is given in [36].

In Sect. 2 we recall the notion of irreducible decomposition of a monomial ideal and prove that the exponents of the variables that occur in the minimal generating set of a monomial ideal I are exactly the exponents of the variables that occur in the minimal generators of the irreducible components of I (Lemma 1). This result indicates that the well known Alexander duality for squarefree monomial ideals could also hold for other families of monomial ideals.

We give algorithms to compute the symbolic powers of monomial ideals using *Macaulay*2 [16] (Lemma 2, Remarks 1 and 5). For a monomial ideal with no embedded primes we classify the normality of its symbolic Rees algebra in terms of the normality of its primary components (Proposition 3).

The normality of a monomial ideal is well understood from the computational point of view. If I is minimally generated by $x^{v_1}, \ldots, x^{v_r}$ and A is the matrix with column vectors $v_1^{\mathrm{t}}, \ldots, v_r^{\mathrm{t}}$, then I is normal if and only if the system $xA \geq \mathbf{1}; x \geq 0$ has the integer rounding property [9, Corollary 2.5]. The normality of I can be determined using the program *Normaliz* [3]. For the normality of monomial ideals of dimension 2 see [6, 12] and the references therein.

To compute the generators of the symbolic Rees algebra of a monomial ideal one can use the algorithm in the proof of [22, Theorem 1.1]. If the primary components of a monomial ideal are normal, we present a procedure that computes the generators of its symbolic Rees algebra using Hilbert bases and *Normaliz* [3] (Proposition 4, Example 4), and give necessary and sufficient conditions for the equality between its ordinary and symbolic powers (Corollary 3).

In Sect. 3 we study edge ideals of weighted oriented graphs. A *directed graph* or *digraph* $\mathscr{D}$ consists of a finite set $V(\mathscr{D})$ of vertices, together with a prescribed collection $E(\mathscr{D})$ of ordered pairs of distinct points called *edges* or *arrows*. An *oriented graph* is a digraph having no oriented cycles of length two. In other words an oriented graph $\mathscr{D}$ is a simple graph G together with an orientation of its edges. We call G the *underlying graph* of $\mathscr{D}$. If a digraph $\mathscr{D}$ is endowed with a function $d\colon V(\mathscr{D}) \to \mathbb{N}_+$, where $\mathbb{N}_+ := \{1, 2, \ldots\}$, we call $\mathscr{D}$ a *vertex-weighted digraph*.

Edge ideals of edge-weighted graphs were introduced and studied by Paulsen and Sather-Wagstaff [33]. In this work we consider edge ideals of graphs which are oriented and have weights on the vertices. In what follows by a weighted oriented graph we shall always mean a vertex-weighted oriented graph.

Let $\mathscr{D}$ be a vertex-weighted digraph with vertex set $V(\mathscr{D}) = \{x_1, \ldots, x_n\}$. The weight $d(x_i)$ of x_i is denoted simply by d_i. The *edge ideal* of $\mathscr{D}$, denoted $I(\mathscr{D})$, is the ideal of R given by

$$I(\mathscr{D}) := (x_i x_j^{d_j} \mid (x_i, x_j) \in E(\mathscr{D})).$$

If a vertex x_i of $\mathscr{D}$ is a *source* (i.e., has only arrows leaving x_i) we shall always assume $d_i = 1$ because in this case the definition of $I(\mathscr{D})$ does not depend on the weight of x_i. In the special case when $d_i = 1$ for all i, we recover the edge ideal of the graph G which has been extensively studied in the literature [7, 11, 13, 17, 20, 30, 38–40, 42]. A vertex-weighted digraph $\mathscr{D}$ is called *Cohen–Macaulay* (over the field K) if $R/I(\mathscr{D})$ is a Cohen–Macaulay ring.

Using a result of [24], we answer a question of Aron Simis and a related question of Antonio Campillo by showing that an oriented graph $\mathscr{D}$ is Cohen–Macaulay if and only if the oriented graph $\mathscr{U}$, obtained from $\mathscr{D}$ by replacing each weight $d_i > 3$ with $d_i = 2$, is Cohen–Macaulay (Corollary 6). Seemingly, this ought to somewhat facilitate the verification of this property.

It turns out that edge ideals of weighted acyclic tournaments are Cohen–Macaulay and satisfy Alexander duality (Corollaries 7 and 8). For transitive weighted oriented graphs it is shown that Alexander duality holds (Theorem 4). Edge ideals of weighted digraphs arose in the theory of Reed-Muller codes as initial ideals of vanishing ideals of projective spaces over finite fields [4, 18, 25].

A major result of Pitones, Reyes and Toledo [34] shows an explicit combinatorial expression for the irredundant decomposition of $I(\mathscr{D})$ as a finite intersection of irreducible monomial ideals (Theorem 2). We will use their result to prove the following explicit combinatorial classification of all Cohen–Macaulay weighted oriented forests.

Theorem 5 *Let $\mathscr{D}$ be a weighted oriented forest without isolated vertices and let G be its underlying forest. The following conditions are equivalent:*

(a) *$\mathscr{D}$ is Cohen–Macaulay.*
(b) *$I(\mathscr{D})$ is unmixed, that is, all its associated primes have the same height.*
(c) *G has a perfect matching $\{x_1, y_1\}, \ldots, \{x_r, y_r\}$ so that $\deg_G(y_i) = 1$ for $i = 1, \ldots, r$ and $d(x_i) = d_i = 1$ if $(x_i, y_i) \in E(\mathscr{D})$.*

All rings considered here are Noetherian. For all unexplained terminology and additional information, we refer to [2] for the theory of digraphs, and [13, 20, 30, 42] for the theory of edge ideals of graphs and monomial ideals.

2 Irreducible Decompositions and Symbolic Powers

In this section we study irreducible representations of monomial ideals and various aspects of symbolic Rees algebras of monomial ideals. Here we continue to employ the notation and definitions used in Sect. 1.

Recall that an ideal L of a Noetherian ring R is called *irreducible* if L cannot be written as an intersection of two ideals of R that properly contain L. Let $R = K[x_1, \ldots, x_n]$ be a polynomial ring over a field K. Up to permutation of variables the irreducible monomial ideals of R are of the form

$$(x_1^{a_1}, \ldots, x_r^{a_r}),$$

where $a_1, \ldots, a_r$ are positive integers. According to [42, Theorem 6.1.17] any monomial ideal I of R has a *unique* irreducible decomposition:

$$I = I_1 \cap \cdots \cap I_m,$$

where $I_1, \ldots, I_m$ are irreducible monomial ideals and $I \neq \cap_{i \neq j} I_i$ for $j = 1, \ldots, m$, that is, this decomposition is irredundant. The ideals $I_1, \ldots, I_m$ are called the *irreducible components* of I.

By [42, Proposition 6.1.7] a monomial ideal $\mathfrak{I}$ is a primary ideal if and only if, after permutation of the variables, it has the form:

$$\mathfrak{I} = (x_1^{a_1}, \ldots, x_r^{a_r}, x^{b_1}, \ldots, x^{b_s}), \tag{1}$$

where $a_i \geq 1$ and $\cup_{i=1}^{s} \mathrm{supp}(x^{b_i}) \subset \{x_1, \ldots, x_r\}$. Thus if $\mathfrak{I}$ is a monomial primary ideal, then $\mathfrak{I}^k$ is a primary ideal for $k \geq 1$. Since irreducible ideals are primary, the irreducible decomposition of I is a primary decomposition of I. Notice that the irreducible decomposition of I is not necessarily a minimal primary decomposition, that is, I_i and I_j could have the same radical for $i \neq j$. If I is a squarefree monomial ideal, its irreducible decomposition is minimal. For edge ideals of weighted oriented graphs one also has that their irreducible decompositions are minimal [34].

Definition 1 An irreducible monomial ideal $L \subset R$ is called a *minimal irreducible* ideal of I if $I \subset L$ and for any irreducible monomial ideal L' such that $I \subset L' \subset L$ one has that $L = L'$.

Proposition 1 *If $I = I_1 \cap \cdots \cap I_m$ is the irreducible decomposition of a monomial ideal I, then $I_1, \ldots, I_m$ are the minimal irreducible monomial ideals of I.*

Proof Let L be an irreducible ideal that contains I. Then $I_i \subset L$ for some i. Indeed if $I_i \not\subset L$ for all i, for each i pick $x_{j_i}^{a_{j_i}}$ in $I_i \setminus L$. Since $I \subset L$, setting $x^a = \mathrm{lcm}\{x_{j_i}^{a_{j_i}}\}_{i=1}^{m}$ and writing $L = (x_{k_1}^{c_{k_1}}, \ldots, x_{k_\ell}^{c_{k_\ell}})$, it follows that x^a is in I and $x_{j_i}^{a_{j_i}}$ is a multiple of $x_{k_t}^{c_{k_t}}$ for some $1 \leq i \leq m$ and $1 \leq t \leq \ell$. Thus $x_{j_i}^{a_{j_i}}$ is in L, a contradiction. Therefore if L is minimal one has $L = I_i$ for some i. To complete the proof notice that I_i is a minimal irreducible monomial ideal of I for all i. This follows from the first part of the proof using that $I = I_1 \cap \cdots \cap I_m$ is an irredundant decomposition. □

The unique minimal set of generators of a monomial ideal I, consisting of monomials, is denoted by $G(I)$. The next result tells us that in certain cases we may have a sort of Alexander duality obtained by switching the roles of minimal generators and irreducible components [42, Theorem 6.3.39] (see Example 7 and Theorem 4).

Lemma 1 *Let I be a monomial ideal of R, with $G(I) = \{x^{v_1}, \ldots, x^{v_r}\}$ and $v_i = (v_{i1}, \ldots, v_{in})$ for $i = 1, \ldots, r$, and let $I = I_1 \cap \cdots \cap I_m$ be its irreducible decomposition. Then*

$$V := \{x_j^{v_{ij}} \mid v_{ij} \geq 1\} = G(I_1) \cup \cdots \cup G(I_m).$$

Proof "⊂": Take $x_j^{v_{ij}}$ in V, without loss of generality we may assume $i = j = 1$. We proceed by contradiction assuming that $x_1^{v_{11}}$ is not in $\cup_{i=1}^m G(I_i)$. Setting $M = x_1^{v_{11}-1} x_2^{v_{12}} \cdots x_n^{v_{1n}}$, notice that M is in I. Indeed for any I_j not containing $x_2^{v_{12}} \cdots x_n^{v_{1n}}$, one has that $x_1^{v_{11}}$ is in I_j because x^{v_1} is in I. Thus there is $x_1^{c_j}$ in $G(I_j)$ such that $v_{11} > c_j \geq 1$ because $x_1^{v_{11}}$ is not in $G(I_j)$. Thus M is in I_j. This proves that M is in I, a contradiction to the minimality of $G(I)$ because this monomial that strictly divides one of the elements of $G(I)$ cannot be in I. Thus $x_1^{v_{11}}$ is in $\cup_{i=1}^m G(I_i)$, as required.

"⊃": Take $x_j^{a_j}$ in $G(I_i)$ for some i, j, without loss of generality we may assume that $i = j = 1$ and $G(I_1) = \{x_1^{a_1}, \ldots, x_\ell^{a_\ell}\}$. We proceed by contradiction assuming that $x_1^{a_1} \notin V$. Setting $L = (x_1^{a_1+1}, x_2^{a_2}, \ldots, x_\ell^{a_\ell})$, notice that $I \subset L$. Indeed take any monomial x^{v_k} in $G(I)$ which is not in $(x_2^{a_2}, \ldots, x_\ell^{a_\ell})$. Then x^{v_k} is a multiple of $x_1^{a_1}$ because $I \subset I_1$. Hence $v_{k1} > a_1$ because $x_1^{a_1} \notin V$. Thus x^{v_k} is in L. This proves that $I \subset L \subsetneq I_1$, a contradiction to the fact that I_1 is a minimal irreducible monomial ideal of I (see Proposition 1). □

Let $I \subset R$ be a monomial ideal. The *Alexander dual* of I, denoted $I^\vee$, is the ideal of R generated by all monomials x^a, with $a = (a_1, \ldots, a_n)$, such that $\{x_i^{a_i} \mid a_i \geq 1\}$ is equal to $G(L)$ for some minimal irreducible ideal L of I. The *dual* of I, denoted I^*, is the intersection of all ideals $(\{x_i^{a_i} \mid a_i \geq 1\})$ such that $x^a \in G(I)$. Thus one has

$$I^\vee = \left(\prod_{f \in G(I_1)} f, \ \ldots, \ \prod_{f \in G(I_m)} f \right) \quad \text{and} \quad I^* = \bigcap_{x^a \in G(I)} (\{x_i^{a_i} \mid a_i \geq 1\}),$$

where $I_1, \ldots, I_m$ are the irreducible components of I. If $I^* = I^\vee$, we say that *Alexander duality* holds for I. There are other related ways introduced by Ezra

Miller [23, 27–29] to define the Alexander dual of a monomial ideal . It is well known that $I^* = I^\vee$ for squarefree monomial ideals [42, Theorem 6.3.39].

Definition 2 Let I be an ideal of a ring R and let $\mathfrak{p}_1, \ldots, \mathfrak{p}_r$ be the minimal primes of I. Given an integer $k \geq 1$, we define the k-th *symbolic power* of I to be the ideal

$$I^{(k)} := \bigcap_{i=1}^{r} \mathfrak{q}_i = \bigcap_{i=1}^{r} (I^k R_{\mathfrak{p}_i} \cap R),$$

where $\mathfrak{q}_i$ is the $\mathfrak{p}_i$-primary component of I^k.

In other words, one has $I^{(k)} = S^{-1} I^k \cap R$, where $S = R \setminus \cup_{i=1}^r \mathfrak{p}_i$. An alternative notion of symbolic power can be introduced using the whole set of associated primes of I instead (see, e.g., [5, 8]):

$$I^{\langle k \rangle} = \bigcap_{\mathfrak{p} \in \mathrm{Ass}(R/I)} (I^k R_{\mathfrak{p}} \cap R) = \bigcap_{\mathfrak{p} \in \mathrm{maxAss}(R/I)} (I^k R_{\mathfrak{p}} \cap R),$$

where $\mathrm{maxAss}(R/I)$ is the set of associated primes which are maximal with respect to inclusion [5, Lemmas 3.1 and 3.2]. Clearly $I^k \subset I^{\langle k \rangle} \subset I^{(k)}$. If I has no embedded primes, e.g. for radical ideals such as squarefree monomial ideals, the two last definitions of symbolic powers coincide. An interesting problem is to give necessary and sufficient conditions for the equality "$I^k = I^{(k)}$ for $k \geq 1$".

For prime ideals the k-th symbolic powers and the k-th usual powers are not always equal. Thus the next lemma does not hold in general but the proof below shows that it will hold for an ideal I in Noetherian ring R under the assumption that $\mathfrak{I}_i^k = \mathfrak{I}_i^{(k)}$ for $i = 1, \ldots, r$. The next lemma is well known for radical monomial ideals [41, Propositions 3.3.24 and 7.3.14].

Lemma 2 *Let $I \subset R$ be a monomial ideal and let $I = \mathfrak{I}_1 \cap \cdots \cap \mathfrak{I}_r \cap \cdots \cap \mathfrak{I}_m$ be an irredundant minimal primary decomposition of I, where $\mathfrak{I}_1, \ldots, \mathfrak{I}_r$ are the primary components associated to the minimal primes of I. Then*

$$I^{(k)} = \mathfrak{I}_1^k \cap \cdots \cap \mathfrak{I}_r^k \textit{ for } k \geq 1.$$

Proof Let $\mathfrak{p}_1, \ldots, \mathfrak{p}_r$ be the minimal primes of I. By [42, Proposition 6.1.7] any power of $\mathfrak{I}_i$ is again a $\mathfrak{p}_i$-primary ideal (see Eq. (1) at the beginning of this section). Thus $\mathfrak{I}_i^k = \mathfrak{I}_i^{(k)}$ for any i, k. Fixing integers $k \geq 1$ and $1 \leq i \leq r$, let

$$I^k = \mathfrak{q}_1 \cap \cdots \cap \mathfrak{q}_r \cap \cdots \cap \mathfrak{q}_s$$

be a primary decomposition of I^k, where $\mathfrak{q}_j$ is $\mathfrak{p}_j$-primary for $j \leq r$. Localizing at $\mathfrak{p}_i$ yields $I^k R_{\mathfrak{p}_i} = \mathfrak{q}_i R_{\mathfrak{p}_i}$ and from $I = \mathfrak{I}_1 \cap \cdots \cap \mathfrak{I}_r \cap \cdots \cap \mathfrak{I}_m$ one obtains:

$$I^k R_{\mathfrak{p}_i} = (I R_{\mathfrak{p}_i})^k = (\mathfrak{I}_i R_{\mathfrak{p}_i})^k = \mathfrak{I}_i^k R_{\mathfrak{p}_i}.$$

Thus $\mathfrak{I}_i^k R_{\mathfrak{p}_i} = \mathfrak{q}_i R_{\mathfrak{p}_i}$ and contracting to R one has $\mathfrak{I}_i^{(k)} = \mathfrak{q}_i$. Therefore

$$I^{(k)} = \mathfrak{I}_1^{(k)} \cap \cdots \cap \mathfrak{I}_r^{(k)} = \mathfrak{I}_1^k \cap \cdots \cap \mathfrak{I}_r^k.$$

□

It was pointed out to us by Ngô Viêt Trung that Lemma 2 is a consequence of [22, Lemma 3.1]. This lemma also follows from [5, Proposition 3.6].

Remark 1 To compute the k-th symbolic power $I^{(k)}$ of a monomial ideal I one can use the following procedure for *Macaulay*2 [16].

```
SPG=(I,k)->intersect(for n from 0 to #minimalPrimes(I)-1
list localize(I^k,(minimalPrimes(I))#n))
```

Example 1 Let I be the ideal $(x_2x_3, x_4x_5, x_3x_4, x_2x_5, x_1^2x_3, x_1x_2^2)$. Using the procedure of Remark 1 we obtain $I^{(2)} = I^2 + (x_1x_2^2x_5, x_1x_2^2x_3)$.

Remark 2 If one uses $\text{Ass}(R/I)$ to define the symbolic powers of a monomial ideal I, the following function for *Macaulay*2 [16] can be used to compute $I^{\langle k \rangle}$.

```
SPA=(I,k)->intersect(for n from 0 to #associatedPrimes(I)-1
list localize(I^k,(associatedPrimes(I))#n))
```

Example 2 Let I be the ideal $(x_1x_2^2, x_3x_1^2, x_2x_3^2)$. Using the procedures of Remarks 1 and 2, we obtain

$$I^{(1)} = I + (x_1x_2x_3) \text{ and } I^{\langle 1 \rangle} = I.$$

Remark 3 The following formula is useful to study the symbolic powers $I^{\langle k \rangle}$ of a monomial ideal I [5, Proposition 3.6]:

$$I^k R_{\mathfrak{p}} \cap R = (I R_{\mathfrak{p}} \cap R)^k \text{ for } \mathfrak{p} \in \text{Ass}(R/I) \text{ and } k \geq 1.$$

Definition 3 An ideal I of a ring R is called *normally torsion-free* if $\text{Ass}(R/I^k)$ is contained in $\text{Ass}(R/I)$ for all $k \geq 1$.

Remark 4 Let I be an ideal of a ring R. If I has no embedded primes, then I is normally torsion-free if and only if $I^k = I^{(k)}$ for all $k \geq 1$.

Lemma 3 ([43, Lemma 5, Appendix 6]) *Let $I \subset R$ be an ideal generated by a regular sequence. Then I^k is unmixed for $k \geq 1$. In particular $I^k = I^{(k)}$ for $k \geq 1$.*

One can also compute the symbolic powers of vanishing ideals of finite sets of reduced projective points using Lemma 2 because these ideals are intersections of finitely many prime ideals that are complete intersections. It is well known that complete intersections are normally torsion-free (Lemma 3).

Remark 5 (Jonathan O'Rourke) If I is a radical ideal of R and all associated primes of I are normally torsion-free, then the k-th symbolic power of I can be computed using the following procedure for *Macaulay*2 [16].

```
SP1 = (I,k) -> (temp = primaryDecomposition I;
temp2 = ((temp_0)^k); for i from 1 to #temp-1 do(temp2 =
    intersect(temp2,(temp_i)^k)); return temp2)
```

Example 3 Let $\mathbb{X}$ be the set $\{[e_1], [e_2], [e_3], [e_4], [(1, 1, 1, 1)]\}$ of 5 points in general linear position in $\mathbb{P}^3$, over the field $\mathbb{Q}$, where e_i is the i-th unit vector, and let $I = I(\mathbb{X})$ be its vanishing ideal. Using *Macaulay2* [16] and Remark 5 we obtain

$$I = (x_2x_4 - x_3x_4, x_1x_4 - x_3x_4, x_2x_3 - x_3x_4, x_1x_3 - x_3x_4, x_1x_2 - x_3x_4),$$

$I^2 = I^{(2)}$, $I^3 \neq I^{(3)}$ and I is a Gorenstein ideal. This example (in greater generality) has been used in [31, proof of Proposition 4.1 and Remark 4.2(2)].

Proposition 2 ([22]) *If $I \subset R$ is a monomial ideal, then the symbolic Rees algebra $R_s(I)$ of I is a finitely generated K-algebra.*

Proof It follows at once from Lemma 2 and [22, Corollary 1.3]. □

To compute the generators of the symbolic Rees algebra of a monomial ideal one can use the procedure given in the proof of [22, Theorem 1.1]. Another method will be presented in this section that works when the primary components are normal.

Remark 6 The symbolic Rees algebra of a monomial ideal I is finitely generated if one uses the associated primes of I to define symbolic powers. This follows from [22, Corollary 1.3] and the following formula [5, Theorem 3.7]:

$$I^{\langle k\rangle} = \bigcap_{\mathfrak{p}\in \operatorname{maxAss}(R/I)} (IR_{\mathfrak{p}} \cap R)^k \text{ for } k \geq 1.$$

Corollary 1 *If I is a monomial ideal, then $R_s(I)$ is Noetherian and there is an integer $k \geq 1$ such that $[I^{(k)}]^i = I^{(ik)}$ for $i \geq 1$.*

Proof It follows at once from [15, p. 80, Lemma 2.1] or by a direct argument using Proposition 2. □

For convenience of notation in what follows we will often assume that monomial ideals have no embedded primes but some of the results can be stated and proved for general monomial ideals.

Proposition 3 *Let $I \subset R$ be a monomial ideal without embedded primes and let $I = \cap_{i=1}^r \mathfrak{I}_i$ be its minimal irredundant primary decomposition. Then $R_s(I)$ is normal if and only if $R[\mathfrak{I}_i t]$ is normal for all i.*

Proof $\Rightarrow$): Since $R_s(I)$ is Noetherian and normal it is a Krull domain by a theorem of Mori and Nagata [26, p. 296]. Therefore, by [37, Lemma 2.5], we get that $R_{\mathfrak{p}_i}[I_{\mathfrak{p}_i}t] = R_{\mathfrak{p}_i}[(\mathfrak{I}_i)_{\mathfrak{p}_i}t]$ is normal. Let $\mathfrak{p}_i$ be the radical of $\mathfrak{I}_i$. Any power of $\mathfrak{I}_i$ is a $\mathfrak{p}_i$-primary ideal. This follows from [42, Proposition 6.1.7] (see Eq. (1) at the beginning of this section). Hence it is seen that $R_{\mathfrak{p}_i}[(\mathfrak{I}_i)_{\mathfrak{p}_i}t] \cap R[t] = R[\mathfrak{I}_i t]$. As $R[t]$ is normal it follows that $R[\mathfrak{I}_i t]$ is normal.

$\Leftarrow$): By Lemma 2 one has $\cap_{i=1}^{r} R[\mathfrak{I}_i t] = R_s(I)$. As $R[\mathfrak{I}_i t]$ and $R_s(I)$ have the same field of quotients it follows that $R_s(I)$ is normal. □

In general, even for monomial ideals without embedded primes, normally torsion-free ideals may not be normal. For instance $I = (x_1^2, x_2^2)$ is normally torsion-free and is not normal. As a consequence of Proposition 3 one recovers the following well known result.

Corollary 2 *Let I be a squarefree monomial ideal. Then $R_s(I)$ is normal and $R[It]$ is normal if I is normally torsion-free.*

Let I be a monomial ideal and let $G(I) = \{x^{v_1}, \ldots, x^{v_m}\}$ be its minimal set of generators. We set

$$\mathscr{A}_I = \{e_1, \ldots, e_n, (v_1, 1), \ldots, (v_m, 1)\},$$

where $e_1, \ldots, e_n$ belong to $\mathbb{Z}^{n+1}$, and denote by $\mathbb{R}_+(I)$ or $\mathbb{R}_+\mathscr{A}_I$ (resp. $\mathbb{N}\mathscr{A}_I$) the cone (resp. semigroup) generated by $\mathscr{A}_I$. The integral closure of $R[It]$ is given by $\overline{R[It]} = K[\mathbb{R}_+(I) \cap \mathbb{Z}^{n+1}]$. Recall that a finite set $\mathscr{H}$ is called a *Hilbert basis* if $\mathbb{N}\mathscr{H} = \mathbb{R}_+\mathscr{H} \cap \mathbb{Z}^{n+1}$, and that $R[It]$ is normal if and only if $\mathscr{A}_I$ is a Hilbert basis [42, Proposition 14.2.3].

Let $C \subset \mathbb{R}^{n+1}$ be a rational polyhedral cone. A finite set $\mathscr{H}$ is called a Hilbert basis of C if $C = \mathbb{R}_+\mathscr{H}$ and $\mathscr{H}$ is a Hilbert basis. A Hilbert basis of C is minimal if it does not strictly contain any other Hilbert basis of C. For pointed cones there is unique minimal Hilbert basis [42, Theorem 1.3.9].

If the primary components of a monomial ideal are normal, the next result gives a simple procedure to compute its symbolic Rees algebra using Hilbert bases.

Proposition 4 *Let I be a monomial ideal without embedded primes and let $I = \cap_{i=1}^{r} \mathfrak{I}_i$ be its minimal irredundant primary decomposition. If $R[\mathfrak{I}_i t]$ is normal for all i and $\mathscr{H}$ is the Hilbert basis of the polyhedral cone $\cap_{i=1}^{r} \mathbb{R}_+(\mathfrak{I}_i)$, then $R_s(I)$ is $K[\mathbb{N}\mathscr{H}]$, the semigroup ring of $\mathbb{N}\mathscr{H}$.*

Proof As $R[\mathfrak{I}_i t] = K[\mathbb{N}\mathscr{A}_{\mathfrak{I}_i}]$ is normal for $i = 1, \ldots, r$, the semigroup $\mathbb{N}\mathscr{A}_{\mathfrak{I}_i}$ is equal to $\mathbb{R}_+(\mathfrak{I}_i) \cap \mathbb{Z}^{n+1}$ for $i = 1, \ldots, r$. Hence, by Lemma 2, we get

$$\begin{aligned} R_s(I) &= \cap_{i=1}^{r} R[\mathfrak{I}_i t] = \cap_{i=1}^{r} K[\mathbb{N}\mathscr{A}_{\mathfrak{I}_i}] = K[\cap_{i=1}^{r} \mathbb{N}\mathscr{A}_{\mathfrak{I}_i}] \\ &= K[\mathbb{R}_+(\mathfrak{I}_1) \cap \cdots \cap \mathbb{R}_+(\mathfrak{I}_r) \cap \mathbb{Z}^{n+1}] = K[\mathbb{N}\mathscr{H}]. \end{aligned}$$

□

Definition 4 The rational polyhedral cone $\cap_{i=1}^{r} \mathbb{R}_+(\mathfrak{I}_i)$ is called the *Simis cone* of I and is denoted by $\mathrm{Cn}(I)$.

For squarefree monomial ideals the Simis cone was introduced in [10]. In particular from Proposition 4 we recover [10, Theorem 3.5].

Example 4 The ideal $I = (x_2x_3,\, x_4x_5,\, x_3x_4,\, x_2x_5,\, x_1^2x_3,\, x_1x_2^2)$ satisfies the hypothesis of Proposition 4. Using *Normaliz* [3] we obtain that the minimal Hilbert basis of the Simis cone is:

```
18 Hilbert basis elements:
 0 0 0 0 1 0     1 2 0 0 0 1
 0 0 0 1 0 0     2 0 1 0 0 1
 0 0 1 0 0 0     1 2 0 0 1 2
 0 1 0 0 0 0     1 2 1 0 0 2
 1 0 0 0 0 0     2 2 1 0 1 3
 0 0 0 1 1 1     2 2 2 0 0 3
 0 0 1 1 0 1     2 4 1 0 2 5
 0 1 0 0 1 1     2 4 2 0 1 5
 0 1 1 0 0 1     2 4 3 0 0 5
```

Hence $R_s(I)$ is generated by the monomials corresponding to these vectors.

Let I be an ideal of R. The equality "$I^k = I^{(k)}$ for $k \geq 1$" holds if and only if I has no embedded primes and is normally torsion-free (see Remark 4). We refer the reader to [8] for a recent survey on symbolic powers of ideals.

In [14, Corollary 3.14] it is shown that a squarefree monomial ideal I is normally torsion-free if and only if the corresponding hypergraph satisfies the max-flow min-cut property. As an application we present a classification of the equality between ordinary and symbolic powers for a family of monomial ideals.

Corollary 3 *Let I be a monomial ideal without embedded primes and let $\mathfrak{I}_1, \ldots, \mathfrak{I}_r$ be its primary components. If $R[\mathfrak{I}_i t]$ is normal for all i, then $I^k = I^{(k)}$ for $k \geq 1$ if and only if* $\mathrm{Cn}(I) = \mathbb{R}_+(I)$ *and $R[It]$ is normal.*

Proof $\Rightarrow$): As $R_s(I) = R[It]$, by Proposition 4, $R[It]$ is normal. Therefore one has

$$K[\mathrm{Cn}(I) \cap \mathbb{Z}^{n+1}] = R_s(I) = R[It] = \overline{R[It]} = K[\mathbb{R}_+(I) \cap \mathbb{Z}^{n+1}].$$

Thus $\mathrm{Cn}(I) = \mathbb{R}_+(I)$.

$\Leftarrow$): By the proof of Proposition 4 one has $R_s(I) = K[\mathrm{Cn}(I) \cap \mathbb{Z}^{n+1}]$. Hence

$$R_s(I) = K[\mathrm{Cn}(I) \cap \mathbb{Z}^{n+1}] = K[\mathbb{R}_+(I) \cap \mathbb{Z}^{n+1}] = \overline{R[It]}.$$

As $R[It]$ is normal, we get $R_s(I) = R[It]$, that is, $I^k = I^{(k)}$ for $k \geq 1$. □

3 Cohen–Macaulay Weighted Oriented Trees

In this section we show that edge ideals of transitive weighted oriented graphs satisfy Alexander duality. It turns out that edge ideals of weighted acyclic tournaments are Cohen–Macaulay and satisfy Alexander duality. Then we classify all Cohen–

Macaulay weighted oriented forests. Here we continue to employ the notation and definitions used in Sects. 1 and 2.

Let G be a graph with vertex set $V(G)$. A subset $C \subset V(G)$ is a *minimal vertex cover* of G if: (i) every edge of G is incident with at least one vertex in C, and (ii) there is no proper subset of C with the first property. If C satisfies condition (i) only, then C is called a *vertex cover* of G.

Let $\mathscr{D}$ be a weighted oriented graph with underlying graph G. Next we recall a combinatorial description of the irreducible decomposition of $I(\mathscr{D})$.

Definition 5 ([34]) Let C be a vertex cover of G. Consider the set $L_1(C)$ of all $x \in C$ such that there is $(x, y) \in E(\mathscr{D})$ with $y \notin C$, the set $L_3(C)$ of all $x \in C$ such that $N_G(x) \subset C$, and the set $L_2(C) = C \setminus (L_1(C) \cup L_3(C))$, where $N_G(x)$ is the *neighbor* set of x consisting of all $y \in V(G)$ such that $\{x, y\}$ is an edge of G. A vertex cover C of G is called a *strong vertex cover* of $\mathscr{D}$ if C is a minimal vertex cover of G or else for all $x \in L_3(C)$ there is $(y, x) \in E(\mathscr{D})$ such that $y \in L_2(C) \cup L_3(C)$ with $d(y) \geq 2$.

Theorem 1 ([34]) *Let $\mathscr{D}$ be a weighted oriented graph. Then L is a minimal irreducible monomial ideal of $I(\mathscr{D})$ if and only if there is a strong vertex cover of $\mathscr{D}$ such that*

$$L = (L_1(C) \cup \{x_i^{d_i} \mid x_i \in L_2(C) \cup L_3(C)\}).$$

Theorem 2 ([34]) *If $\mathscr{D}$ is a weighted oriented graph and $\Upsilon(\mathscr{D})$ is the set of all strong vertex covers of $\mathscr{D}$, then the irreducible decomposition of $I(\mathscr{D})$ is*

$$I(\mathscr{D}) = \bigcap_{C \in \Upsilon(\mathscr{D})} I_C,$$

where $I_C = (L_1(C) \cup \{x_i^{d_i} \mid x_i \in L_2(C) \cup L_3(C)\})$.

Proof This follows at once from Proposition 1 and Theorem 1. □

Corollary 4 ([34]) *Let $\mathscr{D}$ be a weighted oriented graph. Then $\mathfrak{p}$ is an associated prime of $I(\mathscr{D})$ if and only if $\mathfrak{p} = (C)$ for some strong vertex cover C of $\mathscr{D}$.*

Example 5 Let K be the field of rational numbers and let $\mathscr{D}$ be the weighted digraph of Fig. 1 whose edge ideal is $I = I(\mathscr{D}) = (x_1^2x_3,\ x_1x_2^2,\ x_3x_2^2,\ x_3x_4^2,\ x_4^2x_5,\ x_2^2x_5)$. By Theorem 2, the irreducible decomposition of I is

$$I = (x_1^2,\ x_2^2,\ x_4^2) \cap (x_1,\ x_3,\ x_5) \cap (x_2^2,\ x_3,\ x_4^2) \cap (x_2^2,\ x_3,\ x_5).$$

Using *Macaulay*2 [16], we get that I is a Cohen–Macaulay ideal whose Rees algebra is Cohen-Macaulay and whose integral closure is

$$\overline{I} = I + (x_1x_2x_3,\ x_1x_3x_4,\ x_2x_3x_4,\ x_2x_4x_5).$$

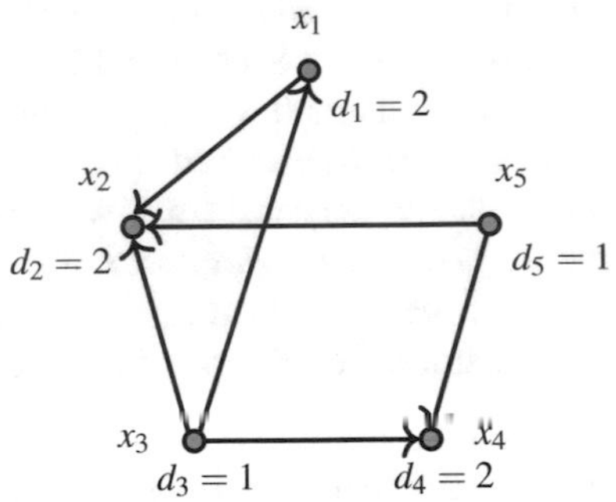

Fig. 1 A Cohen–Macaulay digraph

We note that the Cohen–Macaulayness of both I and its Rees algebra is destroyed (or recovered) by a single stroke of reversing the edge orientation of (x_5, x_2). This also destroys the unmixedness property of I.

In the summer of 2017 Antonio Campillo asked in a seminar at the University of Valladolid if there was anything special if we take an oriented graph $\mathscr{D}$ with underlying graph G and set d_i equal to $\deg_G(x_i)$ for $i = 1, \ldots, n$. It will turn out that in determining the Cohen–Macaulay property of $\mathscr{D}$ one can always make this canonical choice of weights.

Lemma 4 *Let $I \subset R$ be a monomial ideal, let x_i be a variable and let $h_1, \ldots, h_r$ be the monomials of $G(I)$ where x_i occurs. If x_i occurs in h_j with exponent 1 for all j and m is a positive integer, then I is Cohen–Macaulay of height g if and only if $((G(I) \setminus \{h_j\}_{j=1}^r) \cup \{x_i^m h_j\}_{j=1}^r)$ is Cohen–Macaulay of height g.*

Proof It follows at once from [32, Lemmas 3.3 and 3.5]. □

It was pointed out to us by Ngô Viêt Trung that the next proposition follows from the fact that the map $x_i \to y_i^{d_i}$ (replacing x_i by $y_i^{d_i}$) defines a faithfully flat homomorphism from $K[X]$ to $K[Y]$.

Proposition 5 *Let I be a squarefree monomial ideal and let $d_i = d(x_i)$ be a weighting of the variables. If G' is set of monomials obtained from $G(I)$ by replacing each x_i with $x_i^{d_i}$, then I is Cohen–Macaulay if and only if $I' = (G')$ is Cohen–Macaulay.*

Proof It follows applying Lemma 4 to each x_i. □

If a vertex x_i is a *sink* (i.e., has only arrows entering x_i), the next result shows that the Cohen-Macaulay property of $I(\mathscr{D})$ is independent of the weight of x_i.

Corollary 5 *If x_i is a sink of a weighted oriented graph $\mathscr{D}$ and $\mathscr{D}'$ is the digraph obtained from $\mathscr{D}$ by replacing d_i with $d_i = 1$. Then $I(\mathscr{D})$ is Cohen–Macaulay if and only if $I(\mathscr{D}')$ is Cohen–Macaulay.*

That is, to determine whether or not an oriented graph $\mathscr{D}$ is Cohen–Macaulay one may assume that all sources and sinks have weight 1. In particular if all vertices of $\mathscr{D}$ are either sources of sinks and G is its underlying graph, then $I(\mathscr{D})$ is Cohen–Macaulay if and only if $I(G)$ is Cohen–Macaulay.

Let I be a monomial ideal and let x_i be a fixed variable that occurs in $G(I)$. Let q be the maximum of the degrees in x_i of the monomials of $G(I)$ and let $\mathscr{B}_i$ be the set of all monomial of $G(I)$ of degree in x_i equal to q. For use below we set

$$\mathscr{A}_i := \{x^a \mid \deg_{x_i}(x^a) < q\} \cap G(I) = G(I) \setminus \mathscr{B}_i,$$

$p := \max\{\deg_{x_i}(x^a) \mid x^a \in \mathscr{A}_i\}$ and $L := (\{x^a/x_i \mid x^a \in \mathscr{B}_i\} \cup \mathscr{A}_i\})$.

Theorem 3 ([24]) *Let I be a monomial ideal. If $p \geq 1$, and $q - p \geq 2$, then*

$$\operatorname{depth}(R/I) = \operatorname{depth}(R/L).$$

Proof To simplify notation we set $i = 1$. We may assume that $G(I) = \{f_1, \ldots, f_r\}$, where $f_1, \ldots, f_m$ are all the elements of $G(I)$ that contain x_1^q and $f_{m+1}, \ldots, f_s$ are all the elements of $G(I)$ that contain some positive power x_1^ℓ of x_1 for some $1 \leq \ell < q$. Let $X' = \{x_{1,2}, \ldots, x_{1,q-1}\}$ be a set of new variables. If $f = x_1^s f'$ is a monomial with $\gcd(x_1, f') = 1$, we write $f^{\text{pol}} = x_{1,2} \cdots x_{1,t+1} x_1^{s-t} f'$ where $t = \min(q-2, s)$. Making a partial polarization of x_1^q with respect to the new variables $x_{1,2}, \ldots, x_{1,q-1}$ [42, p. 203], gives that f_i polarizes to $f_i^{\text{pol}} = x_{1,2} \cdots x_{1,q-1} x_1^2 f_i'$ for $i = 1, \ldots, m$, where $f_1', \ldots, f_m'$ are monomials that do not contain x_1 and $f_i = x_1^q f_i'$ for $i = 1, \ldots, m$. Hence, using that $q - p \geq 2$, one has the partial polarization

$$I^{\text{pol}} = (x_{1,2} \cdots x_{1,q-1} x_1^2 f_1', \ldots, x_{1,2} \cdots x_{1,q-1} x_1^2 f_m', f_{m+1}^{\text{pol}}, \ldots, f_s^{\text{pol}}, f_{s+1}, \ldots, f_r),$$

where $f_{m+1}^{\text{pol}}, \ldots, f_s^{\text{pol}}$ do not contain x_1 and I^{pol} is an ideal of $R^{\text{pol}} = R[x_{1,2}, \ldots, x_{1,q-1}]$. On the other hand, one has the partial polarization

$$L^{\text{pol}} = (x_{1,2} \cdots x_{1,q-1} x_1 f_1', \ldots, x_{1,2} \cdots x_{1,q-1} x_1 f_m', f_{m+1}^{\text{pol}}, \ldots, f_s^{\text{pol}}, f_{s+1}, \ldots, f_r).$$

By making the substitution $x_1^2 \to x_1$ in each element of $G(I^{\text{pol}})$ this will not affect the depth of $R^{\text{pol}}/I^{\text{pol}}$ (see [32, Lemmas 3.3 and 3.5]). Thus

$$q-2+\operatorname{depth}(R/I) = \operatorname{depth}(R^{\text{pol}}/I^{\text{pol}}) = \operatorname{depth}(R^{\text{pol}}/L^{\text{pol}}) = q-2+\operatorname{depth}(R/L),$$

and consequently $\operatorname{depth}(R/I) = \operatorname{depth}(R/L)$. □

Corollary 6 *Let $I = I(\mathscr{D})$ be the edge ideal of a vertex-weighted oriented graph with vertices $x_1, \ldots, x_n$ and let d_i be the weight of x_i. If $\mathscr{U}$ is the digraph obtained from $\mathscr{D}$ by assigning weight 2 to every vertex x_i with $d_i \geq 2$, then I is Cohen–Macaulay if and only if $I(\mathscr{U})$ is Cohen–Macaulay.*

Proof By applying Theorem 3 to each vertex x_i of $\mathscr{D}$ of weight at least 3, we obtain that depth$(R/I(\mathscr{D}))$ is equal to depth$(R/I(\mathscr{U}))$. Since $I(\mathscr{D})$ and $I(\mathscr{U})$ have the same height, then $I(\mathscr{D})$ is Cohen-Macaulay if and only if $I(\mathscr{U})$ is Cohen–Macaulay. □

Lemma 5 ([19, Theorem 16.3(4),p. 200]) *Let $\mathscr{D}$ be an oriented graph. Then $\mathscr{D}$ is acyclic, i.e., $\mathscr{D}$ has no oriented cycles, if and only if there is a linear ordering of the vertex set $V(\mathscr{D})$ such that all the edges of $\mathscr{D}$ are of the form (x_i, x_j) with $i < j$.*

A complete oriented graph is called a *tournament*. The next result shows that weighted acyclic tournaments are Cohen–Macaulay.

Corollary 7 *Let $\mathscr{D}$ be a weighted oriented graph. If the underlying graph G of $\mathscr{D}$ is a complete graph and $\mathscr{D}$ has no oriented cycles, then $I(\mathscr{D})$ is Cohen–Macaulay.*

Proof By Lemma 5, $\mathscr{D}$ has a source x_i for some i. Hence $\{x_1, \ldots, x_n\}$ is not a strong vertex cover of $\mathscr{D}$ because there is no arrow entering x_i. Thus, by Corollary 4, the maximal ideal $\mathfrak{m} = (x_1, \ldots, x_n)$ cannot be an associated prime of $I(\mathscr{D})$. Therefore $R/I(\mathscr{D})$ has depth at least 1. As $\dim(R/I(\mathscr{D})) = 1$, we get that $R/I(\mathscr{D})$ is Cohen–Macaulay. □

The next result gives an interesting family of digraphs whose edge ideals satisfy Alexander duality. Recall that a digraph $\mathscr{D}$ is called *transitive* if for any two edges (x_i, x_j), (x_j, x_k) in $E(\mathscr{D})$ with i, j, k distinct, we have that $(x_i, x_k) \in E(\mathscr{D})$. Acyclic tournaments are transitive and transitive oriented graphs are acyclic.

Theorem 4 *If $\mathscr{D}$ is a transitive oriented graph and $I = I(\mathscr{D})$ is its edge ideal, then Alexander duality holds, that is, $I^* = I^\vee$.*

Proof "$\supset$": Take $x^a \in G(I^\vee)$. According to Theorem 2, there is a strong vertex cover C of $\mathscr{D}$ such that

$$x^a = \left(\prod_{x_k \in L_1} x_k\right)\left(\prod_{x_k \in L_2 \cup L_3} x_k^{d_k}\right), \tag{2}$$

where $L_i = L_i(C)$ for $i = 1, 2, 3$. Fix a monomial $x_i x_j^{d_j}$ in $G(I(\mathscr{D}))$, that is, $(x_i, x_j) \in E(\mathscr{D})$. It suffices to show that x^a is in the ideal $I_{i,j} := (\{x_i, x_j^{d_j}\})$. If $x_i \in C$, then by Eq. (2) the variable x_i occurs in x^a because C is equal to $L_1 \cup L_2 \cup L_3$. Hence x^a is a multiple of x_i and x^a is in $I_{i,j}$, as required. Thus we may assume that $x_i \notin C$. By Theorem 2 the ideal

$$I_C = (L_1 \cup \{x_k^{d_k} \mid x_k \in L_2 \cup L_3\})$$

is an irreducible component of $I(\mathscr{D})$ and $x_i x_j^{d_j} \in I_C$.

Case (I): $x_ix_j^{d_j} \in (L_1)$. Then $x_ix_j^{d_j} = x_kx^b$ for some $x_k \in L_1$. Hence, as $x_i \notin C$, we get $j = k$. Therefore, as $x_j \in L_1$, there is $x_\ell \notin C$ such that (x_j, x_ℓ) is in $E(\mathscr{D})$. Using that $\mathscr{D}$ is transitive gives $(x_i, x_\ell) \in E(\mathscr{D})$ and $x_ix_\ell^{d_\ell} \in I(\mathscr{D})$. In particular $x_ix_\ell^{d_\ell} \in I_C$, a contradiction because x_i and x_ℓ are not in C. Hence this case cannot occur.

Case (II): $x_ix_j^{d_j} \in (\{x_k^{d_k} | x_k \in L_2 \cup L_3\})$. Then $x_ix_j^{d_j} = x_k^{d_k}x^b$ for some $x_k \in L_2 \cup L_3$. As $x_i \notin C$, we get $j = k$ and by Eq. (2) we obtain $x^a \in I_{i,j}$, as required.

"$\subset$": Take a minimal generator x^α of I^*. By Lemma 1, for each i either $\alpha_i = 1$ or $\alpha_i = d_i$. Consider the set $A = \{x_i | \alpha_i \geq 1\}$. We can write $A = A_1 \cup A_2$, where A_1 (resp. A_2) is the set of all x_i such that $\alpha_i = 1$ (resp. $\alpha_i = d_i \geq 2$). As (A) contains I, from the proof of Proposition 1, and using Theorem 2, there exists a strong vertex cover C of $\mathscr{D}$ contained in A such that the ideal

$$I_C = (L_1(C) \cup \{x_i^{d_i} | x_i \in L_2(C) \cup L_3(C)\})$$

is an irreducible component of $I(\mathscr{D})$. Thus it suffices to show that any monomial of $G(I_C)$ divides x^α because this would give $x^a \in I^\vee$.

Claim (I): If $x_k \in A_1$, then $d_k = 1$ or $x_k \in L_1(A)$. Assume that $d_k \geq 2$. Since x^α is a minimal generator of I^*, the monomial x^α/x_k is not in I^*. Then there is and edge (x_i, x_j) such that x^α/x_k is not in the ideal $I_{i,j} := (\{x_i, x_j^{d_j}\})$. As $x^\alpha \in I^*$ and $d_k \geq 2$, one has that x^α is in $I_{i,j}$ and $i = k$. Notice that x_j is not in A_2 because x^α/x_k is not in $I_{k,j}$. If x_j is not in A_1 the proof is complete because $x_k \in L_1(A)$. Assume that x_k is in A_1. Then $d_j \geq 2$ because x^α/x_k is not in $I_{k,j}$. Setting $k_1 = k$ and $k_2 = j$ and applying the previous argument to x^α/x_{k_2}, there is $x_{k_3} \notin A_2$ such that (x_{k_2}, x_{k_3}) is in $E(\mathscr{D})$. Since $\mathscr{D}$ is transitive, (x_{k_1}, x_{k_3}) is in $E(\mathscr{D})$. If x_{k_3} is not in A_1 the proof is complete. If x_{k_3} is in A_1, then $d_{k_3} \geq 2$ and we can continue using the previous argument. Suppose we have constructed $x_{k_1}, \ldots, x_{k_s}$ for some $s \leq r$ such that $x_{k_s} \notin A_2$, and $(x_{k_1}, x_{k_{s-1}})$ and $(x_{k_{s-1}}, x_{k_s})$ are in $E(\mathscr{D})$. Since $\mathscr{D}$ is transitive, (x_{k_1}, x_{k_s}) is in $E(\mathscr{D})$. If x_{k_s} is not in A_1 the proof is complete. If x_{k_s} is in A_1 and $s < r$, then $d_{k_s} \geq 2$ and we can continue the process. If x_{k_s} is in A_1 and $s = r$, that is, $A_1 = \{x_{k_1}, \ldots, x_{k_r}\}$, then applying the previous argument to x^α/x_{k_r} there is x_{r+1} not in A such that (x_r, x_{r+1}) is in $E(\mathscr{D})$. Thus by transitivity (x_{k_1}, x_{r+1}) is in $E(\mathscr{D})$, that is, x_{k_1} is in $L_1(A)$.

Claim (II): If $x_k \in A_2$, then $x_k \in L_2(A)$. Since $x^\alpha \in G(I^*)$ and $\alpha_k = d_k \geq 2$, there is (x_i, x_k) in $E(\mathscr{D})$ such that x^α/x_k is not in $I_{i,k} = (\{x_i, x_k^{d_k}\})$. In particular x_i is not in A. To prove that x_k is in $L_2(A)$ it suffices to show that x_k is not in $L_1(A)$. If x_k is in $L_1(A)$, there is x_j not in A such that (x_k, x_j) is in $E(\mathscr{D})$. As $\mathscr{D}$ is transitive, we get that (x_i, x_j) is in $E(\mathscr{D})$ and $A \cap \{x_i, x_j\} = \emptyset$, a contradiction because (A) contains I.

Take a monomial $x_k^{a_k}$ of $G(I_C)$.

Case (A): $x_k \in L_1(C)$. Then $a_k = 1$. There is $(x_k, x_j) \in E(\mathscr{D})$ with $x_j \notin C$. Notice $x_k \in A_1$. Indeed if $x_k \in A_2$, then x_k is in $L_2(A)$ because of Claim (II). Then there is (x_i, x_k) in $E(\mathscr{D})$ with $x_i \notin A$. By transitivity $(x_i, x_j) \in E(\mathscr{D})$ and $\{x_i, x_j\} \cap C = \emptyset$, a contradiction because (C) contains I. Thus $x_k \in A_1$, that is, $\alpha_k = 1$. This proves that $x_k^{a_k}$ divides x^α.

Case (B): $x_k \in L_2(C)$. Then $x_k^{a_k} = x_k^{d_k}$. First assume $x_k \in A_1$. Then, by Claim (I), $d_k = 1$ or $x_k \in L_1(A)$. Clearly $x_k \notin L_1(A)$ because $L_1(A) \subset L_1(C)$ and x_k—being in $L_2(C)$—cannot be in $L_1(C)$. Thus $d_k = 1$ and $x_k^{d_k}$ divides x^α. Next assume $x_k \in A_2$. Then, by construction of A_2, $x_k^{d_k}$ divides x^α.

Case (C): $x_k \in L_3(C)$. Then $x_k^{a_k} = x_k^{d_k}$. First assume $x_k \in A_1$. Then, by Claim (I), $d_k = 1$ or $x_k \in L_1(A)$. Clearly $x_k \notin L_1(A)$ because $L_1(A) \subset L_1(C)$ and x_k—being in $L_3(C)$—cannot be in $L_1(C)$. Thus $d_k = 1$ and $x_k^{d_k}$ divides x^α. Next assume $x_k \in A_2$. Then, by construction of A_2, $x_k^{d_k}$ divides x^α. □

Corollary 8 *If $\mathscr{D}$ is a weighted acyclic tournament, then $I(\mathscr{D})^* = I(\mathscr{D})^\vee$, that is, Alexander duality holds.*

Proof The result follows readily from Theorem 4 because acyclic tournaments are transitive. □

Example 6 Let $\mathscr{D}$ be the weighted oriented graph whose edges and weights are

$$(x_2, x_1), (x_3, x_2), (x_3, x_4), (x_3, x_1),$$

and $d_1 = 1, d_2 = 2, d_3 = 1, d_4 = 1$, respectively. This digraph is transitive. Thus $I(\mathscr{D})^* = I(\mathscr{D})^\vee$.

Example 7 The irreducible decomposition of the ideal $I = (x_1x_2^2, x_1x_3^2, x_2x_3^2)$ is

$$I = (x_1, x_2) \cap (x_1, x_3^2) \cap (x_2^2, x_3^2),$$

in this case $I^\vee = (x_1x_2, x_1x_3^2, x_2^2x_3^2) = (x_1, x_2^2) \cap (x_1, x_3^2) \cap (x_2, x_3^2) = I^*$.

Example 8 The irreducible decomposition of the ideal $I = (x_1x_2^2, x_3x_1^2, x_2x_3^2)$ is

$$I = (x_1^2, x_2) \cap (x_1, x_3^2) \cap (x_2^2, x_3) \cap (x_1^2, x_2^2, x_3^2),$$

in this case $I^\vee = (x_1^2x_2, x_1x_3^2, x_2^2x_3) \subsetneq (x_1, x_2^2) \cap (x_3, x_1^2) \cap (x_2, x_3^2) = I^*$.

Example 9 The irreducible decomposition of the ideal $I = (x_1x_2^2, x_1^2x_3)$ is

$$I = (x_1) \cap (x_1^2, x_2^2) \cap (x_3, x_2^2),$$

in this case $I^\vee = (x_1, x_2^2x_3) \supsetneq I^* = (x_1, x_2^2) \cap (x_1^2, x_3) = (x_1^2, x_1x_3, x_2^2x_3)$.

We come to the main result of this section.

Theorem 5 *Let $\mathscr{D}$ be a weighted oriented forest without isolated vertices and let G be its underlying forest. The following conditions are equivalent:*

(a) *$\mathscr{D}$ is Cohen–Macaulay.*
(b) *$I(\mathscr{D})$ is unmixed, that is, all its associated primes have the same height.*
(c) *G has a perfect matching $\{x_1, y_1\}, \ldots, \{x_r, y_r\}$ so that $\deg_G(y_i) = 1$ for $i = 1, \ldots, r$ and $d(x_i) = d_i = 1$ if $(x_i, y_i) \in E(\mathscr{D})$.*

Proof It suffices to show the result when G is connected, that is, when $\mathscr{D}$ is an oriented tree. Indeed $\mathscr{D}$ is Cohen–Macaulay (resp. unmixed) if and only if all connected components of $\mathscr{D}$ are Cohen–Macaulay (resp. unmixed) [34, 40].

$(a) \Rightarrow (b)$: This implication follows from the general fact that Cohen–Macaulay graded ideals are unmixed [42, Corollary 3.1.17].

$(b) \Rightarrow (c)$: According to the results of [40] one has that $|V(G)| = 2r$ and G has a perfect matching $\{x_1, y_1\}, \ldots, \{x_r, y_r\}$ so that $\deg_G(y_i) = 1$ for $i = 1, \ldots, r$. Consider the oriented graph $\mathscr{H}$ with vertex set $V(\mathscr{H}) = \{x_1, \ldots, x_r\}$ whose edges are all (x_i, x_j) such that $(x_i, x_j) \in E(\mathscr{D})$. As $\mathscr{H}$ is acyclic, by Lemma 5, we may assume that the vertices of $\mathscr{H}$ have a "topological" order, that is, if $(x_i, x_j) \in E(\mathscr{H})$, then $i < j$. If $(y_i, x_i) \in E(\mathscr{D})$ for $i = 1, \ldots, r$, there is nothing to prove. Assume that $(x_k, y_k) \in E(\mathscr{D})$ for some k. To complete the proof we need only show that $d(x_k) = d_k = 1$. We proceed by contradiction assuming that $d_k \geq 2$. In particular x_k cannot be a source of $\mathscr{H}$. Setting $X = \{x_1, \ldots, x_r\}$, consider the set of vertices

$$C = (X \setminus N_{\mathscr{H}}^-(x_k)) \cup \{y_i \mid x_i \in N_{\mathscr{H}}^-(x_k)\} \cup \{y_k\},$$

where $N_{\mathscr{H}}^-(x_k)$ is the *in-neighbor* set of x_k consisting of all $y \in V(\mathscr{H})$ such that $(y, x_k) \in E(\mathscr{H})$. Clearly C is a vertex cover of G with $r + 1$ elements because the set $N_{\mathscr{H}}^-(x_k)$ is an independent set of G. Let us show that C is a strong cover of $\mathscr{D}$. The set $N_{\mathscr{H}}^-(x_k)$ is not empty because x_k is not a source of $\mathscr{D}$. Thus x_k is not in $L_3(C)$. Since $L_3(C) \subset \{x_k, y_k\} \subset C$, we get $L_3(C) = \{y_k\}$. There is no arrow of $\mathscr{D}$ with source at x_k and head outside of C, that is, x_k is in $L_2(C)$. Hence (x_k, y_k) is in $E(\mathscr{D})$ with $x_k \in L_2(C)$ and $d(x_k) \geq 2$. This means that C is a strong cover of $\mathscr{D}$. Applying Theorem 4 gives that $\mathfrak{p} = (C)$ is an associated prime of $I(\mathscr{D})$ with $r + 1$ elements, a contradiction because $I(\mathscr{D})$ is an unmixed ideal of height r.

$(c) \Rightarrow (a)$: We proceed by induction on r. The case $r = 1$ is clear because $I(\mathscr{D})$ is a principal ideal, hence Cohen–Macaulay. Let $\mathscr{H}$ be the graph defined in the proof of the previous implication. As before we may assume that the vertices of $\mathscr{H}$ are in topological order and we set $R = K[x_1, \ldots, x_r, y_1, \ldots, y_r]$.

Case (I): Assume that $(y_r, x_r) \in E(\mathscr{D})$. Then x_r is a sink of $\mathscr{D}$ (i.e., has only arrows entering x_r). Using the equalities

$$(I(\mathscr{D})\colon x_r^{d_r}) = (N_G(x_r), I(\mathscr{D} \setminus N_G(x_r))) \text{ and } (I(\mathscr{D}), x_r^{d_r}) = (x_r^{d_r}, I(\mathscr{D} \setminus \{x_r\})),$$

and applying the induction hypothesis to $I(\mathscr{D}\backslash N_G(x_r))$ and $I(\mathscr{D}\backslash\{x_r\})$ we obtain that the ideals $(I(\mathscr{D})\colon x_r^{d_r})$ and $(I(\mathscr{D}), x_r^{d_r})$ are Cohen–Macaulay of dimension r. Therefore, as $I(\mathscr{D})$ has height r, from the exact sequence

$$0 \to R/(I(\mathscr{D})\colon x_r^{d_r})[-d_r] \xrightarrow{x_r^{d_r}} R/I(\mathscr{D}) \to R/(I(\mathscr{D}), x_r^{d_r}) \to 0$$

and using the depth lemma (see [42, Lemma 2.3.9]) we obtain that $I(\mathscr{D})$ is Cohen–Macaulay.

Case (II): Assume that $(x_r, y_r) \in E(\mathscr{D})$. Then $d(x_r) = d_r = 1$ and $x_r y_r^{e_r} \in I(\mathscr{D})$, where $d(y_r) = e_r$. Using the equalities

$$(I(\mathscr{D})\colon x_r) = (N_G(x_r)\backslash\{y_r\}, y_r^{e_r}, I(\mathscr{D}\backslash N_G(x_r))) \text{ and } (I(\mathscr{D}), x_r) = (x_r, I(\mathscr{D}\backslash\{x_r\})),$$

and applying the induction hypothesis to $I(\mathscr{D}\backslash N_G(x_r))$ and $I(\mathscr{D}\backslash\{x_r\})$ we obtain that the ideals $(I(\mathscr{D})\colon x_r)$ and $(I(\mathscr{D}), x_r)$ are Cohen–Macaulay of dimension r. Therefore, as $I(\mathscr{D})$ has height r, from the exact sequence

$$0 \to R/(I(\mathscr{D})\colon x_r)[-1] \xrightarrow{x_r} R/I(\mathscr{D}) \to R/(I(\mathscr{D}), x_r) \to 0$$

and using the depth lemma [42, Lemma 2.3.9] we obtain that $I(\mathscr{D})$ is Cohen–Macaulay. □

The following result was conjectured in a preliminary version of this paper and proved recently in [18] using polarization of monomial ideals.

Theorem 6 ([18, Theorem 3.1]) *Let $\mathscr{D}$ be a weighted oriented graph and let G be its underlying graph. Suppose that G has a perfect matching $\{x_1, y_1\}, \ldots, \{x_r, y_r\}$ where $\deg_G(y_i) = 1$ for each i. The following conditions are equivalent:*

(a) *$\mathscr{D}$ is Cohen–Macaulay.*
(b) *$I(\mathscr{D})$ is unmixed, that is, all its associated primes have the same height.*
(c) *$d(x_i) = 1$ for any edge of $\mathscr{D}$ of the form (x_i, y_i).*

The equivalence between (*b*) and (*c*) was also proved in [34, Theorem 4.16].

Remark 7 If $\mathscr{D}$ is a Cohen–Macaulay weighted oriented graph, then $I(\mathscr{D})$ is unmixed and $\mathrm{rad}(I(\mathscr{D}))$ is Cohen–Macaulay. This follows from the fact that Cohen–Macaulay ideals are unmixed and using a result of Herzog, Takayama and Terai [21, Theorem 2.6] which is valid for any monomial ideal. It is an open question whether the converse is true [34, Conjecture 5.5].

Example 10 The radical of the ideal $I = (x_2x_1, x_3x_2^2, x_3x_4)$ is Cohen–Macaulay and I is not unmixed. The irreducible components of I are (x_1, x_3), (x_2, x_3), (x_1, x_2^2, x_4), (x_2, x_4).

Example 11 (Terai) The ideal $I = (x_1, x_2)^2 \cap (x_2, x_3)^2 \cap (x_3, x_4)^2$ is unmixed, $\mathrm{rad}(I)$ is Cohen-Macaulay, and I is not Cohen–Macaulay.

Acknowledgements We would like to thank Ngô Viêt Trung and the referees for a careful reading of the paper and for the improvements suggested. The first, third and fourth authors were partially supported by the Spanish *Ministerio de Economía y Competitividad* grant MTM2016-78881-P. The second and fourth authors were supported by SNI. The fifth author was supported by a scholarship from CONACYT

References

1. Bahiano, C.: Symbolic powers of edge ideals. J. Algebra **273**(2), 517–537 (2004)
2. Bang-Jensen, J., Gutin, G.: Digraphs: Theory, Algorithms and Applications. Springer Monographs in Mathematics. Springer, London (2006)
3. Bruns, W., Ichim, B., Römer, T., Sieg, R., Söger, C.: Normaliz. Algorithms for rational cones and affine monoids. Available at http://normaliz.uos.de
4. Carvalho, C., Lopez Neumann, V.G., López, H.H.: Projective nested cartesian codes. Bull. Braz. Math. Soc. (N.S.) **48**(2), 283–302 (2017)
5. Cooper, S., Embree, R., Hà, H.T., Hoefel, A.H.: Symbolic powers of monomial ideals. Proc. Edinb. Math. Soc. (2) **60**(1), 39–55 (2017)
6. Crispin Quiñonez, V.: Integral closure and other operations on monomial ideals. J. Commut. Algebra **2**(3), 359–386 (2010)
7. Dao, H., Huneke, C., Schweig, J.: Bounds on the regularity and projective dimension of ideals associated to graphs. J. Algebraic Combin. **38**(1), 37–55 (2013)
8. Dao, H., De Stefani, A., Grifo, E., Huneke, C., Núñez-Betancourt, L.: Symbolic powers of ideals. Preprint, arXiv:1708.03010 (2017)
9. Dupont, L.A., Villarreal, R.H.: Edge ideals of clique clutters of comparability graphs and the normality of monomial ideals. Math. Scand. **106**(1), 88–98 (2010)
10. Escobar, C., Villarreal, R.H., Yoshino, Y.: Torsion freeness and normality of blowup rings of monomial ideals. In: Commutative Algebra. Lecture Notes in Pure and Applied Mathematics, vol. 244, pp. 69–84. Chapman & Hall/CRC, Boca Raton (2006)
11. Francisco, C., Hà, H.T., Mermin, J.: Powers of Square-Free Monomial Ideals and Combinatorics. Commutative Algebra, pp. 373–392. Springer, New York (2013)
12. Gimenez, P., Simis, A., Vasconcelos, W.V., Villarreal, R.H.: On complete monomial ideals. J. Commut. Algebra **8**(2), 207–226 (2016)
13. Gitler, I., Villarreal, R.H.: Graphs, Rings and Polyhedra. Aportaciones Mat. Textos, vol. 35. Sociedad Matemática Mexicana, México (2011)
14. Gitler, I., Valencia, C., Villarreal, R.H.: A note on Rees algebras and the MFMC property. Beiträge Algebra Geom. **48**(1), 141–150 (2007)
15. Goto, S., Nishida, K.: The Cohen–Macaulay and Gorenstein Rees algebras associated to filtrations. Mem. Am. Math. Soc. **110**(526), 1–134 (1994)
16. Grayson, D., Stillman, M.: Macaulay2. Available via anonymous FTP from math.uiuc.edu (1996)
17. Hà, H.T., Morey, S.: Embedded associated primes of powers of square-free monomial ideals. J. Pure Appl. Algebra **214**(4), 301–308 (2010)
18. Hà, H.T., Lin, K.-N., Morey, S., Reyes, E., Villarreal, R.H.: Edge ideals of oriented graphs, Preprint (2017)
19. Harary, F.: Graph Theory. Addison-Wesley, Reading (1972)
20. Herzog, J., Hibi, T.: Monomial Ideals. Graduate Texts in Mathematics, vol. 260. Springer, London/New York (2011)
21. Herzog, J., Takayama, Y., Terai, N.: On the radical of a monomial ideal. Arch. Math. **85**, 397–408 (2005)
22. Herzog, J., Hibi, T., Trung, N.V.: Symbolic powers of monomial ideals and vertex cover algebras. Adv. Math. **210**, 304–322 (2007)

23. Hosten, S., Smith, G.G.: Monomial Ideals. Computations in Algebraic Geometry with Macaulay 2. Algorithms and Computational Mathematics, vol. 8, pp. 73–100. Springer, Berlin (2002)
24. Martínez-Bernal, J., Morey, S., Villarreal, R.H., Vivares, C.E.: Depth and regularity of monomial ideals via polarizations and combinatorial optimization. Preprint (2017)
25. Martínez-Bernal, J., Pitones, Y., Villarreal, R.H.: Minimum distance functions of graded ideals and Reed-Muller-type codes. J. Pure Appl. Algebra **221**, 251–275 (2017)
26. Matsumura, H.: Commutative Algebra. Benjamin-Cummings, Reading (1980)
27. Miller, E.: Alexander duality for monomial ideals and their resolutions. Preprint, arXiv:math/9812095 (1998)
28. Miller, E.: The Alexander duality functors and local duality with monomial support. J. Algebra **231**(1), 180–234 (2000)
29. Miller, E., Sturmfels, B.: Combinatorial Commutative Algebra. Graduate Texts in Mathematics, vol. 227. Springer, New York (2004)
30. Morey, S., Villarreal, R.H.: Edge ideals: algebraic and combinatorial properties. In: Francisco, C., Klingler, L.C., Sather-Wagstaff, S., Vassilev, J.C. (eds.) Progress in Commutative Algebra, Combinatorics and Homology, vol. 1, pp. 85–126. De Gruyter, Berlin (2012)
31. Nejad, A.N., Simis, A., Zaare-Nahandi, R.: The Aluffi algebra of the Jacobian of points in projective space: torsion-freeness. J. Algebra **467**, 268–283 (2016)
32. Neves, J., Vaz Pinto, M., Villarreal, R.H.: Regularity and algebraic properties of certain lattice ideals. Bull. Braz. Math. Soc. (N.S.) **45**, 777–806 (2014)
33. Paulsen, C., Sather-Wagstaff, S.: Edge ideals of weighted graphs. J. Algebra Appl. **12**(5), 1250223, p. 24 (2013)
34. Pitones, Y., Reyes, E., Toledo, J.: Monomial ideals of weighted oriented graphs. Preprint, arXiv:1710.03785 (2017)
35. Simis, A.: Combinatoria Algebrica, XVIII Coloquio Brasileiro de Matematica, IMPA (Apendice. Palimpsesto 2: Potencias simbolicas, 2.1) (1991)
36. Simis, A.: Effective computation of symbolic powers by Jacobian matrices. Commun. Algebra **24**, 3561–3565 (1996)
37. Simis, A., Trung, N.V.: The divisor class group of ordinary and symbolic blow-ups. Math. Z. **198**, 479–491 (1988)
38. Simis, A., Vasconcelos, W.V., Villarreal, R.H.: On the ideal theory of graphs. J. Algebra **167**, 389–416 (1994)
39. Van Tuyl, A.: A beginner's guide to edge and cover ideals. In: Bigatti, A., Gimenez, P., Sáenz-de-Cabezón, E. (eds.) Monomial Ideals, Computations and Applications. Lecture Notes in Mathematics, vol. 2083, pp. 63–94. Springer, Heidelberg (2013)
40. Villarreal, R.H.: Cohen–Macaulay graphs. Manuscripta Math. **66**, 277–293 (1990)
41. Villarreal, R.H.: Monomial Algebras. Monographs and Textbooks in Pure and Applied Mathematics, vol. 238. Chapman and Hall/CRC Press, Boca Raton (2001)
42. Villarreal, R.H.: Monomial Algebras. Monographs and Research Notes in Mathematics, 2nd edn. Chapman and Hall/CRC, Boca Raton (2015)
43. Zariski, O., Samuel, P.: Commutative Algebra, vol. II. Springer, New York (1960)

Asymptotics of Reduced Algebraic Curves Over Finite Fields

J. I. Farrán

Abstract The number $A(q)$ shows the asymptotic behaviour of the quotient of the number of rational points over the genus of non-singular absolutely irreducible curves over $\mathbb{F}_q$. Research on bounds for $A(q)$ is closely connected with the so-called asymptotic main problem in Coding Theory. In this paper, we study some generalizations of this number for non-irreducible curves, their connection with $A(q)$ and their application in Coding Theory. We also discuss the possibility of constructing codes from non-irreducible curves, both from theoretical and practical point of view.

1 Introduction

Let $\mathbb{F}_q$ be the finite field with q elements. Coding Theory is interested in the search of *asymptotically good families* of error-correcting codes, that is, families of codes whose relative parameters have a limit point in the *code domain* over $\mathbb{F}_q$. In such a family, the length of the codes cannot be upper bounded (see [11] for further details). In the case of geometric Goppa codes, that is, codes constructed from non-singular absolutely irreducible algebraic curves X defined over $\mathbb{F}_q$, Tsfasman proved in [13] that this problem is connected with the asymptotic behaviour of the quotient of the number of $\mathbb{F}_q$-rational points over the genus of the curves, which is given by the number

$$A(q) \doteq \limsup_{g \to \infty} \frac{N_q(g)}{g}$$

Partially supported by the project MTM2015-65764-C3-1-P (MINECO/FEDER).

J. I. Farrán (✉)
Universidad de Valladolid, Valladolid, Spain
e-mail: jifarran@eii.uva.es

G.-M. Greuel et al. (eds.), *Singularities, Algebraic Geometry, Commutative Algebra, and Related Topics*, https://doi.org/10.1007/978-3-319-96827-8_22

where $N_q(g) \doteq \max\{N_q(X) \ : \ X$ *non-singular absolutely irreducible algebraic curve defined over* $\mathbb{F}_q$ *with* $g(X) = g\}$ (see [14] for further details). For this type of codes, the length is closely connected with the number of rational points, and so this number cannot be upper bounded. The known results about $A(q)$ can be summarized as follows:

(I) There exists an absolute constant $c > 0$ such that $A(q) \geq c \log_2 q$ for all q (in particular, $A(q) > 0$).
(II) $A(q) \leq \sqrt{q} - 1$ for all q.
(III) If q is a square, then $A(q) = \sqrt{q} - 1$.

Statement (I) was proved by Serre in [12], statement (II) by Drinfeld and Vlăduţ in [7], and statement (III) independently by different people, for example by Tsfasman, Vlăduţ and Zink in [15].

Thus a general upper bound for $A(q)$ is known, but nobody knows its exact value for any non-square q. Moreover, the general lower bound given by Serre is not too explicit, since it is based on class field theory, which is not considered elementary by many authors. Instead, there exist some explicit lower bounds for concrete values of q, like q^3 for example (see [14]). More recently, various people have obtained better lower bounds for non-prime q by using towers of function fields (see for example [8]). In this paper, we give an alternative way to compute $A(q)$ by using non-irreducible curves under certain conditions (cf. Sects. 2, 3 and 6). This may be useful for estimating $A(q)$ and so for the main asymptotic problem in Coding Theory, namely for bounding the code domain.

The main motivation for this work comes from the problem of constructing families of curves given by singular plane models with *prescribed singularities* (see [3, 9]). The asymptotic behaviour of such families is derived from the estimates for the number of rational branches and the genus, depending of the type of singularities which are imposed, but it is typically difficult to decide whether these curves are irreducible or not just from the knowledge of the prescribed singularities (see Sect. 3). In this way, we change the condition of being irreducible by having an upper bound for the number of irreducible components, what still can be expressed as a problem of prescribed singularities. The advantage of using plane models for curves instead of non-singular curves is that they can be treated computationally by means of effective algorithms, vgr. the Brill-Noether algorithm (see [5]), in order to construct geometric Goppa codes.

2 $A(q)$ for Non-irreducible Curves

For each $1 \leq r \in \mathbb{N}$ and each finite field F_q, let $\mathscr{A}_r(q)$ be the set of all possible curves defined over $\mathbb{F}_q$ which are non-singular, *reduced* (without multiple components) and with exactly r disjoint irreducible components over $\overline{\mathbb{F}_q}$ which are all of them defined over $\mathbb{F}_q$. In the same way, let $\mathscr{A}^r(q)$ be the set of all possible curves defined over $\mathbb{F}_q$ which are also non-singular and reduced, but now with *at most* r disjoint irreducible components over $\overline{\mathbb{F}_q}$ which are all of them defined also over $\mathbb{F}_q$. If one writes $X \in \mathscr{A}_r(q)$ as a disjoint union $X = X^1 \cup \ldots \cup X^r$, with $X^i \in \mathscr{A}_1(q)$, one has

$$N_q(X) = \sum_{i=1}^{r} N_q(X^i) \quad \text{and} \quad 2g(X) - 2 = \sum_{i=1}^{r}(2g(X^i) - 2)$$

i.e. both the number of rational points and the *Euler characteristic* are additive. Thus $g(X) = 1 - r + \sum_{i=1}^{r} g(X^i)$ and hence $g(X^i) \leq g(X) + r - 1$ for all $i = 1, \ldots, r$ (see the details in [10]).

Definition 2.1 We define the following subsets of $\mathbb{R}$:

$$\mathfrak{A}_r(q) \doteq \{\limsup_{n\to\infty} \frac{N_q(X_n)}{g(X_n)} \ : \ X_n \in \mathscr{A}_r(q) \text{ and } \lim_{n\to\infty} N_q(X_n) = \infty\}$$

$$\mathfrak{A}^r(q) \doteq \{\limsup_{n\to\infty} \frac{N_q(X_n)}{g(X_n)} \ : \ X_n \in \mathscr{A}^r(q) \text{ and } \lim_{n\to\infty} N_q(X_n) = \infty\}$$

Remark 2.2 Note that the value of $\limsup_{n\to\infty} \frac{N_q(X_n)}{g(X_n)}$ will not change if we either set $g(X_n) - 1$ instead of $g(X_n)$, or remove any irreducible component of X_n either with genus $g = 0, 1$ or without rational points. On the other hand, by using the fact that for any positive real numbers a_i, b_j one has

$$\frac{a_1 + \ldots + a_r}{b_1 + \ldots + b_r} \leq \max\left\{\frac{a_1}{b_1}, \ldots, \frac{a_r}{b_r}\right\}$$

one obtains that for any curve $X = X^1 \cup \ldots \cup X^r$ as above one has

$$\frac{N_q(X)}{g(X) - 1} = \frac{\sum_{i=1}^{r} N_q(X^i)}{\sum_{i=1}^{r}(g(X^i) - 1)} \leq \max\left\{\frac{N_q(X^i)}{g(X^i) - 1} \ : \ i = 1, \ldots, r\right\}$$

In particular, this means that the *rate* N_q/g for a reduced curve X is at most as good as that of one of its components.

Remark 2.3

(a) Notice that $\lim_{n\to\infty} N_q(X_n) = \infty$ implies in both cases that $g(X_n)$ is not upper bounded, by using the Hasse-Weil bound (see [16]). Thus, we can assume that $\lim_{n\to\infty} g(X_n) = \infty$ and that $g(X_n) \neq 0$ for all n.
(b) From this, one has that for each $\alpha \in \mathfrak{A}_r(q), \mathfrak{A}^r(q)$ there exists a sequence $\{X_n\}_{n=1}^{\infty}$ in $\mathscr{A}_r(q), \mathscr{A}^r(q)$ with $\lim_{n\to\infty} N_q(X_n) = \infty$, $\lim_{n\to\infty} g(X_n) = \infty$, $g(X_n) \geq 1$ and

$$\lim_{n\to\infty} \frac{N_q(X_n)}{g(X_n)} = \alpha$$

where we can assume that both $\{N_q(X_n)\}_{n=1}^{\infty}$ and $\{g(X_n)\}_{n=1}^{\infty}$ are increasing.

(c) Also notice that $\mathfrak{A}_r(q), \mathfrak{A}^r(q)$ are bounded subsets of $\mathbb{R}$, since by using the Hasse-Weil bound one has

$$\frac{N_q(X)}{g(X)} \leq r(1+q) + 2r^2\sqrt{q}$$

if $X \in \mathscr{A}_r(q), \mathscr{A}^r(q)$ and $g(X) \geq 1$, and they are lower bounded by 0 by using the previous remarks.

Definition 2.4 Now define the following real numbers

$$A_r(q) \doteq \sup \mathfrak{A}_r(q) \quad \text{and} \quad A^r(q) \doteq \sup \mathfrak{A}^r(q)$$

One can easily check that the sets $\mathfrak{A}_r(q)$ and $\mathfrak{A}^r(q)$ are closed for the usual topology in $\mathbb{R}$ and, as a consequence, one actually has

$$A_r(q) = \max \mathfrak{A}_r(q) \quad \text{and} \quad A^r(q) = \max \mathfrak{A}^r(q)$$

Lemma 2.5 $A^1(q) = A_1(q) = A(q)$.

Proof The first equality is obvious. Now one has

$$A(q) = \lim_{n\to\infty} \frac{N_q(X_n)}{g(X_n)}$$

for a sequence $\{X_n\}_{n=1}^{\infty}$ in $\mathscr{A}_1(q)$ with $N_q(X_n) = N_q(g(X_n))$, $\{g(X_n)\}_{n=1}^{\infty}$ increasing and $\lim_{n\to\infty} g(X_n) = \infty$. But then $\lim_{n\to\infty} N_q(X_n) = \infty$, since $A(q) > 0$. Thus $A(q) \in \mathfrak{A}_1(q)$, and therefore $A(q) \leq A_1(q)$.

Conversely, if one writes $A_1(q) = \lim_{n\to\infty} \frac{N_q(X_n)}{g(X_n)}$ as in Remark 2.3(b), since $N_q(X_n) \leq N_q(g(X_n))$ one has $A_1(q) \leq \lim_{n\to\infty} \frac{N_q(g(X_n))}{g(X_n)} \leq A(q)$.

3 Properties of $A_r(q)$ and $A^r(q)$

The following result shows some elementary properties of the numbers $A_r(q)$ and $A^r(q)$, and the proof is left to the reader.

Lemma 3.1 *For all $r \geq 1$ one has*

i) $A_r(q) \leq A^r(q)$ **iii)** $A^r(q) > 0$
ii) $A^r(q) \leq A^{r+1}(q)$ **iv)** $A_r(q), A^r(q) < \infty$

The following formula is not so elementary, and together with Lemma 3.1(iii), it implies in particular that $A_r(q) > 0$ for all $r \geq 1$.

Proposition 3.2 $A_r(q) = A^r(q)$ *for all* $r \geq 1$.

Proof It suffices to prove that $A^r(q) \leq A_r(q)$. Let $X = X^1 \cup \ldots \cup X^s$ in $\mathscr{A}_s(q)$ with $1 \leq s < r$ and $g(X) \geq 1$, and let $Z \in \mathscr{A}_1(q)$ with $g(Z) = 1$. Then define

$$Y = X^1 \cup \ldots \cup X^r$$

where $X^i \cong Z$ for $i > s$, and so that X^i are all disjoint for $i = 1, \ldots, r$. In this way, for any $X \in \mathscr{A}^r(q)$ we can obtain $Y \in \mathscr{A}_r(q)$ such that $N_q(X) \leq N_q(Y)$ and $g(X) = g(Y) \geq 1$, and hence

$$\frac{N_q(X)}{g(X)} \leq \frac{N_q(Y)}{g(Y)} \tag{1}$$

Now take $A^r(q) \in \mathfrak{A}^r(q)$, write $A^r(q) = \lim_{n\to\infty} \frac{N_q(X_n)}{g(X_n)}$ as in Remark 2.3(b), and proceed as in (1) with $X = X_n$ in order to get a sequence $\{Y_n\}_{n=1}^{\infty}$ in $\mathscr{A}_r(q)$ such that

$$N_q(X_n) \leq N_q(Y_n) \quad \text{and} \quad g(X_n) = g(Y_n) \geq 1$$

Thus one has $\lim_{n\to\infty} N_q(Y_n) = \infty$, $\lim_{n\to\infty} g(Y_n) = \infty$ and

$$\frac{N_q(X_n)}{g(X_n)} \leq \frac{N_q(Y_n)}{g(Y_n)}$$

for all $n \in \mathbb{N}$ and hence

$$A^r(q) = \lim_{n\to\infty} \frac{N_q(X_n)}{g(X_n)} \leq \limsup_{n\to\infty} \frac{N_q(Y_n)}{g(Y_n)} \in \mathfrak{A}_r(q)$$

what proves the inequality.

Now we are able to prove the main result of this paper.

Theorem 3.3 $A_r(q) = A(q)$ *for all* $r \geq 1$.

Proof By using Proposition 3.2 together with Lemmas 2.5 and 3.1, one has just to prove that $A_r(q) \leq A_1(q)$ for $r \geq 1$, and this follows from Remark 2.2.

The above new result is interesting itself from an arithmetic point of view, independently on its possible use in coding theory (for example, for estimating the size of the code domain). Theorem 3.3 shows that the number $A(q)$ can be generalized to the case of non-irreducible (but reduced) curves over $\mathbb{F}_q$, and its value is just the same number as in the irreducible case, whenever the number of components of the considered curves is upper bounded by some integer r. In other words, one does not need irreducibility in order to obtain the number $A(q)$, but just bounding the number of irreducible components. In particular, one has that

$$A_r(q) = A^r(q) = \sqrt{q} - 1$$

if q is a square. On the other hand, if q is non-square, the above result can be used (in theory) for finding lower bounds for the number $A(q)$. In fact, in order to guarantee that there are r components at most, and that all of them are defined over $\mathbb{F}_q$, it suffices to take curves with at most r branches in a certain hyperplane (that of infinity, for example), and so that all those branches are defined over $\mathbb{F}_q$. In this way, we have the following

Corollary 3.4 *Let* $\{Y_n\}_{n=1}^{\infty}$ *be a family of curves, defined over* $\mathbb{F}_q$ *and embedded in* $\mathbb{P}^{t_n} := \mathbb{P}^{t_n}(\mathbb{F}_q)$, *and let r be a positive integer. Assume that for every n there exists a hyperplane* $H_n \subseteq \mathbb{P}^{t_n}$ *(defined also over* $\mathbb{F}_q$*) such that the total number of branches at the points in* $H_n \cap Y_n$ *is at most r, and so that they are all defined over* $\mathbb{F}_q$. *Let* X_n *be the normalization of* Y_n *for every n, and assume that* $\lim_{n\to\infty} N_q(X_n) = \infty$. *Then one has*

$$A(q) \geq \limsup_{n\to\infty} \frac{N_q(X_n)}{g(X_n)}$$

Proof Every irreducible component of Y_n over $\overline{\mathbb{F}_q}$ has at least one point in H_n. Since, by assumption, both the branches at such a point and the hyperplane H_n itself are all defined over $\mathbb{F}_q$, then the irreducible components of Y_n (and so those of X_n) are defined over $\mathbb{F}_q$. Thus, the total number of irreducible components is r at most, and the statement follows from Theorem 3.3.

This corollary suggests us an alternative tool for the (hard) problem of estimating $A(q)$ for the non-square case: "using plane curves with prescribed singularities" (i.e. in the Corollary 3.4 one always takes $t_n = 2$, and H_n is usually taken as the line of infinity). The idea is just trying to find (singular) plane curves passing through certain points and having certain types of singularities at those points. If those prescribed singularities have many branches over $\mathbb{F}_q$, they will produce many rational points in the non-singular model of this curve, and thus the number of rational points will be lower bounded (see for example [3]). On the other hand, we should be able to upper bound the degree of such curves, in order to get an upper bound for the genus (see for example [9] for some recent results in this direction). These observations lead to a lower bound for the quotient $N_q(X_n)/g(X_n)$, and thus we can obtain a lower bound for $A(q)$. The type of singularities imposes local conditions on the corresponding *Newton polytope* of the equation of the plane curve, and for each singularity, the δ-invariant (necessary for the computation of the genus) is also computed from the Newton polytope and the local equation, by means for example of Hamburger-Noether expansions (see [4]). However, the research of concrete asymptotically good examples is not easy at all. As a summary, the **procedure** would be the following:

- Fix points in $\mathbb{P}^2$ and singularity-types at those points.
  In this way we consider:
  - The δ-invariants, to be used in the genus formula.

  - A lower bound for the number of (rational) branches; the key point would be just to find a good balance between the number of branches (places) and the δ-invariants, in order to get a good rate $N_q(X)/g(X)$.
- Upper bound the degree of such a curve X.
  Together with the previous step, this yields:
  - An upper bound for the geometric genus $g(X)$.
  - A lower estimate for $N_q(X)/g(X)$, as a consequence.
- Try to do it "in families" so that the asymptotic behaviour is good, that is, find sequences of curves with "clever enough" prescribed singularities, so that the estimate for $A(q)$ is not trivial (i.e. non-zero).

The conceptual advantage of using $A^r(q)$ instead of $A(q)$ for searching bounds is that it is typically difficult to decide whether the curves are irreducible or not just from the knowledge of the prescribed singularities. The irreducibility could be obtained for example by imposing either strong conditions on the Newton polytope (see [3]), or that there exists just one branch at infinity (see [1]). In this way, we change the condition of being irreducible by just having an upper bound for the number of irreducible components, what is weaker, in principle. Moreover, this last condition is easier to impose in practice from the equation of the curve, since this is still nothing but some kind of problem of prescribed singularities (at infinity), namely having at most r branches at infinity which are defined over $\mathbb{F}_q$. Note that the situation in [1] corresponds to the very particular case when $r = 1$, and then the curve becomes absolutely irreducible.

An additional advantage of using plane curves instead of non-singular curves in higher dimension, is that still one can perform some computations with the aid of some classical algorithms, in order to use $\mathscr{A}^r(q)$ for the construction of asymptotically good sequences of geometric Goppa codes, as we will show in Sect. 5.

4 Geometric Goppa Codes from Non-irreducible Curves

In this section, we will try to take advantage of the results of the previous section about reduced curves in order to construct geometric Goppa codes. In this way, the first idea which arises is why not constructing codes over non-irreducible curves. We discuss this topic now from a theoretical point of view, and we check also what happens with the asymptotic behaviour of the codes. The results are mostly negative, i.e. we show that using curves with several components is not useful in Coding Theory for finding asymptotically good sequences of codes, but nevertheless we do this discussion in order to prevent useless attempts in the future. However, even though the construction does not work fine in an asymptotic sense, it is correct and may be useful for other coding purposes.

Construction R: Assume that the reduced curve χ is a disjoint union of χ_i, for $i = 1, \ldots, r$, and fix rational divisors D_i and G_i on each component. Let then the codes $C_i = C_L(\chi_i, G_i, D_i)$. Consider D as the (formal) sum of the D_i's, G as the sum of the G_i's and $\mathcal{L}(G)$ as the direct sum of the $\mathcal{L}(G_i)$'s, and then define the evaluation of $f \in \mathcal{L}(G_i)$ at a point of the support of D_j by zero if $i \neq j$. One can easily check that the evaluation code $C = C_L(\chi, G, D)$ is the direct sum of the C_i's. First we can easily estimate the dimension of C. In fact, if g, g_i denote respectively the genus of χ, χ_i, one can prove the following version of the Riemann-Roch theorem for the non-irreducible case.

"If $\deg G_i > 2g_i - 2$ for all i, then $\ell(G) = \deg(G) + 1 - g$".

Notice that the hypothesis $\deg G > 2g - 2$ is not sufficient. Moreover, since we are doing componentwise evaluation, in order to obtain injectivity in such evaluation we actually need the extra conditions

$$2g_i - 2 < \deg G_i < n_i = \deg D_i \quad \forall i \tag{2}$$

i.e. we need sufficiently many rational points in all the components we are using for evaluation. This problem can be solved just by forgetting about those irreducible components which do not satisfy this condition, and this does not have any influence on the asymptotic behaviour, if the whole curve χ has a good balance between rational points and genus. However, with the above set of conditions (2), one has that the dimension of C is just $\ell(G)$, which is the sum of the $\ell(G_i)$'s.

Unfortunately, what behaves worse is the minimum distance. In general, for any direct sum of codes one has $d(C) = \min\{d(C_i) \mid i = 1, \ldots, r\}$, and this formula cannot be improved, since a codeword coming from evaluating a function in $\mathcal{L}(G_i)$ will have a zero in all the positions corresponding to those points which do not belong to χ_i. This is bad for the asymptotic problem, since it means that $d(C)$ behaves like the worst of the distances $d_i \doteq d(C_i)$. In particular, note that if in such a sequence of codes one has that $d_i/n_i \to 0$ for some suitable component i, then the same happens to the rate d/n related to the codes constructed from the reduced curves. Hence we need that all the components have a good asymptotic behaviour at the same time, and even in this case the limit seems to be smaller than the components separately (for instance, if all the d_i's are about the same size, then d/n is about d_i/n_i divided by r), what makes impossible to find optimal sequences by using this kind of codes.

Rational functions: We try now to see the above construction from different points of view. In fact, the idea is just trying to generalize the concept of "rational function" over χ and then just define $\mathcal{L}(G) \doteq \{f \mid (f) + G \geq 0\}$. In the sequel, we explore different logical possibilities to define *rational functions* in the non-irreducible case.

With the same notations as in the *construction R*, denote by A_i the (affine) coordinates ring of the component χ_i and by F_i its function field. Since the direct sum of neither the rings A_i nor the fields F_i is not even an integral domain, we have no hope to obtain a function field in this way, and thus we can consider instead the direct product

$$F \doteq F_1 \times \cdots \times F_r \tag{3}$$

This is not a domain either, but its structure is not very far from a field. In this "function ring", functions can be regarded as r-tuples of rational functions f_i (on each component χ_i), and the only reasonable way to do evaluations (for example, if one wants the evaluation to be linear) is to define $f_i(P) = 0$ if $P \in \chi_j$ and $i \neq j$. Thus, when we carry on with the definitions, we essentially fall in the construction R.

In order to try something new, the next possibility is to consider first the "affine coordinates ring" $A \doteq A_1 \times \cdots \times A_r$, denote by S_0 the multiplicative closed system of all the non-zero-divisors (i.e. those r-tuples with all the components different from zero modulo χ_i), and finally define the *total ring of fractions*

$$F \doteq S_0^{-1} A \tag{4}$$

This is a typical concept in Commutative Algebra (see for example [2]). However, one can check again that the constructions (3) and (4) are equivalent, and hence we still are considering essentially the construction R (note that the definition $S_0^{-1}(F_1 \times \cdots \times F_r)$ would not change anything either).

Problem: From a theoretical point of view, an interesting problem for future work arises now: Consider a (possibly singular) plane model for the nonirreducible curve χ, and regarding intersection points between different components as extra singularities, give a description of $\mathscr{L}(G)$ (from the construction of R) in terms of adjoints with sufficiently large degree, in a similar way as in the theory of Brill-Noether for the irreducible case. Note that the resolution of plane curve singularities via symbolic Hamburger-Noether expressions (as explained in [5]) works also for the reduced (possibly non-irreducible) case, so that the work to do is just the generalization of the adjunction theory and the Brill-Noether theorem.

5 Geometric Goppa Codes from Reduced Plane Models

In the sequel, we start again from the results and techniques of the Sect. 3, but now we will try to come back from there to the irreducible case. Our aim is just to give a *computational treatment* of this problem. In fact, assume that we have found an asymptotically good sequence of curves (X_n) in $\mathscr{A}^r(q)$ and we want to construct a good sequence of geometric Goppa codes from this sequence.

Firstly, by using the Remark 2.2, it is easy to check that for each n we can take an irreducible component Y_n of X_n such that

$$\limsup_{n\to\infty} \frac{N_q(Y_n)}{g(Y_n)} = \limsup_{n\to\infty} \frac{N_q(X_n)}{g(X_n)}$$

Thus, the idea is to compute the irreducible components of each X_n and select one, Y_n, having maximum rate N_q/g among all the components. From that moment on, we fall in the irreducible case.

In order to do the computations effective, we will assume that the curves X_n are given by plane models, i.e. by a polynomial F_n in two variables with coefficients in the finite field $\mathbb{F}_q$. In this case, the *procedure* is the following:

Step (1) **Factorization** of F_n over $\mathbb{F}_q$. This is a problem with an effective solution, and it is implemented (for example) in the computer algebra system `Singular` [6]. Note that also for non plane curves the problem is solvable (but more time consuming) with the aid of *primary decomposition* (two different algorithms for computing primary decompositions are also implemented in `Singular`).

Step (2) Computing all the **rational places** (i.e. branches) and the **genus** of each irreducible component of F_n. This can be done effective with the aid of Gröbner bases computations together with the theory of symbolic Hamburger-Noether expressions for plane curves (see [5]), and both kind of computations are very efficient in the system `Singular`. If the curve is non-singular (plane or not) everything can be reduced to Gröbner bases computations, and the genus is not difficult to compute, but if there are singularities, there exists no general method to compute from the equations of the curve neither the rational places nor the genus. There are only some tricky techniques coming usually from the theory of function fields (see for example [8]).

Step (3) Constructing the **geometric Goppa codes** from (irreducible) plane curves is also well-known, and it is also implemented from the release `Singular 2.0`. Again, there are no general methods to calculate, from the equations of the curve, neither the rational places nor the genus, for the case of non-plane curves, since the construction is based on the computation of either the vector space $\mathscr{L}(G_n)$ for some divisor G_n, or the Weierstrass semigroup at P_n with the corresponding rational functions, for some rational point P_n, and such problems do not have a general effective solution apart from the case of plane curves.

Remark 5.1 Notice that, for large input, the complexity of the above procedure has about the same order as that of constructing geometric Goppa codes from irreducible plane curves. In fact, the computations of step (2) are necessary for step (3), and then the computations of both steps can be upper bounded by r times those of steps (2) and (3) for a single component, r being a constant. On the other hand, the complexity of step (1), that is the factorization of one polynomial in two variables over a finite field, is not higher (roughly speaking) than the complexity of computing Gröbner bases of ideals, which is included in step (2). Thus, we do not do essentially worse than in the irreducible case.

Example 5.2 Consider the plane curve over $\mathbb{F}_{16}$ given by

$$F(x, y) = x^{10} + x^5(y^4 + y^2) + (y + 1)(y^5 + y^2)$$

We proceed now with the above algorithm:

Step (1) Factorize the polynomial F:

$$F(x, y) = G(x, y) \cdot H(x, y)$$

where

$$G(x, y) = x^5 + y^4 + y \ , \quad H(x, y) = x^5 + y^2 + y$$

Step (2) Select the best component:

- The curve given by $G(x, y)$ has genus $g = 6$ and $N_{16} = 65$ rational points over $\mathbb{F}_{16}$.
- The curve given by $H(x, y)$ has genus $g = 2$ and $N_{16} = 33$ rational points over $\mathbb{F}_{16}$.
- Then, we select the curve given by $H(x, y)$, since it has better rate N_{16}/g.

Step (3) Compute a suitable geometric Goppa code from the selected curve H (this is implemented in the last version of `Singular` [6], with the library "`brnoeth.lib`", created by J.I. Farrán and Ch. Lossen).

Notice that one could also construct geometric Goppa codes on both components and then consider the direct sum of both, according to the construction R. Since we have 98 rational points in total, the length of such a code could be close to this number.

6 Conclusions

We show now that if the number of irreducible components of the curves which are used is not upper bounded, then the results of Sect. 3 do not hold, and thus we cannot do better than in the above case. In fact, we must now look at the sets

$$\mathscr{A}^{\infty}(q) \doteq \cup_{r=1}^{\infty} \mathscr{A}_r(q)$$

$$\mathfrak{A}^{\infty}(q) \doteq \{ \lim_{n \to \infty} \frac{N_q(X_n)}{g(X_n)} \ : \ X_n \in \mathscr{A}^{\infty}(q), \ g(X_n) \neq 0, \ \lim_{n \to \infty} N_q(X_n) = \infty \}$$

and then define the number $A^{\infty}(q) \doteq \sup \ \mathfrak{A}^{\infty}(q) \in \overline{\mathbb{R}}$.

Firstly, we will see that the condition $\lim_{n \to \infty} g(X_n) = \infty$ is not necessary in $\mathfrak{A}^{\infty}(q)$. In fact, let $\chi \in \mathscr{A}_1(q)$ with $g(\chi) = 1$ and $N_q(\chi) \geq 1$ (note that such a curve exists for all q). Now, for all $n \geq 1$ we get a disjoint union $X_n = X_n^1 \cup \ldots \cup X_n^n$, with $X_n^i \cong \chi$ for all i and n. One obviously has

$$N_q(X_n) \geq n \ \text{ and } \ g(X_n) = 1 \ \text{ for all } \ n$$

Thus $\lim_{n\to\infty} N_q(X_n) = \infty$, even though $\{g(X_n)\}_{n=1}^{\infty}$ is constant. Furthermore, one has

$$\limsup_{n\to\infty} \frac{N_q(X_n)}{g(X_n)} \geq \limsup_{n\to\infty} n = +\infty$$

and hence $A^{\infty}(q) = +\infty$.

In order to avoid this, we may include the hypothesis $\lim_{n\to\infty} g(X_n) = \infty$ in the definition, that is, define

$$\mathfrak{A}_{\infty}(q) \doteq \{\limsup_{n\to\infty} \frac{N_q(X_n)}{g(X_n)} : X_n \in \mathscr{A}^{\infty}(q),\ \lim_{n\to\infty} N_q(X_n) = \lim_{n\to\infty} g(X_n) = \infty\}$$

and $A_{\infty}(q) \doteq \sup\ \mathfrak{A}_{\infty}(q) \in \overline{\mathbb{R}}$. Nevertheless, we will also see that $A_{\infty}(q)$ is not finite.

In fact, take $\Upsilon \in \mathscr{A}_1(q)$ with $g(\Upsilon) = 2$ and $N_q(\Upsilon) \geq 1$ (such a curve also exists for all q) and, for $n \geq 1$, define a disjoint union $X_n = X_n^1 \cup \ldots \cup X_n^n$, with either $X_n^i \cong \chi$ or $X_n^i \cong \Upsilon$ for all i and n (χ as the previous example). Note that $1 \leq g(X_n) \leq 2n + 1 - n = n + 1$ for all n.

Now, since $1 \leq \lfloor \log n \rfloor \leq n + 1$ for $n \gg 0$, we can choose X_n^i in X_n so that $g(X_n) = \lfloor \log n \rfloor$ for $n \gg 0$, namely $X_n^i \cong \Upsilon$ for $1 \leq i \leq \lfloor \log n \rfloor - 1$ and $X_n^i \cong \chi$ for $\lfloor \log n \rfloor \leq i \leq n$. Therefore

$$\limsup_{n\to\infty} \frac{N_q(X_n)}{g(X_n)} \geq \limsup_{n\to\infty} \frac{n}{\log n} = +\infty$$

and then one obtains again that $A_{\infty}(q) = +\infty$.

This shows that, for either calculating $A(q)$ or giving lower bounds for this number, we must necessarily take families of curves whose number of irreducible components is upper bounded. This assumption together with the property of the irreducible components being defined over $\mathbb{F}_q$ can be guaranteed by conditions on the branches at infinity, as we have shown in Sect. 3.

References

1. Abhyankar, S.S.: Irreducibility criterion for germs of analytic functions of two complex variables. Adv. Math. **74**, 190–257 (1989)
2. Atiyah, M.F., MacDonald, I.G.: Introduction to Commutative Algebra. Addison-Wesley Publishing Company, Reading (1969)
3. Beelen, P., Pellikaan, R.: The Newton-polygon of plane curves with many rational points. Des. Codes Crypt. **21**, 41–67 (2000)
4. Campillo, A.: Algebroid Curves in Positive Characteristic. Lecture Notes in Mathematics, vol. 813. Springer, Berlin/New York (1980)
5. Campillo, A., Farrán, J.I.: Symbolic Hamburger-Noether expressions of plane curves and applications to AG codes. Math. Comput. **71**, 1759–1780 (2001)

6. Decker, W., Greuel, G.-M., Pfister, G., Schönemann, H.: "Singular 4-1-0", A computer algebra system for polynomial computations, Centre for Computer Algebra, TU Kaiserslautern (2016). Available via http://www.singular.uni-kl.de/
7. Drinfeld, V.G., Vlăduţ, S.G.: Number of points of an algebraic curve. Funktsional'-nyi Analiz i Ego Prilozhenia **17**, 53–54 (1983)
8. García, A., Stichtenoth, H., Thomas, M.: On towers and composita of towers of function fields over finite fields. Finite Fields Appl. **3**, 257–274 (1997)
9. Greuel, G.-M., Lossen, Ch., Shustin, E.: Plane curves of minimal degree with prescribed singularities. Invent. Math. **133**(3), 539–580 (1998)
10. Hartshorne, R.: Algebraic Geometry. Graduate Texts in Mathematics, vol. 52. Springer, New York (1977)
11. Lachaud, G.: Les Codes Géométriques de Goppa. Sém. Bourbaki, 37ème année **641**, 1984–1985 (1986). Astérisque
12. Serre, J.P.: Sur le nombre des points rationnels d'un courbe algébrique sur un corps fini. C. R. Acad. Sc. Paris **296**, 397–402 (1983)
13. Tsfasman, M.A.: Goppa codes that are better than Varshamov-Gilbert bound. Prob. Peredachi Inform. **18**, 3–6 (1982)
14. Tsfasman, M.A., Vlăduţ, S.G.: Algebraic-Geometric Codes. Mathematics and Its Applications, vol. 58. Kluwer Academic, Amsterdam (1991)
15. Tsfasman, M.A., Vlăduţ, S.G., Zink, Th.: Modular curves, Shimura curves and Goppa codes, better than Varshamov-Gilbert bound. Math. Nachr. **109**, 21–28 (1982)
16. Weil, A.: Basic Number Theory. Grundlehren der Mathematischen Wissenschaften, Bd. 144. Springer, New York (1974)

The Poincaré Polynomial of a Linear Code

Carlos Galindo, Fernando Hernando, Francisco Monserrat, and Ruud Pellikaan

Abstract We introduce the Poincaré polynomial of a linear q-ary code and its relation to the corresponding weight enumerator. The question of whether the Poincaré polynomial is a complete invariant is answered affirmatively for $q = 2, 3$ and negatively for $q \geq 4$. Finally we determine this polynomial for MDS codes and, by means of a recursive formula, for binary Reed-Muller codes.

1 Introduction

In dimension theory within Commutative Algebra, a very useful tool is the so-called Poincaré series [1]. It is associated to a finitely generated graded A-module $M = \bigoplus_{n\geq 0} M_n$, where $A = \bigoplus_{n\geq 0} A_n$ is a Noetherian graded ring with A_0 Artininian, and it is defined as the generating function of the lengths $\ell(M_n)$ of the finite A_0 modules M_n. That is, the Poincaré series is the formal series in $\mathbb{Z}[t]$, $P(M,t) = \sum_{n=0}^{\infty} \ell(M_n)t^n$. This series encodes all the mentioned lengths. In fact it is, by the Hilbert theorem, a rational function.

Using several variables, Poincaré series as generating functions have been extended to other objects with gradings on semigroups or having multi-index filtrations with satisfactory results in singularity theory. In the last 15 years, Antonio

C. Galindo (✉) · F. Hernando
Departamento de Matemáticas, Instituto Universitario de Matemáticas y Aplicaciones de Castellón, Universitat Jaume I, Castelló, Spain
e-mail: galindo@uji.es; carrillf@uji.es

F. Monserrat
Instituto Universitario de Matemática Pura y Aplicada (IUMPA), Universidad Politécnica de Valencia, Valencia, Spain
e-mail: framonde@mat.upv.es

R. Pellikaan
Discrete Mathematics, Technische Universiteit Eindhoven, Eindhoven, The Netherlands
e-mail: g.r.pellikaan@tue.nl

G.-M. Greuel et al. (eds.), *Singularities, Algebraic Geometry, Commutative Algebra, and Related Topics*, https://doi.org/10.1007/978-3-319-96827-8_23

Campillo, together with close colleagues has revitalized this study (see, for instance, [4–10, 12]).

In the above context, Poincaré series usually encode an infinite amount of data. Our resolve of applying similar ideas to error-correcting codes, which are finite sets, is spurred by the fact that the amount of data involved is usually very large and by the subsequent expectation that their grouping according to suitable criteria may lead to interesting results.

Recall that error-correcting codes are used when information is received from some source through a noisy communication channel and one tries to correct (or detect) the produced errors. We will only consider linear (error-correcting) codes which are linear spaces of a vector space $\mathbb{F}_q^n$, where $\mathbb{F}_q$ is the finite field of q elements. For deciding about the goodness of a linear code C, it is customary to consider its parameters $[n, k, d]$, where k is the dimension of the linear space and d the minimum (Hamming) distance of the code, which is the minimum Hamming weight of their non-vanishing codewords. The generating function of the weight distribution of a code is named its weight enumerator [19] and (given the discrete nature of the codes) it is a polynomial in one variable. To determine weight distributions is not easy and the main result to study them is the so-called MacWilliams identity [19] which relates the weight enumerator of a linear code and its dual.

Let us proceed to introduce what we call the Poincaré polynomial of a linear code. This polynomial is also related to the weights of the codewords but contains more information than the weight enumerator. It is a polynomial in several variables and takes into account not only the weight of a codeword, but also the entries contributing to that weight. We will define this polynomial using arrangements of hyperplanes attached to the code. Our first result tells how the weight enumerator can be obtained from the Poincaré polynomial (Theorem 1). Then we show that the Poincaré polynomial and the multivariate Tutte polynomial [17, 20, 22] of the matroid given by the above arrangement determine each other (Theorem 2).

An equivalence φ of $\mathbb{F}_q^n$ is an $\mathbb{F}_q$-linear map $\varphi : \mathbb{F}_q^n \to \mathbb{F}_q^n$ that is a composition of a permutation matrix and diagonal matrix with non zero entries on the diagonal. Now φ is an equivalence of $\mathbb{F}_q^n$ if and only if φ is linear and leaves the Hamming metric *invariant*, that means that $d(\varphi(\mathbf{x}), \varphi(\mathbf{y})) = d(\mathbf{x}, \mathbf{y})$ for all $\mathbf{x}, \mathbf{y} \in \mathbb{F}_q^n$. Let C and D be $\mathbb{F}_q$-linear codes in $\mathbb{F}_q^n$. Then C is called *equivalent* to D if there exists an equivalence φ of $\mathbb{F}_q^n$ such that $\varphi(C) = D$. A map f from the set of all $\mathbb{F}_q$-linear codes to another set is called an *invariant* of $\mathbb{F}_q$-linear codes if $f(C) = f(\varphi(C))$ for every code C in $\mathbb{F}_q^n$ and every equivalence φ of codes. The parameters and the weight enumerator of a code are examples of invariants. A *complete invariant* of $\mathbb{F}_q$-linear codes is an invariant f such that for all $\mathbb{F}_q$-linear codes C and D we have that $f(C) = f(D)$ if and only if C and D are equivalent.

The question of whether the Poincaré polynomial is a complete invariant of $\mathbb{F}_q$-linear codes is answered affirmatively in the binary and ternary cases and negatively when $q \geq 4$ (Remark 1 and Corollary 1). We complete this note by providing an explicit expression (Proposition 6) for the Poincaré polynomial of an MDS code [21], and by giving a recursive formulae (Corollary 2) for computing the Poincaré polynomial of the binary Reed-Muller codes [21].

2 The Poincaré Polynomial

Denote by $\mathbb{F}_q$ the finite field with q elements. For a fixed positive integer k, consider the linear space $\mathbb{F}_q^k$. An arrangement of hyperplanes in $\mathbb{F}_q^k$, $(H_1, H_2, \ldots, H_n)$, is an n-tuple where each H_i, $1 \leq i \leq n$, is the set of solutions in $\mathbb{F}_q^k$ satisfying a linear equation with k variables.

Definition 1 An arrangement of hyperplanes in $\mathbb{F}_q^k$, $(H_1, H_2, \ldots, H_n)$, is called *simple* (respectively, *central*) whenever all the H_i are mutually different (respectively, are linear subspaces of $\mathbb{F}_q^k$). In addition, the arrangement is named *essential* when it is central and $H_1 \cap H_2 \cap \cdots \cap H_n = \{\mathbf{0}\}$.

Central arrangements can be considered in the projective space $\mathbb{P}^{k-1}(\mathbb{F}_q)$, in which case they are essential when $H_1 \cap H_2 \cap \cdots \cap H_n$ is the empty set.

A linear code $C \subseteq \mathbb{F}_q^n$ is *degenerate* when there is an index j such that $x_j = 0$ for all $x \in C$. Arrangements of hyperplanes and linear codes are intimately related to projective systems in the projective space. Indeed, n points $\mathscr{P}$ in the projective space $\mathbb{P}^{k-1}(\mathbb{F}_q)$ spanning $\mathbb{P}^{k-1}(\mathbb{F}_q)$ are called a *projective system* [23, 24]. The $k \times n$ matrix of coordinates of the system has rank k, which is what happens with the generator matrix G of non-degenerate linear $[n, k, d]$ codes over $\mathbb{F}_q$. Specifically we have the following result (see [20, 23, 24]), whose statement uses the concept of equivalence given in the introduction.

Proposition 1 *There exists a bijective map between equivalence classes of essential arrangements of n hyperplanes in $\mathbb{P}^{k-1}(\mathbb{F}_q)$ and equivalence classes of non-degenerate $[n, k, d]$ codes over $\mathbb{F}_q$.*

Recall that two projective systems $\mathscr{P} \subseteq \mathbb{P}$ and $\mathscr{P}' \subseteq \mathbb{P}'$ over projective spaces $\mathbb{P}$ and $\mathbb{P}'$ are called *equivalent* if there is a projective isomorphism between $\mathbb{P}$ and $\mathbb{P}'$ that takes $\mathscr{P}$ to $\mathscr{P}'$. Since projective systems and arrangements of hyperplanes are dual objects, we observe that the bijection mentioned in Proposition 1 comes from and extends to equivalence classes of projective systems of n points in $\mathbb{P}^{k-1}(\mathbb{F}_q)$.

Consider now a non-degenerate $[n, k, d]$ linear code C over $\mathbb{F}_q$ and assume that $G = (g_{ij})_{1 \leq i \leq k;\ 1 \leq j \leq n}$ is a generator matrix. Let H_j be the hyperplane in $\mathbb{F}_q^k$ defined by the equation $g_{1j}X_1 + g_{2j}X_2 + \cdots + g_{kj}X_k = 0$ and denote by $\mathscr{A}_G$ the arrangement of hyperplanes $(H_1, H_2, \ldots, H_n)$. Taking into account that a codeword $\mathbf{c} \in C$ satisfies $\mathbf{c} = \mathbf{x}G$ for some $\mathbf{x} \in \mathbb{F}_q^k$, we have that the jth coordinate of $\mathbf{c}$ satisfies $c_j = \sum_{i=1}^k g_{ij}x_i$ and so $c_j = 0$ if and only if $\mathbf{x}$ lies on the hyplerplane H_j. As a consequence, if we denote by $\mathrm{wt}(\mathbf{c})$ the Hamming weight of the codeword $\mathbf{c} \in C$, then the following result holds (see [23, 24]):

Proposition 2 *With the above notation, if $\mathbf{c} = \mathbf{x}G$ is a codeword of a non-degenerate linear code C over $\mathbb{F}_q$, then the number of hyperplanes of $\mathscr{A}_G$ going through $\mathbf{x}$ is equal to $n - \mathrm{wt}(\mathbf{c})$.*

Next we introduce the object we are interested in.

Definition 2 Let $C \subseteq \mathbb{F}_q^n$ be a non-degenerate linear code of dimension k with generator matrix G. Let $\mathscr{A}_G = (H_1, H_2, \ldots, H_n)$ be the corresponding arrangement of hyperplanes, and for each $\mathbf{x} \in \mathbb{F}_q^k$, define the n-tuple

$$\boldsymbol{\epsilon}(\mathbf{x}) = (\epsilon_1(\mathbf{x}), \epsilon_2(\mathbf{x}), \ldots, \epsilon_n(\mathbf{x})) \in \{0, 1\}^n$$

by $\epsilon_j(\mathbf{x}) = 1$ if $\mathbf{x} \in H_j$ and 0 otherwise.

Then, the *Poincaré polynomial* of the linear code C is defined as

$$P_C(t_1, t_2, \ldots, t_n) = \sum_{\mathbf{x} \in \mathbb{F}_q^k} \underline{t}^{\boldsymbol{\epsilon}(\mathbf{x})} \in \mathbb{Z}[t_1, t_2, \ldots, t_n],$$

where $\underline{t}^{\boldsymbol{\epsilon}(\mathbf{x})} = t_1^{\epsilon_1(\mathbf{x})} t_2^{\epsilon_2(\mathbf{x})} \cdots t_n^{\epsilon_n(\mathbf{x})}$.

For each subset $J \subseteq \{1, \ldots, n\}$, consider the monomial $\underline{t}^J = \prod_{j \in J} t_j$ in $\mathbb{Z}[t_1, \ldots, t_n]$, and define $a(J)$ to be the cardinality of the set of vectors $\mathbf{x} \in \mathbb{F}_q^k$ such that the set of zero coordinates of $\mathbf{x}$ is equal to J, or equivalently, such that $\mathbf{x} \in \cap_{j \in J} H_j$ but, for any $J' \not\supseteq J$, $\mathbf{x} \notin \cap_{j \in J'} H_j$. Then, there exists an alternative expression of the Poincaré polynomial as the following straightforward result states.

Proposition 3 *Let $C \subseteq \mathbb{F}_q^n$ be a non-degenerate linear code. Then, with the above notation, we have that*

$$P_C(t_1, t_2, \ldots, t_n) = \sum_{J \subseteq \{1,2,\ldots,n\}} a(J) \underline{t}^J.$$

As mentioned in the introduction, the weight distribution of a linear code $C \subseteq \mathbb{F}_q^n$ is an important non-complete invariant which provides important information for the structure and practical use of C. Its generating function comes with two equivalent versions. On the one hand, the *weight enumerator of* C, which is defined by $W_C(T) = \sum_{i=0}^n \omega_i T^i \in \mathbb{Z}[T]$, where $\omega_i := \text{card}\{\mathbf{c} \in C \mid \text{wt}(\mathbf{c}) = i\}$ and, on the other hand, the *homogeneous weight enumerator of* C, which is defined as the homogeneous polynomial $W_C(X, Y) = \sum_{i=0}^n \omega_i X^{n-i} Y^i \in \mathbb{Z}[X, Y]$. Now our first result establishes that the Poincaré polynomial of a code contains at least as much information as the weight enumerator.

Theorem 1 *Let $C \subseteq \mathbb{F}_q^n$ be a non-degenerate linear code. With the above notations we have that the homogeneous weight enumerator of C is the homogenization with respect to the variable Y of the polynomial $P_C(X, X, \ldots, X)$. In addition,*

$$P_C(T, T, \ldots, T) = T^n W_C(T^{-1}).$$

Proof It follows from the equality $P_C(X, X, \ldots, X) = \sum_{i=0}^n \omega_i X^{n-i}$. Its homogenization with respect to Y will be $\sum_{i=0}^n \omega_i X^{n-i} Y^i$, that is $W_C(X, Y)$. Now $T^n \left(\sum_{i=0}^n \omega_i T^{i-n}\right) = \sum_{i=0}^n \omega_i T^i = W_C(T)$, which concludes the proof. □

The arrangement $\mathscr{A}_G$ defined by a code C can be regarded as a matroid. Recall that a matroid is a pair $\mathscr{M} := (\mathscr{H}, \mathscr{I})$, where $\mathscr{H}$ is a finite set called the *ground set* of the matroid and $\mathscr{I}$ a family of subsets of $\mathscr{H}$, called the *independent sets* of the matroid, that must satisfy $\emptyset \in \mathscr{I}$ and $J \subseteq I \in \mathscr{I}$ implies $J \in \mathscr{I}$. In addition $(\mathscr{H}, \mathscr{I})$ have to satisfy that if $I, J \in \mathscr{I}$ and $\text{card}(I) > \text{card}(J)$, then there exists $i \in I \setminus J$ such that $J \cup \{i\} \in \mathscr{I}$. The *rank* of a subset J of $\mathscr{H}$ is the number of elements of a maximal independent subset of J.

Definition 3 ([22]) Let $\mathscr{M}$ be a matroid with ground set $\mathscr{H}$, $r_{\mathscr{M}}$ its rank function, and n the cardinality of $\mathscr{H}$. The *multivariate Tutte polynomial* is defined by

$$Z_{\mathscr{M}}(T, t_1, \ldots, t_n) = \sum_{J \subseteq \mathscr{H}} T^{-r_{\mathscr{M}}(J)} \underline{t}^J \in \mathbb{Z}[T^{-1}, t_1, \ldots, t_n].$$

The multivariate Tutte polynomial encodes the full structure of the matroid, contains as a special case the more known two variable Tutte polynomial and also the so-called chromatic polynomial. The Poincaré polynomial of a code and the multivariate Tutte polynomial of the matroid given by the arrangement attached to the code determine each other. For this result we need some results from [17, 20].

Let C be a linear code with generator matrix G. For a subset J of $\{1, 2, \ldots, n\}$ define

$$C(J) = \{\mathbf{c} \in C \mid c_j = 0 \text{ for all } j \in J\} \text{ and } l(J) = \dim C(J).$$

Let $\text{card}(J) = t$ and let G_J be the $k \times t$ submatrix of G consisting of the columns of G indexed by J. Let $r(J)$ be the rank of G_J. Then the map r is equal to $r_{\mathscr{M}}$, the rank of the associated matroid $\mathscr{M}$ of the arrangement of the code. Furthermore the dimension $l(J)$ is equal to $k - r(J)$ by [17] and [20, Lemma 3.2.12]. We have the following result from [20, Proposition 3.2.18].

Proposition 4 *Let C be a linear code of length n and minimum distance d. Assume that $d \leq w \leq n$, then the number of codewords in C of weight w is given by*

$$\omega_w = \sum_{t=n-w}^{n-d} (-1)^{n+w+t} \binom{t}{n-w} \sum_{\text{card}(J)=t} (q^{l(J)} - 1).$$

Furthermore $\omega_0 = 1$ and $\omega_w = 0$ for all $0 < w < d$.

Let $J \subseteq \{1, 2, \ldots, n\}$ consist of the m integers $j_1, j_2, \ldots, j_m$ with

$$1 \leq i_1 < i_2 < \cdots < i_m \leq n.$$

Let $\mathbf{x} \in \mathbb{F}_q^n$. Define

$$\mathbf{x}_J = (x_{j_1}, x_{j_2}, \ldots, x_{j_m}) \in \mathbb{F}_q^m$$

the restriction of $\mathbf{x}$ to the coordinates indexed by J. Let $\overline{J}$ be the relative complement of J in $\{1, 2, \ldots, n\}$. The *shortened* code C^J is obtained from $C(J)$ by deleting the entries that are at positions indexed by J:

$$C^J = \{\mathbf{c}_{\overline{J}} \mid \mathbf{c} \in C \text{ and } \mathbf{c}_J = 0\}.$$

Notice that $C(J)$ is an $\mathbb{F}_q$-linear vector space that is isomorphic with $\cap_{j\in J} H_j$ under the map $\mathbf{x} \mapsto \mathbf{x}G$. By this map we see that $a(J)$ is equal to the number of codewords of the shortened code C^J of maximal weight $n-\text{card}(J)$. The formalism above gives a way to obtain this number.

Proposition 5 *Let C be a linear code of length n, dimension k and minimum distance d. Let $J \subseteq \{1, 2, \ldots, n\}$ and $\text{card}(J) = t$. Then $a(J) = 1$ if $t = n$ and $a(J) = 0$ if $n - d < t < n$. For $0 \leq t \leq n - d$ we have that*

$$a(J) = \sum_{s=0}^{n-t-d} (-1)^s \sum_{I\subseteq \bar{J},\ \text{card}(I)=s} (q^{k-r(J\cup I)} - 1).$$

Proof The shortened code C^J is isomorphic with $C(J)$. By [20, Lemma 3.2.12], it has dimension $l(J) = k - r(J)$ and minimum distance $d(J) \geq d$. We view the codewords of C^J with entries indexed by $\bar{J}$, the complement of J in $\{1, 2, \ldots, n\}$. If $I \subseteq \bar{J}$, then $C^J(I)$ is isomorphic with $C(J\cup I)$. Hence $\dim C^J(I) = k - r(J \cup I)$ by [20, Lemma 3.2.12]. So we better use the notation $k - r(J \cup I)$ instead of $l(I)$ for the dimension of $\dim C^J(I)$, since in the latter notation the dependency on J is not clear. Now $a(J)$ is equal to the number of codewords of the shortened code C^J of maximal weight $n - t$ where $t = \text{card}(J)$ as remarked before. Hence $a(J)$ is given by the formula in Proposition 4 applied to C^J.

Notice that if $s > n - t - d(J)$ we have that in the summation $\sum_{s=0}^{n-t-d}$ the summand $q^{k-r(J\cup I)} - 1$ is zero, since $\text{card}(J \cup I) = s + t > n - d(J)$ and a codeword with $s + t$ zero entries is the all zeros word, so $C^J(I) = \{0\}$. □

Theorem 2 *Let $C \subseteq \mathbb{F}_q^n$ be a non-degenerate linear code with generator matrix G. Let $\mathscr{M}$ be the matroid defined by the arrangement of hyperplanes $\mathscr{A}_G$. Then the Poincaré polynomial $P_C(t_1, t_2, \ldots, t_n)$ and the multivariate Tutte polynomial $Z_{\mathscr{M}}(T, t_1, \ldots, t_n)$ determine each other.*

Proof The multivariate Tutte polynomial of C determines the Poincaré polynomial of C since the latter is determined by the $a(J)$ by Proposition 3, and the $a(J)$ can be computed by means of the rank function $r_{\mathscr{M}}$ by Proposition 5, and the value $r_{\mathscr{M}}(J)$ can be read off from the coefficient $T^{-r_{\mathscr{M}}(J)}$ of $\underline{t}^J$ in $Z_{\mathscr{M}}(T, t_1, \ldots, t_n)$.

Conversely, $a(J)$ is the coefficient of $\underline{t}^J$ in $P_C(t_1, \ldots, t_n)$. Now $a(J)$ is equal to the number of codewords $\mathbf{c}$ in C such that the set of zero coordinates of $\mathbf{c}$ is equal to J. Hence $\sum_{J\subseteq J'} a(J')$ is equal to the number of codewords $\mathbf{c}$ in C such that the zero coordinates of $\mathbf{c}$ are in J, which is equal to card $(C(J)) = l(J)$. Finally, by [20, Lemma 3.2.12], $r_{\mathscr{M}}(J) = k - l(J)$. Hence the Poincaré polynomial of C determines the multivariate Tutte polynomial of C. □

As a consequence of the above result we can state the following one.

Proposition 6 *Let C be an MDS linear code with parameters $[n, k, n-k+1]$ over the field $\mathbb{F}_q$. Then the Poincaré polynomial of C is given by*

$$P_C(t_1, t_2, \ldots, t_n) = t_1 t_2 \cdots t_n + \sum_{t=0}^{k-1} \sum_{\mathrm{card}(J)=t} \sum_{s=0}^{k-1-t} (-1)^s \binom{n-t}{s} (q^{k-t-s} - 1) \underline{t}^J.$$

Proof Since the code C is MDS, then $d(C^{\perp}) = k+1$. So the numbers $l(J)$ and therefore $r(J)$ depend only on the size of J. That is $r(J) = \mathrm{card}(J)$ if $\mathrm{card}(J) \leq k$ and $r(J) = k$ if $\mathrm{card}(J) > k$ by [20, Lemma 3.2.15]. Let $J \subseteq \{1, 2, \ldots, n\}$ and $\mathrm{card}(J) = t$. Then $a(J) = 1$ if $t = n$ and $a(J) = 0$ if $n - d < t < n$. If $0 \leq t \leq n - d$ then $I \subseteq \bar{J}$ and $\mathrm{card}(I) = s$, so $r(J \cup I) = \mathrm{card}(J \cup I) = t + s$ if $t + s \leq k$ and k otherwise. With Proposition 5 we get

$$a(J) = \sum_{s=0}^{k-1-t} (-1)^s \binom{n-t}{s} (q^{k-t-s} - 1).$$

The formula follows now from Proposition 3. □

It is known that the multivariate Tutte polynomial attached to a linear code is not a complete invariant. In the next remark we deduce the same result for the Poincaré polynomial of a q-ary code where $q \geq 5$.

Remark 1 Let $C(a)$ be the $\mathbb{F}_q$-linear code of length 10 and dimension 2 generated by $(1, 0, 0, 1, 1, 1, 1, 1, 1, 1)$ and $(0, 1, 1, 1, 1, 1, a, a, a, a)$ where $a \in \mathbb{F}_q$ and $a \notin \{0, 1\}$. Then $\mathscr{P}(a)$, the corresponding projective system on the projective line consists of the 10-tuple of points P_i with $P_1 = (1 : 0)$, $P_2 = P_3 = (0 : 1)$, $P_4 = P_5 = P_6 = (1 : 1)$ and $P_j = (1 : a)$ for $7 \leq j \leq 10$. That is to say the four points (1:0), (0:1), (1:1) and (1:a) have multiplicity 1, 2, 3 and 4, respectively. All such codes have equivalent matroids and the same multivariate Poincaré polynomial by Theorem 2, but they are not all equivalent. Two projective systems on the projective line are equivalent if and only if the corresponding points are mapped to each other by a fractional transformation. Moreover if three distinct points of the system remain fixed under such a transformation, then the remaining points remain also fixed by [11] and Propositions 5.1.33 and 5.1.34 of [20]. Now suppose that the codes $C(a)$ and $C(b)$ are equivalent with $a, b \in \mathbb{F}_q$ and $a, b \notin \{0, 1\}$, then the projective systems $\mathscr{P}(a)$ and $\mathscr{P}(b)$ are equivalent. Now the points (1:0), (0:1), (1:1) remain fixed under the fractional transformation, since their multiplicities are distinct and should remain the same. Therefore $(1 : a)$ remains also fixed under the transformation. If $q \geq 4$ then there are at least two choices for a giving two inequivalent codes with the same matroid. Hence the Poincaré polynomial is not a complete invariant of $\mathbb{F}_q$-linear codes if $q \geq 4$. The matroid of an $\mathbb{F}_q$-linear code determines the equivalence class of the code if and only if $q = 2$ or $q = 3$. See [3, 18]. That implies by Theorem 2 that the Poincaré polynomial is a complete invariant of $\mathbb{F}_q$-linear codes if and only if $q = 2$ or $q = 3$. In the next section we treat the Poincaré polynomial of binary codes in more detail.

3 The Poincaré Polynomial in the Binary Case

We devote this section to provide some results about the Poincaré polynomial of binary linear codes. However our first result is true for codes over any finite field.

Proposition 7 *Let $C \subseteq \mathbb{F}_q^n$ be a non-degenerate linear code. Then, with the above notation*

$$P_C(t_1, t_2, \ldots, t_n) = \sum_{\mathbf{c} \in C} \underline{t}^{\mathbf{1}-\mathbf{c}^{q-1}},$$

where $\mathbf{1} = (1, \ldots, 1) \in \mathbb{F}_q^n$ and $\mathbf{c}^{q-1} = (c_1^{q-1}, \ldots, c_n^{q-1})$ whenever $\mathbf{c} = (c_1, c_2, \ldots, c_n)$.

Proof The codewords in C are exactly the vectors $\mathbf{x}G$, where $\mathbf{x}$ runs along all vectors in $\mathbb{F}_q^k$ and G is a generator matrix of C. Then the result follows by noting that $\epsilon(\mathbf{x}) = \mathbf{1} - (\mathbf{x}G)^{q-1}$ because $C \subseteq \mathbb{F}_q^n$. □

From now on, our codes will be included in $\mathbb{F}_2^n$. First we state an immediate and interesting consequence of Proposition 7, which also shows that the Poincaré polynomial of a binary code is a complete invariant.

Corollary 1 *Let $C \subseteq \mathbb{F}_2^n$ be a non-degenerate binary linear code, then*

$$P_C(t_1, t_2, \ldots, t_n) = \sum_{\mathbf{c} \in C} \underline{t}^{\mathbf{1}-\mathbf{c}}.$$

Next we explain how the Poincaré polynomial can be obtained for the so-called $(u, u + v)$-construction of binary linear codes. From that result, we will derive a recursive formula for computing the Poincaré polynomial of binary Reed-Muller codes. In our development, we will use the following polynomial, close to the Poincaré one, which is attached to a binary linear code $C \subseteq \mathbb{F}_2^n$:

$$\hat{P}_C(t_1, t_2, \ldots, t_n) = \sum_{\mathbf{c} \in C} \underline{t}^{\mathbf{c}}. \tag{1}$$

The $(u, u + v)$-construction is a particular case of matrix-product code [2] and sometimes it is called the Plotkin sum. Matrix-product codes constitute a natural way to obtain large codes from others previously known (denominated constituent codes) [15]. They admit decoding procedures depending on the decoding methods of the corresponding constituent ones [14, 16], and when the constituent codes are cyclic, their corresponding matrix-product codes are quasi-cyclic codes [13]. Let us show the definition of $(u, u + v)$-construction.

Definition 4 Let C_1 and C_2 be two binary linear codes with parameters $[n, k_1, d_1]$ and $[n, k_2, d_2]$, respectively. The $(C_1, C_1 + C_2)$ code ($(u, u + v)$-construction of C_1 and C_2) is defined to be the following binary linear code

$$(C_1, C_1 + C_2) = \{(\mathbf{c}_1, \mathbf{c}_1 + \mathbf{c}_2) \mid \mathbf{c}_1 \in C_1, \mathbf{c}_2 \in C_2\}.$$

It has parameters $[2n, k_1 + k_2, \min\{2d_1, d_2\}]$.

For the codes on length $2n$ we are going to study, we will distinguish the variables for the Poincaré polynomial setting $t_j = t_{1j}$ and $t_{n+j} = t_{2j}$ for $1 \leq j \leq n$. So we will consider the Poincaré polynomial as an element in the ring

$$\mathbb{Z}[t_{11}, t_{12}, \ldots, t_{1n}, t_{21}, t_{21}, \ldots, t_{2n}].$$

As above, for $i = 1, 2$ the product $\prod_{j=1}^{n} t_{ij}^{c_{ij}}$ is expressed as $\underline{t_i}^{\mathbf{c}_i}$. With this notation, the Poincaré polynomial of a binary $(u, u + v)$ code can be obtained as follows.

Theorem 3 *Let C_1 and C_2 be two binary non-degenerate linear codes both of length n and dimensions k_1 and k_2, respectively. Then.*

$$P_{(C_1,C_1+C_2)}(t_{11}, \ldots, t_{1n}, t_{21}, \ldots, t_{2n}) = \\ P_{(C_1,C_1)}(t_{11}, \ldots, t_{1n}, t_{21}, \ldots, t_{2n}) \hat{P}_{C_2}(t_{21}, \ldots, t_{2n}).$$

Proof Denote by G_1 (respectively, G_2) the generator matrix of C_1 (respectively, C_2). Then, using Corollary 1, the following chain of equalities holds.

$$\begin{aligned} P_{(C_1,C_1+C_2)}(t_{11}, \ldots, t_{1n}, t_{21}, \ldots, t_{2n}) &= \sum_{\mathbf{x}\in\mathbb{F}_2^{k_1},\, \mathbf{y}\in\mathbb{F}_2^{k_2}} \underline{t_1}^{\mathbf{1}-\mathbf{x}G_1} \underline{t_2}^{\mathbf{1}-\mathbf{x}G_1-\mathbf{y}G_2} \\ &= \sum_{\mathbf{x}\in\mathbb{F}_2^{k_1},\, \mathbf{y}\in\mathbb{F}_2^{k_2}} \underline{t_1}^{\mathbf{1}-\mathbf{x}G_1} \underline{t_2}^{\mathbf{1}-\mathbf{x}G_1} \underline{t_2}^{-\mathbf{y}G_2} \\ &= \sum_{\mathbf{x}\in\mathbb{F}_2^{k_1}} \underline{t_1}^{\mathbf{1}-\mathbf{x}G_1} \underline{t_2}^{\mathbf{1}-\mathbf{x}G_1} \sum_{\mathbf{y}\in\mathbb{F}_2^{k_2}} \underline{t_2}^{-\mathbf{y}G_2} \\ &= P_{(C_1,C_1)}(t_{11}, \ldots, t_{1n}, t_{21}, \ldots, t_{2n}) \sum_{\mathbf{y}\in\mathbb{F}_2^{k_2}} \underline{t_2}^{-\mathbf{y}G_2}. \end{aligned}$$

Since we are in the binary field, we have $-\mathbf{y}G_2 = \mathbf{y}G_2$ and the right hand side of the last equality equals

$$P_{(C_1,C_1)}(t_{11},\ldots,t_{1n},t_{21},\ldots,t_{2,n}) \sum_{\mathbf{y}\in\mathbb{F}_2^{k_2}} \underline{t_2}^{\mathbf{y}G_2}.$$

According to Eq. (1), we get $P_{(C_1,C_1)}(t_{11},\ldots,t_{1n},t_{21},\ldots,t_{2n})\hat{P}_{C_2}(t_{21},\ldots,t_{2n})$, which concludes the proof. □

Remark 2 Reasoning as in the proof of Theorem 3, one obtains the following equality of polynomials

$$\hat{P}_{(C_1,C_1+C_2)}(t_{11},\ldots,t_{1n},t_{21},\ldots,t_{2n}) = \hat{P}_{(C_1,C_1)}(t_{11},\ldots,t_{1n},t_{21},\ldots,t_{2n})\hat{P}_{C_2}(t_{21},\ldots,t_{2n}).$$

As an application of Theorem 3, we show recursive formulae for obtaining the Poincaré polynomial of a binary Reed-Muller code. Generally speaking and for two fixed nonnegative integers r and m, the *Reed-Muller code* $\mathrm{RM}_q[r,m]$ is defined to be the linear code obtained as follows:

$$\mathrm{RM}_q[r,m] = \left\{p(\mathbf{a}) \mid p \in \mathbb{F}_q[X_1,X_2,\ldots,X_m],\ \mathbf{a}\in\mathbb{F}_q^m \text{ and } \deg(p)\leq r\right\}.$$

The mentioned recursive formulae apply to binary Reed-Muller codes and are stated in the following result.

Corollary 2 *Let m and r be positive integers such that $0 < r < m$. Then, setting $n = 2^{m-1}$, the following two recursive formulae concerning Poincaré polynomials and polynomials as in Eq.* (1) *hold.*

$$P_{\mathrm{RM}_2[r,m]}(t_{11},\ldots,t_{1n},t_{21},\ldots,t_{2n}) =$$
$$P_{(\mathrm{RM}_2[r,m-1],\mathrm{RM}_2[r,m-1])}(t_{11},\ldots,t_{1n},t_{21},\ldots,t_{2n})\hat{P}_{\mathrm{RM}_2[r-1,m-1]}(t_{21},\ldots,t_{2n}).$$

$$\hat{P}_{\mathrm{RM}_2[r,m]}(t_{11},\ldots,t_{1n},t_{21},\ldots,t_{2n}) =$$
$$\hat{P}_{(\mathrm{RM}_2[r,m-1],\mathrm{RM}_2[r,m-1])}(t_{11},\ldots,t_{1n},t_{21},\ldots,t_{2n})\hat{P}_{\mathrm{RM}_2[r-1,m-1]}(t_{21},\ldots,t_{2n}).$$

Proof It follows from Theorem 3, Remark 2 and the fact that the code $\mathrm{RM}_2[r,m]$ is obtained from the $(u,u+v)$-construction of $\mathrm{RM}_2[r,m-1]$ and $\mathrm{RM}_2[r-1,m-1]$ [21]. □

Acknowledgements We like to thank Rudi Pendavingh for the given information about unique representable matroids.

References

1. Atiyah, M.F., Macdonald, I.G.: Introduction to Commutative Algebra. Addison-Wesley, Reading (1969)
2. Blackmore, T., Norton, G.H.: Matrix-product codes over $\mathbb{F}_q$. Appl. Algebra Eng. Commun. Comput. **12**, 477–500 (2001)
3. Brylawski, T.H., Lucas, D.: Uniquely representable combinatorial geometries. In: Teorie Combinatorie. Proceedings of the 1973 International Colloquium, pp. 83–104. Accademia Nazionale del Lincei, Rome (1976)
4. Campillo, A., Delgado, F., Gusein-Zade S.: Poincaré series of a rational surface singularity. Invent. Math. **155** 45–53 (2004)
5. Campillo, A., Delgado, F., Gusein-Zade S.: Poincaré series of curves on rational surface singularities. Comment. Math. Helvetici **80**, 95–102 (2005)
6. Campillo, A., Delgado, F., Gusein-Zade, S.: Multiindex filtrations and motivic Poincaré series. Monatshefte. Math. **150**, 193–209 (2007)
7. Campillo, A., Delgado, F., Gusein-Zade S., Hernando, F.: Poincaré series of collections of plane valuations. Int. J. Math. **21**, 1461–1473 (2010)
8. Campillo, A., Delgado, F., Gusein-Zade, S.: Equivariant Poincaré series of filtrations. Rev. Mat. Complut. **26**, 241–251 (2013)
9. Campillo, A., Delgado, F., Gusein-Zade, S.: An equivariant Poincaré series of filtrations and monodromy of zeta functions. Rev. Mat. Complut. **28**, 449–467 (2015)
10. Delgado, F., Moyano-Fernández, J.J.: On the relation between the generalized Poincaré series and the Stöhr zeta function. Proc. Am. Math. Soc. **137**, 51–59 (2009)
11. Dür, A.: The automorphism groups of Reed-Solomon codes. J. Comb. Theory Ser. A **44**(1), 69–82 (1987)
12. Galindo, C., Monserrat, F.: The Poincaré series of multiplier ideals of a simple complete ideal in a local ring of a smooth surface. Adv. Math. **225**, 1046–1068 (2010)
13. Hernando, F., Ruano, D.: New linear codes from matrix-product codes with polynomial units. Adv. Math. Commun. **4**, 363–367 (2010)
14. Hernando, F., Ruano, D.: Decoding of matrix-product codes. J. Algebra Appl. **12**, 1250185 (2013)
15. Hernando, F., Lally, K., Ruano, D.: Construction and decoding of matrix-product codes from nested codes. Appl. Algebra Eng. Commun. Comput. **20**, 497–507 (2009)
16. Hernando, F., Høholdt, T., Ruano, D.: List decoding of matrix-product codes from nested codes: an application to quasi-cyclic codes. Adv. Math. Commun. **6**, 259–272 (2012)
17. Jurrius, R., Pellikaan, R.: Codes, arrangements and matroids. In: Martinez-Moro, E. (ed.) Algebraic Geometry Modeling in Information Theory. Series on Coding Theory and Cryptology, vol. 8, pp. 219–325. World Scientific, Hackensack (2013)
18. Kahn, J.: On the uniqueness of matroid representations over GF(4). Bull. Lond. Math. Soc. **20**, 5–10 (1988)
19. MacWilliams, F.J., Sloane, N.J.A.: The Theory of Error-Correcting Codes. North Holland Publishing, Amsterdam/New York (1977)
20. Pellikaan, R., Wu, X.-W., Bulygin, S., Jurrius, R.: Codes, Cryptology and Curves with Computer Algebra. Cambridge University Press, Cambridge (2017)
21. Roman, S.: Coding and Information Theory. Springer, New York (1992)
22. Sokal, A.D.: The multivariate Tutte polynomial (alias Potts model) for graphs and matroids. In: Chapman, R. (ed.) Surveys in Combinatorics. London Mathematical Society Lecture Note Series, vol. 327, pp. 173–226. Cambridge University Press, Cambridge (2005)
23. Tsfasman, M.A., Vlădut, S.G.: Algebraic-Geometric Codes. Kluwer Academic Publishers, Dordrecht (1991)
24. Tsfasman, M.A., Vlădut, S.G., Nogin, D.: Algebraic Geometric Codes: Basic Notions. Mathematical Surveys and Monographs, vol. 139. American Mathematical Society, Providence (2007)

The Metric Structure of Linear Codes

Diego Ruano

Abstract The bilinear form with associated identity matrix is used in coding theory to define the dual code of a linear code, also it endows linear codes with a metric space structure. This metric structure was studied for generalized toric codes and a characteristic decomposition was obtained, which led to several applications as the construction of stabilizer quantum codes and LCD codes. In this work, we use the study of bilinear forms over a finite field to give a decomposition of an arbitrary linear code similar to the one obtained for generalized toric codes. Such a decomposition, called the geometric decomposition of a linear code, can be obtained in a constructive way; it allows us to express easily the dual code of a linear code and provides a method to construct stabilizer quantum codes, LCD codes and in some cases, a method to estimate their minimum distance. The proofs for characteristic 2 are different, but they are developed in parallel.

1 Introduction

Error-correcting codes are used in digital communications in order to recover the information sent through a channel that may corrupt some of the information. The most studied, and in practice used, codes are linear codes [22]. A linear code is a vector subspace of $\mathbb{F}_q^n$, where $\mathbb{F}_q$ is the finite field with q elements. The dual code $\mathscr{C}^\perp \subset \mathbb{F}_q^n$, of a linear code $\mathscr{C} \subset \mathbb{F}_q^n$, is the orthogonal space to $\mathscr{C}$ with respect to the bilinear form $B : \mathbb{F}_q^n \times \mathbb{F}_q^n \to \mathbb{F}_q$, $B(x, y) = \sum x_i y_i$. This bilinear form allows us to consider $\mathbb{F}_q^n$ as a metric space.

Generalized toric codes are an extension of toric codes [17], they are obtained by evaluating polynomials at the algebraic torus $(\mathbb{F}_q^*)^r$. Their metric structure was studied in [27], providing a direct method to compute the dual code of a generalized

D. Ruano (✉)
IMUVA (Mathematics Research Institute), University of Valladolid, Valladolid, Spain

Department of Mathematical Sciences, Aalborg University, Aalborg, Denmark
e-mail: diego.ruano@uva.es

G.-M. Greuel et al. (eds.), *Singularities, Algebraic Geometry, Commutative Algebra, and Related Topics*, https://doi.org/10.1007/978-3-319-96827-8_24

toric code and deduce that there exist no self-dual generalized toric codes. Moreover, J-affine variety codes [14], which include generalized toric codes as a particular case, have a similar metric structure. Stabilizer quantum codes with good parameters [12–15] and new binary and ternary LCD codes [16] were constructed using this characteristic metric structure of J-affine variety codes.

Quantum error-correcting codes are essential for quantum computing since they protect quantum information from decoherence and quantum noise [29]. Although quantum information cannot be cloned, one can construct stabilizer quantum codes from self-orthogonal classical codes [4–6, 20]. A linear code $\mathscr{C}$ is self-orthogonal if $\mathscr{C} \subset \mathscr{C}^{\perp}$.

A linear code $\mathscr{C}$ is called an LCD code (complementary dual code) if $\mathscr{C} \cap \mathscr{C}^{\perp} = \{0\}$ [23]. LCD codes are used in cryptography [7], they play an important role in counter-measures to passive and active side-channel analyses on embedded cryptosystems. LCD codes are also useful for obtaining lattices [19] and in network coding [3]. It has been proved in [8] that q-ary LCD codes are as good as linear codes for $q > 3$. Hence the study of LCD codes is mainly open for binary and ternary fields.

In this paper we give an affirmative answer to the natural question: *May the metric structure and its applications for generalized toric codes or J-affine variety codes be extended for an arbitrary linear code?* To answer this question the classification of bilinear forms on a vector space over a finite field is used [1, 9–11, 18]. We reproduce this classification in Sect. 3 providing constructive proofs. The classification of bilinear forms on vector spaces over finite fields has been already used in coding theory for self-dual and self-orthogonal codes, originally by V. Pless [24–26] and subsequent papers.

For an arbitrary linear code, in Sect. 4, we compute a structure similar to the one of generalized toric codes, called the geometric decomposition of a linear code. The results and their proofs are different for characteristic 2, but they are developed in parallel. The geometric decomposition of a linear code allows us to extend, in Sect. 5, the applications for generalized toric codes: it expresses the dual code of a linear code easily and gives a method to estimate their minimum distance (extending the method in [21]). Moreover, we provide a method for constructing stabilizer quantum codes and LCD codes.

2 Metric Structure of Generalized Toric Codes

Let us introduce generalized toric codes and their metric structure in this section (see [27]), this family of codes motivated this work, as we mentioned in the previous section. They are an extension of toric codes, which are algebraic geometric codes over toric varieties [17]. Let $U \subset H = (\{0, \dots, q-2\})^r$, $T = (\mathbb{F}_q^*)^r$ and the vector space $\mathbb{F}_q[U] = \langle Y^u = Y_1^{u_1} \cdots Y_r^{u_r} \mid u = (u_1, \dots, u_r) \in U \rangle \subset \mathbb{F}_q[Y_1, \dots, Y_r]$. The

generalized toric code $\mathscr{C}_U$ is the image of the $\mathbb{F}_q$-linear map

$$\begin{array}{rcl} \mathrm{ev} : \mathbb{F}_q[U] & \to & \mathbb{F}_q^n \\ f & \mapsto & (f(t))_{t \in T} \end{array}$$

where $n = \#T = (q-1)^r$.

Let $B(x, y) = \sum x_i y_i$ with $x, y \in \mathbb{F}_q^n$. The following result considers the metric structure of a generalized toric code $\mathscr{C}_U \subset \mathbb{F}_q^n$ and computes its dual code.

Theorem 1 ([2, 27]) *With notations as above, if $u, v \in H$, then*

$$B(\mathrm{ev}(Y^u), \mathrm{ev}(Y^v)) = \begin{cases} 0 & \text{if } \overline{u+v} \neq 0, \\ (-1)^r & \text{if } \overline{u+v} = 0, \end{cases}$$

where $u = \overline{u} + b_u$, with $\overline{u} \in H$ and $b_u \in ((q-1)\mathbb{Z})^r$. Let $U \subset H$, $u' = \overline{-u}$ and $U' = \{u' \mid u \in U\}$. Then, $\#U = \#U'$ and the dual code of $\mathscr{C}_U$ is $\mathscr{C}_{U^\perp}$, where $U^\perp = H \setminus U' = (H \setminus U)'$.

Moreover we can order the elements of H in such a way that the matrix of B in the basis $\{\mathrm{ev}(Y^u) \mid u \in H\}$ of $\mathbb{F}_q^n$ has a characteristic form. We consider first in H the pairs of elements u and u', with $u \neq u'$, such that $\overline{u+u'} = 0$ ($u \mapsto u'$ is an involution). Finally, we consider the elements $u \in H$ such that $u = u'$. The matrix N of B in such a basis verifies

$$(-1)^r N = \begin{pmatrix} 0 & 1 & & & & & & \\ 1 & 0 & & & & & & \\ & & \ddots & & & & & \\ & & & 0 & 1 & & & \\ & & & 1 & 0 & & & \\ & & & & & 1 & & \\ & & & & & & \ddots & \\ & & & & & & & 1 \end{pmatrix}.$$

The number of 1's in the main diagonal of the matrix is 2^r if q is odd and 1 if q is even. Therefore, there are no self-dual generalized toric codes. The previous basis and the characteristic matrix N allowed us to obtain stabilizer quantum codes and LCD codes.

3 Bilinear Forms on Vector Spaces Over Finite Fields

In this section, we present an introduction to bilinear forms on vector spaces over finite fields. We refer the reader to [1, 9–11, 18] for a deeper discussion of the definitions and results in this section. All proofs provided are constructive.

A bilinear form over $\mathbb{F}_q^n$ is a bilinear map $B : \mathbb{F}_q^n \times \mathbb{F}_q^n \to \mathbb{F}_q$. It is said to be symmetric if $B(x, y) = B(y, x)\ \forall x, y \in \mathbb{F}_q^n$ and non-degenerate if

$$B(x, y) = 0\ \forall y \in \mathbb{F}_q^n \Rightarrow x = 0,$$
$$B(x, y) = 0\ \forall x \in \mathbb{F}_q^n \Rightarrow y = 0.$$

Let $\mathscr{B} = \{x_1, \ldots, x_n\}$ be a basis of $\mathbb{F}_q^n$; the associated matrix to B in the basis $\mathscr{B}$ is

$$N = \begin{pmatrix} B(x_1, x_1) & \cdots & B(x_1, x_n) \\ \vdots & & \vdots \\ B(x_n, x_1) & \cdots & B(x_n, x_n) \end{pmatrix}.$$

Namely, if $x = (x_1, \ldots, x_n)$, $y = (y_1, \ldots, y_n)$ in the basis $\mathscr{B}$, one has that $B(x, y) = xNy^t$, where y^t is the transpose of y.

From now on, we will consider the metric structure given by the bilinear form $B(x, y) = \sum_{i=1}^n x_i y_i$, which is used to define the dual code of a linear code. Here and subsequently, $\mathbb{F}_q^n$ will be the vector space over $\mathbb{F}_q$ with the non-degenerate symmetric bilinear form B whose associated matrix is the identity matrix. Therefore, B is symmetric and non-degenerate.

Let $x, y \in \mathbb{F}_q^n$, x and y are said to be orthogonal if $B(x, y) = 0$ and we denote it $x \perp y$. Let U, W be two vector subspaces of $\mathbb{F}_q^n$, U and W are said to be orthogonal if $x \perp y$ for all $x \in U$, $y \in W$. Let $U \subset \mathbb{F}_q^n$ be a vector subspace which is direct sum of pairwise orthogonal vector subspaces $U_1, \ldots, U_r$, then we say that $\mathbb{F}_q^n$ is the orthogonal sum of $U_1, \ldots, U_r$ and will denote it by $\mathbb{F}_q^n = U_1 \perp \cdots \perp U_r$. Let $U \subset \mathbb{F}_q^n$ be a vector subspace, the radical of U consists in the vectors of U that are orthogonal to U, that is $\mathrm{rad}(U) = U \cap U^\perp$. Let x, y in $\mathbb{F}_q^n$, they are orthonormal if they are orthogonal and $B(x_1, x_1) = 1$, $B(x_2, x_2) = 1$. A vector $x \in \mathbb{F}_q^n$ is called isotropic if $B(x, x) = 0$, that is, if $\langle x \rangle \subset \mathrm{rad}(\langle x \rangle)$. A vector subspace $U \subset \mathbb{F}_q^n$ is called isotropic if $B(x, y) = 0$ for all $x, y \in U$, that is, if $U \subset \mathrm{rad}(U)$. Every isotropic space U satisfies $\dim(U) \leq \lfloor \frac{n}{2} \rfloor$. An isotropic subspace $U \subset \mathbb{F}_q^n$ is called maximal when it is not strictly contained in any other isotropic subspace. The dimension of all of the maximal isotropic subspaces of a non-singular space U is the same, it is called index of U.

A vector subspace $U \subset \mathbb{F}_q^n$ is said to be non-singular if $\mathrm{rad}(U) = (0)$, and singular otherwise. One has that U is non-singular if and only if the bilinear form restricted to U is non-degenerate. If $U \subset \mathbb{F}_q^n$ is non-singular, then $\mathbb{F}_q^n = U \perp U^\perp$ and $U^\perp$ is non-singular.

Let $H \subset \mathbb{F}_q^n$ be a two-dimensional vector subspace, H is said to be a **hyperbolic plane** if there exist x_1, x_2 generating H such that

$$B(x_1, x_1) = 0,$$
$$B(x_2, x_2) = 0,$$
$$B(x_1, x_2) = 1.$$

hence, H is non-singular. Both ordered generators x_1, x_2 are called **geometric generators** or **geometric basis of** H.

Lemma 1 *Let $\mathbb{F}_q$ have odd characteristic. Then any two-dimensional non-singular subspace of $\mathbb{F}_q^n$ which contains an isotropic vector is a hyperbolic plane.*

Proof Let x_1 be a non-zero isotropic vector. Let y be a vector of the considered two-dimensional subspace linearly independent to x_1 and let $x_2 = \lambda_1 x_1 + \lambda_2 y$, for $\lambda_1, \lambda_2 \in \mathbb{F}_q$. One has that $B(x_1, x_2) = \lambda_2 B(x_1, y)$, moreover, $B(x_1, y) \neq 0$ since a plane is non-singular. Therefore, for $\lambda_2 = B(x_1, y)^{-1} \neq 0$, one has that $B(x_1, x_2) = 1$.

Moreover, $B(x_2, x_2) = 0$ if and only if $2\lambda_1\lambda_2 B(x_1, y) + \lambda_2^2 B(y, y) = 0$. Since $\lambda_2 \neq 0$ and $B(x_1, y) \neq 0$ one has that if

$$\lambda_1 = \frac{-\lambda_2 B(y, y)}{2B(x_1, y)} = \frac{-B(y, y)}{2B(x_1, y)^2}$$

then x_2 is an isotropic vector. □

Note that the previous result does not hold in characteristic 2 as the next example shows.

Example 1 Let $\mathbb{F}_q$ be a field of characteristic 2. Let $x = (x_1, x_2) \in \mathbb{F}_q^2$, x is an isotropic vector if and only if $x_1^2 + x_2^2 = 0$, that is, if and only if $(x_2/x_1)^2 = 1$. Hence, $(1, 1)$ is an isotropic vector, moreover, only the vectors generated by $(1, 1)$ are isotropic, since we have the Frobenius isomorphism. Therefore, $\mathbb{F}_q^2$ contains an isotropic vector but it is not a hyperbolic plane.

We say that a non-singular two-dimensional subvector space $E \subset \mathbb{F}_q^n$ is an **elliptic plane** if it is not a hyperbolic plane and there exist x_1, x_2 generating E and such that

$$B(x_1, x_1) = 0,$$
$$B(x_2, x_2) = 1,$$
$$B(x_1, x_2) = 1.$$

We call x_1, x_2 the **geometric generators** or **geometric basis of** E. For instance $\{(1, 1), (0, 1)\}$ is a geometric basis of the elliptic plane $\mathbb{F}_q^2$, with q even.

3.1 Characteristic Different from 2

One has that -1 is a square element in the field $\mathbb{F}_q$ if and only if $q \equiv 1 \mod 4$. A non-zero vector $x = (x_1, x_2) \in \mathbb{F}_q^2$ is an isotropic vector if and only if $x_1^2 + x_2^2 = 0$, that is, if and only if $(x_2/x_1)^2 = -1$. If -1 is a square element in the field, the previous equation has at least one solution and therefore there exist isotropic vectors. If -1 is non-square element in $\mathbb{F}_q$ there is no isotropic vector and therefore $\mathbb{F}_q^2$ is not a hyperbolic plane (neither an elliptic).

In $\mathbb{F}_q^n$ there exist orthonormal bases for the bilinear form B, for instance the canonical basis. For $x \in \mathbb{F}_q^n$ one can only obtain a linearly dependent vector y of x, with $B(y, y) = 1$, just by multiplying x with the square root of $B(x, x)$, if $B(x, x)$ is a square element in $\mathbb{F}_q$. Therefore, for a linear variety $L = \langle x \rangle$ one has that $B(x, x)$ is equal to a^2 or $a^2 g$ where g is a fixed non-square element in $\mathbb{F}_q$, moreover, multiplying x by a^{-1} we can assume that $B(x, x) = 1$ or $B(x, x) = g$ and then we say that x is a geometric basis of L. From now on we regard g as a fixed non-square element in $\mathbb{F}_q$.

The following result [28, Section 1.7] is used in Proposition 1 and in Lemma 3.

Lemma 2 *Let $a, b, c \in \mathbb{F}_q$ be different from zero. Then the following equation has at least one solution over $\mathbb{F}_q$*

$$aX^2 + bY^2 = c$$

The following result shows whether a non-singular plane is a hyperbolic plane, that is, whether it contains isotropic elements. And, moreover, whether it can be generated by two orthonormal elements, when it is not a hyperbolic plane.

Proposition 1 *Let $P = \langle x_1, x_2 \rangle \subset \mathbb{F}_q^n$ be a non-singular plane with $B(x_1, x_2) = 0$, $B(x_1, x_1) = a$ and $B(x_2, x_2) = b$. If $a = 0$ or $b = 0$ then P is a hyperbolic plane. If $a \neq 0$ and $b \neq 0$, then*

- *For $q \equiv 1 \mod 4$, P is a hyperbolic plane if and only if b/a is a square element. When P is not a hyperbolic plane it cannot be generated by two orthonormal vectors but can be generated by $y_1, y_2 \in \mathbb{F}_q^n$ such that $B(y_2, y_2) = 0$, $B(y_1, y_1) = 1$, $B(y_2, y_2) = g$, where g is a non-square element in $\mathbb{F}_q$.*
- *For $q \equiv 3 \mod 4$, P is a hyperbolic plane if and only if b/a is a non-square element. When P is not a hyperbolic plane it can be generated by two orthonormal vectors.*

Proof If $a = 0$ or $b = 0$ then P is a hyperbolic plane by Lemma 1.

Let $a \neq 0$ and $b \neq 0$. Let $\lambda_1, \lambda_2 \in \mathbb{F}_q$, $B(\lambda_1 x_1 + \lambda_2 x_2, \lambda_1 x_1 + \lambda_2 x_2) = \lambda_1^2 a + \lambda_2^2 b = 0$ if and only if $(\lambda_1/\lambda_2)^2 = -b/a$. Therefore, there are isotropic vector in P (and hence P is a hyperbolic plane by Lemma 1) if and only if $-b/a$ is a square element.

For $q \equiv 1 \mod 4$, one has that $c \in \mathbb{F}_q^*$ is a square element in $\mathbb{F}_q$ if and only if $-c$ is a square element in $\mathbb{F}_q$, since -1 is a square element in $\mathbb{F}_q$. Let $y_1 =$

$\lambda_1 x_1 + \lambda_2 x_2$, $B(y_1, y_1) = \lambda_1^2 a + \lambda_2^2 b$. By Lemma 2 there exist $\lambda_1, \lambda_2 \in \mathbb{F}_q$ such that $B(y_1, y_1) = 1$, since $a \neq 0, b \neq 0$. Let $z \in P$ be non-zero and orthogonal to y_1. One has that $B(\lambda_1 y_1 + \lambda_2 z) = \lambda_1^2 + \lambda_2^2 B(z, z)$. Since there exist no isotropic vectors in P, $-B(z, z)$ is a non-square element in $\mathbb{F}_q$ or, equivalently, $B(z, z)$ is a non-square element in $\mathbb{F}_q$. Therefore, $B(\lambda_2 z, \lambda_2 z) \neq 1$, but for a fixed non-square element g in $\mathbb{F}_q$, there exists $\lambda_2 \in \mathbb{F}_q$ such that for $y_2 = \lambda_2 z$, and one has that $B(y_2, y_2) = g$.

For $q \equiv 3 \mod 4$, one has that $c \in \mathbb{F}_q^*$ is a square element in $\mathbb{F}_q$ if and only if $-c$ is a non-square element in $\mathbb{F}_q$ since -1 is a non-square element in $\mathbb{F}_q$. Let $y_1 = \lambda_1 x_1 + \lambda_2 x_2$, $B(y_1, y_1) = \lambda_1^2 a + \lambda_2^2 b$. By Lemma 2, there exist $\lambda_1, \lambda_2 \in \mathbb{F}_q$ such that $B(y_1, y_1) = 1$, since $a \neq 0, b \neq 0$. Let $z \in P$ be non-zero and orthogonal to y_1. One has that $B(\lambda_1 y_1 + \lambda_2 z) = \lambda_1^2 + \lambda_2^2 B(z, z)$. Since there exist no isotropic vectors in P, $-B(z, z)$ is a non-square element in $\mathbb{F}_q$, or equivalently $B(z, z)$ is a square element in $\mathbb{F}_q$. Therefore, there exists $\lambda_2 \in \mathbb{F}_q$ such that for $y_2 = \lambda_2 z$, one has that $B(y_2, y_2) = 1$. □

The following result computes an isotropic vector in a non-singular space of dimension greater than or equal to 3.

Lemma 3 *Let $U \subset \mathbb{F}_q^n$ be non-singular with dimension greater than or equal to 3, then there exists at least one isotropic non-zero vector in U.*

Proof Let P be a non-singular plane of U and $x_1 \in P^\perp$, assume that $B(x_1, x_1) \neq 0$ (in other case x_1 is isotropic). By Lemma 2 one has that there exists $x_2 \in P$ such that $B(x_2, x_2) = -B(x_1, x_1)$. Therefore $x_1 + x_2 \neq 0$, $B(x_1 + x_2, x_1 + x_2) = 0$ and the result holds. □

Using the previous results one can prove the following proposition.

Proposition 2 *Let $U \subset \mathbb{F}_q^n$ be a non-singular m-dimensional vector space. If q is odd, then one can decompose U in the following way:*

If m is odd, then

(1) $U = H_1 \perp \cdots \perp H_{(m-1)/2} \perp L$, *where each H_i is a hyperbolic plane and L is linear subspace of dimension 1.*

If m is even, then

(2) *If the index of U is $m/2$: $U = H_1 \perp \cdots \perp H_{m/2}$, where each H_i is a hyperbolic plane.*
(3) *If the index of U is $m/2 - 1$: $U = H_1 \perp \cdots \perp H_{(m-2)/2} \perp L_1 \perp L_2$, where each H_i is a hyperbolic plane and L_1 and L_2 are two linear subspaces of dimension 1.*

Proof Let m be odd. Then one can apply Lemmas 1 and 3 to obtain a hyperbolic plane H_1 and therefore one has that $U = H_1 \perp (H_1^\perp \cap U)$. In the same way for $H_1^\perp \cap U$, one obtains another orthogonal hyperbolic planes. Iterating this process, one writes U as the orthogonal sum of $(m - 1)/2$ hyperbolic planes and a linear variety of dimension 1.

In the same way, when m is even, we can apply Lemmas 1 and 3 successively until we compute $(m-2)/2$ pairwise orthogonal hyperbolic planes and a linear variety W of dimension 2. By Lemma 1, we may check whether W contains isotropic vectors and therefore it is a hyperbolic plane and U is decomposed as the orthogonal sum of $m/2$ hyperbolic planes, or on the contrary, it does not contain isotropic vectors and therefore it may be generated by two orthogonal elements and U is decomposed as the orthogonal sum of $m/2-1$ hyperbolic planes and two linear varieties of dimension 1. □

Note that as a corollary of the previous result, one has that the index of an m-dimensional vector subspace is equal to $(m-1)/2$, if m is odd, and $m/2$ or $m/2-1$, if m is even.

3.2 Characteristic 2

In characteristic different from 2, whenever there exists an isotropic vector in a plane, one has a hyperbolic plane. However, as we have seen in Example 1, if q is a power of 2, then $\mathbb{F}_q^2$ is a non-singular plane which contains an isotropic vector but it cannot be generated by two isotropic vectors.

Another important difference between even and odd characteristic is that every element of $\mathbb{F}_q$ is a square element in characteristic 2 (by the Frobenius isomorphism), while this is not the case in odd characteristic. Hence, if x is a non-isotropic vector, then one can always find $y \in \langle x \rangle$ such that $B(y, y) = 1$ since every element in $\mathbb{F}_q^*$ is a square element. Thus, we may say that y is a **geometric basis of** $L = \langle x \rangle$.

The following result allows us to compute a basis of the isotropic vectors in $\mathbb{F}_q^n$.

Proposition 3 *A vector $x \in \mathbb{F}_q^n$, x is isotropic if and only if $\sum_{i=1}^n x_i = 0$. The $n-1$ vectors $y_1 = (1, 1, 0, \dots, 0)$, $y_2 = (0, 1, 1, 0, \dots, 0)$,…, $y_{n-1} = (0, \dots, 0, 1, 1)$ form a basis of the vector space S of isotropic vectors in $\mathbb{F}_q^n$. Furthermore, S is non-singular if n is odd and singular if n is even.*

Proof One has that x is isotropic if and only if $B(x, x) = 0$. That is, $\sum_{i=1}^n x_i^2 = 0$ if and only if $(\sum_{i=1}^n x_i)^2 = 0$ or, equivalently, if $\sum_{i=1}^n x_i = 0$.

The isotropic vectors of $\mathbb{F}_q^n$ form a vector space. Trivially, one has that y_i is isotropic $\forall i$ and that $y_1, \dots, y_{n-1}$ are linearly independent. Let us check that $\{y_1, \dots, y_{n-1}\}$ generates the vector space of isotropic vectors. Let $x = (x_1, \dots, x_n)$ be isotropic, we define then the coefficients of the linear combination

$$\begin{cases} \lambda_1 = x_1, \\ \lambda_2 = x_1 + x_2, \\ \vdots \\ \lambda_{n-1} = x_1 + x_2 + \cdots + x_{n-1}. \end{cases}$$

One has that $\sum_{i=1}^{n-1} \lambda_i y_i = (x_1, \ldots, x_{n-1}, \sum_{i=1}^{n-1} x_i) = (x_1, \ldots, x_{n-1}, x_n)$. The last equality follows from $x_n = \sum_{i=1}^{n-1} x_i$, since x is isotropic. One has that $S^\perp = \langle (1, \ldots, 1) \rangle$ and the result holds because $(1, \ldots, 1) \in S$ if and only if n is even. □

As a corollary of this result, we have that $\mathbb{F}_q^n$ with n even, cannot be decomposed as an orthogonal sum of $n/2$ hyperbolic planes. Note the difference between this case and the one when $\mathbb{F}_q$ is of characteristic different from 2.

Another consequence of the previous result is that, when n is even, $z = (1, \ldots, 1) \in \mathbb{F}_q^n$ is orthogonal to every isotropic vector in $\mathbb{F}_q^n$. Therefore, no plane containing z can be a hyperbolic plane and we will have to consider an elliptic plane for this element. This explains the phenomenum of Example 1.

Although the following result follows from the previous proposition, we present a constructive proof that will allow us to compute a geometric basis.

Lemma 4 *If $P \subset \mathbb{F}_q^n$ is a vector subspace of dimension greater than or equal to 2, then there exists at least one isotropic vector in P.*

Proof Let x_1, x_2 be two linearly independent vectors of P that are non-isotropic. Let $y = \lambda_1 x_1 + \lambda_2 x_2$, with $\lambda_1, \lambda_2 \in \mathbb{F}_q$. One has that $B(y, y) = \lambda_1^2 B(x_1, x_1) + \lambda_2^2 B(x_2, x_2) = 0$, if and only if $(\lambda_1/\lambda_2)^2 = B(x_2, x_2)/B(x_1, x_1)$. Since in a field of characteristic 2 every element is a square, one has that for $\lambda_1 = \sqrt{B(x_2, x_2)}$ and $\lambda_2 = \sqrt{B(x_1, x_1)}$, y is isotropic. □

The following result shows that any non-singular vector space of dimension 2 is either a hyperbolic plane or an elliptic plane.

Proposition 4 *Let $P \subset \mathbb{F}_q^n$ be a two-dimensional non-singular vector subspace. Then P is a hyperbolic plane if and only if $\sum x_i = 0$ for all $x \in P$. If P is not a hyperbolic plane then it is an elliptic plane and it may be generated by two orthonormal elements.*

Proof Let $S \subset \mathbb{F}_q^n$ be the vector space of isotropic vectors, there exist two independent isotropic vectors in P if and only if $P \subset S$. One has that $P \subset S$ if and only if $\sum x_i = 0, \forall x \in P$ by Proposition 3.

Let $P \subset S$, then there exist $x_1, x_2 \in P$ isotropic and linearly independent, therefore $\lambda = B(x_1, x_2)$ is not equal to zero (because B is non-degenerate). Let $y_1 = x_1$, $y_2 = \lambda^{-1} x_2$, one has that y_1, y_2 are the geometric generators of a hyperbolic plane, that is, $B(y_1, y_1) = B(y_2, y_2) = 0$ and $B(y_1, y_2) = 1$.

Let $P \not\subset S$, then there exist $x_1, x_2 \in P$ isotropic and linearly independent, with x_1 isotropic and x_2 non-isotropic. Let $\lambda = B(x_1, x_2)$ and $\mu = B(x_2, x_2) \neq 0$. One has that $y_1 = (\lambda^{-1}\sqrt{\mu}) x_1$ and $y_2 = \sqrt{\mu^{-1}} x_2$ are the geometric generators of an elliptic plane, that is, $B(y_1, y_1) = 0$ and $B(y_1, y_2) = B(y_2, y_2) = 1$.

Let y_1, y_2 be the two generators of the elliptic plane P, then $y_1' = y_1 + y_2$, $y_2' = y_2$. One has that y_1', y_2' form a basis of P, since they are linearly independent. Moreover, $B(y_1', y_1') = 1$ and $B(y_1', y_2') = 0$, therefore P can be generated by two orthonormal elements. □

The following lemma decomposes a non-singular vector space of dimension greater than or equal to 3 as the orthogonal sum of a hyperbolic plane and its orthogonal subspace.

Lemma 5 *Let $U \subset \mathbb{F}_q^n$ be a non-singular vector subspace of dimension greater than or equal to* 3. *Then there exists a hyperbolic plane H such that $U = H \perp U'$ where U' is a non-singular vector subspace.*

Proof By Lemma 4, we can find an isotropic vector $x \in U$ and one has that $U = \langle x \rangle \perp U_1$, where $U_1 = \langle x \rangle^{\perp} \cap U$. Since U_1 is a non-singular vector subspace of dimension greater than or equal to 2, by Lemma 4, there exists an isotropic vector $y \in U$. Therefore, by Proposition 4, $\{x, y\}$ generates a hyperbolic plane H, and $U = H \perp U'$, where $U' = H^{\perp} \cap U$. □

The following result decomposes a non-singular subspace of dimension greater than or equal to 3 as an orthogonal sum of hyperbolic planes and a linear subspace of dimension lower than or equal to 2

Proposition 5 *Let $U \subset \mathbb{F}_q^n$ be an m-dimensional non-singular vector subspace with characteristic of $\mathbb{F}_q$ equal to* 2. *One can decompose U in the following way:*

If m is odd

(1) $U = H_1 \perp \cdots \perp H_{(m-1)/2} \perp L$, *where each H_i is a hyperbolic plane and L is a one-dimensional linear subspace.*

If m is even

(2) $U = H_1 \perp \cdots \perp H_{m/2}$, *where each H_i is a hyperbolic plane.*
(3) $U = H_1 \perp \cdots \perp H_{m/2-1} \perp L_1 \perp L_2$, *where each H_i is a hyperbolic plane and L_1, L_2 are one-dimensional linear subspaces.*

Proof Let m be odd, we can apply Lemma 5 to obtain a hyperbolic plane H_1 and therefore one has that $U = H_1 \perp (H_1^{\perp} \cap U)$. In the same way, we can make further computations in $H_1^{\perp} \cap U$ to obtain more hyperbolic planes pairwise orthogonal. Thus, repeating the process, we write U as the orthogonal sum of $(m-1)/2$ hyperbolic planes and a one-dimensional linear variety.

In the same way, when m is even, we can apply Lemma 5 successively in order to obtain $m/2 - 1$ pairwise orthogonal hyperbolic planes and a two-dimensional linear variety. By Proposition 4, this two-dimensional linear variety is a hyperbolic plane when every element x in it verifies $\sum x_i = 0$. Otherwise it can be generated by two orthonormal vectors. □

Let $U = \mathbb{F}_q^n$ with n even, then U can only have a decomposition of type (3), because in $\mathbb{F}_q^n$ there are just $n-1$ linearly independent isotropic vectors (Proposition 3).

4 Geometric Decompositions of Linear Codes

For an arbitrary linear code, in this section we compute a structure similar to the one of generalized toric codes, called the geometric decomposition of a linear code.

4.1 Characteristic Different from 2

Let $\mathbb{F}_q$ be a finite field of characteristic different from 2. Using the previous results we can write $\mathbb{F}_q^n$ as the orthogonal sum of hyperbolic planes and linear varieties of dimension 1.

We say that $\mathbb{F}_q^n$ has a **geometric decomposition of type** r, s, t if

$$\mathbb{F}_q^n = H_1 \perp \cdots \perp H_r \perp L_1 \perp \cdots \perp L_{s+t}$$

where $H_1, \ldots, H_r$ are hyperbolic planes and $L_1, \ldots, L_{s+t}$ are one-dimensional linear varieties, such that each hyperbolic plane is generated by two geometric generators $H_i = \langle x_{2i-1}, x_{2i} \rangle$, $i = 1, \ldots, r$, each linear variety of dimension 1 is generated by $L_i = \langle x_{2r+i} \rangle$, with $B(x_{2r+i}, x_{2r+i}) = 1$, $i = 1, \ldots, s$, and the other varieties of dimension 1 are generated by $L_i = \langle x_{2r+s+i} \rangle$, with $B(x_{2r+s+i}, x_{2r+s+i}) = g$, $i = 1, \ldots, t$, where g is a fixed non-square element in $\mathbb{F}_q$. One has that $\{x_1, \ldots, x_n\}$ is a basis of $\mathbb{F}_q^n$ and we say that it is a **basis of the geometric decomposition**.

Let M be the matrix whose rows are the elements of the basis of the geometric decomposition, then one has that $MM^t = J_{r,s,t}$. That is, $J_{r,s,t}$ is the matrix of B in such a basis, then we have

$$J_{r,s,t} = \begin{pmatrix} 0 & 1 & & & & & & & & & \\ 1 & 0 & & & & & & & & & \\ & & \ddots & & & & & & & & \\ & & & 0 & 1 & & & & & & \\ & & & 1 & 0 & & & & & & \\ & & & & & 1 & & & & & \\ & & & & & & \ddots & & & & \\ & & & & & & & 1 & & & \\ & & & & & & & & g & & \\ & & & & & & & & & \ddots & \\ & & & & & & & & & & g \end{pmatrix}$$

where g is a fixed non-square element in $\mathbb{F}_q^*$.

A basis $\{x_1, \ldots, x_n\}$ of $\mathbb{F}_q^n$ is said to be a **compatible basis with respect to a decomposition of type** r, s, t if each $\{x_{2i-1}, x_{2i}\}$, with $i = 1, \ldots, r$, is a geometric basis of a hyperbolic plane, and each x_{2r+i}, with $i = 1, \ldots s + t$, generates a one-dimensional linear variety in such a way that all of these subspaces are pairwise orthogonal. Or equivalently, if the matrix of B in such a basis is equal to $J_{r,s,t}$.

Let $\mathscr{C} \subset \mathbb{F}_q^n$ be a linear code, we say that $\mathscr{C}$ is **compatible with a geometric decomposition of type** r, s, t if there exists a basis $\{x_1, \ldots, x_n\}$ of $\mathbb{F}_q^n$ compatible with such a decomposition in such a way that there exists $I \subset \{1, \ldots, n\}$ such that $\{x_i \mid i \in I\}$ is a basis of $\mathscr{C}$.

The following results allows us to compute a geometric decomposition compatible with a given code in characteristic different from 2.

Theorem 2 *Let the characteristic of* $\mathbb{F}_q$ *be different from* 2. *Any linear code* $\mathscr{C} \subset \mathbb{F}_q^n$ *is compatible with at least one geometric decomposition. Furthermore, there is a computable geometric basis, called standard, compatible with* $\mathscr{C}$, *of type* r, s, t *with* $s + t \leq 4$ *and* $t \leq 2$.

Proof Let $\mathscr{C} = \mathrm{rad}(\mathscr{C}) \perp \mathscr{C}_1$, where $\mathrm{rad}(\mathscr{C}) = \langle x_1, \ldots, x_l \rangle$.

We claim that we can compute $x_1', \ldots, x_l' \in \mathbb{F}_q^n$ such that x_i, x_i' are the geometric generators of a hyperbolic plane and, moreover, the hyperbolic planes $H_i = \langle x_i, x_i' \rangle$ and $\mathscr{C}_1$ are pairwise orthogonal. That is, one has that

$$\mathscr{C}' = H_1 \perp \cdots \perp H_l \perp \mathscr{C}_1,$$

where $\mathscr{C}'$ contains $\mathscr{C}$ and is non-singular. We prove the construction of $\mathscr{C}'$ by induction on l (this is Theorem 3.8 in [1]).

For $l = 0$ there is nothing to prove. The subspace $\mathscr{C}_0 = \langle x_1, \ldots, x_{l-1} \rangle \perp \mathscr{C}_1$ is orthogonal to x_l but does not contain it. One has that $x_l \in \mathscr{C}_0^\perp$ but $x_l \notin \mathrm{rad}(\mathscr{C}_0^\perp) = \mathrm{rad}(\mathscr{C}_0)$, therefore there exists $y \in \mathscr{C}_0^\perp$ such that $B(x_l, y) \neq 0$. The plane generated by $\{x_l, y\}$ is non-singular, is contained in $\mathscr{C}_0^\perp$ and by Lemma 1 is generated by a geometric basis $H_l = \langle x_l, x_l' \rangle$. Since $H_l \subset \mathscr{C}_0^\perp$, then $\mathscr{C}_0 \perp H_l$ and $\mathscr{C}_0 \subset H_l^\perp$. As the radical of $\mathscr{C}_0$ has dimension $l - 1$, by inductive hypothesis we can find geometric bases $\{x_i, x_i'\}$ of H_i in $H_l^\perp$, for $i = 1, \ldots, l-1$ such that they are pairwise orthogonal and also to $\mathscr{C}_1$, and since they are orthogonal to H_l and H_l is orthogonal to $\mathscr{C}_1$, the construction of $\mathscr{C}'$ holds.

Therefore, we have $\mathscr{C}' = H_1 \perp \cdots \perp H_l \perp \mathscr{C}_1$, where $H_i = \langle x_i, x_i' \rangle$, with $x_i' \notin \mathscr{C}$. Moreover, $\mathscr{C}'$ is non-singular and one has that $\mathbb{F}_q^n = H_1 \perp \cdots \perp H_l \perp \mathscr{C}_1 \perp \mathscr{C}'^\perp$.

Since $\mathscr{C}_1$ is non-singular, by Proposition 2, we can write $\mathscr{C}_1$ as a sum of hyperbolic planes and a one or two-dimensional linear space W (if the dimension of $\mathscr{C}_1$ is lower than 3 we do not consider any hyperbolic plane and $\mathscr{C}_1 = W$). Hence, we have $\mathscr{C}_1 = H_{l+1} \perp \cdots \perp H_m \perp W$, where $H_{l+i} = \langle x_{l+i}, x_{l+i}' \rangle$.

By Proposition 2, we have three different geometries for W

(a) If $\dim(W) = 1$, we write $W = \langle x \rangle$. Moreover, x is non-isotropic since B is non-degenerate. We can consider $x_{m+1} \in W$ such that $B(x_{m+1}, x_{m+1})$ is equal to 1 (if $B(x_{m+1}, x_{m+1})$ is a square element) or g, where g is a fixed non-square element and $W = L_1 = \langle x_{m+1} \rangle$.

(b) If $\dim(W) = 2$ and W contains some isotropic vector, then W is a hyperbolic plane and $W = H_{m+1} = \langle x_{m+1}, x'_{m+1} \rangle$, by Lemma 1.
(c) If $\dim(W) = 2$ and W does not contain any isotropic vector, then W can be generated by two orthogonal vectors $L_1 = \langle x_{m+1} \rangle$, $L_2 = \langle x_{m+2} \rangle$, where $W = L_1 \perp L_2$, by Proposition 1.

We decompose $\mathscr{C}'^{\perp}$ in the same way as $\mathscr{C}_1$ (using Proposition 2) to obtain

$$\mathscr{C}'^{\perp} = H'_1 \perp \cdots \perp H'_{m'} \perp W'$$

Therefore, with notations as above, we have the geometric decomposition of $\mathbb{F}_q^n$

(a) $\mathbb{F}_q^n = H_1 \perp \cdots \perp H_m \perp L_1 \perp H'_1 \perp \cdots \perp H'_{m'} \perp W'$ and
$\mathscr{C} = \langle x_1, \ldots, x_l, x_{l+1}, x'_{l+1}, \ldots, x_m, x'_m, x_{m+1} \rangle$
(b) $\mathbb{F}_q^n = H_1 \perp \cdots \perp H_m \perp H_{m+1} \perp H'_1 \perp \cdots \perp H'_{m'} \perp W'$ and
$\mathscr{C} = \langle x_1, \ldots, x_l, x_{l+1}, x'_{l+1}, \ldots, x_{m+1}, x'_{m+1} \rangle$
(c) $\mathbb{F}_q^n = H_1 \perp \cdots \perp H_m \perp L_1 \perp L_2 \perp H'_1 \perp \cdots \perp H'_{m'} \perp W'$ and
$\mathscr{C} = \langle x_1, \ldots, x_l, x_{l+1}, x'_{l+1}, \ldots, x_m, x'_m, x_{m+1}, x_{m+2} \rangle$

From the construction of the previous basis of $\mathbb{F}_q^n$ and Proposition 1, it follows that a linear code can be written as the linear subspace generated by a part of a geometric basis of type r, s, t with $s + t \leq 4$, because we have generated hyperbolic planes until their complement (W and W') is a linear subspace of dimension lower than or equal to two. We also have that $t \leq 2$ because in the bases of W and W' there is at most one element x such that $B(x, x) = g$. Note that for $q \equiv 3 \mod 4$ one has that $t = 0$. □

We say that a linear code $\mathscr{C}$ given by the generators $\mathscr{C} = \langle x_1, \ldots, x_k \rangle$ is given in the **standard geometric form** if the matrix of B restricted to $x_1, \ldots, x_k$ is the same matrix as the one of B restricted to the generators of $\mathscr{C}$ of the basis obtained in Theorem 2.

Next example shows a geometric decomposition of $\mathbb{F}_3^{12}$ compatible with the Golay code $\mathscr{G}_{12}$ [22], which is a self-dual code.

Example 2 The Golay code $\mathscr{G}_{12}$ is a self-dual code over $\mathbb{F}_3$ with generator matrix [22]:

$$G = \begin{pmatrix} 1 & 0 & 0 & 0 & 0 & 0 & 1 & 1 & 1 & 1 & 1 & 0 \\ 0 & 1 & 0 & 0 & 0 & 0 & 0 & 1 & 2 & 2 & 1 & 2 \\ 0 & 0 & 1 & 0 & 0 & 0 & 1 & 0 & 1 & 2 & 2 & 2 \\ 0 & 0 & 0 & 1 & 0 & 0 & 2 & 1 & 0 & 1 & 2 & 2 \\ 0 & 0 & 0 & 0 & 1 & 0 & 2 & 2 & 1 & 0 & 1 & 2 \\ 0 & 0 & 0 & 0 & 0 & 1 & 1 & 2 & 2 & 1 & 0 & 2 \end{pmatrix}$$

One has that for $\mathscr{G}_{12} \subset \mathbb{F}_3^{12}$, the standard decomposition of $\mathbb{F}_3^{12}$ compatible with $\mathscr{G}_{12}$ is the orthogonal sum of 6 hyperbolic planes, where the first geometric generator of each one belongs to the code and the second one does not. The matrix M of a standard geometric decomposition is

$$M = \begin{pmatrix} 1 & 0 & 0 & 0 & 0 & 0 & 1 & 1 & 1 & 1 & 1 & 0 \\ 2 & 0 & 0 & 0 & 0 & 0 & 1 & 1 & 1 & 1 & 1 & 0 \\ 0 & 1 & 0 & 0 & 0 & 0 & 0 & 1 & 2 & 2 & 1 & 2 \\ 0 & 2 & 0 & 0 & 0 & 0 & 0 & 1 & 2 & 2 & 1 & 2 \\ 0 & 0 & 1 & 0 & 0 & 0 & 1 & 0 & 1 & 2 & 2 & 2 \\ 0 & 0 & 2 & 0 & 0 & 0 & 1 & 0 & 1 & 2 & 2 & 2 \\ 0 & 0 & 0 & 1 & 0 & 0 & 2 & 1 & 0 & 1 & 2 & 2 \\ 0 & 0 & 0 & 2 & 0 & 0 & 2 & 1 & 0 & 1 & 2 & 2 \\ 0 & 0 & 0 & 0 & 1 & 0 & 2 & 2 & 1 & 0 & 1 & 2 \\ 0 & 0 & 0 & 0 & 2 & 0 & 2 & 2 & 1 & 0 & 1 & 2 \\ 0 & 0 & 0 & 0 & 0 & 1 & 1 & 2 & 2 & 1 & 0 & 2 \\ 0 & 0 & 0 & 0 & 0 & 2 & 1 & 2 & 2 & 1 & 0 & 2 \end{pmatrix}$$

A basis of the code are the rows 1, 3, 5, 7, 9, 11. One has that

$$J_{6,0,0} = MM^t = \begin{pmatrix} 0 & 1 & 0 & 0 & 0 & 0 & 0 & 0 & 0 & 0 & 0 & 0 \\ 1 & 0 & 0 & 0 & 0 & 0 & 0 & 0 & 0 & 0 & 0 & 0 \\ 0 & 0 & 0 & 1 & 0 & 0 & 0 & 0 & 0 & 0 & 0 & 0 \\ 0 & 0 & 1 & 0 & 0 & 0 & 0 & 0 & 0 & 0 & 0 & 0 \\ 0 & 0 & 0 & 0 & 0 & 1 & 0 & 0 & 0 & 0 & 0 & 0 \\ 0 & 0 & 0 & 0 & 1 & 0 & 0 & 0 & 0 & 0 & 0 & 0 \\ 0 & 0 & 0 & 0 & 0 & 0 & 0 & 1 & 0 & 0 & 0 & 0 \\ 0 & 0 & 0 & 0 & 0 & 0 & 1 & 0 & 0 & 0 & 0 & 0 \\ 0 & 0 & 0 & 0 & 0 & 0 & 0 & 0 & 0 & 1 & 0 & 0 \\ 0 & 0 & 0 & 0 & 0 & 0 & 0 & 0 & 1 & 0 & 0 & 0 \\ 0 & 0 & 0 & 0 & 0 & 0 & 0 & 0 & 0 & 0 & 0 & 1 \\ 0 & 0 & 0 & 0 & 0 & 0 & 0 & 0 & 0 & 0 & 1 & 0 \end{pmatrix}$$

4.2 *Characteristic 2*

Now let $\mathbb{F}_q$ be a field of characteristic two. By the results in Sect. 3 we can write $\mathbb{F}_q^n$ as an orthogonal sum of hyperbolic planes, one-dimensional linear varieties and, at most, one elliptic plane.

We say that $\mathbb{F}_q^n$ has a **geometric decomposition of type** r, s, t if

$$\mathbb{F}_q^n = H_1 \perp \cdots \perp H_r \perp L_1 \perp \cdots \perp L_s, \text{ with } t = 0, \text{ or}$$

$$\mathbb{F}_q^n = H_1 \perp \cdots \perp H_r \perp L_1 \perp \cdots \perp L_s \perp E, \text{ with } t = 1$$

where $H_1, \ldots, H_r$ are hyperbolic planes, $L_1, \ldots, L_s$ are non-isotropic one-dimensional linear varieties and E is an elliptic plane. Each hyperbolic plane is generated by two geometric generators $H_i = \langle x_{2i-1}, x_{2i} \rangle$, $i = 1, \ldots, r$, each one-dimensional linear variety is generated by a geometric generator $L_i = \langle x_{2r+i} \rangle$, $i = 1, \ldots, s$ and the elliptic plane is generated by two geometric generators, $E = \langle x_{n-1}, x_n \rangle$ if $t = 1$. One has that $\{x_1, \ldots, x_n\}$ is a basis of $\mathbb{F}_q^n$, called **basis of the geometric decomposition**.

Let M be the matrix whose rows are the elements of the geometric decomposition, then one has that $MM^t = J_{r,s,t}$. That is, $J_{r,s,t}$ is the matrix of B in such a basis.

$$J_{r,s,0} = \begin{pmatrix} 0 & 1 & & & & & & \\ 1 & 0 & & & & & & \\ & & \ddots & & & & & \\ & & & 0 & 1 & & & \\ & & & 1 & 0 & & & \\ & & & & & 1 & & \\ & & & & & & \ddots & \\ & & & & & & & 1 \end{pmatrix}$$

$$J_{r,s,1} = \begin{pmatrix} 0 & 1 & & & & & & & & \\ 1 & 0 & & & & & & & & \\ & & \ddots & & & & & & & \\ & & & 0 & 1 & & & & & \\ & & & 1 & 0 & & & & & \\ & & & & & 1 & & & & \\ & & & & & & \ddots & & & \\ & & & & & & & 1 & & \\ & & & & & & & & 0 & 1 \\ & & & & & & & & 1 & 1 \end{pmatrix}$$

A basis $\{x_1, \ldots, x_n\}$ of $\mathbb{F}_q^n$ is said to be a **compatible basis with respect to a decomposition of type** r, s, t if each $\{x_{2i-1}, x_{2i}\}$, with $i = 1, \ldots, r$, is a geometric basis of a hyperbolic plane; each x_{2r+i}, with $i = 1, \ldots s$, generates a one-dimensional linear variety; $\{x_{2r+s+1}, x_{2r+s+2}\}$ is a geometric basis of an elliptic plane (if $t = 1$) in such a way that all of these subspaces are pairwise orthogonal. Or equivalently, if the matrix of B in such a basis is equal to $J_{r,s,t}$.

Let $\mathscr{C} \subset \mathbb{F}_q^n$ be a linear code, we say that $\mathscr{C}$ is **compatible with a geometric decomposition of type** r, s, t if there exists a basis $\{x_1, \ldots, x_n\}$ of $\mathbb{F}_q^n$ compatible with such a decomposition, in such a way that there exists $I \subset \{1, \ldots, n\}$ such that $\{x_i \mid i \in I\}$ is a basis of $\mathscr{C}$.

The next results allows us to compute a geometric decomposition compatible with a given code in characteristic 2.

Theorem 3 *Let $\mathbb{F}_q$ be of characteristic 2. Any linear code $\mathscr{C} \subset \mathbb{F}_q^n$ is compatible with at least one geometric decomposition. Furthermore, there is a computable geometric basis, called standard, compatible with $\mathscr{C}$, of type r, s, t with $s \leq 4$ and $t = 0$, or, $s \leq 2$ and $t = 1$. A geometric decomposition with an elliptic plane ($s \leq 2$, $t = 1$) is only possible when $(1, \dots, 1) \in \mathrm{rad}(\mathscr{C})$, that is, when $(1, \dots, 1) \in \mathscr{C}$ and all the elements of $\mathscr{C}$ are isotropic.*

Proof One has that $\mathscr{C} = \mathrm{rad}(\mathscr{C}) \perp \mathscr{C}_1$. We shall consider two cases, $z = (1, \dots, 1) \in \mathrm{rad}(\mathscr{C})$ with n even, and the general case. In this particular case ($z \in \mathrm{rad}(\mathscr{C})$ with n even) we consider an elliptic plane because z cannot belong to a hyperbolic plane, since, as we proved in Proposition 3, there exists no isotropic vector in $\mathbb{F}_q^n$ orthogonal to z.

Let $z \in \mathscr{C}$, one has that $z \in \mathrm{rad}(\mathscr{C})$ if and only if all the elements of $\mathscr{C}$ are isotropic, that is, if $\mathscr{C} \subset S$. For instance, for a self-dual code we consider an elliptic plane: let $\mathscr{C}$ be a self-dual code, one has that $z \in \mathscr{C} = \mathrm{rad}(\mathscr{C})$ because otherwise the direct sum of $\mathscr{C}$ and $\langle z \rangle$ would be a vector subspace of index $n/2 + 1$.

First, we prove the general case and, then, the case $(1, \dots, 1) \in \mathrm{rad}(\mathscr{C})$ with n even. Let $\mathscr{C} = \mathrm{rad}(\mathscr{C}) \perp \mathscr{C}_1$, where $\mathrm{rad}(\mathscr{C}) = \langle x_1, \dots, x_l \rangle$. Let S' be equal to S for n odd and to $\langle y_1, \dots y_{n-2} \rangle$ for n even, where $\{y_1, \dots, y_{n-2}, (1, \dots, 1)\}$ is a basis of S. One has that S' is non-singular, has dimension greater than or equal to $n - 2$ and that $\mathrm{rad}(\mathscr{C}) \subset S'$ (by Proposition 3).

We claim that we can compute $x'_1, \dots, x'_l \in \mathbb{F}_q^n$ such that x_i, x'_i are the geometric generators of a hyperbolic plane and, moreover, the hyperbolic planes $H_i = \langle x_i, x'_i \rangle$ and $\mathscr{C}_1$ are pairwise orthogonal. That is, one has that

$$\mathscr{C}' = H_1 \perp \cdots \perp H_l \perp \mathscr{C}_1$$

where $\mathscr{C}'$ contains $\mathscr{C}$ and is non-singular. We prove the construction of $\mathscr{C}'$ by induction on l.

For $l = 0$ there is nothing to prove. The subspace $\mathscr{C}_0 = \langle x_1, \dots, x_{l-1} \rangle \perp \mathscr{C}_1$ is orthogonal to x_l but does not contain it. One has that $x_l \in \mathscr{C}_0^\perp$ but $x_l \notin \mathrm{rad}(\mathscr{C}_0^\perp) = \mathrm{rad}(\mathscr{C}_0)$. Let $\mathscr{C}_0^\perp = \langle x_1, \dots, x_{l-1} \rangle \perp U$. One has that $U \cap S'$ is a non-singular vector space that contains x_l. Therefore there exists $y \in U \cap S'$ such that $B(x_l, y) \neq 0$, since x_l is isotropic. The plane generated by x_l, y is non-singular and is contained in $\mathscr{C}_0^\perp$, so by Proposition 4 it is generated by a geometric basis $H_l = \langle x_l, x'_l \rangle$. Since $H_l \subset \mathscr{C}_0^\perp$, then $\mathscr{C}_0 \perp H_l$ and $\mathscr{C}_0 \subset H_l^\perp$. As the radical of $\mathscr{C}_0$ has dimension $l - 1$, by inductive hypothesis we can compute geometric bases $\{x_i, x'_i\}$ of H_i in $H_l^\perp$, for $i = 1, \dots, l - 1$ such that they are pairwise orthogonal and also to $\mathscr{C}_1$, and since they are orthogonal to H_l and H_l is orthogonal to $\mathscr{C}_1$, the construction of $\mathscr{C}'$ holds.

Therefore, we have $\mathscr{C}' = H_1 \perp \cdots \perp H_l \perp \mathscr{C}_1$, where $H_i = \langle x_i, x'_i \rangle$, with $x'_i \notin \mathscr{C}$. Moreover, $\mathscr{C}'$ is non-singular and one has that $\mathbb{F}_q^n = H_1 \perp \cdots \perp H_l \perp \mathscr{C}_1 \perp \mathscr{C}'^\perp$.

Since $\mathscr{C}_1$ is non-singular, by Proposition 5 we can consider $\mathscr{C}_1$ as a sum of hyperbolic planes and a vector subspace W of dimension 1 or 2 (if the dimension of $\mathscr{C}_1$ is lower than 3 we do not consider any hyperbolic plane and $\mathscr{C}_1 = W$). Hence, we have $\mathscr{C}_1 = H_{l+1} \perp \cdots \perp H_m \perp W$, where $H_{l+i} = \langle x_{l+i}, x'_{l+i} \rangle$.

By Proposition 5 we can have three different geometries for W:

(a) If $\dim(W) = 1$, we write $W = \langle x \rangle$. Moreover, x is non-isotropic since B is non-degenerate. We consider $x_{m+1} \in W$ such that $B(x_{m+1}, x_{m+1})$ is equal to 1 and $W = L_1 = \langle x_{m+1} \rangle$.
(b) If $\dim(W) = 2$ and W contains two linearly independent isotropic vectors (or equivalently $\sum x_i = 0$, for all $x \in W$) then W is a hyperbolic plane, $W = H_{m+1} = \langle x_{m+1}, x'_{m+1} \rangle$, by Proposition 4.
(c) If $\dim(W) = 2$ and W does not contain two lines of isotropic vectors (or equivalently, there exists $x \in W$ with $\sum x_i \neq 0$) then, by Proposition 4, W is an elliptic plane and it can be generated by two orthonormal vectors $L_1 = \langle x_{m+1} \rangle$, $L_2 = \langle x_{m+2} \rangle$, where $W = L_1 \perp L_2$.

We decompose $\mathscr{C}'^{\perp}$ in an analogous way to $\mathscr{C}_1$ (using Proposition 5) and we obtain

$$\mathscr{C}'^{\perp} = H'_1 \perp \cdots \perp H'_{m'} \perp W'$$

With notations as above, we have the following geometric decomposition of $\mathbb{F}_q^n$

(a) $\mathbb{F}_q^n = H_1 \perp \cdots \perp H_m \perp L_1 \perp H'_1 \perp \cdots \perp H'_{m'} \perp W'$ and
$\mathscr{C} = \langle x_1, \ldots, x_l, x_{l+1}, x'_{l+1}, \ldots, x_m, x'_m, x_{m+1} \rangle$
(b) $\mathbb{F}_q^n = H_1 \perp \cdots \perp H_m \perp H_{m+1} \perp H'_1 \perp \cdots \perp H'_{m'} \perp W'$ and
$\mathscr{C} = \langle x_1, \ldots, x_l, x_{l+1}, x'_{l+1}, \ldots, x_{m+1}, x'_{m+1} \rangle$
(c) $\mathbb{F}_q^n = H_1 \perp \cdots \perp H_m \perp L_1 \perp L_2 \perp H'_1 \perp \cdots \perp H'_{m'} \perp W'$ and
$\mathscr{C} = \langle x_1, \ldots, x_l, x_{l+1}, x'_{l+1}, \ldots, x_m, x'_m, x_{m+1}, x_{m+2} \rangle$

From the construction of the previous basis of $\mathbb{F}_q^n$, it follows that a linear code $\mathscr{C}$, such that $(1, \ldots, 1) \notin \mathrm{rad}(\mathscr{C})$ with n even, can be written as the linear subspace generated by a part of a geometric basis of type r, s, t with $s \leq 4$ and $t = 0$, because we have generated hyperbolic planes until their complement (W and W') is a linear subspace of dimension lower than or equal to two which may be decomposed using Proposition 4.

Let us consider $z \in \mathrm{rad}(\mathscr{C})$, when n is even. Let $\mathrm{rad}(\mathscr{C}) = \langle x_1, \ldots, x_l, z \rangle$ and $R = \langle x_1, \ldots, x_l \rangle$. Let $\mathscr{C}_R = R \perp \mathscr{C}_1$, since $z \notin \mathrm{rad}(\mathscr{C}_R) = R$, as in the general case, we can compute $x'_1, \ldots, x'_l \in \mathbb{F}_q^n$ such that x_i, x'_i are the geometric generators of a hyperbolic plane and, moreover, the hyperbolic planes $H_i = \langle x_i, x'_i \rangle$, and $\mathscr{C}_1$ are pairwise orthogonal. That is, one has that

$$\mathscr{C}'_R = H_1 \perp \cdots \perp H_l \perp \mathscr{C}_1$$

where $\mathscr{C}'_R$ contains $\mathscr{C}_R$ and is non-singular.

Since $\mathscr{C}_1$ is non-singular, by Proposition 5 we can consider $\mathscr{C}_1$ as a sum of hyperbolic planes, that is, we have the geometry (b) of the general case since all the elements of $\mathscr{C}_1$ are isotropic. Therefore, we have $\mathscr{C}_1 = H_{l+1} \perp \cdots \perp H_m$, where $H_{l+i} = \langle x_{l+i}, x'_{l+i} \rangle$.

Hence, since all the elements of $\mathscr{C}_R$ are isotropic, one has that U, the direct sum of $\mathscr{C}'_r$ and $\langle z \rangle$ can be written in the following way $U = H_1 \perp \cdots \perp H_m \perp \langle z \rangle$. We claim that in $U^{\perp}$ there exists a non-isotropic vector z' such that z' is orthogonal to $\mathscr{C}'_R$ and $B(z, z') = 1$. Let $E = \langle z, z' \rangle$, we have that E is an elliptic plane. Such vector z' is one solution of the following linear system with at most n equations and n variables

$$\begin{cases} B(x_1, z') = 0, \\ B(x'_1, z') = 0, \\ \quad \vdots \\ B(x_m, z') = 0, \\ B(x'_m, z') = 0, \\ B(z, z') = 1, \\ B(z', z') = 1. \end{cases}$$

Therefore, one has that $U' = H_1 \perp \cdots \perp H_m \perp E$ is non-singular and contains $\mathscr{C}$. We decompose $U'^{\perp}$ using Proposition 5 and obtain

$$U'^{\perp} = H'_1 \perp \cdots \perp H'_{m'} \perp W'$$

With notations as above, we have the following geometric decomposition of $\mathbb{F}_q^n$:

(d) $\mathbb{F}_q^n = H_1 \perp \cdots \perp H_m \perp E \perp H'_1 \perp \cdots \perp H'_{m'} \perp W'$ and
$\mathscr{C} = \langle x_1, \ldots, x_l, x_{l+1}, x'_{l+1}, \ldots, x_{m+1}, x'_{m+1}, z \rangle$

From the construction of the previous basis of $\mathbb{F}_q^n$, it follows that a linear code $\mathscr{C}$, such that $(1, \ldots, 1) \in \mathrm{rad}(\mathscr{C})$ with n even, can be written as the linear subspace generated by a part of a geometric basis of type r, s, t with $s \leq 2$ and $t = 1$, because we have generated hyperbolic planes in $U'^{\perp}$ until W' is a linear subspace of dimension lower than or equal to two that may be decomposed using Proposition 4. □

We say that a linear code $\mathscr{C}$ given by the generators $\mathscr{C} = \langle x_1, \ldots, x_k \rangle$ is given in the **standard geometric form** if the matrix of B restricted to $x_1, \ldots, x_k$ is the same matrix as the one of B restricted to the generators of $\mathscr{C}$ of the basis obtained in Theorem 3.

From Theorem 3 it follows that the standard geometric decomposition of $\mathscr{C} = \mathbb{F}_q^n$ is of type $n/2 - 1, 2, 0$ for n even, and $(n-1)/2, 1, 0$ for n odd. The following example shows the geometric decomposition of a self-dual code in characteristic 2.

Example 3 Let $\mathscr{C} \subset \mathbb{F}_2^6$ be the code with generator matrix

$$G = \begin{pmatrix} 1 & 1 & 0 & 0 & 0 & 0 \\ 1 & 1 & 1 & 0 & 1 & 0 \\ 1 & 1 & 1 & 1 & 1 & 1 \end{pmatrix}$$

One has that $\mathscr{C}$ is a self-dual code because it has dimension $n/2$ and the sum of the coordinates of the generators of the code, that is, the rows of the generator matrix, are 0 (Proposition 3).

Hence, the standard decomposition is given by 2 hyperbolic planes and an elliptic plane. In particular, one has that the matrix M of a standard geometric decomposition is

$$M = \begin{pmatrix} 1 & 1 & 0 & 0 & 0 & 0 \\ 0 & 1 & 1 & 0 & 0 & 0 \\ 1 & 1 & 1 & 0 & 1 & 0 \\ 1 & 1 & 1 & 1 & 0 & 0 \\ 1 & 1 & 1 & 1 & 1 & 1 \\ 0 & 0 & 0 & 0 & 0 & 1 \end{pmatrix}$$

A basis of the code are the rows 1, 3 and 5 of the matrix M, which in this case form the same basis as we have previously considered. The geometric decomposition obtained is $\mathbb{F}_2^6 = H_1 \perp H_2 \perp E$. That is, a geometric decomposition of type 2, 0, 1, hence

$$J_{2,0,1} = \begin{pmatrix} 0 & 1 & 0 & 0 & 0 & 0 \\ 1 & 0 & 0 & 0 & 0 & 0 \\ 0 & 0 & 0 & 1 & 0 & 0 \\ 0 & 0 & 1 & 0 & 0 & 0 \\ 0 & 0 & 0 & 0 & 0 & 1 \\ 0 & 0 & 0 & 0 & 1 & 1 \end{pmatrix}$$

The following example illustrates how to deal with an elliptic plane when $(1, \ldots, 1) \notin \mathrm{rad}(\mathscr{C})$.

Example 4 Let $\mathscr{C}$ be the linear code over $\mathbb{F}_2$ with generator matrix

$$G = \begin{pmatrix} 1 & 1 & 0 & 0 \\ 0 & 0 & 0 & 1 \end{pmatrix}$$

Let $x_1 = (1, 1, 0, 0)$ and $x_2 = (0, 0, 0, 1)$. One has that x_1 is an isotropic vector and that x_2 is non-isotropic. Let $x_1' = (0, 1, 1, 0)$, one has that x_1, x_1' are a geometric basis of an hyperbolic plane $H_1 = \langle x_1, x_1' \rangle$ which is orthogonal to x_2. An orthogonal vector to H_1 and linearly independent to x_2 is $y = (1, 1, 1, 1)$. One has that y is isotropic and y, x_2 form a geometric basis of an elliptic plane. However, we can consider $x_3 = x_2 + y = (1, 1, 1, 0)$ in such a way that $L_2 = \langle x_2 \rangle$ and $L_3 = \langle x_3 \rangle$ are two non-isotropic linear varieties. Therefore, one has a geometric decomposition of $\mathbb{F}_2^4$ compatible with $\mathscr{C}$ of type $1, 2, 0$, given by $\mathbb{F}_2^4 = H_1 \perp L_1 \perp L_2$.

5 Linear Codes and Bilinear Algebra

Since we have proved that a linear code is compatible with a geometric decomposition for arbitrary characteristic, from now on, we will work over an arbitrary positive characteristic.

Let $\{x_1, \ldots, x_n\}$ be a geometric basis of a geometric decomposition of type r, s, t. Let $i \in \{1, \ldots, n\}$. We define i' as

- $i + 1$ if x_i is the first generator of a hyperbolic plane H,
- $i - 1$ if x_i is the second generator of a hyperbolic plane H,
- i if x_i generates a one-dimensional linear space L,
- $i + 1$ if x_i is the first generator of an elliptic plane E.

We do not define i' when x_i is the second geometric generator of an elliptic plane, because we only consider geometric decompositions with at most one elliptic plane E and where only the first generator of E belongs to the code. In the case where both geometric generators of the elliptic plane E belong to the code, by Proposition 4, we consider two orthonormal generators of linear subspaces L (as in Example 4).

For $I \subset \{1, \ldots, n\}$ we define $I' = \{i' \mid i \in I\}$ and $I^\perp = \{1, \ldots, n\} \setminus I'$. In this way we can compute the dual code of a linear code using the following result. Note that this result extends Theorem 1 for an arbitrary linear code.

Theorem 4 *Let $\mathscr{C}$ be a linear code with geometric decomposition of type r, s, t given by the basis $\{x_1, \ldots, x_n\}$ of $\mathbb{F}_q^n$. Let $I \subset \{1, \ldots, n\}$ such that $\mathscr{C} = \langle x_i \mid i \in I \rangle$. Then the dual code of $\mathscr{C}$ is $\mathscr{C}^\perp = \langle x_i \mid i \in I^\perp \rangle$.*

Proof From the matrix $J_{r,s,t}$ of the bilinear form B in the geometric basis it follows that $\langle x_i \rangle^\perp = \langle x_j \mid j \neq i' \rangle$. Therefore, $\mathscr{C}^\perp = \langle x_j \mid j \notin I' \rangle = \langle x_i \mid i \in I^\perp \rangle$. □

Let $\mathscr{C}$ be a linear code of dimension k with a geometric decomposition of type r, s, t given by the basis $\{x_1, \ldots, x_n\}$ of $\mathbb{F}_q^n$ and $I \subset \{1, \ldots, n\}$ such that $\mathscr{C} = \langle x_i \mid i \in I \rangle$. Furthermore, let M be the $n \times n$-matrix whose rows are the elements of the basis $\{x_1, \ldots, x_n\}$, then one has that $MM^t = J_{r,s,t}$. Let $M(I)$ be the $k \times n$-matrix consisting of the k rows given by I, then $M(I)$ is a generator matrix of $\mathscr{C}$. In the same way, $M(I^\perp)$ is a control matrix of $\mathscr{C}$, that is, $M(I^\perp)$ is a generator matrix of the dual code $\mathscr{C}^\perp$ of $\mathscr{C}$.

Example 5 Consider the Matrix M given in Example 3 and the geometric decomposition of type 2,0,1 given by the rows of the matrix M, $\{x_1, \ldots, x_6\}$. One has that $\mathbb{F}_2^6 = H_1 \perp H_2 \perp E$.

Let $I = \{1, 2, 3\}$ and $\mathscr{C} = \langle x_i \mid i \in I\rangle$. By Theorem 4, the dual code of $\mathscr{C}$ is $\langle x_i \mid i \in I^{\perp}\rangle$, where $I' = \{2, 1, 4\}$ and $I^{\perp} = \{1, \ldots, 6\} \setminus I' = \{3, 5, 6\}$.

We have only considered an elliptic plane at the geometric decomposition when the first geometric generator of the elliptic plane belongs to the code and the second one does not. Its motivation rests on the following fact: if x_i is the second generator of an elliptic plane, then $\langle x_i\rangle^{\perp} = \langle x_j \mid j \neq i, i-1\rangle + \langle x_i + x_{i-1}\rangle$, but $x_i + x_{i-1}$ is not an element of the basis of $\mathbb{F}_q^n$ considered.

5.1 Stabilizer Quantum Codes

Stabilizer codes can be constructed from self-orthogonal classical linear codes using the CSS construction (due to Calderbank, Shor and Steane [4, 29]).

Theorem 5 ([4, 20]) *Let $\mathscr{C}$ be a linear $[n, k, d]_q$ error-correcting code such that $\mathscr{C} \subset \mathscr{C}^{\perp}$. Then, there exists an $[[n, n-2k, \geq d^{\perp}]]_q$ stabilizer quantum code, where $d^{\perp}$ denotes the minimum distance of $\mathscr{C}^{\perp}$.*

If we have a geometric decomposition, we can easily check whether a linear code is self-orthogonal and construct a quantum code using the CSS construction.

Theorem 6 *Let $\mathscr{C}$ be a linear $[n, k, d]$ code with geometric decomposition of type r, s, t given by the basis $\{x_1, \ldots, x_n\}$ of $\mathbb{F}_q^n$. Consider $I \subset \{1, \ldots, n\}$ such that $\mathscr{C} = \langle x_i \mid i \in I\rangle$. Let $I \subset I^{\perp}$, then there exists an $[[n, n-2k, \geq d^{\perp}]]_q$ stabilizer quantum code.*

Proof By Theorem 4, the dual code of $\mathscr{C}$ is $\mathscr{C}^{\perp} = \langle x_i \mid i \in I^{\perp}\rangle$. Thus if $I \subset I^{\perp}$, the code $\mathscr{C}$ is self-orthogonal and, by Theorem 5, the result holds. □

Example 6 Consider the Matrix M given in Example 3 and the geometric decomposition of type 2,0,1 given by the rows of the matrix M, $\{x_1, \ldots, x_6\}$. One has that $\mathbb{F}_2^6 = H_1 \perp H_2 \perp E$.

Let $I = \{1, 3\}$ and $\mathscr{C} = \langle x_i \mid i \in I\rangle$. Then $I' = \{2, 4\}$ and $I^{\perp} = \{1, \ldots, 6\} \setminus I' = \{1, 3, 5, 6\}$. By Theorem 6, we can construct a stabilizer quantum code from $\mathscr{C}$ since $I \subset I^{\perp}$.

The technique given in the previous result was used in [12–15] to compute stabilizer quantum codes of J-affine variety codes (and toric codes). Theorem 6 shows which codes, with a geometric decomposition as in Sect. 4, can provide stabilizer quantum codes. That is, one can extend the method in [12–15] for an arbitrary family of codes. Algebraic-geometric codes will be considered in future works. Moreover, an analogous CSS construction also holds for Hermitian duality when the classical code $\mathscr{C}$ is defined over $\mathbb{F}_{q^2}$. The Hermitian metric structure will be studied in future works as well.

5.2 LCD Codes

LCD codes are linear codes whose radical is equal to zero [23], that is, $\mathscr{C}$ is LCD if $\mathscr{C} \cap \mathscr{C}^{\perp} = \{0\}$. If we have a geometric decomposition, we can easily check whether a linear code is LCD.

Theorem 7 *Let $\mathscr{C}$ be a linear code with geometric decomposition of type r, s, t given by the basis $\{x_1, \ldots, x_n\}$ of $\mathbb{F}_q^n$. Let $I \subseteq \{1, \ldots, n\}$ such that $\mathscr{C} = \langle x_i \mid i \in I \rangle$. One has that $\mathscr{C}$ is LCD if and only if $I \cap I^{\perp} = \emptyset$.*

Proof By Theorem 4, the dual code of $\mathscr{C}$ is $\mathscr{C}^{\perp} = \langle x_i \mid i \in I^{\perp} \rangle$. Thus, $I \cap I^{\perp} = \emptyset$ if and only if $\mathscr{C} \cap \mathscr{C}^{\perp} = \{0\}$. □

Example 7 Consider the Matrix M given in Example 3 and the geometric decomposition of type 2, 0, 1 given by the rows of the matrix M, $\{x_1, \ldots, x_6\}$. One has that $\mathbb{F}_2^6 = H_1 \perp H_2 \perp E$.

Let $I = \{1, 2, 3, 4\}$ and $\mathscr{C} = \langle x_i \mid i \in I \rangle$. We have that $I' = \{2, 1, 4, 3\}$ and $I^{\perp} = \{1, \ldots, 6\} \setminus I' = \{5, 6\}$. By Theorem 7, $\mathscr{C}$ is an LCD code since $I \cap I^{\perp} = \emptyset$.

The tecnique given in the previous result was used in [16] to compute new LCD codes from J-affine variety codes (and toric codes). Theorem 7 shows which codes, with a geometric decomposition as in Sect. 4, are LCD. In the same way as for quantum codes, one can extend the method in [16] for an arbitrary family of codes. LCD codes coming from affine variety codes will be considered in future works.

5.3 Minimum Distance of a Linear Code

The following result extends [21, Proposition 1] and [27, Proposition 8] of generalized toric codes for arbitrary linear codes.

Theorem 8 *Let $\mathscr{C}$ be a linear code of dimension k with geometric decomposition of type r, s, t given by the basis $\{x_1, \ldots, x_n\}$ of $\mathbb{F}_q^n$ and $I \subset \{1, \ldots, n\}$ such that $\mathscr{C} = \langle x_i \mid i \in I \rangle$. Let M be the $n \times n$-matrix such that $MM^t = J_{r,s,t}$, where a generator matrix of $\mathscr{C}$ is $M(I)$ and $M(I, J)$ is the submatrix of M corresponding to the rows of I and columns of J, i.e. $M(I, J) = (m_{i,j})_{i \in I, j \in J}$.*

(a) *Let d be the lowest positive integer such that for every set $J \subset \{1, \ldots, n\}$ with $\#J = n - d + 1$ there exists some $K \subset J$ with $\#K = k$ such that* $\det M(I, K) \neq 0$*. Then the minimum distance of $\mathscr{C}$ is d.*
(b) *Let d be the largest positive integer such that for all $J \subset \{1, \ldots, n\}$ with $\#J = d - 1$ there exists $D \subset I^{\perp}$ with $\#D = d - 1$ such that* $\det(D, J) \neq 0$*. Then the minimum distance of $\mathscr{C}$ is d.*

Besides, both previous ways of computing the minimum distance are equivalent.

Proof

(a) One has that the minimum distance of a linear code is d if for any $n-d+1$ columns of a generator matrix there exist k linearly independent columns and there are $n-d$ columns that do not contain k linearly independent columns. A generator matrix of $\mathscr{C}$ is $M(I)$, hence the minimum distance of $\mathscr{C}$ is the greatest positive integer d such that any $n-d+1$ columns of $M(I)$ contain k linearly independent columns, and the result holds.

(b) One has that the minimum distance of a linear code is d if any $d-1$ columns of a control matrix are linearly independent and there exist d linearly independent columns. A control matrix of $\mathscr{C}$ is $M(I^{\perp})$, hence the minimum distance of $\mathscr{C}$ is the largest positive integer d such that any $d-1$ columns of $M(I^{\perp})$ are linearly independent, which is equivalent to the fact that for every $J \subset \{1,\ldots,n\}$, $\#J = d-1$, there exists one minor $M(I^{\perp}, J)$ of size $d-1$ whose determinant is different from 0, and the result holds.

The equivalence between these two results is clear because both compute the minimum distance of a linear code $\mathscr{C}$ and, moreover, both ways of computing the minimum distance are dual. In order to prove it we use Plücker geometry.

Let M be the matrix whose rows are the elements of the basis $\{x_1,\ldots,x_n\}$ of $\mathbb{F}_q^n$, that is, the matrix of the linear transformation from the canonical basis $\{e_1,\ldots,e_n\}$ into $\{x_1,\ldots,x_n\}$, $N = \{1,\ldots,n\}$ and M^* the matrix of the linear transformation from the canonical basis $\{e_1^*,\ldots,e_n^*\}$ into $\{x_1^*,\ldots,x_n^*\}$. Therefore, $x_1 \wedge \cdots \wedge x_k = \sum_{j_i \in N} \det(M(I,K)) e_{j_1} \wedge \cdots \wedge e_{j_k}$, where $K = j_1,\ldots,j_k$. Since $MM^t = J_{r,s,t}$, one has that $M^* = J_{r,s,t}M$.

Let $\zeta(x_1 \wedge \cdots \wedge x_k) = x_{k+1}^* \wedge \cdots \wedge x_n^*$. Then

$$\zeta(x_1 \wedge \cdots \wedge x_k) = \sum_{j_i \in N \setminus K} \det(M^*(N \setminus I, N \setminus K)) e_{j_1}^* \wedge \cdots \wedge e_{j_{n-k}}^*$$

but since ζ is linear, one has that $\zeta(x_1 \wedge \cdots \wedge x_k) =$

$$\sum_{j_i \in K} \det(M(I,K)) \zeta(e_{j_1} \wedge \cdots \wedge e_{j_k}) = \sum_{j_i \in K} \det(M(I,K)) e_{j_1}^* \wedge \cdots \wedge e_{j_k}^*$$

Hence one has that $\det(M(I,K)) = \det(M^*(N \setminus I, N \setminus K)) =$
$\det(J_{r,s,t}M(N \setminus I, N \setminus K)) = \det(M(I \setminus I', N \setminus K)) = \det(M(I^{\perp}, N \setminus K))$. □

In [21, Proposition 1], which is extended by the previous result, the structure of Vandermonde matrix in several variables of the generator matrix of the generalized toric code is used to compute explicitly the minimum distance of two families of codes. For an arbitrary linear code we do not have such an structure and the previous result is not a priori useful. However, the geometric decomposition of a linear code may give rise to the explicit computation of the minimum distance of certain families of linear codes. This will be studied in future works.

Acknowledgements This problem was proposed by Antonio Campillo, I thank him for his many helpful comments. The author gratefully acknowledges the support from RYC-2016-20208 (AEI/FSE/UE), the support from The Danish Council for Independent Research (Grant No. DFF–4002-00367), and the support from the Spanish MINECO/FEDER (Grants No. MTM2015-65764-C2-2-P and MTM2015-69138-REDT).

References

1. Artin, E.: Algèbre Géométrique. Cahiers Scientifiques. Gauthier-Villars, Editeur, Paris (1967)
2. Bras-Amorós, M., O'Sullivan, M.E.: Duality for some families of correction capability optimized evaluation codes. Adv. Math. Commun. **2**(1), 15–33 (2008)
3. Braun, M., Etzion, T., Vardy, A.: Linearity and complements in projective space. Linear Algebra Appl. **430**, 57–70 (2013)
4. Calderbank, A.R., Shor, P.: Good quantum error-correcting codes exist. Phys. Rev. A **54**, 1098–1105 (1996)
5. Calderbank, A.R., Rains, E.M., Shor, P.W., Sloane, N.J.A.: Quantum error correction and orthogonal geometry. Phys. Rev. Lett. **76**, 405–409 (1997)
6. Calderbank, A.R., Rains, E.M., Shor, P.W., Sloane, N.J.A.: Quantum error correction via codes over GF(4). IEEE Trans. Inf. Theory **44**, 1369–1387 (1998)
7. Carlet, C., Guilley, S.: Complementary dual codes for counter-measures to side-channel attacks. Adv. Math. Commun. **10**(1), 131–150 (2016)
8. Carlet, C., Mesnager, S., Tang, C., Qi, Y.: Linear codes over $\mathbb{F}_q$ which are equivalent to LCD codes (2017). arXiv:1703.04346
9. Dickson, L.E.: Linear Groups. With an Exposition of the Galois Field Theory. Dover Publications, Mineola (1958)
10. Dieudonné, J.: La géométrie des groupes classiques (troisième édition). Ergebnisse der Mathematik und ihrer Grenzgebiete, Bd 5. Springer (1971)
11. Dieudonné, J.: Sur les groupes classiques (troisième édition). Publications de L'Institut de Mathématique de L'Université de Strasbourg. Hermann, Paris (1981)
12. Galindo, C., Hernando, F.: Quantum codes from affine variety codes and their subfield subcodes. Des. Codes Crytogr. **76**, 89–100 (2015)
13. Galindo, C., Hernando, F., Ruano, D.: New quantum codes from evaluation and matrix-product codes. Finite Fields Appl. **36**, 98–120 (2015)
14. Galindo, C., Hernando, F., Ruano, D.: Stabilizer quantum codes from J-affine variety codes and a new Steane-like enlargement. Quantum Inf. Process. **14**, 3211–3231 (2015)
15. Galindo, C., Geil, O., Hernando, F., Ruano, D.: On the distance of stabilizer quantum codes from J-affine variety codes. Quantum Inf. Process. **16**, 111 (2017)
16. Galindo, C., Hernando, F., Ruano, D.: New binary and ternary LCD codes (2017). arXiv:1710.00196
17. Hansen, J.P.: Toric varieties Hirzebruch surfaces and error-correcting codes. Appl. Algebra Eng. Commun. Comput. **13**(4), 289–300 (2002)
18. Hirschfeld, J.W.P.: Projective Geometries Over Finite Fields. Oxford Mathematical Monographs, 2nd edn. Oxford University Press, Oxford (1998)
19. Hou, X., Oggier, F.: On LCD codes and lattices. Proc. IEEE Int. Symp. Inf. Theory **2016**, 1501–1505 (2016)
20. Ketkar, A., Klappenecker, A., Kumar, S., Sarvepalli, P.K.: Nonbinary stabilizer codes over finite fields. IEEE Trans. Inf. Theory **52**, 4892–4914 (2006)
21. Little, J., Schwarz, R.: On toric codes and multivariate Vandermonde matrices. Appl. Algebra Eng. Commun. Comput. **18**(4), 349–367 (2007)
22. Macwilliams, F.J., Sloane, N.J.A.: The Theory of Error-Correcting Codes. North-Holland Mathematical Library, vol. 16. North-Holland, Amsterdam/New York (1977)

23. Massey, J.L.: Linear codes with complementary duals. Discrete Math. **106/107**, 337–342 (1992)
24. Pless, V.: On the uniqueness of the Golay codes. J. Comb. Theory **5**, 215–228 (1968)
25. Pless, V.: A classification of self-orthogonal codes over GF(2). Discrete Math. **3**, 209–246 (1972)
26. Pless, V.: Sloane, N.J.A.: On the classification and enumeration of self-dual codes. J. Comb. Theory Ser. A **18**, 313–335 (1975)
27. Ruano, D.: On the structure of generalized toric codes. J. Symbol. Comput. **44**(5), 499–506 (2009)
28. Serre, J.-P.: Cours d'arithmétique. Le Mathématicien, Presses Universitaires de France (1970)
29. Steane, A.M.: Simple quantum error correcting codes. Phys. Rev. Lett. **77**, 793–797 (1996)

On Some Properties of A Inherited by $C_b(X, A)$

Alejandra García, Lourdes Palacios, and Carlos Signoret

Abstract Let X be a completely regular Hausdorff space or a pseudocompact Hausdorff space. We denote by $C(X, A)$ the algebra of all continuous functions on X with values in a complex unital locally pseudo-convex algebra A. Let $C_b(X, A)$ be its subalgebra consisting of all bounded continuous functions endowed with the topology given by the uniform pseudo-seminorms of A on X. In this paper we examine some properties of A that are inherited by $C_b(X, A)$; these properties are projective limit decomposition, inversion, involution, spectral properties and metrizability.

1 Introduction

The importance of the algebras of functions in Functional Analysis and in particular in the general Theory of Topological Algebras is well known. There is a lot of work done on the algebra $C(X)$ of continuous real or complex functions on the topological space X. Some authors have given properties of the algebras of functions with values in a Banach Algebra A or in a more general topological algebra (see [3, 4, 8, 10]). Recently, Arizmendi, Cho and García studied the case when A is a locally convex algebra (see [5]). Throughout this paper we examine the algebra $C_b(X, A)$ of all continuous and bounded functions on X with values in a *locally pseudo-convex algebra* A where X is a completely regular Hausdorff space or a pseudocompact Hausdorff space. This algebra is a subalgebra of $C(X, A)$, the algebra of all continuous functions on X with values in A. Here we examine some properties of A that are inherited by $C_b(X, A)$. Throughout this paper, $\mathbb{F}$ will denote the field of real or complex numbers.

A. García
Facultad de Ciencias UNAM, Ciudad Universitaria México, México City, México
e-mail: a@a.unam.mx

L. Palacios (✉) · C. Signoret
Universidad Autónoma Metropolitana Iztapalapa, México City, México
e-mail: pafa@xanum.uam.mx; casi@xanum.uam.mx

G.-M. Greuel et al. (eds.), *Singularities, Algebraic Geometry, Commutative Algebra, and Related Topics*, https://doi.org/10.1007/978-3-319-96827-8_25

A **topological algebra** A over $\mathbb{F}$ is an algebra over $\mathbb{F}$ whose underlying vector space is a topological Hausdorff vector space with a *jointly continuous* multiplication.

For a topological algebra with unit e and $x \in A$, the **spectrum** of x is the set

$$\sigma(x) = \{\lambda \in \mathbb{F} : x - \lambda e \text{ is not invertible in } A\}$$

and the **spectral radius** of x is the real number[1]

$$r(x) = \sup\{|\lambda| : \lambda \in \sigma(x)\}.$$

A k**-seminorm** on A (with $k \in (0, 1]$), is a function

$$p : A \longrightarrow \mathbb{R}^+ \cup \{0\}$$

such that for each $x, y \in A$ and for $\lambda \in \mathbb{F}$,

$$p(x + y) \leq p(x) + p(y) \quad \text{and} \quad p(\lambda x) = |\lambda|^k \, p(x).$$

If in addition the k-seminorm satisfies

$$p(xy) \leq p(x)p(y),$$

then it is a **submultiplicative** k-seminorm.

For a k-seminorm p (also termed a **pseudo-seminorm**), the number k is called the **homogenity index** of p. A k-seminorm p is a **k-norm** if

$$p(x) = 0 \Rightarrow x = 0$$

A **k-seminormed (k-normed) vector space** (V, p) is a vector space V in which it is defined a k-seminorm (k-norm) p. A topological algebra whose topology is induced by a submultiplicative k-seminorm (submultiplicative k-norm) is called a **k-seminormed (k-normed) algebra**. A k-normed algebra for which the induced topology is complete is a **k-Banach algebra**.

An algebra A is a **locally pseudoconvex algebra** if its underlying vector space is locally pseudoconvex and its multiplication is jointly continuous. The topology of a locally pseudoconvex algebra A can be given by means of a family $\mathscr{P} = \{p_{k_\alpha}\}_{\alpha \in \Lambda}$ of k_α-seminorms, where, for each $\alpha \in \Lambda$, $k_\alpha \in (0, 1]$ is the homogenity index of p_{k_α}.

Moreover, in [6, p.198] it is proved that locally pseudoconvex algebras are precisely those $(A, \mathscr{P})$ where the family $\mathscr{P} = \{p_{k_\alpha}\}_{\alpha \in \Lambda}$ is well-behaved in the following sense:

---

[1] It exists when A is commutative (see [17], p.26).

The family $\mathscr{P} = \{p_{k_\alpha}\}_{\alpha \in \Lambda}$ is **saturated** if

$$p_{k_{\alpha_1}}, p_{k_{\alpha_2}}, \ldots, p_{k_{\alpha_n}} \in \mathscr{P} \Rightarrow \widehat{p} = p_{k_{\alpha_1}} \vee p_{k_{\alpha_2}} \vee \ldots \vee p_{k_{\alpha_n}} \in \mathscr{P}, \tag{1}$$

where $\widehat{p}$ is the β-seminorm on A given by:

$$\widehat{p}(x) = \max(p_{k_{\alpha_1}}^{\beta/k_{\alpha_1}}(x), p_{k_{\alpha_2}}^{\beta/k_{\alpha_2}}(x), \ldots, p_{k_{\alpha_n}}^{\beta/k_{\alpha_n}}(x)) \tag{2}$$

and $\beta = \min(k_{\alpha_1}, k_{\alpha_2}, \ldots, k_{\alpha_n})$.

The family $\mathscr{P} = \{p_{k_\alpha}\}_{\alpha \in \Lambda}$ is **well-behaved** if it is saturated and for each $p_{k_{\alpha_i}} \in \mathscr{P}$, there is a $p_{k_{\alpha_j}} \in \mathscr{P}$ such that

$$p_{k_{\alpha_i}}(x \cdot y) \leq p_{k_{\alpha_j}}^{k_{\alpha_i}/k_{\alpha_j}}(x) \cdot p_{k_{\alpha_j}}^{k_{\alpha_i}/k_{\alpha_j}}(y) \tag{3}$$

for each $x, y \in A$.

If $k_\alpha = 1$ for every $\alpha \in \Lambda$, then the algebra A is a **locally convex algebra**. If $\inf\{k_\alpha : \alpha \in \Lambda\} = k > 0$, then A is a **locally k-convex algebra**, because for this case its topology can be given by a family of pseudo-seminorms with a single index of homogenity k.

A locally pseudoconvex algebra $(A, \{p_{k_\alpha}\}_{\alpha \in \Lambda})$ is a **locally multiplicatively pseudoconvex algebra** (**m-pseudoconvex algebra**, for short) if each k_α-seminorm p_{k_α} is submultiplicative.

A topological algebra with unit A is a **Q-algebra** if the set $G(A)$ of all its invertible elements is an open set.

Recall that a set B in a topological vector space V over $\mathbb{F}$ is **bounded** if for each neighborhood V of zero in V there corresponds an escalar $s > 0$ such that $B \subset tV$ for every $t > s$. (see [16], p.8)

A subset C of a topological vector space is called a **barrel** if it is balanced, convex, closed and absorbing. A locally convex algebra A is **barrelled** if each barrel in A is a neighbourhood of zero in A. A topological algebra is an **F-algebra** if it is metrizable and complete.

A k-Banach algebra A together with an involution $\star$ on A is a **k-$B^\star$-algebra**. The norm has the **C^*-property** if:

$$\|xx^\star\| = \|x\|^2, \text{ for each } x \in A.$$

If A is a $B^\star$-algebra and the norm satisfies the $C^\star$-property, then A is called a **$C^\star$-algebra**. In [9] it was proved that a k-$B^\star$-algebra in which the $C^\star$-property is satisfied, can always be endowed with an equivalent norm with which it is a $C^\star$-algebra.

Recall that a completely regular space is a topological space X for which each closed subset of X and each point not in the set, can be separated by a continuous function $X \to \mathbb{R}$. A pseudocompact space is a completely regular space X for which each continuous function $X \to \mathbb{R}$ is automatically bounded.

Throughout this paper we will denote by (A, τ) the algebra A endowed with the topology τ. Moreover, $(A, \{p_{k_\alpha}\}_{\alpha\in\Lambda})$ will denote a complex Hausdorff commutative locally pseudoconvex algebra with unit. With X we will denote either a completely regular Hausdorff topological space (i.e. a Tychonoff space) or a pseudocompact Hausdorff topological space.

2 $C_b(X, A)$

In this section we present some useful properties of $C_b(X, A)$.

Definition 1 Let $\left(A, \mathscr{P} = \left\{p_{k_\alpha}\right\}_{\alpha\in\Lambda}\right)$ be a commutative locally pseudoconvex algebra and X a completely regular Hausdorff space. Let $C(X, A)$ denote the algebra of all continuous functions from X to A, and $C_b(X, A)$ the subalgebra of $C(X, A)$ of all continuous and bounded functions from X to A.

In $C_b(X, A)$ there can be defined the **uniform k_α-seminorms** $p_{k_\alpha,\infty}$ in the following way:

$$p_{k_\alpha,\infty}(f) = \sup_{x\in X} p_{k_\alpha}(f(x)).$$

Remark 1 Note that if $\alpha \in \Lambda$, then

$$\begin{aligned} p_{k_\alpha,\infty}(\lambda f) &= \sup_{x\in X} p_{k_\alpha}(\lambda f) = \sup_{x\in X} |\lambda|^{k_\alpha} p_{k_\alpha}(f(x) = \\ &= |\lambda|^{k_\alpha} \sup_{x\in X} p_{k_\alpha}(f(x)) = |\lambda|^{k_\alpha} p_{k_\alpha,\infty}(f). \end{aligned}$$

Moreover, it is easily seen that all the properties of a k_α-seminorm are valid. It is also clear that if e is the unit in A, then the map $\widehat{e} : X \longrightarrow A$ given by $\widehat{e}(x) = e$ is bounded and continuous, in fact it is the unit in $(C_b(X, A), \{p_{k_\alpha,\infty}\}_{\alpha\in\Lambda})$. Note also that if p_{k_α} is a submultiplicative k_α-seminorm, then the corresponding k_α-seminorm $p_{k_\alpha,\infty}$ is submultiplicative too.

Concerning the joint continuity of the multiplication in $(C_b(X, A), \{p_{k_\alpha,\infty}\}_{\alpha\in\Lambda})$ it is sufficient to show that the family $\mathscr{P}_\infty = \{p_{k_\alpha,\infty}\}_{\alpha\in\Lambda}$ is well-behaved in the sense defined in (1) and (3).

For, let $p_{k_{\alpha_1},\infty}, p_{k_{\alpha_2},\infty}, \ldots, p_{k_{\alpha_n},\infty} \in \mathscr{P}_\infty$ and let $\widehat{q}$ be defined as in (2). Now consider the corresponding $p_{k_{\alpha_1}}, p_{k_{\alpha_2}}, \ldots, p_{k_{\alpha_n}}$ in $\mathscr{P}$ and $\widehat{p}$ defined also as in (2). It is clear that $\widehat{q} = (\widehat{p})_\infty$ so that $\widehat{q} \in \mathscr{P}_\infty$, since $\mathscr{P}$ is a saturated family.

Now, let $p_{k_{\alpha_i},\infty} \in \mathscr{P}_\infty$ and $f, g \in C_b(X, A)$. Then relation (3) in A implies that

$$p_{k_{\alpha_i},\infty}(f \cdot g) \leq p_{k_{\alpha_j},\infty}^{k_{\alpha_i}/k_{\alpha_j}}(f) \cdot p_{k_{\alpha_j},\infty}^{k_{\alpha_i}/k_{\alpha_j}}(g)$$

so that the family $\mathscr{P}_\infty$ is well behaved.

Hence we have the following:

Proposition 1 *Let $\left(A, \{p_{k_\alpha}\}_{\alpha\in\Lambda}\right)$ be a commutative locally pseudoconvex algebra with unit e, and X be a completely regular Hausdorff space. Then $\left(C_b(X, A), \{p_{k_\alpha,\infty}\}_{\alpha\in\Lambda}\right)$ is a commutative locally pseudoconvex algebra with unit. Moreover, if $\left(A, \{p_{k_\alpha}\}_{\alpha\in\Lambda}\right)$ is an m-pseudoconvex algebra, then $\left(C_b(X, A), \{p_{k_\alpha,\infty}\}_{\alpha\in\Lambda}\right)$ is an m -pseudoconvex algebra too.*

In the case of a k-seminormed algebra (A, p), we get a single k -seminorm p_∞ defined in $C_b(X, A)$ as

$$p_\infty(f) = \sup_{x\in X}\{p(f(x))\};$$

and we have the following:

Corollary 1 *Let (A, p) be a commutative k-seminormed algebra with unit e, and X be a completely regular Hausdorff space. Then*

$$(C_b(X, A), p_\infty)$$

is a commutative k-seminormed algebra with unit. Moreover, if p is a k-norm, then p_∞ is a k-norm too.

Proof The first part follows from Proposition 1; the proof for the second part is straightforward. ■

Concerning the completeness of $(C_b(X, A))$ whenever A is complete, we have the following:

Proposition 2 *Let $\left(A, \{p_{k_\alpha}\}_{\alpha\in\Lambda}\right)$ be a commutative complete locally pseudoconvex algebra with unit e, and X be a completely regular Hausdorff space. Then $\left(C_b(X, A), \{p_{k_\alpha,\infty}\}_{\alpha\in\Lambda}\right)$ is a commutative complete locally pseudoconvex algebra with unit.*

Proof Let $(f_\gamma)_{\gamma\in\Omega}$ be a Cauchy net in $(C_b(X, A), \{p_{k_\alpha,\infty}\}_{\alpha\in\Lambda})$ and let $\varepsilon > 0$ be given; then, for each $\alpha \in \Lambda$, $p_{k_\alpha,\infty}\left(f_\gamma - f_\eta\right) < \varepsilon$ for each $\gamma, \eta \geq \delta_\alpha$ (for some $\delta_\alpha \in \Omega$).

Then $p_{k_\alpha}\left(f_\gamma(x) - f_\eta(x)\right) \leq p_{k_\alpha,\infty}\left(f_\gamma - f_\eta\right) < \varepsilon$ for each $x \in X, \alpha \in \Lambda$ and $\gamma, \eta \geq \delta_\alpha$. This implies that $\left(f_\gamma(x)\right)_{\gamma\in\Omega}$ is a Cauchy net in A. Since $\left(A, \{p_{k_\alpha}\}_{\alpha\in\Lambda}\right)$ is complete by assumption, there exists a well-defined a_x in A such that $\left(f_\gamma(x)\right)_{\gamma\in\Omega}$

converges to a_x in A. Let $f : X \longrightarrow A$ be given by $f(x) = a_x$. We claim that f is a continuous function.

For, let $\alpha \in \Lambda$ and $\varepsilon > 0$. As above, there exists a $\delta_\alpha \in \Omega$ such that $p_{k_\alpha}(f_\gamma(x) - f_\eta(x)) < \varepsilon/6$ for all $x \in X$ and $\eta, \gamma \geq \delta_\alpha$.

Let $x \in X$ be fixed but arbitrary; since $(f_\eta(x))_{\eta\in\Omega}$ converges to $f(x)$, there exists $\rho_{\alpha,x} \in \Omega$ such that $p_{k_\alpha}(f_\eta(x) - f(x)) < \varepsilon/6$ if $\eta \geq \rho_{a,x}$; choose $\rho \in \Omega$ such that $\rho \geq \rho_{\alpha,x}$ and $\rho \geq \delta_\alpha$; so, in particular, $p_{k_\alpha}(f(x) - f_\rho(x)) < \varepsilon/6$ and this implies that $p_{k_\alpha}(f_\gamma(x) - f(x)) < \varepsilon/3$ for any x and $\gamma \geq \delta_\alpha$, because $p_{k_\alpha}(f_\gamma(x) - f(x)) \leq p_{k_\alpha}(f_\gamma(x) - f_\rho(x)) + p_{k_\alpha}(f_\rho(x) - f(x)) < \varepsilon/6 + \varepsilon/6 = \varepsilon/3$.

So we can assume that

$$p_{k_\alpha}(f_\gamma(x) - f(x)) < \varepsilon/3 \text{ for all } x \in X \text{ whenever } \gamma \geq \delta_\alpha. \tag{4}$$

Now take a neighborhood U of x such that $p_{k_\alpha}(f_\gamma(x) - f_\gamma(y)) < \varepsilon/3$ for each $y \in U$. Then, taking any $\gamma \geq \delta_\alpha$, we have $p_{k_\alpha}(f(x) - f(y)) \leq p_{k_\alpha}(f(x) - f_\gamma(x)) + p_{k_\alpha}(f_\gamma(x) - f_\gamma(y)) + p_{k_\alpha}(f_\gamma(y) - f(y)) < \varepsilon/3 + \varepsilon/3 + \varepsilon/3 = \varepsilon$ whenever $y \in U$. This proves that f is a continuous function.

Now we prove that f is a bounded function. For, let us take $\gamma \geq \delta_\alpha$; then, according to (4), $p_{k_\alpha}(f(x)) \leq p_{k_\alpha}(f(x) - f_\gamma(x)) + p_{k_\alpha}(f_\gamma(x)) < \varepsilon/3 + p_{k_\alpha}(f_\gamma(x)) \leq \varepsilon/3 + p_{k_\alpha,\infty}(f_\gamma) < \infty$, because f_γ is a bounded function. Therefore $p_{k_\alpha,\infty}(f) < \infty$.

This completes the proof. ∎

Corollary 2 *Let (A, p) be a commutative k-Banach algebra with unit e, and X be a completely regular Hausdorff space. Then $(C_b(X, A), p_\infty)$ is a commutative k-Banach algebra with unit.*

3 The Generalized Arens-Michael Decomposition

In this section we examine the *generalized Arens-Michael decomposition* for $C_b(X, A)$ in terms of the corresponding one for A.

Let $\left(A, \{p_{k_\alpha}\}_{\alpha\in\Lambda}\right)$ be a complete m-pseudoconvex algebra, where each p_{k_α} is a k_α-seminorm on A. For each $\alpha \in \Lambda$, let $\pi_\alpha : A \twoheadrightarrow A/\ker p_{k_\alpha} \doteq A_\alpha$ be the canonical projection $x \longmapsto \pi_\alpha(x) = x + \ker p_{k_\alpha} \doteq x_\alpha$ and define the k_α-norm $\dot{p}_{k_\alpha}$ on A_α by

$$\dot{p}_{k_\alpha}(x_\alpha) = p_{k_\alpha}(x).$$

Then take $\widetilde{A_\alpha}$ the completion of this k_α-normed algebra and denote its k_α-norm by $\|\cdot\|_{k_\alpha}$. So $(\widetilde{A_\alpha}, \|\cdot\|_{k_\alpha})$ is a k_α-Banach algebra. We denote by $\widetilde{\pi}_\alpha$ the composition map $A \overset{\pi_\alpha}{\twoheadrightarrow} A_\alpha \hookrightarrow \widetilde{A_\alpha}$.

Define a partial ordering $\preceq$ in Λ setting

$$\alpha \preceq \beta \text{ iff } p_{k_\alpha}(x) \leq p_{k_\beta}(x) \text{ for each } x \in A.$$

In this case consider the continuous algebra homomorphism ("connecting homomorphism")

$$\varphi_\alpha^\beta : A_\beta \longrightarrow A_\alpha$$

given by $\varphi_\alpha^\beta(x_\beta) = x_\alpha$; consider also its continuous algebra homomorphism extension

$$\widetilde{\varphi}_\alpha^\beta : \widetilde{A}_\beta \longrightarrow \widetilde{A}_\alpha.$$

Then the k_α-normed algebras projective system

$$(\{A_\alpha, \dot{p}_{k_\alpha}\}_{\alpha\in\Lambda}; \{\varphi_\alpha^\beta : \alpha \preceq \beta\}; \Lambda)$$

and also the k_α-Banach algebras projective system

$$(\{\widetilde{A}_\alpha, \|\cdot\|_{k_\alpha}\}_{\alpha\in\Lambda}; \{\widetilde{\varphi}_\alpha^\beta : \alpha \preceq \beta\}; \Lambda)$$

lead to the inverse limit algebras

$$\varprojlim_\alpha A_\alpha \quad \text{and} \quad \varprojlim_\alpha \widetilde{A}_\alpha.$$

The algebra $\varprojlim_\alpha A_\alpha$ can be realized as the subalgebra of the direct product $\prod_{\alpha\in\Lambda} A_\alpha$ consisting of those elements $(z_\alpha)_{\alpha\in\Lambda}$ for which there exists an $x \in A$ such that $z_\alpha = x_\alpha$ for each $\alpha \in \Lambda$ and $\varphi_\alpha^\beta(z_\beta) = z_\alpha$ whenever $\alpha \preceq \beta$. We denote by

$$\varphi_\beta : \prod_{\alpha\in\Lambda} A_\alpha \longrightarrow A_\beta$$

the β-th canonical projection $\varphi_\beta((z_\alpha)_{\alpha\in\Lambda}) = z_\beta$, and by $\widehat{\varphi}_\beta$ the restriction of φ_β to $\varprojlim_\alpha A_\alpha$. The same considerations are true for the complete algebras $\widetilde{A}_\alpha$.

Recall that the generalized Arens-Michael decomposition (see [1, 2] and [6], Th. 4.5.3, p. 302, see also [11], p. 88) states that, if A is a *complete* m-pseudoconvex algebra, then

$$A \cong \varprojlim_\alpha A_\alpha = \varprojlim_\alpha \widetilde{A}_\alpha$$

within topological algebra isomorphisms. Let us denote by Γ the isomorphism

$$A \to \varprojlim_{\alpha} A_\alpha$$

$$x \longmapsto (x_\alpha)_{\alpha\in\Lambda}$$

Now we turn to the Arens-Michael analysis for the algebra $C_b(X, A)$. First, note that $\ker p_{k_\beta} \subseteq \ker p_{k_\alpha}$ if and only if $\ker p_{k_\beta,\infty} \subseteq \ker p_{k_\alpha,\infty}$, so the partial ordering in Λ given by the family $\{p_{k_\alpha,\infty} : \alpha \in \Lambda\}$ is the same as the one given by the family $\{p_{k_\alpha} : \alpha \in \Lambda\}$.

Second, for the algebra $(C_b(X, A), \{p_{k_\alpha,\infty}\}_{\alpha\in\Lambda})$ let us denote by

$$(C_b(X, A))_\alpha = \left(C_b(X, A)/\ker(p_{k_\alpha,\infty}), \dot{p}_{k_\alpha,\infty}\right),$$

the corresponding k_α-normed algebras in its generalized Arens-Michael analysis; by

$$\Phi_\alpha : C_b(X, A) \to (C_b(X, A))_\alpha,$$

the canonical projection $f \longmapsto f + \ker p_{k_\alpha,\infty} \doteq f_\alpha$ and by

$$\Phi_\alpha^\beta : (C_b(X, A))_\beta \to (C_b(X, A))_\alpha$$

the corresponding connecting homomorphism (whenever $\alpha \preceq \beta$).

So, if A is complete, by virtue of Proposition 2 and the generalized Arens-Michael decomposition, we have that

$$C_b(X, A) \cong \varprojlim_{\alpha} (C_b(X, A))_\alpha$$

Nevertheless, we have another expression of $C_b(X, A)$ as the limit of a projective system naturally associated to the family of seminorms $\{p_{k_\alpha,\infty} : \alpha \in \Lambda\}$. For, let us denote by $\left(\dot{p}_{k_\alpha}\right)_\infty$ the ∞-norm in $C_b(X, A_\alpha)$ and by

$$(\varphi_\alpha^\beta)_\infty : C_b(X, A_\beta) \to C_b(X, A_\alpha)$$

the map given by $(\varphi_\alpha^\beta)_\infty(g) = \varphi_a^\beta \circ g$.

It is easy to check that we have a k_α-normed algebras projective system

$$\left(\{(C_b(X, A_\alpha), \left(\dot{p}_{k_\alpha}\right)_\infty)\}; \{(\varphi_\alpha^\beta)_\infty : \alpha \preceq \beta\}; \Lambda\right)$$

which yields to a projective limit algebra

$$W = \varprojlim_{\alpha} C_b(X, A_\alpha)$$

and maps

$$\widehat{\varrho}_\beta : W \to C_b(X, A_\beta)$$

which are restrictions to W of the canonical projections

$$\varrho_\beta : \prod_{\alpha \in \Lambda} C_b(X, A_\alpha) \longrightarrow C_b(X, A_\beta).$$

Now we can state and prove the next:

Theorem 1 *Let $\left(A, \left\{p_{k_\alpha}\right\}_{\alpha \in \Lambda}\right)$ be a commutative complete locally m-pseudoconvex algebra with unit and X a completely regular Hausdorff space. Then*

$$C_b(X, A) \cong \varprojlim_{\alpha} C_b(X, A_\alpha)$$

within a topological algebras isomorphism.

Proof Consider

$$\varprojlim_{\alpha} A_\alpha \cong A \cong \varprojlim_{\alpha} \widetilde{A_\alpha}$$

the generalized Arens-Michael decomposition for A described above, and let $f \in C_b(X, A)$. First we note that, by the very definition of $p_{k_\alpha,\infty}$, for arbitrary $\alpha \in \Lambda$, $f \in \ker\left(p_{k_\alpha,\infty}\right)$ if and only if $f(x) \in \ker\left(p_{k_\alpha}\right)$ for each $x \in X$.

Consider $\alpha \in \Lambda$ and define the map

$$\begin{aligned} \tau_\alpha \quad &: \quad C_b(X, A) \longrightarrow C_b(X, A_\alpha) \\ f &\longmapsto \tau_\alpha(f) \doteq \pi_\alpha \circ f. \end{aligned}$$

First, since π_α and f are continuous functions, then $\tau_\alpha(f)$ is a continuous function.

Second, note that

$$(\dot{p}_{k_\alpha})_\infty(\tau_\alpha(f)) = (\dot{p}_{k_\alpha})_\infty(\pi_\alpha \circ f) = \sup_{x \in X}\{\dot{p}_{k_\alpha}((\pi_\alpha \circ f)(x))\} =$$

$$= \sup_{x \in X}\{\dot{p}_{k_\alpha}(f(x)_\alpha)\} = \sup_{x \in X}\{p_{k_\alpha}(f(x))\} = p_{k_\alpha,\infty}(f) < \infty, \tag{5}$$

since f is a bounded function. Therefore $\tau_\alpha(f)$ is a bounded function in X.

Due to the previous remarks, the map τ_α is well defined. It is easy to see that it is also linear and multiplicative (since π_α is so).

It is also continuous. For, if $(f_\delta)_{\delta\in\Delta}$ is a net that converges to f in $C_b(X, A)$, then it is clear that $\tau_\alpha(f_\delta) = \pi_\alpha \circ f_\delta$ converges to $\pi_\alpha \circ f$ in $C_b(X, A_\alpha)$, since π_α is a continuous map.

Moreover, if $\alpha \preceq \beta$ in Λ, then

$$((\varphi_\alpha^\beta)_\infty \circ \tau_\beta)(f) = (\varphi_\alpha^\beta)_\infty(\pi_\beta \circ f) = \varphi_\alpha^\beta \circ \pi_\beta \circ f = \pi_\alpha \circ f = \tau_\alpha(f),$$

that is, $(\varphi_\alpha^\beta)_\infty \circ \tau_\beta = \tau_\alpha$.

So, due to the *Universal (Categorical) Property of the Projective Limit* (see [15], p. 231), there exists a topological algebras homomorphism

$$\tau : C_b(X, A) \longrightarrow \varprojlim_\alpha C_b(X, A_\alpha) = W$$

such that $\widehat{\varrho}_\beta \circ \tau = \tau_\beta$ for each $\beta \in \Lambda$. See diagram.

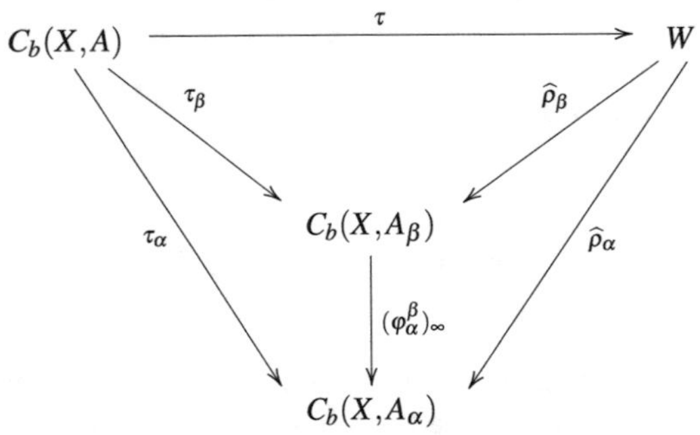

This homomorphism is defined as follows: for $f \in C_b(X, A)$,

$$\tau(f) = ((\tau_\alpha(f))_{\alpha\in\Lambda}).$$

We claim that τ is also a bijection. For, let us suppose that $\tau(f) = 0$; then $\pi_\alpha \circ f = 0$ for each $\alpha \in \Lambda$. Therefore $\pi_\alpha(f(x)) = 0$ for every $\alpha \in \Lambda$ (and for each $x \in X$). Then we have that $p_{k_\alpha}(f(x)) = 0$ for every $\alpha \in \Lambda$ and this implies that $f(x) = 0$. Since this holds for each $x \in X$, we conclude that $f = 0$ and so, τ is injective.

Consider now an arbitrary element $((g_\alpha)_{\alpha\in\Lambda})$ in $W = \varprojlim_\alpha C_b(X, A_\alpha)$. Let us define the function $g : X \longrightarrow A$ as $g(x) = \Gamma^{-1}((g_\alpha(x))_{\alpha\in\Lambda})$.

First, we have to check that $(g_\alpha(x))_{\alpha\in\Lambda} \in \varprojlim_\alpha A_\alpha$. For, take $\alpha \preceq \beta$ in Λ; then $(g_\alpha)_{\alpha\in\Lambda} \in W$ implies $\varphi_\alpha^\beta \circ g_\beta = (\varphi_\alpha^\beta)_\infty(g_\beta) = g_\alpha$ and therefore $\varphi_\alpha^\beta(g_\beta(x)) = g_\alpha(x)$ for each $x \in X$.

Second, g is continuous function. For, take any $\gamma \in \Lambda$ and note that

$$(\widehat{\varphi}_\gamma \circ g)(x) = \widehat{\varphi}_\gamma(((g_\alpha(x))_{\alpha\in\Lambda}) = g_\gamma(x)$$

for any $x \in X$; that is, $\widehat{\varphi}_\gamma \circ g = g_\gamma$, which is a continuous function. Therefore g is continuous.

Third, g is a bounded function. For, let us take any $\gamma \in \Lambda$ and note that

$$p_{k_\gamma,\infty}(g) = \sup_{x\in X}\{p_{k_\gamma}(g(x))\} = \sup_{x\in X}\{\dot{p}_{k_\gamma}(g(x)_\gamma)\} =$$

$$= \sup_{x\in X}\{p_{k_\gamma}(g_\gamma(x))\} = \left((\dot{p}_{k_\gamma})_\infty\right)(g_\gamma) < \infty.$$

Next, note that, for any $\beta \in \Lambda$, and for every $x \in X$, we have

$$(\tau_\beta(g))(x) = (\pi_\beta \circ g)(x) = \pi_\beta((g_\alpha(x))_{\alpha\in\Lambda}) = g_\beta(x).$$

Then $\tau_\beta \circ g = g_\beta$ and this implies that $\tau(g) = ((g_\alpha)_{\alpha\in\Lambda})$ and so, τ is surjective.

Note that relation (5) shows that τ_α is in fact an isometry. Then the isomorphism τ is a homeomorphism too.

This completes the proof. ■

4 Inversion

In this section, X will denote a pseudocompact Hausdorff space. First, we will show that $C_b(X, A)$ has a continuous inversion if the same is true for A and, second, we will show that $C_b(X, A)$ is a spectral algebra if A is so. In order to reach our goal, we prove the following:

Lemma 1 *Let* $(A, (\{p_{k_\alpha}\}_{\alpha\in\Lambda})$ *be a locally pseudoconvex algebra with unit and suppose that A has a continuous inversion; let X be a pseudocompact Hausdorff space. Then for each* $f \in C_b(X, A)$ *we have*

$$\sigma(f) = \bigcup_{x\in X}\{\sigma(f(x))\}.$$

Proof Let $f \in C_b(X, A)$. If $\lambda \notin \sigma(f)$, then $f - \lambda\widehat{e} \in G(C_b(X, A))$. Therefore there exists a $g \in C_b(X, A)$ such that $(f - \lambda\widehat{e})g = \widehat{e}$. Then we have $(f(x) - \lambda e)g(x) = e$

for every $x \in X$, which implies that $(f(x) - \lambda e) \in G(A)$; that is, $\lambda \notin \sigma(f(x))$ for every $x \in X$.

Conversely, suppose that $\lambda \notin \sigma(f(x))$ for every $x \in X$. Then $(f(x) - \lambda e) \in G(A)$ for every $x \in X$. Define $g : X \to A$ by $g(x) = (f(x) - \lambda e)^{-1}$. We claim that $g \in C_b(X, A)$. Since scalar multiplication, sum and inversion (by assumption) are continuous functions on A, then g is a continuous function on A. Therefore, due to the pseudocompactness of X, we have that, for each $\alpha \in \Lambda$,

$$\sup_{x \in X}\{p_{k_\alpha}(g(x)\} < \infty$$

and so g is a bounded function too. This implies that $\lambda \notin \sigma(f)$. This completes the proof. ■

The following Proposition is a generalization of a known fact for m-convex algebras (see for instance [7], 4.8–6).

Proposition 3 *Let* $\left(A, \left(\left\{p_{k_\alpha}\right\}_{\alpha \in \Lambda}\right)\right)$ *be an m-pseudoconvex algebra. Then the inversion in* $\left(A, \left\{p_{k_\alpha}\right\}_{\alpha \in \Lambda}\right)$ *is continuous.*

Proof In fact, the map $G(A) \to A$, $a \mapsto a^{-1}$ is continuous relative to each submultiplicative pseudo-seminorm p_{k_α}; this is a direct generalization of the proof given in the book [14], Th. 1.4.8. ■

Definition 2 We say that a k-seminormed algebra (A, p) is k-**spectral** if

$$r(a)^k \leq p(a)$$

for each $a \in A$.

An algebra A is k-**spectral** if there exists in A a non trivial k-seminorm p such that (A, p) is k-spectral (see [13], p. 233).

Next we will show that if A is k-spectral, then the same is true for $C_b(X, A)$.

Theorem 2 *Let* (A, p) *be a k-seminormed algebra with a continuous inversion and* X *a pseudocompact Hausdorff space. If* (A, p) *is k-spectral, then* $(C_b(X, A), p_\infty)$ *is k-spectral too.*

Proof Since p is k-spectral, then $r(x)^k \leq p(x)$ for each $x \in A$. By hipothesis, one have, on the one hand,

$$\sup\{r(f(x)) : x \in X\} \leq \sup\left\{(p(f(x)))^{1/k} : x \in X\right\}$$
$$= \sup\{(p(f(x)) : x \in X\}^{1/k} = (p_\infty(f))^{1/k}.$$

On the other hand, due to Lemma 1, if $f \in (C_b(X, A), p_\infty)$ and $\lambda \in \sigma(f)$, then there exists $x_0 \in X$ such that $\lambda \in \sigma(f(x_0))$. Therefore

$$|\lambda| \leq r(f(x_0)) \leq \sup_{x \in X} r(f(x)),$$

and so,

$$r(f) \leq \sup_{x \in X} r(f(x))\} \leq (p_\infty(f))^{1/k}.$$

So,

$$r(f)^k \leq p_\infty(f),$$

hence p_∞ is a spectral k-seminorm in $C_b(X, A)$. ■

For m-pseudoconvex algebras we have the following results:

Corollary 3 *Let $\left(A, \left(\{p_{k_\alpha}\}_{\alpha \in \Lambda}\right)\right)$ be an m-pseudoconvex algebra and X a pseudocompact Hausdorff space. If A is k_β-spectral for some $\beta \in \Lambda$, then the algebra $\left(C_b(X, A), \{p_{k_\alpha,\infty}\}_{\alpha \in \Lambda}\right)$ is k_β-espectral relative to the corresponding k_β-seminorm $p_{k_\beta,\infty}$.*

Proof It follows from Proposition 3 and Theorem 2. ■

Corollary 4 *Let $\left(A, \left(\{p_{k_\alpha}\}_{\alpha \in \Lambda}\right)\right)$ be an m-pseudoconvex Q-algebra and X be a pseudocompact Hausdorff space. Then $\left(C_b(X, A), \{p_{k_\alpha,\infty}\}_{\alpha \in \Lambda}\right)$ is a Q-algebra.*

Proof It follows from [12], Th. 3.1. ■

5 Involution

In this section we examine the case where A has an involution $\star$. As already mentioned, any k-$C^\star$-algebra (complete involutive *normed* algebra which satisfies the $C^\star$-property) is in fact a $C^\star$-algebra, so we will focus on $C^\star$-algebras. Nevertheless, we don't know if the same is true for a complete involutive *semi-normed* algebra or even for a k-$B^\star$-algebra which not necessarily satisfies the $C^\star$-property, so we will consider also the case when A is a *locally-pseudo-$C^\star$*-algebra or a k-$B^\star$-algebra.

Let us consider X a topological space and $(A, \star, \tau)$ a topological algebra with a continuous involution. Then we can define in $C_b(X, A)$ a function

$$\star : C_b(X, A) \longrightarrow C_b(X, A)$$

$$f \longmapsto f^\star$$

puting, for $x \in X$,

$$f^\star(x) = (f(x))^\star$$

Since the involution is continuous in A and f is a continuous function, then $f^\star$ is a continuous function $X \to A$.

On the other hand, we claim that $f^\star$ is also a bounded function $X \to A$. Since $f^\star(X) = f(X)^\star$ and $f(X)$ is a bounded set, the claim follows from the fact that a continuous involution is a homeomorphism, and therefore transforms bounded sets into bounded sets. This shows that $f^\star \in C_b(X, A)$ and we have the following:

Proposition 4 *If $(A, \star, \tau)$ is a topological algebra with a continuous involution and X is a topological space, then $(C_b(X, A), \star)$ is an algebra with an involution.*

Proof It is easy to check that all properties of an involution hold for the defined function $\star : C_b(X, A) \to C_b(X, A)$. ∎

Proposition 5 *If $(A, \star, \|\cdot\|)$ is a pseudo-normed algebra with a continuous involution and X is a topological space, then $(C_b(X, A), \star, \|\cdot\|_\infty)$ is a pseudo-normed algebra with a continuous involution.*

Proof Since $\star$ is an additive map, it is sufficient to show that it is continuous at zero. For, due to the continuity of the involution in A, there exists a constant $C > 0$ such that for each $y \in A$ we have that $\|y^*\| \le C\|y\|$ (see [6], Prop 4.3.11, p. 192[2]). Therefore $\|f^*\|_\infty = \sup_{x\in X} \|f(x)^*\| \le C \sup_{x\in X} \|f(x)\| = C\|f\|_\infty$. ∎

Corollary 5 *If $(A, \star, \|\cdot\|)$ is a k-$B^\star$-algebra with a continuous involution and X is a completely regular Hausdorff space, then $(C_b(X, A), \star, \|\cdot\|_\infty)$ is a k-$B^\star$-algebra too.*

Proof It follows from Corollary 2 and Proposition 5. ∎

Corollary 6 *If $(A, \star, \|\cdot\|)$ is a $C^\star$-algebra and X is a completely regular Hausdorff space, then $(C_b(X, A), \star, \|\cdot\|_\infty)$ is a $C^\star$-algebra too.*

Proof By Corollary 5, $(C_b(X, A), \star, \|\cdot\|_\infty)$ is a k-$B^\star$-algebra. The $C^\star$ property $\|aa^\star\| = \|a\|^2$ in A implies that

$$\left\|ff^\star\right\|_\infty = \sup_{x\in X} \left\|f(x)\, f^\star(x)\right\| = \sup_{x\in X} \|f(x)\|^2 = \left(\sup_{x\in X} \|f(x)\|\right)^2 = \|f\|_\infty^2$$

This shows that the norm $\|\cdot\|_\infty$ is a $C^\star$ norm. ∎

Recall that a complete locally pseudoconvex $\star$-algebra $\left(A, \star, \{p_{k_\alpha}\}_{\alpha\in\Lambda}\right)$ is a **locally pseudo-$C^\star$-algebra** if each k_α-seminorm p_{k_α} satisfies $p_{k_\alpha}(aa^\star) = p_{k_\alpha}(a)^2$. Then, by definition of the k_α-seminorm $p_{k_\alpha,\infty}$, we have the following:

Corollary 7 *If $\left(A, \star, \{p_{k_\alpha}\}_{\alpha\in\Lambda}\right)$ is a locally pseudo-$C^\star$-algebra and X is a completely regular Hausdorff space, then $\left(C_b(X, A), \star, \{p_{k_\alpha,\infty}\}_{\alpha\in\Lambda}\right)$ is a locally pseudo-$C^\star$-algebra too.*

[2]The proof of this property involves only real scalars, so there is no problem with the conjugate-homogenity of the involution.

6 Metrizability

In this section X denotes a completely regular Hausdorff space. We will show that if A is an F-algebra, then $C_b(X, A)$ is also an F-algebra.

Proposition 6 *Let A be an F-algebra with unit e and X a completely regular Hausdorff space. Then, $C_b(X, A)$ is an F-algebra too.*

Proof Let d be a compatible metric for the topology on A and define

$$d_\infty(f, g) = \sup_{x \in X} d(f(x), g(x))$$

for each $f, g \in C_b(X, A)$. Then it is straigthforward to prove that d_∞ is a compatible metric for the topology on $C_b(X, A)$. The proof for the completeness of d_∞ is very similar to that of Proposition 2 with d_∞ instead of $p_{k_{\alpha,\infty}}$. ■

Corollary 8 *Let (A, τ) be an F-algebra and X a completely regular Hausdorff space. Then $C_b(X, A)$ is a barrelled algebra.*

We conclude by providing an example.

Example 1 Let $(k_n)_{n \in \mathbb{N}}$ be a sequence of real numbers such that $0 < k_n \leq 1$; consider the algebra $C(\mathbb{R}, \mathbb{C})$ consisting of all $\mathbb{C}$-valued continuous functions on $\mathbb{R}$ endowed with the topology generated by the family of the sub-multiplicative k_n-seminorms $\{\|\cdot\|_n : n \in \mathbb{N}\}$ given by

$$\|x\|_n = \sup_{|t| \leq n} |x(t)|^{k_n}$$

It follows that $\left(C(\mathbb{R}, \mathbb{C}), \{\|\cdot\|_n\}_{n \in \mathbb{N}}\right)$ is an m-pseudoconvex algebra.

If X is a completely regular Hausdorff space, then, due to Proposition 1, the algebra $\left(C_b(X, C(\mathbb{R}, \mathbb{C})), \{\|\cdot\|_{n,\infty}\}_{n \in \mathbb{N}}\right)$ is also an m-pseudoconvex algebra.

Note that $(C_b([-n, n], \mathbb{C}), \|\cdot\|_\infty) = (C([-n, n], \mathbb{C}), \|\cdot\|_\infty) \cong \left(C(\mathbb{R}, \mathbb{C}), \|\cdot\|_n\right)$ and the former algebra is a k_n-Banach algebra; so $(C(\mathbb{R}, \mathbb{C}), \|\cdot\|_n)$ is a k_n-spectral algebra (for each $n \in \mathbb{N}$). It follows that $\left(C_b(X, C(\mathbb{R}, \mathbb{C})), \|\cdot\|_{n,\infty}\right)$ is a k_n-spectral algebra (for each $n \in \mathbb{N}$), and, due to Corollary 3, we have that the m-pseudoconvex algebra $\left(C_b(X, C(\mathbb{R}, \mathbb{C})), \{\|\cdot\|_{n,\infty}\}_{n \in \mathbb{N}}\right)$ is k_n-spectral, for some $n \in \mathbb{N}$. Therefore it is a Q-algebra.

Acknowledgements We want to thank the referee for his useful suggestions. The first author was partially supported by CONACyT Grant 200917, México.

References

1. Abel, M.: Projective limits of topological algebras. Tartu ÜL. Toimetised **836**, 3–27 (1989, in Russian)
2. Abel, M.: Representations of topological algebras by projective limit of Frécher algebras. Commun. Math. Appl. **3**(Nr. 1), 9–15 (2012). RGN Publications
3. Arhippainen, J.: On the ideal structure of algebras of LMC-algebras valued functions. Stud. Math. **101**, 311–318 (1992)
4. Arizmendi, H., Carrillo, A., García, A.: On algebras of Banach algebra-valued bounded continuous functions. Rocky Mt. J. Math. **46**(2), 389–398 (2016)
5. Arizmendi, H., Cho, M., García, A.: On algebras of bounded continuous functions valued in a topological algebra. Comment. Math. **57**(2), 123–129 (2017)
6. Balachandran, V.K.: Topological Algebras. North-Holland Mathematics Studies, vol. 185. Elsevier, Amsterdam (2000)
7. Beckenstein, E., Narici, L., Suffel, C.: Topological Algebras. North-Holland Mathematics Studies, vol. 24. North-Holland, Amsterdam (1977)
8. Buck, C.: Bounded continuous functions on a locally compact space. Mich. Math. J. **5**, 95–104 (1958)
9. El Kinani, A., Ifzarne, A., Oudadess, M.: p -Banach algebras with generalized involution and C^* -structure. Turk. J. Math. **25**, 275–282 (2001). TÜBITAK
10. Hery, W.J.: Maximal ideal in algebras of topological algebra valued functions. Pac. J. Math. **65**, 365–373 (1976)
11. Mallios, A.: Topological Algebras, Selected Topics. North Holland Mathematics Studies, vol. 124. Notas de Matematica 109. North Holland Publishing Co., Amsterdam (1986)
12. Palacios, L., Pérez-Tiscareño, R., Signoret, C.: On Q-algebras and spectral algebras. Poincaré J. Anal. Appl. **1**, 21–28 (2016)
13. Palmer, T.W.: Banach Algebras and the General Theory of *-Algebras. Algebras and Banach Algebras, vol. 1. Cambridge University Press, Cambridge (1994)
14. Rickart, C.E.: General Theory of Banach Algebras. Krieger, Huntington (1974). First published June 1960
15. Rotman, J.J.: An Introduction to Homological Algebra, 2nd edn. Springer, New York (2009)
16. Rudin, W.: Functional Analysis. McGraw-Hill Publishing Company, New York (1974)
17. Żelazko, W.: Selected Topics in Topological Algebras. Lectures Notes Series, No. 31. Mathematik Institut, Aarhus University, Aarhus (1971)

A Fractional Partial Differential Equation for Theta Functions

Rafael G. Campos

Abstract We find that theta functions are solutions of a fractional partial differential equation that generalizes the diffusion equation. This equation is the limit of a sequence of differential equations for the partial sums of theta functions where the fractional derivatives are given as differentiation matrices for trigonometric polynomials in their Fourier representation, i.e., given as similarities of diagonal matrices under the ordinary discrete Fourier transform. This fact enables the fast numerical computation of fractional partial derivatives of theta functions and elliptic integrals.

MSC 14K25, 33E05, 35R11, 42A15, 65T50

1 Introduction

Elliptic functions have a long history and a great number of physical applications (see for example [5, 11, 14, 17, 18]). The doubly-periodic Weierstrass' elliptic functions and Jacobi's elliptic functions can be represented as quotients of the theta functions which are entire and simply-periodic. Theta functions appear in number theory and in the uniformization of complex tori [13, 19]. Many properties of these functions can be found in [6, 7, 15, 16, 20].

Since the partial sums of theta functions are trigonometric polynomials, it is possible to use the differentiation matrices given in [1, 2] to obtain partial derivatives of these incomplete theta functions. The differentiation matrices are diagonalized by the discrete Fourier transform, i.e., they can be written as similar transformations of diagonal matrices, and this fact enables the definition of fractional derivatives in the Fourier representation and a fast numerical calculation of fractional partial derivatives of theta and elliptic functions.

R. G. Campos (✉)
División de Ciencias e Ingeniería, Departamento de Ciencias, Universidad de Quintana Roo, Chetumal, México
e-mail: rafael.gonzalez@uqroo.edu.mx

G.-M. Greuel et al. (eds.), *Singularities, Algebraic Geometry, Commutative Algebra, and Related Topics*, https://doi.org/10.1007/978-3-319-96827-8_26

Using the diagonalized form of the differentiation matrices, a fractional partial differential equation that generalizes the diffusion equation can be obtained as the limit of a sequence of differential equations satisfied by the incomplete theta functions.

2 Differentiation Matrices for Trigonometric Polynomials

This section is based on Refs. [1, 2], therefore the main results are presented as succinctly as possible.

Let $f(z)$ be a one-periodic analytic function with period 2π and let G be a domain of the open strip $0 < \Re(z) < 2\pi$, $-\infty < \Im z < \infty$, containing a closed rectifiable Jordan curve γ. It is possible to find an exact interpolation formula for $f(z)$, in the case in which $f(z)$ is a trigonometric polynomial, by using the Hermite interpolation formula for algebraic polynomials of a complex variable (see [12]). This is due to the fact that any trigonometric polynomial of degree at most m, $\sum_{k=-m}^{m} c_k e^{ikz}$, can be written as $s^{-m}q(s)$, where $q(s)$ is an algebraic polynomial of degree at most $2m$ in s, under the change of variable $s = \varphi(z) = e^{iz}$.

Let us take $N = 2m + 1$ different points $s_k \in \varphi(I(\gamma))$, i.e., $2m + 1$ different complex numbers $z_k \in I(\gamma)$. The set of points s_k, $k = 1, 2, \ldots, N$, define the polynomial $\tilde{\omega}(s) = \prod_{k=1}^{N}(s - s_k)$. The interpolantion $\tilde{p}(s)$ to $\tilde{f}(s) = f(\varphi^{-1}(s))$ corresponding to the set of N points s_k is given by

$$\tilde{p}(s) = \frac{s^{-m}}{2\pi i} \int_{\tilde{\gamma}} \frac{\tilde{f}(\zeta)}{\tilde{\omega}(\zeta)} \frac{s^m \tilde{\omega}(\zeta) - \zeta^m \tilde{\omega}(s)}{\zeta - s} d\zeta,$$

where $\tilde{\gamma} = \varphi(\gamma)$. Since $[s^m \tilde{\omega}(\zeta) - \zeta^m \tilde{\omega}(s)]/(\zeta - s)$ is a polynomial in s of degree $N - 1 = 2m$, $\tilde{p}(s)$ has the required form $s^{-m}q(s)$, where $q(s)$ is a polynomial of degree at most $2m$, to represent a trigonometric polynomial.

To show that $\tilde{f}(s_k) = \tilde{p}(s_k)$, let us consider the residual function $\tilde{R}(s) = \tilde{f}(s) - \tilde{p}(s)$ which is now

$$\tilde{R}(s) = \frac{1}{2\pi i} \frac{\tilde{\omega}(s)}{s^m} \int_{\tilde{\gamma}} \frac{\tilde{f}(\zeta)\zeta^m}{(\zeta - s)\tilde{\omega}(\zeta)} d\zeta.$$

By definition, $\tilde{G} = \varphi(G)$ does not contain points $s_k (\text{mod } 2\pi)$ other than s_k therefore, the integral of the right-hand side of this equation represents an analytic function in $I(\tilde{\gamma})$ and we have that $\tilde{R}(s_k) = 0$. Since $s^m \tilde{\omega}(\zeta) - \zeta^m \tilde{\omega}(s)$ is divisible by $\zeta - s$, the poles of the integrand are simple and located at s_k. The residue theorem yields

$$\tilde{p}(s) = \sum_{j=1}^{N} \tilde{f}(s_j) \left(\frac{s_j}{s}\right)^m \frac{\prod_{k \neq j}^{N}(s - s_k)}{\prod_{k \neq j}^{N}(s_j - s_k)}, \qquad s \in \tilde{G}$$

and the trigonometric polynomial of degree $m = (N-1)/2$ interpolating to $f(z)$ is

$$p(z) = \sum_{j=1}^{N} f(z_j) e^{i(N-1)(z_j - z)/2} \frac{\prod_{k \neq j}^{N} (e^{iz} - e^{iz_k})}{\prod_{k \neq j}^{N} (e^{iz_j} - e^{iz_k})}, \qquad z \in G,$$

which can be written as the Gauss interpolation formula

$$p(z) = \sum_{k=1}^{N} f(z_k) \frac{s_k(z)}{s_k(z_k)}, \tag{1}$$

where $s_k(z) = \prod_{j \neq k}^{N} \sin\left(\frac{z - z_j}{2}\right)$. The differentiation matrix D for trigonometric polynomials can be obtained by writing the derivative of $p(z)$ at z_j as

$$\frac{dp(z_j)}{dz} = \sum_{k=1}^{N} D_{jk} f(z_k). \tag{2}$$

Thus, the elements of the matrix D are given by

$$D_{jk} = \begin{cases} \dfrac{1}{2} \displaystyle\sum_{l \neq j}^{N} \cot\left(\dfrac{z_j - z_l}{2}\right), & j = k, \\[2ex] \dfrac{1}{2} \dfrac{s'(z_j)}{s'(z_k)} \csc\left(\dfrac{z_j - z_k}{2}\right), & j \neq k, \end{cases} \tag{3}$$

where $\{z_1, z_2, \cdots, , z_N\}$ are $N = 2m + 1$ different points contained in G, and

$$s(z) = \prod_{k=1}^{N} \sin\left(\frac{z - z_k}{2}\right).$$

This matrix is a projection of d/dz in the subspace of trigonometric polynomials of degree at most $(N-1)/2$. If $f(z)$ is given identically by $p(z)$, Eq. (2) gives the derivative of $f(z)$ at z_j and it can be written simply as $f' = Df$, and in general,

$$f^{(n)} = D^n f, \quad n = 0, 1, 2, \cdots, \tag{4}$$

where, $f^{(n)}$ stands for the vector of entries $d^n f(z_k)/dz^n$.

Now consider the trigonometric polynomial

$$p_k(z) = e^{ikz}, \quad |k| \le \frac{N-1}{2}, \quad k \in \mathbb{Z}.$$

Then, $p_k = (e^{ikz_1}, e^{ikz_2}, \cdots, e^{ikz_N})^T$ is an unnormalized eigenvector of D with eigenvalue ik, i.e.,

$$Dp_k = ikp_k, \tag{5}$$

where $k = -(N-1)/2, \cdots, (N-1)/2$. The matrix D takes a simpler form if the points z_l are selected according to

$$z_{l+1} = \frac{2\pi l}{N} + iy, \quad l = 0, 1, \cdots, N-1, \tag{6}$$

for a fixed y. Then we have that

$$\frac{s_j(z_j)}{s_k(z_k)} = \frac{s'(z_j)}{s'(z_k)} = (-1)^{j+k}.$$

Since $\sum_{k\neq j}^{N} \cot \frac{(j-k)\pi}{N} = 0$, the elements of D become

$$D_{jk} = \begin{cases} 0, & j = k, \\ \dfrac{(-1)^{j+k}}{2\sin\frac{(j-k)\pi}{N}} & j \neq k. \end{cases} \tag{7}$$

In this form, iD is a Hermitian matrix with real eigenvalues and orthogonal eigenvectors. The jth component of the kth eigenvector p_k is now $e^{ikz_{j+1}} = e^{i2\pi jk/N-ky}$, where $j = 0, 1, \cdots, N-1$ and $k = -(N-1)/2, \cdots, (N-1)/2$. Since the product ky is a fixed number for a given k, the normalized kth eigenvector $\hat{p}_k$ has components

$$(\hat{p}_k)_j = \frac{1}{\sqrt{N}} e^{-i\pi\frac{N-1}{N}j} e^{i2\pi jk/N}, \quad j, k = 0, 1, \cdots, N-1.$$

Let F be the unitary matrix containing the eigenvector $\hat{p}_k$ in the kth column. Then, F can be written in terms of the ordinary discrete Fourier transform as

$$F = SF_0, \tag{8}$$

where F_0 stands for the ordinary discrete Fourier transform and S is a diagonal matrix containing the elements $e^{-i\pi\frac{N-1}{N}j}$ along the diagonal. Note that (5) and (8) imply that F diagonalizes D as given by (7), i.e.,

$$F^{-1}DF = i\Lambda, \tag{9}$$

where Λ is the diagonal matrix whose diagonal elements are given by the set of integers

$$\{-\frac{N-1}{2}, \cdots, 0, \cdots, \frac{N-1}{2}\}. \tag{10}$$

Consider now an even number $N = 2m + 2$ of points z_k in (6) and the 2π-antiperiodic function

$$q(z) = \sin(z/2)p(z), \tag{11}$$

where $p(z)$ is a 2π-periodic trigonometric polynomial of degree at most $(N-2)/2$. For simplicity consider $y = 0$ in (6). Take the nonzero points $z_2, z_3, \cdots, z_N$. This is an odd number of complex numbers at which the polynomial $p(z)$ can be interpolated to yield

$$q(z) = \sum_{k=2}^{N} p(z_k)\frac{\sin(z/2)s_k(z)}{s_k(z_k)} = \sum_{k=2}^{N} q(z_k)\frac{\sin(z/2)s_k(z)}{\sin(z_k/2)s_k(z_k)},$$

and taking into account that $z_1 = 0$, we have that

$$q(z) = \sum_{k=1}^{N} q(z_k)\frac{\sin(z/2)s_k(z)}{\sin(z_k/2)s_k(z_k)}. \tag{12}$$

The previous arguments also hold for functions of the form $\cos(z/2)p(z)$, since π is contained in the set of points z_k. Therefore, (12) is an interpolation formula for functions of the form

$$q(z) = (ae^{iz/2} + be^{-iz/2})p(z), \tag{13}$$

at the $N = 2m + 2$ points $\{0, z_2, \cdots, \pi, \cdots, z_N\}$. Here, a and b are constants. In this case, the differentiation matrix D is also given by (7) and yields the exact derivative of anti-periodic polynomials. By considering a polynomial of the form

$$q_k(z) = e^{i(k\pm 1/2)z}, \quad k = -N/2 + 1, \cdots, N/2 - 1,$$

one can see that Eq. (9) is also valid for this case in which N is even, where Λ is now the diagonal matrix whose diagonal elements are given by the set of semi-integers

$$\{-\frac{N-1}{2}, \cdots, -1/2, 1/2, \cdots, \frac{N-1}{2}\}, \tag{14}$$

Thus we see that, whether N is odd or not, the matrix F given by (8), is the kernel matrix of the periodic discrete Fourier transformation that diagonalizes the differentiation matrix which gives derivatives of 2π-periodic trigonometric polynomials if N is odd, or derivatives of 2π-antiperiodic polynomials if N is even.

Note that the nth-order derivative can be written as

$$D^n = F(i\Lambda)^n F^{-1}, \quad n = 0, 1, 2, \cdots.$$

This expression leads us to the following definition of fractional derivative

$$D^\alpha = F(i\Lambda)^\alpha F^{-1}, \quad \alpha \geq 0,$$

where the diagonal entries of Λ are given by (14) or (10) according to the (anti)periodicity of the trigonometric polynomial and a convenient branch of the power function is selected.

Since D is nonsingular for N even, we can still extend this result to the differentiation/integration matrix

$$D^\alpha = F(i\Lambda)^\alpha F^{-1}, \quad \alpha \in \mathbb{R}, \tag{15}$$

where the diagonal entries of Λ are only given by (14). A formula similar to (15) for the differentiation matrix for square-integrable functions has been given in the context of one-dimensional boundary value problems in the real line [3].

Equation (15) is an important consequence of the similarity between the differentiation matrix D and the diagonal matrix Λ under the discrete Fourier transform. But in general, if $f(z)$ is an analytic function in an open set of the imaginary axis containing the integers (10) (or the semi-integers (14)), the matrix function $f(D)$ can be defined [8] by

$$f(D) = Ff(i\Lambda)F^{-1}, \tag{16}$$

where $f(\Lambda)$ is the diagonal matrix with diagonal entries given by

$$f(-(N+1)/2 + k), \quad k = 1, 2, \cdots, N.$$

3 A Fractional Equation for Theta Functions

We begin by writing the theta functions in the form [7]

$$\begin{aligned}
\vartheta_1(z,q) &= i\sum_{n=-\infty}^{\infty}(-1)^n q^{(n+1/2)^2}e^{i(n+1/2)z},\\
\vartheta_2(z,q) &= \sum_{n=-\infty}^{\infty} q^{(n+1/2)^2}e^{i(n+1/2)z},\\
\vartheta_3(z,q) &= \sum_{n=-\infty}^{\infty} q^{n^2}e^{inz},\\
\vartheta_4(z,q) &= \sum_{n=-\infty}^{\infty}(-1)^n q^{n^2}e^{inz},
\end{aligned} \tag{17}$$

for $|q| < 1$. Now we define for $N = 2m+2$, the 2π-antiperiodic trigonometric polynomials $\vartheta_1^N(z,q)$ and $\vartheta_2^N(z,q)$ in the variable z as the partial sums

$$\begin{aligned}
\vartheta_1^N(z,q) &= i\sum_{n=-(m+1)}^{m}(-1)^n q^{(n+1/2)^2}e^{i(n+1/2)z},\\
\vartheta_2^N(z,q) &= \sum_{n=-(m+1)}^{m} q^{(n+1/2)^2}e^{i(n+1/2)z},
\end{aligned} \tag{18}$$

and for $N = 2m+1$, the 2π-periodic trigonometric polynomials $\vartheta_3^N(z,q)$ and $\vartheta_4^N(z,q)$ as

$$\begin{aligned}
\vartheta_3^N(z,q) &= \sum_{n=-m}^{m} q^{n^2}e^{inz},\\
\vartheta_4^N(z,q) &= \sum_{n=-m}^{m}(-1)^n q^{n^2}e^{inz},
\end{aligned} \tag{19}$$

Let us consider $\vartheta_3^N(z,q)$ as a sample function. Since $\vartheta_3^N(z,q)$ is a 2π-periodic trigonometric polynomial in z, its derivatives can be given in terms of differentiation matrices. More generally, we can define a differential operator through (16) and apply it to $\vartheta_3^N(z,q)$, for fixed q. Note that $-D^2$ is a positive-semidefinite matrix for N odd, and a positive-definite matrix por N even, and that the function

$$f(z) = \frac{\Gamma(z+1)}{\Gamma(z+1-\alpha)}$$

is analytic for $\Re(z) > -1$, therefore, according to (16), the differential operator $f(-D^2)$ is defined as

$$\frac{\Gamma(-D^2+I)}{\Gamma(-D^2+I-\alpha I)} = F\frac{\Gamma(\Lambda^2+I)}{\Gamma(\Lambda^2+I-\alpha I)}F^{-1},$$

and, acting on $\vartheta_3^N(z,q)$, yields

$$\frac{\Gamma(-D^2+I)}{\Gamma(-D^2+I-\alpha I)}\vartheta_3^N(z,q) = \sum_{n=-m}^{m}\frac{\Gamma(n^2+1)}{\Gamma(n^2+1-\alpha)}q^{n^2}e^{inz}. \tag{20}$$

Here, I stands for the identity matrix. The term $\Gamma(n^2+1)q^{n^2}/\Gamma(n^2+1-\alpha)$ appearing in the right-hand side of (20) can be split in two as

$$q^{\alpha}\frac{\Gamma(n^2+1)}{\Gamma(n^2+1-\alpha)}q^{n^2-\alpha}.$$

Note that $\Gamma(n^2+1)q^{n^2-\alpha}/\Gamma(n^2+1-\alpha)$ can be identified with some adequate fractional derivative of the power function q^{n^2}. Taking into account that the sum of the right-hand side of (20) contains a non-null term for $n=0$ and $\Re(\alpha) < 1$, the Riemann-Liouville fractional derivative D_q^{α} with the lower terminal at 0, is the fractional derivative to be considered here. We remind the reader that the Riemann-Liouville derivative of a constant does not vanish, and the derivative of a power function q^m is given by [10]

$$D_q^{\alpha}q^m = \frac{\Gamma(m+1)}{\Gamma(m+1-\alpha)}q^{m-\alpha}, \quad m > -1, \quad \Re(\alpha) \geq 0. \tag{21}$$

Therefore, we have that Eq.(20) can be written in the form

$$\frac{\Gamma(-D_z^2+I)}{\Gamma(-D_z^2+I-\alpha I)}\vartheta_3^N(z,q) - q^{\alpha}\, D_q^{\alpha}\vartheta_3^N(z,q) = 0, \quad 0 \leq \Re(\alpha) < 1, \tag{22}$$

where D_z stands for the differentiation matrix with respect to z for odd N. The function $\vartheta_4^N(z,q)$ also satisfies an equation like this. In the case of $\vartheta_1^N(z,q)$ and $\vartheta_2^N(z,q)$, the differential equation has the same form, with D_z standing for the differentiation matrix with respect to z for even N. As it can be proved easily, (22) holds for $\alpha = 1$. This case gives rise to the diffusion equation. A similar situation occurs for negative values of α. Note that Eq. (21) (and Eq. (22)) remains valid for $\Re(\alpha) < 0$, but taking into account that higher derivatives (or integrals) can be composed with derivatives (integrals) of integer order, it is enough to consider $-1 \leq \Re(\alpha) < 1$. According to the standard notation, the symbol $D_q^{-\alpha}$, denoting now an integral, should be replaced by I_q^α. However, we will maintain the use of D_q^α to denote differentiation for positive α or integration for negative α.

Since D_z is a representation of the ordinary derivative with respect to z in the subspaces of (anti)periodic trigonometric polynomials, the asymptotic form of Eq. (22) for large N is

$$\frac{\Gamma\left(-\frac{\partial^2}{\partial z^2}+1\right)}{\Gamma\left(-\frac{\partial^2}{\partial z^2}+1-\alpha\right)}\vartheta(z,q) - q^\alpha\, D_q^\alpha \vartheta(z,q) = 0, \quad -1 \le \Re(\alpha) \le 1. \tag{23}$$

where $\vartheta(z,q)$ can be any of the four theta functions (17).

As it has been noted above, this equation is a fractional partial integral equation for theta functions for negative values of α. Note that for $\alpha = -1$, Eq. (23) can be written as

$$\left(-\frac{\partial^2}{\partial z^2}+1\right)^{-1}\vartheta(z,q) - \frac{1}{q}\int_0^q \vartheta(z,r)dr = 0, \quad q \ne 0.$$

On the other hand, the differential operator $\Gamma(D_z^2+I)/\Gamma(D_z^2+I-\alpha I)$ can be implemented numerically in terms of the fast Fourier transform via Eq. (16). This fact allows us to test numerically Eq. (23) through the simple relation given by Eq. (20). As a sample case, we take 10 points in the unit circle of the q-plane according to $q_j = x_j + i x_j^3$, $x_j = -1 + 2j/11$, $j = 1, 2, .., 10$. At each q_j-point we compute the vector f_j of components $f_{kj} = \vartheta_3^{61}(z_k, q_j)$, with

$$z_j = \frac{2\pi k}{N_z} + i\frac{\pi}{4}, \quad N_z = 2^{10}+1,$$

and then, we compute

$$f_j^{(\alpha)} = \frac{\Gamma(-D_z^2+I)}{\Gamma(-D_z^2+I-\alpha I)} f_j, \quad j = 1, 2, \cdots, 10,$$

through Eq. (16) for $\alpha = j/10$, $j = -10, \cdots, 10$. For given q_j and α, the vector $g_j^{(\alpha)}$ of components $g_{kj}^{(\alpha)} = q_j^\alpha\, D_q^\alpha \vartheta(z_k, q_j)$ can be computed by using the sum of the right-hand side of (20). If the norm $\max_{j,k} |f_{kj}^{(\alpha)}|$ is denoted by $\|f^{(\alpha)}\|$, we find that the numerical residue $\|f^{(\alpha)} - g^{(\alpha)}\|$ at the values of α can be plotted as in Fig. 1.

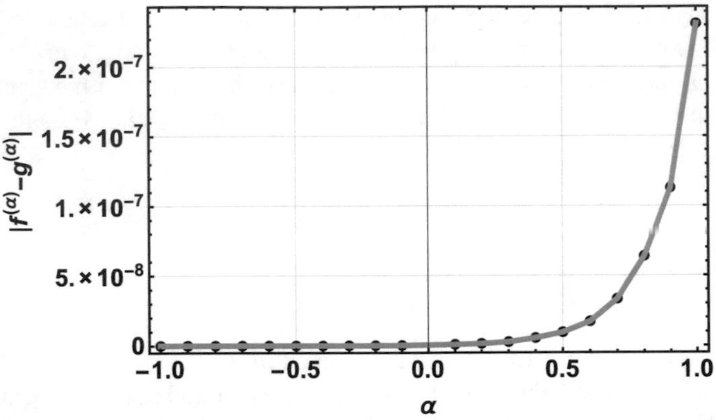

Fig. 1 Numerical computation of $\|f^{(\alpha)} - g^{(\alpha)}\|$ at $\alpha = m/10$, $m = \pm 1, \pm 2, \cdots, \pm 10$. The minimum value of $\|f^{(\alpha)} - g^{(\alpha)}\|$ is 1.0×10^{-12} and it occurs at $\alpha = -1$. The maximum value is 2.3×10^{-7} and it occurs at $\alpha = 1$

4 Computing Jacobi Elliptic Functions

The above method, which is based in differentiation matrices for trigonometric polynomials, can also be applied to compute elliptic functions by computing elliptic integrals. Just as a sample case, consider the Jacobi elliptic functions which can be computed in terms of the amplitude φ given in

$$F(\varphi, \kappa) = \int_0^{\varphi} \frac{d\theta}{\sqrt{1 - \kappa^2 \sin^2 \theta}}.$$

Let $f(\theta, \kappa)$ denote the integrand of this integral. This is a π-periodic function that can be differentiated as an antiperiodic function if we subtract the mean value of the minimum and maximum values of $f(\theta, \kappa)$ as a function of θ and change the period to $\pi/2$. Thus, the integral can be computed by using (15) with $\alpha = -1$. The subtracted terms are not significant for this computation and can be discarded. Let us consider the $N = 2m$ points of $[0, \pi)$

$$\theta_{j+1} = \pi \frac{j}{N}, \quad j = 0, 1, \cdots, N-1,$$

and let $f(\kappa)$ be the vector of entries $f_j(\kappa) = f(\theta_j, \kappa)$, $j = 1, \cdots, N$. Thus, taking the change of period as a change of variable, we obtain that the vector

$$u(\kappa) = \frac{1}{2} D^{-1} f(\kappa) \tag{24}$$

gives an approximation to $F(\theta_j, \kappa)$ up to a constant of integration. Since $F(0, \kappa) = 0$, this constant is found to be the vector whose entries are all equal to $u_1(\kappa)$, and therefore, the vector $F(\kappa)$ of entries $F_j(\kappa) = u_j(\kappa) - u_1(\kappa)$ is the approximation that we are looking for. Thus, the set of points $\left(\theta_j, F_j(\kappa)\right)$ can be interpolated to yield a continuous curve to approximate $F(\varphi, \kappa)$. Therefore, the Jacobi amplitude is given by the inverse function, the one interpolating the set of points

$$\left(F_j(\kappa), \theta_j\right), \quad j = 1, 2, \cdots, N.$$

In other words, the values of the Jacobi amplitude at the non-evenly spaced abscissas $F_j(\kappa)$ are given by θ_j for $j = 1, 2, \cdots, N$. Once the amplitude is obtained, N accurate values of Jacobi elliptic functions can be computed in $\mathcal{O}(N \log N)$ operations. A comparison between this procedure and some of the well-established algorithms for computing elliptic functions can be done. Just to give an idea of the performance of this approach, we compare it with the `sn` function of the Elliptic R package [9] and the `JacobiSN` function of Wolfram Mathematica v10 running on a 3.4 GHz Intel Core i5 processor using standard machine-number precision. We compute $N = 2^{20}$ values $\mathrm{sn}(u_j, \kappa)$, $j = 1, 2, \cdots, N$, of this Jacobi elliptic function for $\kappa = 99{,}999/100{,}000$. In the case of the R package, we show the user cpu time computed with the `system.time` function. In the case of Mathematica, we show the cpu time obtained with the `Timing` function.

Only N values of the standard sine function are required in our procedure, i.e., $\sin\theta_j$, however, the N values $F_j(\kappa)$, i.e., the $u_j(\kappa)$-values, are also required. Therefore, the cpu time for computing the $2N$ values $u_j(\kappa)$ and $\sin\theta_j$ is shown in this case. We compare two cases: the real case, i.e., the case in which

$$\theta_j = \frac{\pi}{2}\left(-1 + \frac{2j}{N}\right), \quad j = 1, \cdots, N,$$

and the complex case, where θ_j is given by

$$\theta_j = \frac{\pi}{2}\left(-1 + \frac{2j}{N}\right) + \frac{i}{2}, \quad j = 1, \cdots, N.$$

Please note that, according to this method, a proper integration constant has to be added to (24). The results are displayed in Tables 1 and 2.

Table 1 Computing 2^{20} values of $\mathrm{sn}(u, 0.99999)$ with standard machine-number precision: comparison between this approach and the function `sn` of the elliptic R package. The error corresponds to the maximum absolute value of the difference between outputs

	User time (secs) real case	User time (secs) complex case
Elliptic R package	1.5	1.6
This approach	0.4	0.4
Error	3×10^{-4}	4×10^{-6}

Table 2 Computing 2^{20} values of $\mathrm{sn}(u, 0.99999)$ with standard machine-number precision: comparison between this approach and the function `JacobiSN` of Wolfram Mathematica. The error corresponds to the maximum absolute value of the difference between outputs

	Timing time (secs) real case	Timing time (secs) complex case
Mathematica 10	5.1	19.2
This approach	7.3	8.0
Error	4×10^{-4}	3×10^{-15}

5 Final Remark

The main advantage of the method presented above is that it gives a finite-dimensional scheme to differentiate or integrate trigonometric polynomials with exact results. This fact generates sequences of matrix operators that represent continuum operators defined in a flat torus, providing room for studying this kind of problems from other perspective. This is the case of Eq. (23). Further examples can be found in Refs. [2, 4].

References

1. Campos, R.G., Meneses, C.: Differentiation matrices for meromorphic functions. Bol. Soc. Mat. Mexicana **12**, 121–132 (2006)
2. Campos, R.G., Pimentel, L.O.: A finite-dimensional representation of the quantum angular momentum operator. Il Nuovo Cimento **116B**, 31–45 (2001)
3. Campos, R.G., Ruiz, R.G.: Fast integration of one-dimensional boundary value problems. Int. J. Mod. Phys. C **24** (2013). https://doi.org/10.1142/S0129183113500824
4. Campos, R.G., López-López, J.L., Vera, R.: Lattice calculations on the spectrum of Dirac and Dirac-Kähler operators. Int. J. Mod. Phys. **23**, 1029–1038 (2008)
5. Chapront, J., Simon, J.L.: Planetary theories with the aid of the expansions of elliptic functions. Celest. Mech. Dyn. Astron. **63**, 171–188 (1996)
6. Dubrovin, B.A.: Theta functions and non-linear equations. Russ. Math. Surv. **36**, 11–92 (1981)
7. Erdélyi, A. (ed.): Higher Transcendental Functions, vol. II. McGraw Hill, New York (1953)
8. Golub, G.H., Van Loan, C.F.: Matrix Computations. The Johns Hopkins University Press, Baltimore (1996)
9. Hankin, R.K.S.: Introducing elliptic, an R package for elliptic and modular functions. J. Stat. Softw. **15**, 1–22 (2006)
10. Kilbas, A.A., Srivastava, H.M., Trujillo, J.J.: Theory and Applications of Fractional Differential Equations. Elsevier, Amsterdam (2006)
11. Lawden, D.F.: Elliptic Functions and Applications. Springer Science+Business Media, New York (1989)
12. Markushevich, A.I.: Theory of Functions of a Complex Variable. AMS Chelsea Publishing, Rhode Island (2011)
13. McKean, H., Moll, V.: Elliptic Curves. Cambridge University Press, Cambridge (1999)
14. Meyer, K.R.: Jacobi elliptic functions from a dynamical systems point of view. Am. Math. Mon. **108**, 729–737 (2001)

15. Mumford, D.: Tata Lectures on Theta II. Jacobian Theta Functions and Differential Equations. Birkäuser, Boston (1984)
16. Mumford, D.: Tata Lectures on Theta I. Birkäuser, Boston (2007)
17. Petrović, N.Z., Bohra, M.: General Jacobi elliptic function expansion method applied to the generalized (3+1)-dimensional nonlinear Schödinger equation. Opt. Quant. Electron. (2016). https://doi.org/10.1007/s11082-016-0522-1
18. Rodríguez, C.M.: Orbits in General Relativity: The Jacobian Elliptic Functions. Il Nuovo Cimento **98B**, 87–96 (1987)
19. Siegel, C.L.: Topics in Complex Function Theory. Volume I: Elliptic Functions and Uniformization Theory. Wiley, New York (1988)
20. Zudilin, V.V.: Thetanulls and differential equations. Sbornik Mathematics **191**, 1–45 (2000)

On Continued Fractions

Gerardo Gonzalez Sprinberg

Para Antonio Campillo en sus 65 vueltas alrededor del sol

Abstract This paper on the geometry, algebra and arithmetics of continued fractions is based on a lecture for students, teachers and a non-specialist audience, beginning with the history of the golden number and Fibonacci sequence, continued fractions of rational and irrational numbers, Lagrange theorem on periodicity of continued fractions for quadratic irrationals, Klein's geometric interpretation of the convergents as integer points, Jung-Hirzebruch continued fractions with negative signs and two dimensional singularities, higher dimensional generalizations, and ending with a result on a periodic generalized 3-dimensional continued fraction for a cubic irrational.

1 Introduction: The Golden Number

Probably one of the main emblematic examples is the golden number

$$\varphi = \frac{1+\sqrt{5}}{2} = 1.618\,033\,988\,749\,894\,848\,204\ldots$$

The three suspension points represent an infinity of digits without any known periodicity or regularity, and it would take an indefinite time and space to write them down explicitly!

The golden number is usually denoted by the Greek letter φ in honor of the sculptor and architect Phidias who built the Parthenon in Athens, or by τ, from

G. Gonzalez Sprinberg (✉)
Facultad de Ciencias, Centro de Matemática, UdelaR, Montevideo, Uruguay

Université Grenoble Alpes, Grenoble, France
e-mail: gerardogs@cmat.edu.uy

G.-M. Greuel et al. (eds.), *Singularities, Algebraic Geometry, Commutative Algebra, and Related Topics*, https://doi.org/10.1007/978-3-319-96827-8_27

the Greek word $\tau o\mu\eta$, section. It is already known by the Pythagoreans (in the -6 century) in its geometric form. Studied later by Plato (see [8]) and Eudoxus of Cnidus it is presented geometrically in the Book VI, definition 3 of Elements of Euclide (in the -3 century), see [5]:

A straight segment is said to be cut in extreme and mean ratio when the whole is to the greater segment as the greater segment is to the smaller one.

$$\frac{a+b}{a} = \frac{a}{b}$$

This quotient which is independent of the length of the segment, is the *golden number*. In the Proposition 30 of Book VI, Euclid gives a construction method of the extreme and mean section of a segment.

It can be proved that the golden number φ is not the quotient of two integer numbers, i.e. it is an irrational number. This fact seems to have created panic and political crisis among the Pythagoreans, since for them everything was representable by the numbers they knew, namely the rational numbers (this may be the origin of the name "irrational" for these numbers). The number φ is a quadratic irrational, i.e. an algebraic irrational number, root of a polynomial of degree 2 with integer coefficients. Denoting by x the quotient $\frac{a}{b}$, the equality $\frac{a+b}{a} = \frac{a}{b}$ is equivalent to $1 + \frac{1}{x} = x$. Multiplying this equality by x one gets the polynomial equation $x^2 - x - 1 = 0$ and φ is the positive root of this equation. The other (conjugate) root is $-1/\varphi$, since the product of the two roots is -1. Note that $1/\varphi = \varphi - 1$ is the fractionary part of φ.

A segment of length φ is obtained by using the theorem of Pythagoras (Euclid, Book IV.10):

A right triangle with sides of length 1 and $1/2$ has an hypotenuse of length $\sqrt{5}/2$. Then the segment AB has length φ if oB is $\sqrt{5}/2$, where o is the midpoint of AC of length 1.

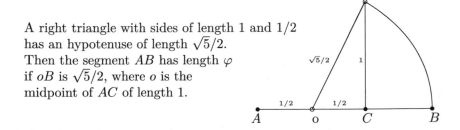

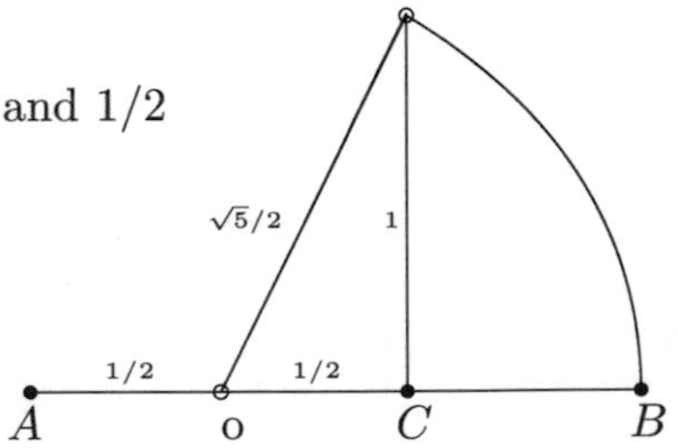

1.1 On Relations Between φ, Fibonacci, Polygones and Polyhedra

By iterating the equality $\varphi = 1 + 1/\varphi$, we obtain $\varphi = 1 + 1/(1 + 1/\varphi)$ and so on. As a consequence, there is an expression of φ as a so-called *continued fraction*:

$$\varphi = 1 + \cfrac{1}{1 + \cfrac{1}{\ddots + \cfrac{1}{1 + \cfrac{1}{\ddots}}}}$$

Or more concisely $\varphi = 1 + \frac{1|}{|1} + \cdots + \frac{1|}{|1} + \cdots$ Note that all the coefficients are equal to 1, with a regularity contrasting with the decimal expression.

Keeping only a finite beginning part of the continued fraction one gets a rational number called a "*convergent*".

Denote by F_n/F_{n-1} the convergent obtained with the first n terms of the continued fraction of φ. Then $F_1/F_0 = 1$, $F_2/F_1 = 1 + 1/1$, $F_3/F_2 = 1 + 1/(1 + 1/1)$, $\cdots$, $F_{n+1}/F_n = 1 + 1/(1 + F_n/F_{n-1})$, so $F_0 = 1$, $F_1 = 1$, $F_2 = 2$, $\cdots$, $F_{n+1} = F_n + F_{n-1}$ if $n \geq 1$.

This sequence of integers $1, 1, 2, 3, 5, 8, 13, 21, \ldots$, where each term is the sum of the two preceding ones, was introduced in the arithmetic treatise "Liber Abaci" (Book of calculations, see [6]) in 1202 by Leonardo de Pisa, called *Fibonacci* (son of Bonacci), as solution of the famous problem on the number of couples of rabbits giving birth to a new couple each month. Remark that each new couple of rabbits are brother and sister, but the church did not mind this incestuous mathematical problem, even if Galileo was excomunicated for more than three centuries for "e pur si muove". Fibonacci also popularized the indo-arabic notation of numbers used until today. The quotients F_n/F_{n-1} : $(1,\ 2,\ 3/2 = 1.5, 5/3 = 1,\overline{6}, 8/5 = 1.6,\ 13/8 = 1.625, 21/13 = 1.615384615384615,\ \cdots)$ converge towards $\varphi = 1.618\cdots$ with values alternatively smaller and bigger.

Let us collect here some properties of the Fibonacci sequence:

$$(F_n, F_{n+1}) = 1 \quad \text{for} \quad n \geq 0 \quad ; \quad F_n^2 - F_{n-1}F_{n+1} = (-1)^n \quad \text{for} \quad n \geq 1$$

$$|F_{n+1}/F_n - \varphi| \leq 1/F_n^2 \quad ; \quad F_n = \frac{1}{\sqrt{5}}(\varphi^{n+1} - \varphi'^{n+1}) \text{ with } \varphi' = \frac{1-\sqrt{5}}{2}$$

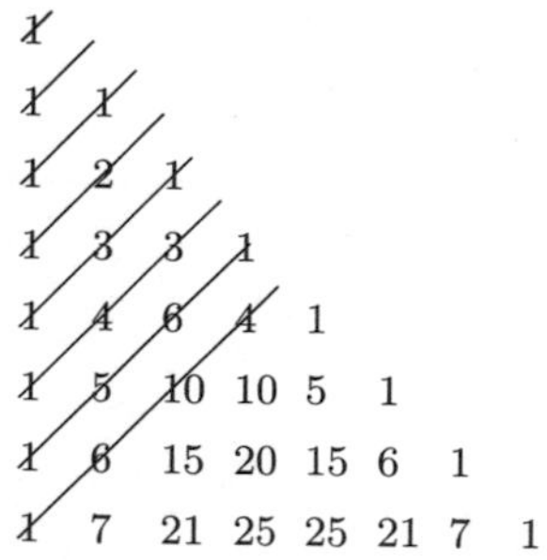

The sums of numbers in each indicated diagonal of the "chinese" Pascal triangle are consecutive terms of the Fibonacci sequence.

There is a relation between φ and π given by elementary trigonometric functions: $\varphi = 2\cos\pi/5, \quad 1/\varphi = 2\sin\pi/10, \quad \varphi = 1 + 1/\varphi = 1 + 2\sin\pi/10$. Therefore there are relations between φ, the regular pentagon, decagon and polyhedra. The quotient between the lengths of a diagonal and of a side of a regular pentagon is φ. Thus a regular pentagon is obtained from a "golden triangle" ACE (see Fig. 1) whose angles are $\widehat{ACE} = \pi/5, \widehat{CAE} = \widehat{CEA} = 2\pi/5$.

A regular pentagon may be obtained as a "golden knot" from a rectangular paper (see Fig. 2); this is proved knowing the angles in the pentagon.

The interior intersections of the diagonals are the five vertices of a new regular pentagon and so on indefinitely inward or outward (see Fig. 3).

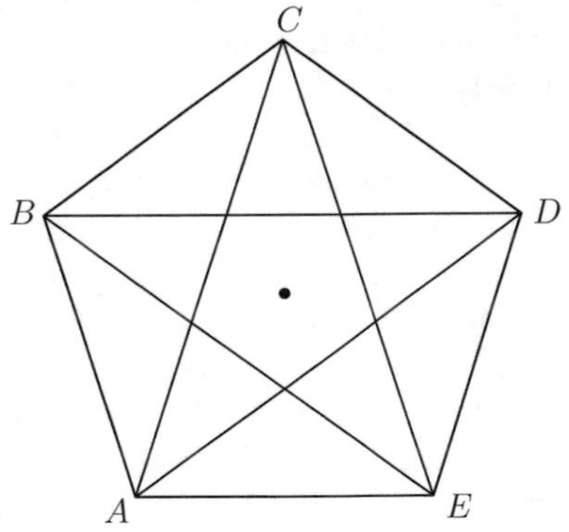

Fig. 1 The pentagram or pentagonal star formed by the five diagonals of the pentagon was used by Pithagoreans as a symbol of recognition and brotherhood

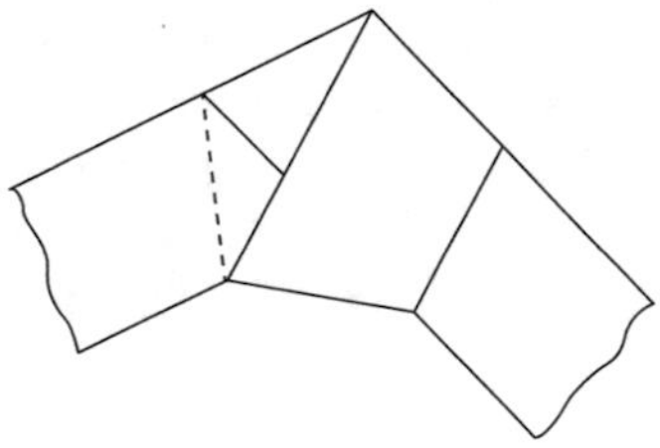

Fig. 2 A "golden knot"

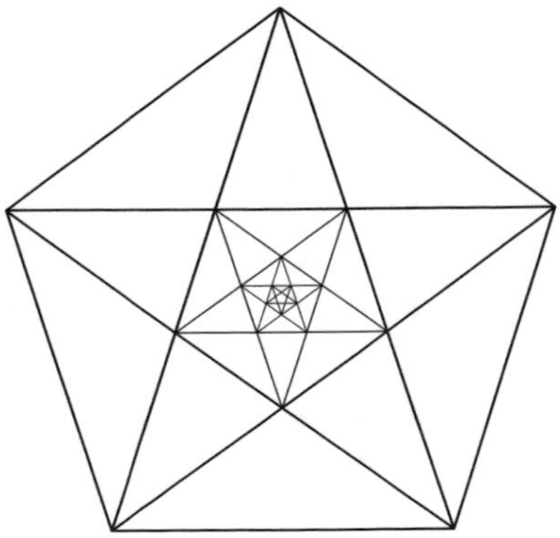

Fig. 3 Iterated diagonals of a regular pentagon

A regular decagon is also easily obtained, since the quotient between the radius and the side is φ.

Knowing φ one obtains also a regular dodecahedron (see [3]), with 12 regular pentagons as faces, 3 of them meeting at each vertex, since

$$(\pm 1, \pm 1, \pm 1), (0, \pm\varphi, \pm 1/\varphi), (\pm 1/\varphi, 0, \pm\varphi), (\pm\varphi, \pm 1/\varphi, 0)$$

are the coordinates of the 20 vertices of a regular dodecahedron.

The 12 centers of the faces of the dodecahedron are the vertices of an icosahedron, dual of the dodecahedron, with 20 equilateral triangles as faces, 5 of them meeting at each vertex.

Each pair of opposite parallel edges of the icosahedron limit a *golden rectangle* with φ as the quotient of sides. There are 30 edges, 15 rectangles and 5 trios of such rectangles which are orthogonal (see Fig. 4).

With coordinates with respect to the planes of such an orthogonal trio of rectangles the points $(0, \pm\varphi, \pm 1)$, $(\pm 1, 0, \pm\varphi)$, $(\pm\varphi, \pm 1, 0)$ are the vertices of the icosahedron.

These points divide in the proportion $\varphi : 1$ the 12 edges of the regular octahedron with vertices $(\pm\varphi^2, 0, 0)$, $(0, \pm\varphi^2, 0)$, $(0, 0, \pm\varphi^2)$.

The number φ and the Fibonacci sequence appear endlessly in nature, for example in phyllotaxis, i.e. the distribution of leaves along plant stems (see [4]), or in the logarithmic spiral of polar equation $r = \varphi^{2\theta/\pi}$ on variables (r, θ) in the Nautilus mollusk. They also appear in art (e.g. in paintings and drawings by Botticelli and Leonardo da Vinci, in sculptures by Phidias), and in music (e.g. in Pythagoras cycle of fifths, in proportions of Stradivarius violins).

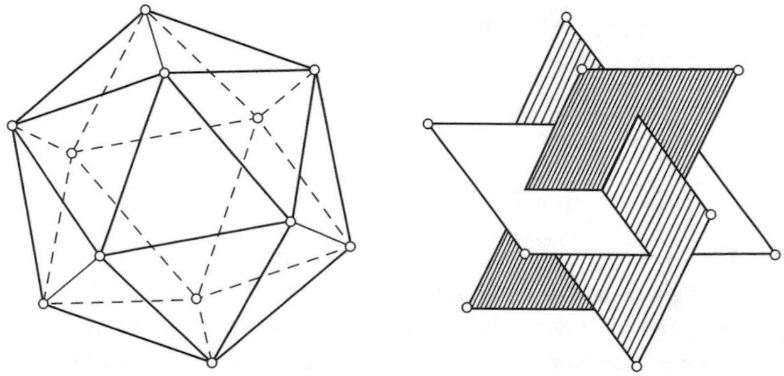

Fig. 4 Icosahedron and orthogonal golden rectangles

2 Continued Fractions of Rational and Irrational Numbers

The continued fractions of rational numbers result from the classical euclidian algorithm. Given a rational a/b, with a and b integers (with no common factor), $b > 0$, the integer division $a = bc_0 + r_0$, with c_0 the quotient and r_0 the remainder, $0 \leq r_0 < b$, may be iterated (finitely) till the last remainder obtained is zero : $a/b = c_0 + 1/(b/r_0) = c_0 + 1/(c_1 + 1/(r_0/r_1)) =$

Remark The last non zero remainder is $\gcd(a, b)$. This is the original geometric presentation of Euclid, as the common measure between two commensurable segments, Book 7,2.

A number is rational if and only if its continued fraction is finite.

The decimal expression may be infinite, like $1/7 = 0,\overline{142857}$ with period overlined. The continued fraction of a rational is unique (with last term > 1 if it has at least 2 terms), but its decimal expression is not unique, e.g. $1,2 = 1,1999\cdots = 1,1\overline{9}$.

This algorithm may be generalized, remarking that the integer quotient is the *integer part* of the rational, i.e. the greatest integer less or equal the given number r, notation : $\lfloor r \rfloor$ the "floor" of r.

If r is a real irrational number, let $\{r\} = r - \lfloor r \rfloor$ be its fractional part. Then again, by iterating, a continued fraction is obtained:

$$r = \lfloor r \rfloor + \frac{1}{1/\{r\}} = \lfloor r \rfloor + \cfrac{1}{\lfloor 1/\{r\} \rfloor + \cfrac{1}{1/\{1/\{r\}\}}} = \cdots =$$

$$= c_0 + \cfrac{1}{c_1 + \cfrac{1}{\ddots + \cfrac{1}{c_n + \cfrac{1}{\ddots}}}}$$

$= c_0 + 1|c_1 + \cdots + 1|c_n + \cdots$, usual short notation $[c_0, c_1, \cdots, c_n, \cdots]$.

It is an infinite continued fraction, since otherwise r would be a rational number. Conversely, *any infinte sequence of integers, positive starting from the second one, defines a continued fraction converging towards an irrational number.*

We have seen the remarkable example: $\varphi = (1 + \sqrt{5})/2 = [1, 1, \cdots, 1, \cdots]$

It is a particular case of the positive root of $x^2 = ax + 1$ for $a = 1$. For any integer $a > 0$ its positive root has the continued fraction $[a, a, \cdots, a, \cdots]$.

For example if $a = 2$, the positive root is $1 + \sqrt{2} = [2, 2, \cdots, 2, \cdots]$. Therefore $\sqrt{2} = [1, 2, \cdots, 2, \cdots]$.

Note: This number is used for the standard formats since if the ratio of the sides of a rectangular paper sheet is $\sqrt{2}$, then if folded the ratio is the same. From a sheet of 1 m^2, $A_0 = 84,1 \times 118,9$ cm, folding four times gives the usual $A_4 = 21 \times 29,7$ cm.

There are also important results on continued fractions of *trascendental* irrationals e.g. $e = [2, 1, 2, 1, 1, 4, 1, 1, 6, 1, 1, 8, 1, 1, 10, 1, \cdots]$ $(Euler)$
or examples of *generalized* continued fractions (i.e. with numerators $\neq 1$).
$e = 2 + 1\overline{|1} + 1\overline{|2} + 2\overline{|3} + 3\overline{|4} + \cdots$ $(Euler, Cesàro)$
$\pi = 3 + 1^2\overline{|6} + 3^2\overline{|6} + 5^2\overline{|6} + 7^2\overline{|6} + \cdots$ $(Wallis)$
An infinite continued fraction $[c_0, c_1, \cdots, c_n, \cdots]$ is called *periodic* if it reproduces periodically from a certain rank, i.e. $[c_0, \cdots, c_{k-1}, \overline{c_k, \cdots, c_{k+n}}]$.
It is *pure periodic* if $k = 0$, i.e. of the form $[\overline{c_0, \cdots, c_n}]$.

A remarkable theorem of Joseph-Louis Lagrange proves that *any quadratic irrational has a periodic continued fraction.* The converse proposition, easier to prove, was known by Leonhard Euler, so there is an equivalence.

Evariste Galois proved in particular: *a continued fraction is pure periodic if and only if it is a quadratic irrational* > 1 *with conjugate between* 0 *and* 1.

3 Geometric Interpretation of Continued Fractions

Let r be a positive real number. Consider the ray L with equation $y = rx$ of slope r in the first quadrant of the plane. Consider the continued fraction of r with convergents $p_n/q_n = [c_0, \cdots, c_n]$.

Felix Klein proposed the following geometric interpretation. *If a thread along L moves away from L in both sides then the first points with integer coordinates it meets have coordinates* (q_n, p_n) *given by the convergents. These points are the vertices of two polygonal lines approaching L, one above and one below* (see Fig. 5 and the paper [14]).

3.1 Some Proofs and Corollaries

The following recurrence relations for $k \geq 2$ are proved by induction (see [2, 15, 17, 19]):

$$p_k = c_k p_{k-1} + p_{k-2} \tag{1}$$

$$q_k = c_k q_{k-1} + q_{k-2} \tag{2}$$

If we put $p_{-1} := 1$ and $q_{-1} := 0$, then they are true also for $k = 1$.

The difference $[(2)p_{k-1} - (1)q_{k-1}]$ eliminates c_k and then
$q_k p_{k-1} - p_k q_{k-1} = -(q_{k-1} p_{k-2} - p_{k-1} q_{k-2})$. As $q_0 p_{-1} - p_0 q_{-1} = 1$ it follows

$$q_k p_{k-1} - p_k q_{k-1} = (-1)^k \tag{3}$$

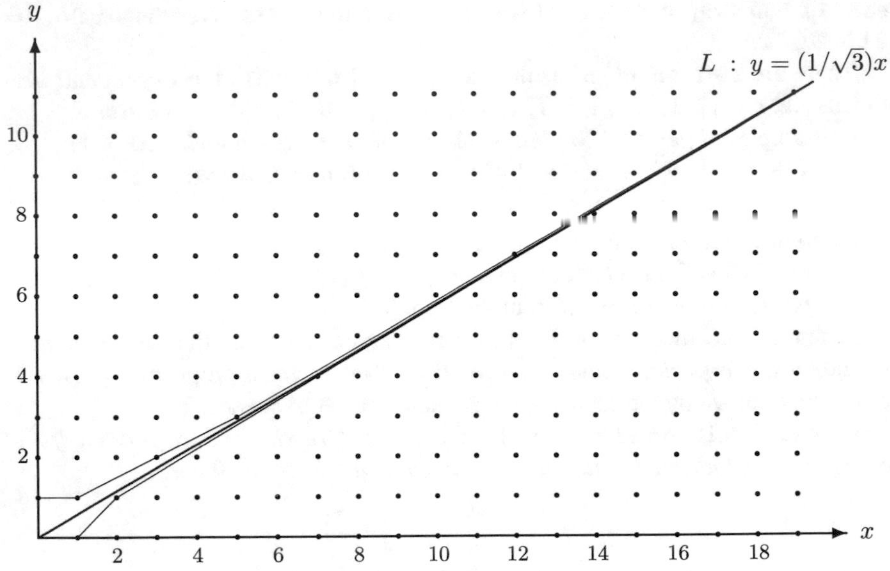

Fig. 5 *Example*. Consider the ray L in the first quadrant with slope $1/\sqrt{3}$. The continued fraction of $1/\sqrt{3}$ is $[0, 1, \overline{1, 2}]$ and the first six integer points near L given by the convergents are $(1, 0)$, $(1, 1)$, $(2, 1)$, $(5, 3)$, $(7, 4)$, $(19, 11)$. We can see in this picture the two polygonal lines near L with vertices these points, one below and the other above L, according to Klein's geometric interpretation of the convergents

It results from (3) that the representative p_n/q_n of each convergent is irreducible, since any divisor of p_n and q_n is a divisor of $(-1)^k$.

Considering the difference $[(2)p_{k-2} - -(1)q_{k-2}]$ one obtains, using the equality (3), that:

$$q_k p_{k-2} - p_k q_{k-2} = c_k(q_{k-1}p_{k-2} - p_{k-1}q_{k-2}) = (-1)^{k-1}c_k. \quad \text{Therefore:}$$

$$p_{k-2}/q_{k-2} - p_k/q_k = (-1)^{k-1}c_k/q_k q_{k-2} \tag{4}$$

As a consequence of (4), *if r is irrational therefore the convergents of even (resp. odd) order form an increasing (resp.decreasing) sequence.*

$$\frac{p_0}{q_0} < \frac{p_2}{q_2} < \frac{p_4}{q_4} < \frac{p_6}{q_6} < \cdots < r < \cdots < \frac{p_7}{q_7} < \frac{p_5}{q_5} < \frac{p_3}{q_3} < \frac{p_1}{q_1}$$

The points V_k of coordinates (q_k, p_k) in the plane are below the line L if k is even, and above it if k is odd.

As $q_k = c_k q_{k-1} + q_{k-2} \geq q_{k-1} + q_{k-2} \geq 2q_{k-2}$ it follows that $q_k \geq 2^{(k-1)/2}$, and by (3), we obtain:

The two sequences converge towards the same limit r. If r is rational, both are finite and one of them arrives to r. The points with coordinates given by the convergents tend monotonously from below and from above towards the line L.

The left-hand side of the equality (3) $q_k p_{k-1} - p_k q_{k-1} = (-1)^k$ may be seen as the determinant of the matrix with column vectors

$$V_k = \begin{pmatrix} q_k \\ p_k \end{pmatrix} \quad \textit{and} \quad V_{k-1} = \begin{pmatrix} q_{k-1} \\ p_{k-1} \end{pmatrix}$$

Therefore the equality (3) implies $|\det(V_k, V_{k-1})| = 1$ and it follows that (V_k, V_{k-1}) is a basis of $\mathbb{Z}^2$. As a consequence, the vertices of the triangle obtained as the convex hull of the set $\{0, V_k, V_{k-1}\}$ are its only integer points.

This proves that V_k and V_{k-1} are the integer points nearest to the ray L in this triangle, and we have proved the Klein's geometric interpretation of the convergents of a continued fraction.

The relations (1) and (2) may be interpreted geometrically:

$$V_k - V_{k-2} = c_k V_{k-1}$$

which means that the segment $\overline{(V_{k-2}, V_k)}$ is parallel to V_{k-1} and contains it c_k times. Then the number of integer points in $\overline{(V_{k-2}, V_k)}$ is $c_k + 1$.

4 Jung-Hirzebruch Continued Fractions

A continued fraction where minus signs replace the plus ones is obtained by considering for r rational the excess euclidean division, or equivalently for any real r the least integer $\geq r$ ("roof" of r), denoted $\lceil r \rceil$.

If r is not an integer then $\lceil r \rceil = \lfloor r \rfloor + 1$, $r = \lceil r \rceil - (1 - \{r\}) =$
$\lceil r \rceil - 1/(1/(1 - \{r\})) = \lceil r \rceil - 1/(\lceil 1/(1 - \{r\}) \rceil - (1 - \{1/(1 - \{r\})\})) = \cdots$

$$= d_1 - 1\lfloor d_2 - \cdots - 1\lfloor d_n - \cdots =: [d_1, d_2 \cdots, d_n, \cdots]^-$$

$$d_k \geq 2 \text{ if } k > 1$$

This is by definition a so called Jung-Hirzebruch continued fraction.

If $0 < a < b$ and b/a is an irreducible rational, then its Jung-Hirzebruch continued fraction $[d_1, d_2, \cdots]^-$ allows to compute all the nearest integer points in the positive quadrant below the ray of equation $y = (b/a)x$ as follows:

$$Q_0 = (1, 0), Q_1 = (1, 1), \quad Q_{i+1} = d_i Q_i - Q_{i-1}, i \geq 1$$

The points Q_{i-1}, Q_i, Q_{i+1} are aligned if $d_i = 2$. There is a relation with the ordinary continued fraction, from which are obtained first the vertices and then all the integer points in each segment of the polygonal line $Q_0Q_1Q_2\ldots$ (see [10, 18]).

The Jung-Hirzebruch continued fraction of φ is more difficult to find than its ordinary continued fraction:

$$\varphi = [2, \overline{3}]^- = 2 - 1\rfloor 3 - \cdots - 1\rfloor 3 - \cdots$$

Historical note: Jung used this kind of continued fractions in his method of resolution of surfaces, published in [13]. The singularity at the origin in the surface with equation $x^n = yz^p$ after being normalized may be described as a toric singularity (see [7]), defined by the algebra associated to the semigroup of integer points of a cone as we consider in next section. Its desingularization is obtained as a fan of cones defined by the nearest integer points of the line defining the toric surface, or equivalently by the Hilbert basis of the semigroup. These singularities are essential in the Jung's method of desingularization of surfaces. The minimal resolution has a chain of rational curves in its exceptional divisor, with self-intersections given by (nowadays called) Jung-Hirzebruch continued fraction of n/p (see [11, 13, 16]).

Periodic Jung-Hirzebruch continued fractions of quadratic irrationals appear in the description of the resolution of cuspidal singularities of Hilbert-Blumenthal by Hirzebruch. Cycles of exceptional divisors are obtained for these elliptic singularities ($h_1 = 1$, i.e. the first cohomology group has dimension 1) (see [9, 12]).

5 Continued Fractions in Higher Dimension

Let C be a simplicial cone in euclidean space $\mathbb{R}^n$ and $(e_1, \cdots, e_n)$ the canonical basis of $\mathbb{R}^n$. The convex hull of the non-zero points of $C \cap \mathbb{Z}^n$ is called the *Klein polyhedron of C*.

The boundary of a Klein polyhedron is a polyhedral variety called "*sail*" (see [1]). If r is a positive real number, by the geometric interpretation its continued fraction is equivalent to the data of the sails of the cones $\langle e_1, (1, r)\rangle$ and $\langle (1, r), e_2\rangle$, whose vertices or angular points correspond to the convergents.

This allows a generalization of continued fractions in higher dimension. If L is a ray in the positive octant of $\mathbb{R}^3$, the nearest integer coordinates points to L are those in the polyhedral sails of the cones $C_1 = \langle L, e_2, e_3\rangle$, $C_2 = \langle e_1, L, e_3\rangle$, $C_3 = \langle e_1, e_2, L\rangle$.

One may also give a matrix description of the continued fraction which is generalizable in higher dimension. Recurrence relations (1) and (2) are expressed as a product in $GL(2, \mathbb{Z})$:

$$\begin{pmatrix} p_k & p_{k-1} \\ q_k & q_{k-1} \end{pmatrix} = \begin{pmatrix} c_0 & 1 \\ 1 & 0 \end{pmatrix} \begin{pmatrix} c_1 & 1 \\ 1 & 0 \end{pmatrix} \cdots \begin{pmatrix} c_k & 1 \\ 1 & 0 \end{pmatrix}$$

5.1 Some Results in Dimension 3

Let $L_1 = (x_1, y_1, z_1)$ be a point in $\mathbb{R}^3$ with positive coordinates defining a ray OL_1 in the positive octant of $\mathbb{R}^3$. Let P_1 be the first integer point nearest to the ray OL_1. If P_1 belongs to the cone C_i replace e_i by P_1 in the canonical basis and call T_1 the base change matrix. Let $L_2 = T_1^{-1}L_1$ and iterate the procedure with respect to the ray OL_2 and the new basis and so forth. The sequence of base change matrices T_k belong in fact to $GL(3, \mathbb{Z})$ and it is the matrix equivalent of the continued fraction. The matrices $M_k := T_1T_2 \cdots T_k$ play the role of the convergents and one has $M_{k-1}L_k = L_1$ for $k \geq 2$ (article in preparation).

5.2 Example: The Cubic Root of 2

$\sqrt[3]{2} = 1.25992104989\cdots$ is a pretty mysterious cubic irrational number. Its continued fraction (in dimension 2) is not periodic by Lagrange's theorem since it is not a quadratic irrational, and there is no regularity

$$\sqrt[3]{2} = [1, 3, 1, 5, 1, 1, 4, 1, 1, 8, 1, 14, 1, 10, 2, 1, 4, 12, 2, 3, 2, 1, 3, 4, 1, 1, 2, 14, 3, 12, 1, 15, 3, 1, 4, 534, 1, 1, 5, 1, 1, 121, 1, 2, 2, 4, 10, \cdots]$$

Result: Let $r = \sqrt[3]{2}$. For the ray given by $L_1 = (1, r, r^2)$ (with coordinates being a natural basis of $\mathbb{Q}[r]$ over $\mathbb{Q}$) the above procedure gives a periodic continued fraction, in the sense that the sequence of matrices T_k is periodic with a period of length 3 from the third term:

$$\left[\begin{pmatrix}1&0&0\\1&1&0\\1&0&1\end{pmatrix}, \begin{pmatrix}1&3&0\\0&1&0\\0&2&1\end{pmatrix}, \overline{\begin{pmatrix}1&0&3\\0&1&3\\0&0&1\end{pmatrix}, \begin{pmatrix}1&0&0\\3&1&0\\3&0&1\end{pmatrix}, \begin{pmatrix}1&3&0\\0&1&0\\0&3&1\end{pmatrix}}\right]$$

The sequence of first integer points nearest to the ray OL_i in each new basis is given by the columns of the matrices T_k which do not belong to the canonical basis. It is also periodic with same kind of period:

$$\left[\begin{pmatrix}1\\1\\1\end{pmatrix}, \begin{pmatrix}3\\1\\2\end{pmatrix}, \overline{\begin{pmatrix}3\\3\\1\end{pmatrix}, \begin{pmatrix}1\\3\\3\end{pmatrix}, \begin{pmatrix}3\\1\\3\end{pmatrix}}\right]$$

The integer points P_i nearest to the ray in the (original) canonic basis, are columns of the matrices M_k. They have coordinates that grow very fast, as can be noted in the following first nine points:

$$\begin{pmatrix}1\\1\\1\end{pmatrix}, \begin{pmatrix}3\\4\\5\end{pmatrix}, \begin{pmatrix}12\\15\\19\end{pmatrix}, \begin{pmatrix}46\\58\\73\end{pmatrix}, \begin{pmatrix}177\\223\\281\end{pmatrix}, \begin{pmatrix}681\\858\\1081\end{pmatrix}, \begin{pmatrix}2620\\3301\\4159\end{pmatrix}, \begin{pmatrix}10080\\12700\\16001\end{pmatrix}, \begin{pmatrix}38781\\48861\\61561\end{pmatrix}$$

The quotient of the second over the first (or of the third over the second) coordinate converge fast towards r.

The inverse of the minimum coordinate of L_i may be written in the basis $(r^2, r, 1)$ of $Q[r]$ over Q. It turns out that the coefficients of this linear combination are the integer coordinates of the point P_{i-1}, for $i > 1$.

The cone given by three consecutive points P_{i-1}, P_i, P_{i+1} contains the initial ray and has determinant 1, so that the tetrahedron with vertices these points and the origin contains only the vertices as integer points.

References

1. Arnold, V.I.: Higher dimensional continued fractions. Regul. Chaotic Dyn **3**, 10–17 (1998)
2. Beskin, N.: Fracciones Maravillosas. Ediciones MIR, Moscú (1987)
3. Coxeter, H.S.M.: Regular Polytopes. Dover Publications, New York (1973)
4. Coxeter, H.S.M.: The role of intermediate convergents in Tait's explanation for phyllotaxis. J. Algebra **20**, 167–175 (1972)
5. Euclid: The Thirteen Books of the Elements (translated by Sir T. L. Heath). Dover Publications, New York (1956)
6. Fibonacci: Liber Abaci. Springer, New York (2002)
7. Fulton, W.: Introduction to Toric Varieties. Annals of Mathematics Studies, vol. 131. Princeton University Press, Princeton (1993)
8. Fowler, D.H.: The Mathematics of Plato's Academy. A New Reconstruction, 2nd ed. Clarendon Press, Oxford (1999)
9. Giraud, J.: Surfaces d'Hilbert-Blumenthal (d'après Hirzebruch, etc ...). In: Séminaire de Géométrie Algébrique d'Orsay, LNM 868. Springer, Berlin (1981)
10. Gonzalez Sprinberg, G.: Éventails en dimension deux et transformé de Nash. Secrétariat Mathémathique de l'E.N.S., Paris (1977)
11. Hirzebruch, F.: Über vierdimensionale Riemannsche Flächen Mehrdeutiger analytischer Funktionen von zwei komplexen Veränderlichen. Math. Ann. **126**, 1–22 (1953)
12. Hirzebruch, F.: Hilbert Modular Surfaces, vol. 19, no. 2, pp. 183–281. Enseignement Mathématique, Genève (1973)
13. Jung, H.W.E.: Darstellung der Funktionen eines algebraischen Körpers zweier unabhängigen Veränderlichen x,y in der Umgebung einer Stelle x = a, y = b. J. Reine Angew. Math. **133**, 289–314 (1908)
14. Klein, F.: Über eine geometrische Auffassung der gewöhnlichen Kettenbruchentwicklung. Nachr. Ges. Wiss. Göttingen. Math.-Phys. Kl. **3**, 357–359 (1895). French translation: Sur une représentation géométrique du développement en fraction continue ordinaire. Nouvelles Annales de Mathématiques **15**, 327–331 (1896)
15. Khinchin, A.Ya.: Continued Fractions. Dover Publications, New York (1997)
16. Lipman, J.: Introduction to Resolution of Singularities. Proceedings of Symposia in Pure Mathematics, vol. 29, pp. 187–230. American Mathematical Society, Providence (1975)
17. Perron, O.: Die Lehre von den Kettenbrüchen. Teubner, Germany (1954)
18. Popescu-Pampu, P.: The geometry of continued fractions and the topology of surface singularities. In: Brasselet, J.-P., Suwa, T. (eds.) Singularities in Geometry and Topology 2004. Advanced Studies in Pure Mathematics, vol. 46, pp. 119–195. Mathematical Society of Japan, Tokyo (2007)
19. Stark, H.M.: An Introduction to Number Theory. Markham Publishing Company, Chicago (1970); MIT Press, Cambridge (1998)

Printed in the United States
By Bookmasters